Quantum Optics

In the past decade many important advances have taken place in the field of quantum optics, with numerous potential applications. This textbook provides an up-to-date account of the basic principles of the subject, and is ideal for graduate courses.

Focusing on applications of quantum optics, the textbook covers recent developments such as engineering of quantum states, quantum optics on a chip, nano-mechanical mirrors, quantum entanglement, quantum metrology, spin squeezing, control of decoherence, and many other key topics. Readers are guided through the principles of quantum optics and their uses in a wide variety of areas including quantum information science and quantum mechanics.

The textbook features over 150 end-of-chapter exercises with solutions available for instructors at www.cambridge.org/9781107006409. It is invaluable to both graduate students and researchers in physics and photonics, quantum information science, and quantum communications.

Girish S. Agarwal is Noble Foundation Chair and Regents Professor at Oklahoma State University. A recognized leader in the field of theoretical quantum optics, he is a Fellow of the Royal Society and has won several awards, including the Max-Born Prize from the Optical Society of America and the Humboldt Research Award.

Quantum Optics

GIRISH S. AGARWAL
Oklahoma State University

CAMBRIDGE
UNIVERSITY PRESS

University Printing House, Cambridge CB2 8BS, United Kingdom

One Liberty Plaza, 20th Floor, New York, NY 10006, USA

477 Williamstown Road, Port Melbourne, VIC 3207, Australia

4843/24, 2nd Floor, Ansari Road, Daryaganj, Delhi - 110002, India

79 Anson Road, #06-04/06, Singapore 079906

Cambridge University Press is part of the University of Cambridge.

It furthers the University's mission by disseminating knowledge in the pursuit of education, learning and research at the highest international levels of excellence.

www.cambridge.org
Information on this title: www.cambridge.org/9781107006409

First published 2013
Reprinted 2015

A catalogue record for this publication is available from the British Library

Library of Congress Cataloging in Publication data
Agarwal, G. S. (Girish S.), 1946–
Quantum optics / Girish S. Agarwal.
pages cm
Includes bibliographical references and index.
ISBN 978-1-107-00640-9
1. Quantum optics – Textbooks. I. Title.
QC446.2.A33 2012
535′.15 – dc23 2012025069

ISBN 978-1-107-00640-9 Hardback

Additional resources for this publication at www.cambridge.org/9781107006409

Dedicated to the memory of my wife Sneh who had hoped that the book would be completed and to my daughters Anjali and Mranjali who have been a constant source of inspiration.

Contents

Preface

The development of new sources of radiation that produce nonclassical and entangled light has changed the landscape of quantum optics. The production, characterization, and detection of single photons is important not only in understanding fundamental issues but also in the transfer of quantum information. Entangled light and matter sources as well as ones possessing squeezing are used for precision interferometry and for implementing quantum communication protocols. Furthermore, quantum optics is making inroads in a number of interdisciplinary areas, such as quantum information science and nano systems.

These new developments require a book which covers both the basic principles and the many emerging applications. We therefore emphasize fundamental concepts and illustrate many of the ideas with typical applications. We make every possible attempt to indicate the experimental work if an idea has already been tested. Other applications are left as exercises which contain enough guidance so that the reader can easily work them out. Important references are given, although the bibliography is hardly complete. Thus students and postdocs can use the material in the book to do independent research. We have presented the material in a self-contained manner. The book can be used for a two-semester course in quantum optics after the students have covered quantum mechanics and classical electrodynamics at a level taught in the first year of graduate courses. Some advanced topics in the book, such as exact non-Markovian dynamics of open systems, quantum walks, and nano-mechanical mirrors, can be used for seminars in quantum optics.

The material in the book is broadly divided into two parts. The first part deals with many old and emerging aspects of the quantized radiation fields, such as the engineering and characterization of quantum states and the generation of entanglement. The working of an interferometer using one or a few photons is extensively treated. A chapter is devoted to quantum optics with fields carrying orbital angular momentum. Many applications of entangled fields are given. A thorough discussion of quantum noise in amplification and attenuation is also given. The second part deals with the interaction of radiation with matter. Coherent, squeezed, and cat states of atoms are treated. Dissipative processes are treated from a microscopic approach. This part includes a discussion of electromagnetically induced transparency and a host of applications. Special emphasis is placed on quantum interference and entanglement. It is shown how measurement can produce entanglement. Furthermore, the deterministic production of entanglement is discussed. Many relatively newer aspects of cavity QED, such as single photon switches, photon blockade, and anti-Jaynes–Cummings interaction are presented. The book concludes with a look at developments such as quantum optics on a chip, quantum optical effects arising from radiation pressure and mechanical motion, quantum walks, control of decoherence, and disentanglement.

The focus in the book is on emerging areas in quantum optics and therefore important topics like the quantum theory of lasers, micromasers, optical multi-stability, self-induced transparency, etc. have been left out as these are well covered in earlier textbooks like those of Mandel and Wolf (*Optical Coherence and Quantum Optics*) and Scully and Zubairy (*Fundamentals of Quantum Optics*). Other topics like polarization are treated from the new perspective of quantum fluctuations in Stokes parameters. Clearly it is impossible to present in a single volume all that has happened since the discovery of the laser and the pioneering quantum optical works in the early 1960s. We have emphasized aspects that we consider essential for the newer directions in which quantum optics is moving.

This book is the outcome of teaching courses in Quantum Optics at the University of Hyderabad and the Oklahoma State University and extensive lecturing at the International Center for Theoretical Physics, Trieste, and at many scientific schools in India and elsewhere. Research carried out over several decades and my earlier writings, as well as interaction with students, postdocs, and collaborators, has shaped the book. A part of the chapters dealing with states of the radiation field was evolved while I spent half a year in 1992–93 at the Max-Planck Institute for Quantum Optics, Garching, Germany. My collaborator Subhash Chaturvedi contributed by refining some of this material.

The book would never have been completed except for the tireless efforts of my student Sumei Huang who extensively worked on it. I am grateful to her. I also acknowledge considerable assistance from my student Kenan Qu.

I thank my collaborators Jay Banerji, Subhash Chaturvedi, Tarak Dey, Jacques Perk, Gautam Vemuri, and Joachim von Zanthier for reading several chapters and for providing useful input. I thank Ravi Puri, Surya Tewari, Subhasish Dutta Gupta, and a large number of students whose works have been used for the writing of the book. I thank Mustansir Barma for the hospitality at TIFR, Mumbai where I continued to work on my book.

Over the years I have learnt a lot through interactions with a large number of physicists. These interactions have had a deep influence in the writing of the book and I thank especially Bob Boyd, Jinx Cooper, Joe Eberly, Marlan Scully, Herman Haken, Sudhansu Jha, Peter Knight, Emil Wolf, the late Len Mandel, and Herbert Walther.

I am grateful to the Oklahoma State University, especially Dean Peter Sherwood, and the Noble Foundation for supporting my work and for providing ideal facilities.

I thank Dr. John Fowler who was instrumental in getting the book project going and Dr. Simon Capelin for his deep interest in the project. Finally, I thank the very supportive staff at Cambridge University Press, and in particular Ms. Lindsay Barnes, who always came up with a solution to my problems.

1 Quantized electromagnetic field and coherent state representations

The foundations of the quantum theory of radiation were laid by the work of Planck, Einstein, Dirac, Bose, Wigner, and many others. Historically Planck's [1] work on black body radiation is the foundation of any work on the quantum theory of radiation. Einstein's [2] work on the photoelectric effect established the particle nature of the radiation field. These particles were named as photons by Lewis [3]. Einstein [4] also introduced the *A* and *B* coefficients to describe the interaction of radiation and matter. He characterized stimulated emission using the *B* coefficient. Using thermodynamic arguments, he could also extract the *A* coefficient describing spontaneous emission which is at the heart of the origin of all spectral lines. This was quite a remarkable achievement. Dirac [5] implemented the quantization of the electromagnetic field and showed how Einstein's *A* coefficient emerges naturally from the quantization of the radiation field. It should be remembered that stimulated emission is the key to the working of any laser system. Following Dirac's quantization of the radiation field, Weisskopf and Wigner [6] were able explain in a very fundamental way the decay of the excited states of a system and hence derive the remarkable law of exponential decay. Bose [7] discovered a quantitative explanation for Planck's law. He introduced a new way of counting statistics relevant to quantum particles with zero mass. This was the beginning of quantum statistics. Bose's work was followed by Einstein [8] who produced a counting statistics for particles with finite mass (now known as Bosons). Fürth [9] studied the fluctuations in the energy distribution of black body radiation and thus shed light on the wave–particle duality of light. Einstein found that the fluctuations had two types of contributions – one can be interpreted in terms of the particle characteristics and the other in terms of the wave characteristics. Dirac also investigated the question of a proper phase operator for a quantized radiation field. This is because in classical physics the phase is extensively used to characterize coherent fields. He introduced one but also worried about its unitary nature. The question posed by Dirac on the phase operator became a subject of intense activity during the 1990s [10], but it is still not fully resolved. In this chapter we will discuss the quantization of the radiation field. We present important states of the field and give details of the phase-space distributions for the radiation field.

1.1 Quantization of the electromagnetic field

In this section we introduce the key features associated with the quantization of the electromagnetic field. As we will see in subsequent chapters, quantization is essential in order to understand a very wide variety of quantum optical phenomena. Let us first consider a plane electromagnetic wave of frequency ω_k that is propagating in the direction $\vec{k}$ in free space. The electric $\vec{E}$ and magnetic $\vec{B}$ fields associated with such a plane wave are given by [11]

$$\begin{aligned} \vec{E}(\vec{r},t) &= \vec{\epsilon}E_0\mathrm{e}^{\mathrm{i}\vec{k}\cdot\vec{r}-\mathrm{i}\omega_k t} + c.c., \\ \vec{B}(\vec{r},t) &= \tfrac{\vec{k}\times\vec{\epsilon}}{k}E_0\mathrm{e}^{\mathrm{i}\vec{k}\cdot\vec{r}-\mathrm{i}\omega_k t} + c.c., \end{aligned} \tag{1.1}$$

for $k = |\vec{k}| = \omega_k/c$, where c is the velocity of light and $\vec{\epsilon}$ denotes the polarization vector for the electromagnetic field. The symbol $c.c.$ stands for the complex conjugate. We will use CGS units throughout the book. We choose especially form (1.1) as this would immediately correspond to the form used in quantum theory. The polarization vector is orthogonal to the direction of propagation and lies in a plane perpendicular to $\vec{k}$. There are two orthogonal polarizations. Let us denote these by $\vec{\epsilon}_{\vec{k}s}$, with s taking two values 1, 2. For a wave in the direction z we can write $\vec{\epsilon}$ in terms of the unit vectors along the x and y axes as

$$\vec{\epsilon}_{\vec{k}s} = \epsilon_{xs}\hat{x} + \epsilon_{ys}\hat{y}, \quad |\epsilon_{xs}|^2 + |\epsilon_{ys}|^2 = 1. \tag{1.2}$$

For real ϵ_{xs} and ϵ_{ys} we have linearly polarized light; for

$$\vec{\epsilon}_{\vec{k}s} = \frac{1}{\sqrt{2}}(\hat{x} \pm \mathrm{i}\hat{y}), \tag{1.3}$$

we have circularly polarized light. Note that the vectors $\vec{E}$, $\vec{B}$, and $\vec{k}$ form a right-handed orthogonal coordinate system. The energy of the electromagnetic field, contained in volume V, is given by

$$U \equiv \frac{1}{8\pi}\int_V [E^2(\vec{r},t) + B^2(\vec{r},t)]\,\mathrm{d}^3r, \tag{1.4}$$

which for a plane wave reduces to

$$U \equiv \frac{1}{2\pi}|E_0|^2 V. \tag{1.5}$$

The Poynting vector $\vec{S}$, defined by

$$\vec{S} \equiv \frac{c}{4\pi}\vec{E}\times\vec{B}, \tag{1.6}$$

reduces for the plane wave to

$$\vec{S} \equiv \frac{c}{2\pi}|E_0|^2\left(\frac{\vec{k}}{k}\right). \tag{1.7}$$

In deriving (1.5) and (1.7) we have dropped the fast-oscillating terms at frequency $2\omega_k$.

In many problems it is more convenient to work with potentials, such as the vector potential $\vec{A}(\vec{r},t)$. Since the current book is devoted to problems in the nonrelativistic domain, we adopt the Coulomb gauge or transverse gauge in which the scalar potential is set to zero and $\vec{A}$ satisfies

$$\mathrm{div}\vec{A}(\vec{r},t) = 0. \tag{1.8}$$

The electric and magnetic fields are related to $\vec{A}$ via

$$\vec{E} = -\frac{1}{c}\frac{\partial\vec{A}}{\partial t}, \quad \vec{B} = \vec{\nabla}\times\vec{A}. \tag{1.9}$$

For the case of a plane wave we can write

$$\vec{A}(\vec{r},t) = \vec{\epsilon} A_0 \mathrm{e}^{\mathrm{i}\vec{k}\cdot\vec{r}-\mathrm{i}\omega_k t} + c.c.; \quad E_0 = \frac{\mathrm{i}\omega_k}{c} A_0. \tag{1.10}$$

In quantum theory the vector potential is more fundamental than the electric and magnetic fields.

Consider next an electromagnetic field confined to a box with volume V, we expand the field into a complete set of plane waves. The complete set can be obtained by imposing boundary conditions at the walls of the box. For convenience let us take the box to be a cube with volume L^3. Then imposing periodic boundary conditions, the allowed values of $\vec{k}$ are

$$\vec{k} \equiv 2\pi \frac{\vec{n}}{L}, \quad \vec{n}_i \equiv 0, \pm 1, \pm 2, \ldots. \tag{1.11}$$

Here each component of $\vec{n}$ is an integer with all possible values. Thus we write the vector potential in the form

$$\vec{A}(\vec{r},t) = \sum_{\vec{k},s} \frac{A_{\vec{k}s}}{\sqrt{V}} \vec{\epsilon}_{\vec{k}s} \mathrm{e}^{\mathrm{i}\vec{k}\cdot\vec{r}-\mathrm{i}\omega_k t} + c.c.. \tag{1.12}$$

The summation in (1.12) is over all allowed values of $\vec{k}$. It should be borne in mind that for each $\vec{k}$, s takes two values. The coefficients $A_{\vec{k}s}$ are arbitrary. The specific form of $A_{\vec{k}s}$ would be determined by the electromagnetic field at hand. Using (1.12) and (1.9) and the orthogonality of the plane waves for $\vec{n} = (n_x, n_y, n_z)$ and $\vec{n}' = (n'_x, n'_y, n'_z)$,

$$\frac{1}{V} \int_V \mathrm{e}^{\mathrm{i}\vec{k}\cdot\vec{r}-\mathrm{i}\vec{k}'\cdot\vec{r}} \mathrm{d}^3 r \equiv \delta_{n_x n'_x} \delta_{n_y n'_y} \delta_{n_z n'_z}, \tag{1.13}$$

we obtain from (1.4) the expression for the energy

$$U = \frac{1}{2\pi} \sum_{\vec{k},s} \left(\frac{\omega_k^2}{c^2}\right) |A_{\vec{k}s}|^2, \quad \omega_k = kc. \tag{1.14}$$

The energy has been expressed as a sum over modes – each mode is a plane wave with a given polarization.

We would now proceed with the quantization of the field. Clearly $\hbar\omega_k$ is the quantum of energy associated with a single mode. Let $n_{\vec{k}s}$ be the number of quanta associated with each mode. Therefore the total energy would be

$$U = \sum_{\vec{k},s} \hbar\omega_k n_{\vec{k}s}. \tag{1.15}$$

On comparison with (1.14) we can thus identify

$$\frac{\omega_k}{2\pi\hbar c^2} |A_{\vec{k}s}|^2 \leftrightarrow n_{\vec{k}s}. \tag{1.16}$$

In quantum theory all fields $\vec{E}$, $\vec{B}$, and $\vec{A}$ become operators. The energy U becomes the Hamiltonian operator. The number becomes the number operator. Each mode of the electromagnetic field can be identified with a photon. It was demonstrated by Bose [7] that photons obey what is now called Bose statistics and thus each mode can be occupied by an arbitrary number of photons. Planck had already established that for a black body at temperature T, the average occupation number $n_{\vec{k}s}$ is

$$\langle n_{\vec{k}s}\rangle = \frac{1}{\exp(\frac{\hbar\omega_k}{K_{\mathrm{B}}T}) - 1}, \tag{1.17}$$

where K_{B} is the Boltzmann constant. Depending on the temperature and ω_k, $\langle n_{\vec{k}s}\rangle$ can take arbitrary values. In quantum theory the number operator $n_{\vec{k}s}$ is a positive definite Hermitian operator, with eigenvalues 0, 1, 2, ... It can thus be written in terms of the non-Hermitian operators $a_{\vec{k}s}$ and $a^{\dagger}_{\vec{k}s}$ as

$$n_{\vec{k}s} \equiv a^{\dagger}_{\vec{k}s}a_{\vec{k}s}. \tag{1.18}$$

The operators $a_{\vec{k}s}$ and $a^{\dagger}_{\vec{k}s}$ obey the Boson commutation relations

$$[a_{\vec{k}s}, a^{\dagger}_{\vec{k}'s'}] = \delta_{\vec{k}\vec{k}'}\delta_{ss'}, \tag{1.19}$$

$$[a_{\vec{k}s}, a_{\vec{k}'s'}] = 0. \tag{1.20}$$

The noncommutativity of a and $a^{\dagger}$ adds a new dimension to the electromagnetic field. This is because all field operators are linear in a and $a^{\dagger}$, whereas the energy is quadratic in a and $a^{\dagger}$. Thus the energy can be nonzero even if there are no quanta in the field. This can be seen explicitly by using $A_{\vec{k}s} \to \sqrt{\frac{2\pi\hbar c^2}{\omega_k}}a_{\vec{k}s}$ on the basis of (1.16). All quantum fields can be expressed as [12]

$$\vec{A}(\vec{r}, t) = \sum_{\vec{k},s}\sqrt{\frac{2\pi\hbar c^2}{\omega_k V}}\vec{\epsilon}_{\vec{k}s}a_{\vec{k}s}\mathrm{e}^{\mathrm{i}\vec{k}\cdot\vec{r}-\mathrm{i}\omega_k t} + H.c., \tag{1.21}$$

$$\vec{E}(\vec{r}, t) = \mathrm{i}\sum_{\vec{k},s}\sqrt{\frac{2\pi\hbar\omega_k}{V}}\vec{\epsilon}_{\vec{k}s}a_{\vec{k}s}\mathrm{e}^{\mathrm{i}\vec{k}\cdot\vec{r}-\mathrm{i}\omega_k t} + H.c., \tag{1.22}$$

$$\vec{B}(\vec{r}, t) = \mathrm{i}\sum_{\vec{k},s}\sqrt{\frac{2\pi\hbar\omega_k}{V}}\frac{\vec{k}\times\vec{\epsilon}_{\vec{k}s}}{k}a_{\vec{k}s}\mathrm{e}^{\mathrm{i}\vec{k}\cdot\vec{r}-\mathrm{i}\omega_k t} + H.c.. \tag{1.23}$$

Here $H.c.$ stands for Hermitian conjugate because we are now dealing with quantum fields, which have to be Hermitian. The expression (1.4), after we replace energy U by the corresponding operator Hamiltonian H, becomes

$$H = \sum_{\vec{k},s}\frac{1}{2}\hbar\omega_k(a^{\dagger}_{\vec{k}s}a_{\vec{k}s} + a_{\vec{k}s}a^{\dagger}_{\vec{k}s}) \tag{1.24}$$

$$= \sum_{\vec{k},s}\hbar\omega_k\left(n_{\vec{k}s} + \frac{1}{2}\right). \tag{1.25}$$

The contribution $\sum_{\vec{k},s}\hbar\omega_k/2$ is called the zero point energy of the electromagnetic field. Furthermore, in quantum theory one writes all Hermitian field operators $\vec{F} = (\vec{A}, \vec{E}, \vec{B})$ as

$$\vec{F}(\vec{r}, t) = \vec{F}^{(+)}(\vec{r}, t) + \vec{F}^{(-)}(\vec{r}, t), \tag{1.26}$$

where $F^{(-)}$ is the adjoint of $F^{(+)}$ and $F^{(+)}$ contains only the positive frequencies ω_k. $F^{(+)}$ is called as the positive frequency part of F and consists of only the annihilation operators $a_{\vec{k}s}$. The decomposition (1.26) is related to how the photon detectors respond to the electromagnetic field (see Section 8.1). Finally, note that in dealing with free space we would take the limit $V \to \infty$ at the end of the calculation, i.e. after we have calculated the physical observables.

1.2 State space for the electromagnetic field – Fock space and Fock states

In quantum theory all observables are represented by Hermitian operators. The expectation values of such operators in the state of the quantum system gives the measureable quantities. We thus need to specify the appropriate state space for the electromagnetic field. The quantization of a system of bosons with finite mass was done by Fock. The corresponding space is called the Fock space and the basis states are called Fock states. Since all the fields (Eqs. (1.21)–(1.23)) are written as superpositions of all the independent modes, we can construct states for each mode and from this obtain states for the multimode field.

1.2.1 State space for a single mode of the radiation field

For brevity we drop the subscript $\vec{k}s$ and denote the single-mode operators as a, $a^\dagger$, and n, with

$$[a, a^\dagger] = 1. \tag{1.27}$$

The smallest eigenvalue of the number operator is zero. Let us denote the states of the number operator as $|n\rangle$

$$a^\dagger a|n\rangle = n|n\rangle; \quad n = 0, 1, 2, \ldots, \infty. \tag{1.28}$$

The states $|n\rangle$ are called Fock states. For $n = 0$, $a^\dagger a|0\rangle = 0$, therefore we can define the state $|0\rangle$ by

$$a|0\rangle = 0. \tag{1.29}$$

The state $|0\rangle$ is called the vacuum state as it contains no quanta of the radiation field. Now using the commutation relation between a and $a^\dagger$ and applying it to the vacuum state we get $(aa^\dagger - a^\dagger a)|0\rangle = |0\rangle \Rightarrow a(a^\dagger|0\rangle) = |0\rangle$, which can be rewritten as

$$a^\dagger a(a^\dagger|0\rangle) = a^\dagger|0\rangle, \tag{1.30}$$

and hence $a^\dagger|0\rangle$ is an eigenstate of $a^\dagger a$ with eigenvalue 1, i.e.

$$|1\rangle = a^\dagger|0\rangle. \tag{1.31}$$

The state is called a single-photon state. One can continue this process and obtain the n photon state $|n\rangle$ by repeated application of the operator $a^\dagger$ on $|0\rangle$. The state $|n\rangle$ is found to be

$$|n\rangle = \frac{(a^\dagger)^n}{\sqrt{n!}}|0\rangle. \tag{1.32}$$

The factor $1/\sqrt{n!}$ leads to the correct normalization of the state. The set of states $|n\rangle$ are orthogonal and complete

$$\sum_{n=0}^{\infty} |n\rangle\langle n| = 1, \quad \langle n|m\rangle = \delta_{nm}, \tag{1.33}$$

and further have the important property

$$a|n\rangle = \sqrt{n}|n-1\rangle, \quad a^\dagger|n\rangle = \sqrt{n+1}|n+1\rangle. \tag{1.34}$$

The property (1.34) is proved using (1.32) and by using the commutator (1.27). Note that the operators a and $a^\dagger$ are respectively called the annihilation and creation operators. This is because the action of a $[a^\dagger]$ on the n photon state $|n\rangle$ yields the $(n-1)$ $[(n+1)]$ photon state.

For a multimode field we would write the states as $|\{n_{\vec{k}s}\}\rangle$, which means that the mode $\vec{k}s$ has $n_{\vec{k}s}$ photons. These states are a product of the states for each mode

$$|\{n_{\vec{k}s}\}\rangle = \prod_{\vec{k}s} |n_{\vec{k}s}\rangle, \tag{1.35}$$

and have the properties

$$\begin{aligned} a_{\vec{k}s}|\{n_{\vec{k}s}\}\rangle &= \sqrt{n_{\vec{k}s}}|n_{\vec{k}s}-1\rangle \prod_{\vec{k}'s'\neq\vec{k}s} |n_{\vec{k}'s'}\rangle, \\ a^\dagger_{\vec{k}s}|\{n_{\vec{k}s}\}\rangle &= \sqrt{n_{\vec{k}s}+1}|n_{\vec{k}s}+1\rangle \prod_{\vec{k}'s'\neq\vec{k}s} |n_{\vec{k}'s'}\rangle. \end{aligned} \tag{1.36}$$

1.3 Quadratures of the field

For a plane-wave field (1.1), the amplitude E_0 is a complex number. Thus E_0 has a phase which can be measured by using an interferometer. One can thus obtain information on both the real and imaginary parts of E_0. These are called the in-phase and out-of-phase quadratures of the field. The well-known homodyne measurement can directly yield these quadratures. In quantum theory E_0 gets replaced by the non-Hermitian annihilation operator a. We can then define the two Hermitian quadrature operators X and Y as

$$a = \frac{X + \mathrm{i}Y}{\sqrt{2}}, \quad X = \frac{a + a^\dagger}{\sqrt{2}}, \quad Y = \frac{a - a^\dagger}{\sqrt{2}\mathrm{i}}. \tag{1.37}$$

In view of (1.27) we now get

$$[X, Y] = \mathrm{i}. \tag{1.38}$$

Note that this commutation relation for the quadratures is similar to the commutation relation for the position and momentum operators of a particle with mass. The operator X should not be confused with the position of the photon. We can, however, introduce a quadrature space by representing Y as $Y = -\mathrm{i}\frac{\partial}{\partial X}$, which follows from the commutation relation (1.38). This enables us to write the Fock states in quadrature space as

$$\Psi_n(X) = \langle X|n\rangle = (2^n n!\sqrt{\pi})^{-1/2} H_n(X) \mathrm{e}^{-X^2/2}, \tag{1.39}$$

where $H_n(X)$ is the Hermite polynomial of degree n. Furthermore, the Heisenberg uncertainty relation would give

$$\Delta X \Delta Y \geq \frac{1}{2}, \quad (\Delta X)^2 = \langle X^2\rangle - \langle X\rangle^2, \quad (\Delta Y)^2 = \langle Y^2\rangle - \langle Y\rangle^2. \tag{1.40}$$

For Fock states the relation (1.40) reduces to

$$\Delta X \Delta Y = n + \frac{1}{2}, \quad (\Delta X)^2 = (\Delta Y)^2 = n + \frac{1}{2}. \tag{1.41}$$

Clearly the quadratures carry the phase-dependent information on the field and are important in the context of the characterization and detection of the squeezed states of the field.

1.4 Coherent states

In classical theory, one can have a very well-defined electromagnetic field, i.e. a field E_0 with well-defined amplitude and phase. We have seen that in quantum theory E_0 is replaced by the annihilation operator a. Thus one would clearly identify E_0 with the expectation value of the quantum field a in a given quantum-mechanical state of the field. However, for the Fock states $|n\rangle$ of the field the mean values of a and the quadrature operators vanish

$$\langle n|a|n\rangle = \langle n|X|n\rangle = \langle n|Y|n\rangle = 0. \tag{1.42}$$

Thus Fock states could not represent fields with well-defined amplitudes and phase at a classical level. We know that a field produced by a single-mode laser is coherent, i.e. has a well-defined amplitude and phase. So the question is – what is the appropriate state of the radiation field that would represent such a coherent field? Glauber [13, 14] gave the answer to such a question and derived a new class of states that he called coherent states, which are usually denoted by the symbol $|\alpha\rangle$. If the eigenvalue equation

$$a|\alpha\rangle = \alpha|\alpha\rangle, \quad \alpha \text{ complex number}, \tag{1.43}$$

has a normalizable solution $|\alpha\rangle$, then the field and its quadratures would have nonzero values

$$\langle\alpha|a|\alpha\rangle = \alpha; \quad \langle\alpha|X|\alpha\rangle = \frac{\mathrm{Re}\{\alpha\}}{\sqrt{2}}, \quad \langle\alpha|Y|\alpha\rangle = \frac{\mathrm{Im}\{\alpha\}}{\sqrt{2}}. \tag{1.44}$$

Thus the state $|\alpha\rangle$ would correspond to a classical field with well-defined amplitude and phase and hence the name coherent states is used for such states. The intensity, which is proportional to $\langle a^\dagger a\rangle$, would be

$$\langle\alpha|a^\dagger a|\alpha\rangle = |\alpha|^2 = |\langle\alpha|a|\alpha\rangle|^2. \tag{1.45}$$

This is again like a coherent classical field as then the intensity of the field is the modulus of the square of the mean amplitude of the field.

1.4.1 Solution to the eigenvalue equation (1.43)

We can expand $|\alpha\rangle$ in terms of the Fock states as these form a complete set

$$|\alpha\rangle = \sum_{n=0}^{\infty} c_n|n\rangle, \tag{1.46}$$

which on substituting in (1.43) yields the recursion relation

$$c_{n+1} = \frac{\alpha}{\sqrt{n+1}}c_n, \tag{1.47}$$

whose solution is

$$c_n = \frac{\alpha^n}{\sqrt{n!}}c_0, \tag{1.48}$$

with c_0 fixed by the normalization condition $\langle\alpha|\alpha\rangle = 1 \Rightarrow c_0 = \exp(-\frac{1}{2}|\alpha|^2)$. Thus for all complex values of α we have the solution

$$|\alpha\rangle = \mathrm{e}^{-\frac{1}{2}|\alpha|^2}\sum_{n=0}^{\infty}\frac{\alpha^n}{\sqrt{n!}}|n\rangle. \tag{1.49}$$

The coherent states are superpositions of Fock states. The probability p_n of finding the system in the state $|n\rangle$ is then given by

$$p_n = |c_n|^2 = \mathrm{e}^{-|\alpha|^2}\frac{(|\alpha|^2)^n}{n!}. \tag{1.50}$$

The probability of finding n photons in a coherent state is given by the Poisson distribution (1.50) with mean $|\alpha|^2$. This distribution is shown in Figure 1.1 and has the property that its variance is equal to the mean

$$\langle n^2\rangle - \langle n\rangle^2 = \langle n\rangle; \quad \langle n\rangle = |\alpha|^2, \tag{1.51}$$

where n is the number operator $a^\dagger a$.

1.4.2 Properties of coherent states

Next we present some important properties of coherent states. The states form a complete set

$$\frac{1}{\pi}\int \mathrm{d}^2\alpha|\alpha\rangle\langle\alpha| = 1, \quad \alpha = x + \mathrm{i}y, \quad \mathrm{d}^2\alpha = \mathrm{d}x\,\mathrm{d}y, \quad -\infty \le x, y \le +\infty. \tag{1.52}$$

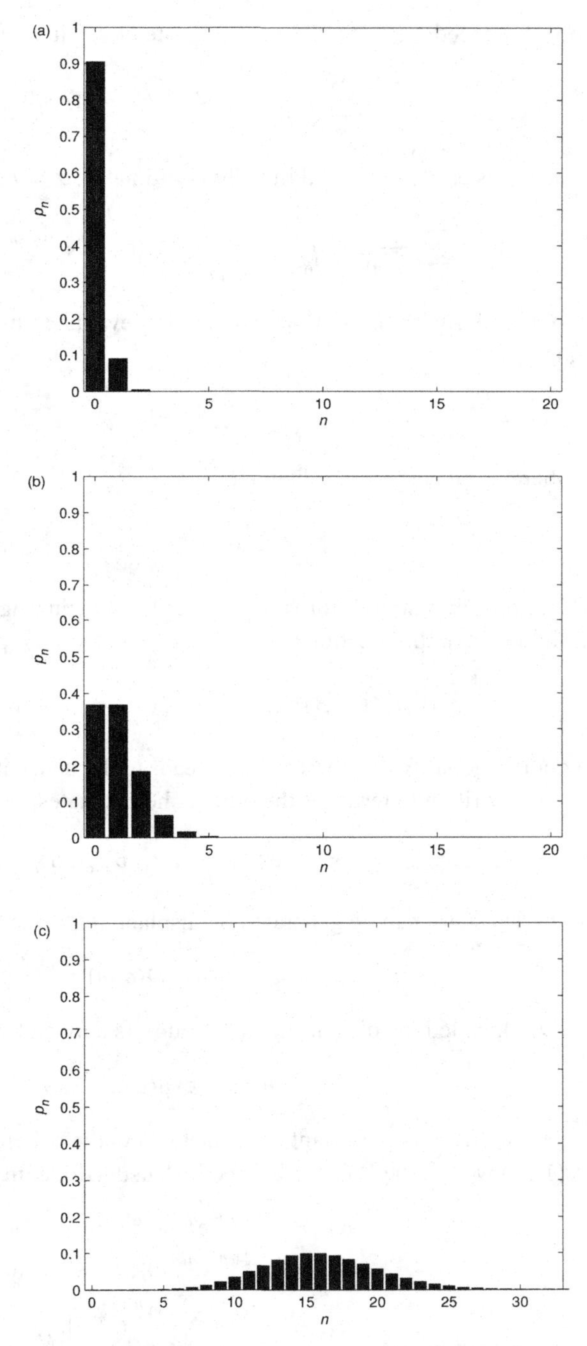

Fig. 1.1 The photon distribution p_n for a coherent state with different values of $|\alpha|^2$. (a) $|\alpha|^2 = 0.1$; (b) $|\alpha|^2 = 1$; and (c) $|\alpha|^2 = 16$.

The proof proceeds as follows – we substitute (1.49) into (1.52), hence the LHS becomes

$$\sum_{n,m} \frac{1}{\sqrt{n!m!}} \int \mathrm{d}^2\alpha\, \alpha^n \alpha^{*m} |n\rangle\langle m| \mathrm{e}^{-|\alpha|^2}.$$

The integral is easily evaluated in polar coordinates, $\alpha = r\mathrm{e}^{\mathrm{i}\theta}$,

$$\sum_{n,m} \frac{1}{\sqrt{n!m!}} \int_0^\infty r\mathrm{d}r \int_0^{2\pi} \mathrm{d}\theta\, (r)^{n+m} \mathrm{e}^{i\theta(n-m)} |n\rangle\langle m| \mathrm{e}^{-r^2}.$$

The θ integral gives δ_{nm}, and the r integral is evaluated in terms of the gamma function, yielding

$$\sum_n |n\rangle\langle n| = 1. \qquad \square$$

The coherent states are nonorthogonal

$$\langle\alpha|\beta\rangle \equiv \exp\left(\alpha^*\beta - \frac{1}{2}|\alpha|^2 - \frac{1}{2}|\beta|^2\right), \tag{1.53}$$

which follows in a straightforward manner by substituting (1.49) into (1.53). In view of (1.52) into (1.53), the function (1.53) is an example of a reproducing kernel $K(\alpha, \beta)$

$$\int \mathrm{d}^2\beta\, K(\alpha, \beta)K(\beta, \gamma) = K(\alpha, \gamma), \quad K(\alpha, \beta) = \langle\alpha|\beta\rangle/\pi. \tag{1.54}$$

The nonorthogonality of coherent states leads to the unusual property that a given coherent state can be written in terms of the other coherent states

$$|\alpha\rangle = \int K(\alpha, \beta)|\beta\rangle\, \mathrm{d}^2\beta. \tag{1.55}$$

The coherent states can be generated by displacing the vacuum

$$|\alpha\rangle = D(\alpha)|0\rangle, \tag{1.56}$$

where $D(\alpha)$, called the displacement operator, is defined by

$$D(\alpha) = \exp(\alpha a^\dagger - \alpha^* a). \tag{1.57}$$

The relation (1.56) is important as it shows how the coherent states can be generated. To prove (1.56) we use the Baker–Campbell–Hausdorff identity [15] to write (1.56) as

$$\begin{aligned} |\alpha\rangle &= \mathrm{e}^{-\frac{1}{2}|\alpha|^2} \mathrm{e}^{\alpha a^\dagger} \mathrm{e}^{-\alpha^* a}|0\rangle \\ &= \mathrm{e}^{-\frac{1}{2}|\alpha|^2} \mathrm{e}^{a^\dagger \alpha}|0\rangle, \quad \text{as} \quad a^n|0\rangle = 0 \\ &= \mathrm{e}^{-\frac{1}{2}|\alpha|^2} \sum_{n=0}^\infty \frac{\alpha^n}{n!}(a^\dagger)^n|0\rangle, \end{aligned} \tag{1.58}$$

which leads to (1.49) by using the definition (1.32). The displacement operators are quite important in many calculations with coherent states. We list in Table 1.1 many of their mathematical properties.

Table 1.1 Important properties of the displacement operator $D(\alpha) = \exp(\alpha a^\dagger - \alpha^* a)$.

$$D(\alpha) = \mathrm{e}^{-\frac{1}{2}|\alpha|^2}\mathrm{e}^{\alpha a^\dagger}\mathrm{e}^{-\alpha^* a} = \mathrm{e}^{\frac{1}{2}|\alpha|^2}\mathrm{e}^{-\alpha^* a}\mathrm{e}^{\alpha a^\dagger} \text{(BCH relation).} \quad (1)$$

$$D^{-1}(\alpha) = D^\dagger(\alpha) = D(-\alpha). \quad (2)$$

$$D^\dagger(\alpha)G(a, a^\dagger)D(\alpha) = G(a+\alpha, a^\dagger + \alpha^*),\ D^\dagger(\alpha)aD(\alpha) = a + \alpha. \quad (3)$$

$$D(\alpha)D(\beta) = D(\alpha+\beta)\exp\left[\tfrac{1}{2}(\alpha\beta^* - \alpha^*\beta)\right],\ [D(\alpha), D(\beta)] \neq 0. \quad (4)$$

$$\mathrm{Tr}D(\alpha) = \pi\delta^{(2)}(\alpha) = \pi\delta(\mathrm{Re}\{\alpha\})\delta(\mathrm{Im}\{\alpha\}). \quad (5)$$

$$\mathrm{Tr}[D(\alpha)D^\dagger(\beta)] = \pi\delta^{(2)}(\alpha - \beta). \quad (6)$$

$$D(\alpha)|\beta\rangle = |\alpha+\beta\rangle\exp\left[\tfrac{1}{2}(\alpha\beta^* - \alpha^*\beta)\right]. \quad (7)$$

$$\langle\alpha|D(\gamma)|\beta\rangle = \langle\alpha|\beta\rangle\exp\left(\gamma\alpha^* - \gamma^*\beta - \tfrac{1}{2}|\gamma|^2\right). \quad (8)$$

$$\langle n|D(\gamma)|m\rangle = \sqrt{\tfrac{m!}{n!}}\mathrm{e}^{-|\alpha|^2/2}(\alpha)^{n-m}\mathrm{L}_m^{(n-m)}(|\alpha|^2), \quad n \geq m. \quad (9)$$

$$= \sqrt{\tfrac{n!}{m!}}\mathrm{e}^{-|\alpha|^2/2}(-\alpha^*)^{m-n}\mathrm{L}_m^{(m-n)}(|\alpha|^2), \quad n \leq m. \quad (10)$$

where $\mathrm{L}_n^{(k)}(x)$ are the associated Laguerre polynomials [19]

$$\mathrm{L}_n^{(k)}(x) = \sum_{m=0}^{n}(-1)^m\binom{n+k}{m+k}\frac{x^m}{m!}. \quad (11)$$

The coherent states are known as the minimum uncertainty states as [16–18]

$$\Delta X^2 = \Delta Y^2 = \frac{1}{2}, \quad \Delta X \Delta Y = \frac{1}{2}, \quad \forall\,\alpha. \tag{1.59}$$

These relations can be proved by writing, say, X in terms of a and $a^\dagger$ and using $\langle\alpha|a^2|\alpha\rangle = \alpha^2 = \langle\alpha|a^{\dagger 2}|\alpha\rangle^*$; $\langle\alpha|a^\dagger a|\alpha\rangle = |\alpha|^2$. The coherent states, besides being minimum uncertainty states, also have the important characteristic that the uncertainties in the two quadratures are equal. The coherent states have the most important property that these do not spread in time unlike the wavepackets for a free particle. Schrödinger discovered coherent states from this very requirement. In order to see this we examine the time evolution of coherent states under the free Hamiltonian (1.25) for a single mode:

$$\begin{aligned}
\mathrm{e}^{-\mathrm{i}Ht/\hbar}|\alpha\rangle &= \mathrm{e}^{-\mathrm{i}\omega t/2}\mathrm{e}^{-\mathrm{i}\omega t a^\dagger a}|\alpha\rangle \\
&= \mathrm{e}^{-\mathrm{i}\omega t/2}\mathrm{e}^{-\mathrm{i}\omega t a^\dagger a}\sum\frac{\alpha^n}{\sqrt{n!}}|n\rangle\mathrm{e}^{-\frac{1}{2}|\alpha|^2} \\
&= \sum\frac{\alpha^n\mathrm{e}^{-\mathrm{i}\omega t n}}{\sqrt{n!}}|n\rangle\mathrm{e}^{-\frac{1}{2}|\alpha|^2}\mathrm{e}^{-\mathrm{i}\omega t/2} \\
&= |\alpha\mathrm{e}^{-\mathrm{i}\omega t}\rangle\mathrm{e}^{-\mathrm{i}\omega t/2}.
\end{aligned} \tag{1.60}$$

Hence a coherent state evolves into another coherent state with a different phase. Since the uncertainties ΔX and ΔY do not depend on the amplitude of the coherent state, and hence we have no spreading of coherent wavepackets [16, 17], i.e.

$$\Delta X^2(t) = \Delta Y^2(t) = \frac{1}{2}, \quad \forall\, t. \tag{1.61}$$

Finally, we examine the structure of coherent states in terms of the quadrature spaces

$$\Psi(X) = \langle X|\alpha\rangle, \quad \Psi(Y) = \langle Y|\alpha\rangle. \tag{1.62}$$

Using (1.49) and (1.39) we get

$$\Psi(X) = \mathrm{e}^{-\frac{1}{2}|\alpha|^2} \sum_{n=0}^{\infty} \frac{\alpha^n}{\sqrt{n!}} \cdot \frac{1}{\sqrt{2^n n! \sqrt{\pi}}} \mathrm{e}^{-X^2/2} H_n(X). \tag{1.63}$$

The sum in (1.63) can be carried out by using the generating function for the Hermite polynomials

$$\sum_{n=0}^{\infty} \frac{H_n(X)}{n!} S^n = \exp(-S^2 + 2SX), \tag{1.64}$$

leading to

$$\Psi(X) = \frac{1}{\pi^{1/4}} \exp\left(-\frac{1}{2}X^2 - \frac{1}{2}\alpha^2 - \frac{1}{2}|\alpha|^2 + \sqrt{2}\alpha X\right). \tag{1.65}$$

The quadrature distribution would be

$$|\Psi(X)|^2 = \frac{1}{\sqrt{\pi}} \exp\left\{-\left[X - \frac{1}{\sqrt{2}}(\alpha + \alpha^*)\right]^2\right\}. \tag{1.66}$$

The quadrature distribution is Gaussian whose center is located at the value $\frac{1}{\sqrt{2}}(\alpha + \alpha^*) = \langle X \rangle$. We can similarly work out the quadrature distribution for Y with the results

$$\Psi(Y) = \frac{1}{\pi^{1/4}} \exp\left(-\frac{1}{2}Y^2 + \frac{1}{2}\alpha^2 - \frac{1}{2}|\alpha|^2 - \sqrt{2}\mathrm{i}\alpha Y\right), \tag{1.67}$$

$$|\Psi(Y)|^2 = \frac{1}{\sqrt{\pi}} \exp\left\{-\left[Y - \frac{1}{\sqrt{2}\mathrm{i}}(\alpha - \alpha^*)\right]^2\right\}. \tag{1.68}$$

The quadrature distribution for Y is also Gaussian. We will see in Section 5.18 that these quadrature distributions can be measured using a homodyne technique.

1.5 Mixed states of the radiation field

In the previous sections we introduced two different classes of states of the radiation field. The Fock states $|n\rangle$ and the coherent states $|\alpha\rangle$ have very different physical properties. One could construct a variety of other types of states using superpositions of either Fock states or coherent states. The black body case belongs to a newer class – this was the case that led to the birth of the quantum theory of radiation. A single mode of the black body radiation

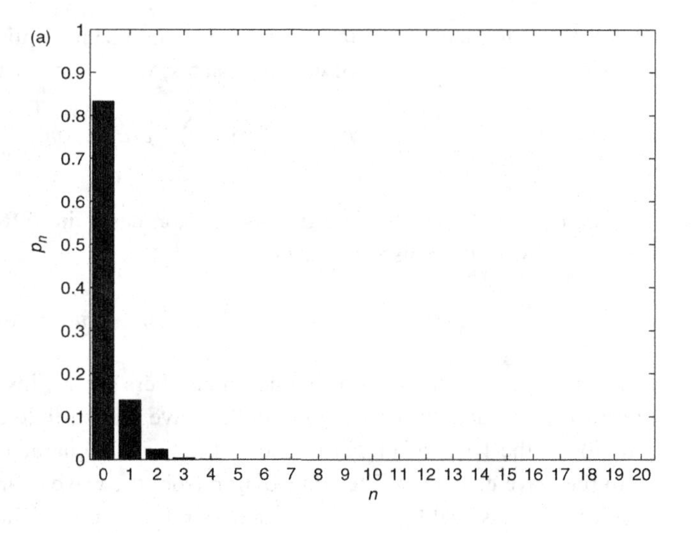

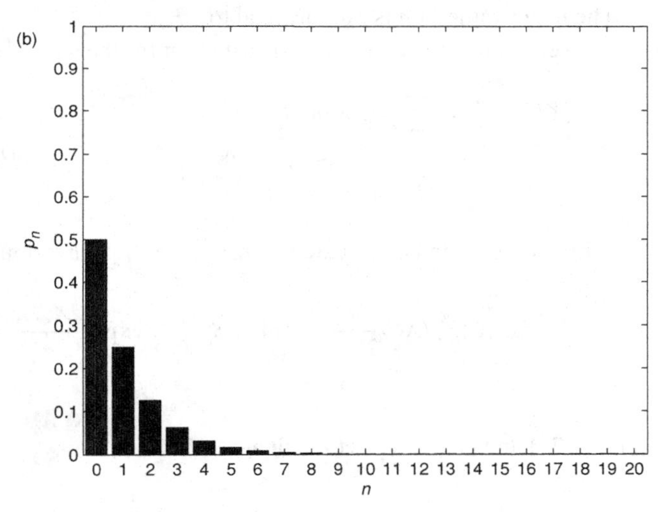

Fig. 1.2 The Bose–Einstein distribution p_n for different mean number of photons $\langle n \rangle$. (a) $\langle n \rangle = 0.2$; (b) $\langle n \rangle = 1$.

corresponds to a mixed state of the radiation field. This mixed state is described in terms of the density matrix ρ given by

$$\rho_{\mathrm{T}} = \exp(-\beta\hbar\omega a^{\dagger}a)/\mathrm{Tr}[\exp(-\beta\hbar\omega a^{\dagger}a)], \quad \beta = 1/(K_{\mathrm{B}}T), \tag{1.69}$$

which can be expressed in terms of Fock states as

$$\rho_{\mathrm{T}} = \sum_n p_n |n\rangle\langle n|; \quad p_n = \frac{\langle n \rangle^n}{(1 + \langle n \rangle)^{n+1}}, \quad \langle n \rangle = 1/[\exp(\beta\hbar\omega) - 1]. \tag{1.70}$$

The distribution p_n is called the Bose–Einstein (BE) distribution and depends on the mean number of photons $\langle n \rangle$, which in turn depends on the temperature T and the frequency ω. This distribution is shown in Figure 1.2. Note that there are no off-diagonal terms in ρ_{T}, the state (1.70) is an incoherent state of the field. The mean amplitude $\langle a \rangle$ of the field is zero. This is in contrast to the case of coherent states, which have density matrix ρ_c given by

$$\rho_c = |\alpha\rangle\langle\alpha| = \sum_{n,m} c_n c_m^* |n\rangle\langle m|, \tag{1.71}$$

where c_n is given by (1.48). The BE distribution has quite different fluctuation properties as compared with the Poisson distribution

$$\langle n^2 \rangle - \langle n \rangle^2 = \langle n \rangle^2 + \langle n \rangle. \tag{1.72}$$

As $\langle n \rangle$ increases the fluctuations in photon numbers grow. Einstein [8,9] interpreted (1.72) in terms of the fluctuations due to both the wave and particle characteristics of light. The term $\langle n \rangle$ on the RHS of (1.72) is due to the particle character whereas the term $\langle n \rangle^2$ is due to the wave character. The contribution from the wave character can be understood by considering a classical light wave E_0 Gaussian fluctuations. The details are left as Exercise 1.3. The mean value of n is proportional to $\langle |E_0|^2 \rangle$.

Next we examine the quadrature distribution for the state (1.70)

$$\begin{aligned}\langle X|\rho_T|X'\rangle &= \sum p_n \langle X|n\rangle\langle n|X'\rangle \\ &= \frac{1}{1+\langle n\rangle} \sum \left(\frac{\langle n\rangle}{1+\langle n\rangle}\right)^n \cdot \mathrm{e}^{-(X^2+X'^2)/2} \frac{H_n(X)H_n(X')}{2^n n! \sqrt{\pi}}.\end{aligned} \tag{1.73}$$

The sum in (1.73) can be evaluated by using the Mehler formula for Hermite polynomials

$$\begin{aligned}\sum H_n(X)H_n(X')\frac{S^n}{2^n n!} &= (1-S^2)^{-1/2} \exp\left[\frac{2XX'S - (X^2+X'^2)S^2}{1-S^2}\right], \\ |S| &< 1.\end{aligned} \tag{1.74}$$

Using (1.74) and (1.73) we get the final result

$$\begin{aligned}\langle X|\rho_T|X'\rangle &= \frac{1}{\sqrt{\pi(1+2\langle n\rangle)}} \\ &\times \exp\left\{-\frac{[2\langle n\rangle(\langle n\rangle+1)+1](X^2+X'^2) - 4XX'\langle n\rangle(1+\langle n\rangle)}{2(1+2\langle n\rangle)}\right\}.\end{aligned} \tag{1.75}$$

The quadrature distribution is Gaussian

$$\langle X|\rho_T|X\rangle = \frac{1}{\sqrt{\pi(1+2\langle n\rangle)}} \exp\left(-\frac{X^2}{1+2\langle n\rangle}\right), \tag{1.76}$$

with $\langle X \rangle = 0$, $\langle X^2 \rangle = \langle n \rangle + \frac{1}{2}$. The uncertainty product for these states is

$$\Delta X \Delta Y = \langle n \rangle + \frac{1}{2}.$$

1.6 Diagonal coherent state representation for electromagnetic fields – P-representation

One of the most remarkable developments in quantum optics has been the possibility of expressing any state of the radiation field as a diagonal sum over coherent states even though such states are nonorthogonal

$$\rho = \int P(\alpha)|\alpha\rangle\langle\alpha|\,\mathrm{d}^2\alpha, \tag{1.77}$$

with

$$\int P(\alpha)\,\mathrm{d}^2\alpha = 1. \tag{1.78}$$

The representation (1.77) was introduced first in the context of thermal fields as in Section 1.5 by Glauber [13] who named it the P-representation. Sudarshan [20] realized the very general nature of the representation (1.77) and argued that it is valid for all radiation fields with the proviso that $P(\alpha)$ need not have the properties of a probability distribution. Sudarshan allowed $P(\alpha)$ to be singular and argued that it may not even exist. The mathematical complexities of $P(\alpha)$ are discussed extensively in the book by Klauder and Sudarshan [21].

A formal solution for $P(\alpha)$ can be written down if we use the properties of the delta function, such as $x^n\delta(x) = 0$ and

$$\int \mathrm{d}x \left(\frac{\mathrm{d}^n}{\mathrm{d}x^n}\delta(x)\right) x^m = n!(-1)^n\delta_{mn}. \tag{1.79}$$

We need the two-dimensional version of (1.79)

$$\int \mathrm{d}^2\alpha \frac{1}{n!m!}(-1)^{n+m}\left(\frac{\partial^{n+m}}{\partial\alpha^n\partial\alpha^{*m}}\delta^{(2)}(\alpha)\right)\alpha^p\alpha^{*q} = \delta_{np}\delta_{qm}. \tag{1.80}$$

Note that $\delta^{(2)}(\alpha) = \delta(x)\delta(y)$ since $\alpha = x + \mathrm{i}y$. Using (1.80) and the expansion of the coherent states in terms of Fock states we obtain

$$P(\alpha) = \sum_{nm} \rho_{nm}(-1)^{n+m}\mathrm{e}^{|\alpha|^2}\frac{1}{\sqrt{n!m!}}\frac{\partial^{n+m}}{\partial\alpha^n\partial\alpha^{*m}}\delta^{(2)}(\alpha). \tag{1.81}$$

This formal form shows the singular nature of the P-function in general. The structure (1.81) does not necessarily imply that $P(\alpha)$ is always singular. For a Fock state it is

$$\rho = |n\rangle\langle n| \rightarrow P_n(\alpha) = \mathrm{e}^{|\alpha|^2}\frac{1}{n!}\frac{\partial^{2n}}{\partial\alpha^n\partial\alpha^{*n}}\delta^{(2)}(\alpha). \tag{1.82}$$

For a coherent state it has the expected form

$$\rho = |\alpha_0\rangle\langle\alpha_0| \rightarrow P_c(\alpha) = \delta^{(2)}(\alpha - \alpha_0). \tag{1.83}$$

For a thermal state (Section 1.5)

$$\begin{aligned} \rho = \rho_T &= \exp(-\beta\hbar\omega a^\dagger a)/\mathrm{Tr}[\exp(-\beta\hbar\omega a^\dagger a)] \\ &\rightarrow P_T(\alpha) = \frac{1}{\pi\langle n\rangle}\exp\left(-\frac{|\alpha|^2}{\langle n\rangle}\right), \end{aligned} \tag{1.84}$$

which is like a classical probability distribution. Many methods can be used to derive (1.84). Using (1.70) in (1.81) we find

$$\begin{aligned} P_{\mathrm{T}}(\alpha) &= \frac{1}{1+\langle n\rangle}\mathrm{e}^{|\alpha|^2}\exp\left(\frac{\langle n\rangle}{1+\langle n\rangle}\frac{\partial^2}{\partial\alpha\partial\alpha^*}\right)\delta^{(2)}(\alpha) \\ &= \frac{1}{\pi^2(1+\langle n\rangle)}\mathrm{e}^{|\alpha|^2}\exp\left(\frac{\langle n\rangle}{1+\langle n\rangle}\frac{\partial^2}{\partial\alpha\partial\alpha^*}\right)\int \mathrm{e}^{\beta^*\alpha-\beta\alpha^*}\,\mathrm{d}^2\beta \\ &= \frac{1}{\pi^2(1+\langle n\rangle)}\mathrm{e}^{|\alpha|^2}\int \mathrm{e}^{-\frac{\langle n\rangle}{1+\langle n\rangle}|\beta|^2}\mathrm{e}^{\beta^*\alpha-\beta\alpha^*}\,\mathrm{d}^2\beta. \end{aligned} \tag{1.85}$$

The expression (1.85) can be simplified by using the integral

$$\frac{1}{\pi}\int \mathrm{e}^{-\gamma|\beta|^2}\mathrm{e}^{\beta^*\alpha-\beta\alpha^*}\,\mathrm{d}^2\beta = \frac{1}{\gamma}\mathrm{e}^{-|\alpha|^2/\gamma}. \tag{1.86}$$

On combining (1.85) and (1.86) we obtain (1.84). It is interesting to note that a singular-looking expression like (1.81) for a thermal field leads to a nice Gaussian function (1.84).

A useful method to compute $P(\alpha)$ was given by Mehta [22]

$$P(\alpha) = \frac{1}{\pi^2}\mathrm{e}^{|\alpha|^2}\int \langle-\beta|\rho|\beta\rangle \mathrm{e}^{|\beta|^2-(\beta\alpha^*-\beta^*\alpha)}\,\mathrm{d}^2\beta. \tag{1.87}$$

To prove (1.87) we use (1.77) and (1.53) in (1.87)

$$\begin{aligned} P(\alpha) &= \int \mathrm{d}^2\gamma\, P(\gamma)\frac{1}{\pi^2}\mathrm{e}^{|\alpha|^2}\int \langle-\beta|\gamma\rangle\langle\gamma|\beta\rangle \mathrm{e}^{|\beta|^2-(\beta\alpha^*-\beta^*\alpha)}\,\mathrm{d}^2\beta \\ &= \mathrm{e}^{|\alpha|^2}\int \mathrm{d}^2\gamma P(\gamma)\mathrm{e}^{-|\gamma|^2}\frac{1}{\pi^2}\int \mathrm{e}^{\beta^*(\alpha-\gamma)-\beta(\alpha-\gamma)^*}\mathrm{d}^2\beta \\ &= \mathrm{e}^{|\alpha|^2}\int \mathrm{d}^2\gamma P(\gamma)\mathrm{e}^{-|\gamma|^2}\delta^{(2)}(\alpha-\gamma) = P(\alpha). \end{aligned} \tag{1.88}$$

It should be noted that the representation (1.77) also holds for any operator G and not just the density operator. Consider G written as an antinormally ordered monomial of the form

$$G = a^n a^{\dagger m}. \tag{1.89}$$

We can easily write it in the form (1.77) by inserting the resolution (1.52) of the identity in the middle of a and $a^\dagger$

$$G = a^n\frac{1}{\pi}\int \mathrm{d}^2\alpha|\alpha\rangle\langle\alpha|a^{\dagger m}, \tag{1.90}$$

which on using the fact that $|\alpha\rangle$ ($\langle\alpha|$) is an eigenstate of a ($a^\dagger$) reduces to

$$G = \frac{1}{\pi}\int \mathrm{d}^2\alpha(\alpha^n\alpha^{*m})|\alpha\rangle\langle\alpha|. \tag{1.91}$$

Thus the P-representation can be easily obtained for those operators which are already in the antinormal ordered form – just replace a and $a^\dagger$ by α and α^* and divide the result by π, i.e. if

$$G = \sum G_{nm}a^n a^{\dagger m}, \quad \text{then} \quad P_G(\alpha) = \frac{1}{\pi}\sum G_{nm}\alpha^n\alpha^{*m}. \tag{1.92}$$

The expectation values of the normally ordered moments of the field can be expressed as

$$\langle a^{\dagger m} a^n \rangle = \int P(\alpha)\alpha^{*m}\alpha^n \mathrm{d}^2\alpha. \tag{1.93}$$

In Chapter 8 we will see that the measurements of the characteristics of the field by a photon detector, which works by the absorption of photons [23, 24], are related to the normally ordered moments $\langle a^{\dagger m} a^n \rangle$. Thus the measurable quantities can be directly obtained in an almost classical manner, i.e. by averaging over the distribution $P(\alpha)$. However, it must be borne in mind that unlike the classical case the function $P(\alpha)$ can be singular and need not exist. This is the content of Sudarshan's optical equivalence theorem [20]. In view of (1.93) $P(\alpha)$ is like a phase-space distribution for a quantum system. In the literature, functions like $P(\alpha)$ are also called quasiprobabilities.

A closely related phase-space distribution $Q(\alpha, \alpha^*)$ is obtained [25–27] by taking the diagonal matrix elements of ρ

$$Q(\alpha) = \frac{1}{\pi}\langle \alpha|\rho|\alpha \rangle. \tag{1.94}$$

It is easily seen that, unlike the P-function, the Q-function has all the properties of a classical probability distribution. These properties make the Q-function quite attractive for calculations and for the visualization of physical systems, particularly in situations where $P(\alpha)$ does not exist. To give a simple example, the vacuum state of the radiation field, $Q_{\mathrm{v}}(\alpha) = \frac{1}{\pi}\exp(-|\alpha|^2)$, which is a distribution with finite width. Thus it reflects in a sense the zero point fluctuations. We will see the utility of $Q(\alpha)$ in many applications in subsequent chapters. However, unlike the P-function, the Q-function is not directly related to the measurements in the sense of (1.93). The functions $P(\alpha)$ and $Q(\alpha)$ can be related to each other by using (1.77) and (1.94). On taking the diagonal matrix element of (1.77) and using the expression for $\langle \alpha|\beta \rangle$ we obtain

$$Q(\alpha) = \frac{1}{\pi}\int P(\beta)\exp(-|\alpha-\beta|^2)\mathrm{d}^2\beta. \tag{1.95}$$

The relation (1.95) can be inverted in terms of the Widder transform [28]

$$P(\alpha) = \exp\left(-\frac{\partial^2}{\partial\alpha\partial\alpha^*}\right) Q(\alpha, \alpha^*). \tag{1.96}$$

For the thermal state (1.69), the Q-function is

$$\begin{aligned} Q_{\mathrm{T}}(\alpha) &= \frac{1}{\pi}\sum_{n=0}^{\infty}\frac{\langle n\rangle^n}{(1+\langle n\rangle)^{n+1}}\langle \alpha|n\rangle\langle n|\alpha\rangle \\ &= \frac{1}{\pi}\sum_{n=0}^{\infty}\frac{\langle n\rangle^n}{(1+\langle n\rangle)^{n+1}}\frac{|\alpha|^{2n}}{n!}\mathrm{e}^{-|\alpha|^2} \\ &= \frac{1}{\pi(1+\langle n\rangle)}\exp\left[-\frac{|\alpha|^2}{1+\langle n\rangle}\right], \end{aligned} \tag{1.97}$$

which is Gaussian. Note the presence of the factor 1 in the width of the Gaussian. This is again a signature of the vacuum noise.

1.7 The Wigner function for the electromagnetic field

Wigner [29] in 1932 showed the interesting possibility of a phase-space distribution for a quantum system. He constructed a plausible distribution which yielded the correct marginal distributions for a quantum system and can be used to compute quantum-mechanical expectation values by using a prescription. However, Wigner also found that while constructing the possible phase-space distribution he had to give up the positivity property of the distribution. Thus the Wigner distribution $W(\alpha)$ could be negative. Nevertheless this function has proved to be very useful in a variety of fields as it always exists. The Wigner function has turned out to be remarkably useful in quantum optics, particularly in the characterization and visualization of nonclassical fields. We next introduce the Wigner function using the coherent state framework. A comprehensive discussion of the phase space in quantum optics is given by Schleich [30]. It is well known in probability theory that a distribution is the Fourier transform of the characteristic function. Using this we define the Wigner function

$$W(\alpha) = \frac{1}{\pi^2} \int \mathrm{d}^2\beta \, \mathrm{Tr}[\rho D(\beta)] \mathrm{e}^{-(\beta\alpha^* - \beta^*\alpha)}, \tag{1.98}$$

as a two-dimensional Fourier transform of the quantum-mechanical characteristic function

$$\mathrm{Tr}[\rho \exp(\beta a^\dagger - \beta^* a)] = \langle \exp(\beta a^\dagger - \beta^* a) \rangle. \tag{1.99}$$

Clearly

$$\int W(\alpha) \mathrm{d}^2\alpha = \int \mathrm{d}^2\beta \, \mathrm{Tr}[\rho D(\beta)] \delta^{(2)}(\beta) = \mathrm{Tr}\rho = 1. \tag{1.100}$$

Let us first evaluate the Wigner function for some standard states. For a coherent state $|\gamma\rangle$, $\mathrm{Tr}[\rho D(\beta)] = \langle\gamma|D(\beta)|\gamma\rangle = \mathrm{e}^{-\frac{1}{2}|\beta|^2} \mathrm{e}^{\beta\gamma^* - \beta^*\gamma}$, which on using in (1.98) and the integral (1.86) gives

$$W_{\mathrm{c}}(\alpha) = \frac{2}{\pi} \exp(-2|\alpha - \gamma|^2). \tag{1.101}$$

Thus the coherent state is represented by a Gaussian with a width $\frac{1}{2}$, which in fact is just the variance

$$\langle\gamma|\frac{a^\dagger a + a a^\dagger}{2}|\gamma\rangle - \langle\gamma|a|\gamma\rangle\langle\gamma|a^\dagger|\gamma\rangle \equiv \langle\gamma|a^\dagger a|\gamma\rangle - |\langle\gamma|a|\gamma\rangle|^2 + \frac{1}{2}. \tag{1.102}$$

Clearly the Wigner function carries implicitly part of the vacuum fluctuation. We can in fact use (1.101) and (1.77) to obtain a relation between the P-function and the Wigner function

$$W(\alpha) = \frac{2}{\pi} \int P(\gamma) \exp(-2|\alpha - \gamma|^2) \mathrm{d}^2\gamma, \tag{1.103}$$

which can also be inverted in terms of the Widder transform

$$P(\alpha) = \exp\left[-\frac{1}{2} \frac{\partial^2}{\partial\alpha\partial\alpha^*}\right] W(\alpha). \tag{1.104}$$

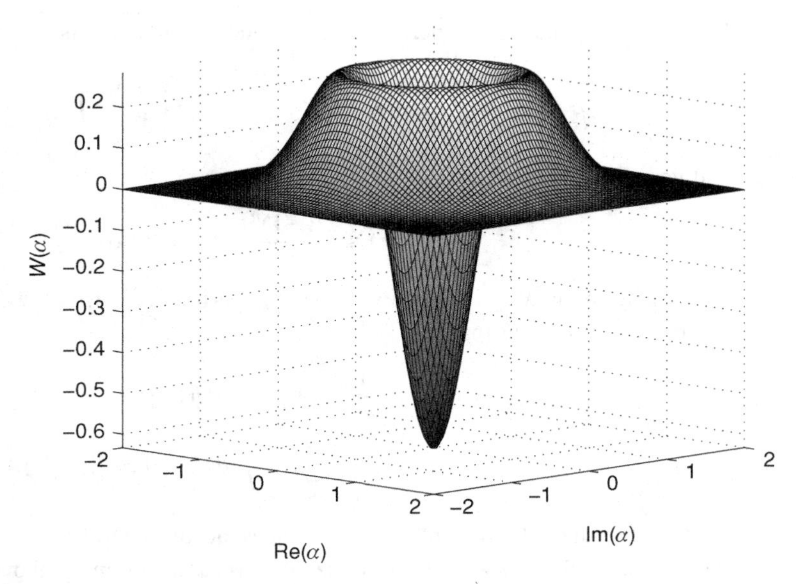

Fig. 1.3 The Wigner function $W(\alpha)$ as a function of Re(α) and Im(α).

For a thermal state we can use (1.84) and the integral (1.86) in (1.103) to obtain

$$W_{\mathrm{T}}(\alpha) = \frac{1}{\pi\left(\langle n\rangle + \frac{1}{2}\right)} \exp\left[-\frac{|\alpha|^2}{\langle n\rangle + \frac{1}{2}}\right]. \tag{1.105}$$

The evaluation of the Wigner function for a Fock state $|n\rangle$ is more complicated and is found to be

$$W_n(\alpha) = \frac{2}{\pi}(-1)^n \mathrm{e}^{-2|\alpha|^2} \mathrm{L}_n(4|\alpha|^2). \tag{1.106}$$

Clearly the Wigner function can be negative. At $\alpha = 0$, $W(0) = \frac{2}{\pi}(-1)^n < 0$ if n is odd. For a single photon state $n = 1$

$$W(\alpha) = \frac{2}{\pi}\mathrm{e}^{-2|\alpha|^2}(4|\alpha|^2 - 1), \tag{1.107}$$

and thus the Wigner function takes negative values in the region $|\alpha|^2 < 1/4$. The function (1.107) is plotted in Figure 1.3 to exhibit the negativity of the Wigner function.

To derive (1.106) it should be noted that the density matrix for the thermal field has the structure (1.70) and thus the Wigner function (1.105) can be considered as the generating function for the Wigner function $W_n(\alpha)$ of the Fock state, i.e.

$$W_{\mathrm{T}}(\alpha) = \sum S^n W_n(\alpha)(1 - S), \quad S = \frac{\langle n\rangle}{1 + \langle n\rangle}. \tag{1.108}$$

On substituting (1.105) in (1.108) we get

$$\sum S^n W_n(\alpha) = \frac{2}{\pi(1+S)} \exp\left(\frac{4|\alpha|^2 S}{1+S}\right) \exp(-2|\alpha|^2). \tag{1.109}$$

We recall the generating function for the Laguerre polynomials

$$\sum_{n=0}^{\infty} \mathrm{L}_n(x)t^n = \frac{1}{1-t}\exp\left(\frac{-xt}{1-t}\right), \tag{1.110}$$

and hence (1.109) leads to

$$W_n(\alpha) \equiv \frac{2(-1)^n}{\pi}\mathrm{e}^{-2|\alpha|^2}\mathrm{L}_n(4|\alpha|^2). \tag{1.111}$$

The value of the Wigner function at the origin is important in many tests of nonclassicality. From (1.98) it is clear that

$$\begin{aligned} W(0) &\equiv \frac{1}{\pi^2}\int \mathrm{d}^2\beta\, \mathrm{Tr}[\rho D(\beta)], \\ &= \frac{1}{\pi^2}\sum_n \rho_{nn}\int \mathrm{d}^2\beta\, \mathrm{Tr}[|n\rangle\langle n|D(\beta)]. \end{aligned} \tag{1.112}$$

All the off-diagonal terms in (1.112) vanish as, according to Eqs. (9) and (10) in Table 1.1, the off-diagonal terms would have nonzero phase and their integral over the phase vanishes. We can rewrite (1.112) as

$$W(0) = \sum_n \rho_{nn} W_n(0) = \frac{2}{\pi}\sum_n \rho_{nn}(-1)^n, \tag{1.113}$$

which can also be written in terms of the parity operator $\exp(\mathrm{i}\pi a^\dagger a)$

$$W(0) = \frac{2}{\pi}\langle \exp(\mathrm{i}\pi a^\dagger a)\rangle. \tag{1.114}$$

Clearly a density matrix, for which $\rho_{nn} \neq 0$ only for odd n, would have a negative Wigner function at the origin. The relations (1.113) and (1.98) can be used to obtain a bound on the Wigner function as $|W(\alpha)| \leq |W(0)| \leq \frac{2}{\pi}$.

We next present the expectation values as phase-space averages. Let G be the operator corresponding to a physical observable. The displacement operators form a complete set and hence any operator can be expanded as

$$G = \frac{1}{\pi}\int \mathrm{d}^2\gamma\, \mathrm{Tr}[GD^\dagger(\gamma)]D(\gamma). \tag{1.115}$$

Thus the expectation value of $\langle G\rangle$ is

$$\langle G\rangle = \mathrm{Tr}(\rho G) = \frac{1}{\pi}\int \mathrm{d}^2\gamma\, \mathrm{Tr}[GD^\dagger(\gamma)]\mathrm{Tr}[\rho D(\gamma)]. \tag{1.116}$$

On inverting (1.98) we have $\mathrm{Tr}[\rho D(\gamma)] = \int \mathrm{d}^2\alpha W(\alpha)\mathrm{e}^{\gamma\alpha^* - \gamma^*\alpha}$, and hence

$$\begin{aligned} \langle G\rangle &= \int \mathrm{d}^2\alpha W(\alpha)\frac{1}{\pi}\int \mathrm{d}^2\gamma\, \mathrm{Tr}[GD^\dagger(\gamma)]\mathrm{e}^{\gamma\alpha^* - \gamma^*\alpha} \\ &= \pi\int \mathrm{d}^2\alpha W(\alpha)W_G(\alpha), \end{aligned} \tag{1.117}$$

Table 1.2 Quantum phase-space distributions for some typical density operators.

(1) Fock state: $\rho = |m\rangle\langle m|$.

$$P(\alpha) = \frac{1}{m!}\mathrm{e}^{|\alpha|^2}\left(\frac{\partial^2}{\partial\alpha\partial\alpha^*}\right)^m \delta^{(2)}(\alpha). \qquad (1)$$

$$W(\alpha) = \frac{2(-1)^m}{\pi}\mathrm{e}^{-2|\alpha|^2}\mathrm{L}_m(4|\alpha|^2). \qquad (2)$$

$$Q(\alpha) = \frac{1}{\pi m!}\mathrm{e}^{-|\alpha|^2}(|\alpha|^2)^m. \qquad (3)$$

(2) Coherent state: $\rho = |\alpha_o\rangle\langle\alpha_o|$.

$$P(\alpha) = \delta^{(2)}(\alpha - \alpha_o). \qquad (4)$$

$$W(\alpha) = \frac{2}{\pi}\mathrm{e}^{-2|\alpha-\alpha_o|^2}. \qquad (5)$$

$$Q(\alpha) = \frac{1}{\pi}\mathrm{e}^{-|\alpha-\alpha_o|^2}. \qquad (6)$$

(3) Thermal state: $\rho = \mathrm{e}^{-\beta\hbar\omega a^\dagger a}/\mathrm{Tr}(\mathrm{e}^{-\beta\hbar\omega a^\dagger a})$.

$$P(\alpha) = \frac{1}{\pi\langle n\rangle}\exp\left[-\frac{|\alpha|^2}{\langle n\rangle}\right], \ \langle n\rangle = \frac{1}{\mathrm{e}^{\beta\hbar\omega}-1}. \qquad (7)$$

$$W(\alpha) = \frac{1}{\pi(\langle n\rangle + \frac{1}{2})}\exp\left(-\frac{|\alpha|^2}{\langle n\rangle + \frac{1}{2}}\right). \qquad (8)$$

$$Q(\alpha) = \frac{1}{\pi(\langle n\rangle + 1)}\exp\left(-\frac{|\alpha|^2}{\langle n\rangle + 1}\right). \qquad (9)$$

where $W_G(\alpha)$ is the Wigner function associated with the operator G defined by (1.98) with $\rho \to G$. The relation (1.117) can be used to prove that $W(\alpha)$ is a square integrable function by replacing G by ρ

$$\mathrm{Tr}\rho^2 = \pi\int \mathrm{d}^2\alpha W^2(\alpha) \leq 1. \qquad (1.118)$$

The last inequality follows from the property of the density matrix, $\mathrm{Tr}\rho^2 \leq \mathrm{Tr}\rho = 1$.

The important properties of the Wigner function can be summarized as follows:

1. It always exists but is not necessarily positive.
2. It is square integrable (Eq. (1.118)).
3. It is bounded $|W(\alpha)| \leq |W(0)| \leq 2/\pi$.
4. For a single photon state $W(0) = -2/\pi$.

So far we have discussed three different types of phase-space distributions for quantum systems. We summarize these for later use in Table 1.2. There are a whole host of phase-space distributions that have been extensively discussed in the literature [31,32]. However, $P(\alpha)$, $Q(\alpha)$, and $W(\alpha)$ are the only ones that we use in this book. In all the coherent

Table 1.3 Some important properties of Gaussians.

(1) n-variable Gaussian distribution, $A = n \times n$ matrix:

$$P(x_1 \cdots x_n) = \frac{(\det A)^{1/2}}{(2\pi)^{n/2}} \exp\left[-\frac{1}{2}\sum_{ij}(x_i - \bar{x}_i)A_{ij}(x_j - \bar{x}_j)\right].$$

$\bar{x}_i$ is the mean value of the random variable x_i.

(2) $(A^{-1})_{ij}$ is called the covariance matrix defined by

$$\int P(x_1 \cdots x_n)(x_i - \bar{x}_i)(x_j - \bar{x}_j)\mathrm{d}x_1 \cdots \mathrm{d}x_n = (A^{-1})_{ij}.$$

(3) The Fourier transform of P:

$$\int P(x_1 \cdots x_n) \exp\left(\mathrm{i}\sum k_i x_i\right) \mathrm{d}x_1 \cdots \mathrm{d}x_n$$

$$= \exp\left[\mathrm{i}\sum k_i \bar{x}_i - \frac{1}{2}\sum k_i (A^{-1})_{ij} k_j\right].$$

(4) A two-dimensional Fourier transform in complex coordinates:

$$\frac{1}{\pi}\int \exp(-\mu\alpha^2 - \nu\alpha^{*2} - z^*\alpha + z\alpha^* - |\alpha|^2)\mathrm{d}^2\alpha$$

$$= \frac{1}{\tau}\exp[-(\mu z^2 + \nu z^{*2} + zz^*)/\tau^2], \quad \tau = \sqrt{1 - 4\mu\nu}, \quad \mu + \nu < 1.$$

(5) $\frac{1}{\pi^2}\int \mathrm{d}^2\alpha \exp(-z^*\alpha + z\alpha^*) = \delta^2(z).$

state representations we encounter Gaussian distributions all the time. Thus for the sake of convenience we list some important properties of Gaussians in Table 1.3.

1.8 Bosonic systems with finite mass – coherent states and phase-space representations

So far we have discussed coherent states and various representations in the context of the electromagnetic fields. Our discussion applies to all types of harmonic oscillators satisfying Bosonic algebra. Consider a harmonic oscillator with mass m and frequency ω

$$H = \frac{p^2}{2m} + \frac{1}{2}m\omega^2 x^2, \quad [x, p] = \mathrm{i}\hbar, \tag{1.119}$$

where x and p are the coordinate and momentum operators. The Hamiltonian can be written as

$$\begin{gathered} H = \hbar\omega\left(a^\dagger a + \frac{1}{2}\right), \quad a = \frac{X + \mathrm{i}Y}{\sqrt{2}}, \quad a^\dagger = \frac{X - \mathrm{i}Y}{\sqrt{2}}, \\ [a, a^\dagger] = 1, \quad [X, Y] = \mathrm{i}, \quad X = \sqrt{\frac{m\omega}{\hbar}}x, \quad Y = \frac{p}{\sqrt{m\hbar\omega}}. \end{gathered} \tag{1.120}$$

In view of (1.120), all that we have developed in Sections 1.2–1.4, 1.6, and 1.7, applies to the massive harmonic oscillator. This will also be the case for other states for Bosonic operators that we develop in subsequent chapters. Various quantum states for the massive oscillator would be useful, say, in connection with the quantum nature of a nano-mechanical oscillator.

Since the Wigner function plays an important role in the description of physical systems, we also present how the Wigner function can be directly calculated by using the wave-functions in coordinate space for a quantum system. We can define the Wigner function in analogy to (1.98) by writing the characteristic function using position and momentum operators x and p

$$W(u,v)=\frac{1}{(2\pi)^2}\int_{-\infty}^{+\infty}\mathrm{d}\xi\int_{-\infty}^{+\infty}\mathrm{d}\eta\,\mathrm{Tr}[\rho\exp(\mathrm{i}x\xi+\mathrm{i}p\eta)]\exp(-\mathrm{i}u\xi-\mathrm{i}v\eta). \quad (1.121)$$

Here x and p are operators obeying the commutation relation $[x,p]=\mathrm{i}\hbar$ and u,v are numbers (phase-space variables). We require

$$\iint W(u,v)\mathrm{d}u\mathrm{d}v\equiv\mathrm{Tr}\rho. \quad (1.122)$$

Writing ρ in coordinate-space representation as

$$\rho=\iint\mathrm{d}x\mathrm{d}x'|x\rangle\langle x'|\rho(x,x'),\qquad \rho(x,x')=\langle x|\rho|x'\rangle, \quad (1.123)$$

and using the Baker–Campbell–Hausdorff identity

$$\exp(\mathrm{i}x\xi+\mathrm{i}p\eta)=\exp(\mathrm{i}x\xi)\exp(\mathrm{i}p\eta)\exp\left(\frac{1}{2}\mathrm{i}\hbar\xi\eta\right), \quad (1.124)$$

we get

$$W(u,v)=\frac{1}{(2\pi)^2}\int\mathrm{d}\xi\int\mathrm{d}\eta\int\mathrm{d}x\int\mathrm{d}x'\rho(x,x')\langle x'|\mathrm{e}^{\mathrm{i}x\xi}\mathrm{e}^{\mathrm{i}p\eta}|x\rangle\mathrm{e}^{\frac{1}{2}\mathrm{i}\hbar\xi\eta}\mathrm{e}^{-\mathrm{i}u\xi-\mathrm{i}v\eta}. \quad (1.125)$$

Noting that

$$\langle x'|\mathrm{e}^{\mathrm{i}x\xi}\mathrm{e}^{\mathrm{i}p\eta}|x\rangle=\mathrm{e}^{\mathrm{i}x'\xi}\delta(x'-x+\hbar\eta), \quad (1.126)$$

the ξ integral gives $2\pi\delta(x'+\frac{1}{2}\hbar\eta-u)$, then (1.125) can be simplified to

$$W(u,v)=\frac{1}{2\pi}\int\mathrm{d}\eta\,\rho\left(u+\frac{1}{2}\hbar\eta;u-\frac{1}{2}\hbar\eta\right)\mathrm{e}^{-\mathrm{i}v\eta}. \quad (1.127)$$

This is the celebrated form of the Wigner function [29]. If W_G is the Wigner function for any operator G obtained from (1.127) by replacing ρ by G, then

$$\int\mathrm{d}u\int\mathrm{d}vW(u,v)W_G(u,v)=\frac{1}{2\pi\hbar}\mathrm{Tr}(\rho G). \quad (1.128)$$

This is the standard formula for the expectation values in terms of the Wigner functions. For the nth excited state of the harmonic oscillator

$$\Psi_n(u)=\frac{\exp(-\frac{1}{2}\frac{m\omega}{\hbar}u^2)H_n(\sqrt{\frac{m\omega}{\hbar}}u)}{2^{n/2}\sqrt{n!\sqrt{\pi}}}, \quad (1.129)$$

the Wigner function is

$$W(u,v) = \frac{(-1)^n}{2\pi\hbar}\exp\left[-\frac{2}{\hbar\omega}H(u,v)\right]\mathrm{L}_n\left[\frac{4}{\hbar\omega}H(u,v)\right],$$

$$H(u,v) = \frac{v^2}{2m} + \frac{1}{2}m\omega^2 u^2, \tag{1.130}$$

which exhibits complete dependence on mass, frequency, and the Planck constant. For the first excited state the nonclassical region is given by

$$\frac{4}{\hbar\omega}H(u,v) < 1. \tag{1.131}$$

In the literature there has been considerable amount of work [30,33] on the limit of (1.130) as $\hbar \to 0$. Clearly we should also take the limit $n \to \infty$ such that $n\hbar \to$ constant.

The quantum states that we develop in this book would also be useful to describe a system of cold Bosons in connection with atom optics. Consider a system of N Bosons in a confining potential, which we take to be harmonic and interacting via two-body interaction, then the Hamiltonian can be written as

$$H = \hbar\omega\sum_i a_i^\dagger a_i + \sum v_{ij} a_i^\dagger a_j^\dagger a_i a_j, \tag{1.132}$$

where ω is the trapping frequency. The term v_{ii} corresponds to self-interaction. The operators a_i satisfy Bosonic algebra

$$[a_i, a_j^\dagger] = \delta_{ij}, \quad [a_i, a_j] = [a_i^\dagger, a_j^\dagger] = 0. \tag{1.133}$$

In view of the algebra (1.133), we are allowed to use the states and representations developed in previous sections and in subsequent chapters.

Exercises

1.1 Follow the same procedure that we used in calculating the eigenstate of a, to show that the eigenvalue equation

$$a^\dagger|\Psi\rangle = \beta|\Psi\rangle$$

has no normalizable solutions.

1.2 Using (1.46) prove (1.51).

1.3 Consider a classical field with complex amplitude E_0 characterized by the probability distribution $p(E_0) = \exp(-|E_0|^2/\sigma^2)/(\pi\sigma^2)$. Then show that $\langle|E_0|^4\rangle - \langle|E_0|^2\rangle^2 = \sigma^4; \sigma^2 = \langle|E_0|^2\rangle$.

1.4 For a thermal field, using (1.93) and (1.84) show that

$$\langle a^{\dagger n}a^n\rangle = \mathrm{Tr}(\rho_{\mathrm{T}} a^{\dagger n}a^n) = n!\langle a^\dagger a\rangle^n = n!\langle n\rangle^n,$$

whereas for a coherent state $\langle a^{\dagger n}a^n\rangle = \langle n\rangle^n$.

1.5 If the state ρ has the P-function $P(\alpha)$ then find the P-function $\tilde{P}$ associated with the state $\tilde{\rho} = D^\dagger(\beta)\rho D(\beta)$. The result should be

$$\tilde{P}(\alpha) = P(\alpha + \beta).$$

1.6 Repeat Exercise 1.5 for the Wigner function, i.e. show that

$$\tilde{W}(\alpha) \equiv W(\alpha + \beta), \quad \text{if} \quad \tilde{\rho} = D^\dagger(\beta)\rho D(\beta).$$

1.7 Find the effect of displacement on a coherent state $|\alpha\rangle$ shown in the figure

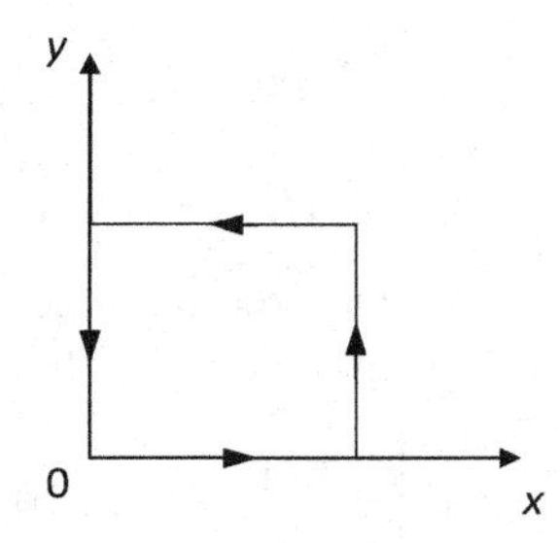

i.e.

$$D(-\mathrm{i}y)D(-x)D(\mathrm{i}y)D(x)|\alpha\rangle,$$

and show that the phase of $|\alpha\rangle$ changes. Does the final result have anything to do with the area of the rectangle? This change in phase is related to the well-known Berry phase [34] and was calculated in the context of coherent states by Chaturvedi *et al.* [35].

1.8 Using the BCH relation (Table 1.1, Eq. (1)) show that

$$\langle\alpha|D(\gamma)|\beta\rangle = \langle\alpha|\beta\rangle \exp\left(\gamma\alpha^* - \gamma^*\beta - \frac{1}{2}|\gamma|^2\right).$$

1.9 The state $D(\alpha)|m\rangle$ is called the displaced Fock state. We can clearly expand $D(\alpha)|m\rangle$ in terms of the Fock states

$$D(\alpha)|m\rangle = \sum \langle n|D(\alpha)|m\rangle|n\rangle.$$

Show that

$$\langle n|D(\alpha)|m\rangle = \mathrm{e}^{-|\alpha|^2/2}\mathrm{L}_m^{|n-m|}(|\alpha|^2) \times \begin{cases} \sqrt{\frac{m!}{n!}}\alpha^{n-m}, & n \geq m, \\ \sqrt{\frac{n!}{m!}}(-\alpha^*)^{m-n}, & n \leq m, \end{cases}$$

where $\mathrm{L}_n^{(k)}$ is the associated Laguerre polynormial defined by

$$\mathrm{L}_n^{(k)}(x) = \sum_{m=0}^{n}(-1)^m \binom{n+k}{m+k}\frac{x^m}{m!}.$$

To find the result use the Baker–Campbell–Hausdorff (BCH) identity to write $D(\alpha)$ as $\mathrm{e}^{-|\alpha|^2/2}\mathrm{e}^{a^\dagger\alpha}\mathrm{e}^{-a\alpha^*}$, then expand the action of the exponential on the Fock state, i.e.

$e^{-a\alpha^*}|m\rangle$ in a series of Fock states. Write the final result in terms of the associated Laguerre polynomials.

1.10 Consider a harmonic oscillator under the influence of a time-dependent force. The Hamiltonian is then

$$H(t) = \hbar\omega a^\dagger a + \hbar f(t)a + \hbar f^*(t)a^\dagger.$$

Let the initial state of the oscillator be a vacuum $|0\rangle$, and find the state at time t. It is useful to work in the interaction picture. The unitary evolution operator in the interaction picture is

$$\begin{aligned} U(t) &= T\exp\left[-\frac{\mathrm{i}}{\hbar}\int_0^t \tilde{H}(t')\mathrm{d}t'\right], \\ \tilde{H}(t') &= \hbar f(t')a\mathrm{e}^{-\mathrm{i}\omega t'} + H.c., \end{aligned}$$

where T is the time-ordering operator. Using the fact that the commutator $[\tilde{H}(t_1), \tilde{H}(t_2)]$ is a c number, show that

$$U(t) = \exp\left\{-\frac{\mathrm{i}}{\hbar}\int_0^t \tilde{H}(t')\mathrm{d}(t') - \frac{1}{2\hbar^2}\int_0^t \mathrm{d}t'\int_0^{t'} \mathrm{d}t''[\tilde{H}(t'), \tilde{H}(t'')]\right\}.$$

Use this evolution to obtain the explicit form of the generated coherent state, i.e. find $U(t)|0\rangle$. (Details can be found in Glauber [36].)

1.11 In classical electrodynamics, Green's functions or propagators are important as these determine the fields radiated by, say, dipoles [37]. An especially useful propagator is the dyadic Green's function

$$\overleftrightarrow{G}(\vec{r}, \vec{r}\,', \omega) = \left(\overleftrightarrow{1} + \frac{\vec{\nabla}\vec{\nabla}}{k^2}\right)\frac{\mathrm{e}^{\mathrm{i}k|\vec{r}-\vec{r}\,'|}}{|\vec{r}-\vec{r}\,'|}, \qquad k = \frac{\omega}{c}.$$

It turns out that the commutators of the quantized fields (1.21)–(1.23) are related to the propagators. Using, for example, (1.22), prove that

$$G_{ij}(\vec{r}, \vec{r}\,', \omega) = \frac{\mathrm{i}c^2}{\hbar\omega^2}\int_0^\infty \mathrm{d}t\,\mathrm{e}^{\mathrm{i}\omega t}\langle[E_i(\vec{r}, t), E_j(\vec{r}\,', 0)]\rangle.$$

The evaluation of the commutator can be found in [38].

References

[1] M. Planck, *Verh. Dtsch. Phys. Ges. Berlin* **2**, 202 (1990); *ibid.* **2**, 237 (1990).
[2] A. Einstein, *Ann. Physik* **17**, 132 (1905).
[3] G. N. Lewis, *Nature (London)* **118**, 874 (1926).
[4] A. Einstein, *Z. Phys.* **18**, 121 (1917).
[5] P. A. M. Dirac, *Proc. Roy. Soc. London A* **114**, 243 (1927).
[6] V. Weisskopf and E. Wigner, *Z. Phys.* **63**, 54 (1930); *ibid.* **65**, 18 (1930).
[7] S. N. Bose, *Z. Phys.* **26**, 178 (1924).

[8] A. Einstein, *Ber. d. Berl. Akad.* p; 3 (1925).
[9] R. Fürth, *Z. Phys.* **48**, 323 (1928); *ibid.* **50**, 310 (1928).
[10] J. W. Noh, A. Fougères, and L. Mandel, *Phys. Rev. Lett.* **67**, 1426 (1991).
[11] J. D. Jackson, *Classical Electrodynamics*, 3rd edn. (New York: Wiley, 1999).
[12] R. Loudon, *The Quantum Theory of Light*, 3rd edn. (New York: Oxford University Press, 2000).
[13] R. J. Glauber, *Phys. Rev. Lett.* **10**, 84 (1963).
[14] R. J. Glauber, *Phys. Rev.* **131**, 2766 (1963).
[15] R. M. Wilcox, *J. Math. Phys.* **8**, 962 (1967).
[16] L. I. Schiff, *Quantum Mechanics*, 3rd edn. (New York: McGraw-Hill, 1968).
[17] E. Schrödinger, *Naturwissenschaften* **14**, 664 (1926).
[18] C. C. Gerry and P. L. Knight, *Introductory Quantum Optics* (Cambridge: Cambridge Univerity Press, 2005).
[19] M. Abramowitz and I. A. Stegun, *NIST Handbook of Mathematical Functions* (Cambridge: Cambridge Univerity Press, 2010).
[20] E. C. G. Sudarshan, *Phys. Rev. Lett.* **10**, 277 (1963).
[21] J. R. Klauder and E. C. G. Sudarshan, *Fundamentals of Quantum Optics* (New York: Benjamin, 1968).
[22] C. L. Mehta, *Phys. Rev. Lett.* **18**, 752 (1967).
[23] R. J. Glauber, *Phys. Rev.* **130**, 2529 (1963).
[24] P. L. Kelley and W. H. Kleiner, *Phys. Rev.* **136**, A316 (1964).
[25] C. L. Mehta and E. C. G. Sudarshan, *Phys. Rev.* **138**, B274 (1965).
[26] Y. Kano, *J. Phys. Soc. Japan* **19**, 1555 (1964); *J. Math. Phys.* **6**, 1913 (1965).
[27] K. Hushimi, *Proc. Phys. Math. Soc. Japan* **22**, 264 (1940).
[28] I. I. Hirschman and D. V. Widder, in *The Convolution Transform* (Princeton, NJ: Princeton University Press, 1955), Chap. VIII.
[29] E. Wigner, *Phys. Rev.* **40**, 749 (1932).
[30] W. P. Schleich, *Quantum Optics in Phase Space* (Berlin: Wiley-VCH, 2001).
[31] G. S. Agarwal and E. Wolf, *Phys. Rev. D* **2**, 2161, 2187 (1970).
[32] K. E. Cahill and R. J. Glauber, *Phys. Rev.* **177**, 1857, 1882 (1969).
[33] M. V. Berry, *Phil. Trans. Roy. Soc. London A* **287**, 237 (1977).
[34] M. V. Berry, *Proc. Roy. Soc. London A* **392**, 45 (1984).
[35] S. Chaturvedi, M. S. Sriram, and V. Srinivasan. *J. Phys. A* **20**, L1071 (1987).
[36] R. J. Glauber, in *Quantum Optics and Electronics*, edited by C. deWitt, A. Blanden, and C. Cohen-Tannoudji (New York: Gordon and Breach, 1965), p. 65.
[37] C.-T Tai, *Dyadic Green's Functions in Electromagnetic Theory* (Intext Educational Publishers, 1971).
[38] W. Heitler, *The Quantum Theory of Radiation* (New York: Oxford University Press, 1954), pp. 76–8.

2 Nonclassicality of radiation fields

In the previous chapter we discussed two important classes of the states of the radiation field, viz coherent states and Fock states. The coherent states were like classical states with well-defined amplitudes and phase. In addition, the relative fluctuation in photon number in a coherent state,

$$(\langle n^2 \rangle - \langle n \rangle^2)/\langle n \rangle^2 = 1/\langle n \rangle, \tag{2.1}$$

becomes smaller and smaller as $\langle n \rangle$ increases. Furthermore, all measurements with photodetectors could not differentiate between the results for a quantum field described by a coherent state and by a classical field. All this followed from the fact that the P-function for the coherent field was like a classical probability distribution. It should be borne in mind that classical probability theory allows for delta function distributions.

The Fock states of the radiation field, which contain a precise number of photons, are quite different from coherent fields. The average field proportional to $\langle a \rangle$ is zero. The P-function for the n-photon state is highly singular. Even the Wigner function has regions of negativity. Thus the n-photon state can be considered as a state with properties far from classical fields.

We thus consider a field to be nonclassical if the underlying P-function does not possess properties of a classical probability distribution – we adopt a characterization of nonclassicality in terms of the P-function as this is the relevant phase-space distribution as far as the measurements are made with photo detectors. Clearly thermal states are not nonclassical as the P-function for the thermal states has all the features of a classical probability distribution. The mean number of photons in the thermal state, however, depends on the Planck constant $\hbar$. Since for nonclassical fields the P-function can not be directly measured, we next formulate the experimentally accessible measures of nonclassicality [1–5].

2.1 The Mandel $Q_{\rm M}$ parameter

A very useful parameter to characterize nonclassicality was introduced by Mandel. He observed that the photon number distribution for a coherent field is Poissonian and therefore any distribution which is narrower than Poissonian must correspond to a nonclassical field. He introduced the parameter [5]

$$Q_{\rm M} = \frac{\langle (a^\dagger a)^2 \rangle - \langle a^\dagger a \rangle^2 - \langle a^\dagger a \rangle}{\langle a^\dagger a \rangle}. \tag{2.2}$$

Clearly $Q_M = 0$ for a coherent field. For a Fock state $Q_M = -1$. The negativity of Q_M is a sufficient condition for the field to be nonclassical. If $Q_M > 0$, no conclusion can be drawn about nonclassicality. Note that for a thermal field $Q_M = \langle a^\dagger a\rangle$ is always positive. We next prove how the negativity of Q_M implies nonclassicality. For these purpose we only need to consider the numerator in (2.2)

$$\begin{aligned} f &= \langle (a^\dagger a)^2\rangle - \langle a^\dagger a\rangle^2 - \langle a^\dagger a\rangle \\ &= \langle a^{\dagger 2} a^2\rangle - \langle a^\dagger a\rangle^2, \end{aligned} \tag{2.3}$$

where we used the commutator $[a, a^\dagger] = 1$. In terms of the P-function

$$\begin{aligned} \langle a^{\dagger 2} a^2\rangle &= \int P(\alpha)\alpha^{*2}\alpha^2 \mathrm{d}^2\alpha = \langle \alpha^{*2}\alpha^2\rangle_P, \\ \langle a^\dagger a\rangle &= \int P(\alpha)\alpha^*\alpha \mathrm{d}^2\alpha = \langle \alpha^*\alpha\rangle_P, \end{aligned} \tag{2.4}$$

where $\langle\ \rangle_P$ denotes the average with respect to $P(\alpha)$. Thus (2.3) can be written as

$$f = \langle (\alpha^*\alpha - \langle \alpha^*\alpha\rangle_P)^2\rangle_P. \tag{2.5}$$

The quantity (2.5) would always be positive if $P(\alpha)$ were a classical probability distribution. Thus if one were to find that $f < 0$, then $P(\alpha)$ must be nonclassical. Therefore we have established that the negativity of Q_M is definitely a measure of the nonclassicality of the field. Note that the parameter Q_M is in terms of the photon number operator and thus does not carry any information on the phase characteristics of the field. Furthermore, Q_M does not exhaust the possible measures of nonclassicality. We also note in passing that the lowest value that Q_M can have is -1. This is because the smallest value for $\langle (a^\dagger a)^2\rangle - \langle a^\dagger a\rangle^2$ is zero.

2.2 Phase-dependent measure of nonclassicality – squeezing parameter S

As with fluctuations in the number of photons, we can also measure fluctuations in the quadratures X and Y of the field. We will show in Chapter 5 how such fluctuations can be measured by homodyne schemes. We consider here a more general quadrature X_θ, which is a linear combination of X and Y (cf. Eq. (1.37))

$$X_\theta = \frac{a e^{-\mathrm{i}\theta} + a^\dagger e^{\mathrm{i}\theta}}{\sqrt{2}}, \qquad [X_\theta, X_{\theta+\pi/2}] = \mathrm{i}, \tag{2.6}$$

which for $\theta = 0\ [\pi/2]$ reduces to $X\ [Y]$. We next introduce the parameter [6, 7]

$$S_\theta = \langle : (X_\theta)^2 : \rangle - \langle X_\theta\rangle^2, \tag{2.7}$$

where : : stands for the normal ordering of operators defined by $: aa^\dagger := a^\dagger a$, whereas $aa^\dagger = a^\dagger a + 1$. Note further that for a coherent state $|\alpha\rangle$

$$S_\theta = 0; \quad \langle X_\theta \rangle = \frac{\alpha e^{-i\theta} + \alpha^* e^{i\theta}}{\sqrt{2}},$$
$$\langle : X_\theta^2 : \rangle = (\alpha^2 e^{-2i\theta} + \alpha^{*2} e^{2i\theta} + 2\alpha^* \alpha)/2. \tag{2.8}$$

For a Fock state and for a thermal state

$$S_{\theta \mathrm{F}} = n, \quad S_{\theta \mathrm{T}} = \langle n \rangle. \tag{2.9}$$

We next prove that negative values of S_θ imply the nonclassicality of the underlying field [1–4]. To prove this we write (2.7) in terms of the averages over the P-function

$$\begin{aligned} S_\theta &= \int P(\alpha) \left(\frac{\alpha e^{-i\theta} + \alpha^* e^{i\theta}}{\sqrt{2}} \right)^2 \mathrm{d}^2\alpha - \left[\int P(\alpha) \left(\frac{\alpha e^{-i\theta} + \alpha^* e^{i\theta}}{\sqrt{2}} \right) \mathrm{d}^2\alpha \right]^2 \\ &= \left\langle \left(\frac{\alpha e^{-i\theta} + \alpha^* e^{i\theta}}{\sqrt{2}} \right)^2 \right\rangle_P - \left\langle \frac{\alpha e^{-i\theta} + \alpha^* e^{i\theta}}{\sqrt{2}} \right\rangle_P^2, \end{aligned} \tag{2.10}$$

which can be negative only if $P(\alpha)$ is nonclassical. Thus we have another sufficient condition for the nonclassicality of the field. Note that S_θ is closely related to the variance of X_θ:

$$S_\theta = (\langle X_\theta^2 \rangle - \langle X_\theta \rangle^2) - \frac{1}{2} = (\Delta X_\theta)^2 - \frac{1}{2}. \tag{2.11}$$

Since $(\Delta X_\theta)^2 \geq 0$, the parameter S_θ has a lower bound

$$\min[S_\theta] = -\frac{1}{2} \quad \text{or} \quad S_\theta \geq -\frac{1}{2}. \tag{2.12}$$

The quadratures X and Y do not commute, hence ΔX and ΔY obey the uncertainty relation

$$\Delta X \Delta Y \geq \frac{1}{2}, \quad \Delta X^2 = S_X + \frac{1}{2}, \quad \Delta Y^2 = S_Y + \frac{1}{2}. \tag{2.13}$$

Clearly if $\Delta X < 1/\sqrt{2}$, then $\Delta Y > 1/\sqrt{2}$ and vice versa, as shown in Figure 2.1. For a coherent state $\Delta X = \Delta Y = 1/\sqrt{2}$. We call the property $\Delta X_\theta < \frac{1}{\sqrt{2}}$ the squeezing property of the field because the width of the quadrature distribution $|\Psi(X)|^2$ is narrower than that for a coherent state (Eq. (1.66)).

We note here that experimentalists give the amount of squeezing in terms of decibels (dB). The dB scale is obtained from $-10 \log_{10}(2\Delta X_\theta^2) \Rightarrow$ dB. Thus 3 dB squeezing corresponds to $\Delta X_\theta^2 = \frac{1}{4}$ or 50% squeezing.

We will discuss the squeezed states of the field in what follows and in later chapters. These states show that $S_\theta < 0$ for certain value of θ. It may happen that the two experimental measures of nonclassicality Q_{M} and S_θ fail to test the nonclassicality of a given field. In such cases one needs alternate measures as we discuss later. It should also be borne in mind that a nonclassical field may have both Q_{M} and S_θ negative, only one negative, or even both positive.

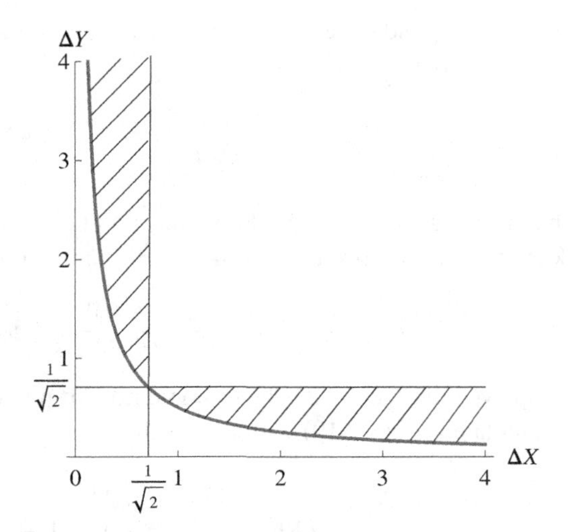

Fig. 2.1 The uncertainty parabola; the shaded region shows the region where squeezing occurs.

2.3 Single-mode squeezed states – squeezed vacuum

In Section 1.4 we saw that the coherent states $|\alpha\rangle$ can be generated from the vacuum state by the application of the displacement operator. Following Yuen [2,6,7], we now define the squeezed state $|\xi\rangle$ via the application of the unitary operator

$$S(\xi) = \exp\left(\frac{1}{2}\xi a^{\dagger 2} - \frac{1}{2}\xi^* a^2\right), \quad \xi = r\mathrm{e}^{\mathrm{i}\varphi}, \tag{2.14}$$

on the vacuum state

$$|\xi\rangle = S(\xi)|0\rangle. \tag{2.15}$$

We can write $|\xi\rangle$ as a superposition of Fock states by decomposing the unitary operator $S(\xi)$ as

$$S(\xi) = \exp\left(\mathrm{e}^{\mathrm{i}\varphi}\tanh r\,\frac{a^{\dagger 2}}{2}\right)\exp\left[-(\ln\cosh r)\left(a^\dagger a + \frac{1}{2}\right)\right] \times \exp\left(-\mathrm{e}^{-\mathrm{i}\varphi}\tanh r\,\frac{a^2}{2}\right). \tag{2.16}$$

The decomposition (2.16) is the BCH formula for the SU(1,1) group. The advantage of the form (2.16) is that the action of the last exponential on the vacuum would yield unity as $a|0\rangle = 0$. The middle exponential on the vacuum would lead to $\exp[-(\ln\cosh r)\frac{1}{2}]|0\rangle$ and thus

$$|\xi\rangle = \frac{1}{\sqrt{\cosh r}}\exp\left(\mathrm{e}^{\mathrm{i}\varphi}\tanh r\,\frac{a^{\dagger 2}}{2}\right)|0\rangle. \tag{2.17}$$

Finally we can expand the exponential in (2.17) in infinite series to write the squeezed state $|\xi\rangle$ in terms of Fock states

$$|\xi\rangle = \frac{1}{\sqrt{\cosh r}} \sum_{n=0}^{\infty} \mathrm{e}^{in\varphi} (\tanh r)^n \frac{\sqrt{(2n)!}}{n! 2^n} |2n\rangle. \tag{2.18}$$

Note that all the states with an odd number of photons are missing from (2.18). The state $|\xi\rangle$ is known as the squeezed vacuum. The photon number distribution is

$$p_{2n} = \frac{(\tanh r)^{2n}}{\cosh r} \frac{(2n)!}{(n!)^2 2^{2n}}, \quad p_{2n+1} = 0, \tag{2.19}$$

and is displayed in Figure 2.2. The mean number of photons and the Mandel Q_{M} parameter can be calculated from (2.19) with the results

$$\begin{aligned} \langle a^\dagger a\rangle &= \sinh^2 r, \\ Q_{\mathrm{M}} &= 2\sinh^2 r + 1 = 1 + 2\langle a^\dagger a\rangle. \end{aligned} \tag{2.20}$$

The number fluctuations are super-Poissonian and in the limit of a large photon number, $Q_{\mathrm{M}} \to 2\langle a^\dagger a\rangle$. Such a result has been verified [8] for the twin beams produced by a downconverter by using two-photon counting in a semiconductor.

The unitary operator (2.14) has some interesting transformation properties that enable us to understand the squeezing character of the state. To derive these we write $S(\xi)$ as

$$S(\xi) = \exp(rh), \quad h = \frac{1}{2}\mathrm{e}^{\mathrm{i}\varphi} a^{\dagger 2} - H.c.. \tag{2.21}$$

Let

$$a^\dagger(\xi) = S^\dagger(\xi) a^\dagger S(\xi), \quad a(\xi) = S^\dagger(\xi) a S(\xi), \tag{2.22}$$

then it is easily seen by taking the derivatives with respect to r

$$\frac{\mathrm{d}}{\mathrm{d}r} a^\dagger = \mathrm{e}^{-\mathrm{i}\varphi} a, \quad \frac{\mathrm{d}}{\mathrm{d}r} a = \mathrm{e}^{\mathrm{i}\varphi} a^\dagger. \tag{2.23}$$

The solution to (2.23) is straightforward

$$a(\xi) = a\cosh r + a^\dagger \mathrm{e}^{\mathrm{i}\varphi} \sinh r. \tag{2.24}$$

The $a^\dagger(\xi)$ can be obtained by taking adjoint of (2.24). The transformation (2.24) is known as the Bogoliubov transformation. Relations like (2.24) are useful in the calculation of expectation values. We illustrate this with an example

$$\begin{aligned} \langle a^\dagger a\rangle &= \langle \xi | a^\dagger a | \xi\rangle = \langle 0 | S^\dagger(\xi) a^\dagger a S(\xi) | 0\rangle \\ &= \langle 0 | S^\dagger(\xi) a^\dagger S(\xi) S^\dagger(\xi) a S(\xi) | 0\rangle \\ &= \langle 0 | a^\dagger(\xi) a(\xi) | 0\rangle. \end{aligned} \tag{2.25}$$

If we use (2.24), then the calculation of expectation values in the squeezed vacuum reduces to the calculation of vacuum expectation values. By substituting relations like (2.24) in (2.25) we immediately obtain (2.20). From the definition (2.22) it is also seen that the

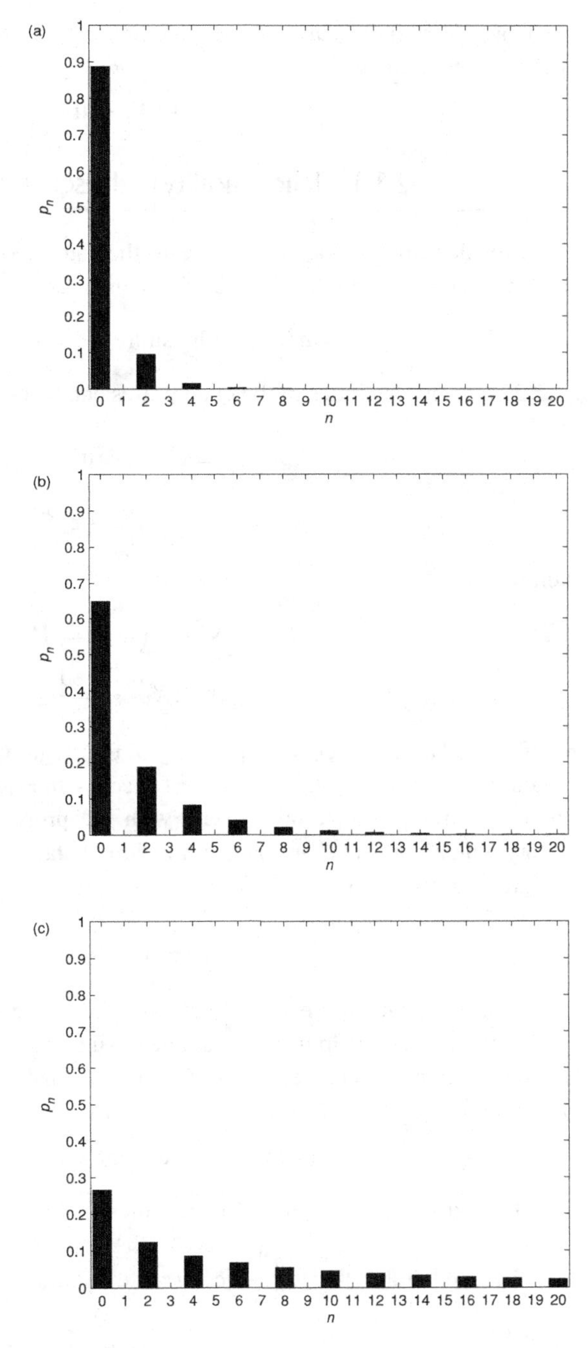

Fig. 2.2 The photon number distribution p_n as a function of n for the squeezed vacuum state. (a) $r = 0.5$; (b) $r = 1$; and (c) $r = 2$.

squeezed vacuum is the vacuum of the Bogoliubov transformed operator. More explicitly, the eigenvalue equation is

$$a(-\xi)|\xi\rangle = 0. \tag{2.26}$$

2.3.1 Nonclassicality of the squeezed vacuum

We next consider what is nonclassical about the squeezed state. For this purpose we first consider the quadratures in the state (2.15). Using (2.24), a simple calculation leads to

$$\langle a^2\rangle = \cosh r \sinh r\, \mathrm{e}^{\mathrm{i}\varphi} = \langle a^{\dagger 2}\rangle^*. \tag{2.27}$$

Using (2.27) the uncertainties in the quadratures can be evaluated

$$\begin{aligned} \langle\xi|X^2_{\frac{\varphi}{2}}|\xi\rangle &= \mathrm{e}^{2r}\langle 0|X^2_{\frac{\varphi}{2}}|0\rangle = \frac{1}{2}\mathrm{e}^{2r}, \\ \langle\xi|X^2_{\frac{\varphi}{2}+\frac{\pi}{2}}|\xi\rangle &= \frac{1}{2}\mathrm{e}^{-2r}, \end{aligned} \tag{2.28}$$

and therefore

$$\begin{aligned} \left(\Delta X_{\frac{\varphi}{2}}\right)^2 = \frac{1}{2}\mathrm{e}^{2r}, \quad \left(\Delta X_{\frac{\varphi}{2}+\frac{\pi}{2}}\right)^2 &= \frac{1}{2}\mathrm{e}^{-2r}, \\ \Delta X_{\frac{\varphi}{2}}\Delta X_{\frac{\varphi}{2}+\frac{\pi}{2}} &= \frac{1}{2}. \end{aligned} \tag{2.29}$$

We therefore find a very interesting property of the Bogoliubov transformation: it stretches the quadrature $X_{\frac{\varphi}{2}}$ and at the same time squeezes the quadrature $X_{\frac{\varphi}{2}+\frac{\pi}{2}}$. The squeezed vacuum is a minimum uncertainty state with the property that one quadrature has an uncertainty which is lower than that for a coherent state. The squeezing parameter (2.11) for the squeezed vacuum is

$$S_{\frac{\varphi}{2}+\frac{\pi}{2}} = -\frac{1}{2}(1-\mathrm{e}^{-2r}) < 0, \tag{2.30}$$

which is a definite signature of the nonclassicality of the squeezed vacuum. It should be borne in mind that the quadratures are defined using the phase of ξ.

We will now give the representation of $|\xi\rangle$ in the quadrature-X space. To do this we use (2.26) to obtain

$$(a\cosh r - a^\dagger \mathrm{e}^{\mathrm{i}\varphi}\sinh r)|\xi\rangle = 0. \tag{2.31}$$

Using (1.37) and $Y = -\mathrm{i}\frac{\partial}{\partial X}$, Eq. (2.31) becomes

$$X(\cosh r - \mathrm{e}^{\mathrm{i}\varphi}\sinh r)\Psi_\xi + \frac{\partial\Psi_\xi}{\partial X}(\cosh r + \mathrm{e}^{\mathrm{i}\varphi}\sinh r) = 0, \;\; \Psi_\xi(X) = \langle X|\xi\rangle, \tag{2.32}$$

$$\Psi_\xi(X) = \Psi^{(0)}\exp\left(-\frac{1}{2}X^2\frac{\cosh r - \mathrm{e}^{\mathrm{i}\varphi}\sinh r}{\cosh r + \mathrm{e}^{\mathrm{i}\varphi}\sinh r}\right), \tag{2.33}$$

leading to a Gaussian wavefunction with r-dependent width and chirp. The constant $\Psi^{(0)} = \langle X = 0|\xi\rangle$ can be easily determined using (2.18).

We next examine whether the squeezed state would spread under free-time evolution. From (2.15) we get

$$
\begin{aligned}
e^{-iHt/\hbar}|\xi\rangle &= e^{-\frac{1}{2}i\omega t}e^{-i\omega t a^\dagger a}\exp\left(\frac{1}{2}\xi a^{\dagger 2}-\frac{1}{2}\xi^* a^2\right)|0\rangle \\
&= e^{-\frac{1}{2}i\omega t}e^{-i\omega t a^\dagger a}\exp\left(\frac{1}{2}\xi a^{\dagger 2}-\frac{1}{2}\xi^* a^2\right)e^{i\omega t a^\dagger a}e^{-i\omega t a^\dagger a}|0\rangle \\
&= e^{-\frac{1}{2}i\omega t}\exp\left(\frac{1}{2}\xi' a^{\dagger 2}-\frac{1}{2}\xi'^* a^2\right)|0\rangle, \qquad \xi' = \xi e^{-2i\omega t} \\
&= e^{-\frac{1}{2}i\omega t}|\xi e^{-2i\omega t}\rangle.
\end{aligned} \tag{2.34}
$$

Thus a squeezed state transforms into another squeezed state; nevertheless, the phase of ξ evolves with time. Hence we conclude that the squeezing remains invariant under free-time evolution. However, the direction of the squeezing ellipse rotates as t changes. We recover the relations (2.29) but with $\varphi \to (\varphi - 2\omega t)$.

2.3.2 Quantum phase-space distributions

The P-function for the squeezed state does not exist. The Q-function can be obtained by using (2.17) and noting that $\langle\alpha|a^\dagger = \alpha^*\langle\alpha|$,

$$
\begin{aligned}
Q_\xi(\alpha) &= \frac{1}{\pi}|\langle\alpha|\xi\rangle|^2 \\
&= \frac{1}{\pi\cosh r}\exp\left[\frac{\tanh r}{2}(e^{i\varphi}\alpha^{*2}+e^{-i\varphi}\alpha^2)-|\alpha|^2\right].
\end{aligned} \tag{2.35}
$$

The function is a Gaussian distribution in two-dimensional α space. The quadratic form in the exponent implies that Q_ξ is a constant along the ellipse whose major and minor axes are proportional to $(1-\tanh r)^{-1/2}$, $(1+\tanh r)^{-1/2}$, with the major axis making an angle $-\varphi/2$ with the x axis. To see this we write $\alpha e^{-i\varphi/2} = \frac{u+iv}{\sqrt{2}}$, hence the exponent becomes $-\frac{1}{2}[u^2(1-\tanh r)+v^2(1+\tanh r)]$.

Next we obtain the Wigner function. For the calculation of the Wigner function the Bogoliubov transformation (2.24) is quite useful. We use the definition (1.98) for the Wigner function

$$
W_\xi(\alpha) = \frac{1}{\pi^2}\int d^2\beta\,\langle 0|S^\dagger(\xi)D(\beta)S(\xi)|0\rangle e^{-(\beta\alpha^*-\beta^*\alpha)}, \tag{2.36}
$$

and using (2.24)

$$
\begin{aligned}
S^\dagger(\xi)D(\beta)S(\xi) &= \exp\{(a^\dagger\cosh r + ae^{-i\varphi}\sinh r)\beta - H.c.\} = D(\beta'), \\
\beta' &= \beta\cosh r - \beta^* e^{i\varphi}\sinh r,
\end{aligned} \tag{2.37}
$$

we can then change the integration variable in (2.36) to β', leading to

$$
\begin{aligned}
W_\xi(\alpha) &= \frac{1}{\pi^2}\int d^2\beta'\,\langle 0|D(\beta')|0\rangle e^{-(\beta'\alpha'^*-\beta'^*\alpha')}, \\
\alpha' &= \alpha\cosh r - \alpha^* e^{i\varphi}\sinh r.
\end{aligned} \tag{2.38}
$$

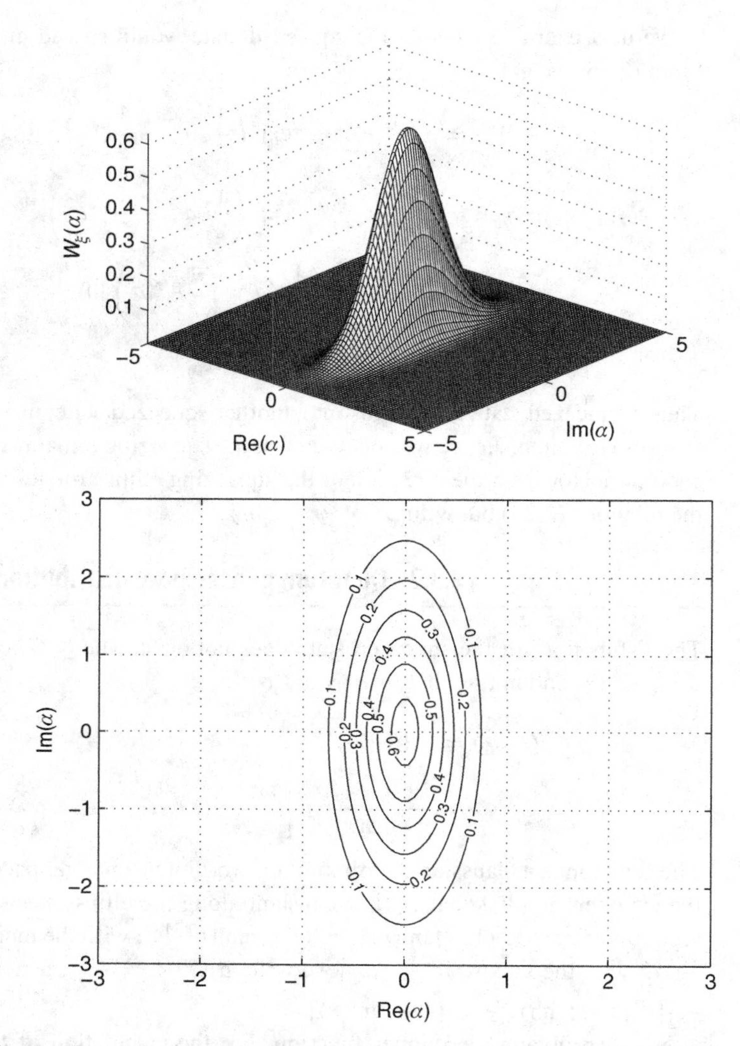

Fig. 2.3 The Wigner function $W_\xi(\alpha)$ for the squeezed vacuum state for $r = 0.6, \varphi = \pi$. The lower plot shows the contours along which the Wigner function is constant.

The relation (2.38) shows that the Wigner function for the squeezed vacuum can be obtained from the Wigner function for the vacuum state by changing α to α'

$$W_\xi(\alpha) = \frac{2}{\pi} \exp(-2|\alpha \cosh r - \alpha^* e^{i\varphi} \sinh r|^2). \tag{2.39}$$

The Wigner function is Gaussian even though the state has nonclassical properties. It is constant along ellipses in α space with the major and minor axes proportional to $\frac{1}{\sqrt{2}}e^r$, $\frac{1}{\sqrt{2}}e^{-r}$, and with the major axis making an angle $-\frac{\varphi}{2}$ with the x axis. This can be seen by expressing (2.39) in terms of the u and v variables as we did following Eq. (2.35). A plot of the Wigner function is shown in Figure 2.3.

2.4 Squeezed coherent state

The squeezed coherent states are defined by [2, 6, 7]

$$|\xi, \beta\rangle = D(\beta)S(\xi)|0\rangle. \tag{2.40}$$

The properties of these states can be obtained since we have already studied the detailed properties of $D(\beta)$ and $S(\xi)$. For example, using (2.24) and Eq. (3) of Table 1.1, we obtain

$$a(\xi, \beta) = S^\dagger(\xi)D^\dagger(\beta)aD(\beta)S(\xi) = a\cosh r + a^\dagger e^{i\varphi}\sinh r + \beta. \tag{2.41}$$

It should be borne in mind that the operators $D(\beta)$ and $S(\xi)$ do not commute. Thus many expectation values can be obtained by using (2.41) and the results for expectation values in the vacuum as

$$\begin{aligned}\langle\xi, \beta|G(a, a^\dagger)|\xi, \beta\rangle &= \langle 0|S^\dagger(\xi)D^\dagger(\beta)GD(\beta)S(\xi)|0\rangle \\ &= \langle 0|G(a(\xi, \beta), a^\dagger(\xi, \beta))|0\rangle.\end{aligned} \tag{2.42}$$

Using (2.41) and (2.42) we obtain some important expectation values

$$\begin{gathered}\langle a\rangle = \beta, \quad \langle a^\dagger a\rangle = |\beta|^2 + \sinh^2 r, \\ \langle a^2\rangle = \beta^2 + \frac{1}{2}\sinh 2r\, e^{i\varphi}, \quad \langle X_\theta\rangle = \frac{1}{\sqrt{2}}(\beta e^{-i\theta} + \beta^* e^{i\theta}), \\ \left(\Delta X_{\frac{\varphi}{2}}\right)^2 = \frac{1}{2}e^{2r}, \quad \left(\Delta X_{\frac{\varphi}{2}+\frac{\pi}{2}}\right)^2 = \frac{1}{2}e^{-2r}, \quad \Delta X_{\frac{\varphi}{2}}\Delta X_{\frac{\varphi}{2}+\frac{\pi}{2}} = \frac{1}{2}.\end{gathered} \tag{2.43}$$

As one would have expected, the squeezing properties of the squeezed coherent states are identical to the squeezing properties of the squeezed vacuum. The displacement of the state can not squeeze the state. However, one has the possibility that the Mandel Q_M parameter can be negative. Thus there are regions of (r, β) for which the photon number distribution can be sub-Poissonian. Using (2.42) we find for the special case $\varphi = \pi$, and real β, that

$$Q_M = \frac{\sinh^4 r + 2\beta^2\sinh^2 r + \sinh^2 r\cosh^2 r - 2\beta^2\sinh r\cosh r}{\beta^2 + \sinh^2 r}. \tag{2.44}$$

A plot of the parameter Q_M as a function of β is shown in Figure 2.4, which clearly shows the regions where the photon statistics become sub-Poissonian.

The phase-space distributions $Q_{\xi\beta}$ and $W_{\xi\beta}$ can be obtained from those for the squeezed vacuum. Since the state (2.40) is obtained by displacing $|\xi\rangle$,

$$\begin{aligned}Q_{\xi\beta}(\alpha) &= Q_\xi(\alpha - \beta), \\ W_{\xi\beta}(\alpha) &= W_\xi(\alpha - \beta),\end{aligned} \tag{2.45}$$

where Q_ξ and W_ξ are given by (2.35) and (2.39), respectively. Thus both $Q_{\xi\beta}$ and $W_{\xi\beta}$ are Gaussians. A plot for $W_{\xi\beta}$ can be obtained from Figure 2.3 by displacing the origin of the coordinate system, and is shown in Figure 2.5.

The photon number distribution for squeezed coherent states has attracted considerable attention. We give its derivation [6], although it has a complicated structure. Let

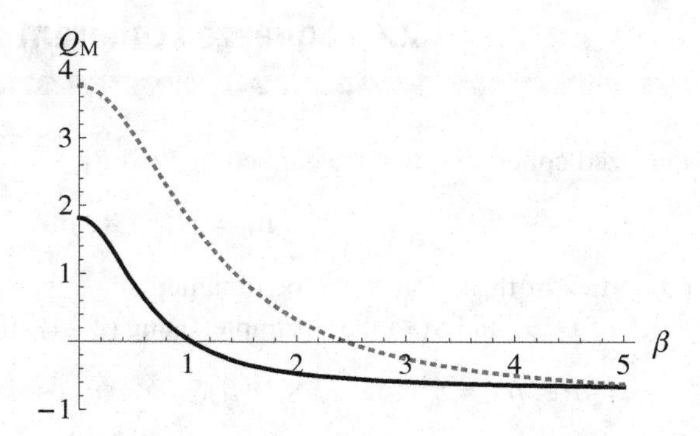

Fig. 2.4 The Mandel Q_M parameter as a function of β for the squeezed coherent state. $r = 0.6$ (solid), 1 (dotted).

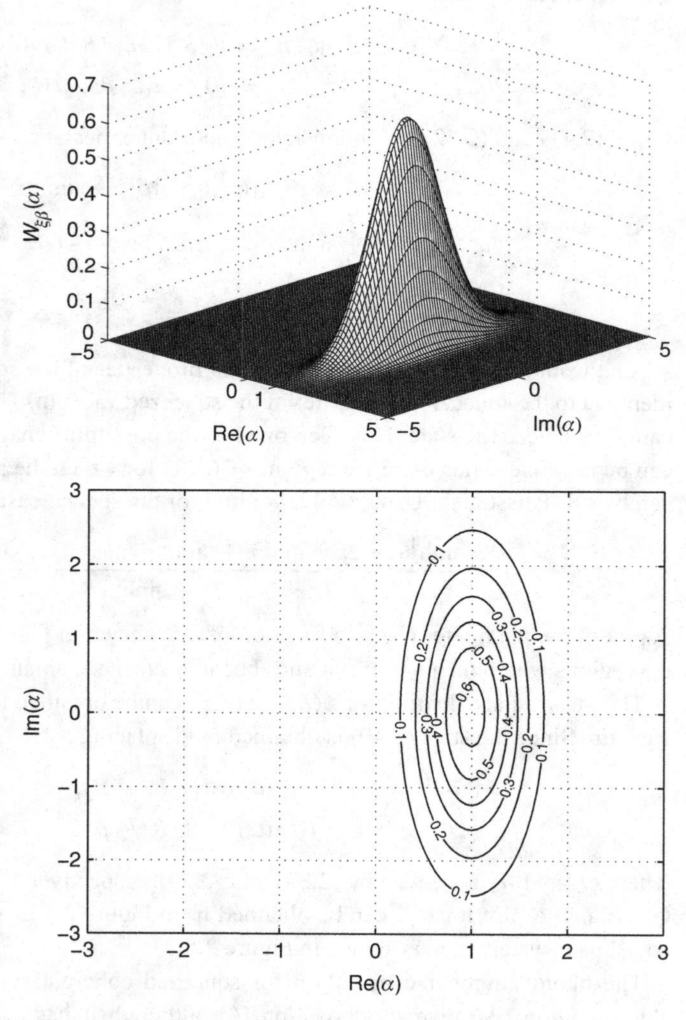

Fig. 2.5 The Wigner function $W_{\xi\beta}(\alpha)$ for the squeezed coherent state for $r = 0.6, \varphi = \pi$, and $\beta = 1$.

$c_n = \langle n|\xi, \beta\rangle$, then using the definition of coherent states we find the relation

$$\sum_{n=0}^{\infty} \frac{\alpha^{*n}}{\sqrt{n!}} c_n = \mathrm{e}^{\frac{1}{2}|\alpha|^2} \langle \alpha|\xi, \beta\rangle. \tag{2.46}$$

Clearly the function on the right-hand side is the generating function for the coefficients c_n. The right-hand side of (2.46) is evaluated by using the relation (2.17), Table 1.1 (Eq. (7)). The important steps are given

$$\begin{aligned}\langle \alpha|\xi, \beta\rangle &= \langle \alpha|D(\beta)S(\xi)|0\rangle \\ &= \frac{1}{\sqrt{\cosh r}} \langle \alpha - \beta| \exp\left(\mathrm{e}^{\mathrm{i}\varphi} \frac{1}{2} a^{\dagger 2} \tanh r\right) |0\rangle \exp\left[-\frac{1}{2}(\alpha\beta^* - \alpha^*\beta)\right] \\ &= \frac{1}{\sqrt{\cosh r}} \exp\left[\mathrm{e}^{\mathrm{i}\varphi} \frac{(\alpha^* - \beta^*)^2}{2} \tanh r - \frac{1}{2}|\alpha - \beta|^2 - \frac{1}{2}(\alpha\beta^* - \alpha^*\beta)\right],\end{aligned}$$

and thus the function on the right-hand side of (2.46) becomes

$$\begin{aligned}\mathrm{e}^{\frac{1}{2}|\alpha|^2} \langle \alpha|\xi, \beta\rangle &= \frac{1}{\sqrt{\cosh r}} \exp\left(-\frac{1}{2}|\beta|^2 + \beta^{*2} \frac{\mathrm{e}^{\mathrm{i}\varphi}}{2} \tanh r\right) \\ &\times \exp\left[\alpha^{*2} \frac{\mathrm{e}^{\mathrm{i}\varphi}}{2} \tanh r + \alpha^*(\beta - \mathrm{e}^{\mathrm{i}\varphi} \tanh r\, \beta^*)\right].\end{aligned} \tag{2.47}$$

In order to obtain c_n we need to expand the second exponential in (2.47) in terms of Hermite polynomials by using the generating function (1.64) for these. This leads to

$$\begin{aligned}\langle n|\xi, \beta\rangle &= \exp\left(-\frac{1}{2}|\beta|^2 + \beta^{*2} \frac{\mathrm{e}^{\mathrm{i}\varphi}}{2} \tanh r\right) \mathrm{i}^n \sqrt{\frac{\mathrm{e}^{\mathrm{i}n\varphi/2}}{n! \cosh r}} \left(\frac{\tanh r}{2}\right)^{n/2} \\ &\times H_n\left[-\frac{\mathrm{i}}{2} \mathrm{e}^{-\mathrm{i}\varphi/2} \sqrt{\frac{2}{\tanh r}} (\beta - \mathrm{e}^{\mathrm{i}\varphi} \tanh r\, \beta^*)\right].\end{aligned} \tag{2.48}$$

Note the appearance of Hermite polynomials with complex argument. We show the photon number distribution in Figure 2.6, which is oscillatory. These oscillations have been interpreted by Schleich and Wheeler [9] as due to interference in phase space.

2.5 Other measures of nonclassicality

If both the Mandel Q_{M} parameter and the squeezing parameter S_θ are positive, then no conclusion can be drawn on the nonclassical nature of the radiation field. A negative Wigner function is surely a signature of nonclassicality. However, no conclusion can be drawn on the nonclassicality of the field if W is positive. A criterion has also been developed in terms of the Q-function, which states that the zeroes of the Q-function are signatures of nonclassicality [11]. On the other hand, if the Q-function has no zeroes then no conclusions on nonclassicality can be drawn. This follows by using the relation (1.95) between the Q-function and the P-function

$$Q(\alpha) = \frac{1}{\pi} \int P(\beta) \mathrm{e}^{-|\alpha - \beta|^2} \mathrm{d}^2\beta. \tag{2.49}$$

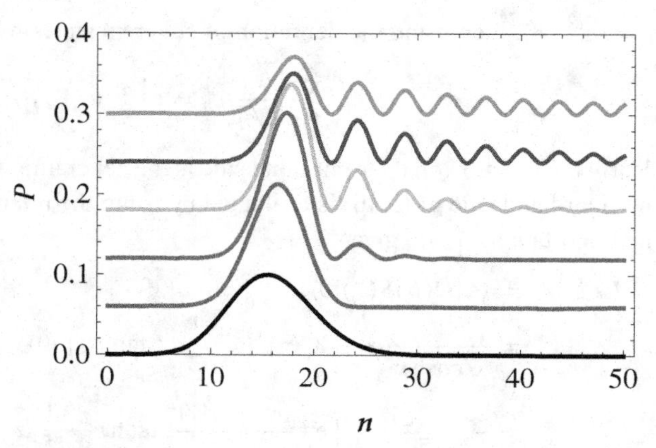

Fig. 2.6 The photon number distribution p_n as a function of n for the squeezed coherent state, $\beta = |\beta|e^{i(\varphi+\pi)/2}$, $|\beta| = 4$. The curves from bottom to top are for the squeezing parameter $r = 0, 0.5, 1, 1.5, 2, 2.5$, respectively. For clarity, different curves are displaced, i.e. $P = p_n + 0.12r$, after [10].

If $Q(\alpha) = 0$ for some α, then $P(\beta)$ must be nonclassical, for if $P(\beta)$ were like classical probability, then the right-hand side of (2.49) could not be zero. In Section 4.1, we will see that the "CAT" state leads to zeroes of $Q(\alpha)$. We have already seen that for squeezed states, which have nonclassical properties, the Q-function is Gaussian and has no zeroes in the complex α plane, $|\alpha| < \infty$.

Phase-space distributions like $Q(\alpha)$ and $W(\alpha)$ can be extracted from tomographic data. We now show that there are other experimental measures of nonclassicality besides the parameters Q_{M} and S_θ. The parameter S_θ [Q_{M}] depends on second moments of the quadrature operator [number operator]. Clearly we need to introduce parameters which would depend on higher moments of either quadrature operators or number operators, or even combinations of both. Most subsequent developments consider the positivity of expectation values of the quadratic forms if $P(\alpha)$ were classical. The negativity of the quadratic form is then taken as a sufficient condition for nonclassicality. We now briefly describe some of these developments.

2.5.1 Nonclassicality in terms of higher-order moments of quadrature operators – phase-sensitive criterion

Consider the nth-order moments of the quadrature operators [12]

$$X_n(\theta) = \left\langle : \left(\frac{a e^{-i\theta} + a^\dagger e^{i\theta}}{\sqrt{2}} \right)^n : \right\rangle$$

$$= \int P(\alpha) \left(\frac{\alpha e^{-i\theta} + \alpha^* e^{i\theta}}{\sqrt{2}} \right)^n d^2\alpha \tag{2.50}$$

$$= \left\langle \left(\frac{\alpha e^{-i\theta} + \alpha^* e^{i\theta}}{\sqrt{2}} \right)^n \right\rangle_P . \tag{2.51}$$

Consider the quadratic form

$$|f(\alpha,\alpha^*)|^2 = \left|\sum_{j=1}^{n} c_j \left(\frac{\alpha e^{-i\theta} + \alpha^* e^{i\theta}}{\sqrt{2}}\right)^{j-1}\right|^2, \qquad n = 2, 3, \ldots. \tag{2.52}$$

Clearly, the positivity of $P(\alpha)$ would lead to

$$\langle |f(\alpha,\alpha^*)|^2 \rangle_P \geq 0. \tag{2.53}$$

Using the well-known results for the quadratic forms, we then find that the classicality of $P(\alpha)$ would imply that the $n \times n$ matrix $X^{(n)}(\theta)$ whose jl element is

$$X_{jl}^{(n)}(\theta) = X_{j+l-2}(\theta) \tag{2.54}$$

must be positive definite, which in turn requires that all its principal minors be positive

$$\det X^{(k)}(\theta) \geq 0, \quad 1 \leq k \leq n. \tag{2.55}$$

A violation of (2.55) would be a signature of the nonclassicality of $P(\alpha)$. The negativity of the matrix

$$X^{(2)}(\theta) = \begin{pmatrix} 1 & X_1(\theta) \\ X_1(\theta) & X_2(\theta) \end{pmatrix}$$

is equivalent to the existence of squeezing, i.e. $S_\theta < 0$.

For example, consider a Fock state $|n\rangle$, then

$$X_{2l} = \binom{2l}{l}\binom{n}{l}\frac{l!}{2^l}, \qquad \theta = 0, \tag{2.56}$$

which leads to

$$\det X^{(2)}(0) = n; \quad \det X^{(3)}(0) = \frac{n^2}{2}(n-3) < 0, \qquad \text{if } n < 3. \tag{2.57}$$

It is remarkable that the nonclassicality of the Fock state with two photons can be tested via higher-order phase-sensitive moments.

2.5.2 Nonclassicality in terms of higher-order moments of the number operator

We next follow a procedure similar to the one that led to (2.54) by using moments of the number operator

$$\begin{aligned} M_n &= \langle : (a^\dagger a)^n : \rangle \\ &= \int P(\alpha)\alpha^{*n}\alpha^n \mathrm{d}^2\alpha = \langle (\alpha^*\alpha)^n \rangle_P. \end{aligned} \tag{2.58}$$

We consider the quadratic form

$$|f(\alpha,\alpha^*)|^2 = \left|\sum_{j=1}^{n} c_j (\alpha^*\alpha)^{j-1}\right|^2. \tag{2.59}$$

Clearly, the positivity of $P(\alpha)$ would imply $\langle |f(\alpha,\alpha^*)|^2\rangle_P \geq 0$, leading to the positivity of the $n \times n$ matrices $M^{(n)}$ with elements

$$M_{jl}^{(n)} = M_{j+l}. \tag{2.60}$$

Thus if any of the minors of $M^{(n)}$ is negative, then we conclude that the underlying P-function is nonclassical [13]. The condition

$$\det M^{(2)} = \det \begin{pmatrix} 1 & \langle a^\dagger a\rangle \\ \langle a^\dagger a\rangle & \langle a^{\dagger 2} a^2\rangle \end{pmatrix} < 0$$

is equivalent to the existence of sub-Poissonian statistics, i.e. $Q_{\mathrm{M}} < 0$. We have seen that $Q_{\mathrm{M}} \geq -1$. Thus it is useful to define a normalized parameter using higher-order moments. Let $\mu^{(n)}$ be the matrix obtained from $M^{(n)}$ by replacing $\langle : (a^\dagger a)^n : \rangle$ by $\langle (a^\dagger a)^n\rangle$. Note that $\mu^{(n)}$ is positive definite since it involves moments of the number operator and further for Fock states $\det\mu^{(n)} = 0$. Thus a useful normalized parameter would be [13]

$$A_n = \det M^{(n)}/(\det\mu^{(n)} - \det M^{(n)}). \tag{2.61}$$

For coherent states $\det M^{(n)} = 0$. Clearly $P(\alpha)$ would be nonclassical if

$$0 \geq A_n \geq -1. \tag{2.62}$$

The upper and lower bounds correspond respectively to coherent states and Fock states. The relation (2.62) will be applied to several non-Gaussian nonclassical states in Chapter 4.

2.5.3 A necessary and sufficient condition for nonclassicality

A necessary and sufficient condition for nonclassicality [14] has been formulated by using the Bochner theorem from probability theory. This theorem states that a continuous function $C(\beta)$ obeying the condition $C(0) = 1$ is a classical characteristic function if and only if it is positive semidefinite. Let us introduce a characteristic function (cf. (1.98)) via

$$C_P(\beta) = \int P(\alpha)\, e^{\beta\alpha^* - \beta^*\alpha} d^2\alpha, \tag{2.63}$$

thus $P(\alpha)$ has all the properties of a probability distribution if and only if $C_P(\beta)$ satisfies the positivity condition

$$\sum_{ij=1}^{n} C_P(\beta_i - \beta_j) c_i^* c_j \geq 0, \tag{2.64}$$

for any complex numbers $c_i'^s$ and set of points $\beta_1 \cdots \beta_n$ in complex number space. Thus the matrix $C^{(n)}$ with elements $C_P(\beta_i - \beta_j)$ would be positive definite. The negativity of $C^{(n)}$, i.e.

$$\det C^{(k)} \leq 0, \tag{2.65}$$

would imply nonclassicality of $P(\alpha)$ and vice versa. This is a powerful tool as it is both necessary and sufficient. An application of these conditions would be considered in the context of non-Gaussian states.

In the special case of $n = 2$, we get

$$C^{(2)} = \begin{pmatrix} 1 & C_P(\beta) \\ C_P(-\beta) & 1 \end{pmatrix},$$

and hence det$C^{(2)}$ is nonpositive if

$$|C_P(\beta)| > 1, \tag{2.66}$$

and this becomes a necessary and sufficient condition for nonclassicality. The characteristic function $C_P(\beta)$ is related to the characteristic function $C_W(\beta)$ for the Wigner function via

$$\begin{aligned} C_W(\beta) = \mathrm{Tr}[\rho D(\beta)], \quad C_W(\beta) = \mathrm{Tr}(\rho \mathrm{e}^{\beta a^\dagger} \mathrm{e}^{-\beta^* a}) \mathrm{e}^{-\frac{1}{2}|\beta|^2}, \\ C_P(\beta) = \mathrm{e}^{\frac{1}{2}|\beta|^2} C_W(\beta). \end{aligned} \tag{2.67}$$

The characteristic function is bounded $|C_W(\beta)| < 1$; however, $|C_P(\beta)|$ could be greater than 1. As a simple application of (2.67) we check the nonclassicality of the squeezed vacuum. The value of C_W is already evaluated in (2.37)

$$\begin{aligned} C_W(\beta) &= \langle 0|D(\beta')0\rangle, \quad \beta' = \beta \cosh r - \beta^* \mathrm{e}^{\mathrm{i}\varphi} \sinh r, \\ &= \exp\left(-\frac{1}{2}|\beta \cosh r - \beta^* \mathrm{e}^{\mathrm{i}\varphi} \sinh r|^2\right), \end{aligned} \tag{2.68}$$

and therefore

$$C_P(\beta) = \exp\left(\frac{1}{2}|\beta|^2 - \frac{1}{2}|\beta \cosh r - \beta^* \mathrm{e}^{\mathrm{i}\varphi} \sinh r|^2\right), \tag{2.69}$$

which for real β, $\varphi = 0$ becomes

$$C_P(\beta) = \exp\left[\frac{1}{2}\beta^2(1 - \mathrm{e}^{-2r})\right] > 1, \text{ if } r \neq 0. \tag{2.70}$$

Thus this criterion is the same as what we found earlier using S_θ except that it is both necessary and sufficient.

2.6 Mixed nonclassical states – degradation in squeezing

Physical systems, in general, are subject to a number of dissipative processes. For example, the field produced by a laser arises as a consequence of not only stimulated processes but also spontaneous processes. Similarly the output of a cavity containing a nonlinear crystal and pumped by lasers is determined by the quality factor of the cavity. Dissipative processes generally lead to the production of mixed states. Therefore we examine mixed nonclassical states. The simplest mixed state can be obtained by the action of the squeezing operator on the thermal state (1.69)

$$\rho = S(\xi)\rho_{\mathrm{T}} S^\dagger(\xi), \tag{2.71}$$

where $S(\xi)$ is defined by (2.14). The number state representation for the state is complicated [10]. However, the Wigner function associated with (2.71) is relatively simple.

This can be obtained from what we proved using (2.36). The function $W_\rho(\alpha)$ is related to $W_{\rm T}(\alpha)$ via

$$W_\rho(\alpha) = W_{\rm T}(\alpha\cosh r - \alpha^* e^{i\varphi}\sinh r), \tag{2.72}$$

which on using (1.105) becomes

$$W_\rho(\alpha) = \frac{2}{\pi(2\langle n\rangle + 1)} \exp\left\{-\frac{2}{2\langle n\rangle+1}\left|\alpha\cosh r - \alpha^* e^{i\varphi}\sinh r\right|^2\right\}. \tag{2.73}$$

In order to understand the meaning of the Gaussian in (2.73), we examine the mean values of a^2 and $a^\dagger a$. According to our discussion following (2.24), these mean values can be obtained by using the Bogoliubov transformations

$$\langle a^2\rangle = {\rm Tr}\rho_{\rm T}[a(\xi)]^2, \quad \langle a^\dagger a\rangle = {\rm Tr}[\rho_{\rm T} a^\dagger(\xi) a(\xi)], \tag{2.74}$$

where $a(\xi)$ is defined by (2.24). Note that in the thermal state ${\rm Tr}(\rho_{\rm T} a^2) = 0$. Thus

$$\begin{aligned}\langle a^2\rangle &= \cosh r\sinh r\, e^{i\varphi}{\rm Tr}\rho_{\rm T}(aa^\dagger + a^\dagger a)\\ &= \frac{1}{2}(1+2\langle n\rangle)\sinh 2r e^{i\varphi},\end{aligned} \tag{2.75}$$

$$\begin{aligned}\langle a^\dagger a\rangle &= \cosh^2 r{\rm Tr}(\rho_{\rm T}a^\dagger a) + \sinh^2 r{\rm Tr}(\rho_{\rm T}aa^\dagger)\\ &= \frac{1}{2}(2\langle n\rangle+1)(\cosh^2 r + \sinh^2 r) - \frac{1}{2}\\ &= \frac{1}{2}(2\langle n\rangle+1)\cosh 2r - \frac{1}{2}.\end{aligned} \tag{2.76}$$

The Wigner function for the mixed nonclassical state (2.71) can be written in standard two-dimensional Gaussian form as

$$W_\rho = \frac{1}{\pi(\tau^2 - 4|\mu|^2)^{1/2}} \exp\left(-\frac{\mu\alpha^2 + \mu^*\alpha^{*2} + \tau|\alpha|^2}{\tau^2 - 4|\mu|^2}\right). \tag{2.77}$$

The parameters μ and τ are related to the second moments

$$\langle a^2\rangle = -2\mu^*, \quad \langle a^\dagger a\rangle = \tau - \frac{1}{2}. \tag{2.78}$$

Such a Gaussian Wigner function was first introduced in [15] in connection with studies on the nonequilibrium dynamics of a chain of coupled harmonic oscillators. Its connection with squeezing was realized later. Such Wigner functions were also found to be useful in the theory of interferometers [16]. The states (2.71), for obvious reasons, are known as squeezed thermal states [17, 18].

It is now clear that if we are given a two-dimensional Gaussian Wigner function, then we can immediately relate to the form (2.71) by comparing (2.78) with (2.75) and (2.76). Most physical systems used for producing squeezed light, such as nonlinear crystals in cavities, produce a state of the form (2.77), where the parameters μ and τ depend on the physical system.

It should be borne in mind that every two-dimensional Gaussian is not a Wigner function [19]. The Gaussian must be consistent with the uncertainty relation. Consider the mean value $\langle (c_1^* a^\dagger + c_2^* a)(c_1 a + c_2 a^\dagger) \rangle$ for arbitrary coefficients c_1, c_2. Since the density matrix is positive definite, this mean value must be ≥ 0, i.e.

$$|c_1|^2 \langle a^\dagger a \rangle + |c_2|^2 \langle a a^\dagger \rangle + c_1^* c_2 \langle a^{\dagger 2} \rangle + c_1 c_2^* \langle a^2 \rangle \geq 0, \tag{2.79}$$

for all values of c_1 and c_2. Therefore

$$\begin{vmatrix} \langle a^\dagger a \rangle & \langle a^{\dagger 2} \rangle \\ \langle a^2 \rangle & \langle a a^\dagger \rangle \end{vmatrix} \geq 0, \tag{2.80}$$

i.e.

$$\langle a^\dagger a \rangle \langle a a^\dagger \rangle \geq \langle a^2 \rangle \langle a^{\dagger 2} \rangle. \tag{2.81}$$

Hence the parameters μ and τ in (2.77) are subject to [15]

$$\tau^2 - 4|\mu|^2 \geq \frac{1}{4}. \tag{2.82}$$

The values given by (2.75) and (2.76) are clearly consistent with the requirement (2.81). Note that (2.77) is well defined as a Gaussian if $\tau^2 - 4|\mu|^2 \geq 0$, which is a much weaker condition than what follows (2.81) from the uncertainty principle or from the positivity of the density matrix.

The mixed state (2.71) would be less nonclassical than the pure state $|\xi\rangle$. This can, for example, be seen by calculating the squeezing parameter S. Using (2.75) and (2.76), we find that (2.29) is modified to

$$\left(\Delta X_{\frac{\varphi}{2}+\frac{\pi}{2}}\right)^2 = \frac{1}{2} e^{-2r} (1 + 2\langle n \rangle). \tag{2.83}$$

Clearly the amount of squeezing is reduced by the factor $(1 + 2\langle n \rangle)$, and the squeezing is lost if $(1 + 2\langle n \rangle)$ exceeds e^{2r}.

Exercises

2.1 Calculate the Mandel parameter Q_M for the displaced Fock state $D(\alpha_0)|n\rangle$. Note that relations like $D^\dagger(\alpha_0) a D(\alpha_0) = a + \alpha_0$ would simplify the calculations.

2.2 Repeat the calculation of Exercise 2.1 for displaced thermal states with density matrix given by

$$D(\alpha_0) \rho_T D^\dagger(\alpha_0),$$

where ρ_T is defined by Eq. (1.69).

2.3 Using the form of $P(\alpha)$ for both coherent state $|\alpha_0\rangle$ and thermal state ρ_T, evaluate the characteristic function $C_P(\beta)$ defined by $\mathrm{Tr}(\rho e^{a^\dagger \beta} e^{-a\beta^*})$ and show that these are well behaved.

2.4 Calculate the P-function for the displaced thermal state defined in Exercise 2.2.

2.5 Show by direct calculation that the squeezed vacuum is an eigenstate of $(a \cosh r - a^\dagger e^{i\varphi} \sinh r)$ with zero eigenvalue.

2.6 Diagonalize the Hamiltonian

$$H = \omega a^\dagger a + g a^2 + g^* a^{\dagger 2},$$

by using $a = b\cosh r + b^\dagger \mathrm{e}^{-\mathrm{i}\theta} \sinh r$ so that $H = \Omega b^\dagger b + \eta$. Show that Ω, r, θ, η are given by

$$\Omega = \sqrt{\omega^2 - 4|g|^2}, \quad r = \frac{1}{4}\ln\frac{\omega - 2\sqrt{|g|^2}}{\omega + 2\sqrt{|g|^2}}, \quad \theta = \mathrm{i}\ln\sqrt{\frac{g^*}{g}}, \quad \eta = \frac{\Omega - \omega}{2}.$$

Note that a and b are Boson operators.

2.7 The purpose of this exercise is to show how nonclassical states can be obtained from minimum uncertainty relations. Let the quadratures X and Y defined by Eq. (2.6) for $\theta = 0$ and $\pi/2$ satisfy the minimum uncertainty relation

$$\Delta X \Delta Y = \frac{1}{2}, \quad [X, Y] = \mathrm{i}.$$

Show that the derivation of the uncertainty relation implies that the underlying state must satisfy the relation

$$(X - \langle X \rangle)|\Psi\rangle = -\mathrm{i}\lambda(Y - \langle Y \rangle)|\Psi\rangle, \quad \lambda = \text{complex}.$$

Establish that the solution of this equation is identical to the solution of

$$a(\xi, \beta)|\Psi\rangle = 0,$$

where the annihilation operator $a(\xi, \beta)$ is defined by (2.41). Find ξ, β in terms of $\langle X \rangle$, $\langle Y \rangle$, and λ. Clearly the squeezed coherent state (2.40) can be obtained by the requirement that the uncertainty product be minimum. An elementary discussion of a minimum uncertainty wavepacket is given in [20]. A detailed review of minimum uncertainty states in quantum optical system is given in [21].

2.8 Solve the eigenvalue problem

$$\frac{1}{1 + a^\dagger a} a^2 |\Psi\rangle = \lambda |\Psi\rangle.$$

Show that $|\Psi\rangle$ goes over to the squeezed vacuum if we further impose the condition $\lim_{\lambda \to 0} |\Psi\rangle = |0\rangle$ (for details see [22]). What would be the solution to the eigenvalue problem if we impose the condition $\lim_{\lambda \to 0} |\Psi\rangle = |1\rangle$?

2.9 Show that it is possible to represent the squeezed vacuum

$$|\xi\rangle = \exp\left[\frac{1}{2} r (a^{\dagger 2} - a^2)\right] |0\rangle$$

in terms of superpositions involving coherent states on a line

$$|\xi\rangle = \left(\frac{1}{2\pi \sinh r}\right)^{1/2} \int_{-\infty}^{+\infty} \mathrm{d}\alpha_x \exp\left(-\frac{\alpha_x^2 \mathrm{e}^{-r}}{2 \sinh r}\right) |\alpha_x\rangle,$$

where α_x is a coherent state on a line with real values of α_x. A proof of this is given in [23]. A general discussion of representations in terms of coherent states on a line can be found in [24]. Two-mode extensions are given in [25].

2.10 The purpose of this exercise is to demonstrate how squeezed states can be generated by a sudden change in the frequency of oscillator. Consider an harmonic oscillator with

$$H = \frac{p^2}{2m} + \frac{1}{2}m\omega^2 x^2, \quad [x, p] = \mathrm{i}\hbar.$$

Let the oscillator be in its ground state

$$\Psi_0(x) = \left(\frac{\alpha}{\sqrt{\pi}}\right)^{1/2} \exp\left(-\frac{1}{2}\alpha^2 x^2\right), \quad \alpha^2 = \frac{m\omega}{\hbar}.$$

Let the frequency of the oscillator be changed to ω' suddenly at $t = 0$. Find the wave function of the oscillator at $t > 0$ in terms of the eigenfunctions of H. Show that this Ψ is the same as the squeezed vacuum. The integral $\int_{-\infty}^{+\infty} \mathrm{e}^{-x^2} H_n(x\eta)\mathrm{d}x = \delta_{n,2m}\sqrt{\pi}\,\frac{(2m)!}{m!}(\eta^2-1)^2$ would be useful. The production of squeezed states by changes in the frequency of the oscillator has been extensively treated [26–29], and has been experimentally realized [30].

References

[1] D. N. Klyshko, *Phys. Lett. A* **213**, 7 (1996); *Sov. Phys. Usp.* **39**, 573 (1996).
[2] R. Loudon and P. L. Knight, *J. Mod. Opt.* **34**, 709 (1987).
[3] D. F. Walls, *Nature (London)* **306**, 141 (1983).
[4] V. Buzek and P. L. Knight, *Progress in Optics*, ed. E. Wolf (Amsterdam: Elsevier) **34**, 1 (1995).
[5] L. Mandel, *Opt. Lett.* **4**, 205 (1979).
[6] H. P. Yuen, *Phys. Rev. A* **13**, 2226 (1976).
[7] C. M. Caves and B. L. Schumaker, *Phys. Rev. A* **31**, 3093 (1985).
[8] F. Boitier, A. Godard, N. Dubreuil, P. Delaye, C. Fabre, and E. Rosencher, *Nature Communications* **2**, 425 (2011).
[9] W. Schleich and J. A. Wheeler, *Nature (London)* **326**, 574 (1987).
[10] G. S. Agarwal and G. Adam, *Phys. Rev. A* **38**, 750 (1988).
[11] N. Lütkenhaus and S. M. Barnett, *Phys. Rev. A* **51**, 3340 (1995).
[12] G. S. Agarwal, *Opt. Commun.* **95**, 109 (1993).
[13] G. S. Agarwal and K. Tara, *Phys. Rev. A* **46**, 485 (1992).
[14] Th. Richter and W. Vogel, *Phys. Rev. Lett.* **89**, 283601 (2002).
[15] G. S. Agarwal, *Phys. Rev. A* **3**, 828 (1971).
[16] G. S. Agarwal, *J. Mod. Opt.* **34**, 909 (1987).
[17] S. Chaturvedi and V. Srinivasan, *Phys. Rev. A* **40**, 6095 (1989).
[18] S. M. Barnett and P. L. Knight, *J. Opt. Soc. Am. B* **2**, 467 (1985).
[19] R. Simon, E. C. G. Sudarshan, and N. Mukunda, *Phys. Lett. A* **124**, 223 (1987).
[20] L. I. Schiff, *Quantum Mechanics*, 3rd ed. (New York: McGraw-Hill, 1968), Sec. 12.
[21] G. S. Agarwal, *Fortschritte der Physik* **50**, 575 (2002).
[22] C. L. Mehta, A. K. Roy, and G. M. Saxena, *Phys. Rev. A* **46**, 1565 (1992).

[23] G. S. Agarwal and R. Simon, *Opt. Commun.* **92**, 105 (1992).
[24] P. Adam, I. Földesi, and J. Janszky, *Phys. Rev. A* **49**, 1281 (1994).
[25] H. Fan, N. Jiang, and H. Lu, *Opt. Commun.* **234**, 277 (2004).
[26] J. Janszky and Y. Y. Yushin, *Opt. Commun.* **59**, 151 (1986).
[27] R. Graham, *J. Mod. Opt.* **34**, 873 (1987).
[28] G. S. Agarwal and S. A. Kumar, *Phys. Rev. Lett.* **67**, 3665 (1991).
[29] V. V. Dodonov and V. I. Man'ko, in *Invariants and the Evolution of Nonstationary Quantum Systems*, edited by M. A. Markov (Moscow: Lebedev Institute, 1990).
[30] G. A. Garrett, A. G. Rojo, A. K. Sood, J. F. Whitaker, and R. Merlin, *Science* **275**, 1638 (1997).

3 Two-mode squeezed states and quantum entanglement

In Chapter 2, we introduced the idea of nonclassicality and discussed at length the single-mode squeezed state. In this chapter we start with two-mode squeezed states. This discussion would naturally take us to the idea of quantum entanglement between different modes of the field. Let us denote the two modes by a and b with commutators

$$[a, a^\dagger] = [b, b^\dagger] = 1, \qquad [a, b^\dagger] = 0, \text{ etc.} \tag{3.1}$$

The two modes can correspond to different frequencies, for example to signal and idler modes in type-I parametric down-conversion or to different polarizations in type-II parametric down-conversion discussed in Section 3.10. The two modes can also correspond to two different states of orbital angular momentum.

3.1 The two-mode squeezed states

We generate two-mode squeezed states by applying the two-mode squeezing operator [1]

$$S(\xi) = \exp(\xi a^\dagger b^\dagger - \xi^* ab), \qquad \xi = r\mathrm{e}^{\mathrm{i}\varphi}, \tag{3.2}$$

on the two-mode vacuum state $|0, 0\rangle$

$$|\xi\rangle = S(\xi)|0, 0\rangle; \qquad a|0, 0\rangle = b|0, 0\rangle = 0. \tag{3.3}$$

Although we use the same notation as for the one-mode squeezed vacuum, the context would make it clear which squeezed state is implied. The relation analogous to (2.16) is

$$\begin{aligned} S(\xi) = {} & \exp\left(\mathrm{e}^{\mathrm{i}\varphi} \tanh r\, a^\dagger b^\dagger\right) \exp\left[-(\ln\cosh r)(a^\dagger a + b^\dagger b + 1)\right] \\ & \times \exp(-\mathrm{e}^{\mathrm{i}\varphi} \tanh r\, ab). \end{aligned} \tag{3.4}$$

On using (3.4) in (3.3) and the properties of the vacuum, we have

$$S(\xi)|0, 0\rangle = \frac{1}{\cosh r} \exp\left(\mathrm{e}^{\mathrm{i}\varphi} \tanh r\, a^\dagger b^\dagger\right) |0, 0\rangle, \tag{3.5}$$

which on expanding the exponential gives the explicit result for the two-mode squeezed vacuum:

$$|\xi\rangle = \frac{1}{\cosh r} \sum_{n=0}^{\infty} \mathrm{e}^{\mathrm{i}n\varphi} (\tanh r)^n |n, n\rangle. \tag{3.6}$$

We now have a very interesting result on the probability $P_{n,m}$ of finding n photons in the mode a and m photons in the mode b

$$P_{n,m} = \delta_{nm} \frac{1}{\cosh^2 r} (\tanh r)^{2n}. \tag{3.7}$$

Thus if one knew that the mode a has n photons, then one can conclude that the mode b must also have n photons even if no experiment is carried out to detect the b mode.

We next consider the two-mode Bogoliubov transformations and the computation of the mean values. Defining

$$a(\xi) = S^\dagger(\xi) a S(\xi), \qquad b(\xi) = S^\dagger(\xi) b S(\xi), \tag{3.8}$$

and writing the squeezing operator as

$$S(\xi) = \exp(rh), \qquad h = \mathrm{e}^{\mathrm{i}\varphi} a^\dagger b^\dagger - H.c., \tag{3.9}$$

we can derive

$$\frac{\mathrm{d}}{\mathrm{d}r} a = \mathrm{e}^{\mathrm{i}\varphi} b^\dagger, \qquad \frac{\mathrm{d}}{\mathrm{d}r} b = \mathrm{e}^{\mathrm{i}\varphi} a^\dagger. \tag{3.10}$$

These can be integrated with the result

$$\begin{aligned} a(\xi) &= a \cosh r + b^\dagger \mathrm{e}^{\mathrm{i}\varphi} \sinh r, \\ b(\xi) &= b \cosh r + a^\dagger \mathrm{e}^{\mathrm{i}\varphi} \sinh r, \end{aligned} \tag{3.11}$$

which is the two-mode Bogoliubov transformation. The expectation values can be computed using a formula analogous to (2.25)

$$\langle \xi | G(a, b) | \xi \rangle = \langle 00 | G(a(\xi), b(\xi)) | 00 \rangle, \tag{3.12}$$

where G is a function of the operators a, b, $a^\dagger$, $b^\dagger$. For example, the mean number of photons in any mode is

$$\langle a^\dagger a \rangle = \langle b^\dagger b \rangle = \sinh^2 r. \tag{3.13}$$

The correlation between two modes is given by

$$\langle ab \rangle = \cosh r \sinh r \, \mathrm{e}^{\mathrm{i}\varphi}, \qquad \langle a^\dagger b \rangle = 0. \tag{3.14}$$

Note also the properties $\langle a^2 \rangle = \langle b^2 \rangle = 0$, $\langle a \rangle = \langle b \rangle = 0$.

3.2 Nonclassicality of the two-mode squeezed vacuum

The correlation $\langle ab \rangle$ between the two modes leads to nonclassical properties such as the squeezing of the two-mode state (3.3). In view of the fact that $\langle a^2 \rangle = \langle b^2 \rangle = 0$, we can show that the squeezing parameters for the modes a and b as defined by (2.11) and (2.6) are positive. Thus the modes a and b themselves are not squeezed. The nonvanishing of $\langle ab \rangle$

suggests that we must consider quadratures of the operators which are linear combinations of a and b, for instance

$$d = \frac{a+b}{\sqrt{2}}, \qquad [d, d^\dagger] = 1. \tag{3.15}$$

The second-order moments for d are found using (3.13) and (3.14),

$$\langle d^\dagger d \rangle = \sinh^2 r, \qquad \langle d^2 \rangle = \cosh r \sinh r \mathrm{e}^{\mathrm{i}\varphi}. \tag{3.16}$$

Thus the operator d for the two-mode squeezed vacuum behaves like the operator a for the single-mode squeezed vacuum (Eq. (2.27)), and hence on defining

$$X_\theta^{(d)} = \frac{d\mathrm{e}^{-\mathrm{i}\theta} + d^\dagger \mathrm{e}^{\mathrm{i}\theta}}{\sqrt{2}}, \tag{3.17}$$

we would get the relations (2.29), (2.30) with X_θ replaced by $X_\theta^{(d)}$, i.e. for the two-mode squeezed vacuum we have the key result

$$S^{(d)}_{\frac{\varphi}{2}+\frac{\pi}{2}} = -\frac{1}{2}(1 - \mathrm{e}^{-2r}) < 0. \tag{3.18}$$

All this suggests a possible relationship between the two-mode squeezed vacuum and the single-mode squeezed vacuum. To see this let us introduce the operator c, which commutes with d

$$c = \frac{a-b}{\sqrt{2}}, \qquad [c, c^\dagger] = 1, \qquad [c, d^\dagger] = 0. \tag{3.19}$$

Thus starting with two modes (a, b) we construct two new ones (c, d). Now note

$$\frac{1}{2}(d^2 - c^2) = ab, \tag{3.20}$$

and hence the two-mode squeezed vacuum can be written as

$$\exp\left(\xi a^\dagger b^\dagger - \xi^* ab\right) = \exp\left(\frac{\xi}{2} d^{\dagger 2} - \frac{\xi^*}{2} d^2\right) \exp\left(-\frac{\xi}{2} c^{\dagger 2} + \frac{\xi^*}{2} c^2\right). \tag{3.21}$$

Thus the squeezing operator for two modes can be written as a product of two-single mode squeezing operators. The transformations (3.15) and (3.19) can be realized by 50-50 beam splitters (see Section 5.1).

Finally we note that we can also define two-mode squeezed coherent states by displacing the state (3.3)

$$|\xi, \alpha_0, \beta_0\rangle = D(\alpha_0) D(\beta_0) S(\xi) |0, 0\rangle, \tag{3.22}$$

where $D(\alpha_0)$ $[D(\beta_0)]$ is the displacement operator for the a-mode [b-mode]. We will not discuss this any further as its properties can be obtained by using the algebraic properties of both the displacement and squeezing operators.

3.3 Quantum phase-space distributions and quadrature distributions

The Q-function is easy to calculate by using (3.5)

$$
\begin{aligned}
Q_\xi(\alpha,\beta) &= \frac{1}{\pi}|\langle\alpha,\beta|\xi\rangle|^2 = \frac{1}{\pi\cosh^2 r}\left|\langle\alpha,\beta|00\rangle\exp\left(\mathrm{e}^{\mathrm{i}\varphi}\tanh r\,\alpha^*\beta^*\right)\right|^2 \\
&= \frac{1}{\pi\cosh^2 r}\exp[(\mathrm{e}^{\mathrm{i}\varphi}\tanh r\,\alpha^*\beta^* + c.c) - |\alpha|^2 - |\beta|^2], \qquad (3.23)
\end{aligned}
$$

which is a Gaussian distribution in four-dimensional space (Re$\{\alpha\}$, Re$\{\beta\}$, Im$\{\alpha\}$, Im$\{\beta\}$). The Wigner function can be evaluated by following the same procedure as in the context of the single-mode squeezed vacuum (2.37). Let us consider two density matrices ρ and $\tilde{\rho}$ related by the transformation (3.3)

$$\rho = S(\xi)\tilde{\rho}S^\dagger(\xi), \qquad (3.24)$$

then the Wigner functions are related by

$$W_\rho(\alpha,\beta) = W_{\tilde{\rho}}(\tilde{\alpha},\tilde{\beta}), \qquad (3.25)$$

where

$$\begin{pmatrix}\tilde{\alpha}\\ \tilde{\beta}^*\end{pmatrix} = \begin{pmatrix}\cosh r & -\sinh r\,\mathrm{e}^{\mathrm{i}\varphi}\\ -\sinh r\,\mathrm{e}^{-\mathrm{i}\varphi} & \cosh r\end{pmatrix}\begin{pmatrix}\alpha\\ \beta^*\end{pmatrix}. \qquad (3.26)$$

Choosing $\tilde{\rho}$ as $|00\rangle\langle 00|$, then

$$W_{\tilde{\rho}}(\tilde{\alpha},\tilde{\beta}) = \frac{4}{\pi^2}\exp\left[-2\left(|\tilde{\alpha}|^2 + |\tilde{\beta}|^2\right)\right], \qquad (3.27)$$

and hence the Wigner function for the two-mode squeezed vacuum is

$$
\begin{aligned}
W_\xi(\alpha,\beta) = \frac{4}{\pi^2}&\exp(-2|\alpha\cosh r - \beta^*\sinh r\,\mathrm{e}^{\mathrm{i}\varphi}|^2 \\
&-2|-\alpha^*\sinh r\,\mathrm{e}^{\mathrm{i}\varphi} + \beta\cosh r|^2). \qquad (3.28)
\end{aligned}
$$

Finally we derive the quadrature distribution associated with (3.6). Let X_a and X_b be the two quadratures for the modes a and b. Then

$$
\begin{aligned}
\Psi(X_a,X_b) &= \langle X_a, X_b|\xi\rangle \\
&= \frac{1}{\cosh r}\sum_{n=0}\mathrm{e}^{\mathrm{i}n\varphi}(\tanh r)^n\frac{1}{2^n n!\sqrt{\pi}}H_n(X_a)H_n(X_b)\mathrm{e}^{-\frac{X_a^2}{2}-\frac{X_b^2}{2}}, \qquad (3.29)
\end{aligned}
$$

where we have used (1.39). The series in (3.29) can be summed up using Mehler's formula for Hermite polynomials

$$\sum_{n=0}^{\infty}H_n(X)H_n(Y)\frac{\eta^n}{2^n n!} = (1-\eta^2)^{-1/2}\exp\left[\frac{2XY\eta - (X^2+Y^2)\eta^2}{1-\eta^2}\right], \ |\eta| < 1, \qquad (3.30)$$

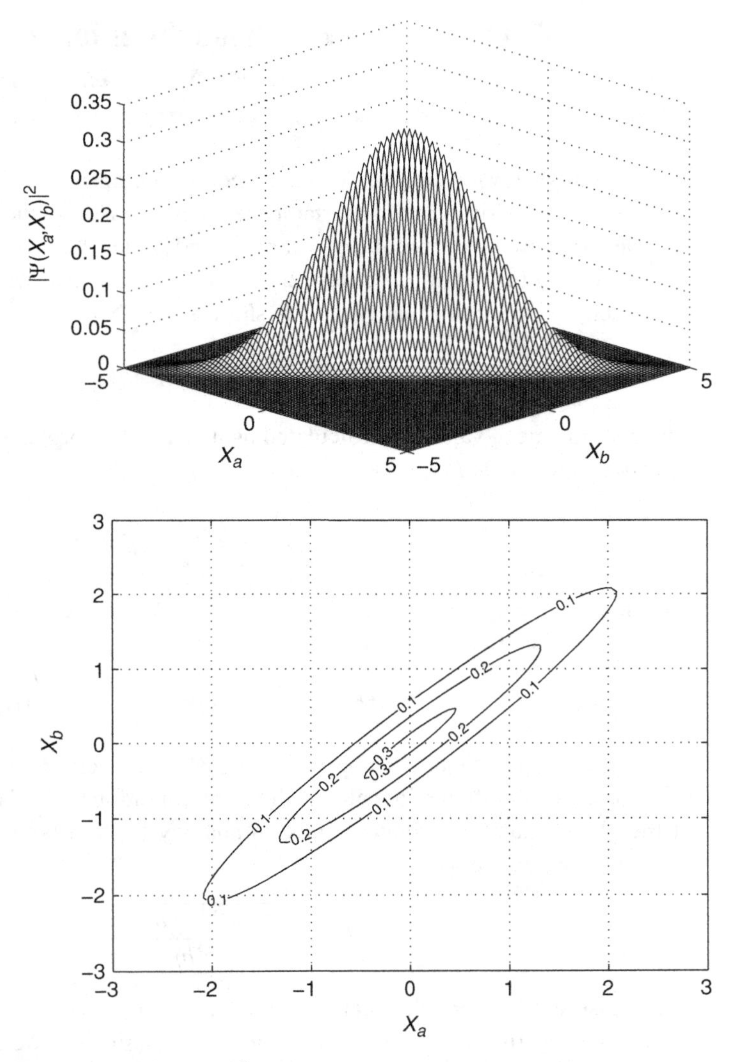

Fig. 3.1 The probability distribution $|\Psi(X_a, X_b)|^2$ as a function of X_a and X_b for $r = 1$ and $\varphi = 0$.

with the result

$$\Psi(X_a, X_b) = \frac{1}{\sqrt{(1-\eta^2)\pi \cosh^2 r}} \exp\left[\frac{2X_aX_b\eta - (X_a^2 + X_b^2)\eta^2}{1-\eta^2} - \frac{1}{2}(X_a^2 + X_b^2)\right],$$
$$\eta = \mathrm{e}^{\mathrm{i}\varphi} \tanh r. \tag{3.31}$$

The quadrature distribution is a two-dimensional Gaussian in (X_a, X_b). The behavior of the probability distribution $|\Psi(X_a, X_b)|^2$ is shown in Figure 3.1. This figure clearly shows the entangled character of $\Psi(X_a, X_b)$ discussed in Section 3.6.

3.4 Cauchy–Schwarz inequalities for nonclassicality in two-mode states

The Cauchy–Schwarz inequalities have been important in the development of quantum mechanics. The derivation of uncertainty relations is based on these inequalities. We would now discuss how a version of these can be used in developing experimental measures of nonclassicality for two-mode states [2, 3]. Let f and g be two well-defined functions, then the Cauchy–Schwarz inequality (CSI in short) states that

$$\langle f^* f\rangle\langle g^* g\rangle \geq |\langle f^* g\rangle|^2, \tag{3.32}$$

where all the mean values are calculated using a positive probability distribution. We now consider a two-mode P-representation

$$\rho = \int P(\alpha,\beta)|\alpha,\beta\rangle\langle\alpha,\beta|\mathrm{d}^2\alpha\mathrm{d}^2\beta, \tag{3.33}$$

and choose $f = f^* = \alpha^*\alpha$, $g = g^* = \beta^*\beta$. Assuming that $P(\alpha, \beta)$ is a probability distribution, then (3.32) becomes

$$\langle a^{\dagger 2}a^2\rangle\langle b^{\dagger 2}b^2\rangle \geq |\langle a^\dagger a\, b^\dagger b\rangle|^2, \qquad \langle G\rangle = \mathrm{Tr}(\rho G). \tag{3.34}$$

This is the CSI that should be obeyed if $P(\alpha, \beta)$ is classical. Any violation of (3.34) would be a sufficient condition of nonclassicality. The condition (3.34) is in terms of the moments of the photon number operator. Thus, in analogy to the Mandel Q_{M} parameter, we can introduce a parameter [4]

$$I = \frac{\sqrt{\langle a^{\dagger 2}a^2\rangle\langle b^{\dagger 2}b^2\rangle}}{\langle a^\dagger a\, b^\dagger b\rangle} - 1, \tag{3.35}$$

as a measure of the nonclassicality of the two-mode field. Note that $I = 0$ for coherent states. The negative values of I would imply nonclassicality. We can now apply this to the two-mode nonclassical state (3.6). Note that (3.6) has a structure such that

$$\langle a^{\dagger 2}a^2\rangle = \langle b^{\dagger 2}b^2\rangle = \langle(a^\dagger a)^2\rangle - \langle a^\dagger a\rangle, \qquad \langle a^\dagger a\, b^\dagger b\rangle = \langle(a^\dagger a)^2\rangle, \tag{3.36}$$

and hence

$$I = -\frac{\langle a^\dagger a\rangle}{\langle(a^\dagger a)^2\rangle} = -\mathrm{sech}2r < 0. \tag{3.37}$$

The violation of the Cauchy–Schwarz inequality for different values of the squeezing parameter r is shown in Figure 3.2. Its smallest value is -1. We have given two distinct criteria for the nonclassicality of the state (3.6) – one based on the study of the phase-sensitive properties (3.18) and the other on the moments (3.37) of the photon number distributions.

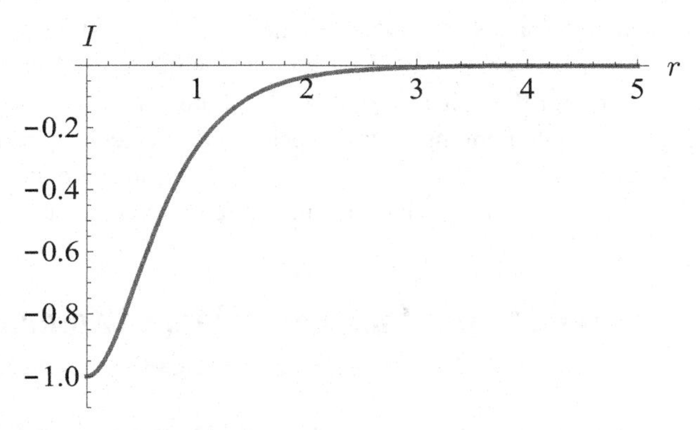

Fig. 3.2 The parameter I as a function of the squeezing parameter r.

3.5 Conditional measurements on the two-mode squeezed vacuum

The two-mode squeezed vacuum has important nonclassical properties like squeezing. A number of other important states can be generated by conditional measurements on it. Let us first consider the case when, say, the b mode is not measured at all. In this case the reduced state of the a mode can be obtained by tracing over the b mode, i.e.

$$\rho^{(a)} = \mathrm{Tr}_b|\xi\rangle\langle\xi|, \tag{3.38}$$

which on using (3.6) becomes

$$\begin{aligned}\rho^{(a)} &= \mathrm{Tr}_b\frac{1}{\cosh^2 r}\sum_{n,m}\mathrm{e}^{in\varphi}\mathrm{e}^{-im\varphi}(\tanh r)^{n+m}|n,n\rangle\langle m,m| \\ &= \frac{1}{\cosh^2 r}\sum_{n,m}\mathrm{e}^{\mathrm{i}(n-m)\varphi}(\tanh r)^{n+m}|n\rangle\langle m|\delta_{nm} \\ &= \frac{1}{\cosh^2 r}\sum_{n}(\tanh r)^{2n}|n\rangle\langle n|.\end{aligned} \tag{3.39}$$

The state (3.39) is just the thermal state (1.70) with $\langle n\rangle = \sinh^2 r$. Thus by doing measurements on just one mode we lose all the nonclassical properties of the field. The result (3.39) also implies that a mixed state (3.39) for one mode is a pure state (3.6) in higher-dimensional Hilbert space. This observation has been used to develop the thermal field representation of mixed states in terms of pure states [5].

Next we consider the conditional state of the mode a for certain specific measurements on the mode b. Let the measurements on the mode b be represented by the projection operator $\wp_b$, then the conditional state $\rho^{(a,c)}$ of the a mode would be

$$\rho^{(a,c)} = \mathrm{Tr}_b\,\wp_b\rho\,\wp_b/\mathrm{Tr}\,\wp_b\rho\,\wp_b. \tag{3.40}$$

Suppose we measure precisely n photons in the b mode, then $\wp_b = |n\rangle\langle n|$, and hence

$$\rho^{(a,c)} = |n\rangle\langle n|, \tag{3.41}$$

i.e. the conditional state of the a mode consists of just n photons. This gives us a practical method for the generation of the n-photon state of a radiation field. This method is extensively used for the generation of single-photon states [6–8]. This is the so-called heralded source of single photons. Many experiments, particularly on interference with single photons, have been performed using such a technique to generate single-photon Fock states (Chapter 5). The method has also been used to generate two-photon Fock states [9].

3.6 Quantum entanglement in the two-mode squeezed vacuum

We have found that the state $|\xi\rangle$ in the quadrature space (Eq. (3.31)) has a Gaussian structure with the property

$$\Psi(X_a, X_b) \neq f(X_a)g(X_b), \qquad \text{if } r \neq 0, \tag{3.42}$$

where f [g] is a function of X_a [X_b] alone. This is also the case in the Fock space representation (3.6), i.e.

$$|\xi\rangle \neq |\Psi_a\rangle|\Psi_b\rangle, \tag{3.43}$$

where $|\Psi_a\rangle$ [$|\Psi_b\rangle$] is a state of the a mode [b mode] alone. We found a strong correlation between the modes a and b (Eq. (3.14)), which is nonzero if $r \neq 0$. Clearly the state of the two modes can not be written as a factorized product of the states for the individual modes. We rather have a form (3.6) which in literature has been called the Schmidt form [10]. Such a state is called an entangled state. The entanglement in the state depends on the parameter r. In order to characterize the amount of entanglement in the state, we have to introduce quantitative measures of entanglement. We next consider more important measures of entanglement.

3.7 Peres–Horodecki separability criterion for continuous variable systems

Consider a system of two particles. Let ρ be the density matrix for a composite system of two particles. Let $\rho^{(a)}$ [$\rho^{(b)}$] be the density matrices for the particle a or mode a [b] alone. The state ρ is said to be separable if it has the structure

$$\rho = \sum_i p_i \rho^{(a)i} \rho^{(b)i}, \qquad p_i \geq 0, \tag{3.44}$$

where $\rho^{(a)i}$, $\rho^{(b)i}$ are the single-particle density matrices. All other states which do not have the structure (3.44) are entangled. Clearly the state (3.6) does not have the structure (3.44) and hence is entangled. Peres [11] developed an elegant criterion for separability for two particles. His condition is both necessary and sufficient in (2×2)- and (2×3)-dimensional cases. However, it is not sufficient in higher dimensions [12]. In Section 3.1 we considered two modes of the radiation field which is a system in Hilbert space of infinite dimensions. We have continuous variables, as quadratures X and Y are continuous. Simon and Duan *et al.* [13,14] used the Peres criterion to obtain experimental measures to test the separability of continuous variable systems.

The Peres–Horodecki separability criterion states that a separable density operator under partial transpose operation goes into a genuine density operator. The partial transpose of ρ is obtained by taking the transpose of the operators associated with the particle b. Thus the Peres–Horodecki criterion for entanglement is that the partial transpose ρ^{PT} of ρ is not a semi-positive definite matrix. This is a sufficient condition for entanglement. Thus, in principle, one can test the entanglement of the state by examining the eigenvalues of the partial transpose of ρ. Subsequent works have converted the Peres–Horodecki criterion to quantities which can be related directly to measurements. A very useful measure is the logarithmic negativity defined by

$$E_{\mathcal{N}}(\rho) = \log_2[1 + 2\mathcal{N}(\rho)], \tag{3.45}$$

where $\mathcal{N}(\rho)$ is the absolute value of the sum of the negative eigenvalues of ρ^{PT} [15,16].

We consider the calculation of the parameter $E_{\mathcal{N}}(\rho)$ for the state (3.6), which we write in the form

$$\rho = \sum_{nm} c_n c_m^* |nn\rangle\langle mm|. \tag{3.46}$$

Clearly the partial transpose is

$$\rho^{\mathrm{PT}} = \sum_{nm} c_n c_m^* |nm\rangle\langle mn|. \tag{3.47}$$

For a given n and m, let us consider the Hermitian combination

$$c_n c_m^* |nm\rangle\langle mn| + c_m c_n^* |mn\rangle\langle nm|,$$

which on using $c_n c_m^* = |c_n c_m| \mathrm{e}^{\mathrm{i}\theta_{nm}}$ can be written as

$$\begin{aligned} &|c_n c_m| (\mathrm{e}^{\mathrm{i}\theta_{nm}} |nm\rangle\langle mn| + \mathrm{e}^{-\mathrm{i}\theta_{nm}} |mn\rangle\langle nm|) \\ &= |c_n c_m| \left[\left(\frac{|nm\rangle + \mathrm{e}^{-\mathrm{i}\theta_{nm}} |mn\rangle}{\sqrt{2}} \right) \left(\frac{\langle nm| + \mathrm{e}^{\mathrm{i}\theta_{nm}} \langle mn|}{\sqrt{2}} \right) \right. \\ &\quad \left. - \left(\frac{|nm\rangle - \mathrm{e}^{-\mathrm{i}\theta_{nm}} |mn\rangle}{\sqrt{2}} \right) \left(\frac{\langle nm| - \mathrm{e}^{\mathrm{i}\theta_{nm}} \langle mn|}{\sqrt{2}} \right) \right]. \end{aligned} \tag{3.48}$$

This gives the diagonal form of ρ^{PT}. The negative eigenvalues are given by

$$-|c_n c_m|, \tag{3.49}$$

and therefore

$$E_{\mathcal{N}}(\rho) = \log_2 \left(1 + \sum_{n \neq m} |c_n c_m| \right) \tag{3.50}$$

$$= \log_2 \left(\sum_n |c_n| \right)^2, \tag{3.51}$$

where we used the condition $\sum_n |c_n|^2 = 1$. Thus we have a compact formula for the log negativity of a class of entangled states defined by (3.46). In the special case of a two-mode squeezed vacuum $|c_n| = (\tanh r)^n / \cosh r$, we get

$$E_{\mathcal{N}}(\rho) = \log_2(e^{2r}), \tag{3.52}$$

showing the clear connection between the log negativity and the squeezing parameter.

Let us consider another example of a two-mode state such that N photons can be found either in the mode a or in the mode b. Such a state is called the NOON state

$$|\Psi\rangle = \frac{1}{\sqrt{2}}(|N,0\rangle + |0,N\rangle). \tag{3.53}$$

For $N = 2$, this state was first discussed by Hong, Ou, and Mandel [17] in the context of the photons produced by a down-converter. Hong, Ou, and Mandel showed how such a state can be produced by two single photons on the two sides of a 50-50 beam splitter. This then led to the Hong–Ou–Mandel dip in two-photon interferometry, which is now extremely important in any tests of the single-photon nature of the source. The state for arbitrary N was introduced by [18] who discussed the importance of such states in the context of superresolution. The partial transpose for the state (3.53) is

$$\rho^{\text{PT}} = \frac{1}{2}(|N,0\rangle\langle N,0| + |0,N\rangle\langle 0,N| + |N,N\rangle\langle 0,0| + |0,0\rangle\langle N,N|). \tag{3.54}$$

The last two terms in (3.54) are diagonalized as

$$\frac{1}{2}(|N,N\rangle + |0,0\rangle)(\langle N,N| + \langle 0,0|) - \frac{1}{2}(|N,N\rangle - |0,0\rangle)(\langle N,N| - \langle 0,0|),$$

showing that ρ^{PT} has one negative eigenvalue $-\frac{1}{2}$. The log negativity parameter for the NOON state is

$$E_{\mathcal{N}}(\rho) = \log_2(2) = 1. \tag{3.55}$$

We will present a more complete description of quantum entanglement as well as applications to mixed states in Section 3.13.

3.8 Generation of two-mode nonclassical and entangled states – optical parametric down-conversion

The squeezing operator introduced in Section 3.2 is unitary and bears a resemblance to the unitary evolution operator in quantum mechanics. In fact, (3.2) can be written in the form

$$S(\xi) = \exp\left(-\frac{i}{\hbar}Ht\right), \tag{3.56}$$

where

$$H = i\hbar g\left(e^{i\varphi}a^\dagger b^\dagger - e^{-i\varphi}ab\right), \qquad gt = r. \tag{3.57}$$

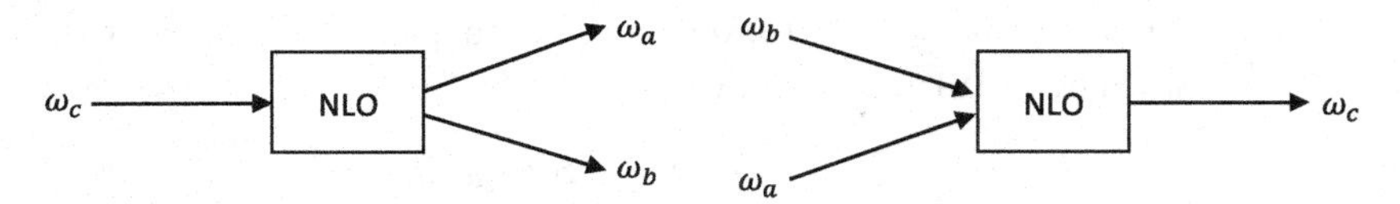

Fig. 3.3 Down-conversion and its inverse.

From (3.56) it is clear that the squeezing operation is realized by the Hamiltonian (3.57). We now connect the Hamiltonian (3.57) to the well-known three-wave interaction in nonlinear crystals lacking inversion symmetry [19, 20]. Such crystals are characterized by the nonlinear susceptibility $\chi^{(2)}_{ijk}$. Let us consider a three-wave interaction of the form

$$H = \hbar(Gca^\dagger b^\dagger + H.c.), \tag{3.58}$$

such that $\omega_c = \omega_a + \omega_b$. The coupling constant G is related to $\chi^{(2)}$ and the geometrical factors of the crystal. The mode c is converted into two modes and vice versa. The physical processes are represented by Figure 3.3.

Thus a photon of frequency ω_c is converted into two photons of lower frequencies ω_a and ω_b and vice versa. This process is called down-conversion. The photons ω_a and ω_b are generated from a vacuum of the modes a and b. This is the process of spontaneous generation. The reverse process $\omega_a + \omega_b \rightarrow \omega_c$ is generally unimportant unless one is dealing with down-conversion in a doubly resonant cavity. If the pumping field is intense, then it is safe to replace its mode operator in (3.58) by a number and rewrite (3.58) in the form (3.57). The parameter g is proportional to the amplitude of the pump at ω_c. The factor φ in (3.57) is related to the phase of the pump. The modes a and b are typically referred to as the signal and idler modes. The directions $\vec{k}_a$ and $\vec{k}_b$ of the signal and idler fields are determined by the phase-matching conditions

$$\vec{k}_c = \vec{k}_a + \vec{k}_b, \tag{3.59}$$

which is also the condition for conservation of momentum. All the wave vectors in (3.59) are inside the nonlinear crystal. In the degenerate case $\vec{k}_a = \vec{k}_b$, $\omega_a = \omega_b$, the Hamiltonian reduces to

$$H = \hbar(Gca^{\dagger 2} + H.c.), \tag{3.60}$$

which under the assumption that the pump field remains constant would generate a unitary evolution operator identical to the squeezing operator (2.14) for a single-mode field.

The Heisenberg operators associated with (3.57) can be written down using (3.11)

$$\begin{aligned} a(t) &= a\cosh gt + b^\dagger e^{i\varphi}\sinh gt, \\ b(t) &= b\cosh gt + a^\dagger e^{i\varphi}\sinh gt. \end{aligned} \tag{3.61}$$

We note that, for a steady-state traveling wave through the crystal, t is generally to be replaced by L/v_g, where L is the length of the crystal and v_g is the group velocity.

The two equations in (3.61) can be decoupled by using the mode operators c and d defined by (3.19) and (3.15), with the result

$$d(t) = d \cosh gt + d^\dagger e^{i\varphi} \sinh gt,$$
$$c(t) = c \cosh gt - c^\dagger e^{i\varphi} \sinh gt, \tag{3.62}$$

which lead to very simple results for the quadratures if we set $\varphi = 0$,

$$\frac{d(t) \pm d^\dagger(t)}{\sqrt{2}} = e^{\pm r} \frac{d \pm d^\dagger}{\sqrt{2}}, \qquad \frac{c(t) \pm c^\dagger(t)}{\sqrt{2}} = e^{\mp r} \frac{c \pm c^\dagger}{\sqrt{2}}. \tag{3.63}$$

The operator form of the relations (3.63) is noteworthy. The operators in (3.63) at time t are obtained by scaling the ones at time $t = 0$. No quantum noise terms are added. This property is extensively used in the construction of phase-sensitive amplifiers.

3.9 Parametric amplification of signals

So far we have considered the case when the fields are generated, starting from no inputs at the frequencies of the modes a and b. We have concentrated on the spontaneous generation of signals. Using the solution (3.61) we can also study the amplification of the signals. Let us suppose that the mode a is initially in a coherent state $|\alpha\rangle$ and that the mode b is in a vacuum state. Then using (3.61) we obtain

$$\begin{aligned} \langle a^\dagger(t)a(t)\rangle &= \cosh^2 r \langle a^\dagger a\rangle + \sinh^2 r \langle bb^\dagger\rangle + (\cosh r \sinh r e^{i\varphi} \langle a^\dagger b^\dagger\rangle + c.c.) \\ &= \cosh^2 r\, |\alpha|^2 + \sinh^2 r \\ &\rightarrow \frac{|\alpha|^2}{4} e^{2r}, \qquad \text{for larger } r. \end{aligned} \tag{3.64}$$

The input signal at ω_a grows, i.e. we have amplification. The factor e^{2r} (more precisely $\frac{1}{4}e^{2r}$) essentially gives the gain of the amplifier. At the same time the idler mode also grows

$$\langle b^\dagger(t)b(t)\rangle = \frac{|\alpha|^2}{4} e^{2r}, \qquad r \gg 1. \tag{3.65}$$

We note that the spontaneous emission, for example the $\sinh r$ term in (3.64), always contributes to the signals. However, the quadratures $X_d = (d + d^\dagger)/\sqrt{2}$, $Y_d = (d - d^\dagger)/\sqrt{2}\,i$ at the output are related to the quadratures at the input (3.63) in a very simple way

$$X_d(t) = e^r X_d, \qquad Y_d(t) = e^{-r} Y_d. \tag{3.66}$$

This is a remarkable feature of parametric amplifiers: one quadrature is amplified whereas the other quadrature is deamplified. The process of parametric amplification is phase-sensitive in view of (3.66) and such amplifiers are called phase-sensitive amplifiers. In view of (3.66) the amplification is also called noise free as we have, for example,

$$\langle X_d(t)\rangle = e^r \langle X_d\rangle, \qquad \langle X_d^2(t)\rangle - \langle X_d(t)\rangle^2 = e^{2r}(\langle X_d^2\rangle - \langle X_d\rangle^2), \tag{3.67}$$

and hence the signal-to-noise ratio

$$\frac{\langle X_d(t)\rangle}{\sqrt{\langle X_d^2(t)\rangle - \langle X_d(t)\rangle^2}} = \frac{\langle X_d\rangle}{\sqrt{\langle X_d^2\rangle - \langle X_d\rangle^2}} \tag{3.68}$$

does not depend on the gain of the amplifier. We will discuss details of optical amplifiers in much more detail in Chapter 10.

3.10 Type-II optical parametric down-conversion – production of entangled photons

We have seen in the previous section that in the process of down-conversion the conservation of energy and momentum gives

$$\omega_c = \omega_a + \omega_b, \qquad \vec{k}_c = \vec{k}_a + \vec{k}_b. \tag{3.69}$$

These conservation laws allow the possibility of signal and idler modes to have large widths and large momentum spread. Typically a state containing two photons can be written as

$$|\Psi\rangle \sim \int \Phi(\nu)\left|\frac{\omega_c}{2} + \nu; \frac{\omega_c}{2} - \nu\right\rangle \mathrm{d}\nu, \tag{3.70}$$

which is clearly an example of an entangled state. The function $\Phi(\nu)$ depends on the band width considerations, which in turn depend on phase-matching conditions. In practice one makes a selection by using narrow pinholes and by using frequency filters. In addition, typical nonlinear crystals like $LiNbO_3$ are uniaxial. Therefore one has the possibility of choosing signal or idler to be either an extraordinary wave or an ordinary wave. These two waves are orthogonally polarized. We could then choose, for example, signal and idler to have the same frequency but different polarization. A state containing one signal photon and one idler photon would have the structure $|\Psi\rangle \sim (|e, o\rangle + \mathrm{e}^{\mathrm{i}\theta}|o, e\rangle)/\sqrt{2}$, since the signal (or idler) could either be an e-wave or an o-wave. This is an example of a two-photon state which is entangled in polarization.

In view of the foregoing, and the fact that one can produce a variety of entangled states using down-conversion, we present the general theory for down-conversion. We concentrate on type-II generation as these are the ones most used these days in studies on quantum entanglement. The process is of type I (II) if signal and idler photons have identical (orthogonal) polarizations. It is collinear (noncollinear) if signal and idler photons travel in the same (different) directions. The quantum theory of type-II downconverters is discussed extensively in [21,22], which we follow closely.

The interaction Hamiltonian in the interaction picture can be written down as

$$H_\mathrm{I} = \chi \int_V \mathrm{d}^3r\, E_p^{(+)}(\vec{r})E_o^{(-)}(\vec{r})E_e^{(-)}(\vec{r}) + H.c.. \tag{3.71}$$

The mode c is the field of the pump laser E_p, which we treat classically and write $E_p^{(+)}$ as

$$E_p^{(+)} = \varepsilon_p \mathrm{e}^{\mathrm{i}\vec{k}_p\cdot\vec{r} - \mathrm{i}\omega_p t}. \tag{3.72}$$

The quantum fields corresponding to ordinary and extraordinary fields would have the form (1.22), i.e.

$$E_j^{(+)} = \sum_k \mathrm{i}\sqrt{\frac{2\pi\hbar\omega_{jk}}{n_{jk}^2 V}}\, a_{j\vec{k}} \mathrm{e}^{\mathrm{i}\vec{k}\cdot\vec{r}-\mathrm{i}\omega_{jk}t}, \qquad j = o, e, \qquad |\vec{k}| = \frac{\omega_{jk} n_{jk}}{c}. \tag{3.73}$$

Here n_{jk} is the refractive index at the frequency ω_{jk}. We assume that the coupling is small enough, which is the case unless the crystal is pumped by a high-power pulsed laser. Then the wave function of the field at the output can be evaluated to the lowest order in the interaction H_1. For a spontaneous down-conversion the first-order correction to the wave function is

$$|\Psi^{(1)}\rangle \cong \sum_{\vec{k},\vec{k}'} a^\dagger_{o\vec{k}} a^\dagger_{e\vec{k}'} F_{\vec{k}\vec{k}'} |0\rangle, \tag{3.74}$$

where

$$F_{\vec{k}\vec{k}'} = \varepsilon_p \chi \mathrm{i} \int_V \mathrm{d}^3 r\, \mathrm{e}^{\mathrm{i}\vec{k}_p\cdot\vec{r}-\mathrm{i}\vec{k}\cdot\vec{r}-\mathrm{i}\vec{k}'\cdot\vec{r}} \int \mathrm{d}t\, \mathrm{e}^{-\mathrm{i}\omega_p t+\mathrm{i}\omega_{ok}t+\mathrm{i}\omega_{ek'}t} \frac{2\pi\sqrt{\omega_{ok}\omega_{ek'}}}{V n_{ok} n_{ek'}}. \tag{3.75}$$

The form of the function $F_{\vec{k}\vec{k}'}$ depends on the phase-matching conditions. It also depends on whether we consider collinear or noncollinear signal and idler propagation.

The case of collinear propagation is much simpler. Let us assume that the pump is propagating in the z direction $\vec{k}_p = \hat{z}k_p$ with large transverse dimensions (T) and area A of the crystal, where T stands for the transverse components. The integration over x and y in (3.75) would yield $\delta_{\vec{k}_T+\vec{k}'_T}$. In the limit of large T, the time integration would give $2\pi\delta(\omega_p - \omega_{ok} - \omega_{ek'})$. The integration over z is then the usual phase-matching integral

$$h(x) = \frac{-1+\mathrm{e}^{+\mathrm{i}x}}{\mathrm{i}x}, \qquad x = (k_p - k_{oz} - k'_{ez})L. \tag{3.76}$$

The function $F_{\vec{k}\vec{k}'}$ reduces to

$$F_{\vec{k}\vec{k}'} = \frac{4\pi^2 \mathrm{i}\chi\varepsilon_p\sqrt{\omega_{ok}\omega_{ek'}}}{n_{ok}n_{ek'}} \delta_{\vec{k}_T+\vec{k}'_T} \delta(\omega_p - \omega_{ok} - \omega_{ek'}) h[(k_p - k_{oz} - k'_{ez})L]. \tag{3.77}$$

The structure of the function $F_{\vec{k}\vec{k}'}$ leads to entanglement both in frequency and momentum. The delta function in (3.77) implies that both ω_{ok} and $\omega_{ek'}$ photons are entangled. The nonfactorizable form of the function h leads to entanglement in momentum. As mentioned earlier, the selection of transverse momentum and frequency is made by using pinholes and frequency filters.

The down-conversion is optimized if

$$\omega_p = \omega_{ok} + \omega_{ek'}, \qquad k_p = k_{oz} + k'_{ez}. \tag{3.78}$$

Note that the refractive index of the e wave depends on the angle ψ of the optic axis with the z axis [23]

$$1 = n_e^2(\omega_e, \psi)\left[\frac{\cos^2\psi}{n_o^2(\omega_e)} + \frac{\sin^2\psi}{n_e^2(\omega_e)}\right]. \tag{3.79}$$

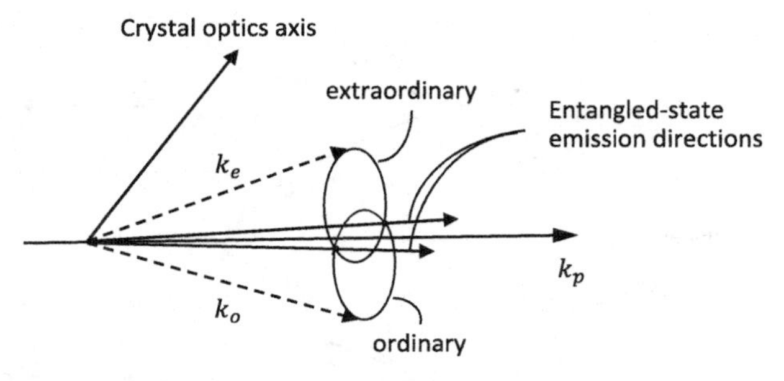

Fig. 3.4 Spontaneous down-conversion cones present with type-II phase matching.

Many crystals such as BBO used for down-conversion are negative uniaxial so that $n_o > n_e$ (for $\psi = 0$). The phase-matching conditions [24] can be satisfied depending on the frequency dependence of the refractive indices and angle ψ.

The phase-matching condition for the noncollinear propagation is much more complicated and a discussion of it would take us deep into crystal optics. We state some facts [25]: let ψ be the angle between the optics axis and the direction of propagation of the pump. Then for degenerate emission the down-converted photons are emitted on two cones. The polarizations of the two photons are orthogonal to each other since one cone is for the ordinary ray and the other cone is for the extraordinary ray as shown in Figure 3.4. The two cones intersect along two lines. Along each intersection the polarization of the photon could be either ordinary or extraordinary. This leads to a polarization entangled two-photon state

$$|\Psi\rangle = (|H_1, V_2\rangle + e^{i\chi}|V_1, H_2\rangle)/\sqrt{2}, \tag{3.80}$$

where the phase factor χ depends on the properties of the nonlinear crystal. Here H stands for the horizontal polarization of the extraordinary ray and V stands for the vertical polarization of the ordinary ray. By the use of waveplates, entangled states of the form

$$|\Phi\rangle = (|H_1, H_2\rangle + e^{i\chi}|V_1, V_2\rangle)/\sqrt{2} \tag{3.81}$$

can be generated.

3.11 Four-photon entanglement using optical parametric down-conversion

The parametric Hamiltonian (3.57) leads to the generation of photons in the modes a and b. For the noncollinear case of type-II phase matching we have to generalize (3.57) so that it has four annihilation and four creation operators. The reason is that we have two directions and in each direction we have two orthogonal polarizations. Let us then consider the Hamiltonian

$$H = i\hbar g(e^{i\varphi_1}a_H^\dagger b_V^\dagger + e^{i\varphi_2}a_V^\dagger b_H^\dagger - H.c.), \tag{3.82}$$

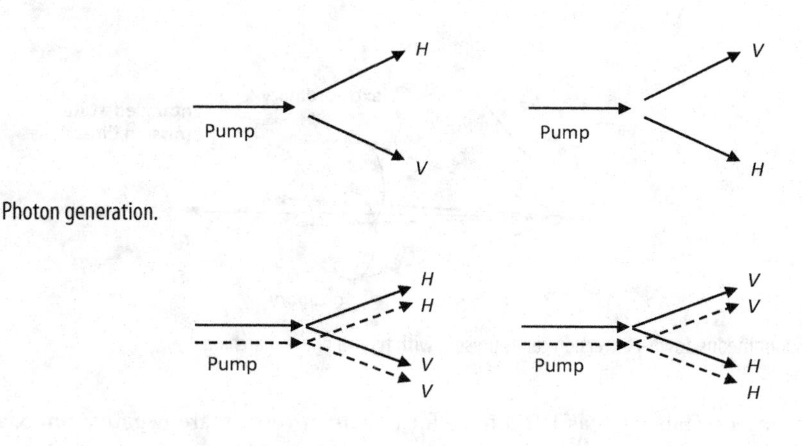

Fig. 3.5 Photon generation.

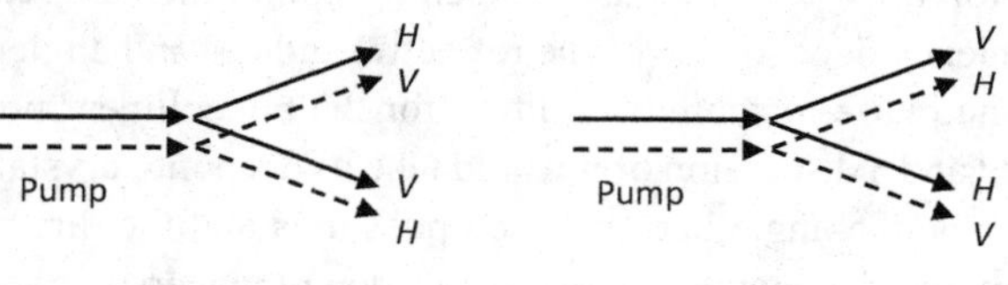

Fig. 3.6 Four-photon entangled state.

where a and b refer to two directions, and H and V refer to two orthogonal polarizations in each direction. The elementary processes of photon generation are shown in Figure 3.5.

The state of the field at time t would be

$$
\begin{aligned}
|\Psi\rangle &= \mathrm{e}^{-\mathrm{i}Ht/\hbar}|0\rangle \\
&= |0\rangle - \frac{\mathrm{i}Ht}{\hbar}|0\rangle + \left(\frac{\mathrm{i}}{\hbar}\right)^2 \frac{t^2}{2} H^2|0\rangle + \cdots .
\end{aligned} \tag{3.83}
$$

The first-order contribution to $|\Psi\rangle$ is

$$
|\Psi\rangle^{(1)} = -\frac{\mathrm{i}Ht}{\hbar}|0\rangle = gt(\mathrm{e}^{\mathrm{i}\varphi_1}|H, V\rangle + \mathrm{e}^{\mathrm{i}\varphi_2}|V, H\rangle), \tag{3.84}
$$

which is the two-photon polarization entangled state considered in Section 3.10. The next-order contribution to the wave function is

$$
\begin{aligned}
|\Psi\rangle^{(2)} = &\frac{g^2t^2}{2}\left(\mathrm{e}^{\mathrm{i}\varphi_1} a_H^\dagger b_V^\dagger + \mathrm{e}^{\mathrm{i}\varphi_2} a_V^\dagger b_H^\dagger\right)^2 |0\rangle \\
&+ \text{contributions proportional to } |0\rangle.
\end{aligned} \tag{3.85}
$$

The first term in (3.85) is the four-photon entangled state

$$
\begin{aligned}
\frac{g^2t^2}{2}(&\mathrm{e}^{2\mathrm{i}\varphi_1} 2|HHVV\rangle + \mathrm{e}^{2\mathrm{i}\varphi_2} 2|VVHH\rangle \\
&+ \mathrm{e}^{\mathrm{i}(\varphi_1+\varphi_2)}|HVHV\rangle + \mathrm{e}^{\mathrm{i}(\varphi_1+\varphi_2)}|VHVH\rangle).
\end{aligned} \tag{3.86}
$$

The four terms clearly correspond to the four elementary processes shown in Figure 3.6. The factor 2 in (3.86) arises from the Bosonic nature of photons expressed mathematically as $a_H^{\dagger 2}|0\rangle = 2|HH\rangle$. The existence of this factor 2 makes the state (3.86) entangled. It can not be written as a product of, say, entangled two-photon states. Several groups have used four-photon entanglement in producing a variety of other types of entangled states which one can utilize in quantum information processing [26,27].

3.12 Two-mode mixed nonclassical states

We next introduce a class of mixed nonclassical states. As explained in Section 2.6, dissipative effects are inherent to physical systems and hence any attempt to produce a nonclassical state would not produce a pure state but a mixed state. We can define a class of mixed states for the two-mode case by applying the squeezing operator $S(\xi)$ (Eq. (3.2)) on a two-mode thermal state

$$\begin{aligned} \rho &= S(\xi)\rho_{\mathrm{T}}S^{\dagger}(\xi), \\ \rho_{\mathrm{T}} &= \exp[-\beta\hbar\omega(a^{\dagger}a + b^{\dagger}b)]/\mathrm{Tr}\{\exp[-\beta\hbar\omega(a^{\dagger}a + b^{\dagger}b)]\}. \end{aligned} \tag{3.87}$$

The Wigner function for the mixed state can be obtained by using (3.25) and (3.26)

$$W_{\rho}(\alpha,\beta) = W_{\mathrm{T}}(\tilde{\alpha},\tilde{\beta}), \tag{3.88}$$

$$\begin{pmatrix} \tilde{\alpha} \\ \tilde{\beta}^{*} \end{pmatrix} = \begin{pmatrix} \cosh r & -\sinh r\, \mathrm{e}^{\mathrm{i}\varphi} \\ -\sinh r\, \mathrm{e}^{-\mathrm{i}\varphi} & \cosh r \end{pmatrix} \begin{pmatrix} \alpha \\ \beta^{*} \end{pmatrix}, \tag{3.89}$$

$$W_{\mathrm{T}} = \frac{4}{\pi^2(2\langle n\rangle + 1)^2}\exp\left[-\frac{2(|\alpha|^2 + |\beta|^2)}{2\langle n\rangle + 1}\right]. \tag{3.90}$$

The expectation values in the state (3.87) can be obtained by using the Bogoliubov transformations (3.11)

$$\begin{aligned} \langle ab\rangle &= \langle a(\xi)b(\xi)\rangle_{\mathrm{T}} = \sinh 2r\, \mathrm{e}^{\mathrm{i}\varphi}\left(\langle n\rangle + \frac{1}{2}\right), \\ \langle a^{\dagger}a\rangle &= \langle a^{\dagger}(\xi)a(\xi)\rangle_{\mathrm{T}} = \left(\langle n\rangle + \frac{1}{2}\right)\cosh 2r - \frac{1}{2} = \langle b^{\dagger}b\rangle. \end{aligned} \tag{3.91}$$

We could also start with a class of Gaussian Wigner functions for two modes of the form [28]

$$W_{\rho} = \frac{1}{\pi^2(\tau^2 - 4|\mu|^2)}\exp\left[-\frac{2\mu\alpha\beta + 2\mu^{*}\alpha^{*}\beta^{*} + \tau(|\alpha|^2 + |\beta|^2)}{\tau^2 - 4|\mu|^2}\right], \tag{3.92}$$

under the following condition, which follows from the positivity of the density matrix,

$$\tau^2 - 4|\mu|^2 > \frac{1}{4}. \tag{3.93}$$

The parameters τ and μ are given by

$$\langle ab\rangle = -2\mu^{*}, \qquad \langle a^{\dagger}a\rangle = \langle b^{\dagger}b\rangle = \tau - \frac{1}{2}. \tag{3.94}$$

On comparison of (3.91) and (3.94), we can obtain τ and μ in terms of r and $\langle n\rangle$. All the physical properties of the field in the mixed state (3.87) can be calculated either by using Bogoliubov transformations or via the Wigner function (3.92).

3.13 Entanglement in two-mode mixed Gaussian states

Gaussian states are used extensively in both quantum optics and quantum information science. The reason for their widespread use is the ease with which such states can be used. Entanglement criteria for Gaussian states have been extensively developed [13, 14, 29]. These are both necessary and sufficient. For Gaussian states the covariance matrix is the key quantity. For example, we already saw that the Gaussian state (3.92) is completely determined in terms of the second-order moments of a and b (Eq. (3.94)). Let us write the Gaussian two-mode state in terms of the quadratures x_a, y_a, x_b, and y_b defined as $x_j = (a_j + a_j^\dagger)/\sqrt{2}, y_j = (a_j - a_j^\dagger)/(\sqrt{2}\mathrm{i}), j = a, b$. Let X be the row vector with elements (x_a, y_a, x_b, y_b) and let σ be the covariance matrix

$$\sigma_{ij} = \frac{1}{2}\langle X_iX_j + X_jX_i\rangle - \langle X_i\rangle\langle X_j\rangle. \tag{3.95}$$

This is a real, positive symmetric matrix. A normalized two-mode Gaussian distribution has the form

$$W(X) = \frac{\exp\left[-(X - \langle X\rangle)\sigma^{-1}(X - \langle X\rangle)^T/2\right]}{(2\pi)^n\sqrt{\det(\sigma)}}. \tag{3.96}$$

In the two-mode case, the covariance matrix has the form

$$\sigma = \begin{pmatrix} \alpha & \gamma \\ \gamma^T & \beta \end{pmatrix}. \tag{3.97}$$

Given the covariance matrix for a two-mode state ρ, the two symplectic eigenvalues of the covariance matrix associated with its "partial transpose" $\tilde{\rho}$ are explicitly given by

$$\tilde{\nu}_{\pm} = \sqrt{\frac{\tilde{\Delta}(\sigma) \pm \sqrt{\tilde{\Delta}(\sigma)^2 - 4\det(\sigma)}}{2}}, \tag{3.98}$$

where $\tilde{\Delta}(\sigma) = \det(\alpha) + \det(\beta) - 2\det(\gamma)$. If $\tilde{\nu}_<$ is taken to denote the smaller of the two symplectic eigenvalues, then the necessary and sufficient conditions for ρ to be an entangled state can be expressed as

$$\tilde{\nu}_< < \frac{1}{2}. \tag{3.99}$$

The corresponding logarithmic negativity $E_N(\rho)$ is given by [29]

$$E_N(\rho) = \max[0, -\ln(2\tilde{\nu}_<)]. \tag{3.100}$$

Here a logarithm, in contrast to (3.45), to the base e is used. For the special case when σ has the form

$$\sigma = \begin{pmatrix} A & 0 & B & C \\ 0 & A & C & -B \\ B & C & A' & 0 \\ C & -B & 0 & A' \end{pmatrix}, \tag{3.101}$$

the two symplectic eigenvalues $\tilde{\nu}_\pm$ turn out to be

$$\tilde{\nu}_\pm \equiv \frac{1}{2}\left[(A + A') \pm \sqrt{(A - A')^2 + 4(B^2 + C^2)}\right]. \tag{3.102}$$

For the two-mode squeezed vacuum

$$\begin{aligned} \rho &= S(\xi)|0,0\rangle\langle 0,0|S^\dagger(\xi), \\ S(\xi) &= \exp(\xi a^\dagger b^\dagger - \xi^* ab), \qquad \xi = re^{i\varphi}, \end{aligned} \tag{3.103}$$

which belongs to the class of Gaussian entangled states, the two symplectic eignevalues are $\tilde{\nu}_\pm = e^{\pm 2r}/2$ and thus the quantum entanglement in this state is directly linked to the squeezing parameter r as the logarithmic negativity $E_N = 2r$. As $r \to 0$, $\tilde{\nu}_< \to 1/2$ and the entanglement disappears. The result (3.99) is a very powerful criteria for entanglement in a two-mode Gaussian state which could even be a mixed state. We will use it in Section 10.4 for the study of degradation in entanglement by the phase-sensitive amplifier.

3.14 Application of entanglement to the teleportation of a quantum state

As an interesting application of entangled state we describe the teleportation of an unknown quantum state by using an entangled pair of particles. The word teleportation implies the disappearance of an object from one place and its reappearance somewhere else. The scheme was discovered by Bennett *et al.* [30]. It has been extensively implemented using entangled photons produced by parametric down-converters [31] as well as entangled atomic pairs [32, 33]. In the first demonstration [31], a single photon in an arbitrary polarization state was teleported from Alice to Bob. The teleportation idea has been extended to systems with continuous variables, such as coherent states [34], and even to the teleportation of nonclassical states like cat states has been achieved [35].

Let us consider two-level systems such as spins, which are two polarization states of a single photon. Let $|\Phi\rangle_1$ be the unknown state of a particle with Alice labeled as 1 and suppose that we need to teleport it to a different location, i.e. to Bob. We can write the state $|\Phi\rangle_1$ as a superposition of two states

$$|\Phi\rangle_1 = \alpha|+\rangle_1 + \beta|-\rangle_1. \tag{3.104}$$

The idea is to use an EPR pair of particles 2 and 3 prepared, say, in a singlet state

$$|\Psi\rangle_{2,3} = \frac{1}{\sqrt{2}}(|+\rangle_2|-\rangle_3 - |+\rangle_3|-\rangle_2), \tag{3.105}$$

and to send particle 2 to Alice and particle 3 to Bob. Alice and Bob together have particles in the state $|\Phi\rangle_1|\Psi\rangle_{2,3}$. Alice then carries out a measurement on the particles 1 and 2 in the Bell basis and communicates over a classical channel the result of the Bell measurement. We can rewrite the wave function $|\Phi\rangle_1|\Psi\rangle_{2,3}$ in terms of the Bell states of particles 1 and 2

$$|\Psi^{\pm}\rangle_{1,2} = \frac{1}{\sqrt{2}}(|+\rangle_1|-\rangle_2 \pm |+\rangle_2|-\rangle_1),$$
$$|\Phi^{\pm}\rangle_{1,2} = \frac{1}{\sqrt{2}}(|+\rangle_1|+\rangle_2 \pm |-\rangle_2|-\rangle_1), \tag{3.106}$$

with the result

$$\begin{aligned}|\Phi\rangle_1|\Psi\rangle_{2,3} = &-\frac{1}{2}|\Psi^-\rangle_{1,2}|\Phi\rangle_3\\ &+\frac{1}{2}|\Psi^+\rangle_{1,2}(-\alpha|+\rangle_3 + \beta|-\rangle_3)\\ &+\frac{1}{2}|\Phi^-\rangle_{1,2}(\alpha|-\rangle_3 + \beta|+\rangle_3)\\ &+\frac{1}{2}|\Phi^+\rangle_{1,2}(\alpha|-\rangle_3 - \beta|+\rangle_3).\end{aligned} \tag{3.107}$$

Thus each Bell-state measurement by Alice with a success probability of $\frac{1}{4}$ reduces the state of particle 3 with Bob according to:

(i) $|\Psi^-\rangle_{1,2} \to -|\Phi\rangle_3$,

(ii) $|\Psi^+\rangle_{1,2} \to -\begin{pmatrix} 1 & 0 \\ 0 & -1 \end{pmatrix}|\Phi\rangle_3$,

(iii) $|\Phi^-\rangle_{1,2} \to \begin{pmatrix} 0 & 1 \\ 1 & 0 \end{pmatrix}|\Phi\rangle_3$,

(iv) $|\Phi^+\rangle_{1,2} \to \begin{pmatrix} 0 & -1 \\ 1 & 0 \end{pmatrix}|\Phi\rangle_3$.

Thus in 25% of the cases, corresponding to the Bell measurement $|\Psi^-\rangle_{1,2}$, the state of particle 3 with Bob becomes, except for a sign, the state of particle 1 with Alice. In other cases one needs to make a unitary transformation to recover $|\Phi\rangle_1$. Thus, as remarked by Bennett *et al.* [30], an accurate teleportation can be achieved in all cases by having Alice communicate over a classical channel to Bob the outcome of her measurement, after which Bob applies the required unitary operation to transform the state of his particle to the state of particle 1. Note that, in this protocol, one need not know the state that one is teleporting. In the first experimental confirmation of teleportation, Bouwmeester *et al.* [31] used for the EPR pair polarization-entangled photons from a down-converter for different settings of the polarization of the single photon. Teleportation of continuous variables has been studied by Kimble and collaborators [34]. Lee *et al.* [35] demonstrated teleportation of a cat state. The Wigner function of the teleported state exhibited negative values at the origin, though the negativity was reduced. For continuous variables we are limited by the amount of entanglement in the EPR state used for transportation. Thus the negativity [36] of the teleported Wigner function depends on the squeezing parameter of the two-mode entangled state (Eq. (3.52)).

3.15 Nonclassical fields in optical fibers

In Section 3.10 we showed how parametric processes arising from second-order nonlinearities lead to amplification of signals as well as the generation of nonclassical states of light. One also has the possibility of parametric processes using the third-order nonlinearities where two photons of the pump are converted into two new photons by the process of four-wave mixing

$$\omega_1 + \omega_2 \rightarrow \omega_3 + \omega_4. \tag{3.108}$$

The generation is subject to the phase-matching conditions in a nonlinear medium. Such four-wave mixing processes have been extensively studied in optical fibers [37] and are of a great deal of interest in connection with the generation of nonclassical fields at telecommunication wavelengths [38–40]. Let us consider the special case (as in fiber) that all four waves are propagating in the same direction z and let each wave be x-polarized. The nonlinear polarization has the form

$$\vec{P}_{\mathrm{NL}} = \chi^{(3)} \cdot \vec{E}\vec{E}\vec{E}, \tag{3.109}$$

where $\chi^{(3)}$ is a tensor of rank four and

$$\vec{E} = \hat{x}\sum_{j=1}^{4} E_j \exp\left[\mathrm{i}\left(n_j\frac{z}{c} - t\right)\omega_j\right] + c.c.. \tag{3.110}$$

On substituting (3.110) in (3.109) we can find the components of $\vec{P}_{\mathrm{NL}}$ at ω_3 and ω_4. The contribution at ω_3, for example, can arise from terms $|E_1|^2E_3$, $|E_2|^2E_3$, $E_1E_2E_4^*$, $|E_3|^2E_3$, etc. We assume that the pump waves ω_1 and ω_2 are intense so that the effect of the generated fields ω_3 and ω_4 on the pump fields is insignificant. The phase-matching condition would be

$$\Delta k = n_3\omega_3 + n_4\omega_4 - n_1\omega_1 - n_2\omega_2 = 0. \tag{3.111}$$

Assuming continuous wave operation, the field ω_j would propagate in the jth mode u_j of the fiber. The mode would have some (x, y) dependence. Thus each E_j in (3.110) can be written as

$$E_j = u_j(x, y)A_j(z). \tag{3.112}$$

We drop any (x, y) changes in E_j as it propagates. The wave equation at a frequency ω is

$$\left(\nabla^2 + \frac{\omega^2}{c^2}n^2\right)E(\omega) = -4\pi P_{\mathrm{NL}}(\omega), \tag{3.113}$$

where $P_{\mathrm{NL}}(\omega)$ is the component of nonlinear polarization at ω. The wave equation (3.113) is simplified in the standard slowly varying envelope approximation $\frac{\mathrm{d}^2E}{\mathrm{d}z^2} \ll \frac{\omega}{c}\frac{\mathrm{d}E}{\mathrm{d}z}$. As indicated earlier, if the action back of ω_3 and ω_4 on ω_1 and ω_2 is dropped (i.e. the pump is undepleted), then the z dependence of the two pump fields is given by

$$A_1(z) = A_1 \exp[i\gamma(P_1 + 2P_2)z],$$
$$A_2(z) = A_2 \exp[i\gamma(P_2 + 2P_1)z],$$
$$P_i = |A_i(0)|^2, \tag{3.114}$$

i.e. P_i is the power of the ith pump at the input. In a lossless medium, the propagation of the pump field is affected by the self- and cross-phase modulation. The parameter γ is the nonlinearity coefficient and is approximately equal to

$$\gamma \sim \frac{n_2' \omega}{cA_{eff}}, \tag{3.115}$$

where A_{eff} is related to the overlap coefficient between the modes. For simplicity this can be considered to be mode independent. The corresponding equations for the generated fields are

$$\frac{dA_3}{dz} = 2i\gamma\left[(P_1 + P_2)A_3 + (P_1P_2)^{1/2}\exp(i\theta)A_4^*\right],$$
$$\frac{dA_4^*}{dz} = -2i\gamma\left[(P_1 + P_2)A_4^* + (P_1P_2)^{1/2}\exp(-i\theta)A_3\right], \tag{3.116}$$

where

$$\theta = [\Delta k + 3\gamma(P_1 + P_2)]z + (\varphi_1 + \varphi_2). \tag{3.117}$$

Here φ_i is the phase of the ith pump at $z = 0$, and $A_i(0) = \sqrt{P_i}e^{i\varphi_i}$. In writing (3.116) we dropped third-order terms in A_3, A_4, etc. These are the key equations giving the generation of ω_3 and ω_4. These are similar to the equations (3.10) for the parametric process in a down-converter, though more complicated. Note that in fibers we consider propagation over long distances – which could be of the order of several hundred meters or more. Thus in fibers the γ term in θ is very important. For a single pump, we set $P_1 = P_2 = P/2$ and replace $3\gamma(P_1 + P_2)$ by $2\gamma P$. This is because $3\gamma(P_1 + P_2)$ comes from both self-phase modulation and cross-phase modulation.

So far we have treated all fields classically. We can now quantize the fields. We use the quantization rules for the single-mode fields

$$[A_i(z), A_j^\dagger(z)] = \delta_{ij}, \quad [A_i(z), A_j(z)] = 0, \quad i, j = 3, 4. \tag{3.118}$$

The mean value $\langle A_3^\dagger A_3 \rangle$ represents the number of photons in the field ω_3 with power defined as $P_3 = \hbar\omega\langle A_3^\dagger A_3\rangle c$ (units 10^{-7} watts/cm^2). Thus in quantum theory Eqs. (3.116) are to be considered as Heisenberg equations and are to be solved subject to the commutation relations (3.118).

From now onwards, we treat A_3 and A_4 as operators and relabel these as a and b and the frequencies as ω_a and ω_b, respectively. Equations (3.116) can be solved by matrix methods and we write the solutions as

$$a(z) = \mu a(0) + \nu b^\dagger(0), \qquad b(z) = \mu b(0) + \nu a^\dagger(0), \tag{3.119}$$

where for a single input pump the values of μ and ν are

$$\mu = \mathrm{e}^{\frac{\mathrm{i}}{2}[\Delta k+2\gamma P]z}\left[\cosh(gz) - \frac{\mathrm{i}}{2g}(\Delta k - 2\gamma P)\sinh(gz)\right],$$

$$\nu = \mathrm{i}\mathrm{e}^{\frac{\mathrm{i}}{2}[\Delta k+2\gamma P]z}\frac{\gamma P}{g}\mathrm{e}^{2\mathrm{i}\varphi}\sinh(gz), \qquad |\mu|^2 - |\nu|^2 = 1, \tag{3.120}$$

where the parameter g is called the gain coefficient and is defined by

$$g = \sqrt{(\gamma P)^2 - \frac{1}{4}(\Delta k - 2\gamma P)^2}. \tag{3.121}$$

We note that because of the phase modulation of the pump inside the fiber we have an effective phase-mismatch parameter $\Delta\tilde{k}$ given by

$$\Delta\tilde{k} = \Delta k - 2\gamma P. \tag{3.122}$$

The parameter Δk can be simplified if we expand the refractive index at the frequencies $\omega_a (= \omega_3)$ and $\omega_b (= \omega_4)$ in terms of the index at the pump frequency $\omega_p = \omega_1 = \omega_2$

$$\Delta k \approx -\beta_2(\omega_a - \omega_p)^2, \qquad \beta_2 \approx \frac{\omega_p}{c}\frac{\mathrm{d}^2}{\mathrm{d}\omega_p^2}n(\omega_p). \tag{3.123}$$

The solutions for the parametric processes in an optical fiber are just the Bogoliubov transformation discussed in detail in Sections 3.1–3.7. For $\Delta\tilde{k} \to 0$, $g \to \gamma P$ and

$$\mu \to \mathrm{e}^{\mathrm{i}\Delta kz}\cosh(\gamma Pz), \qquad \nu \to \mathrm{e}^{\mathrm{i}\Delta kz}\mathrm{i}\mathrm{e}^{2\mathrm{i}\varphi}\sinh(\gamma Pz). \tag{3.124}$$

The solution (3.119) is almost identical to (3.11) and hence all the discussion in Chapter 3 on the generation of nonclassical and entangled states would apply to parametric processes in an optical fiber.

The non-phase-matched case is somewhat different. However, all the calculations can be done using the Bogoliubov transformation (3.119). One can define four different quadratures, such as

$$X_\pm = \left(\frac{a \pm b}{\sqrt{2}} + \frac{a^\dagger \pm b^\dagger}{\sqrt{2}}\right)\Big/\sqrt{2}, \qquad Y_\mp = \left(\frac{a \mp b}{\sqrt{2}} - \frac{a^\dagger \mp b^\dagger}{\sqrt{2}}\right)\Big/\sqrt{2}\mathrm{i}. \tag{3.125}$$

It is seen from (3.119) that

$$\Delta X_\pm^2 = \frac{1}{2}(1 + 2|\nu|^2 \pm \mu\nu \pm \mu^*\nu^*). \tag{3.126}$$

At the same time for coherent inputs, i.e. a and b in coherent states with equal amplitudes α, we get the gain

$$G_a \cong \frac{\langle a^\dagger a\rangle}{|\alpha|^2} = |\mu|^2 + |\nu|^2 + 2|\mu||\nu|\cos(\theta_a + \theta_b + \theta_\mu - \theta_\nu), \tag{3.127}$$

where θ_μ and θ_ν are the phases of μ and ν. The vacuum fluctuation term is dropped in (3.127). The variance $\Delta X_\mp^2$ will be minimum if $\cos(\theta_\mu + \theta_\nu) = -1$. We can thus study the variance and gain as a function of the phase of the pump. Since gain and fluctuation depend in different ways on θ_μ and θ_ν, we can have minimum quadrature variance and significant gain at the same time by a judicious choice of parameters. An example is shown in Figure 3.7

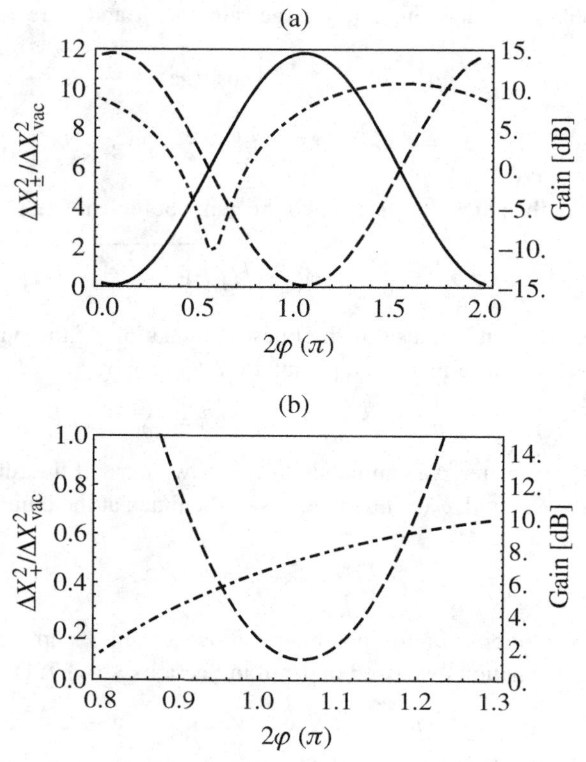

Fig. 3.7 (a) The normalized variances $\Delta X^2_\pm/\Delta X^2_{\text{vac}}$ and the gain (dB) as a function of 2φ, where ΔX^2_{vac} is the variance of the vacuum state and the gain (dB) is defined as $10\log_{10} G_a$. The dashed curve represents the normalized variance $\Delta X^2_+/\Delta X^2_{\text{vac}}$, the solid curve represents the normalized variance $\Delta X^2_-/\Delta X^2_{\text{vac}}$, and the dot-dashed curve gives the gain (dB). (b) The plot zooms the squeezed variance ΔX^2_+ and the corresponding gain (dB). Parameters: $\theta_a + \theta_b = 0$, $\Delta\beta = 1.8$/km, $z = 1$ km, $\gamma = 2.5$/W·km, and $P = 0.5$ W.

for the non-phase-matched case. This shows that at the point of perfect squeezing, the gain is about 7 dB. We have presented the basic equations for nonclassical and entangled fields in fibers. The main results are the Bogoliubov transformations (3.119) which can be used to calculate all the relevant physical quantities. Even the entanglement in the two modes can be calculated using the results of Section 3.13. The covariance matrix (3.95) can be calculated using Eq. (3.119). We remind the reader that since the input modes a and b are in coherent states the generated fields $a(z)$ and $b(z)$ would have a Gaussian Wigner function.

Exercises

3.1 Find the scalar product $\langle\xi_1|\xi_2\rangle$ of two different two-mode squeezed states (3.6), where $\xi_i = r_i e^{i\varphi_i}$, $i = 1, 2$.

3.2 Show that the correlation $\langle a^\dagger a b^\dagger b\rangle - \langle a^\dagger a\rangle\langle b^\dagger b\rangle$ between the modes a and b for modes in the state (3.6) is $\sinh^2 r + \sinh^4 r$.

3.3 Show that the operators S_-, S_+, and S_3 defined by

$$S_- = ab^\dagger, \qquad S_+ = a^\dagger b, \qquad S_3 = \frac{1}{2}(a^\dagger a - b^\dagger b),$$

satisfy SU(2) algebra

$$[S_-, S_+] = -2S_3, \qquad [S_3, S_-] = -S_-, \qquad [S_3, S_+] = S_+.$$

Show also that the operator $(a^\dagger a + b^\dagger b)$ commutes with all S_i. This representation would be useful in the context of polarization of light (Chapter 6), and is called the Schwinger boson representation of angular momentum algebra [41].

3.4 Show that the operators K_-, K_+ and K_3, $K_- = ab$, $K_+ = a^\dagger b^\dagger$, $K_3 = \frac{1}{2}(a^\dagger a + b^\dagger b + 1)$, constructed out of two-mode Bosonic operators satisfy SU(1,1) algebra [41]:

$$[K_-, K_+] = 2K_3, \qquad [K_3, K_+] = K_+, \qquad [K_3, K_-] = -K_-.$$

3.5 Prove the relation (3.25) for the transformation (3.24).

3.6 Use the Bogoliubov transformation for two modes

$$a = c\cosh r + d^\dagger e^{-i\theta}\sinh r,$$
$$b = d\cosh r + c^\dagger e^{-i\theta}\sinh r,$$

to diagonalize the Hamiltonian

$$H = \omega a^\dagger a + \omega b^\dagger b + g a^\dagger b^\dagger + g^* ab$$

to $\Omega(c^\dagger c + d^\dagger d) + \alpha$. Then find Ω, r, θ, α.
[Solution: $\Omega = \sqrt{\omega^2 - |g|^2}$, $r = \frac{1}{4}\ln\frac{\omega - \sqrt{|g|^2}}{\omega + \sqrt{|g|^2}}$, $\theta = i\ln\sqrt{\frac{g^*}{g}}$, $\alpha = \sqrt{\omega^2 - |g|^2} - \omega$.]

3.7 Obtain a 2×2 matrix representation of the squeezing operator $S(\xi)$

$$\begin{pmatrix} \cosh r & \sinh r e^{i\varphi} \\ \sinh r e^{-i\varphi} & \cosh r \end{pmatrix}$$

by using the nonunitary realization of SU(1,1) algebra

$$a^\dagger b^\dagger = \begin{pmatrix} 0 & 1 \\ 0 & 0 \end{pmatrix}, \quad -ab = \begin{pmatrix} 0 & 0 \\ 1 & 0 \end{pmatrix}, \quad \frac{a^\dagger a + b^\dagger b + 1}{2} = \begin{pmatrix} \frac{1}{2} & 0 \\ 0 & -\frac{1}{2} \end{pmatrix}.$$

This nonunitary representation is useful in calculations, for example it can easily yield the decomposition (3.4) [42, 43].

3.8 Using the 2×2 matrix form of $S(\xi)$ from Exercise 3.7, prove the product theorem for two squeezing operators

$$S(\xi_1)S(\xi_2) = S(\xi_3)\exp\left[\frac{i}{2}\Phi(\xi_1, \xi_2)(a^\dagger a + b^\dagger b + 1)\right],$$

$$\Phi(\xi_1, \xi_2) = \frac{1}{i}\ln\left(\frac{1 + \zeta_1\zeta_2^*}{1 + \zeta_1^*\zeta_2}\right),$$

$$(\tanh r_3)e^{i\varphi_3} = \frac{\zeta_1 + \zeta_2}{1 + \zeta_1^*\zeta_2},$$

$$\zeta_i = (\tanh r_i)e^{i\varphi_i}, \qquad \xi_i = r_i e^{i\varphi_i}.$$

Such a product theorem is useful in analyzing the effect of several down-converters in a cascade arrangement as in the context of SU(1,1) interferometers.

3.9 Solve the eigenvalue problem

$$\frac{1}{\sqrt{(1+a^\dagger a)(1+b^\dagger b)}}\, ab|\Psi_{\lambda q}\rangle = \lambda|\Psi_{\lambda q}\rangle,$$

$$(a^\dagger a - b^\dagger b)|\Psi_{\lambda q}\rangle = q|\Psi_{\lambda q}\rangle,$$

where q is an integer. Show that, in the limit, $q \to 0$, $|\Psi_{\lambda q}\rangle$ is identical to the two-mode squeezed vacuum (for details see [44]).

References

[1] B. L. Schumaker and C. M. Caves, *Phys. Rev. A* **31**, 3093 (1985).
[2] R. Loudon, *Rep. Prog. Phys.* **43**, 913 (1980).
[3] M. D. Reid and D. F. Walls, *Phys. Rev. A* **34**, 1260 (1986).
[4] G. S. Agarwal, *J. Opt. Soc. Am. B* **5**, 1940 (1988).
[5] H. Umezawa, H. Matsumoto, and M. Tachiki, *Thermo Field Dynamics and Condensed States* (Amsterdam: North-Holland, 1982).
[6] A. Zavatta, S. Viciani, and M. Bellini, *Science* **306**, 660 (2004).
[7] G. Bertocchi, O. Alibart, D. B. Ostrowsky, S. Tanzilli, and P. Baldi, *J. Phys. B* **39**, 1011 (2006).
[8] Special issue, *J. Mod. Opt.* **56**, Nos. 2,3, (2009).
[9] A. Ourjoumtsev, R. Tualle-Brouri, and P. Grangier, *Phys. Rev. Lett.* **96**, 213601 (2006).
[10] C. K. Law, I. A. Walmsley, and J. H. Eberly, *Phys. Rev. Lett.* **84**, 5304 (2000).
[11] A. Peres, *Phys. Rev. Lett.* **77**, 1413 (1996).
[12] P. Horodecki, *Phys. Lett. A* **232**, 333 (1997).
[13] R. Simon, *Phys. Rev. Lett.* **84**, 2726 (2000).
[14] Lu-Ming Duan, G. Giedke, J. I. Cirac, and P. Zoller, *Phys. Rev. Lett.* **84**, 2722 (2000).
[15] G. Vidal and R. F. Werner, *Phys. Rev. A* **65**, 032314 (2002).
[16] M. B. Plenio, *Phys. Rev. Lett.* **95**, 090503 (2005).
[17] C. K. Hong, Z. Y. Ou, and L. Mandel, *Phys. Rev. Lett.* **59**, 2044 (1987).
[18] J. P. Dowling, *Contemp. Phys.* **49**, 125 (2008).
[19] L. Mandel and E. Wolf, *Optical Coherence and Quantum Optics* (New York: Cambridge University Press, 1995), Chap. 22.
[20] R. Boyd, *Nonlinear Optics*, 2nd ed. (San Diego, CA: Academic Press, 2003).
[21] M. H. Rubin, *Phys. Rev. A* **54**, 5349 (1996).
[22] M. H. Rubin, D. N. Klyshko, Y. H. Shih, and A. V. Sergienko, *Phys. Rev. A* **50**, 5122 (1994).
[23] F. A. Hopf and G. I. Stegeman, *Applied Classical Electrodynamics, Vol. 2: Nonlinear Optics* (New York: Wiley-Interscience, 1986), Chap. 7.

[24] F. A. Hopf and G. I. Stegeman, *Applied Classical Electrodynamics, Vol. 2: Nonlinear Optics* (New York: Wiley-Interscience, 1986), Chap. 15.

[25] P. G. Kwiat, K. Mattle, H. Weinfurter, A. Zeilinger, A. V. Sergienko, and Y. Shih, *Phys. Rev. Lett.* **75**, 4337 (1995).

[26] S. Gaertner, M. Bourennane, M. Eibl, C. Kurtsiefer, and H. Weinfurter, *Appl. Phys. B* **77**, 803 (2003).

[27] H. Weinfurter and M. Żukowski, *Phys. Rev. A* **64**, 010102(R) (2001).

[28] G. S. Agarwal, *Quantum Opt.* **2**, 1 (1990).

[29] M. B. Plenio, *Phys. Rev. Lett.* **95**, 090503 (2005); G. Adesso, A. Serafini, and F. Illuminati, *Phys. Rev. A* **70**, 022318 (2004).

[30] C. H. Bennett, G. Brassard, C. Crépeau, R. Jozsa, A. Peres, and W. K. Wootters, *Phys. Rev. Lett.* **70**, 1895 (1993).

[31] D. Bouwmeester, J. W. Pan, K. Mattle, M. Eibl, H. Weinfurter, and A. Zeilinger, *Nature (London)* **390**, 575 (1997).

[32] M. Riebe, H. Häffner, C. F. Roos *et al.*, *Nature (London)* **429**, 734 (2004).

[33] M. D. Barrett, J. Chiaverini, T. Schaetz *et al.*, *Nature (London)* **429**, 737 (2004).

[34] A. Furusawa, J. L. Sørensen, S. L. Braunstein, C. A. Fuchs, H. J. Kimble, and E. S. Polzik, *Science* **282**, 706 (1998).

[35] N. Lee, H. Benichi, Y. Takeno *et al.*, *Science* **332**, 330 (2011).

[36] L. Mišta, Jr., R. Filip, and A. Furusawa, *Phys. Rev. A* **82**, 012322 (2010).

[37] G. P. Agrawal, *Nonlinear Fiber Optics* (New York: Academic Press, 1995), Chap. 10.

[38] J. Hansryd, P. A. Andrekson, M. Westland, J. Li, and P. O. Hedekvist, *IEEE J. Sel. Top. Quantum Electron.* **8**, 506 (2002).

[39] C. J. McKinstrie, M. Yu, M. G. Raymer, and S. Radic, *Opt. Express* **13**, 4986 (2005).

[40] J. E. Sharpening, M. Fiorentino, and P. Kumar, *Opt. Lett.* **26**, 367 (2001).

[41] R. R. Puri, *Mathematical Methods of Quantum Optics* (Berlin: Springer-Verlag, 2001).

[42] R. A. Fisher, M. M. Nieto, and V. D. Sandberg, *Phys. Rev. D* **29**, 1107 (1984).

[43] Ibid., Ref. 1, Appendix B.

[44] G. S. Agarwal, *Opt. Commun.* **100**, 479 (1993).

4 Non-Gaussian nonclassical states

Most nonclassical states that we discussed in Chapters 2 and 3 are Gaussian in the sense that the Wigner function for such states was Gaussian. The only exception to the Gaussian states is the n photon Fock state for which the Wigner function is non-Gaussian and negative for certain domains in the phase space. In this chapter we introduce many other classes of nonclassical states which typically are non-Gaussian.

4.1 Schrödinger cat state and the cat paradox

Schrödinger, in 1935, proposed a thought experiment which shows how quantum-mechanical description can be in contradiction to our everyday experience [1]. He considered a closed box containing a cat, which is a macroscopic object, and a radioactive atom. The box is also supposed to contain a vial of cyanide. When the radioactive atom has decayed, the cyanide is released and the cat dies. Let $|e\rangle$ and $|g\rangle$ be the excited and ground states of the radioactive atom and let τ be the life time of the excited state $|e\rangle$. A consistent quantum-mechanical description of what happens in the box would require us to treat a macroscopic object like the cat as a quantum system. Clearly, after a time $\tau/2$ the combined state of atom and cat would be

$$|\text{atom–cat}\rangle = \frac{1}{\sqrt{2}}(|e\rangle|\text{live cat}\rangle + |g\rangle|\text{dead cat}\rangle), \tag{4.1}$$

which is an entangled state of a microscopic and a macroscopic system. Thus when we look there is a 50% probability of finding the cat dead or alive. This is clearly in contradiction to our everyday experience. This is called the cat paradox and has led to a lively debate over the years about the completeness of quantum mechanics in the description of reality [2].

The quantum optics community has taken a great interest in these issues because the coherent states describe naturally the states of macroscopic systems if the excitation amplitude is large. We need to consider the possibility of superpositions of such states. Mathematically speaking, we are dealing with superpositions of Gaussian wavepackets and hence such superpositions are examples of non-Gaussian states. Generally in quantum mechanics we use the states of a Hamiltonian to construct the superposition of states. The coherent states and squeezed states are superpositions of harmonic oscillator states. However, the cat states are formed from a superposition [3–5] of two different coherent states

$$|\alpha_0\rangle_c = \mathcal{N}^{-1}(|\alpha_0\rangle + e^{i\varphi}|-\alpha_0\rangle), \tag{4.2}$$

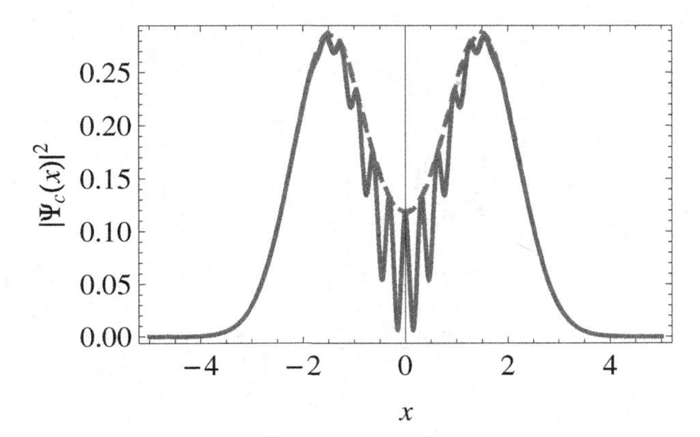

Fig. 4.1 The quadrature distribution $|\Psi_c(x)|^2$ as a function of x for $\varphi = 0$, $x_0 = 1.5$, and $y_0 = 0$ (dashed), 10 (solid).

where $\mathcal{N}$ is the normalization constant

$$\begin{aligned} {}_c\langle\alpha_0|\alpha_0\rangle_c = 1 &= \mathcal{N}^{-2}(\langle\alpha_0| + e^{-i\varphi}\langle-\alpha_0|)(|\alpha_0\rangle + e^{i\varphi}|-\alpha_0\rangle) \\ &= \mathcal{N}^{-2}(2 + \langle\alpha_0|-\alpha_0\rangle e^{i\varphi} + e^{-i\varphi}\langle-\alpha_0|\alpha_0\rangle), \end{aligned} \tag{4.3}$$

and therefore

$$\mathcal{N} = \left(2 + 2e^{-2|\alpha_0|^2}\cos\varphi\right)^{1/2}. \tag{4.4}$$

The two parts of (4.2) are nonorthogonal $\langle\alpha_0|-\alpha_0\rangle \neq 0$. In order to understand the meaning of (4.2), we examine it in the quadrature space

$$\Psi_c(x) = \langle x|\alpha_0\rangle_c = \mathcal{N}^{-1}(\langle x|\alpha_0\rangle + e^{i\varphi}\langle x|-\alpha_0\rangle), \tag{4.5}$$

which on using (1.65) and writing $\alpha_0 = \frac{x_0+iy_0}{\sqrt{2}}$, becomes

$$\Psi_c(x) = \mathcal{N}^{-1}e^{-i\frac{x_0y_0}{2}}\pi^{-1/4}\left[e^{iy_0x-\frac{1}{2}(x-x_0)^2} + e^{i\varphi}e^{-iy_0x}e^{-\frac{1}{2}(x+x_0)^2}\right]. \tag{4.6}$$

The quadrature distribution is then

$$|\Psi_c(x)|^2 = \mathcal{N}^{-2}\pi^{-1/2}\left[e^{-(x-x_0)^2} + e^{-(x+x_0)^2} + 2e^{-(x^2+x_0^2)}\cos(\varphi - 2y_0x)\right]. \tag{4.7}$$

This is a superposition of two Gaussian wavepackets located at $\pm x_0$ and an interference term whose oscillatory character depends on the imaginary part of α_0. The magnitude of the interference term depends on the real part of α_0. We display the behavior of the quadrature distribution (4.7) in Figure 4.1 for $\varphi = 0$. There is no interference for $y_0 = 0$. However, in this case a well-defined interference appears in a complementary space, i.e. in the quadrature space y. Following the procedure that led to (4.7), it can be shown that for $y_0 = 0$

$$|\Psi_c(y)|^2 = 2\mathcal{N}^{-2}\pi^{-1/2}e^{-y^2}[1 + \cos(\varphi + 2x_0y)]. \tag{4.8}$$

The quadrature distribution $|\Psi_c(y)|^2$ shows modulations as a function of y if $x_0 \neq 0$. This is shown in Figure 4.2.

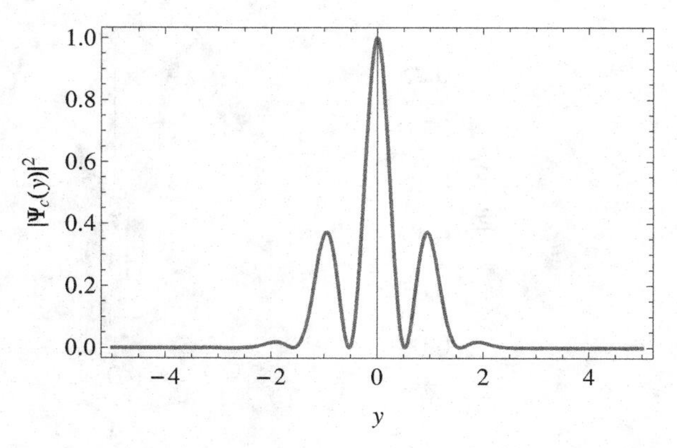

Fig. 4.2 The quadrature distribution $|\Psi_c(y)|^2$ as a function of y for $\varphi = 0$, $x_0 = 3$, and $y_0 = 0$.

This simple exercise suggests that if interferences are lost in one space, then they can be seen in the complementary space [6]. This is a consequence of the Heisenberg uncertainty relation for the quadratures x and y. This also suggests that we need a formulation where the totality of the interference effects can be seen. This is where the Wigner function is useful because it displays the totality of interferences (Eq. (4.15)). Interferences in complementary spaces were first seen in neutron interferometry [7], where the neutron flux after the interferometer can be measured either in coordinate space or in momentum space. In Chapter 8, we will give another example of interferences in complementary spaces involving the time domain and spectral measurements.

We next examine the nonclassical properties of (4.1). The P-function for the state (4.2) does not exist. The Q-function is easily calculated to be

$$Q(\alpha) = |\langle\alpha|\alpha_0\rangle_c|^2 = 4\mathcal{N}^{-2}\mathrm{e}^{-|\alpha|^2-|\alpha_0|^2}\left|\cosh\left(\alpha^*\alpha_0 - \mathrm{i}\frac{\varphi}{2}\right)\right|^2. \tag{4.9}$$

The Q-function has zeroes in the complex α plane. In order to see this, let us set α_0 real, $\alpha^* = -\mathrm{i}\mathrm{Im}\{\alpha\}$. Then the zeroes are given by

$$\cos\left(\alpha_0\mathrm{Im}\{\alpha\} + \frac{\varphi}{2}\right) = 0, \tag{4.10}$$

which lie on the imaginary α axis. Hence the state (4.2) has nonclassical properties since the zeroes in the Q-function signify the nonclassical nature of the underlying state [8].

The Wigner function for the state (4.2) can take negative values. The calculation of the Wigner function is complicated as the density matrix associated with (4.2) is

$$\rho_c = \mathcal{N}^{-2}(|\alpha_0\rangle\langle\alpha_0| + |-\alpha_0\rangle\langle-\alpha_0| + |\alpha_0\rangle\langle-\alpha_0|\mathrm{e}^{-\mathrm{i}\varphi} + |-\alpha_0\rangle\langle\alpha_0|\mathrm{e}^{\mathrm{i}\varphi}). \tag{4.11}$$

The Wigner function for the first two terms is simple and is obtained from (1.101). Using (1.98) the contribution ϑ of $|\alpha_0\rangle\langle-\alpha_0|$ to the Wigner function can be obtained as follows:

$$\begin{aligned}\vartheta &= \frac{1}{\pi^2}\int \mathrm{d}^2\beta\langle-\alpha_0|D(\beta)|\alpha_0\rangle\mathrm{e}^{-(\beta\alpha^*-\beta^*\alpha)} \\ &= \frac{1}{\pi^2}\int \mathrm{d}^2\beta\langle-\alpha_0|\mathrm{e}^{a^\dagger\beta}\mathrm{e}^{-a\beta^*}|\alpha_0\rangle\mathrm{e}^{-\frac{1}{2}|\beta|^2}\mathrm{e}^{-\beta\alpha^*+\beta^*\alpha},\end{aligned} \tag{4.12}$$

where we used the normally ordered form of the displacement operator (see Table 1.1, Eq. (1)). The evaluation of the matrix element in (4.12) is straightforward if we use $a|\alpha_0\rangle = \alpha_0|\alpha_0\rangle$, $\langle\alpha_0|a^\dagger = \alpha_0^*\langle\alpha_0|$; leading to

$$\vartheta = \frac{1}{\pi^2}\int d^2\beta e^{-\frac{1}{2}|\beta+2\alpha_0|^2-\beta\alpha^*+\beta^*\alpha}. \tag{4.13}$$

The integral can be done using (1.86) with the result

$$\vartheta = \frac{2}{\pi}e^{-2|\alpha|^2}e^{2\alpha_0\alpha^*-2\alpha_0^*\alpha}. \tag{4.14}$$

Using now (1.101) and (4.14), the Wigner function for the cat state is

$$\begin{aligned} W_c(\alpha) = {} & \frac{2\mathcal{N}^{-2}}{\pi}\{\exp(-2|\alpha-\alpha_0|^2)+\exp(-2|\alpha+\alpha_0|^2) \\ & +2e^{-2|\alpha|^2}\cos[\varphi+4\text{Im}\{\alpha_0^*\alpha\}]\}. \end{aligned} \tag{4.15}$$

We show the behavior of the Wigner function in Figures 4.3 and 4.4 for $\varphi = 0, \pi$. We also show the contours of the Wigner functions. The noteworthy features are (i) the negativity of the Wigner function, and (ii) the oscillatory character arising from the interference due to the superposition of two coherent states. The oscillations occur in the region around $|\alpha| \sim 0$ as then the coefficient of the cosine term in (4.15) is close to unity.

We can now characterize the nonclassicality of the cat state in terms of sub-Poissonian statistics and squeezing. We note the unusual property of the cat state

$$a^2|\alpha_0\rangle_c = \alpha_0^2|\alpha_0\rangle_c, \tag{4.16}$$

and therefore

$$\langle a^{\dagger 2}a^2\rangle = |\alpha_0|^4, \quad \langle a^2\rangle = \alpha_0^2. \tag{4.17}$$

The mean number of photons is

$$\langle a^\dagger a\rangle = 2|\alpha_0|^2(1-e^{-2|\alpha_0|^2}\cos\varphi)\mathcal{N}^{-2}. \tag{4.18}$$

The Mandel Q_M parameter is found to be

$$\begin{aligned} Q_M &= \frac{\langle a^{\dagger 2}a^2\rangle - \langle a^\dagger a\rangle^2}{\langle a^\dagger a\rangle} = 4|\alpha_0|^2\cos\varphi\, e^{-2|\alpha_0|^2}/(1-\cos^2\varphi\, e^{-4|\alpha_0|^2}) \\ &< 0 \quad \text{if} \quad \cos\varphi < 0. \end{aligned} \tag{4.19}$$

The cat states have sub-Poissonian statistics if the relative phase in the coherent superposition lies between $\frac{\pi}{2}$ and $\frac{3\pi}{2}$.

The squeezing parameter is more complex due to the number of phase angles. The mean amplitude of a is

$$\begin{aligned} \langle a\rangle &= \mathcal{N}^{-2}\alpha_0(\langle\alpha_0| + e^{-i\varphi}\langle-\alpha_0|)(|\alpha_0\rangle - e^{i\varphi}|-\alpha_0\rangle) \\ &= -2\mathcal{N}^{-2}\alpha_0 i e^{-2|\alpha_0|^2}\sin\varphi \neq 0 \quad \text{if} \quad \varphi \neq n\pi. \end{aligned} \tag{4.20}$$

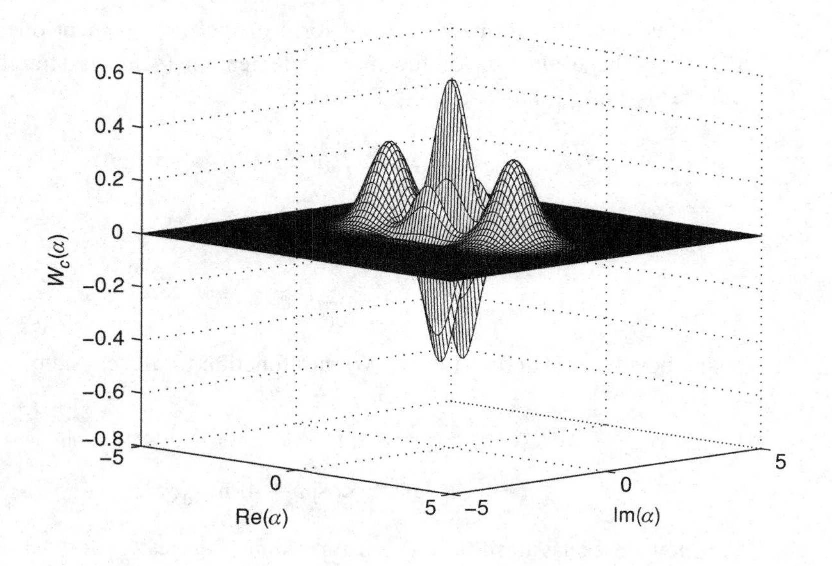

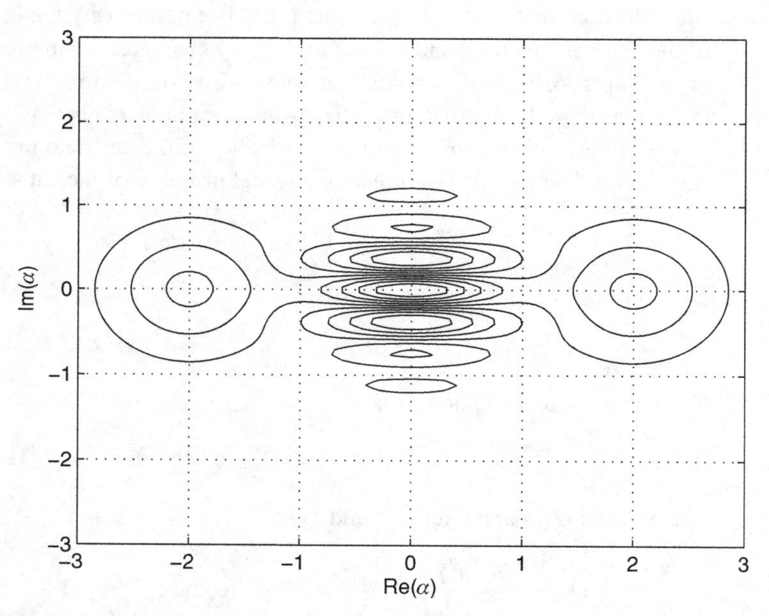

Fig. 4.3 The Wigner function $W_c(\alpha)$ for the cat state for $\alpha_0 = 2$, $\varphi = 0$.

All the quantities needed for the evaluation of the squeezing parameter are given by (4.17), (4.18), and (4.20). For real α_0, a simple algebraic calculation yields

$$S_\theta = \frac{|\alpha_0|^2}{(1+e^{-2|\alpha_0|^2}\cos\varphi)^2}\left[\cos 2\theta\,(1+e^{-2|\alpha_0|^2}\cos\varphi)^2 + 1 - e^{-4|\alpha_0|^2} + \cos^2\theta\,\sin^2\varphi\, e^{-4|\alpha_0|^2}\right]. \tag{4.21}$$

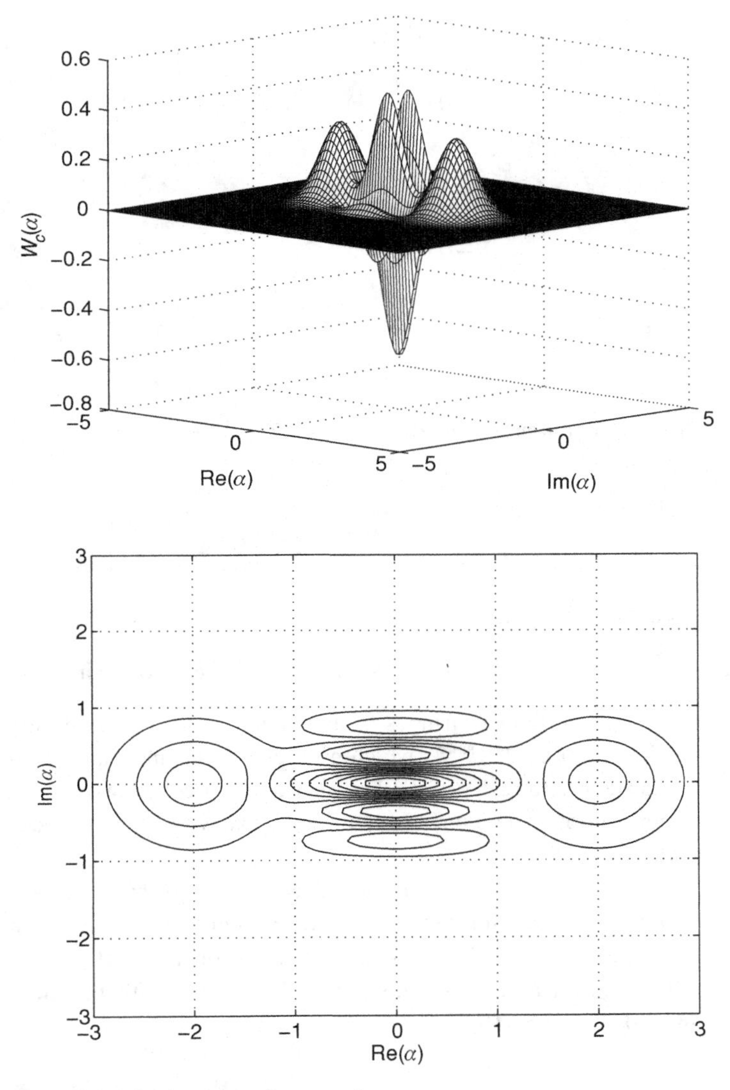

Fig. 4.4 The Wigner function $W_c(\alpha)$ for the cat state for $\alpha_0 = 2, \varphi = \pi$.

For fixed $|\alpha_0|$, the behavior of S_θ is shown in Figure 4.5. This figure shows the regions where the cat state has squeezing.

The cat states have been realized experimentally for several Bosonic systems, using (a) Rydberg atoms in high-quality dispersive cavities [9], and (b) ion traps where motional degrees of freedom are prepared in cat states [10]. The observation of the cat states is rather indirect as one studies experimentally the population of the atomic states or electronic states of the ion. Here the observed quantity is directly the norm of the cat state. Consider

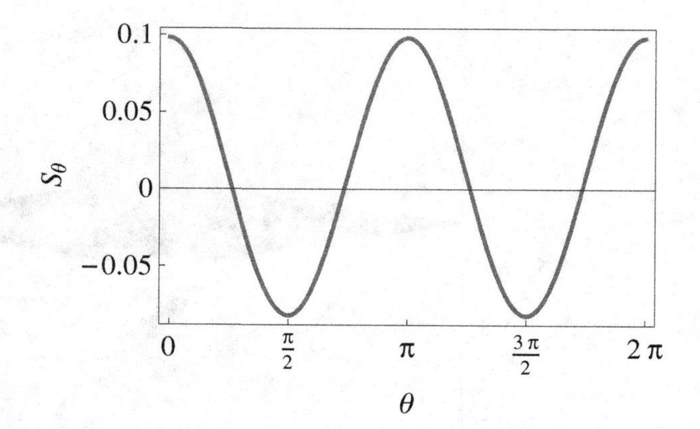

Fig. 4.5 The squeezing parameter $S(\theta)$ as a function of θ for $|\alpha_0| = 0.3$ and $\varphi = 0$.

a slightly more general state (4.2)

$$|\alpha_0\rangle_c = \mathcal{N}^{-1}(|\alpha_0 e^{i\theta}\rangle + e^{i\varphi}|\alpha_0 e^{-i\theta}\rangle), \tag{4.22}$$

then the norm of the state is

$$\mathcal{N}^2 = 2 + 2e^{-2|\alpha_0|^2 \sin^2\theta} \cos(|\alpha_0|^2 \sin 2\theta - \varphi). \tag{4.23}$$

One has interference fringes as a function of φ. The visibility of these fringes is given by the factor $e^{-2|\alpha_0|^2 \sin^2\theta}$, which goes down as θ increases. Note that the visibility is just the distance $|\langle\alpha_0 e^{-i\theta}|\alpha_0 e^{i\theta}\rangle|$ between the two coherent states $|\alpha_0 e^{i\theta}\rangle$ and $|\alpha_0 e^{-i\theta}\rangle$ in the superposition. The interference effects go down as the distance between the two coherent states increases.

A direct observation [11–13] of cat-like states has also been made by the technique of photon subtraction using the states produced by a down-converter – this is discussed in Section 4.6. Finally, an important question is: what is the life time of the coherent superposition as the systems generally interact with the environment? This subject is treated in Section 10.7.

4.2 Photon-added and -subtracted states

A number of new states can be produced by the process of addition or subtraction of photons. This is also the route to non-Gaussian states from Gaussian states. Furthermore, the process of addition/subtraction changes the quantum statistics of the fields. We could then inquire if it is possible to produce nonclassical fields starting from classical fields. We will show that the process of subtraction can not make a classical field nonclassical. Let ρ be the density matrix of a given field. If we subtract one photon, then the state becomes

$$\rho^{(s)} = a\rho a^\dagger/\mathrm{Tr}(a\rho a^\dagger) = a\rho a^\dagger/\langle a^\dagger a\rangle, \tag{4.24}$$

where all averages with respect to original ρ are denoted by $\langle\ \rangle$. Furthermore, we denote all averages w.r.t. $\rho^{(s)}$ by $\langle\ \rangle_s$. Let us find the P-function for the state $\rho^{(s)}$. Using (4.24) and the property of coherent states $a|\alpha\rangle = \alpha|\alpha\rangle$ we obtain

$$\rho^{(s)} = \int P(\alpha)|\alpha|^2|\alpha\rangle\langle\alpha|\mathrm{d}^2\alpha/\langle a^\dagger a\rangle = \int P_s(\alpha)|\alpha\rangle\langle\alpha|\mathrm{d}^2\alpha, \tag{4.25}$$

where

$$P_s(\alpha) = |\alpha|^2 P(\alpha)/\langle a^\dagger a\rangle. \tag{4.26}$$

From (4.26) we conclude that if $P(\alpha)$ is classical, then $P_s(\alpha)$ can not be nonclassical. Thus $P_s(\alpha)$ can exhibit nonclassical properties only if the original $P(\alpha)$ is nonclassical.

Let us now consider the process of addition of photons [14]. This is expected to produce nonclassical states out of classical states. Consider the vacuum state. Addition of one photon produces the state $|1\rangle$, which is highly nonclassical. More generally, we examine the properties of the photon-added state

$$\rho^{(a)} = a^\dagger\rho a/\mathrm{Tr}(a^\dagger\rho a) = a^\dagger\rho a/\langle aa^\dagger\rangle. \tag{4.27}$$

The Q-function associated with (4.27) is

$$Q^{(a)}(\alpha) = \langle\alpha|a^\dagger\rho a|\alpha\rangle/\langle aa^\dagger\rangle = |\alpha|^2 Q(\alpha)/\langle aa^\dagger\rangle. \tag{4.28}$$

The function $Q^{(a)}(\alpha)$ can have a zero even if $Q(\alpha)$ did not have any zero, which leads to the conclusion that $\rho^{(a)}$ can be nonclassical even if ρ were classical. Let us then evaluate the P-function for the state $\rho^{(a)}$

$$\rho^{(a)} = \int P(\alpha)a^\dagger|\alpha\rangle\langle\alpha|a\,\mathrm{d}^2\alpha/\langle aa^\dagger\rangle. \tag{4.29}$$

To simplify (4.29) we use the expansion of coherent states in terms of Fock states

$$\begin{aligned}
a^\dagger|\alpha\rangle\langle\alpha|a &= \sum_{nm} \mathrm{e}^{-|\alpha|^2}\frac{\alpha^n}{\sqrt{n!}}a^\dagger|n\rangle\langle m|a\frac{\alpha^{*m}}{\sqrt{m!}} \\
&= \sum_{nm} \mathrm{e}^{-|\alpha|^2}\frac{\alpha^n(n+1)}{\sqrt{(n+1)!}}|n+1\rangle\langle m+1|\frac{\alpha^{*m}(m+1)}{\sqrt{(m+1)!}} \\
&= \mathrm{e}^{-|\alpha|^2}\frac{\partial^2}{\partial\alpha\partial\alpha^*}\sum_{nm}\frac{\alpha^{n+1}(\alpha^*)^{m+1}}{\sqrt{(n+1)!(m+1)!}}|n+1\rangle\langle m+1| \\
&= \mathrm{e}^{-|\alpha|^2}\frac{\partial^2}{\partial\alpha\partial\alpha^*}\mathrm{e}^{|\alpha|^2}|\alpha\rangle\langle\alpha|.
\end{aligned} \tag{4.30}$$

On substituting (4.30) in (4.29) and on integration by parts we get

$$P^{(a)}(\alpha) = \mathrm{e}^{|\alpha|^2}\frac{\partial^2}{\partial\alpha\partial\alpha^*}P(\alpha)\mathrm{e}^{-|\alpha|^2}/\langle aa^\dagger\rangle. \tag{4.31}$$

This is an important relation and shows how the state $\rho^{(a)}$ can be nonclassical even if ρ were classical.

4.3 Single-photon-added coherent and thermal states

Consider two important cases where nonclassical states can be obtained from well-known classical states. We examine the photon-added coherent state defined by [14]

$$\rho_c^{(a)} = a^\dagger |\alpha_0\rangle\langle\alpha_0| a/(1+|\alpha_0|^2) \tag{4.32}$$

and the photon-added thermal state defined by [15]

$$\rho_{\mathrm{T}}^{(a)} = a^\dagger \mathrm{e}^{-\beta\hbar\omega a^\dagger a} a/\mathrm{Tr}(\mathrm{e}^{-\beta\hbar\omega a^\dagger a} a a^\dagger). \tag{4.33}$$

Note that in the limit $\alpha_0 \to 0$, (4.32) goes over to the highly nonclassical single-photon state. For large α_0, the addition of a single photon is expected to be insignificant. The state (4.32) is thus remarkable in the sense that it can allow us to study quantum to classical transition by changing the value of α_0. The nonclassical properties are reflected in the P-function for the states (4.32) and (4.33). Using (4.31) and

$$P_c(\alpha) = \delta^{(2)}(\alpha - \alpha_0), \qquad P_{\mathrm{T}}(\alpha) = \frac{1}{\pi\bar{n}} \mathrm{e}^{-|\alpha|^2/\bar{n}}, \tag{4.34}$$

we get the P-function for the single-photon-added states

$$P_c^{(a)}(\alpha) = \mathrm{e}^{|\alpha|^2} \frac{\partial^2}{\partial\alpha\partial\alpha^*} \delta^{(2)}(\alpha - \alpha_0) \mathrm{e}^{-|\alpha|^2}/(1+|\alpha_0|^2), \tag{4.35}$$

$$P_{\mathrm{T}}^{(a)}(\alpha) = \frac{1+\langle n\rangle}{\langle n\rangle^2} \left[|\alpha|^2 - \frac{\langle n\rangle}{1+\langle n\rangle} \right] P_{\mathrm{T}}(\alpha), \tag{4.36}$$

showing that

$$P_{\mathrm{T}}^{(a)}(\alpha) < 0, \qquad \text{if} \qquad |\alpha|^2 < \frac{\langle n\rangle}{1+\langle n\rangle}. \tag{4.37}$$

The P-function associated with the single-photon-added thermal state $\rho_{\mathrm{T}}^{(a)}$ can be negative. The negative region depends on the average number of photons in the thermal field. Here we perhaps have the unique case where $P_{\mathrm{T}}^{(a)}$ exists but is negative. Most known cases of nonclassical P involve P-functions which do not even exist.

Since the P-function for the photon-added coherent state is singular, it is advantageous to study its nonclassical properties in terms of the Wigner function. To do this it is useful to find the relationship between the Wigner functions for $\rho^{(a)}$ and ρ. Using the definition (1.98), the Wigner function for $\rho^{(a)}$ will be

$$W^{(a)}(\alpha) = \frac{1}{\pi^2} \int \mathrm{d}^2\beta \, \mathrm{Tr}[\rho a D(\beta) a^\dagger] \mathrm{e}^{-(\beta\alpha^* - \beta^*\alpha)}/\langle a a^\dagger\rangle. \tag{4.38}$$

To simplify (4.38) we write $D(\beta)$ in antinormally ordered form

$$\begin{aligned} \langle a D(\beta) a^\dagger\rangle &= \mathrm{e}^{\frac{1}{2}|\beta|^2} \langle a \mathrm{e}^{-\beta^* a} \mathrm{e}^{\beta a^\dagger} a^\dagger\rangle \\ &= \mathrm{e}^{\frac{1}{2}|\beta|^2} \frac{\partial^2}{\partial(-\beta^*)\partial\beta} \langle \mathrm{e}^{-\beta^* a} \mathrm{e}^{\beta a^\dagger}\rangle \\ &= \mathrm{e}^{\frac{1}{2}|\beta|^2} \frac{\partial^2}{\partial\beta\partial(-\beta^*)} \langle D(\beta)\rangle \mathrm{e}^{-\frac{1}{2}|\beta|^2}. \end{aligned} \tag{4.39}$$

Substituting in (4.38) and integration by parts leads to

$$W^{(a)}(\alpha) = \frac{1}{\pi^2}\int \mathrm{d}^2\beta \langle D(\beta)\rangle \mathrm{e}^{-\frac{1}{2}|\beta|^2} \frac{\partial^2}{\partial\beta\partial(-\beta^*)} \mathrm{e}^{\frac{1}{2}|\beta|^2-\beta\alpha^*+\beta^*\alpha}/\langle aa^\dagger\rangle, \tag{4.40}$$

which is easily shown to be equivalent to

$$W^{(a)}(\alpha) = \frac{1}{\langle aa^\dagger\rangle}\left(\alpha^* - \frac{1}{2}\frac{\partial}{\partial\alpha}\right)\left(\alpha - \frac{1}{2}\frac{\partial}{\partial\alpha^*}\right) W(\alpha). \tag{4.41}$$

This is the desired relation between the Wigner functions of $\rho^{(a)}$ and ρ. After carrying out the necessary differentiations we can obtain $W^{(a)}$ for both single-photon-added coherents and thermal states. The final results are:

I.
$$\begin{aligned} W_c^{(a)}(\alpha) &= \frac{1}{\langle aa^\dagger\rangle}\left(\alpha^* - \frac{1}{2}\frac{\partial}{\partial\alpha}\right)\left(\alpha - \frac{1}{2}\frac{\partial}{\partial\alpha^*}\right)\frac{2}{\pi}\mathrm{e}^{-2|\alpha-\alpha_0|^2}, \\ &= \frac{2}{\pi(1+|\alpha_0|^2)}\left[4|\alpha|^2 - 2(\alpha^*\alpha_0 + \alpha\alpha_0^*) - 1 + |\alpha_0|^2\right]\mathrm{e}^{-2|\alpha-\alpha_0|^2}, \end{aligned} \tag{4.42}$$

which is negative for

$$4|\alpha|^2 - 2(\alpha^*\alpha_0 + \alpha\alpha_0^*) - 1 + |\alpha_0|^2 < 0. \tag{4.43}$$

II.
$$\begin{aligned} W_{\mathrm{T}}^{(a)}(\alpha) &= \frac{1}{\langle aa^\dagger\rangle}\left(\alpha^* - \frac{1}{2}\frac{\partial}{\partial\alpha}\right)\left(\alpha - \frac{1}{2}\frac{\partial}{\partial\alpha^*}\right)\frac{1}{\pi(\langle n\rangle + \frac{1}{2})}\mathrm{e}^{-|\alpha|^2/(\langle n\rangle+\frac{1}{2})} \\ &= \left[|\alpha|^2\left(\frac{2\langle n\rangle+2}{2\langle n\rangle+1}\right) - \frac{1}{2}\right]\frac{\mathrm{e}^{-2|\alpha|^2/(1+2\langle n\rangle)}}{\pi(\langle n\rangle+\frac{1}{2})^2}, \end{aligned} \tag{4.44}$$

which is also negative for

$$|\alpha|^2 < \frac{2\langle n\rangle + 1}{4(\langle n\rangle + 1)}. \tag{4.45}$$

This analysis has shown that both single-photon-added coherent and thermal states are nonclassical. The negative regions can be explored by tomographic reconstruction of the Wigner function. The quadrature distribution for the single-photon-added coherent state can be evaluated as follows:

$$\Psi_c^{(a)}(x) = \langle x|a^\dagger|\alpha\rangle/\sqrt{1+|\alpha|^2} = \frac{1}{\sqrt{2}}\left(x - \frac{\partial}{\partial x}\right)\Psi_c(x)/\sqrt{1+|\alpha|^2}, \tag{4.46}$$

where we have used the representation of $a^\dagger$ in terms of the quadrature x. Using (1.65) the expression (4.46) reduces to

$$\Psi_c^{(a)}(x) = \frac{1}{\sqrt{2}}\frac{2x - \sqrt{2}\alpha}{\sqrt{1+|\alpha|^2}}\Psi_c(x), \tag{4.47}$$

and hence

$$|\Psi_c^{(a)}(x)|^2 = 2\frac{\left|x - \frac{\alpha}{\sqrt{2}}\right|^2}{(1+|\alpha|^2)\sqrt{\pi}}\exp\left\{-\left[x - \frac{1}{\sqrt{2}}(\alpha+\alpha^*)\right]^2\right\}. \tag{4.48}$$

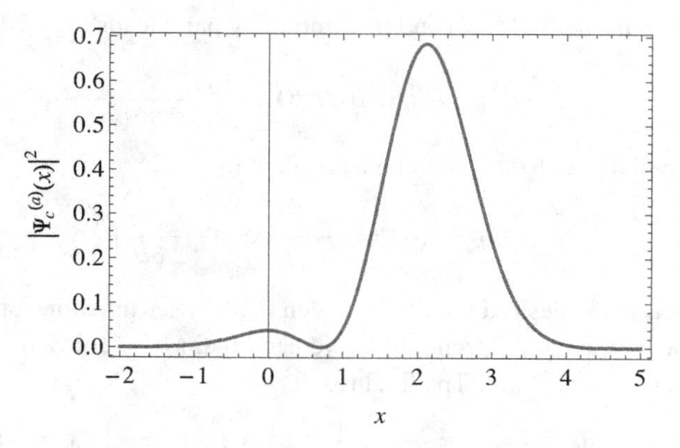

Fig. 4.6 The non-Gaussian quadrature distribution $|\Psi_c^{(a)}(x)|^2$ as a function of x for real $\alpha = 1$.

This has a zero at $x = \frac{\alpha}{\sqrt{2}}$ if α is real. The behavior of $|\Psi_c^{(a)}(x)|^2$ is shown in Figure 4.6 and the distribution is non-Gaussian.

4.4 Squeezing and sub-Poissonian properties of single-photon-added states

So far we have considered the nonclassicality of the state obtained by the addition of a photon using various phase-space distributions $P(\alpha)$, $Q(\alpha)$, and $W(\alpha)$. In order to have direct access to nonclassicality one can evaluate the Mandel Q_M parameter and the squeezing parameter S_θ.

The photon-added thermal state has a P-function that depends on the modulus of $|\alpha|$ and hence S_θ can not be negative. For the photon-added coherent state, the relevant expectation values can be computed by the repeated use of $[a, a^\dagger] = 1$

$$\begin{aligned}\langle a\rangle_c^{(a)} &= \langle\alpha_0|aaa^\dagger|\alpha_0\rangle/(1+|\alpha_0|^2)\\ &= \alpha_0(2+|\alpha_0|^2)/(1+|\alpha_0|^2),\end{aligned} \tag{4.49}$$

$$\begin{aligned}\langle a^2\rangle_c^{(a)} &= \langle\alpha_0|aa^2a^\dagger|\alpha_0\rangle/(1+|\alpha_0|^2)\\ &= \alpha_0^2(3+|\alpha_0|^2)/(1+|\alpha_0|^2),\end{aligned} \tag{4.50}$$

$$\begin{aligned}\langle a^\dagger a\rangle_c^{(a)} &= \langle\alpha_0|aa^\dagger aa^\dagger|\alpha_0\rangle/(1+|\alpha_0|^2)\\ &= (|\alpha_0|^4+3|\alpha_0|^2+1)/(1+|\alpha_0|^2).\end{aligned} \tag{4.51}$$

Hence the parameter S_θ defined by (2.7) becomes

$$S_\theta = \frac{1}{(1+|\alpha_0|^2)^2}[1-|\alpha_0|^2\cos(2\theta+2\varphi)]; \qquad \alpha_0 = |\alpha_0|e^{-i\varphi}. \tag{4.52}$$

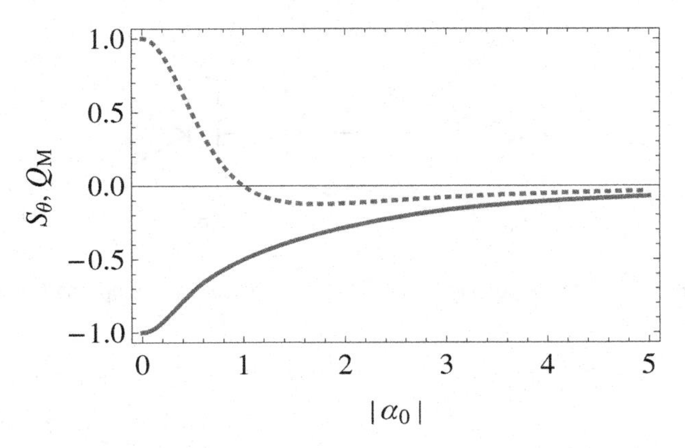

Fig. 4.7 The squeezing parameter S_θ (dotted) for $\theta + \varphi = 0$ and Mandel Q_M parameter (solid) as a function of $|\alpha_0|$.

The single-photon-added coherent state has squeezing if $|\alpha_0|$ and phases are such that

$$|\alpha_0|^2 \cos(2\theta + 2\varphi) > 1. \tag{4.53}$$

Next we calculate the Mandel Q_M parameter to check if the photon-added states can possess sub-Poissonian statistics. A useful relation obtained again by the use of the commutation relations is

$$aa^{\dagger 2}a^2 a^\dagger = a^{\dagger 3}a^3 + 5a^{\dagger 2}a^2 + 4a^\dagger a, \tag{4.54}$$

which leads to

$$\begin{aligned} \langle a^{\dagger 2}a^2\rangle_c^{(a)} &= \langle\alpha_0|aa^{\dagger 2}a^2a^\dagger|\alpha_0\rangle/(1+|\alpha_0|^2) \\ &= |\alpha_0|^2(|\alpha_0|^4 + 5|\alpha_0|^2 + 4)/(1+|\alpha_0|^2). \end{aligned} \tag{4.55}$$

Using (4.51) and (4.55), calculation of Q_M (defined by (2.2)) is straightforward

$$Q_M = -(2|\alpha_0|^4 + 2|\alpha_0|^2 + 1)/[(1+|\alpha_0|^2)(|\alpha_0|^4 + 3|\alpha_0|^2 + 1)]. \tag{4.56}$$

Hence there is a range of $|\alpha_0|$ where the single-photon-added coherent state can have sub-Poissonian statistics. Figure 4.7 shows the squeezing and sub-Poissonian properties of the single-photon-added coherent state.

The single-photon-added thermal state is found to have sub-Poissonian statistics. From (4.54) and using the property of the thermal state

$$\langle a^{\dagger n}a^n\rangle = n!\bar{n}^n, \tag{4.57}$$

we obtain

$$\begin{aligned} \langle a^{\dagger 2}a^2\rangle_T^{(a)} &= \langle a^{\dagger 3}a^3 + 5a^{\dagger 2}a^2 + 4a^\dagger a\rangle/\langle aa^\dagger\rangle \\ &= (6\bar{n}^3 + 10\bar{n}^2 + 4\bar{n})/(1+\bar{n}), \end{aligned} \tag{4.58}$$

$$\begin{aligned} \langle a^\dagger a\rangle_T^{(a)} &= \langle aa^\dagger aa^\dagger\rangle/(1+\bar{n}) \\ &= (2\bar{n}^2 + 3\bar{n} + 1)/(1+\bar{n}), \end{aligned} \tag{4.59}$$

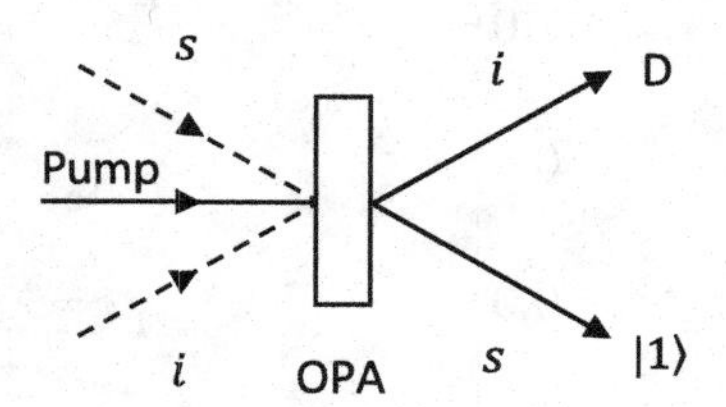

Fig. 4.8 Heralded generation of single-photon states of the signal. At the input port both signal and idler are in the vacuum state.

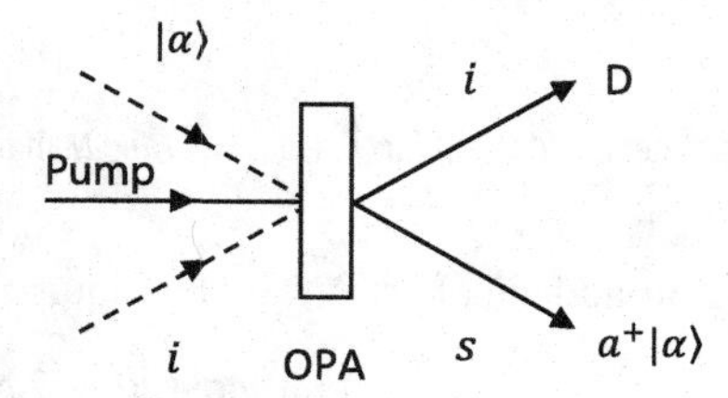

Fig. 4.9 The single-photon-added coherent state is generated by replacing the vacuum port of the signal by a coherent state.

and therefore the single-photon-added thermal state is sub-Poissonian if

$$Q_{\mathrm{M}} = \frac{2\bar{n}^2 - 1}{2\bar{n} + 1} < 0, \qquad \text{i.e. if} \qquad \bar{n} < \frac{1}{\sqrt{2}}. \tag{4.60}$$

4.5 Experimental realization of photon-added nonclassical non-Gaussian states

In Section 3.5 we discussed the production of single-photon states by using the process of parametric down-conversion. Furthermore, we have the standard methods of generating coherent and thermal states. A combination of these two techniques would yield the single-photon-added coherent and thermal states [16–18]. We depict this diagramatically in Figures 4.8–4.10.

In order to justify the generation of single-photon-added states, consider the unitary evolution operator (3.56), which to first order in the interaction can be written as

$$S(\xi) \cong 1 - \frac{\mathrm{i}t}{\hbar}\mathrm{i}\hbar g(\mathrm{e}^{\mathrm{i}\varphi}a^\dagger b^\dagger - \mathrm{e}^{-\mathrm{i}\varphi}ab). \tag{4.61}$$

The joint state of signal and idler can be written if their initial state is $|\Psi_a, 0\rangle$

$$|\Psi_a(t), \Psi_b(t)\rangle = |\Psi_a, 0\rangle + gt\mathrm{e}^{\mathrm{i}\varphi}a^\dagger|\Psi_a\rangle|1_b\rangle. \tag{4.62}$$

Here we keep the input state of the signal as arbitrary whereas the idler port is taken in the vacuum state. The detection of one photon at the idler port would lead to a conditional state

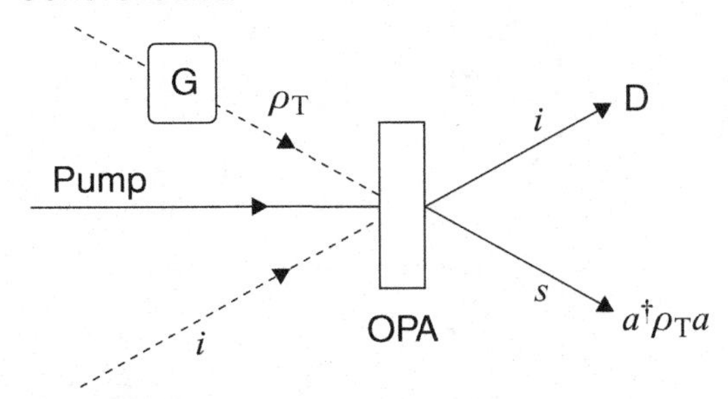

Fig. 4.10 The single-photon-added thermal state is generated by replacing the vacuum of the signal by a thermal state. A thermal state is usually generated from a coherent beam by the use of a ground glass plate marked as G.

of the signal as

$$|\Psi_a(t)\rangle \sim gt e^{i\varphi} a^\dagger |\Psi_a\rangle, \tag{4.63}$$

which on normalization can be written as

$$\rho^{(a)} = a^\dagger \rho a / \mathrm{Tr}(a^\dagger \rho a), \qquad \rho = |\Psi_a\rangle\langle\Psi_a|. \tag{4.64}$$

Choosing $|\Psi_a\rangle$ as a coherent (thermal) state results in the production of the single-photon-added coherent (thermal) state. It should be borne in mind that for a thermal state, ρ in (4.64) is to be replaced by the mixed state.

4.6 Single-photon-subtracted states

We saw earlier that the process of subtraction on a classical state can not produce nonclassical states. However, the process of subtraction on a nonclassical state can produce a very interesting nonclassical state [11–13, 19–21]. Consider the squeezed state defined by the equation

$$|\xi\rangle = S(\xi)|0\rangle = \exp\left(\frac{\xi}{2} a^{\dagger 2} - \frac{\xi^*}{2} a^2\right)|0\rangle. \tag{4.65}$$

Let us define the single-photon-subtracted state by

$$|\xi\rangle^{(s)} = a|\xi\rangle / \sqrt{\langle\xi|a^\dagger a|\xi\rangle} = a|\xi\rangle / \sinh r. \tag{4.66}$$

The state (4.66) can be rewritten by using the properties (2.24) of the unitary squeezing transformation

$$\begin{aligned} a|\xi\rangle &= aS(\xi)|0\rangle = S(\xi)S^\dagger(\xi)aS(\xi)|0\rangle \\ &= S(\xi)(a\cosh r + a^\dagger \sinh r\, e^{i\varphi})|0\rangle \\ &= S(\xi)\, e^{i\varphi}|1\rangle \sinh r, \end{aligned} \tag{4.67}$$

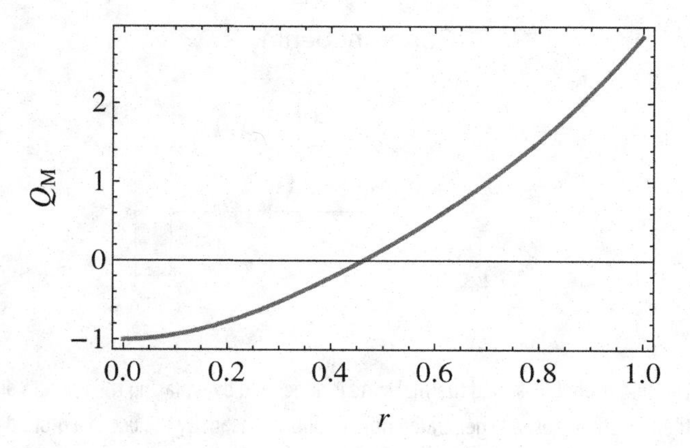

Fig. 4.11 The Mandel Q_M parameter as a function of the squeezing parameter r, after [21].

and hence

$$|\xi\rangle^{(s)} = e^{i\varphi} S(\xi)|1\rangle, \tag{4.68}$$

where the overall phase factor is unimportant. The state $|\xi\rangle^{(s)}$ involves only Fock states with odd numbers of photons. This is clear from the expansion (Eq. (2.18)) of the squeezed vacuum

$$\begin{aligned} |\xi\rangle^{(s)} &= \frac{a}{\sinh r} \frac{1}{\sqrt{\cosh r}} \sum_{n=0}^{\infty} e^{in\varphi} (\tanh r)^n \frac{\sqrt{(2n)!}}{n! 2^n} |2n\rangle \\ &= \sum_{n=0}^{\infty} c_{2n+1} |2n+1\rangle, \end{aligned} \tag{4.69}$$

where

$$c_{2n+1} = \frac{e^{i(n+1)\varphi} (\tanh r)^n \sqrt{(2n+1)!}}{(\cosh r)^{3/2} n! 2^n}. \tag{4.70}$$

The photon-number distribution clearly is

$$p_m = \delta_{m,2n+1} \frac{(\tanh r)^{2n} (2n+1)!}{(\cosh r)^3 (n!)^2 2^{2n}}. \tag{4.71}$$

The photon-subtracted squeezed state has sub-Poissonian statistics for squeezing parameter r less than an upper bound. An elementary calculation using (4.71) shows that

$$\langle a^{\dagger 2} a^2 \rangle = \frac{3 \tanh^2 r (3 + 2 \tanh^2 r)}{(1 - \tanh^2 r)^2}, \qquad \langle a^\dagger a \rangle = \frac{1 + 2 \tanh^2 r}{1 - \tanh^2 r}. \tag{4.72}$$

We show a plot of the Mandel Q_M parameter in Figure 4.11 which also shows that Q_M becomes positive if $r \geq 0.46$. One thus finds that the single-photon-subtracted squeezed vacuum can exhibit sub-Poissonian statistics even though the original state does not have sub-Poissonian statistics [21].

We next demonstrate that the Wigner function for the state $|\xi\rangle^{(s)}$ becomes negative in certain regions of the phase space. For this purpose it is useful to work with (4.68), which shows the connection to the Fock state $|1\rangle$. From this connection one would expect the Wigner function to be negative as the Fock state has a negative Wigner function. The density matrix for $|\xi\rangle^{(s)}$ is

$$\rho^{(s)} = S(\xi)|1\rangle\langle 1|S^{\dagger}(\xi). \tag{4.73}$$

In Section 2.3.2 we proved the relation between the Wigner functions for the two density matrices related by the transformation (4.73). Using this relation and the Wigner function (1.107) for the single-photon Fock state we find

$$W(\alpha) = \frac{2}{\pi}(4|\tilde{\alpha}|^2 - 1)e^{-2|\tilde{\alpha}|^2}, \tag{4.74}$$

$$\tilde{\alpha} = \alpha \cosh r - e^{i\varphi}\alpha^* \sinh r. \tag{4.75}$$

We show in Figure 4.12 the Wigner function for the state $|\xi\rangle^{(s)}$. The new property is the negativity in the region given by

$$C = 4|\tilde{\alpha}|^2 - 1 < 0. \tag{4.76}$$

The regions of negativity are separately displayed in Figure 4.13. The squeezing property is clearly evident from the plots in Figures 4.12 and 4.13. The single-photon-subtracted squeezed vacuum has some similarities to the cat state (4.2) for small α [11–13, 19–21]. In order to see this we compare the Wigner function of the state $|\xi\rangle^{(s)}$ with the Wigner function of the state $(|\alpha\rangle - |-\alpha\rangle)$. The similarities of the two can be seen. Finally, we note that the negativity of the Wigner function of the state $|\xi\rangle^{(s)}$ survives on phase averaging, i.e. the state

$$\bar{\rho}^{(s)} = \frac{1}{2\pi}\int |\xi\rangle^{(s)(s)}\langle\xi| d\varphi \tag{4.77}$$

has a negative Wigner function (Exercise 4.2).

4.6.1 Realization of photon subtraction by a beam splitter

Consider the scheme shown in Figure 4.14. The state $|\Psi\rangle$ from which the photon is to be subtracted is incident on a beam splitter with very low reflectivity. The vacuum enters from the other part. We prove below that the detection of $|1\rangle$ photon at the port where $|0\rangle$ would be transmitted in the absence of the beam splitter would produce a photon-subtracted state $a|\Psi\rangle$. To show this we recall that the action of a symmetric beam splitter on the input states is described by the unitary operator

$$U = e^{i\theta(a^{\dagger}b + ab^{\dagger})}, \tag{4.78}$$

where $\sin^2\theta$ is the reflectivity of the beam splitter. The outgoing state would be

$$\begin{aligned} |\Psi_{ab}\rangle_{\text{out}} &= e^{i\theta(a^{\dagger}b+ab^{\dagger})}|\Psi\rangle|0\rangle \\ &\cong |\Psi\rangle|0\rangle + i\theta(a^{\dagger}b + ab^{\dagger})|\Psi\rangle|0\rangle \\ &= |\Psi\rangle|0\rangle + i\theta a|\Psi\rangle|1\rangle, \end{aligned} \tag{4.79}$$

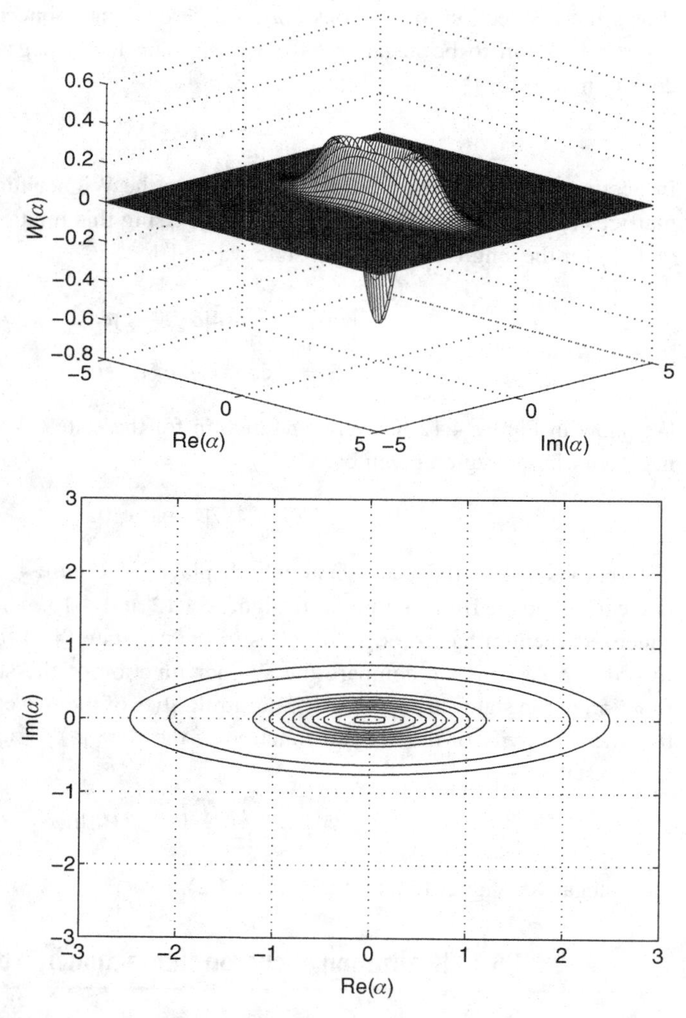

Fig. 4.12 The Wigner function $W(\alpha)$ for the single-photon-subtracted state with $r = 0.6$ and $\varphi = 0$.

if the reflectivity is very small. The state of the a mode after detection of one photon at the b-port leads to the single-photon-subtracted state

$$|\Psi_a\rangle_{\text{out}} \sim a|\Psi\rangle = |\Psi\rangle^{(s)}. \tag{4.80}$$

Ourjoumtsev *et al.* [11] (and also others [12, 13, 19]) produced the single-photon-subtracted squeezed vacuum by the down-conversion of frequency doubled laser pulses. They employed a beam splitter with about 10% reflectivity ($\theta \approx 5^o$) and the single photon was detected by APD (avlanche photo diode). Since the beam splitter reflectivity is small, the probability that two photons strike APD is very small. The detection at the APD heralds the generation of the state $|\xi\rangle^{(s)}$. The homodyne measurements on $|\xi\rangle^{(s)}$ allowed one to do tomography and thereby obtain the Wigner function of $|\xi\rangle^{(s)}$. Ourjoumtsev *et al.* confirmed the generation of $|\xi\rangle^{(s)}$ for several values of $|\xi|$.

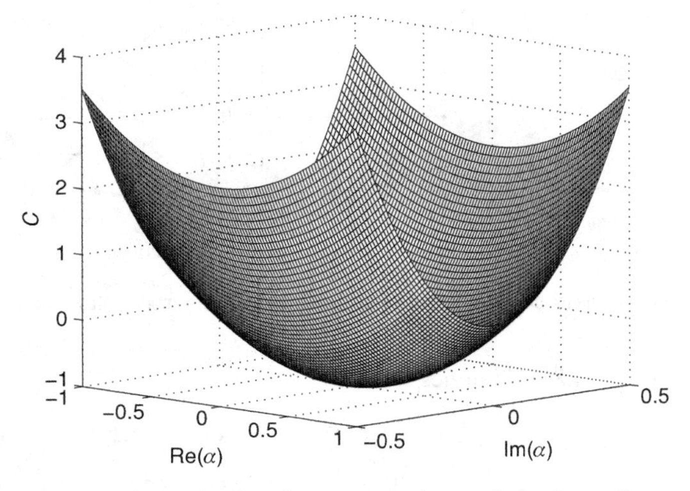

Fig. 4.13 The parameter C as a function of the real and imaginary parts of α for $r = 0.6$ and $\varphi = 0$.

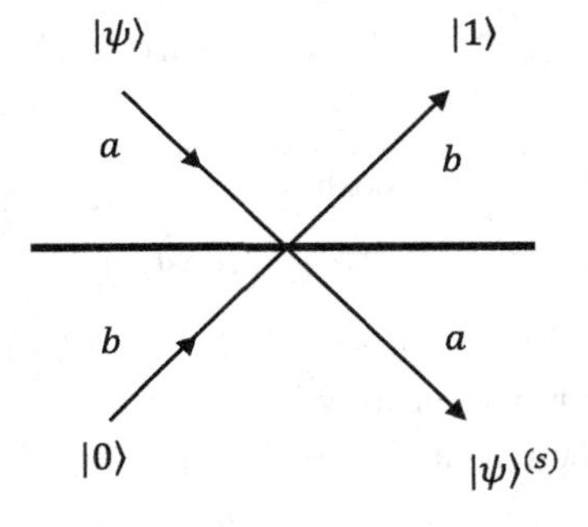

Fig. 4.14 The scheme for realization of photon subtraction by a beam splitter.

4.7 Single-photon-subtracted two-mode states with vortex structure

Next we consider the generalization of the results of Section 4.6 to two modes. We can add or subtract a photon to the two-mode state. Let us start with the two-mode state of Section 3.1 and subtract a photon, say, from the idler mode. The resulting state would be

$$|\xi\rangle^{(s)} = \mathcal{N} b \exp\left(\xi a^\dagger b^\dagger - \xi^* ab\right)|0,0\rangle, \ \mathcal{N} = \langle\xi|b^\dagger b|\xi\rangle^{-1/2} = 1/\sinh r. \tag{4.81}$$

This state can be produced by starting from a nondegenerate down-conversion process and then subtracting a photon from the idler mode using the beam splitter method as outlined in Section 4.6.1. The process is displayed in Figure 4.15. Using the Bogoliubov transformation

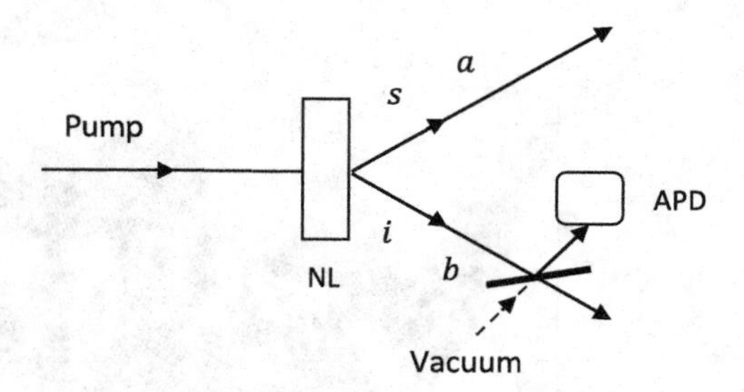

Fig. 4.15 The scheme for production of single-photon-subtracted two-mode states.

(3.11), (4.81) simplifies to

$$|\xi\rangle^{(s)} = \mathcal{N}\exp\left(\xi a^\dagger b^\dagger - \xi^* ab\right)\left(b\cosh r + a^\dagger \sinh r e^{i\varphi}\right)|0,0\rangle$$
$$= \exp\left(\xi a^\dagger b^\dagger - \xi^* ab\right)|1,0\rangle e^{i\varphi}. \tag{4.82}$$

The state has many nonclassical properties which arise from the presence of both Fock state $|1\rangle$ and the squeezing operator. Using (3.4), the expression (4.82) can be simplified to

$$|\xi\rangle^{(s)} = e^{i\varphi}\exp[-2(\ln\cosh r)]\exp\left(e^{i\varphi}\tanh r\, a^\dagger b^\dagger\right)|1,0\rangle$$
$$= \frac{e^{i\varphi}}{\cosh^2 r}\sum_{n=0}^{\infty} e^{ni\varphi}(\tanh r)^n\sqrt{n+1}|n+1,n\rangle, \tag{4.83}$$
$$= e^{i\varphi}a^\dagger|\xi\rangle/(\cosh r). \tag{4.84}$$

The state (4.83) has the property that the difference between the number of photons in the signal and idler modes is unity. The Wigner function for the state $|\xi\rangle^{(s)}$ has negative values. The Wigner function can be obtained by using (3.25) and the Wigner function for the Fock state $|1,0\rangle$

$$W^{(s)}(\alpha,\beta) = \frac{4}{\pi^2}(4|\tilde\alpha|^2 - 1)\exp[-2(|\tilde\alpha|^2 + |\tilde\beta|^2)], \tag{4.85}$$

where

$$\begin{pmatrix}\tilde\alpha\\ \tilde\beta^*\end{pmatrix} = \begin{pmatrix}\cosh r & -\sinh r\, e^{i\varphi}\\ -\sinh r\, e^{-i\varphi} & \cosh r\end{pmatrix}\begin{pmatrix}\alpha\\ \beta^*\end{pmatrix}. \tag{4.86}$$

The negative region of the Wigner function is given by

$$4|\alpha\cosh r - \beta^*\sinh r e^{i\varphi}|^2 < 1. \tag{4.87}$$

The state (4.83) has an unusual property. It has the possibility of a vortex structure. The quadrature distribution $\Psi^{(s)}(x_a, x_b)$ can be calculated using (4.84) and the representation of $a^\dagger$ in terms of x_a

$$a^\dagger = \frac{1}{\sqrt{2}}\left(x_a - \frac{\partial}{\partial x_a}\right), \tag{4.88}$$

$$\Psi^{(s)}(x_a, x_b) = \frac{e^{i\varphi}}{\sqrt{2}\cosh r}\left(x_a - \frac{\partial}{\partial x_a}\right)\Psi(x_a, x_b), \tag{4.89}$$

where $\Psi(x_a, x_b)$ is the quadrature distribution (3.31) for the two-mode squeezed vacuum. On simplification (4.89) becomes

$$\Psi^{(s)}(x_a, x_b) = \frac{\sqrt{2}e^{i\varphi}(x_a - \eta x_b)}{\cosh^2 r(1-\eta^2)^{3/2}\pi^{1/2}} \exp\left[\frac{2x_a x_b \eta - (x_a^2 + x_b^2)\eta^2}{1-\eta^2} - \frac{1}{2}(x_a^2 + x_b^2)\right]. \tag{4.90}$$

For $\eta = i|\eta|$ this has the structure of a vortex. The contours of constant $|\Psi^{(s)}(x_a, x_b)|^2$ are shown in Figures 4.16 and 4.17. The vortex structure is clear. Details of a variety of states with vortex structure are given in [22]. For real η, the state is entangled and has the structure shown in Figure 4.18. Thus a variety of entangled states can be engineered by changing the parameter η or by changing the phase of the pump field driving the down-converter.

We next consider quantitatively the entanglement in the vortex state by computing the log negativity parameter defined by [23, 24]

$$\mathcal{E} = \log_2(1 + 2\mathcal{N}), \tag{4.91}$$

where $\mathcal{N}$ is the modulus of the sum of all the negative eigenvalues associated with the partial transpose [25, 26] ρ^{PT} of the density matrix $|\xi\rangle^{(s)(s)}\langle\xi|$. From Eq. (4.83), the partial transpose is

$$\rho^{\mathrm{PT}} = \sum_{n,m} c_n c_m |n+1, m\rangle\langle m+1, n| e^{i(n-m)\varphi}, \tag{4.92}$$

$$c_n = (\tanh r)^n \sqrt{1+n}/\cosh^2 r. \tag{4.93}$$

The diagonal elements in (4.92) are all positive. The terms $n \neq m$ in (4.92) have the form

$$|n+1, m\rangle\langle m+1, n| e^{i(n-m)\varphi} + |m+1, n\rangle\langle n+1, m| e^{i(m-n)\varphi},$$

which can be written in diagonal form as

$$\begin{aligned}
&\frac{1}{2}(e^{in\varphi}|n+1, m\rangle + e^{im\varphi}|m+1, n\rangle) \\
&\quad \times (e^{-in\varphi}\langle n+1, m| + e^{-im\varphi}\langle m+1, n|) \\
&\quad - \frac{1}{2}(e^{in\varphi}|n+1, m\rangle - e^{im\varphi}|m+1, n\rangle) \\
&\quad \times (e^{-in\varphi}\langle n+1, m| - e^{-im\varphi}\langle m+1, n|).
\end{aligned} \tag{4.94}$$

Thus all the negative eigenvalues are $-c_n c_m$ and hence the log negativity parameter (4.91) becomes

$$\begin{aligned}
\mathcal{E} &= \log_2\left(1 + \sum_{n \neq m} c_n c_m\right) \\
&= \log_2\left(\sum_n c_n\right)^2, \quad \text{since} \sum_n c_n^2 = 1.
\end{aligned} \tag{4.95}$$

The sum in (4.95) can be evaluated numerically. A similar calculation for the two-mode squeezed vacuum (where the corresponding $c_n^{\prime s}$ are given by $(\tanh r)^n/\cosh r$) gives the

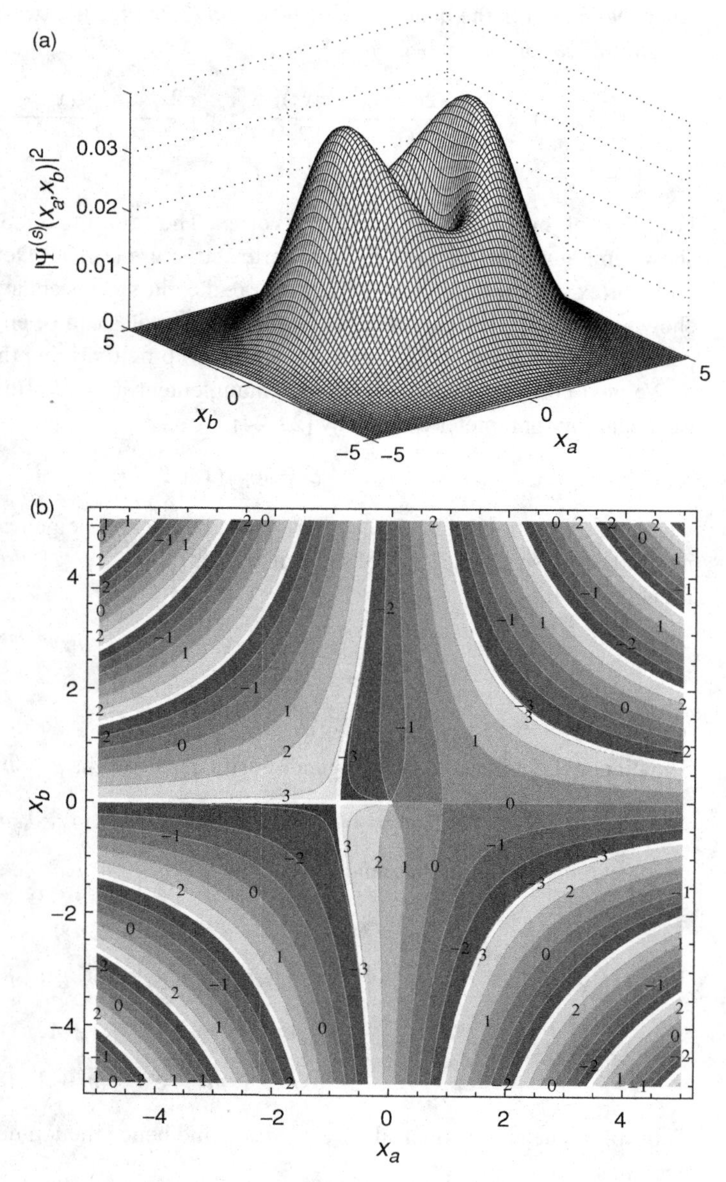

Fig. 4.16 (a) The intensity $|\Psi^{(s)}(x_a, x_b)|^2$ and (b) the phase of $\Psi^{(s)}(x_a, x_b)$ for $r = 1$ and $\eta = \mathrm{i}\tanh r$, after [22].

result $\log_2(\mathrm{e}^{2r})$. Thus entanglement in the vortex state can be compared with that in the state $|\xi\rangle$ by studying the ratio

$$\tilde{\mathcal{E}} = \left(\sum_n c_n \mathrm{e}^{-r}\right)^2. \tag{4.96}$$

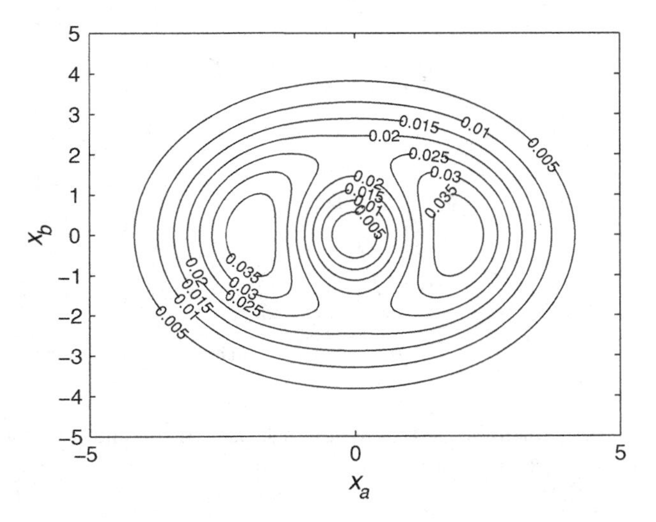

Fig. 4.17 The contour of $|\Psi^{(s)}(x_a, x_b)|^2$ as a function of x_a and x_b for $r = 1$ and $\eta = \mathrm{i}\tanh r$, after [22].

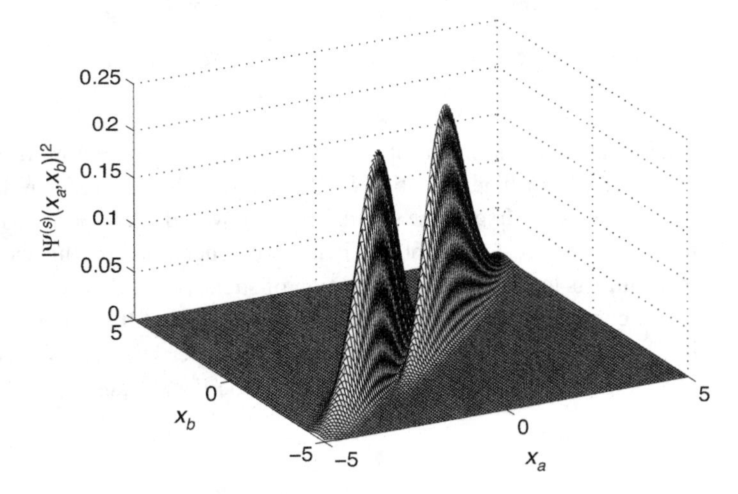

Fig. 4.18 $|\Psi^{(s)}(x_a, x_b)|^2$ as a function of x_a and x_b for $r = 1$ and $\eta = \tanh r$.

This ratio is shown in Figure 4.19 over a range of values of the parameter r. It is to be noted that $\tilde{\mathcal{E}} > 1$ and hence in a sense the vortex state $|\xi\rangle^{(s)}$ is more entangled than the two-mode squeezed vacuum $|\xi\rangle$.

4.8 Pair-coherent states

A different class of two-mode nonclassical non-Gaussian states are the pair-coherent states. The non-Gaussian states in the previous sections were obtained by addition and subtraction

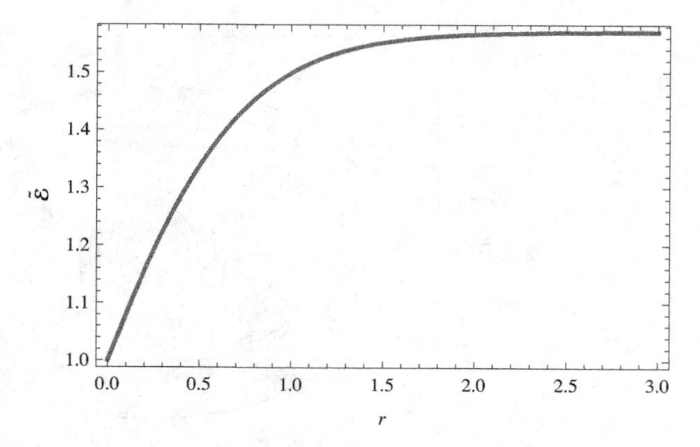

Fig. 4.19 The ratio $\tilde{\mathcal{E}}$, which gives the entanglement in the vortex state relative to the entanglement in the two-mode squeezed vacuum, as a function of r, after [22].

of photons on Gaussian states. The pair-coherent states (PCS in short) are produced by a different mechanism. These states are defined by the eigenvalue problem [27–29]

$$ab|\zeta\rangle = \zeta|\zeta\rangle, \tag{4.97}$$

$$(a^\dagger a - b^\dagger b)|\zeta\rangle = q|\zeta\rangle. \tag{4.98}$$

Thus simultaneous eigenstates of the annihilation operators a and b are subject to the condition (4.98). The condition (4.98) says that the difference in the number of photons in the two modes is fixed. The symmetric case would correspond to $q = 0$. In most situations this would be most relevant and hence we concentrate on this case. The possible physical mechanisms leading to the production of such states are discussed in [27, 30, 31]. Because of the constraint (4.98), the PCS are not products of the coherent states that we discussed in Chapter 1. Many important nonclassical properties of PCS are due to (4.98). In view of (4.98), the PCS can be written in terms of the Fock states as

$$|\zeta\rangle = \sum_{n=0}^{\infty} c_n|n, n\rangle. \tag{4.99}$$

The coefficients can be obtained by using (4.97):

$$ab|\zeta\rangle = \sum_{n=0}^{\infty} c_n n|n-1, n-1\rangle = \zeta \sum c_n|n, n\rangle, \tag{4.100}$$

and hence

$$\zeta c_n = c_{n+1}(n+1), \qquad c_n = c_0 \frac{\zeta^n}{n!}. \tag{4.101}$$

The coefficient c_0 is determined by the normalization condition

$$\sum_n |c_n|^2 = 1, \qquad \text{i.e.} \qquad |c_0|^2 \sum \frac{|\zeta|^{2n}}{(n!)^2} = 1, \tag{4.102}$$

which can be written in terms of the modified Bessel function

$$I_0(z) = \sum_{n=0}^{\infty} \frac{(z^2/4)^n}{(n!)^2}, \qquad c_0 = [I_0(2|\zeta|)]^{-1/2}. \tag{4.103}$$

The final expression for PCS is

$$|\zeta\rangle = c_0 \sum_{n=0}^{\infty} \frac{\zeta^n}{n!} |n, n\rangle. \tag{4.104}$$

The structure of (4.104) is similar to that of the two-mode squeezed vacuum (Eq. (3.6)). However, the presence of the factor $1/n!$ gives the state $|\zeta\rangle$ some nonclassical properties, such as sub-Poissonian statistics. The probability of finding n photons in either a or b mode is

$$p_n = c_0^2 \frac{|\zeta|^{2n}}{(n!)^2}. \tag{4.105}$$

The mean number of photons is

$$\langle a^\dagger a\rangle \equiv \bar{n} = c_0^2 \sum \frac{n|\zeta|^{2n}}{(n!)^2} = c_0^2 \sum_{n=0}^{\infty} \frac{|\zeta|^{2n}}{n!(n-1)!}. \tag{4.106}$$

The second-order moment can be obtained using (4.97) and (4.98):

$$\begin{aligned}
\langle a^{\dagger 2} a^2\rangle &= \langle a^\dagger a a^\dagger a\rangle - \langle a^\dagger a\rangle \\
&= \langle (a^\dagger a - b^\dagger b) a^\dagger a\rangle + \langle b^\dagger a^\dagger b a\rangle - \langle a^\dagger a\rangle, \qquad (4.107)\\
&= |\zeta|^2 - \bar{n}. \qquad (4.108)
\end{aligned}$$

The very first term in (4.107) is zero due to (4.98), the second term gives $|\zeta|^2$ due to (4.97). On combining (4.106) and (4.108) we obtain the Mandel Q_M parameter

$$Q_\mathrm{M} = |\zeta|^2/\bar{n} - 1 - \bar{n}. \tag{4.109}$$

A plot of Q_M as a function of $|\zeta|$ is shown in Figure 4.20. We can prove analytically that $Q_\mathrm{M} < 0$. We use the operator inequality

$$|\langle AB\rangle|^2 \le \langle AA^\dagger\rangle\langle B^\dagger B\rangle. \tag{4.110}$$

Choosing $A = a$, $B = b$,

$$|\langle ab\rangle|^2 \le \langle aa^\dagger\rangle\langle b^\dagger b\rangle, \tag{4.111}$$

and hence

$$(\bar{n} + 1)\bar{n} \ge |\zeta|^2, \tag{4.112}$$

which leads to the sub-Poissonian property of the PCS, i.e. $Q_\mathrm{M} < 0$.

In Section 3.4 we introduced the violation of the Cauchy–Schwarz inequality as a measure of nonclassicality. The negative values of I (Eq. (3.35)) imply nonclassicality. In view of the eigenvalue problem (4.97), (4.98) it is seen that

$$I = -\bar{n}/|\zeta|^2. \tag{4.113}$$

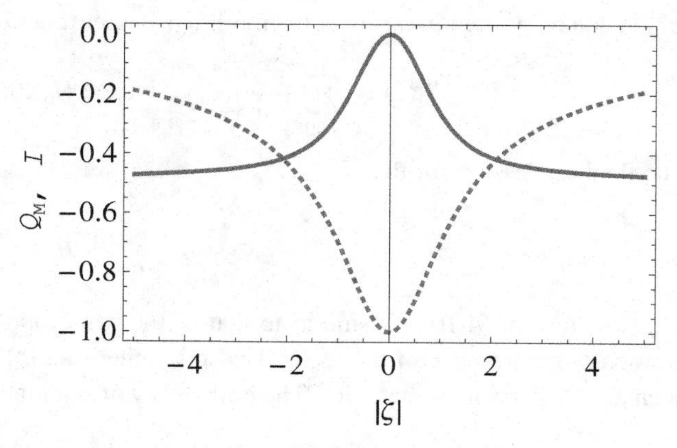

Fig. 4.20 The Mandel Q_{M} parameter (solid) and the parameter I (dotted) as a function of $|\zeta|$.

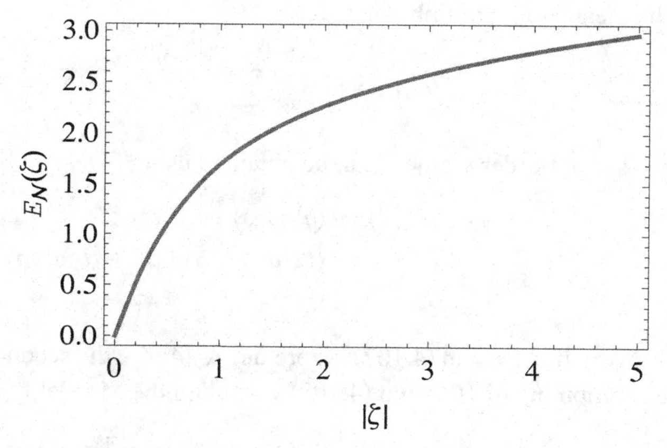

Fig. 4.21 The entanglement in the pair coherent state as given by the logarithmic negativity parameter $E_{\mathcal{N}}(\zeta)$ as a function of $|\zeta|$.

The violation of the CS inequality is shown in Figure 4.20. The violation of the CS inequality can be seen in the two-photon interference experiment as explained in Chapter 5.

The PCS has strong entanglement between the two modes. This should be evident from the form (4.104) as this is the Schmidt form. We derived a general formula characterizing entanglement in Section 3.7. This is given by (3.51), i.e. the logarithmic negativity parameter is

$$\begin{aligned} E_{\mathcal{N}}(\zeta) &= 2\log_2\left(c_0\sum_n \frac{|\zeta|^n}{n!}\right) \\ &= 2\log_2 c_0 + 2\log_2 \mathrm{e}^{|\zeta|}. \end{aligned} \tag{4.114}$$

The behavior of (4.114) is shown in Figure 4.21. The increase with $|\zeta|$ is expected in view of (4.104), as then more and more states in (4.104) become important.

Exercises

4.1 Show that the single-photon-added squeezed vacuum state is identical to the single-photon-subtracted squeezed state, i.e. show that $a^\dagger|\xi\rangle$ and $a|\xi\rangle$ differ only by trivial scaling factors.

4.2 Show that the Wigner function of the state (4.77) is given by

$$\bar{W}^{(s)}(\alpha) = \frac{2}{\pi}\exp[-2|\alpha|^2\cosh 2r][(4|\alpha|^2\cosh 2r - 1)I_0(2|\alpha|^2\sinh 2r) - 4|\alpha|^2\sinh 2r\, I_1(2|\alpha|^2\sinh 2r)],$$

where I_0 and $I_1(z)$ are the modified Bessel functions

$$I_p(z) = \frac{1}{\pi}\int_0^\pi e^{z\cos\theta}(\cos\theta)^p d\theta.$$

Confirm the negativity by plotting $\bar{W}^{(s)}(\alpha)$ [21].

4.3 Conditional generation of the state $f|0\rangle + g|1\rangle$ with $|f| \sim |g|$. The method used for the generation of the single-photon-added coherent state can be used to generate the superposition $f|0\rangle + g|1\rangle$. One should now use the initial state where the idler is in a weak coherent state $|\beta\rangle$, $|\beta| \ll 1$, and the signal is in a vacuum state. Prove this using Eq. (3.2). Display the two processes that lead to one photon at the idler port. Such a generation was successfully demonstrated in [32] (see also for an alternate procedure [33]).

4.4 Show that the solution to the eigenvalue problem (4.97) and (4.98) for nonzero q is given by

$$|\zeta\rangle = c_q \sum \frac{\zeta^n}{\sqrt{n!(n+q)!}}|n+q, n\rangle,$$

$$c_q = [|\zeta|^{-q} I_q(2|\zeta|)]^{-1/2},$$

where $I_q(z)$ is the modified Bessel function of order q.

4.5 Show that the pair-coherent states can also have quadrature squeezing. Construct the quadrature (cf. Eq. (3.17)) $d = (be^{-i\theta/2} + a^\dagger e^{i\theta/2} + H.c.)/2$, then obtain $S_\theta = \langle : d^2 :\rangle - \langle d\rangle^2$. Find the range of θ, ζ for which $S_\theta < 0$.

4.6 The purpose of this exercise is to test nonclassicality using the parameter A_3 (Eq. (2.61)) when the parameters Q_M and S_θ fail. Calculate the nonclassicality parameter A_3 for the photon-added thermal state (Eq. (4.32)) as a function of $\bar{n}$. Show that there is a region where A_3 continues to be negative even when Q_M becomes positive.

4.7 Repeat Exercise 4.6 for a cat state $\sim |\alpha e^{i\theta/2}\rangle + |\alpha e^{-i\theta/2}\rangle$ with $\alpha = 3$. Find A_3 and Q_M as a function of θ. Show that there are regions where Q_M is positive and A_3 is negative.

4.8 Find the evolution of the coherent state under a nonlinear Hamiltonian $H = g\hbar a^{\dagger 2}a^2$, i.e. find $e^{-igta^{\dagger 2}a^2}|\alpha\rangle$. Show that at a time given by $gt = \pi/2$, the evolved coherent state is a cat state of the form

$$c_+|\alpha e^{i\pi/2}\rangle + c_-|\alpha e^{-i\pi/2}\rangle.$$

Find c_+ and c_- [3].

References

[1] E. Schrödinger, *Naturwissenschaften* **23**, 807–812, 823–828, 844–849 (1935); *Proc. Am. Philos. Soc.* **124**, 323–338 (1980).
[2] A. Einstein, B. Podolsky, and N. Rosen, *Phys. Rev.* **41**, 777 (1935).
[3] B. Yurke and D. Stoler, *Phys. Rev. Lett.* **57**, 13 (1986).
[4] G. J. Milburn and C. A. Holmes, *Phys. Rev. Lett.* **56**, 2237 (1986).
[5] V. Bužek, A. Vidiella-Barranco, and P. L. Knight, *Phys. Rev. A* **45**, 6570 (1992).
[6] G. S. Agarwal, *Foundations of Physics*, **25**, 219 (1995).
[7] D. L. Jacobson, S. A. Werner, and H. Rauch, *Phys. Rev. A* **49**, 3196 (1994).
[8] N. Lütkenhaus and S. M. Barnett, *Phys. Rev. A* **51**, 3340 (1995).
[9] M. Brune, E. Hagley, J. Dreyer *et al.*, *Phys. Rev. Lett.* **77**, 4887 (1996).
[10] C. Monroe, D. M. Meekhof, B. E. King, and D. J. Wineland, *Science* **272**, 1131 (1996).
[11] A. Ourjoumtsev, R. Tualle-Brouri, J. Laurat, and P. Grangier, *Science* **312**, 83 (2006).
[12] T. Gerrits, S. Glancy, T. S. Clement *et al.*, *Phys. Rev. A* **82**, 031802(R) (2010).
[13] N. Namekata, Y. Takahashi, G. Fujii, D. Fukuda, S. Kurimura, and S. Inoue, *Nature Photon.* **4**, 655 (2010).
[14] G. S. Agarwal and K. Tara, *Phys. Rev. A* **43**, 492 (1991).
[15] G. S. Agarwal and K. Tara, *Phys. Rev. A* **46**, 485 (1992).
[16] A. Zavatta, S. Viciani, and M. Bellini, *Science* **306**, 660 (2004).
[17] A. Zavatta, V. Parigi, and M. Bellini, *Phys. Rev. A* **75**, 052106 (2007).
[18] A. Zavatta, V. Parigi, M. S. Kim, and M. Bellini, *New J. Phys.* **10** 123006 (2008).
[19] J. S. Neergaard-Nielsen, B. M. Nielsen, C. Hettich, K. Mølmer, and E. S. Polzik, *Phys. Rev. Lett.* **97**, 083604 (2006).
[20] M. S. Kim, E. Park, P. L. Knight, and H. Jeong, *Phys. Rev. A* **71**, 043805 (2005).
[21] A. Biswas and G. S. Agarwal, *Phys. Rev. A* **75**, 032104 (2007).
[22] G. S. Agarwal, *New J. Phys.* **13**, 073008 (2011).
[23] G. Vidal and R. F. Werner, *Phys. Rev. A* **65**, 032314 (2002).
[24] M. B. Plenio, *Phys. Rev. Lett.* **95**, 090503 (2005).
[25] A. Peres, *Phys. Rev. Lett.* **77**, 1413 (1996).
[26] P. Horodecki, *Phys. Lett. A* **232**, 333 (1997).
[27] G. S. Agarwal, *J. Opt. Soc. Am. B* **5**, 1940 (1988).
[28] A. O. Barut and L. Girardello, *Commun. Math. Phys.* **21**, 41 (1971).
[29] D. Bhaumik, K. Bhaumik, and B. Dutta-Roy, *J. Phys. A* **9**, 1507 (1976).
[30] M. D. Reid and L. Krippner, *Phys. Rev. A* **47**, 552 (1993).
[31] S. Gou, J. Steinbach, and P. L. Knight, *Phys. Rev. A* **54**, R1014 (1996).
[32] K. J. Resch, J. S. Lundeen, and A. M. Steinberg, *Phys. Rev. Lett.* **88**, 113601 (2002).
[33] A. I. Lvovsky and J. Mlynek, *Phys. Rev. Lett.* **88**, 250401 (2002).

5 Optical interferometry with single photons and nonclassical light

Having discussed in the previous chapters many different aspects of single photons and nonclassical light, we are now ready to discuss interferometry with single photons. We first discuss traditional interferometers and their performance if classical light beams are replaced by quantum fields.

The earliest interferometer is the Young's double slit interferometer. Young's work on interference confirmed the wave nature of light and was a turning point in optics. A complete description of Young's double slit is more complicated as it involves the propagation of wavefronts through the slits and hence we will take it up in Chapter 8. Michelson designed an interferometer which he very successfully used in measurements of spectral lines and the diameter of stars. However, in this chapter we focus on the Mach–Zehnder and Sagnac interferometers which are currently used extensively. We note that all interferometers use the interference between light beams arising from at least two paths. In what follows we assume that the beams arising from the two paths are coherent with respect to each other. This would be the case if the path difference was short compared to the coherence length of the light sources at the input ports of the interferometer. In Chapter 8, we will consider more general cases, which will allow us to relax this assumption somewhat.

All optical interferometers use optical devices such as beam splitters, mirrors, and phase shifters. The action of all such optical devices is very well understood for classical beams of light. Therefore, before we discuss interferometers, we must first understand how nonclassical beams of light behave with different types of optical devices [1–4]; in particular, how the optical devices transform the photon statistics of the incoming beams of light.

5.1 Transformation of quantized light fields at beam splitters

Consider first a classical beam of light with amplitude E. Upon incidence on a beam splitter the transmitted and reflected beams are clearly given by

$$E_t = t_1 E, \qquad E_r = r_1 E, \tag{5.1}$$

where r_1 and t_1 are, respectively, the reflection and transmission amplitudes, as shown in Figure 5.1. In quantum theory the field amplitudes are replaced by Heisenberg operators $E \to a$, $E_t \to c$, $E_r \to d$, then

$$d = r_1 a, \qquad c = t_1 a. \tag{5.2}$$

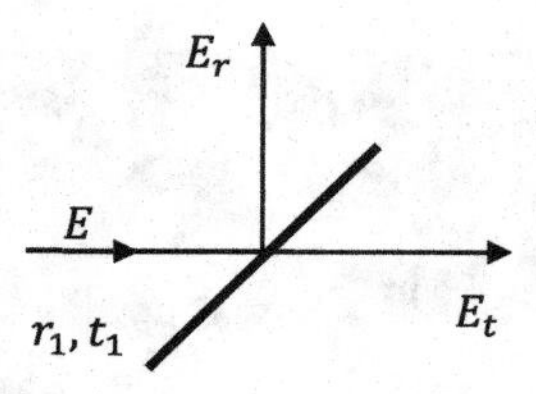

Fig. 5.1 The beam splitter's action on a classical field.

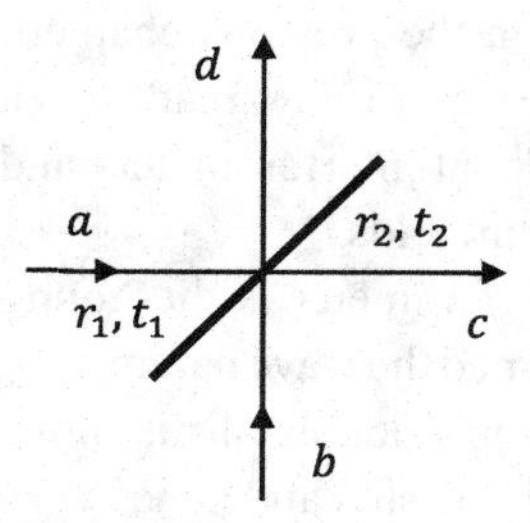

Fig. 5.2 The beam splitter in quantum theory.

Clearly (5.2) does not satisfy the commutation relation $[a, a^\dagger] = [d, d^\dagger] = 1$. Hence we need to modify this simple description. In quantum theory vacuum fields are always present. Hence even if no field is sent from the other port of the beam splitter, the vacuum field enters from this port. Thus in quantum theory we need to have the description shown in Figure 5.2, where r_i, t_i are the reflection and transmission amplitudes from the two sides of the beam splitter.

The Heisenberg operators c and d of the output fields are related to the Heisenberg operators of the input fields by

$$\begin{pmatrix} c \\ d \end{pmatrix} = \begin{pmatrix} t_1 & r_2 \\ r_1 & t_2 \end{pmatrix} \begin{pmatrix} a \\ b \end{pmatrix} = U \begin{pmatrix} a \\ b \end{pmatrix}. \tag{5.3}$$

The transformation matrix U has to be unitary to ensure the validity of the commutation relations

$$\begin{aligned} &[c, c^\dagger] = [d, d^\dagger] = 1, \qquad [c, d^\dagger] = 0, \qquad [c, d] = 0, \\ &\text{if } [a, a^\dagger] = [b, b^\dagger] = 1, \qquad [a, b^\dagger] = 0. \end{aligned} \tag{5.4}$$

The unitary requirement is equivalent to

$$|t_1|^2 + |r_1|^2 = |t_2|^2 + |r_2|^2 = 1, \qquad t_1^* r_2 + r_1^* t_2 = 0. \tag{5.5}$$

These requirements are also equivalent to the conservation of energy at a lossless beam splitter. Assuming $|t_1| = |t_2|$, $|r_1| = |r_2|$, the unitary matrix can be written as

$$U = \begin{pmatrix} \cos\theta & \mathrm{e}^{\mathrm{i}\varphi} \sin\theta \\ -\sin\theta\, \mathrm{e}^{-\mathrm{i}\varphi} & \cos\theta \end{pmatrix}. \tag{5.6}$$

Clearly $\cos^2\theta$ ($\sin^2\theta$) gives the transmissivity (reflectivity) of the beam splitter. For a symmetric beam splitter, $t_1 = t_2, r_1 = r_2$, we choose $\varphi = \pi/2$, then

$$U = \begin{pmatrix} \cos\theta & \mathrm{i}\sin\theta \\ \mathrm{i}\sin\theta & \cos\theta \end{pmatrix} \tag{5.7}$$

$$\rightarrow \frac{1}{\sqrt{2}}\begin{pmatrix} 1 & \mathrm{i} \\ \mathrm{i} & 1 \end{pmatrix} \quad \text{for a 50-50 beam splitter.} \tag{5.8}$$

Note that other forms of U corresponding to $\varphi = 0$ or π are also used in the literature. For a 100% mirror, we say that the phase of the field changes by π. In this case the outgoing field d is $-a$. Using the transformation property (5.3), all the quantum statistical properties of the output fields can be calculated in terms of the properties of the input fields.

5.2 Beam splitter transformation equivalent to evolution under a Hamiltonian

Since the beam splitter transformation (5.3) is unitary, we expect that there must be an effective Hamiltonian description for it. In order to see this, we write (5.3) and (5.7) as

$$\begin{aligned} c &= a(\theta) = a\cos\theta + \mathrm{i}b\sin\theta, \\ d &= b(\theta) = b\cos\theta + \mathrm{i}a\sin\theta. \end{aligned} \tag{5.9}$$

These equations are equivalent to the differential equations

$$\frac{\mathrm{d}}{\mathrm{d}\theta}a(\theta) = \mathrm{i}b(\theta), \qquad \frac{\mathrm{d}}{\mathrm{d}\theta}b(\theta) = \mathrm{i}a(\theta); \qquad a(0) = a, \qquad b(0) = b, \tag{5.10}$$

which are like Heisenberg equations for a and b with the evolution given by the effective Hamiltonian

$$H = -\hbar(a^\dagger b + ab^\dagger), \tag{5.11}$$

and the operators defined by

$$\begin{aligned} a(\theta) &= u^\dagger(\theta)au(\theta), & b(\theta) &= u^\dagger(\theta)bu(\theta), \\ u(\theta) &= \exp\left(-\frac{\mathrm{i}}{\hbar}\theta H\right), & \frac{\mathrm{d}}{\mathrm{d}\theta}a(\theta) &= -\frac{\mathrm{i}}{\hbar}[a(\theta), H]. \end{aligned} \tag{5.12}$$

The effective Hamiltonian (5.11) is like a mode-mixing Hamiltonian, where the parameter θ (the equivalent of time) is related to the reflectivity of the beam splitter.

5.3 Transformation of states by the beam splitter

Given the effective Hamiltonian description, it is easy to transform the state. Let ρ be the density operator of the input fields. It would be a function of a, b, $a^\dagger$, and $b^\dagger$. The output density operator can be obtained by using the standard rules of quantum mechanics

$$\rho_{\text{out}} = u(\theta)\rho u^\dagger(\theta) = \rho(uau^\dagger, ubu^\dagger) = \rho(a(-\theta), b(-\theta)). \tag{5.13}$$

As an application of (5.13) we find how the coherent states are transformed by the beam splitters:

$$\begin{aligned}
|\alpha, \beta\rangle_{\text{out}} &= u(\theta)|\alpha, \beta\rangle = u(\theta)\mathrm{e}^{a^\dagger\alpha - a\alpha^* + b^\dagger\beta - b\beta^*}|0, 0\rangle \\
&= \exp\left[a^\dagger(-\theta)\alpha - a(-\theta)\alpha^* + b^\dagger(-\theta)\beta - b(-\theta)\beta^*\right]u(\theta)|0, 0\rangle \\
&= \exp\left[a^\dagger\alpha(\theta) - a\alpha^*(\theta) + b^\dagger\beta(\theta) - b\beta^*(\theta)\right]|0, 0\rangle \\
&= |\alpha(\theta), \beta(\theta)\rangle.
\end{aligned} \tag{5.14}$$

Here we used the property $u(\theta)|0, 0\rangle = |0, 0\rangle$, and $\alpha(\theta)$, and $\beta(\theta)$ are given by

$$\begin{pmatrix}\alpha(\theta) \\ \beta(\theta)\end{pmatrix} = U\begin{pmatrix}\alpha \\ \beta\end{pmatrix}. \tag{5.15}$$

Thus the fields at the output ports are coherent. This is an important property of the beam splitter where coherent fields are transformed into coherent fields. The transformation of coherent amplitudes (5.15) is the same as in the classical case. Coherent fields at a beam splitter behave like classical fields.

5.4 Transformation of photon number states by a beam splitter

Consider next how fields in states with a fixed number of photons behave at a beam splitter. Let the input state be $|n, m\rangle$. The transformation laws for the state $|n, m\rangle$ can be obtained from (5.13) or more directly. We expect that the total number of photons would be conserved. This is clearly seen from (5.11), which yields

$$a^\dagger a + b^\dagger b = \text{constant}. \tag{5.16}$$

Hence we expect that the state $|n, m\rangle$ would go over to

$$|n, m\rangle \to \sum_{p=0}^{n+m} f_p\,|p, n+m-p\rangle. \tag{5.17}$$

To derive the explicit form of f_p, we proceed as in (5.14):

$$\begin{aligned}
|n, m\rangle_{\text{out}} &= u(\theta)|n, m\rangle = u(\theta)\frac{a^{\dagger n}b^{\dagger m}}{\sqrt{n!m!}}|0, 0\rangle \\
&= \frac{[a^\dagger(-\theta)]^n[b^\dagger(-\theta)]^m}{\sqrt{n!m!}}|0, 0\rangle \\
&= \frac{(a^\dagger\cos\theta + \mathrm{i}\sin\theta\, b^\dagger)^n(b^\dagger\cos\theta + \mathrm{i}\sin\theta\, a^\dagger)^m}{\sqrt{n!m!}}|0, 0\rangle.
\end{aligned} \tag{5.18}$$

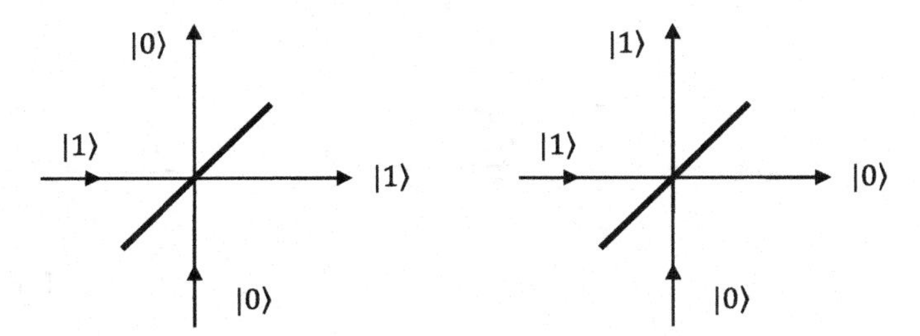

Fig. 5.3 The two paths when single photons enter the port a, and no light enters from the port b.

A binomial expansion yields

$$|n, m\rangle_{\text{out}} = \sum_{q,q'} \binom{n}{q} \binom{m}{q'} \sqrt{\frac{(q+q')!(n+m-q-q')!}{n!m!}} \times (\cos\theta)^{m+q-q'} (\mathrm{i} \sin\theta)^{n-q+q'} |q+q', n+m-q-q'\rangle, \tag{5.19}$$

which has precisely the form (5.17) with

$$f_p = \sum_{q=0}^{n} \sum_{q'=0}^{m} \delta_{p,q+q'} \binom{n}{q} \binom{m}{q'} \sqrt{\frac{(q+q')!(n+m-q-q')!}{n!m!}} \times (\cos\theta)^{m+q-q'} (\mathrm{i} \sin\theta)^{n-q+q'}. \tag{5.20}$$

The formula (5.19) can be used to understand the behavior of single photons at a beam splitter. As mentioned earlier, all interferometers with single photons use beam splitters and thus it is important to understand how single photons behave at beam splitters.

5.5 Single photons at beam splitters

Let us consider the simplest case when no light enters from the port b and single photons enter the port a. The input state is $|1, 0\rangle$. The output state is then (cf. Eq. (5.18))

$$|1, 0\rangle_{\text{out}} = \cos\theta |1, 0\rangle + \mathrm{i} \sin\theta |0, 1\rangle. \tag{5.21}$$

The production of the output state (5.21) can be thought of as due to the two paths as displayed in Figure 5.3. The single photon can take two paths – it either gets transmitted or is reflected. The transmission (reflection) amplitude is $\cos\theta$ ($\mathrm{i}\sin\theta$). It should be borne in mind that the photon is an indivisible particle. The state (5.21) produced by a single photon on a beam splitter has the form of an entangled state; more precisely it is a path-entangled state. The entanglement is maximal for a 50-50 beam splitter ($\theta = \pi/4$).

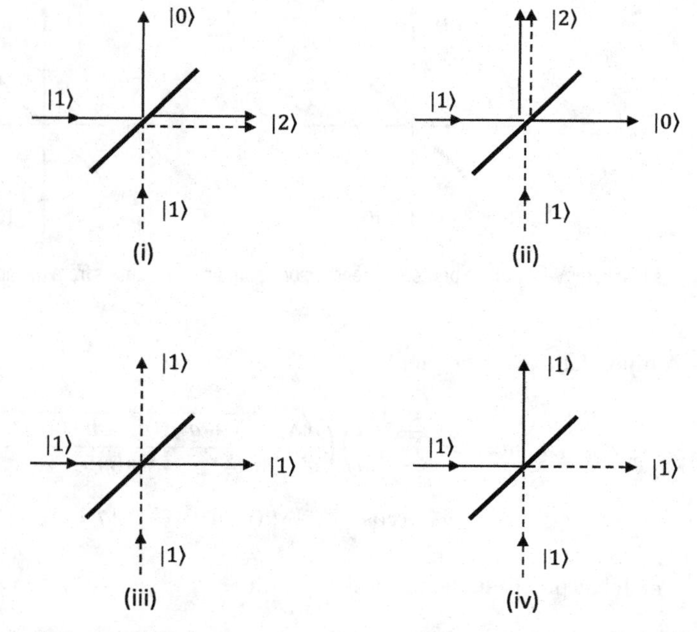

Fig. 5.4 Four elementary paths for two single photons with one photon incident from each side of the beam splitter.

5.6 Pairs of single photons at beam splitters

We next consider two single photons with one photon incident from each side of the beam splitter, i.e. the input state is $|1, 1\rangle$. The output state is easily evaluated from (5.18):

$$|1, 1\rangle_{\text{out}} = (\cos^2\theta - \sin^2\theta)|1, 1\rangle + \sqrt{2}\mathrm{i}\sin\theta\cos\theta(|2, 0\rangle + |0, 2\rangle). \tag{5.22}$$

The state (5.22) results from the four elementary paths shown in Figure 5.4. The path (iii) involves two transmissions whereas the path (iv) involves two reflections. This leads to an amplitude $(-\sin^2\theta)$ for the path (iv). The minus sign comes from two reflections as each reflection gives a factor i. Hence there is a possibility of quantum interference between the paths (iii) and (iv). This interference is complete at a 50-50 beam splitter ($\theta = \pi/4$) when the state (5.22) reduces to

$$|1, 1\rangle_{\text{out}} = \frac{\mathrm{i}}{\sqrt{2}}(|2, 0\rangle + |0, 2\rangle). \tag{5.23}$$

The possibility of complete cancellation of the contribution from the paths (iii) and (iv) was first discovered by Hong, Ou, and Mandel [1] and is now known as Hong–Ou–Mandel interference. The state (5.23) is a maximally entangled state of two photons. The consequences and applications of this will be discussed later in this chapter.

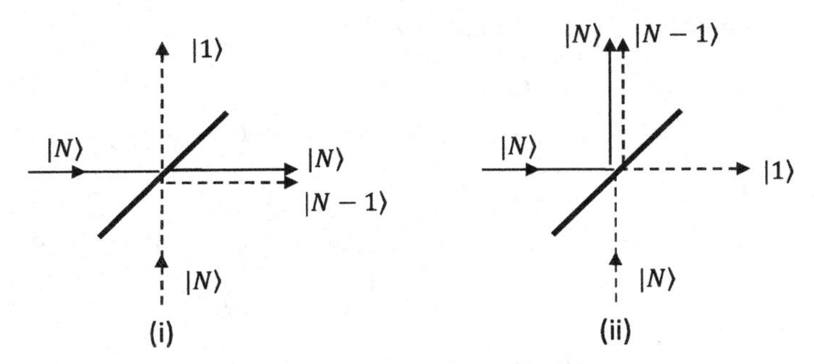

Fig. 5.5 The two pathways for the output state $|N, N\rangle$ that can lead to one photon at any of the output ports.

5.7 Generalization of the Hong–Ou–Mandel interference to N photons from both ports of the beam splitter

Let us now consider a more general case involving larger number of photons, i.e. the input state $|N, N\rangle$. This is of interest in quantum metrology. For simplicity we first consider the case of a 50-50 beam splitter. Then from (5.18) the output state will be

$$
\begin{aligned}
|N, N\rangle_{\text{out}} &= \frac{(a^\dagger + \mathrm{i}b^\dagger)^N (b^\dagger + \mathrm{i}a^\dagger)^N}{N! 2^N} |0, 0\rangle \\
&= \frac{\mathrm{i}^N}{N! 2^N} (a^{\dagger 2} + b^{\dagger 2})^N |0, 0\rangle \\
&= \frac{\mathrm{i}^N}{N! 2^N} \sum_{p=0}^{N} \binom{N}{p} \sqrt{(2p)!(2N - 2p)!} |2p, 2N - 2p\rangle .
\end{aligned} \tag{5.24}
$$

The result (5.24) has the very interesting property that one never finds an odd number of photons on any port. This is again due to quantum interference between different pathways. To illustrate this point, consider the possibility of finding one photon at one of the output ports and the remaining $(2N - 1)$ photons at the other port. Two elementary paths as shown in Figure 5.5 contribute to this possibility.

The number of reflections involved in the path (i) is $(N - 1)$ whereas in the path (ii) it is $(N + 1)$. The relative phase would be $(\mathrm{i})^{N+1}/(\mathrm{i})^{N-1} = \mathrm{i}^2 = -1$. Therefore the paths (i) and (ii) interfere destructively at a 50-50 beam splitter, leading to the nonappearance of one photon on one side and $(2N - 1)$ photons on the other side.

5.8 Transformation of a two-mode squeezed state by a 50-50 beam splitter

We next consider the transformation of a two-mode squeezed state by a beam splitter. Such states are employed in interferometry and metrology. Consider the state (3.3) incident on a beam splitter. The output state will be

$$\begin{aligned}|\xi\rangle_{\text{out}} &= u(\theta)\exp(\xi a^\dagger b^\dagger - \xi^* ab)|0,0\rangle \\ &= \exp[\xi a^\dagger(-\theta)b^\dagger(-\theta) - \xi^* a(-\theta)b(-\theta)]|0,0\rangle. \end{aligned} \tag{5.25}$$

For $\theta = \pi/4$,

$$\begin{aligned} a^\dagger(-\theta)b^\dagger(-\theta) &= (a^\dagger \cos\theta + \mathrm{i}\sin\theta\, b^\dagger)(b^\dagger\cos\theta + \mathrm{i}\sin\theta\, a^\dagger) \\ &= \frac{1}{2}\mathrm{i}(a^{\dagger 2} + b^{\dagger 2}), \end{aligned} \tag{5.26}$$

and hence (5.25) reduces to

$$|\xi\rangle_{\text{out}} = \exp\left[\frac{1}{2}\mathrm{i}\,(\xi a^{\dagger 2} + \xi^* a^2)\right]|0\rangle \exp\left[\frac{1}{2}\mathrm{i}\,(\xi b^{\dagger 2} + \xi^* b^2)\right]|0\rangle. \tag{5.27}$$

The output state is a product of two identical single-mode squeezed states. The beam splitter works like a disentangler for the two-mode squeezed state. One can similarly consider the transformation of single-mode squeezed states or even thermal squeezed states, although the results are more complicated [4].

5.9 Generation of two-mode entangled states by the interference of coherent fields and single photons

As a final example we consider the possibility of generating more general entangled states. We have already seen how a single photon can produce a path-entangled state (5.21). Let us replace the vacuum on the port b by a coherent state $|\beta\rangle$ (Figure 5.6). Then the output state would be

$$\begin{aligned} |1,\beta\rangle_{\text{out}} &= u(\theta)|1,\beta\rangle \\ &= u(\theta)a^\dagger D(\beta)|0,0\rangle \\ &= a^\dagger(-\theta)u(\theta)D(\beta)u^\dagger(\theta)|0,0\rangle \\ &= a^\dagger(-\theta)\exp[\beta b^\dagger(-\theta) - \beta^* b(\theta)]|0,0\rangle \\ &= (a^\dagger\cos\theta + \mathrm{i}\sin\theta\, b^\dagger)\exp[\beta(b^\dagger\cos\theta + \mathrm{i}\sin\theta\, a^\dagger) - c.c.]|0,0\rangle \\ &= (a^\dagger\cos\theta + \mathrm{i}\sin\theta\, b^\dagger)|\mathrm{i}\beta\sin\theta, \beta\cos\theta\rangle \\ &= \cos\theta\, a^\dagger|\mathrm{i}\beta\sin\theta, \beta\cos\theta\rangle + \mathrm{i}\sin\theta\, b^\dagger|\mathrm{i}\beta\sin\theta, \beta\cos\theta\rangle. \end{aligned} \tag{5.28}$$

Thus at the output port we have a combination of photon-added coherent states where a photon is added either on the port a or on the port b.

Beam splitters can be used to generate new types of heralded quantum states. Consider, for example, the state that can be produced after a single photon is detected at the output port c as shown in Figure 5.6. This can be obtained by taking the projection of (5.28) $\langle 1|1,\beta\rangle_{\text{out}}$. The generated state up to a normalization factor will be

$$|\Psi_b\rangle \sim \cos\theta|\beta\cos\theta\rangle - \beta\sin^2\theta\, b^\dagger|\beta\cos\theta\rangle. \tag{5.29}$$

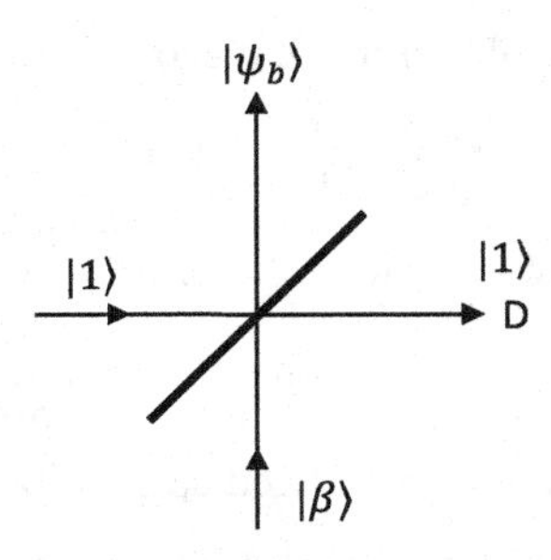

Fig. 5.6 Generation of the superposition of vacuum and one photon state by the interference of a field in a coherent state with single photons.

If we now assume that the coherent state amplitude β and the transmission $\cos\theta$ are small, then to lowest order in small quantities we get

$$|\Psi_b\rangle \sim (\cos\theta|0\rangle - \beta|1\rangle). \tag{5.30}$$

We can obviously make $\beta \sim \cos\theta$. In this case we have generated a coherent superposition of vacuum and one photon states. This was experimentally implemented by Lvovsky and Mlynek [5].

5.10 Beam splitter as an attenuator

The beam splitter can be considered as a model for the attenuation of light beams. For classical light beams the transmitted field is tE, where $|t| < 1$. Thus if observation is made only on the transmitted beam, then the beam splitter is just an attenuator of the light beam. For the quantum case, we have seen that (5.15)

$$|\alpha, \beta\rangle \to |\alpha(\theta), \beta(\theta)\rangle, \qquad \begin{pmatrix} \alpha(\theta) \\ \beta(\theta) \end{pmatrix} = \begin{pmatrix} \cos\theta & e^{i\varphi}\sin\theta \\ -e^{-i\varphi}\sin\theta & \cos\theta \end{pmatrix} \begin{pmatrix} \alpha \\ \beta \end{pmatrix}. \tag{5.31}$$

More generally, if the input state is given by

$$\rho = \int P(\alpha, \beta)|\alpha, \beta\rangle\langle\alpha, \beta|d^2\alpha d^2\beta, \tag{5.32}$$

then the output state will be

$$\rho_{\text{out}} = \int P(\alpha, \beta)|\alpha(\theta), \beta(\theta)\rangle\langle\alpha(\theta), \beta(\theta)|d^2\alpha d^2\beta. \tag{5.33}$$

If there is no input from the port b, then

$$P(\alpha, \beta) = P(\alpha)\delta^{(2)}(\beta), \tag{5.34}$$

and then (5.33) reduces to

$$\rho_{\text{out}} = \int P(\alpha)|\cos\theta\,\alpha, -\sin\theta\, e^{-i\varphi}\alpha\rangle\langle\cos\theta\,\alpha, -\sin\theta\, e^{-i\varphi}\alpha|d^2\alpha. \tag{5.35}$$

The reduced state $\rho^{(a)}$ of mode a is obtained by tracing over mode b leading to

$$\rho_{\text{out}}^{(a)} = \int P(\alpha)|\alpha\cos\theta\rangle\langle\alpha\cos\theta|\mathrm{d}^2\alpha, \tag{5.36}$$

and therefore the P-function of the field at the transmitted port is

$$P_{\text{out}}^{(a)}(\alpha) = \frac{1}{\cos^2\theta}P(\alpha/\cos\theta). \tag{5.37}$$

From (5.36) the normally ordered correlations are related to the input via

$$\langle a^{\dagger m}a^n\rangle_{\text{out}} = (\cos\theta)^{m+n}\langle a^{\dagger m}a^n\rangle. \tag{5.38}$$

Thus all the normally ordered moments of order (m, n) are attenuated by the factor $(\cos\theta)^{m+n}$. The beam splitter becomes a model of an attenuator if no observations are made on the reflected port. Mathematically this is the process of taking a trace over mode b. For an n-photon input state, the density matrix of the transmitted field can be obtained from (5.17) by setting $m = 0$

$$\rho^{(a)} = \sum_{p=0}^{n} |f_p|^2|p\rangle\langle p|, \tag{5.39}$$

where f_p can be obtained from (5.20) by setting $m = 0$, $q' = 0$, $q = p$,

$$|f_p|^2 = \binom{n}{p}\left(\cos^2\theta\right)^p\left(\sin^2\theta\right)^{n-p}. \tag{5.40}$$

The transmitted field has a binomial distribution. Generally the transmitted field would be a mixed state even if the input state was a pure state. This is due to the loss of the reflected beam. In deriving (5.36)–(5.40) we assumed no input from the port b. A more general description of the attenuator can be obtained by replacing (5.34) by $P(\alpha)\exp(-|\beta|^2/\bar{n})/(\pi\bar{n})$ if some incoherent noise enters from the port b.

5.11 Transformation of quantized light fields by phase shifters

Consider next the action of a phase shifter. For classical light beams, the amplitude of the field changes by

$$E \longrightarrow E\mathrm{e}^{-\mathrm{i}\varphi}. \tag{5.41}$$

For quantized fields, we can characterize the action of a phase shifter by the unitary operator

$$u_p(\varphi) = \exp(-\mathrm{i}a^\dagger a\,\varphi). \tag{5.42}$$

Clearly the operator a would transform as

$$a \to u_p^\dagger(\varphi)au_p(\varphi) = a\mathrm{e}^{-\mathrm{i}\varphi}. \tag{5.43}$$

Any state of the field would transform as

$$|\Psi\rangle \to u_p(\varphi)|\Psi\rangle. \tag{5.44}$$

For example, a phase shifter would transform the coherent state as follows

$$\begin{aligned}|\alpha\rangle &\to u_p(\varphi)D(\alpha)|0\rangle \\ &= u_p(\varphi)D(\alpha)u_p^\dagger(\varphi)u_p(\varphi)|0\rangle \\ &= \exp\left(a^\dagger e^{-i\varphi}\alpha - a e^{i\varphi}\alpha^*\right)|0\rangle \\ &= D(\alpha e^{-i\varphi})|0\rangle \\ &= |\alpha e^{-i\varphi}\rangle. \end{aligned} \tag{5.45}$$

Thus the coherent state with amplitude α goes over to a coherent state with amplitude $\alpha e^{-i\varphi}$.

For a Fock state, the transformation is rather simple

$$|n\rangle \to u_p(\varphi)|n\rangle = e^{-in\varphi}|n\rangle. \tag{5.46}$$

Thus the n-photon state picks up the phase factor $n\varphi$ rather than φ. This is very significant in quantum metrology and points to a definite advantage of using Fock states in quantum metrology.

Finally, for a squeezed vacuum the phase shifter changes the direction of squeezing. For a single-mode squeezed vacuum

$$\begin{aligned}|\xi\rangle &= \exp\left(\frac{\xi}{2}a^{\dagger 2} - \frac{\xi^*}{2}a^2\right)|0\rangle \longrightarrow \\ u_p(\varphi)|\xi\rangle &= \exp\left(\frac{\xi}{2}e^{-2i\varphi}a^{\dagger 2} - \frac{\xi^*}{2}e^{2i\varphi}a^2\right)|0\rangle = |\xi e^{-2i\varphi}\rangle. \end{aligned} \tag{5.47}$$

Similarly for a two-mode squeezed vacuum, we obtain

$$|\xi\rangle \to |\xi e^{-i(\varphi_a+\varphi_b)}\rangle, \tag{5.48}$$

where φ_a, φ_b are respectively the phase shifts for modes a and b.

5.12 The Mach–Zehnder interferometer

One of the most versatile interferometers is the Mach–Zehnder interferometer shown in Figure 5.7. Here BS1 and BS2 are two 50-50 beam splitters, and M1 and M2 are two 100% mirrors. We can detect an interference pattern either at the port 1 or at the port 2, i.e. we can detect the intensities with the detectors 1 and 2. The detectors typically are avalanche photodiodes. Ideally one would need photon number resolving detectors and such detectors are beginning to become available and experiments have started appearing [6, 7]. We assume that, in one of the paths, there is an object which shifts the phase of the field in that arm. We can assume for simplicity that the path lengths of the two arms are same. Even if the path lengths are different, the difference can be absorbed in the phase φ. If the interferometer is operated with classical light, then there is no input from the side b. In quantum interferometry, there could be an input from the side b as well. In any case we

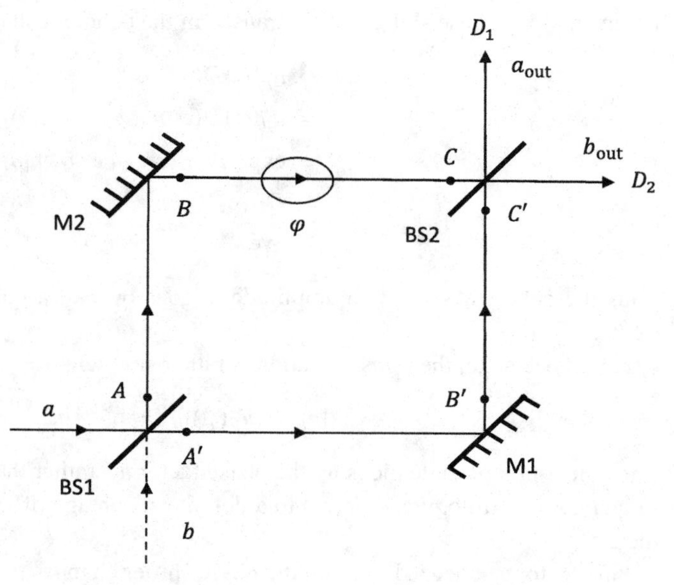

Fig. 5.7 The Mach–Zehnder interferometer.

have to account for the vacuum entering from the port b. We will assume symmetric beam splitters.

It is simplest to discuss the working of the interferometer in terms of the transformations of the field operators. We would use the results obtained in Section 5.1. The operators immediately after BS1, i.e. at the points A and A', are given by $\frac{b+ia}{\sqrt{2}}, \frac{a+ib}{\sqrt{2}}$. The mirrors M1 and M2 impart a phase shift of π so that the operators at B and B' are $-\frac{b+ia}{\sqrt{2}}, -\frac{a+ib}{\sqrt{2}}$, which become $-\mathrm{e}^{-\mathrm{i}\varphi}\frac{b+ia}{\sqrt{2}}, -\frac{a+ib}{\sqrt{2}}$ at the points C and C'. The beam splitter BS2 transforms these so that the output fields at D_1 and D_2 are given by

$$\begin{aligned} a_{\text{out}}(D_1) &= \frac{1}{\sqrt{2}}\left(-\frac{a+ib}{\sqrt{2}} - \mathrm{i}\mathrm{e}^{-\mathrm{i}\varphi}\frac{b+\mathrm{i}a}{\sqrt{2}}\right), \\ b_{\text{out}}(D_2) &= \frac{1}{\sqrt{2}}\left(-\mathrm{e}^{-\mathrm{i}\varphi}\frac{b+\mathrm{i}a}{\sqrt{2}} - \mathrm{i}\frac{a+\mathrm{i}b}{\sqrt{2}}\right). \end{aligned} \tag{5.49}$$

We have adopted the convention so that in the absence of beam splitters and $\varphi = 0$; $b_{\text{out}} = -b$, $a_{\text{out}} = -a$. It is easy to write these in terms of matrices

$$\begin{pmatrix} a_{\text{out}} \\ b_{\text{out}} \end{pmatrix} = \frac{1}{\sqrt{2}}\begin{pmatrix} 1 & \mathrm{i} \\ \mathrm{i} & 1 \end{pmatrix}\begin{pmatrix} 1 & 0 \\ 0 & \mathrm{e}^{-\mathrm{i}\varphi} \end{pmatrix}\begin{pmatrix} -1 & 0 \\ 0 & -1 \end{pmatrix}\frac{1}{\sqrt{2}}\begin{pmatrix} 1 & \mathrm{i} \\ \mathrm{i} & 1 \end{pmatrix}\begin{pmatrix} a \\ b \end{pmatrix}. \tag{5.50}$$

Given the output fields (5.49), one can calculate all the features in terms of the input fields. If the field b is in the vacuum state, then from output fields (5.49) it can be shown that

$$\begin{aligned} \langle a^\dagger_{\text{out}} a_{\text{out}} \rangle &= \sin^2\frac{\varphi}{2}\,\langle a^\dagger a\rangle, \\ \langle b^\dagger_{\text{out}} b_{\text{out}} \rangle &= \cos^2\frac{\varphi}{2}\,\langle a^\dagger a\rangle. \end{aligned} \tag{5.51}$$

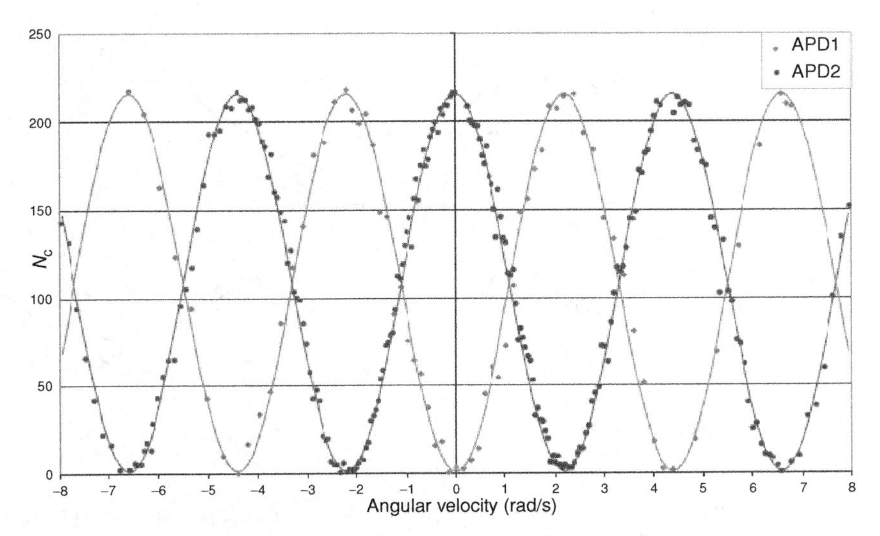

Fig. 5.8 The interference pattern shows the Sagnac effect with visibility up to $(99.2 \pm 0.4)\%$. Each point corresponds to the photon number count N_c for a 300 ms integration, after [10].

Thus the intensity patterns at both D_1 and D_2 show fringes as a function of φ. The maxima at one detector coincide with the minima at another detector. The result (5.51) is the same as one finds in classical fields. Thus for no field from the port b, the results for intensities coincide for classical and quantum fields. The interferometer is said to be balanced if $\varphi = 0$.

Long before the advent of the theory of quantized fields, the question of whether interference could be obtained with very weak beams of light was being investigated. According to (5.51) interference will occur even if the average number $\langle a^\dagger a\rangle$ is much smaller than unity. The earliest experiment was performed by Taylor [8] following a suggestion of Sir J. J. Thompson. Taylor observed the interference, although the exposure time was 2000 hours. A series of interference experiments at single-photon level were performed by Hariharan and coworkers [9].

In recent times several experiments with single photons have been carried out. For example, reference [10] uses the Sagnac interferometer [11, 12] in measurements. In these experiments one uses a heralded source of single photons – we have described such a source in Section 3.5. It is clear from (5.51) that one would get an interference pattern with single photons $\langle a^\dagger a\rangle = 1$. The experimental output would be as shown in Figure 5.8. Physically one would expect that if a click was registered by detector 1, then no click would be registered by detector 2 at the same time, i.e. the detectors 1 and 2 can not click simultaneously. This is confirmed by examining the quantity $\langle a^\dagger_{\text{out}} b^\dagger_{\text{out}} a_{\text{out}} b_{\text{out}}\rangle$. From (5.49), a simple algebra shows that

$$\langle a^\dagger_{\text{out}} b^\dagger_{\text{out}} a_{\text{out}} b_{\text{out}}\rangle \propto \langle a^{\dagger 2} a^2\rangle = 0, \tag{5.52}$$

if the input state of the field is a single-photon Fock state. Dirac [13] noted that a single photon at a beam splitter would follow either the path ABC or $A'B'C'$. The experiments have shown that in order to build up the interference pattern one has to count for a sufficient

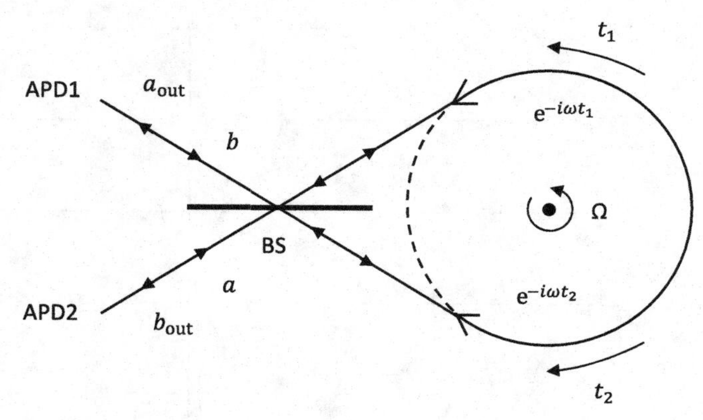

Fig. 5.9 The Sagnac interferometer.

amount of time otherwise one would be recording just a small number of clicks arising from individual photons hitting the detector. For example, in the experimental data shown in Figure 5.8 the integration time was 300 ms for each point.

The theory of the Sagnac interferometer is very similar to that of the Mach–Zehnder interferometer. We note that the Sagnac interferometer (Figure 5.9) is easily realized by using optical fibers. It works in a way that is very similar to the Mach–Zehnder interferometer as shown in Figure 5.7. The interferometer rotates with an angular velocity Ω in the counterclockwise direction. The light going clockwise (counterclockwise) takes a longer (smaller) time through the interferometer:

$$t_1 = \frac{2\pi R}{c + R\Omega}, \qquad t_2 = \frac{2\pi R}{c - R\Omega}, \qquad t_1 < t_2, \tag{5.53}$$

where R is the radius of the interferometer. Using the transformation laws for the beam splitter, we can now write the output fields in terms of the input fields as

$$\begin{pmatrix} a_{out} \\ b_{out} \end{pmatrix} = \frac{1}{\sqrt{2}} \begin{pmatrix} 1 & \mathrm{i} \\ \mathrm{i} & 1 \end{pmatrix} \begin{pmatrix} e^{-i\omega t_2} & 0 \\ 0 & e^{-i\omega t_1} \end{pmatrix} \frac{1}{\sqrt{2}} \begin{pmatrix} 1 & \mathrm{i} \\ \mathrm{i} & 1 \end{pmatrix} \begin{pmatrix} a \\ b \end{pmatrix}, \tag{5.54}$$

where the middle matrix in (5.54) corresponds to the phase shifts for light going clockwise/counterclockwise through the interferometer. The transformation (5.54) is almost the same as (5.50). If the input field b is the vacuum, then clearly

$$\begin{aligned} \langle a^\dagger_{out} a_{out} \rangle &= \sin^2 \frac{\varphi}{2} \langle a^\dagger a \rangle, \\ \langle b^\dagger_{out} b_{out} \rangle &= \cos^2 \frac{\varphi}{2} \langle a^\dagger a \rangle, \qquad \varphi = \omega(t_2 - t_1). \end{aligned} \tag{5.55}$$

In view of the similarity of (5.54) and (5.50), all the results for the Mach–Zehnder interferometer would also apply to the Sagnac interferometer. In particular, this would also be the case if nonclassical fields were used as inputs to the interferometer.

5.13 Wheeler's delayed choice gedanken experiment

Let us consider the case if the beam splitter 2 in Figure 5.7 were removed after a single photon had entered the interferometer. In (5.50), we drop the very first matrix and obtain

$$a_{\text{out}} = -\frac{a + ib}{\sqrt{2}}, \qquad b_{\text{out}} = -\mathrm{e}^{-\mathrm{i}\varphi} \frac{b + ia}{\sqrt{2}}, \tag{5.56}$$

which for no light from the port b leads to

$$\langle a_{\text{out}}^{\dagger} a_{\text{out}} \rangle = \langle b_{\text{out}}^{\dagger} b_{\text{out}} \rangle = \frac{1}{2}. \tag{5.57}$$

The interference pattern disappears. On the one hand, if the photon is transmitted, then it continues on the path $A'B'C'$ and appears at detector D_1. On the other hand, if the photon at the beam splitter BS1 is reflected, then it appears at detector D_2. The photon has an equal probability of appearing either on D_1 or D_2, although in a single shot it would appear either on D_1 or D_2. The existence of the interference pattern given by (5.51) implies that the photon has arrived by both routes [14]. One has here a realization of Bohr's complementarity principle since the two experimental setups, with and without BS2, are mutually exclusive. In one setup (with BS2 present) one observes the wave character of light, whereas with BS2 removed one has a realization of the particle characteristic of light. In Wheeler's gedanken experiment, the choice of wave or particle characteristic is made after the photon has entered the interferometer via BS1. In order to verify Wheeler's gedanken delayed choice experiments one needs a source of single photons, and these are now available. Furthermore, one has to build up an interferometer in which both BS2 and BS1 have relativistic space-like separation. That is, one should be able to quickly move BS2 back and forth so that the information about the presence or absence of BS2 does not reach the photon that has already entered the interferometer. All these conditions were realized by Jacques *et al.* [15]. In their experiment with a single photon source the time taken for the photon to travel from BS1 to BS2 was larger than the time taken to move BS2 in and out. Their data unambiguously verified (I) the result (5.51) (BS2 in) when the photon traveled by both paths, (II) the result (5.57) (BS2 out) when the photon traveled by one path or the other.

5.14 Interaction-free measurements

Let us consider the situation when the phase shifter φ in Figure 5.7 is replaced by a strong scatterer. We first discuss the case that the scatterer completely scatters the photon if it travels along path AB. In that case, no field arrives at the beam splitter BS2 from the path ABC. The situation then becomes as shown in Figure 5.10. In quantum theory, we now have

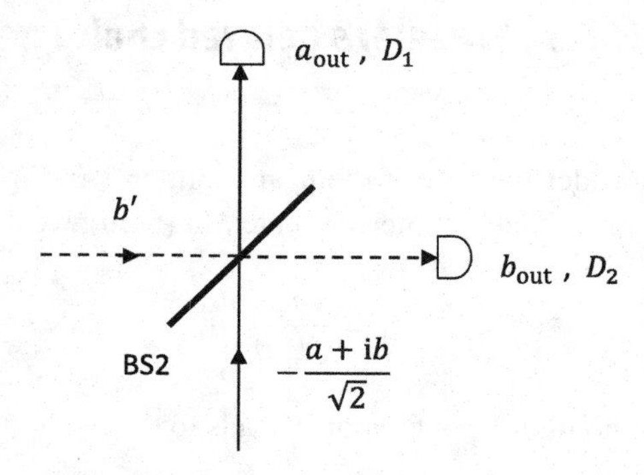

Fig. 5.10 The fields at BS2 in a Mach–Zehnder interferometer in which the phase shifter φ is replaced by a strong scatterer; b' is the vacuum field.

to include the vacuum mode b'. The output field would be

$$\begin{aligned} a_{\text{out}} &= \frac{1}{\sqrt{2}}\left(-\frac{a+ib}{\sqrt{2}} + \mathrm{i}b'\right), \\ b_{\text{out}} &= \frac{1}{\sqrt{2}}\left(b' - \mathrm{i}\frac{a+ib}{\sqrt{2}}\right). \end{aligned} \tag{5.58}$$

Noting that both b and b' represent vacuum fields, the detected signals would be

$$\langle a_{\text{out}}^\dagger a_{\text{out}}\rangle = \frac{1}{4}\langle a^\dagger a\rangle, \qquad \langle b_{\text{out}}^\dagger b_{\text{out}}\rangle = \frac{1}{4}\langle a^\dagger a\rangle. \tag{5.59}$$

If the strong scatterer were absent, then the detected signals were (Eq. (5.51), $\varphi = 0$)

$$\langle a_{\text{out}}^\dagger a_{\text{out}}\rangle = 0, \qquad \langle b_{\text{out}}^\dagger b_{\text{out}}\rangle = \langle a^\dagger a\rangle. \tag{5.60}$$

The port a_{out} becomes bright when the scatterer is present, due to a photon traveling on the path $A'B'C'$. Furthermore, none of the detectors would click if the photon were to travel by the path ABC. One thus has a remarkable result that the presence of the scatterer can be sensed even when the photon did not interact with the scatterer. These situations are known as interaction-free measurements [16]. Penrose discussed the possibility of using quantum mechanics to detect a bomb without actually exploding it [17].

Instead of a strong scatterer which completely removes the photon if the photon goes on the path AB, we can consider a more general situation which can easily be implemented in the laboratory [18]. We replace the scatterer by another beam splitter BS3 with transmission coefficient $\cos\theta'$ $(0 \le \theta' \le \frac{\pi}{2})$. This is sketched in Figure 5.11. The fields at the points C and C' are now equal to

$$-\frac{b+\mathrm{i}a}{\sqrt{2}}\cos\theta' + \mathrm{i}b'\sin\theta', \qquad -\frac{a+\mathrm{i}b}{\sqrt{2}},$$

and hence the output field say at the port a_{out} will be

$$a_{\text{out}} = \frac{1}{\sqrt{2}}\left[-\frac{a+\mathrm{i}b}{\sqrt{2}} + \mathrm{i}\left(-\frac{b+\mathrm{i}a}{\sqrt{2}}\cos\theta' + \mathrm{i}b'\sin\theta'\right)\right]. \tag{5.61}$$

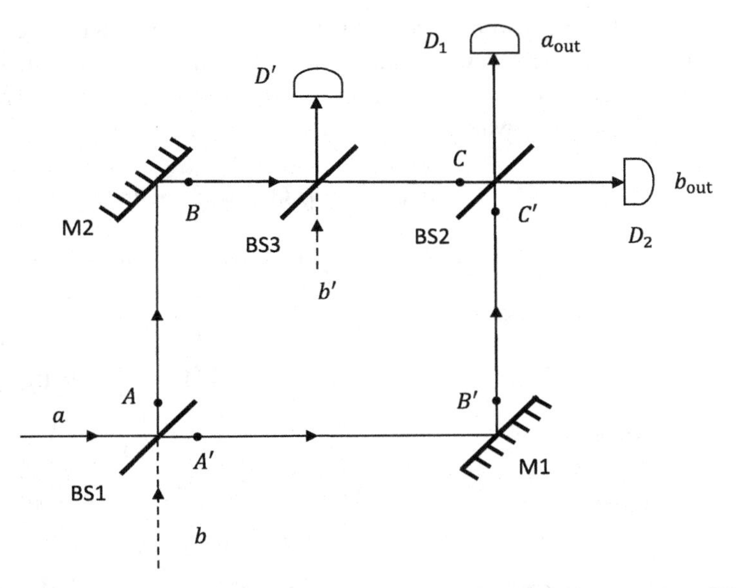

Fig. 5.11 The Mach–Zehnder interferometer in which the phase shifter φ is replaced by another beam splitter BS3 with b' representing the vacuum at the open port.

The detected signal at D_1 will be

$$\langle a_{\text{out}}^{\dagger} a_{\text{out}} \rangle = \sin^4\left(\frac{\theta'}{2}\right) \langle a^{\dagger} a \rangle, \tag{5.62}$$

which for $\theta' = \pi/2$ reduces to the previous result (5.59). An auxiliary detector D' would detect the photon reflected by BS3. The signal at D' will be

$$\langle a_{D'}^{\dagger} a_{D'} \rangle = \frac{1}{2} \sin^2 \theta' \langle a^{\dagger} a \rangle, \tag{5.63}$$

the factor $\frac{1}{2}$ in (5.63) arises as the field at B is $\frac{b+ia}{\sqrt{2}}$. In the experiment of Kwiat *et al.* [18], the single photon was from a heralded source of photons using a downconverter. They used a Michelson interferometer instead of a Mach–Zehnder interferometer. Furthermore, they measured and compared signals like (5.62) and (5.63) for different reflectivities of the beam splitter BS3.

5.15 Two-photon Mach–Zehnder interferometer

So far we have described various aspects of the single-photon Mach–Zehnder interferometers. Let us now consider a two-photon interferometer in which a pair of photons is incident at BS1 (Figure 5.7) such that from each side we have one photon. The state of the input field is $|1, 1\rangle$, i.e. each of the modes a and b contains exactly one photon. As discussed in Chapter 3, such a pair can be obtained from a parametric downconverter. The relations

(5.49) for the field operators at the output of the interferometer hold for all states of the input. Using (5.49) we evaluate the probability $p(1, 1)$ of simultaneously detecting one photon at the detector D_1 and one photon at the detector D_2. This probability is equal to $\langle a^\dagger_{\text{out}} b^\dagger_{\text{out}} a_{\text{out}} b_{\text{out}} \rangle$ where

$$\begin{aligned} a_{\text{out}} &= -\frac{1}{2}\left[a(1 - \mathrm{e}^{-\mathrm{i}\varphi}) + \mathrm{i}b(1 + \mathrm{e}^{-\mathrm{i}\varphi})\right], \\ b_{\text{out}} &= -\frac{1}{2}\left[\mathrm{i}a(1 + \mathrm{e}^{-\mathrm{i}\varphi}) - b(1 - \mathrm{e}^{-\mathrm{i}\varphi})\right]. \end{aligned} \tag{5.64}$$

From (5.64) it is seen that

$$a_{\text{out}} b_{\text{out}} |1, 1\rangle = -\frac{1}{2}(1 + \mathrm{e}^{-2\mathrm{i}\varphi})|0, 0\rangle, \tag{5.65}$$

and hence

$$P(1, 1) = \frac{1}{4}|1 + \mathrm{e}^{-2\mathrm{i}\varphi}|^2 = \cos^2 \varphi, \tag{5.66}$$

$$\rightarrow 0 \quad \text{for} \quad \varphi = \pi/2. \tag{5.67}$$

The result (5.67) is a consequence of the Hong–Ou–Mandel interference that we discussed in Section 5.6. The equation (5.23) shows that for a 50-50 beam splitter the output after BS1 is $\frac{1}{\sqrt{2}}(|2, 0\rangle + |0, 2\rangle)$. The phase shifter for $\varphi = \pi/2$ changes this symmetric state into an antisymmetric state $\frac{1}{\sqrt{2}}(|2, 0\rangle - |0, 2\rangle)$. The second beam splitter in the Mach–Zehnder interferometer has no effect on the antisymmetric state $\frac{1}{\sqrt{2}}(|2, 0\rangle - |0, 2\rangle)$. This can be verified using (5.18):

$$\begin{aligned} \frac{1}{\sqrt{2}}(|2, 0\rangle - |0, 2\rangle) &\rightarrow \frac{1}{2\sqrt{2}}(a^{\dagger 2} - b^{\dagger 2})|0, 0\rangle \\ &\rightarrow \frac{1}{2\sqrt{2}}[(a^\dagger \cos\theta + \mathrm{i}\sin\theta b^\dagger)^2 - (b^\dagger \cos\theta + \mathrm{i}\sin\theta a^\dagger)^2]|0, 0\rangle \\ &= \frac{1}{2\sqrt{2}}(a^{\dagger 2} - b^{\dagger 2})|0, 0\rangle \\ &= \frac{1}{\sqrt{2}}(|2, 0\rangle - |0, 2\rangle). \end{aligned} \tag{5.68}$$

In the output one sees either two photons at the detector 1 or at the detector 2. For $\varphi \neq \pi/2$, the input state at BS2 is a linear combination of both symmetric and antisymmetric states $\frac{1}{\sqrt{2}}(|2, 0\rangle \pm |0, 2\rangle)$. The symmetric part is transformed to $|1, 1\rangle$ leading to nonzero $p(1, 1)$. In the landmark experiment of Hong, Ou, and Mandel, pairs of photons from a downconverter were used. The downconverter produces pairs over a finite bandwidth, where one can select the bandwidth using appropriate filters. Their filters had a bandwidth $\Delta\omega$ of about 10^{13} Hz. Their interferometeric setup is shown in Figure 5.12.

The beam splitter is translated up and down so that the time of arrival of the signal and idler photons at the beam splitter is somewhat different. The simultaneous clicking of the detectors D_1 and D_2 was studied as a function of the time difference τ in the arrival times. One needs to generalize our discussion of the beam splitter in Section 5.17 to many modes.

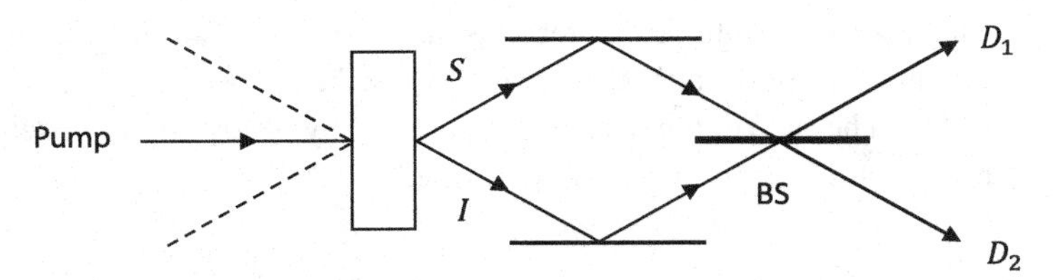

Fig. 5.12 The Hong–Ou–Mandel interferometer; the BS can be translated up or down.

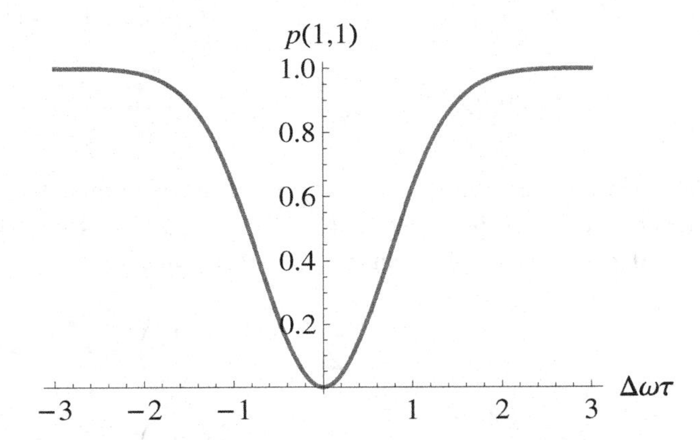

Fig. 5.13 The coincidence probability $p(1, 1)$ as a function of the delay between the arrival of the signal and idler photons at the beam splitter.

The final result for the coincidence probability is proportional to

$$p(1, 1) \propto 1 - \exp(-\Delta\omega^2\tau^2), \tag{5.69}$$

which leads to the famous Hong–Ou–Mandel dip at $\tau = 0$. We reproduce their experimental result in Figure 5.13. With this technique they could measure correlation times in the sub-picosecond regime. The Hong–Ou–Mandel interference is now at the heart of quantum optics and quantum information science. It is used as a tool to test the single photon nature of the source.

5.16 Multiphoton interference and engineering of quantum states

We have seen in the previous section that the Hong–Ou–Mandel interference can be used to produce an entangled state $\frac{1}{\sqrt{2}}(|2, 0\rangle + |0, 2\rangle)$ starting from a separable state $|1, 1\rangle$. These ideas can be extended to multiphoton interference and towards the production of a variety of entangled states. We have seen in Chapter 3 that the down-conversion can produce a two-mode squeezed vacuum (Eq. (3.6)), which shows the production of paired states of the form $|n, n\rangle$. Thus at higher pump powers we can produce not only $|1, 1\rangle$ but also $|2, 2\rangle$, etc. In the following we specifically consider launching of the field in the state $|2, 2\rangle$ through a

Mach–Zehnder interferometer. It is instructive to examine the behavior of the state $|2,2\rangle$ in the interferometer, although we can study the properties at the output by using the relation (5.50). The state of the field at AA' (denoted by $|\Psi\rangle_{AA'}$) immediately after the beam splitter 1 can be obtained from (5.24) by setting $N = 2$:

$$|\Psi\rangle_{AA'} = -\left[\frac{1}{2}|2,2\rangle + \sqrt{\frac{3}{8}}(|4,0\rangle + |0,4\rangle)\right]. \tag{5.70}$$

The mirrors $M1$ and $M2$ do not change it. The phase shifter changes it to

$$|\Psi\rangle_{CC'} = -\left[\frac{1}{2}\mathrm{e}^{-2i\varphi}|2,2\rangle + \sqrt{\frac{3}{8}}(\mathrm{e}^{-4i\varphi}|4,0\rangle + |0,4\rangle)\right]. \tag{5.71}$$

We next transform the state (5.71) by the beam splitter BS2. The state $|2,2\rangle$ would be transformed to a superposition of $|2,2\rangle$, $|4,0\rangle$, and $|0,4\rangle$ (as in (5.70)). To transform $|4,0\rangle$ and $|0,4\rangle$ we use the formula (5.17), (5.20) with $\theta = \pi/4$.

$$\begin{aligned}
|4,0\rangle &\to \sum_{p=0}^{4}\sqrt{\binom{4}{p}}\left(\frac{1}{\sqrt{2}}\right)^4 \cdot (\mathrm{i})^{4-p}|p,4-p\rangle, \\
|0,4\rangle &\to \sum_{p=0}^{4}\sqrt{\binom{4}{p}}\left(\frac{1}{\sqrt{2}}\right)^4 \cdot (\mathrm{i})^{p}|p,4-p\rangle.
\end{aligned} \tag{5.72}$$

The output state of the Mach–Zehnder interferometer when fed with the state $|2,2\rangle$ can be obtained by combining (5.70) and (5.72). We focus on the probability of detecting three photons at one port and one photon at the other output port [19,20]. The state $|2,2\rangle$ in (5.71) would not contribute to such a probability. The state $|4,0\rangle$, $|0,4\rangle$ would give contributions (coefficients of the state $|3,1\rangle$ in (5.72))

$$|4,0\rangle \Rightarrow \frac{1}{2}\mathrm{i}|3,1\rangle, \qquad |0,4\rangle \Rightarrow -\frac{1}{2}\mathrm{i}|3,1\rangle, \tag{5.73}$$

and the probability $p(3,1)$ will be

$$p(3,1) = \frac{3}{8}\sin^2 2\varphi. \tag{5.74}$$

This can be compared with the probability $\cos^2\varphi$ (Eq. (5.66)) of detecting one photon at each of the output ports when the interferometer is fed with light in the state $|1,1\rangle$. In a given interval the signal (5.74) shows twice as many fringes than the signal (5.66).

The multiphoton interference has been discussed by several authors using beam splitters with variable reflectivities [21,22]. Wang and Kobayashi [21] discuss the interferometer with the input state $|2,1\rangle$ and for each BS with reflectivity $2/3$. They show (see Exercise 5.10) that the fringes would occur three times as often than with classical light, provided the detected signal corresponds to the detection of two photons at one port and one photon at the other port.

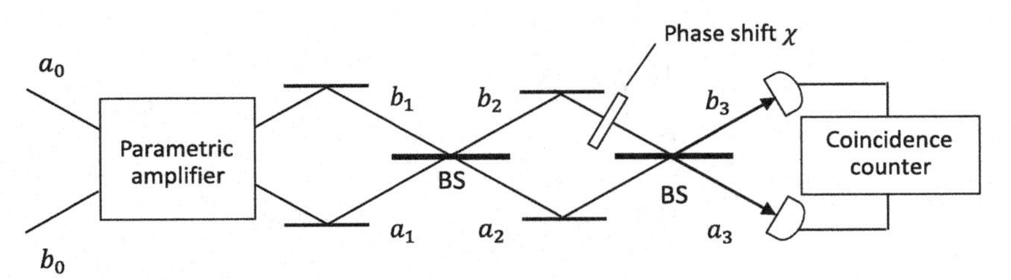

Fig. 5.14 Mach–Zehnder interferometer with two-mode squeezed vacuum as input, after [23].

5.17 Mach–Zehnder interferometer with two-mode squeezed vacuum as input

We have discussed the behavior of the Mach-Zehnder interferometer using pairs of photons produced by a downconverter. The natural question is now what happens if the downconverter is working in the region of high gain. What is the nature of the interference pattern? Let us then consider the setup [23] shown in Figure 5.14. The output operators (a_3, b_3) can be written in terms of the input operators (a_1, b_1) as

$$\begin{pmatrix} a_3 \\ b_3 \end{pmatrix} = \frac{1}{\sqrt{2}} \begin{pmatrix} 1 & \mathrm{i} \\ \mathrm{i} & 1 \end{pmatrix} \begin{pmatrix} 1 & 0 \\ 0 & \mathrm{e}^{-\mathrm{i}\chi} \end{pmatrix} \frac{1}{\sqrt{2}} \begin{pmatrix} 1 & \mathrm{i} \\ \mathrm{i} & 1 \end{pmatrix} \begin{pmatrix} a_1 \\ b_1 \end{pmatrix}, \tag{5.75}$$

where we have dropped the matrix -1 for the two 100% mirrors. This is really redundant for photon detectors. The fields a_1 and b_1 are produced by the parametric amplifier (Eq. (3.61))

$$\begin{aligned} a_1 &= a_0 \cosh r + b_0^\dagger \mathrm{e}^{\mathrm{i}\varphi} \sinh r, \\ b_1 &= b_0 \cosh r + a_0^\dagger \mathrm{e}^{\mathrm{i}\varphi} \sinh r, \end{aligned} \tag{5.76}$$

where a_0 and b_0 are the modes in the vacuum state. From (5.75) we have

$$\begin{aligned} a_3 &= \frac{1}{2}\left[a_1(1 - \mathrm{e}^{-\mathrm{i}\chi}) + \mathrm{i}b_1(1 + \mathrm{e}^{-\mathrm{i}\chi})\right], \\ b_3 &= \frac{1}{2}\left[\mathrm{i}a_1(1 + \mathrm{e}^{-\mathrm{i}\chi}) - b_1(1 - \mathrm{e}^{-\mathrm{i}\chi})\right], \end{aligned} \tag{5.77}$$

leading to

$$a_3 b_3 = \frac{1}{4}\left[a_1^2(1 - \mathrm{e}^{-2\mathrm{i}\chi})\mathrm{i} - b_1^2(1 - \mathrm{e}^{-2\mathrm{i}\chi})\mathrm{i} - 2a_1 b_1(1 + \mathrm{e}^{-2\mathrm{i}\chi})\right]. \tag{5.78}$$

Furthermore, from (5.76) on using the properties of the vacuum state we get

$$\begin{aligned} a_1^2|0,0\rangle &= \sqrt{2}\mathrm{e}^{2\mathrm{i}\varphi} \sinh^2 r|0,2\rangle, \\ b_1^2|0,0\rangle &= \sqrt{2}\mathrm{e}^{2\mathrm{i}\varphi} \sinh^2 r|2,0\rangle, \\ a_1 b_1|0,0\rangle &= \mathrm{e}^{2\mathrm{i}\varphi} \sinh^2 r|1,1\rangle + \mathrm{e}^{\mathrm{i}\varphi} \sinh r \cosh r|0,0\rangle, \end{aligned} \tag{5.79}$$

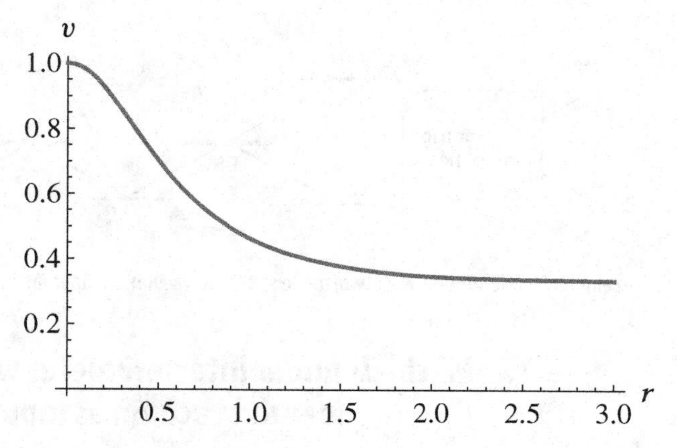

Fig. 5.15 The behavior of the visibility v as a function of r, after [23].

which on using in (5.78) gives

$$a_3b_3|0,0\rangle = \frac{i\sqrt{2}}{4}(1-e^{-2i\chi})e^{2i\varphi}\sinh^2 r(|0,2\rangle - |2,0\rangle) \\ -\frac{1}{2}(1+e^{-2i\chi})\left(e^{2i\varphi}\sinh^2 r|1,1\rangle + e^{i\varphi}\sinh r\cosh r|0,0\rangle\right). \quad (5.80)$$

From (5.80) we derive the coincidence signal

$$c(a,b) = \langle a_3^\dagger b_3^\dagger a_3 b_3\rangle = \sinh^2 r\left[\cos^2\chi + \sinh^2 r(1+\cos^2\chi)\right]. \quad (5.81)$$

In the limit of small r, $c(a,b) \propto \cos^2\chi$ and the visibility of the fringes would be 100%. The visibility drops as r increases. For arbitrary r the visibility is

$$v = (1+\sinh^2 r)/(1+3\sinh^2 r). \quad (5.82)$$

The behavior of the visibility as a function of r (which is also called the single-pass gain) is given in Figure 5.15. For large gain of the amplifier the visibility saturates to 1/3.

We note that on the experimental side we have several options. We could study (i) the coincidence detection of one photon on each side of BS2, or (ii) the detection of two photons on one side of BS2. Such experiments have been performed by Sciarrino *et al.* [24] with results in agreement with theoretical predictions [23, 25, 26]. One also has the possibility of detecting more than two photons. For example, we could detect three photons on one side and one photon on the other side. We have presented such a calculation in the previous section for the low-gain case. The results for the high-gain case are given in [27], which show that one can design detection schemes depending on the number of photons detected on each side so that 100% visibility can be obtained.

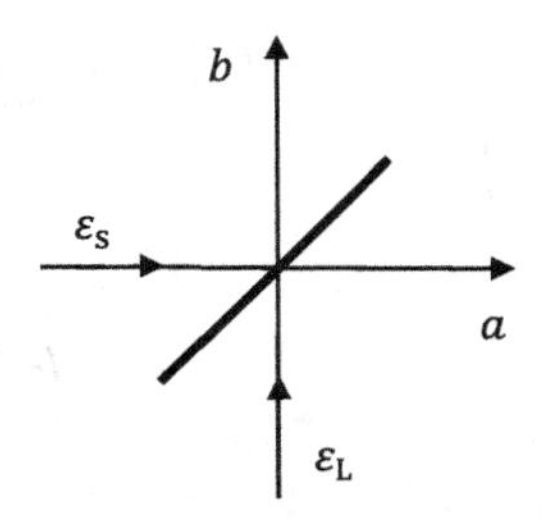

Fig. 5.16 Homodyne technique.

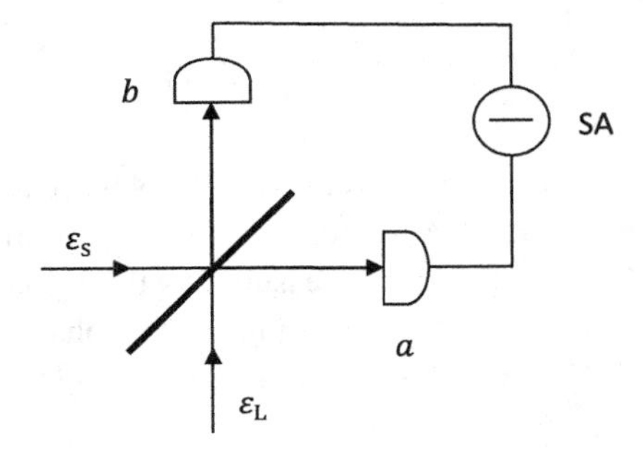

Fig. 5.17 Balanced homodyne interferometer.

5.18 Balanced homodyne interferometers for measuring the squeezing of light

The phase of an unknown signal is measured by mixing it with a field called the local oscillator with well-defined amplitude and phase using a beam splitter (see Figure 5.16). Thus if ε_S is the signal and ε_L is the local oscillator, then the intensity at the port a behind the beam splitter would be

$$I_a = \frac{|\varepsilon_S + i\varepsilon_L|^2}{2} = \frac{|\varepsilon_S|^2}{2} + \frac{|\varepsilon_L|^2}{2} + \frac{1}{2}(i\varepsilon_L\varepsilon_S^* + c.c.), \tag{5.83}$$

which under the condition $|\varepsilon_L|^2 \gg |\varepsilon_S|^2$ reduces to

$$I_a = \frac{|\varepsilon_L|^2}{2} + \frac{1}{2}(i\varepsilon_L\varepsilon_S^* + c.c.). \tag{5.84}$$

Clearly the measurement of I_a yields the phase of ε_S. It is desirable to remove the $|\varepsilon_L|^2$ term from (5.84). This can be done using the arrangement shown in Figure 5.17. This is called balanced homodyne detection [28]. Here the outputs on the ports a and b are subtracted electronically and one measures the signal S. More generally, for fields with temporal fluctuations, which arise from interactions with other systems, the environment, etc., one measures the spectrum with a spectrum analyzer (SA). In analogy to (5.84), we obtain

$$I_b = \frac{|\varepsilon_L + i\varepsilon_s|^2}{2} \approx \frac{1}{2}|\varepsilon_L|^2 - \frac{1}{2}(i\varepsilon_L\varepsilon_s^* + c.c.), \tag{5.85}$$

and therefore

$$S = I_a - I_b \approx i\varepsilon_L\varepsilon_s^* + c.c.. \tag{5.86}$$

So far we have taken all the fields to be classical. The field to be measured ε_s could be a quantum field or it could be a nonclassical field. Let a and $a^\dagger$ be the annihilation and creation operators for such a field, then the signal at the SA will be proportional to

$$S \approx i\varepsilon_L a^\dagger - i\varepsilon_L^* a. \tag{5.87}$$

The local oscillator is treated classically. Note that (5.87) does not depend on the optical frequency as $\varepsilon_L \propto e^{-i\omega_L t}$, $a^\dagger \propto e^{i\omega_L t}$. The equation (5.87) shows that the two quadratures of the field X and Y can be measured as the signal that is being fed to the SA becomes proportional to X or Y, depending on the phase of the local oscillator. Recalling the definition (1.37) $a = \frac{X+iY}{\sqrt{2}}$ and writing $\varepsilon_L = |\varepsilon_L|e^{-i\varphi_L}$, Eq. (5.87) becomes

$$S = \sqrt{2}|\varepsilon_L|(X \sin\varphi_L + Y \cos\varphi_L). \tag{5.88}$$

A study of $\langle S\rangle$ and $\langle S^2\rangle$ would yield the squeezing properties of the field. The homodyne method has been very successfully used in studies of the spectrum of squeezing.

5.19 Manipulation of quantum states by homodyning and feed-forward

Many of the ideas from classical control theory have been extended to control of quantum states of light. One possibility is, of course, the conditional detection on one part of the system to control the quantum state of the other part, which we have already discussed in Section 3.5. The other possibility is to precisely monitor certain quantum variables and use the values so obtained to feedback and control another part of the system. A very popular and successful scheme to do this is schematically shown in Figure 5.18. The input at the port 2 is an auxiliary state. We use this state, a homodyne measurement, and feed-forward to control the state at the output port 1. A detailed review of the theory and the range of experiments is given in [29]. Here we describe the key elements of the theory. We have already discussed the transformation laws for a beam splitter (Section 5.1). The input field operators a and b are transformed into

$$a(\theta) = a\cos\theta + ib\sin\theta, \qquad b(\theta) = b\cos\theta + ia\sin\theta, \tag{5.89}$$

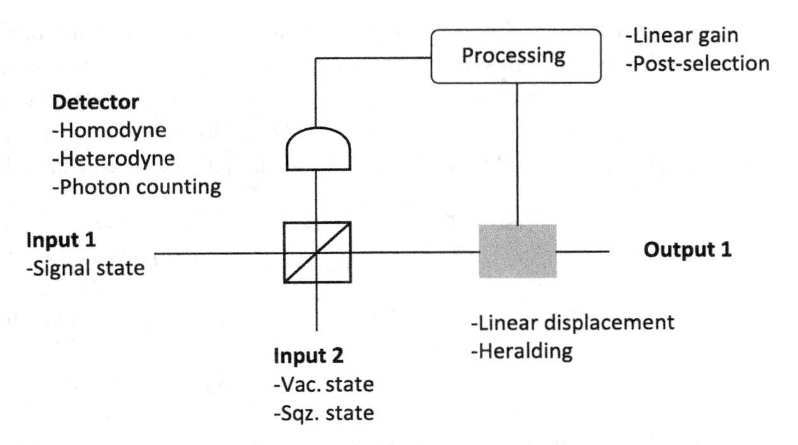

Fig. 5.18 Schematic of the simple feed-forward setup [29], with many different possibilities of detection followed by feed-forward.

where $\cos^2\theta$ is the transmissivity of the mirror. Using these we can obtain equations for the quadratures $X_i(\theta)$, $Y_i(\theta)$; $i = a, b$:

$$\begin{aligned} X_a(\theta) &= X_a\cos\theta - Y_b\sin\theta, \\ Y_a(\theta) &= Y_a\cos\theta + X_b\sin\theta, \\ X_b(\theta) &= X_b\cos\theta - Y_a\sin\theta, \\ Y_b(\theta) &= Y_b\cos\theta + X_a\sin\theta. \end{aligned} \tag{5.90}$$

Now suppose that the quadrature Y_b is measured precisely and is scaled by a factor g and then fed to the output of the port 1. In that case the quadrature $X_a(\theta)$ and $Y_a(\theta)$ become

$$X_a(\theta) = X_a(\cos\theta + g\sin\theta) + Y_b(g\cos\theta - \sin\theta), \tag{5.91}$$

$$Y_a(\theta) = Y_a\cos\theta + X_b\sin\theta. \tag{5.92}$$

Clearly if g is chosen such that

$$g\cos\theta = \sin\theta, \tag{5.93}$$

then

$$X_a(\theta) = X_a/\cos\theta, \tag{5.94}$$

$$Y_a(\theta) = Y_a\cos\theta + X_b\sin\theta. \tag{5.95}$$

Note that if the input on the port 2 is a perfect squeezed state in the X-quadrature, then $X_b \to 0$ and then

$$X_a(\theta) = X_a/\cos\theta, \qquad Y_a(\theta) = Y_a\cos\theta, \tag{5.96}$$

which is a perfect squeezing transformation. This was experimentally verified by Yoshikawa *et al.* [30] where the input coherent states on the port 1 were converted into squeezed states. We also note from (5.94) and (5.95) that a measurement of the output X-quadrature would imply that it is amplified without addition of noise even if the input at the port 2 were a vacuum state. Noiseless amplification of a quadrature was demonstrated in this manner [31].

In the foregoing we have not explicitly written down the states at the output ports. These can always be obtained by using the transformation laws (5.91) and (5.92). Using the result of Exercise 5.1, the Wigner function immediately after the beam splitter is

$$\begin{aligned} W_{\text{out}}(\alpha,\beta) &= W_{\text{in}}(\alpha\cos\theta - \mathrm{i}\beta\sin\theta, \beta\cos\theta - \mathrm{i}\alpha\sin\theta) \\ &\equiv W_{\text{in}}[X_a(\theta), Y_a(\theta), X_b(\theta), Y_b(\theta)], \end{aligned} \tag{5.97}$$

which after the homodyne measurement of the quadrature Y_b becomes

$$\int W_{\text{in}}\left[X_a(\theta), Y_a(\theta), X_b(\theta), \bar{Y}_b(\theta)\right] \mathrm{d}X_b, \tag{5.98}$$

where $\bar{Y}_b$ means the measured value. This value is used to displace X_a, then the final Wigner function for the field at the port 1 is

$$W^{(1)}(X_a, Y_a) \equiv \int \mathrm{d}Y_b \int \mathrm{d}X_b W_{\text{in}}[X_a(\theta) + GY_b(\theta), Y_a(\theta); X_b(\theta), Y_b(\theta)]. \tag{5.99}$$

Note that if the input fields have a Gaussian Wigner function, then the $W^{(1)}(X_a, Y_a)$ would also be Gaussian. The covariance matrix for the output at the port 1 is easily computed from (5.91) and (5.92):

$$\begin{aligned} \Delta X_a^2(\theta) &= \Delta X_a^2(\cos\theta + g\sin\theta)^2 + \Delta Y_b^2(g\cos\theta - \sin\theta)^2, \\ \Delta Y_a^2(\theta) &= \Delta Y_a^2\cos^2\theta + \Delta X_b^2\sin^2\theta, \end{aligned} \tag{5.100}$$

if a and b modes at the input are uncorrelated.

5.20 Quantum state tomography

In this section we give a method to recover the quantum state of a system by homodyne measurements. Vogel and Risken [32] showed how the data on the quadrature measurements of the field can be used to reconstruct the Wigner function. Once the Wigner function is known, then we can construct the density matrix of the system. Let us consider a single-mode field and let us assume that the quadrature X_θ defined by Eq. (2.6), i.e. $X_\theta = (a\mathrm{e}^{-\mathrm{i}\theta} + a^\dagger\mathrm{e}^{\mathrm{i}\theta})/\sqrt{2}$, has been measured via homodyne measurements. Let us then assume that the homodyne measurements yield the probability distribution $P(X,\theta)$ defined by

$$P(X,\theta) = \frac{1}{2\pi}\int \mathrm{e}^{-\mathrm{i}\eta X}\mathrm{d}\eta\, \mathrm{Tr}\left[\rho\exp\left(\mathrm{i}\eta\frac{a\mathrm{e}^{-\mathrm{i}\theta} + a^\dagger\mathrm{e}^{\mathrm{i}\theta}}{\sqrt{2}}\right)\right]. \tag{5.101}$$

On inverting (5.101) we have

$$\left\langle \exp\left[\frac{\mathrm{i}\eta}{\sqrt{2}}\left(a\mathrm{e}^{-\mathrm{i}\theta} + a^\dagger\mathrm{e}^{\mathrm{i}\theta}\right)\right]\right\rangle = \int_{-\infty}^{+\infty} P(X,\theta)\mathrm{e}^{\mathrm{i}\eta X}\mathrm{d}X. \tag{5.102}$$

Now from the definition (1.98) of the Wigner function

$$W(\alpha) = \frac{1}{\pi^2}\int d^2\beta\, e^{-(\beta\alpha^*-\beta^*\alpha)}\langle \exp(\beta a^\dagger - \beta^* a)\rangle,$$
$$= \frac{1}{\pi^2}\int d^2 v\, \langle \exp(iva^\dagger + iv^* a)\rangle \exp[-i(v\alpha^* + v^*\alpha)], \tag{5.103}$$

where we changed $\beta \to iv$. We can rewrite Eq. (5.103) by first changing the integration to polar coordinates and then reducing the θ integral from 0 to 2π to 0 to π

$$W(\alpha) = \frac{1}{\pi^2}\int_{-\infty}^{+\infty} dr\,|r| \int_0^{\pi} d\theta\, \langle \exp[ir(a^\dagger e^{i\theta} + a e^{-i\theta})]\rangle \exp[-ir(\alpha^* e^{i\theta} + \alpha e^{-i\theta})], \tag{5.104}$$

which on using (5.102) becomes

$$W(\alpha) = \frac{1}{\pi^2}\int_{-\infty}^{+\infty} dr\,|r| \int_0^{\pi} d\theta \int_{-\infty}^{+\infty} P(X,\theta)\, e^{i\sqrt{2}rX}\, e^{-ir(\alpha^* e^{i\theta} + \alpha e^{-i\theta})} dX. \tag{5.105}$$

On writing

$$\alpha = \frac{Q + iP}{\sqrt{2}},$$
$$W(\alpha)\, d^2\alpha = W(Q,P)\, dQ\, dP,$$
$$W(Q,P) = \frac{W(\alpha)}{2},$$

Eq. (5.105) leads to

$$W(Q,P) = \frac{1}{2\pi^2}\int_{-\infty}^{+\infty}\int_0^{\pi} d\theta P(X,\theta) K[(Q\cos\theta + P\sin\theta) - X]\, dX, \tag{5.106}$$

where

$$K(\eta) = \frac{1}{2}\int_{-\infty}^{+\infty} dr\,|r|\, e^{ir\eta} = -\frac{P}{\eta^2}. \tag{5.107}$$

The result (5.106) is the main result obtained by Vogel and Risken for the reconstruction of the Wigner function from the homodyne measurements of the probability distribution for the quadratures. The kernel K is well known in the theory of Radon transforms [33] and is infinite at $\eta = 0$. In practical implementation one puts a cut-off such that no unusual artifacts of the cut-off appear. The first implementation of the Vogel–Risken construction was done by Smithey *et al.* [34], who reconstructed the Wigner functions for the vacuum and squeezed vacuum states of a single mode of the electromagnetic field. There are many refinements to the above scheme and a detailed review is given by Lvovsky and Raymer [35]. The quantum state tomography is now a fairly standard tool for reconstructing density matrix elements from tomographic data.

5.21 Sensitivity of an optical interferometer

We now discuss the question of how precisely the phase φ can be measured by an optical interferometer. This question has been thoroughly investigated and the answer depends on

(a) the input states to the interferometer, (b) the detection scheme, and (c) quantitative measures used to characterize the sensitivity of the interferometer.

We first consider the simplest possible situation when the Mach–Zehnder interferometer in Figure 5.7 is fed by a coherent beam of light, i.e. the input state for the mode a is a coherent state $|\alpha\rangle$. From (5.51), the measured signal, i.e. the number of photons N_1 at the output port D_1, is

$$N_1 = \sin^2 \frac{\varphi}{2} |\alpha|^2. \tag{5.108}$$

We have seen in Eq. (5.15) that a beam splitter transforms a coherent state into coherent states at the output. Thus the fields at the output ports D_1 and D_2 would be in coherent states as the interferometer is a combination of two beam splitters. We also learnt in Chapter 1 that the photon number fluctuations in a coherent state are Poissonian. Hence the fluctuation in the signal would be

$$\Delta N_1 = \sqrt{N_1}. \tag{5.109}$$

Following [36] we define the phase sensitivity by

$$\Delta\varphi = \Delta S / |\partial S / \partial\varphi|, \tag{5.110}$$

where S is the detected signal and ΔS is the fluctuation in the signal. For Eq. (5.108), $S \propto N_1$ and hence on using (5.108) and (5.109), (5.110) becomes

$$\Delta\varphi \simeq \frac{1}{|\alpha|} \cdot \frac{1}{|\cos\frac{\varphi}{2}|} \sim \frac{1}{|\alpha|}, \tag{5.111}$$

if φ is close to zero. The phase sensitivity of the interferometer is proportional to the square root of the number of photons fed into the interferometer. This result is also consistent with the number phase uncertainty relation for any coherent signal

$$\Delta N \Delta\varphi \geq 1. \tag{5.112}$$

The uncertainty relation (5.112) suggests that one should be able to get a phase sensitivity better than (5.111), since ΔN could be of the order of N or even larger:

$$\Delta\varphi \geq \frac{1}{N}. \tag{5.113}$$

This limit is called the Heisenberg limit and a lot of effort has been devoted to achieving this limit (see Section 12.9).

Another popular measure for studying phase sensitivity is the Cramer–Rao bound (CRB), which is based on the Fisher information [37]. Let $p(n_1, n_2)$ be the joint probability of detecting n_1 (n_2) photons at the port 1 (2). This probability is a function of the phase φ. The Fisher information $F(\varphi)$ is defined by

$$F(\varphi) = \sum_{n_1,n_2} \frac{1}{p(n_1,n_2)} \left(\frac{\partial p(n_1,n_2)}{\partial\varphi} \right)^2. \tag{5.114}$$

According to CRB, the largest phase sensitivity is given by

$$(\Delta\varphi)_{\mathrm{CRB}} = 1/\sqrt{pF(\varphi)}, \tag{5.115}$$

where p is the number of measurements done to estimate the phase φ.

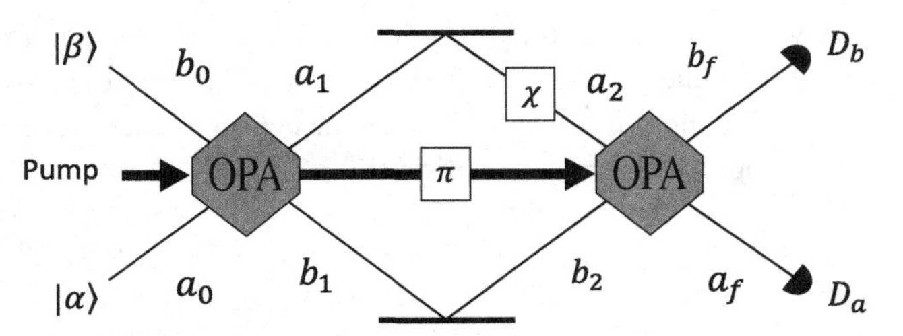

Fig. 5.19 A strong laser beam pumps the first OPA. The beam (which is assumed to be undepleted) undergoes a π-phase shift and then pumps the second OPA. The input modes of the first OPA are fed with coherent light. After the first OPA, one of the outputs interacts with the phase to be probed. Both outputs are then brought back together as the inputs for the second OPA. Measurements are taken on the second OPA's outputs [39].

Let us check (5.115) for the single-photon interferometer for which according to (5.51)

$$p(1,0) = \sin^2\frac{\varphi}{2}, \qquad p(0,1) = \cos^2\frac{\varphi}{2}. \tag{5.116}$$

This yields $(\Delta\varphi)_{\mathrm{CRB}} = 1/\sqrt{p}$, which is the expected outcome. For a coherent input, the joint distribution will be

$$p(n_1,n_2) = p(n_1)p(n_2), \qquad p(n_i) = \frac{\bar{n}_i^{n_i}\mathrm{e}^{-\bar{n}_i}}{n_i!}, \qquad i = 1,2, \tag{5.117}$$

where

$$\bar{n}_1 = \sin^2\frac{\varphi}{2}|\alpha|^2, \qquad \bar{n}_2 = \cos^2\frac{\varphi}{2}|\alpha|^2. \tag{5.118}$$

The Fisher information and $(\Delta\varphi)_{\mathrm{CRB}}$ can be computed and studied as a function of the input photon number.

The sensitivity depends on the type of signal that one measures. For example, one can think of a setup [36, 38] where the signal is the difference of the number of photons at the ports 1 and 2, i.e. $S = \langle n_1 - n_2\rangle$, $\Delta S = \langle (n_1-n_2)^2\rangle - \langle n_1 - n_2\rangle^2$. Clearly $\Delta\varphi$ is dependent on the setup.

5.22 Heisenberg limited sensitivity of interferometers based on parametric amplifiers or four-wave mixers

Yurke *et al.* [36] introduced a new class of interferometers where the two beam splitters of the Mach–Zehnder interferometer are replaced by parametric amplifiers or four-wave mixers. The great advantage of such interferometers is the possibility of achieving Heisenberg limited measurements of phase. We consider the interferometric setup shown in Figure 5.19. Here we also consider both signal and idler ports to be pumped by coherent fields – in contrast to Yurke *et al.* who considered signal and idler fields in the vacuum state. The second OPA is pumped by a pump which is phase-shifted by π from the laser pumping the first OPA. This arrangement would make the output fields identical to the input fields if there was no object in the interferometer. Thus any change in the output fields would be a signature of the presence of the object. We can write down the output fields a_f and b_f in terms of the input fields by following the action of each element in the setup. The first OPA does a Bogoliubov transformation (Eq. (3.11))

$$a_1 = \mu a_0 + \nu b_0^\dagger, \quad b_1 = \mu b_0 + \nu a_0^\dagger,$$
$$\mu = \cosh r, \quad \nu = \mathrm{e}^{\mathrm{i}\varphi} \sinh r. \tag{5.119}$$

The object transforms the operators as

$$a_2 = \mathrm{e}^{\mathrm{i}\chi} a_1, \quad b_2 = b_1. \tag{5.120}$$

The second OPA transforms a_2 and b_2 as

$$a_f = \mu a_2 - \nu b_2^\dagger,$$
$$b_f = \mu b_2 - \nu a_2^\dagger. \tag{5.121}$$

The sign of ν is different because the field pumping the second OPA has been shifted by π. The fields a_0 and b_0 are in coherent states $|\alpha\rangle$ and $|\beta\rangle$, respectively. If we think of a measurement where we measure the total photon number at the output, then the signal S would be

$$S = \langle a_f^\dagger a_f + b_f^\dagger b_f \rangle, \tag{5.122}$$

and the fluctuation would be

$$\Delta S^2 = \langle (a_f^\dagger a_f + b_f^\dagger b_f)^2 \rangle - S^2. \tag{5.123}$$

The calculations of S and ΔS can be done by using the transformation equations (5.119)–(5.121). The expressions for the final field operators are simple

$$a_f = \mathrm{e}^{\mathrm{i}\chi}(A a_0 + B b_0^\dagger),$$
$$b_f = A b_0 + a_0^\dagger B,$$
$$A = \mu^2 - |\nu|^2 \mathrm{e}^{-\mathrm{i}\chi}, \quad B = \mu\nu(1 - \mathrm{e}^{-\mathrm{i}\chi}). \tag{5.124}$$

Clearly for $\chi \to 0$, $a_f \to a_0$, $b_f \to b_0$ as $\mu^2 - |\nu|^2 = 1$. Furthermore, $|A|^2 - |B|^2 = 1$ so that the expected commutation relations for a_f and b_f hold. We have thus obtained a very interesting result: the output fields are related to the input fields via a Bogoliubov transformation. The parameters of the Bogoliubov transformation depend on the object χ.

We first discuss the case originally considered by Yurke *et al.* [36]. The fields a_0 and b_0 are in a vacuum state, then

$$\langle a_f^\dagger a_f \rangle = |B|^2, \quad \langle b_f^\dagger b_f \rangle = |B|^2, \tag{5.125}$$

and the signal is

$$S = 8|\mu|^2|\nu|^2 \sin^2\left(\frac{\chi}{2}\right). \tag{5.126}$$

To simplify (5.123) we use the property of the Bogoliubov transformation $a_f^\dagger a_f - b_f^\dagger b_f =$ constant $= 0$,

$$\begin{aligned}\Delta S^2 &= 4\langle a_f^{\dagger 2} a_f^2\rangle + 4\langle a_f^\dagger a_f\rangle - S^2,\\ &= 8|B|^4 + 4|B|^2 - 4|B|^4,\\ &= 4|B|^2 + 4|B|^4.\end{aligned} \tag{5.127}$$

Using (5.126) and (5.127) we find the sensitivity for $\chi = 0$

$$\Delta\chi = \sqrt{\frac{(\Delta S)^2}{\left|\frac{\partial S}{\partial \chi}\right|^2}} = \frac{1}{2|\mu||\nu|}, \tag{5.128}$$

which clearly scales as the inverse of the photon number. Each OPA produces $|\nu|^2$ photons and for large $|\nu|$, $|\mu||\nu| \approx |\nu|^2$. Thus the interferometer of Figure 5.19 with $\alpha = \beta = 0$ can yield Heisenberg limited measurements of phase.

Calculations for the case when α and β are nonzero get algebraically more complicated. We cite the result of Plick *et al.* [39] for the sensitivity near $\chi \approx 0$, $\alpha = \beta = |\alpha|e^{-i\pi/4}$,

$$\Delta\chi^2 = \frac{1}{(2|\alpha|^2)(4\sinh^2 r\cosh^2 r)}. \tag{5.129}$$

Thus when the OPAs are seeded we have the advantage of both coherent pumping and parametric interactions. The interferometers of the type discussed above have been realized using parametric amplifiers [40] and phase-sensitive fiber amplifiers.

5.23 The quantum statistics of fields at the output ports

In the previous section we saw that the characterization of the phase sensitivity of the interferometer in terms of CRB requires knowledge of the photon statistics of the output fields. For this purpose, and in view of the very special place that the Wigner function has in the context of nonclassical fields, we can calculate the Wigner functions of the fields at the output in terms of the Wigner function of the input fields. We note that the transformation (5.50) for the Mach–Zehnder interferometer is linear

$$\begin{pmatrix} a_{\text{out}} \\ b_{\text{out}} \end{pmatrix} = m\begin{pmatrix} a \\ b \end{pmatrix}, \qquad m^\dagger m = 1. \tag{5.130}$$

This linearity enables us to write the Wigner function of the fields at the output ports (see Exercise 5.1) as

$$W_{\text{out}}(\alpha,\beta) = W_{\text{in}}\left((m^{-1})_{11}\alpha + (m^{-1})_{12}\beta;\ (m^{-1})_{21}\alpha + (m^{-1})_{22}\beta\right). \tag{5.131}$$

Given the explicit form of the Wigner function we can calculate the photon statistics of the fields at the output port. Note that if $W_{\text{in}}(\alpha,\beta)$ is Gaussian, then $W_{\text{out}}(\alpha,\beta)$ is also Gaussian. This covers coherent states, single and two-mode squeezed vacuums, thermal states, etc.

The Wigner function (5.131) would have information on joint correlations between the fields at the ports D_1 and D_2. In particular, the counting distribution can be obtained using (1.117) and (1.111)

$$p(n, m) = \int d^2\alpha \int d^2\beta \, 4e^{-2|\alpha|^2-2|\beta|^2} L_n(4|\alpha|^2) L_m(4|\beta|^2)(-1)^{n+m} W_{\text{out}}(\alpha, \beta), \tag{5.132}$$

which, though involved, can be computed numerically.

Exercises

5.1 Show that under the transformations (5.3) and (5.6), the P, Q, and W functions transform as follows:

$$F_{\text{out}}(\alpha, \beta) = F_{\text{in}}(\alpha', \beta'); \qquad \begin{pmatrix} \alpha' \\ \beta' \end{pmatrix} = U^{-1} \begin{pmatrix} \alpha \\ \beta \end{pmatrix},$$

where F stands for any of the functions P, Q, and W.

5.2 Show that under the transformation (5.43), the P, Q, and W functions transform as

$$F_{\text{out}}(\alpha) = F_{\text{in}}(\alpha e^{i\varphi}),$$

where F stands for any of the functions P, Q, and W.

5.3 Prove the conservation law (5.16) directly by showing that $[a^\dagger a + b^\dagger b, H] = 0$, where H is defined by (5.11).

5.4 Use (5.18) to find the state $|2, 0\rangle_{\text{out}}$ for $\theta = \pi/4$. Then calculate the quadrature function $\Psi(x, y) = \langle x, y|2, 0\rangle_{\text{out}}$. Sketch the phase and modulus of $\Psi(x, y)$. Is this a vortex? Find the relation of $\Psi(x, y)$ to Laguerre–Gaussian functions defined by

$$u_{mn}(r, \varphi) = \sqrt{\frac{2}{\pi\omega^2} \frac{(-1)^p p!}{m!n!}} e^{-i\varphi(m-n)} \left(\frac{\sqrt{2}r}{\omega}\right)^{|m-n|} L_p^{|m-n|}\left(\frac{2r^2}{\omega^2}\right) e^{-r^2/\omega^2},$$

where $r^2 = x^2 + y^2$, $\varphi = \arctan\left(\frac{y}{x}\right)$, $p = \min(n, m)$, $\omega = \frac{1}{\sqrt{2}}$ and $L_p^l(x)$ is generalized Laguerre polynomial. (For details see [41].)

5.5 Consider an array of identical beam splitters, each with a transmission amplitude of $\cos\theta$. The fields $b_1, b_2, \ldots, b_n$ are in the vacuum state. Find the reduced density matrix of the output field c if a single photon is incident as shown in the figure.

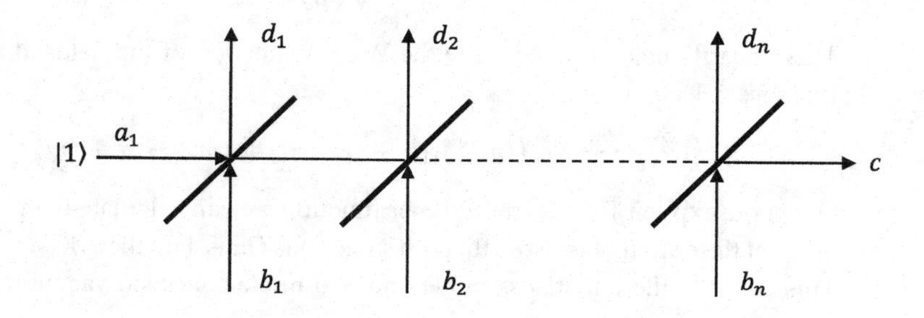

Show that the probability of finding one photon at the output c is $\cos^{2n}\theta$. What is the limiting value if $\theta = \pi/2n$ and n is large? This system is relevant to purely optical realizations of quantum zeno effect [42–44].

5.6 Consider a field in the state $|1, 3\rangle$ incident on a 50-50 beam splitter as shown in the figure. Show that the quantum interference leads to nonappearance of the output field in the state $|2, 2\rangle$. Display the two paths which lead to such a destructive interference.

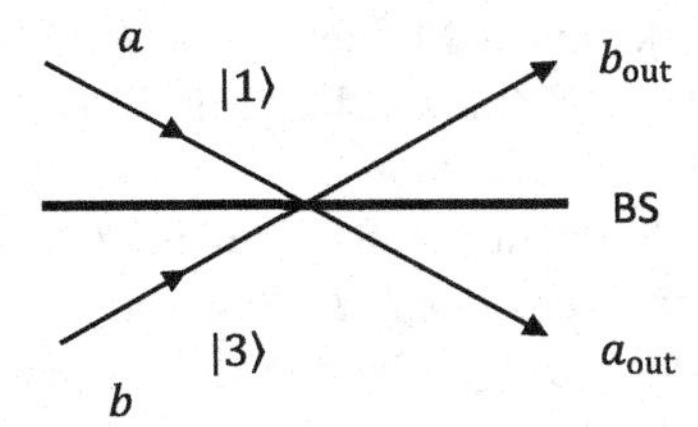

5.7 Because of the Pauli exclusion principle a Fermionic interferometer behaves differently. We characterize the input state of an electron as $|1_{\vec{k}s}\rangle$, where $\vec{k}$ is the direction of propagation and s is the spin direction. Find the possible output states for the configuration shown in the figure.

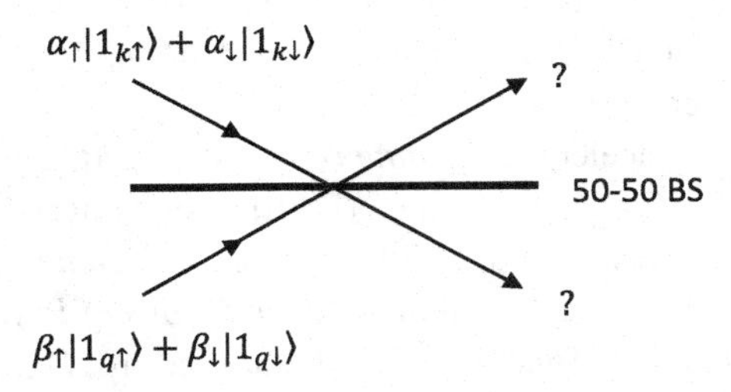

The electrons from two sides of the beam splitter come in superposition states.

5.8 For the Mach–Zehnder interferometer fed with the two-mode squeezed vacuum, calculate, following the procedure of Section 5.17, the probability of detecting two photons at one port, i.e. $\langle a_3^{\dagger 2} a_3^2\rangle$. Find the visibility of the signal and its asymptotic value. (For detail see [25].)

5.9 Consider the beam splitter with arbitrary reflection coefficient $\cos\theta$. Let a pair $|2, 2\rangle$ be launched at the input. Using the formula (5.18) find the wave function of the field at the output, find the condition on the reflectivity of the beam splitter so that at the output the state $|2, 2\rangle$ is absent [22]. Sketch the basic processes in terms of transmission and reflection of the individual photons that contribute to $|2, 2\rangle$ and how these processes interfere leading to null amplitude for the state $|2, 2\rangle$ at the output.

5.10 Consider the interferometer of Figure 5.7 where the input state is $|2, 1\rangle$ and where both BS1 and BS2 have reflectivities 2/3 ($\sin^2\theta = 2/3$), calculate $\langle a^{\dagger}_{\text{out}} b^{\dagger 2}_{\text{out}} a_{\text{out}} b^2_{\text{out}}\rangle$, which is the probability of detecting two photons at the port D_2 and one photon at the port D_1. Find the frequency of the fringes and compare with the frequency of the fringes (see Eq. (5.51)) at the output when one uses the input as $|1, 0\rangle$ [21].

5.11 Using (5.117) and (5.108) calculate and plot $(\Delta\varphi)_{\text{CRB}}$ as a function of $|\alpha|^2$.

References

[1] C. K. Hong, Z. Y. Ou, and L. Mandel, *Phys. Rev. Lett.* **59**, 2044 (1987).
[2] G. S. Agarwal, *J. Mod. Opt.* **34**, 909 (1987).
[3] R. A. Campos, B. E. A. Saleh, and M. C. Teich, *Phys. Rev. A* **40**, 1371 (1989).
[4] H. Huang and G. S. Agarwal, *Phys. Rev. A* **49**, 52 (1994).
[5] A. I. Lvovsky and J. Mlynek, *Phys. Rev. Lett.* **88**, 250401 (2002).
[6] N. Namekata, Y. Takahashi, G. Fuji, D. Fukuda, S. Kurimura, and S. Inoue, *Nature Photon.* **4**, 655 (2010).
[7] T. Gerrits, S. Glancy, T. S. Clement *et al.*, *Phys. Rev. A* **82**, 031802(R)(2010).
[8] G. I. Taylor, *Proc. Camb. Philos. Soc.* **15**, 114 (1909).
[9] P. Hariharan and B. C. Sanders, in *Progress in Optics*, edited by E. Wolf (Amsterdam: North-Holland, 1996), **36**, p. 49.
[10] G. Bertocchi, O. Alibart, D. B. Ostrowsky, S. Tanzilli, and P. Baldi, *J. Phys. B: At. Mol. Opt. Phys.* **39**, 1011 (2006).
[11] G. Sagnac, *C. R. Acad. Sci.* **95**, 708 (1913).
[12] W. W. Chow, J. Gea-Banacloche, L. M. Pedrotti, V. E. Sanders, W. Schleich, and M. O. Scully, *Rev. Mod. Phys.* **57**, 61 (1985).
[13] P. A. M. Dirac, *The Principles of Quantum Mechanics*, 4th edition (New York: Oxford University Press, 1982).
[14] J. A. Wheeler, in *Quantum Theory and Measurement*, edited by J. A. Wheeler and W. H. Zurek (Princeton, NJ: Princeton University Press, 1984), pp. 182–213.
[15] V. Jacques, E. Wu, F. Grosshans *et al.*, *Science* **315**, 966 (2007).
[16] A. Elitzur and L. Vaidman, *Foundations of Physics* **23**, 987 (1993).
[17] R. Penrose, *Shadows of the Mind* (Oxford: Oxford University Press, 1994).
[18] P. Kwiat, H. Weinfurter, T. Herzog, A. Zeilinger, and M. Kasevich, *Phys. Rev. Lett.* **74**, 4763 (1995).
[19] A. Kolkiran and G. S. Agarwal, *Opt. Express* **15**, 6798 (2007).
[20] T. Nagata, R. Okamoto, J. L. O'Brien, K. Sasaki, and S. Takeuchi, *Science* **316**, 726 (2007).
[21] H. Wang and T. Kobayashi, *Phys. Rev. A* **71**, 021802(R) (2005).
[22] B. H. Liu, F. W. Sun, Y. X. Gong, Y. F. Huang, G. C. Guo, and Z. Y. Ou. *Opt. Lett.* **32**, 1320 (2007).
[23] E. M. Nagasako, S. J. Bentley, R. W. Boyd, and G. S. Agarwal, *Phys. Rev. A* **64**, 043802 (2001).
[24] F. Sciarrino, C. Vitelli, F. D. Martini, R. Glasser, H. Cable, and J. P. Dowling, *Phys. Rev. A* **77**, 012324 (2008).
[25] G. S. Agarwal, R. W. Boyd, E. M. Nagasako, and S. J. Bentley, *Phys. Rev. Lett.* **86**, 1389 (2001).
[26] G. S. Agarwal, K. W. Chan, R. W. Boyd, H. Cable, and J. P. Dowling, *J. Opt. Soc. Am. B* **24**, 270 (2007).
[27] A. Kolkiran and G. S. Agarwal, *Opt. Express* **16**, 6479 (2008).

[28] H. P. Yuen and V. W. S. Chan, *Opt. Lett.* **8**, 177 (1983).
[29] U. L. Andersen and R. Filip, in *Progress in Optics*, edited by E. Wolf (Amsterdam: Elsevier, 2009), **53**, p. 365.
[30] J. I. Yoshikawa, T. Hayashi, T. Akiyama *et al.*, *Phys. Rev. A* **76**, 060301(R) (2007).
[31] P. K. Lam, T. C. Ralph, E. H. Huntington, and H.-A. Bachor, *Phys. Rev. Lett.* **79**, 1471 (1997).
[32] K. Vogel and H. Risken, *Phys. Rev. A* **40**, 2847 (1989).
[33] M. Born and E. Wolf, *Principles of Optics*, 7th ed. (Cambridge: Cambridge University Press, 2002), Sections 4.11.4 and 4.11.5.
[34] D. T. Smithey, M. Beck, and M. G. Raymer, *Phys. Rev. Lett.* **70**, 1244 (1993).
[35] A. I. Lvovsky and M. G. Raymer, *Rev. Mod. Phys.* **81**, 299 (2009).
[36] B. Yurke, S. L. McCall, and J. R. Klauder, *Phys. Rev. A* **33**, 4033 (1986).
[37] C. W. Helstrom, *Quantum Detection and Estimation Theory* (New York: Academic Press, 1976).
[38] M. J. Holland and K. Burnett, *Phys. Rev. Lett.* **71**, 1355 (1993).
[39] W. N. Plick, J. P. Dowling, and G. S. Agarwal, *New J. Phys.* **12**, 083014 (2010).
[40] J. T. Jing, C. J. Liu, Z. F. Zhou, Z. Y. Ou, and W. P. Zhang, *Appl. Phys. Lett.* **99**, 011110 (2011).
[41] G. S. Agarwal and J. Banerji, *J. Phys. A* **39**, 11503 (2006).
[42] G. S. Agarwal and S. P. Tewari, *Phys. Lett. A* **185**, 139 (1994).
[43] B. Misra and E. C. G. Sudarshan, *J. Math. Phys.* **18**, 756 (1977).
[44] A. Peres, *Am. J. Phys.* **48**, 931 (1980).

6 Polarization and orbital angular momentum of quantum fields

Electromagnetic fields have several important characteristics which we can control externally. This is important in studying matter, particularly when electromagnetic probes are used. We can change, for example, both the intensity and frequency of the field. Another important property is its polarization, which is usually exploited to study the Zeeman states of atomic and molecular systems. Simple optical devices exist which change the polarization of light. The polarization is intimately related to the spin angular momentum in quantum mechanics. Clearly, light beams also have orbital angular momentum. In this chapter we discuss both the polarization and orbital angular momentum of quantum fields. We first briefly present the polarization of classical beams of light, the details of which can be found in standard books, for example [1].

6.1 Characterization of the polarization properties of quantized fields

As already mentioned in Section 1.1, a plane wave propagating in the z direction can be written as

$$\vec{E}(z,t) = \vec{E}_0 e^{ikz - i\omega t} + c.c., \tag{6.1}$$

where

$$\vec{E}_0 = \alpha\hat{x} + \beta\hat{y}. \tag{6.2}$$

In the notation of Section 1.1, $\alpha = E_0\epsilon_{xs}$, $\beta = E_0\epsilon_{ys}$. We describe next the Poincaré sphere representation of the field (6.2). It should be borne in mind that both α and β are complex. The field is, in general, elliptically polarized. The intensity of light is given by $|\alpha|^2 + |\beta|^2$. The measurable quantities are, in general, quadratic in α and β. It is useful then to work with the Stokes parameters defined by

$$\begin{aligned} s_0 &= |\alpha|^2 + |\beta|^2, & s_1 &= |\alpha|^2 - |\beta|^2, \\ s_2 &= \alpha^*\beta + \alpha\beta^*, & s_3 &= i(\alpha^*\beta - \alpha\beta^*), \end{aligned} \tag{6.3}$$

which are real. Furthermore, the Stokes parameters obey

$$\sum_{j=1}^{3} s_j^2 = s_0^2, \tag{6.4}$$

which is easily shown by using (6.3). Hence the vector $\vec{s} = (s_1, s_2, s_3)$ for all values of α and β traces out a sphere with radius s_0. This sphere is called the Poincaré sphere. Note that as α and β change, the polarization of light changes. Thus each state of polarization is represented by a point on the sphere. Let us introduce the angles θ and φ by

$$|\alpha| = l_0 \cos\theta, \qquad |\beta| = l_0 \sin\theta, \qquad \alpha\beta^* = l_0^2 \sin\theta \cos\theta \, e^{+2i\varphi}, \tag{6.5}$$

where $l_0 = \sqrt{s_0}$, and $0 < \theta < \pi/2$, then

$$s_1 = s_0 \cos 2\theta, \qquad s_2 = s_0 \sin 2\theta \cos 2\varphi, \qquad s_3 = s_0 \sin 2\theta \sin 2\varphi. \tag{6.6}$$

The relative phase between α and β is 2φ. For linearly polarized light $\varphi = 0$ and therefore the Stokes vector $\vec{s}$ lies in the xy plane. For circularly polarized light $2\varphi = \pm\pi/2$, $2\theta = \pi/2$, leading to $s_1 = s_2 = 0$, $s_3 = \pm s_0$. For left (right) circular polarization

$$\alpha = \frac{l_0}{\sqrt{2}}, \qquad \beta = \frac{il_0}{\sqrt{2}}\left(-\frac{il_0}{\sqrt{2}}\right), \qquad 2\varphi = -\frac{\pi}{2}\left(+\frac{\pi}{2}\right). \tag{6.7}$$

The right (left) circular polarization is represented by a north (south) pole. Born and Wolf [1] define

$$s_3 = s_0 \sin 2\chi, \qquad s_1 = s_0 \cos 2\chi \cos 2\psi, \qquad s_2 = s_0 \cos 2\chi \sin 2\psi. \tag{6.8}$$

Here the angle ψ $(0 \le \psi < \pi)$ specifies the orientation of the polarization ellipse and the angle χ $(-\pi/4 \le \chi \le \pi/4)$ characterizes the ellipticity and the sense in which the ellipse is being described. We do not follow this further as all the details of the ellipse can be obtained by using (6.5), i.e. by writing the field as

$$\vec{E}_0 = l_0(\cos\theta\,\hat{x} + \sin\theta\, e^{-2i\varphi}\hat{y})e^{ikz - i\omega t} + c.c.. \tag{6.9}$$

6.2 Polarization of quantized fields – Stokes operators

For a quantum field, $\vec{E}_0$ in (6.1) becomes an operator and we write (6.2) as

$$\vec{E}_0 = \varepsilon_0(a\hat{x} + b\hat{y}), \tag{6.10}$$

where the annihilation operators a and b satisfy the commutation algebra

$$[a, a^\dagger] = [b, b^\dagger] = 1, \qquad [a, b^\dagger] = 0, \qquad \text{etc.} \tag{6.11}$$

The quantity ε_0 is a scale factor, which is irrelevant for our discussion below. In quantum theory we introduce the Stokes operators [2, 3] whose expectation values will be related to the Stokes parameters

$$\begin{aligned} S_0 &= a^\dagger a + b^\dagger b, \qquad & S_1 &= a^\dagger a - b^\dagger b, \\ S_2 &= a^\dagger b + ab^\dagger, \qquad & S_3 &= -i(a^\dagger b - ab^\dagger). \end{aligned} \tag{6.12}$$

We also have to specify the state of the quantum field. The state could be a pure state or a mixed state. The mean value of S_0 is the total number of photons. The operators S as defined above satisfy SU(2) algebra

$$[S_i, S_j] = 2\mathrm{i}\epsilon_{ijk}S_k, \qquad i, j, k = 1, 2, 3,$$
$$S_0(S_0 + 2) = S_1^2 + S_2^2 + S_3^2. \tag{6.13}$$

The operator S_0 commutes with S_1, S_2, and S_3. The factor 2 on the left-hand side arises from the noncommutativity of the operators a, $a^\dagger$, etc. If the field is in a coherent state $|\alpha, \beta\rangle$, then

$$\langle S_0\rangle = \alpha^*\alpha + \beta^*\beta, \qquad \langle S_1\rangle = \alpha^*\alpha - \beta^*\beta,$$
$$\langle S_2\rangle = \alpha^*\beta + \alpha\beta^*, \qquad \langle S_3\rangle = -\mathrm{i}(\alpha^*\beta - \alpha\beta^*), \tag{6.14}$$

which are just the Stokes parameters (6.3) (except for $\langle S_3\rangle$ which differs by a minus sign – which is just a matter of convention).

The degree of polarization P can be defined in terms of the matrix J_P, which is called the coherency matrix [1],

$$J_P = \frac{1}{2}\begin{pmatrix} \langle S_0 + S_1\rangle & \langle S_2 + \mathrm{i}S_3\rangle \\ \langle S_2 - \mathrm{i}S_3\rangle & \langle S_0 - S_1\rangle \end{pmatrix}, \tag{6.15}$$

$$P = \sqrt{1 - \frac{4\det J_P}{(\mathrm{Tr}J_P)^2}}. \tag{6.16}$$

For the case (6.14), $\det J_P = 0$ and hence $P = 1$, i.e. the light is fully polarized. The unpolarized light has been defined traditionally as one for which the relative phase between α and β is random and the intensities of the two components are equal. Thus for unpolarized light $\langle S_i\rangle = 0$ for $i = 1, 2, 3$; $4\det J_P = (\mathrm{Tr}J_P)^2$ and hence $P = 0$. More general criteria for unpolarized light are developed in Section 6.4. These are based on the transformation properties of the density matrix for the field under rotations. A field in thermal equilibrium with the density matrix

$$\rho = \exp[-\beta\hbar\omega(a^\dagger a + b^\dagger b)]/\mathrm{Tr}\left[\mathrm{e}^{-\beta\hbar\omega(a^\dagger a + b^\dagger b)}\right] \tag{6.17}$$

is unpolarized as all $\langle S_i\rangle = 0$.

For a complete characterization of the polarization properties of a quantum field, a description based on Stokes parameters is not adequate, and one has to study the fluctuations in Stokes parameters. In order to appreciate this, let us examine fields with one and two photons.

For a field with a single photon, the state can be $\alpha|1, 0\rangle + \beta|0, 1\rangle$, $|\alpha|^2 + |\beta|^2 = 1$. In this case, the Stokes parameters are given by (6.14). A knowledge of the Stokes parameters would yield α and β. Next, consider a field with two photons with a state given by $\alpha|2, 0\rangle + \beta|1, 1\rangle + \gamma|0, 2\rangle$, $|\alpha|^2 + |\beta|^2 + |\gamma|^2 = 1$. Now the Stokes parameters are

$$\langle S_0\rangle = 2, \qquad \langle S_1\rangle = 2(|\alpha|^2 - |\gamma|^2),$$
$$\langle S_2\rangle = \sqrt{2}(\gamma\beta^* + \alpha^*\beta) + c.c., \qquad \langle S_3\rangle = -\mathrm{i}\sqrt{2}[(\gamma\beta^* + \alpha^*\beta) - c.c.]. \tag{6.18}$$

Clearly, a knowledge of the Stokes parameters is not enough to determine the complete state of the field. We note that the degree of polarization P is not equal to unity even though the state is pure. This is because of the occurrence of quantum fluctuations in any pure state. For special values of α, β, and γ, $P = 1$ (see Exercise 6.4). We will discuss in Section 6.5 a procedure to determine the fluctuations in the Stokes parameters.

The subject of polarization continues to attract a great deal of attention both in classical optics [4] and quantum optics [3]. The basic question is: what is an appropriate measure of polarization for fields that are not in coherent states? It appears that more than a single measure like (6.16) is required to characterize the polarization properties.

6.3 Action of polarizing devices on quantized fields

We have to understand first the action of polarizing devices like half-wave plates (HWP) and quarter-wave plates (QWP) before we can treat the problem of determination of Stokes parameters and fluctuations in Stokes operators. An optical system which carries out linear transformations must be a 2×2 matrix. Thus we write the transformation as

$$\begin{pmatrix} a_p \\ b_p \end{pmatrix} = J \begin{pmatrix} a \\ b \end{pmatrix}, \tag{6.19}$$

where J is called the Jones matrix of the system. The Jones matrix must be unitary if the optical system is lossless. A birefringent plate [1] has slow and fast axes which are orthogonal. Light has larger refractive index along the slow axis and hence compared to the fast axis, it acquires an extra phase. If we assume that the slow axis is along the x direction, then the Jones matrix is then given by

$$J = \mathrm{e}^{\mathrm{i}\delta} \begin{pmatrix} \mathrm{e}^{\mathrm{i}\eta/2} & 0 \\ 0 & \mathrm{e}^{-\mathrm{i}\eta/2} \end{pmatrix}, \tag{6.20}$$

where δ is an overall phase factor. For $\eta = \pi/2$ (π) we have a quarter-wave plate (half-wave plate). A QWP converts a linearly polarized light to circular polarized light. If the slow axis of the wave plate is at an angle φ with the x axis, then introducing the rotation matrix

$$R(\varphi) = \begin{pmatrix} \cos\varphi & -\sin\varphi \\ \sin\varphi & \cos\varphi \end{pmatrix}, \tag{6.21}$$

the Jones matrix is

$$\begin{aligned} J &= \mathrm{e}^{\mathrm{i}\delta} R(\varphi) \begin{pmatrix} \mathrm{e}^{\mathrm{i}\frac{\eta}{2}} & 0 \\ 0 & \mathrm{e}^{-\mathrm{i}\frac{\eta}{2}} \end{pmatrix} R^{-1}(\varphi) \\ &= \mathrm{e}^{\mathrm{i}\delta} \begin{pmatrix} \cos\frac{\eta}{2} + \mathrm{i}\sin\frac{\eta}{2}\cos 2\varphi & \mathrm{i}\sin\frac{\eta}{2}\sin 2\varphi \\ \mathrm{i}\sin\frac{\eta}{2}\sin 2\varphi & \cos\frac{\eta}{2} - \mathrm{i}\sin\frac{\eta}{2}\cos 2\varphi \end{pmatrix}. \end{aligned} \tag{6.22}$$

We also note here that $R(\varphi)$ is the Jones matrix for an optically active medium.

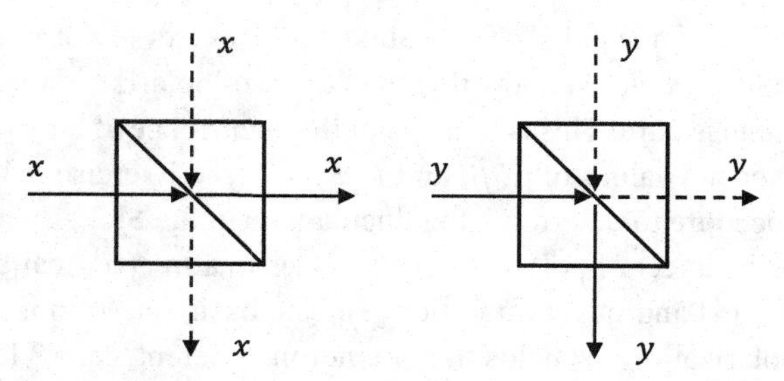

Fig. 6.1 The polarizing beam splitter.

The transformation (6.19) of the Heisenberg operators is sufficient to obtain the transformation of the states. We can also give an effective Hamiltonian description for the transformation (6.22), as we did in connection with the beam splitters in Section 5.2. It is clear that (6.20) can be generated by (dropping δ)

$$H = a^\dagger a - b^\dagger b, \qquad \begin{pmatrix} a_p \\ b_p \end{pmatrix} = \mathrm{e}^{-\mathrm{i}H\eta/2} \begin{pmatrix} a \\ b \end{pmatrix} \mathrm{e}^{\mathrm{i}H\eta/2}. \tag{6.23}$$

More generally, for generating the Jones matrix (6.22), we have [5]

$$\begin{gathered} H = \cos 2\varphi (a^\dagger a - b^\dagger b) + \sin 2\varphi (a^\dagger b + a b^\dagger), \\ \begin{pmatrix} a_p \\ b_p \end{pmatrix} = u_p^\dagger \begin{pmatrix} a \\ b \end{pmatrix} u_p = J \begin{pmatrix} a \\ b \end{pmatrix}, \qquad u_p = \mathrm{e}^{\mathrm{i}H\eta/2}. \end{gathered} \tag{6.24}$$

For $\eta \to 0$, $u_p \to 1$. The result (6.24) is easily proved by using standard methods of quantum mechanics. For instance, one can write differential equations for a_p and b_p:

$$\begin{pmatrix} a'_p \\ b'_p \end{pmatrix} = -\mathrm{i} \begin{pmatrix} \cos 2\varphi & \sin 2\varphi \\ \sin 2\varphi & -\cos 2\varphi \end{pmatrix} \begin{pmatrix} a_p \\ b_p \end{pmatrix}; \qquad a'_p = -\frac{\partial}{\partial(\eta/2)} a_p. \tag{6.25}$$

Integration of (6.25) is straightforward using the condition $\binom{a_p}{b_p} \to \binom{a}{b}$ for $\eta \to 0$. Having given an effective Hamiltonian description of the polarization elements, we can now transform states as follows

$$|\Psi\rangle_{\text{out}} = u_p |\Psi\rangle_{\text{in}}, \tag{6.26}$$

with u_p defined by (6.24).

Another very useful optical device is the polarizing beam splitter (PBS). A polarizing beam splitter has the property depicted in Figure 6.1. We have two directions of incidence which we label as h and v. We need two modes, a_i and b_i, for each direction $i = h, v$. In each direction we can have either x or y polarization. Let a stand for x polarization and b stand for y polarization. Clearly the transformation law for PBS is

$$\begin{pmatrix} a_h \\ b_h \\ a_v \\ b_v \end{pmatrix} \to \begin{pmatrix} 1 & 0 & 0 & 0 \\ 0 & 0 & 0 & 1 \\ 0 & 0 & 1 & 0 \\ 0 & 1 & 0 & 0 \end{pmatrix} \begin{pmatrix} a_h \\ b_h \\ a_v \\ b_v \end{pmatrix}. \tag{6.27}$$

For any initial state, the properties of the output can be worked out from (6.27). The PBS is important in applications as it can separate two orthogonal polarizations, because x polarization is transmitted and y polarization is reflected.

Finally, the question is: how do we describe the transformation of a quantized field by a device which changes the degree of polarization of light? A partially polarizing device involves a loss of intensity. Thus such a transformation can not be represented by an unitary operator. This is to be treated in the framework of the theory of dissipation (see Chapter 9).

6.4 Description of unpolarized light beyond Stokes parameters

In this section we find the density matrix of the unpolarized light using the invariance properties [6–9]. We consider a rotation of the x and y axes by an angle φ around the z axis. Under this rotation the operators transform as

$$\begin{pmatrix} a \\ b \end{pmatrix} \to \begin{pmatrix} a' \\ b' \end{pmatrix} = \begin{pmatrix} \cos\varphi & -\sin\varphi \\ \sin\varphi & \cos\varphi \end{pmatrix} \begin{pmatrix} a \\ b \end{pmatrix}. \tag{6.28}$$

This transformation is generated by

$$\exp\{\mathrm{i}\varphi[-\mathrm{i}(a^\dagger b - ab^\dagger)]\}\ a\ \exp\{-\mathrm{i}\varphi[-\mathrm{i}(a^\dagger b - ab^\dagger)]\} = a'. \tag{6.29}$$

If the observable properties of light are invariant under (6.28), then its state ρ must satisfy

$$[\rho, a^\dagger b - ab^\dagger] = 0. \tag{6.30}$$

Next consider the invariance of light under arbitrary relative phase change between a and b. This is the transformation given by (6.20) and (6.23). Hence ρ must satisfy

$$[\rho, a^\dagger a - b^\dagger b] = 0. \tag{6.31}$$

We have also noticed in Eq. (6.13) that both $a^\dagger b - ab^\dagger$ and $a^\dagger a - b^\dagger b$ are noncommuting elements of the SU(2) algebra. The only operator that commutes these with two elements of the SU(2) algebra is $(a^\dagger a + b^\dagger b)$. Hence the relations (6.30) and (6.31) imply that the density matrix of the unpolarized light must be a function of $(a^\dagger a + b^\dagger b)$ only and, in terms of the Fock states, it has the form

$$\rho = \sum_{N=0}^{\infty} \sum_{p=0}^{N} C_N |N-p, p\rangle\langle N-p, p|. \tag{6.32}$$

It is easily seen that for the state (6.32) $\langle S_1\rangle = \langle S_2\rangle = \langle S_3\rangle$ and $P = 0$. Lehner *et al.* [9, 10] considered more general type of unpolarized light by relaxing the property (6.31). Tsegaye *et al.* [11] extended these ideas to polarized light by considering their invariance under rotations (Eq. (6.30)).

6.5 Stokes operator tomography

In this section we discuss how the polarization characteristics of the quantized fields can be obtained by studies on the properties of the Stokes operators [12–14]. We first need to do a number of measurements to obtain mean values of the Stokes operators and then examine fluctuations in the mean values. For this purpose, we adopt a method outlined in Born and Wolf [1] and then show how it can be generalized to obtain fluctuations in the Stokes parameters. Consider a general SU(2) transformation

$$\begin{pmatrix} a(\theta,\varphi) \\ b(\theta,\varphi) \end{pmatrix} = J^{(s)} \begin{pmatrix} a \\ b \end{pmatrix}, \qquad J^{(s)} = \begin{pmatrix} \cos\theta & \sin\theta\, \mathrm{e}^{\mathrm{i}\varphi} \\ -\sin\theta\, \mathrm{e}^{-\mathrm{i}\varphi} & \cos\theta \end{pmatrix}. \tag{6.33}$$

The intensity in the output mode $a(\theta,\varphi)$ is

$$\begin{aligned} I(\theta,\varphi) &= \langle a^\dagger(\theta,\varphi) a(\theta,\varphi)\rangle \\ &= \cos^2\theta \langle a^\dagger a\rangle + \sin^2\theta \langle b^\dagger b\rangle + \mathrm{e}^{\mathrm{i}\varphi} \sin\theta\cos\theta \langle a^\dagger b\rangle \\ &\quad + \mathrm{e}^{-\mathrm{i}\varphi}\sin\theta\cos\theta\langle ab^\dagger\rangle. \end{aligned} \tag{6.34}$$

The unknowns $\langle a^\dagger a\rangle$, $\langle b^\dagger b\rangle$, $\langle a^\dagger b\rangle$, and $\langle ab^\dagger\rangle$ can be obtained from four measurements $I(0,0), I(\pi/2,0), I(\pi/4,0), I(\pi/4,\pi/2)$ as the following equations show

$$\begin{aligned} I(0,0) &= \langle a^\dagger a\rangle, \qquad I(\pi/2,0) = \langle b^\dagger b\rangle, \\ I(\pi/4,0) &= \frac{1}{2}(\langle a^\dagger a\rangle + \langle b^\dagger b\rangle) + \frac{1}{2}(\langle a^\dagger b\rangle + \langle ab^\dagger\rangle), \\ I(\pi/4,\pi/2) &= \frac{1}{2}(\langle a^\dagger a\rangle + \langle b^\dagger b\rangle) + \frac{1}{2}\mathrm{i}(\langle a^\dagger b\rangle - \langle ab^\dagger\rangle). \end{aligned} \tag{6.35}$$

Thus all the Stokes parameters $\langle S_i\rangle$ (Eq. (6.12)) can be obtained from these four measurements.

Next, we consider the fluctuations in Stokes parameters $\langle S_iS_j\rangle - \langle S_i\rangle\langle S_j\rangle$. Clearly, $\langle S_iS_j\rangle$ would involve expectation values of the products involving a total of four operators a, $a^\dagger$, b, and $b^\dagger$. Therefore, we propose measurements of photon–photon correlations of the form $\langle (a^\dagger(\theta,\varphi))^2 (a(\theta,\varphi))^2\rangle$, $\langle a^\dagger(\theta,\varphi)a(\theta,\varphi)b^\dagger(\theta,\varphi)b(\theta,\varphi)\rangle$ for the determination of quantities like $\langle S_iS_j\rangle$. Following Mukunda and Jordan [15], we consider measurements for different values of θ and φ:

(I) $\varphi = 0$

$$\begin{aligned} \langle [a^\dagger(\theta,0)]^2[a(\theta,0)]^2\rangle &= \langle a^\dagger a^\dagger aa\rangle\cos^4\theta + \langle b^\dagger b^\dagger bb\rangle \sin^4\theta \\ &\quad + (4\langle a^\dagger b^\dagger ab\rangle + \langle a^\dagger a^\dagger bb\rangle + \langle b^\dagger b^\dagger aa\rangle)\cos^2\theta\sin^2\theta \\ &\quad + 2(\langle a^\dagger a^\dagger ab\rangle + \langle a^\dagger b^\dagger aa\rangle)\cos^3\theta\sin\theta \\ &\quad + 2(\langle a^\dagger b^\dagger bb\rangle + \langle b^\dagger b^\dagger ab\rangle)\cos\theta\sin^3\theta, \end{aligned} \tag{6.36}$$

(II) $\varphi = \pi/2$

$$
\begin{aligned}
\langle [a^\dagger(\theta,\pi/2)]^2[a(\theta,\pi/2)]^2\rangle &= \langle a^\dagger a^\dagger aa\rangle \cos^4\theta + \langle b^\dagger b^\dagger bb\rangle \sin^4\theta \\
&+ (4\langle a^\dagger b^\dagger ab\rangle - \langle a^\dagger a^\dagger bb\rangle - \langle b^\dagger b^\dagger aa\rangle)\cos^2\theta\sin^2\theta \\
&+ 2\mathrm{i}(\langle a^\dagger a^\dagger ab\rangle - \langle a^\dagger b^\dagger aa\rangle)\cos^3\theta\sin\theta \\
&+ 2\mathrm{i}(\langle a^\dagger b^\dagger bb\rangle - \langle b^\dagger b^\dagger ab\rangle)\cos\theta\sin^3\theta,
\end{aligned} \tag{6.37}
$$

(III) $\theta = \pi/4,\ \varphi = \pi/4$

$$
\begin{aligned}
&\langle a^\dagger(\pi/4,\pi/4)b^\dagger(\pi/4,\pi/4)a(\pi/4,\pi/4)b(\pi/4,\pi/4)\rangle \\
&= \frac{1}{4}\langle a^\dagger a^\dagger aa\rangle + \frac{1}{4}\langle b^\dagger b^\dagger bb\rangle - \frac{\mathrm{i}}{4}(\langle a^\dagger a^\dagger bb\rangle - \langle b^\dagger b^\dagger aa\rangle).
\end{aligned} \tag{6.38}
$$

Equation (6.36), for five different of values θ, would yield $\langle a^{\dagger 2}a^2\rangle$, $\langle b^{\dagger 2}b^2\rangle$, $(\langle a^{\dagger 2}ab\rangle + \langle a^\dagger b^\dagger a^2\rangle)$, $(\langle a^\dagger b^\dagger b^2\rangle + \langle b^{\dagger 2}ab\rangle)$, $(4\langle a^\dagger b^\dagger ab\rangle + \langle a^{\dagger 2}b^2\rangle + \langle b^{\dagger 2}a^2\rangle)$. Equation (6.37), for three different values of θ, would yield $(4\langle a^\dagger b^\dagger ab\rangle - \langle a^{\dagger 2}b^2\rangle - \langle b^{\dagger 2}a^2\rangle)$, $(\langle a^{\dagger 2}ab\rangle - \langle a^\dagger b^\dagger a^2\rangle)$,$(\langle a^\dagger b^\dagger b^2\rangle - \langle b^{\dagger 2}ab\rangle)$. Thus these eight measurements and the measurement (6.38) enable us to determine all the required fourth-order correlations $\langle a^{\dagger 2}a^2\rangle$, $\langle b^{\dagger 2}b^2\rangle$, $(\langle a^{\dagger 2}ab\rangle$, $\langle a^\dagger b^\dagger a^2\rangle)$, $(\langle a^\dagger b^\dagger b^2\rangle$, $\langle b^{\dagger 2}ab\rangle)$, $\langle a^\dagger b^\dagger ab\rangle$, $\langle a^{\dagger 2}b^2\rangle$, $\langle b^{\dagger 2}a^2\rangle$ needed for the determination of $\langle S_iS_j\rangle$.

We next give the arrangements [16] of quarter-wave plates and half-wave plates to realize (6.33). The Jones matrix for any wave plate is given by (6.22). For clarity, let us write it as $J(\eta,\varphi)$. For $\eta = \pi/2$ (π), it is a quarter-wave plate (half-wave plate). The result is (6.33), i.e.

$$
\begin{aligned}
J^{(s)}(\theta,0) &= Q_{\pi/4}Q_{\pi/4}H_{-\pi/4+\theta/2}, \\
J^{(s)}(\theta,\pi/2) &= Q_{\pi/2}Q_{\theta+\pi/2}H_{\theta/2}, \\
J^{(s)}(\pi/4,\pi/4) &= Q_{\pi/4+(\arctan 2^{1/2})/2}Q_{5\pi/12+(\arctan 2^{1/2})/2}H_{\pi/12},
\end{aligned} \tag{6.39}
$$

where the subscripts to Q and H give the angles by which the quarter-wave and half-wave plates have to be rotated around the z axis. The matrices for the half-wave and quarter-wave plates are given by

$$
H_\phi = \mathrm{i}\begin{pmatrix} \cos 2\phi & \sin 2\phi \\ \sin 2\phi & -\cos 2\phi \end{pmatrix}, \tag{6.40}
$$

$$
Q_\phi = \frac{\mathrm{i}}{\sqrt{2}}\begin{pmatrix} \cos 2\phi - \mathrm{i} & \sin 2\phi \\ \sin 2\phi & -\cos 2\phi - \mathrm{i} \end{pmatrix}. \tag{6.41}
$$

The scheme outlined above is quite general and can be extended to higher-order correlations of the Stokes operators [17]. It can also be used in the tomography of entangled states. The case of two-photon entangled states has been discussed in detail [18]. Figure 6.2 and the foregoing discussion makes it clear how the scheme would work for two-photon entangled states.

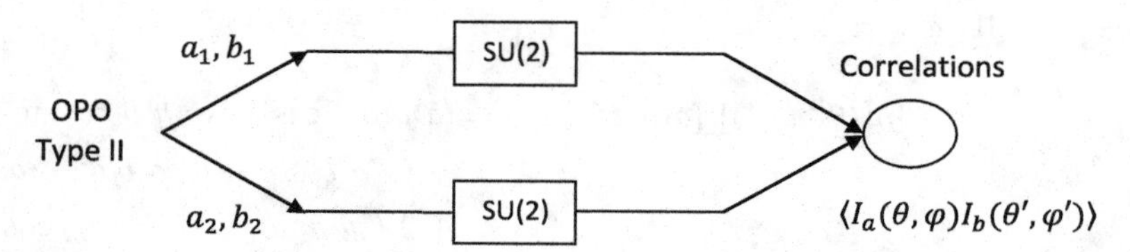

Fig. 6.2 The OPO produces an entangled state with a total of two photons, one in the direction labeled by the subscript 1 and one in the direction 2. The intensity–intensity correlations are measured after carrying out SU(2) transformations.

6.6 Orbital angular momentum of fields – HG and LG modes

In the previous sections we examined the polarization properties of light. Another important property of light is its orbital angular momentum if we are dealing with two-dimensional fields. We would concentrate on paraxial beams, i.e. light beams with propagation vectors that are close to the axis of the beam. More precisely, the transverse components of the momentum must be much smaller than the longitudinal or axial component. It was demonstrated by Beth [19] that the interaction of a circularly polarized beam with a birefringent plate can produce a mechanical torque [20]. In paraxial optics two types of beams have been extensively used; assuming a time dependence of the form $e^{-i\omega t}$, these are:

(A) Hermite–Gauss (HG) modes

$$u_{nm}(x, y, w) = \sqrt{\frac{2}{\pi}} \left(\frac{1}{2^{n+m} w^2 n! m!} \right)^{1/2} \mathrm{H}_n \left(\frac{\sqrt{2}x}{w} \right) \mathrm{H}_m \left(\frac{\sqrt{2}y}{w} \right) \mathrm{e}^{-(x^2+y^2)/w^2},$$

$$\iint |u_{nm}(x, y, w)|^2 \, \mathrm{d}x \, \mathrm{d}y = 1, \tag{6.42}$$

where w is the beam waist. The amplitude (6.42) is in the waist plane and the integers n, m satisfy $0 \leq n, m \leq \infty$. These modes are also called $\mathrm{TEM}_{n,m}$ modes.

(B) Laguerre–Gaussian (LG) modes
Let ρ and θ be the polar coordinates in two dimensions $\rho = \sqrt{x^2 + y^2}$, $\tan\theta = y/x$:

$$v_{n,m}(\rho, \theta) = \mathrm{e}^{\mathrm{i}(n-m)\theta} \mathrm{e}^{-\rho^2/w^2} (-1)^{\min(n,m)} \left(\frac{\rho\sqrt{2}}{w} \right)^{|n-m|} \sqrt{\frac{2}{\pi n! m! w^2}}$$

$$\times \mathrm{L}_{\min(n,m)}^{|n-m|} \left(\frac{2\rho^2}{w^2} \right) (\min(n, m))!,$$

$$\iint |v_{nm}(\rho, \theta)|^2 \, \mathrm{d}x \, \mathrm{d}y = 1, \tag{6.43}$$

where $\mathrm{L}_p^l(x)$ is the generalized Laguerre polynomial

$$\mathrm{L}_p^l(x) = \sum_{m=0}^{p} (-1)^m \binom{p+l}{p-m} \frac{x^m}{m!}, \qquad l > -1. \tag{6.44}$$

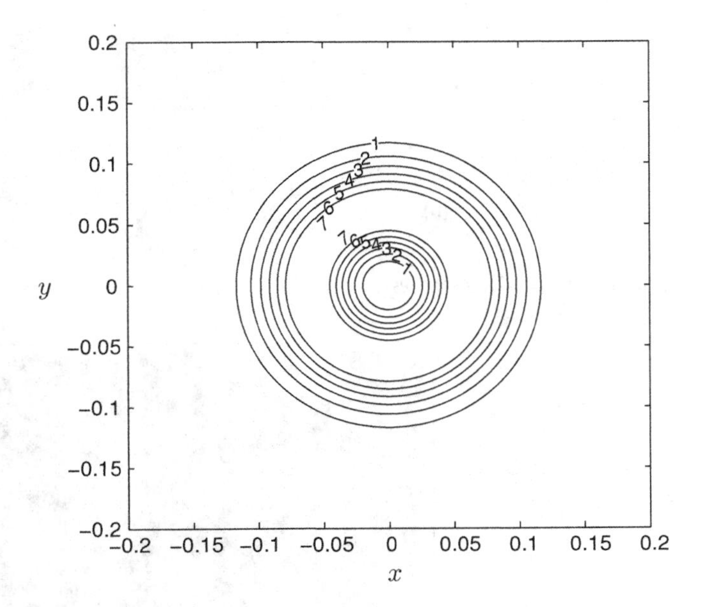

Fig. 6.3 Intensity distribution for the mode v_{30}.

Here $N = n + m$ is called the order of the mode, $l = n - m$ is the azimuthal index, and $p = \min(n, m)$ is called the radial index. Allen *et al.* [21] found that LG modes carry an orbital angular momentum $\hbar l$ per photon. For reviews on orbital angular momentum see [22].

For LG modes with $n \neq m$, the intensity is zero at $\rho = 0$. These modes are generally called vortices because they carry definite angular momentum. The name "donut mode" is also used because of their intensity pattern. The mode v_{30} is displayed in Figure 6.3.

For $n = m = 0$, u_{00} is the Gaussian beam, similarly $v_{00}(\rho, \theta)$ is also a Gaussian beam. The simplest case where the two modes differ corresponds to $n + m = N = 1$, $l = \pm 1$ and $p = 0$. Here, for the HG case

$$u_{10} \propto x\mathrm{e}^{-(x^2+y^2)/w^2}, \qquad u_{01} \propto y\mathrm{e}^{-(x^2+y^2)/w^2}, \tag{6.45}$$

whereas

$$\begin{aligned} v_{10} &\propto \rho \mathrm{e}^{\mathrm{i}\theta} \mathrm{e}^{-\rho^2/w^2} = (x + \mathrm{i}y)\mathrm{e}^{-(x^2+y^2)/w^2}, \\ v_{01} &\propto \rho \mathrm{e}^{-\mathrm{i}\theta} \mathrm{e}^{-\rho^2/w^2} = (x - \mathrm{i}y)\mathrm{e}^{-(x^2+y^2)/w^2}. \end{aligned} \tag{6.46}$$

Thus the LG modes are linear combinations of HG modes for the same value of $N = n+m$. The relation between (6.45) and (6.46) is pictorially shown in Figure 6.4 [23].

A very general relationship between the LG and HG modes for arbitrary order can be obtained using operator techniques [24] and has been utilized by Allen and coworkers [25] to find an experimental method to produce LG modes.

Experimentally, the vortices can be studied by examining the interferogram obtained by using a reference beam of the same amplitude. For example, let us consider the intensity

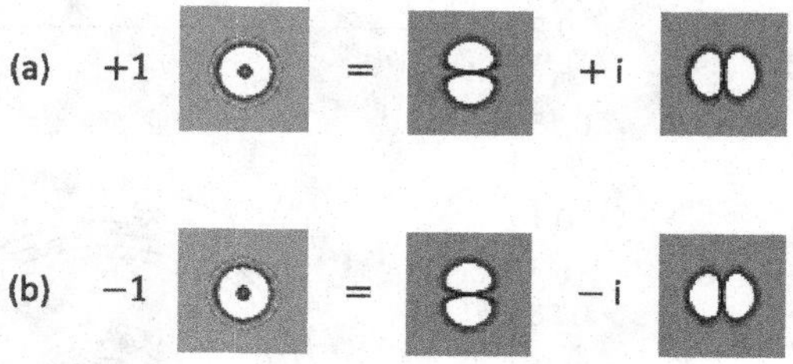

(a) +1 = +i

(b) −1 = −i

Fig. 6.4 Intensity distribution for the modes given by (a) $v_{10} = u_{10} + \mathrm{i}u_{01}$, (b) $v_{01} = u_{10} - \mathrm{i}u_{01}$. (Redrawn after [23].)

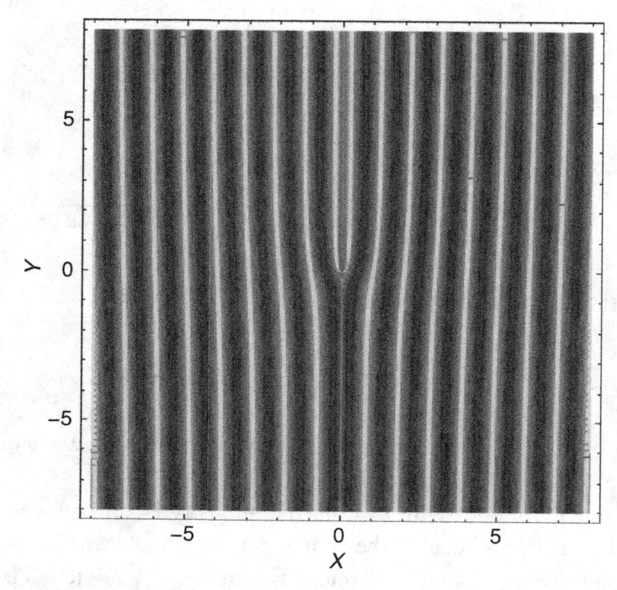

Fig. 6.5 Interferogram of a single point vortex of charge $m = 1$, in which $X = x/\lambda$ and $Y = y/\lambda$. The vortex core is located at the fork of the equiphase lines. (Redrawn after [26].)

pattern obtained by superposition of a vortex with $l = 1$, $A_0 e^{\mathrm{i}\theta}$, where $\theta = \arctan(y/x)$, and a plane wave $A_0 e^{2\pi \mathrm{i}x/\lambda}$. The intensity in the plane $z = 0$ will be

$$I = 2A_0^2 \left[1 + \cos\left(\frac{2\pi x}{\lambda} - \theta \right) \right]. \tag{6.47}$$

Here λ is the spatial period of the reference wave in the xy plane. The resulting interferogram (6.47) is shown in Figure 6.5 [26]. This interferogram has nearly the form of a parallel grating except for a forking pattern near the core $x, y = 0$.

Having defined the modes, we discuss briefly how LG modes can be produced from HG modes. Several methods are known – one can use cylindrical lenses, diffractive optics elements, spatial light modulators [27]. A relative phase shift of $\pi/2$ or π between the HG modes u_{01} and u_{10} can be introduced by a combination of two cylindrical lenses each of focal length f and separated by $\sqrt{2}f$ and $2f$ respectively. This has been demonstrated by Beijersbergen *et al.* [25]. The method is based on different amounts of Gouy phase shifts if

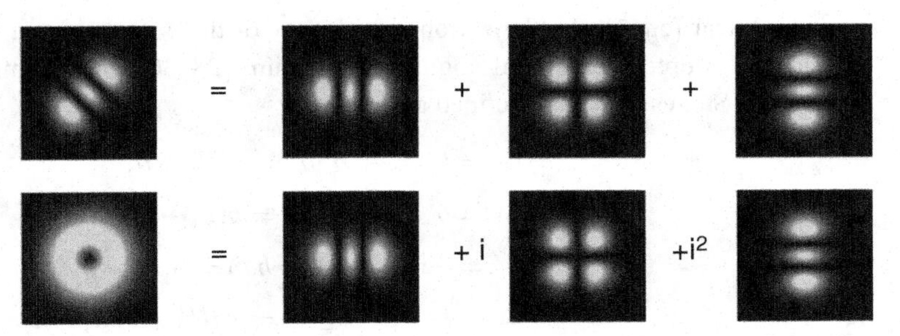

Fig. 6.6 The decomposition of a TEM_{02} Hermite–Gaussian mode at 45° into a set of Hermite–Gaussian modes and the decomposition of a Laguerre–Gaussian mode into the same with different phase factors, after [25].

the HG beams in the x and y directions have different Rayleigh range $\pi w^2/\lambda$. From (6.46) it then follows that a π phase shift would change v_{01} to v_{10} and vice versa. Similarly, HG modes, whose principal axes make an angle of 45° with the x axis, would be changed to LG modes by a $\pi/2$ phase shift, as shown in Figure 6.6. A popular technique for generating vortex beams is to use computer-generated holograms. For example, we can generate the pattern given in Figure 6.5 and launch a Gaussian beam to generate LG modes v_{01} and v_{10} (see Exercise 6.7).

6.7 Orbital Stokes operators and the Poincaré sphere

Let us consider a special case of a linearly polarized quantum field with orbital angular momentum which can be written as a superposition of v_{10} and v_{01} modes in the waist plane

$$E(x, y) = v_{10}(x, y)b_+ + v_{01}(x, y)b_-. \tag{6.48}$$

We will continue to assume that the field is monochromatic. The operators $b_\pm$ and their adjoints $b_\pm^\dagger$ satisfy Bosonic communtation relations

$$[b_+, b_+^\dagger] = [b_-, b_-^\dagger] = 1, \qquad [b_+, b_-^\dagger] = 0. \tag{6.49}$$

The field (6.48) can also be written in terms of HG modes

$$E(x, y) = u_{10}(x, y)b_1 + u_{01}(x, y)b_2, \tag{6.50}$$

where

$$b_1 = \frac{b_+ + b_-}{\sqrt{2}}, \qquad b_2 = \frac{\mathrm{i}(b_+ - b_-)}{\sqrt{2}}, \qquad b_\pm = \frac{b_1 \mp \mathrm{i}b_2}{\sqrt{2}}. \tag{6.51}$$

The operator b_1 and b_2 also satisfy Bosonic algebra

$$[b_\alpha, b_\beta^\dagger] = \delta_{\alpha\beta}. \tag{6.52}$$

The nonclassical character of the light with OAM arises from the states of the field (b_+, b_-). Note the similarity of (6.48) and (6.50) to equations (6.10) in the description of the polarization of light. The representation (6.50) [(6.48)] is analogous to the representation in terms of a linear [circular] polarization. Thus many of the concepts used for polarization optics can be adopted for orbital angular momentum [28–30]. For example, we can introduce orbital Stokes operators defined as

$$
\begin{aligned}
\mathcal{L}_0 &= b_+^\dagger b_+ + b_-^\dagger b_- = b_1^\dagger b_1 + b_2^\dagger b_2, \\
\mathcal{L}_1 &= b_+^\dagger b_- + b_-^\dagger b_+ = b_1^\dagger b_1 - b_2^\dagger b_2, \\
\mathcal{L}_2 &= -\mathrm{i}(b_+^\dagger b_- - b_-^\dagger b_+) = b_1^\dagger b_2 + b_2^\dagger b_1, \\
\mathcal{L}_3 &= b_+^\dagger b_+ - b_-^\dagger b_- = -\mathrm{i}(b_1^\dagger b_2 - b_2^\dagger b_1).
\end{aligned}
\tag{6.53}
$$

These operators also satisfy the SU(2) algebra

$$
\begin{aligned}
[\mathcal{L}_i, \mathcal{L}_j] &= 2\mathrm{i}\epsilon_{ijk}\mathcal{L}_k, \qquad 1 \le i, j, k \le 3, \\
\mathcal{L}_0(\mathcal{L}_0 + 2) &= \mathcal{L}_1^2 + \mathcal{L}_2^2 + \mathcal{L}_3^2.
\end{aligned}
\tag{6.54}
$$

The operator $\mathcal{L}_0$ commutes with the rest of the operators. For a field in a coherent state $|\beta_+, \beta_-\rangle$,

$$
b_\pm|\beta_+, \beta_-\rangle = \beta_\pm|\beta_+, \beta_-\rangle. \tag{6.55}
$$

$$
\begin{aligned}
\langle\mathcal{L}_0\rangle &= |\beta_+|^2 + |\beta_-|^2, \qquad \langle\mathcal{L}_3\rangle = |\beta_+|^2 - |\beta_-|^2, \\
\langle\mathcal{L}_1\rangle &= \beta_+^*\beta_- + c.c., \qquad \langle\mathcal{L}_2\rangle = -\mathrm{i}(\beta_+^*\beta_- - c.c.).
\end{aligned}
\tag{6.56}
$$

In analogy to the polarization case these would be orbital Stokes parameters. For classical fields

$$
\langle\mathcal{L}_1\rangle^2 + \langle\mathcal{L}_2\rangle^2 + \langle\mathcal{L}_3\rangle^2 = \langle\mathcal{L}_0\rangle^2, \tag{6.57}
$$

and hence the collection of states of a light beam with OAM can be represented as points on a sphere [30] which we can call the orbital Poincaré sphere. If either β_+ or β_- are zero, i.e. we have states with either mode v_{10} or v_{01}, then $\langle\mathcal{L}_1\rangle$ and $\langle\mathcal{L}_2\rangle$ are zero. Furthermore, if the relative phase between β_+ and β_- is zero (multiple of π), then the orbital Stokes vector lies in the xz plane. In this case, the mean value of the OAM is nonzero. The north (south) pole on the orbital Poincaré sphere represents the LG mode $(p = 0, l = 1)$ $[(p = 0, l = -1)]$. For vectors in the equatorial plane $|\beta_+| = |\beta_-|$, writing $\beta_+ = \beta_0 \mathrm{e}^{-\mathrm{i}\varphi/2}$, $\beta_- = \beta_0 \mathrm{e}^{\mathrm{i}\varphi/2}$, we find from (6.51), $\langle b_1\rangle = \sqrt{2}\beta_0 \cos\varphi/2$, $\langle b_2\rangle = \sqrt{2}\beta_0 \sin\varphi/2$. Thus the mean value of the field is

$$
\langle E\rangle = \sqrt{2}\beta_0(u_{10}\cos\varphi/2 + u_{01}\sin\varphi/2). \tag{6.58}
$$

In analogy to linearly polarized states lying in the equatorial plane of the Poincaré sphere, the states with linear combinations of u_{01} and u_{10} and that are in-phase lie in the equatorial plane.

Since the coherent states of light are like classical states, we now discuss how single-photon states can be represented on the orbital Poincaré sphere. Consider the most general pure state containing one photon

$$
|\Psi\rangle = \beta_+|1, 0\rangle + \beta_-|0, 1\rangle, \qquad |\beta_+|^2 + |\beta_-|^2 = 1. \tag{6.59}
$$

Here $|1, 0\rangle$ $[|0, 1\rangle]$ means one photon in the mode v_{10} $[v_{01}]$. It is easily shown that

$$
\langle\mathcal{L}_1 + \mathrm{i}\mathcal{L}_2\rangle = 2\beta_+^*\beta_-, \qquad \langle\mathcal{L}_3\rangle = |\beta_+|^2 - |\beta_-|^2, \tag{6.60}
$$

and

$$|\langle\vec{\mathcal{L}}\rangle|^2 = (|\beta_+|^2 + |\beta_-|^2)^2 \qquad \forall\, \beta_+, \beta_-. \tag{6.61}$$

Therefore most general one-photon states can be represented by points on the orbital Poincaré sphere.

We have thus given effectively a two-mode description of the field carrying OAM. We could thus consider squeezed states as well as the entangled states of these two modes as described in Chapters 1–3. The discussion there was based on two modes satisfying Bosonic commutation relations. Thus these modes can be taken as the modes as defined in Eq. (6.48).

6.8 Mixed states of orbital angular momentum

We can also consider mixed states of orbital angular momentum for which $\langle\mathcal{L}_1\rangle^2 + \langle\mathcal{L}_2\rangle^2 + \langle\mathcal{L}_3\rangle^2 \neq \langle\mathcal{L}_0\rangle^2$. To describe these we introduce, in analogy to (6.15), the matrix

$$V = \begin{pmatrix} \langle b_+^\dagger b_+\rangle & \langle b_+^\dagger b_-\rangle \\ \langle b_-^\dagger b_+\rangle & \langle b_-^\dagger b_-\rangle \end{pmatrix} = \frac{1}{2}\begin{pmatrix} \langle\mathcal{L}_0 + \mathcal{L}_3\rangle & \langle\mathcal{L}_1 + i\mathcal{L}_2\rangle \\ \langle\mathcal{L}_1 - i\mathcal{L}_2\rangle & \langle\mathcal{L}_0 - \mathcal{L}_3\rangle \end{pmatrix}. \tag{6.62}$$

One can also define the degree of vorticity by

$$\mathcal{V} = \frac{\lambda_1 - \lambda_2}{\lambda_1 + \lambda_2} = \left[1 - \frac{4\det V}{(\mathrm{Tr}V)^2}\right]^{1/2}, \tag{6.63}$$

where λ_1 and λ_2 are the eigenvalues of V. States of zero vorticity (in analogy to unpolarized light) would correspond to $\lambda_1 = \lambda_2$ or $4\det V = (\mathrm{Tr}V)^2$. States like (6.55) and (6.59) correspond to $\det V = 0$. These are the states for which we have the orbital Poincaré sphere representation.

For LG modes of higher order we need to consider the SU(N) group, as instead of (6.48) we need to write

$$E(x,y) = \sum_{n+m=N} v_{n,m}(x,y) b_l, \qquad l = n - m. \tag{6.64}$$

Thus, for a fixed N, the number of field operators needed is $N+1$. The Bosonic operators $b_\alpha^\dagger b_\beta$ can be used to construct the SU($N+1$) algebra. All the nonclassical properties of the higher-order LG modes can be studied in terms of the expectation values involving the products of the operators b_{n-m}, $b_{n-m}^\dagger$, etc.

6.9 Entangled states of the orbital angular momentum

The entangled states of OAM can be produced by using parametric down-conversion [31, 36]. We rewrite the interaction Hamiltonian (3.71) in terms of the signal and idler fields as

$$H_1 = \chi \int_V \mathrm{d}^3 r\, E_p^{(+)}(\vec{r}, t) E_s^{(-)}(\vec{r}, t) E_i^{(-)}(\vec{r}, t) + c.c.. \tag{6.65}$$

We assume as before that the pump field is strong and treat it classically. Furthermore, the pump field is taken to be the lowest-order Gaussian mode u_{00}, i.e.

$$E_p^{(+)}(\vec{r}, t) = A(\omega_p) \mathrm{e}^{-\mathrm{i}\omega_p t} u_{00}(\vec{r}). \tag{6.66}$$

We expand the signal and idler fields in terms of the LG mode v_{mn} or v_p^l. Here l is the azimuthal index $(n - m)$ and p the radial index $\min(m, n)$. For the purpose of this section we write v_p^l as

$$v_p^l(\vec{r}) = R_p(\rho, z) \mathrm{e}^{\mathrm{i}l\theta} \Big/ \sqrt{2\pi}, \tag{6.67}$$

where $R_p^{\prime s}$ are radial functions. We assume that these are orthogonal and normalized as

$$\begin{aligned} \sum_p \rho R_p^*(\rho, z) R_p(\rho', z) &= \delta(\rho - \rho'), \\ \int \rho\, \mathrm{d}\rho\, R_p^*(\rho, z) R_{p'}(\rho, z) &= \delta_{pp'}. \end{aligned} \tag{6.68}$$

We would consider a detection system where only the angular part of (6.67) is measured. Next we write the signal and idler fields as

$$\begin{aligned} E_s^{(-)}(\vec{r}, t) &= \sum_{l_s, p_s} b_{l_s p_s}^\dagger(\omega_s) v_{p_s}^{l_s *}(\vec{r}) \mathrm{e}^{\mathrm{i}\omega_s t}, \\ E_i^{(-)}(\vec{r}, t) &= \sum_{l_i, p_i} b_{l_i p_i}^\dagger(\omega_i) v_{p_i}^{l_i *}(\vec{r}) \mathrm{e}^{\mathrm{i}\omega_i t}. \end{aligned} \tag{6.69}$$

There is no summation over frequencies as we assume that signal and idler are monochromatic fields. Following the argument that led to (3.74) we can write the first-order wave function as

$$|\Psi^{(i)}\rangle \approx \sum_{l, p_s, p_i} b_{l p_s}^\dagger b_{-l p_i}^\dagger |0, 0\rangle F_{l p_s p_i}, \tag{6.70}$$

where

$$F_{l p_s p_i} = \frac{\chi A}{\mathrm{i}\hbar} \iint \rho\, \mathrm{d}\rho\, \mathrm{d}z\, R_{p_s}^*(\rho, z) R_{p_i}^*(\rho, z) R_{p_o}(\rho, z). \tag{6.71}$$

Clearly F gives the probability amplitude for the production of signal and idler photons in the OAM states $|l, p_s\rangle$ and $|-l, p_i\rangle$ respectively. If now consider a bucket detection system that is insensitive to radial degrees of freedom, then the state (6.70) effectively becomes

$$|\Psi^{(1)}\rangle \approx \sum \sqrt{P_l} |l\rangle_s |-l\rangle_i, \tag{6.72}$$

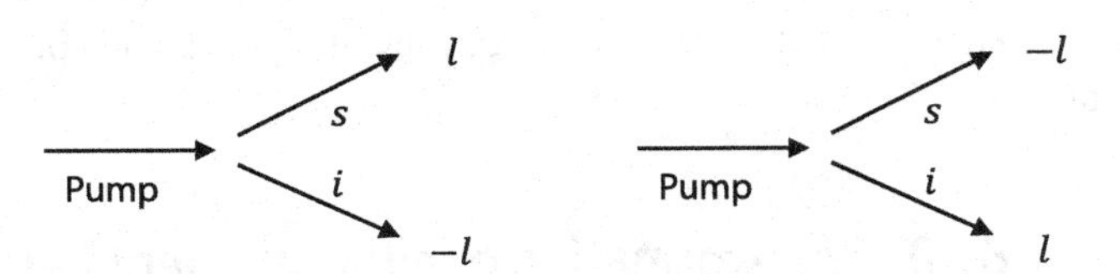

Fig. 6.7 Pairs of photons.

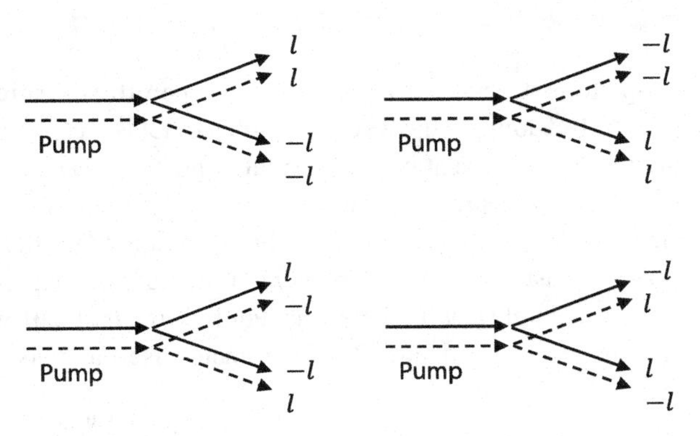

Fig. 6.8 Generation of the four-photon entangled states of OAM.

where

$$P_l = \sum_{p_s, p_i} |F_{l,p_s,p_i}|^2 \tag{6.73}$$

is the probability of producing a pair of photons, one with OAM l and the other with OAM $-l$. We have thus produced an entangled state of OAM degrees of freedom. The pump has no orbital angular momentum and it produces pairs of photons $(l, -l)$. For a given l we have an EPR-like state

$$|\Psi_{EPR}^{OAM}\rangle = \frac{1}{\sqrt{2}}(|l\rangle_s|-l\rangle_i + |-l\rangle_s|l\rangle_i), \tag{6.74}$$

as shown in Figure 6.7. We can also produce four-photon entangled states of OAM as we did in Section 3.11 for the polarization degrees of freedom. This is shown in Figure 6.8.

The four-photon entangled state of OAM for bucket detection would have the form

$$|\Psi^{(2)}\rangle = \sum_{ll'} \sqrt{P_{ll'}}|l, l'\rangle_s|-l, -l'\rangle_i, \tag{6.75}$$

$$P_{ll'} = \sum_{p_s p_i} \sum_{p'_s p'_i} |F_{lp_s p_i}|^2 |F_{l' p'_s p'_i}|^2. \tag{6.76}$$

Having generated entangled states of the orbital angular momentum, one can now do a variety of interferometric measurements as well as use these in supersensitive measurements of angular displacements [31]. The Hong–Ou–Mandel dip has been studied using the entangled states of the orbital angular momentum [32]. In the original experiment, Hong,

Ou, and Mandel used the finite frequency of the entangled photons. For OAM, one uses the finite width inherent in the state (6.72).

6.10 Transformation of entanglement between polarization and orbital angular momentum q-plates

The q-plate is a device which can couple the orbital degrees of freedom to polarization degrees of freedom. At the level of quantized fields it enables us to produce entangled states involving both polarization and orbital angular momentum degrees of freedom [33, 34]. Consider, for example, a half-wave plate made from a birefringent medium, i.e. a uniaxial crystal, as discussed in Section 6.3. Let us assume that the plate lies in the xy plane with its slow and fast axes in the xy plane. Let the light be incident normally on it. Let the optic axis make an angle α with the x axis. For the standard half-wave plates α is a constant. Let us now make α dependent on the coordinates itself and take it to be of the form

$$\alpha(\theta) = q\theta + \alpha_0. \tag{6.77}$$

Such wave plates can be realized by using nematic liquid crystals. The cases when q is a half integer or an integer are of special interest. The Jones matrix (6.22) using (6.77) and $\eta = \pi$ is

$$J = \mathrm{i}\mathrm{e}^{\mathrm{i}\delta}\begin{pmatrix} \cos 2(q\theta + \alpha_0) & \sin 2(q\theta + \alpha_0) \\ \sin 2(q\theta + \alpha_0) & -\cos 2(q\theta + \alpha_0) \end{pmatrix}. \tag{6.78}$$

The overall phase factor can be dropped. Thus a left-hand polarized circular wave $\binom{1}{\mathrm{i}}$ will be transformed to

$$\begin{pmatrix} 1 \\ \mathrm{i} \end{pmatrix} \longrightarrow \begin{pmatrix} 1 \\ -\mathrm{i} \end{pmatrix} \mathrm{e}^{2\mathrm{i}q\theta + 2\mathrm{i}\alpha_0}. \tag{6.79}$$

The left-hand polarized circular wave transforms into right-hand polarized circular wave. In addition, the outgoing wave acquires orbital angular momentum if $2q$ is an integer.

We will now show by a simple example how the entangled states can be produced by using a q-plate with $q = 1$. Let us consider an input state of a single photon with horizontal (H) polarization and zero OAM. The H polarization is a combination of right- and left-circular polarizations and thus we have the transformation

$$\begin{pmatrix} 1 \\ 0 \end{pmatrix} = \frac{1}{2}\left[\begin{pmatrix} 1 \\ \mathrm{i} \end{pmatrix} + \begin{pmatrix} 1 \\ -\mathrm{i} \end{pmatrix}\right] \xrightarrow{\text{plate}} \frac{1}{2}\begin{pmatrix} 1 \\ -\mathrm{i} \end{pmatrix}\mathrm{e}^{2\mathrm{i}\theta + 2\mathrm{i}\alpha_0} + \frac{1}{2}\begin{pmatrix} 1 \\ \mathrm{i} \end{pmatrix}\mathrm{e}^{-2\mathrm{i}\theta - 2\mathrm{i}\alpha_0}, \tag{6.80}$$

which is an entangled state consisting of a superposition of a photon in the state with left-circular polarization and with $l = -2$ and a state with right-circular polarization and $l = +2$. For the preparation and detection of such states, see [34]. Clearly, for two photons, starting with, say, polarization-entangled states one can produce a variety of entangled states by passing each beam through different q-plates.

The q-plate clearly mixes the spin and orbital angular momenta. It is thus an example of spin–orbit coupling in optical systems. Such a coupling is often referred to as the Hall effect of light [35]. The θ-dependent optic axis naturally provides the basic ingredient for spin–orbit coupling. Once we know the transformations by a q-plate one can discuss the behavior of nonclassical fields. Besides the Jones matrix for the transformation, it is also useful to have an equivalent description in terms of an effective Hamiltonian. Let us write the Bosonic operator $a_{s,l}$, where the index s stands for the polarization (left-circular or right-circular) and l for the orbital angular momentum. Clearly the q-plate for integer values of $2q$ can be described by an effective Hamiltonian of the form

$$H = \sum \left(c_{Ll} a^{\dagger}_{Ll} a_{Rl+2q} + c_{Rl} a^{\dagger}_{Rl} a_{Ll-2q} + H.c. \right), \qquad 2q = \text{integer}. \tag{6.81}$$

The coefficients c can be easily fixed from the Jones matrix (6.78). The effective Hamiltonian is especially useful in studying the transformation of nonclassical fields by systems having spin–orbit coupling.

Exercises

6.1 Solve (6.25) to show that J is indeed equal to (6.22) ($\delta = 0$).

6.2 If a device shifts the phase of the fields in a circular basis then find the Jones matrix J, i.e. find J for the transformation $(a \pm ib) \to (a \pm ib)\mathrm{e}^{-\mathrm{i}\varphi_{\pm}}$. This is important in the context of magneto-optical rotations.

6.3 Consider the two-photon states defined in the circular polarization basis as

$$|++\rangle = \frac{a_{+}^{\dagger 2}}{\sqrt{2}}|00\rangle, \qquad |+-\rangle = a_{+}^{\dagger} a_{-}^{\dagger}|00\rangle, \qquad |--\rangle = \frac{a_{-}^{\dagger 2}}{\sqrt{2}}|00\rangle,$$

where $a_{\pm} = \frac{a \pm ib}{\sqrt{2}}$, $[a_+, a_+^{\dagger}] = [a_-, a_-^{\dagger}] = 1$, $[a_+, a_-^{\dagger}] = 0$. Show that all these states are invariant under rotation, i.e. $[\rho, a^{\dagger}b - ab^{\dagger}] = 0$, where ρ stands for the density matrix of any of the above three states. The wave function, however, would acquire a phase.

6.4 Show that $\det J_P = 0$ for all pure states is defined by $\alpha|2,0\rangle + \beta|1,1\rangle + \gamma|0,2\rangle$ with

$$\begin{pmatrix} \alpha \\ \beta \\ \gamma \end{pmatrix} = \exp\left(-\mathrm{i}\theta \vec{S} \cdot \vec{n}\right) \begin{pmatrix} 1 \\ 0 \\ 0 \end{pmatrix},$$

where J_P is defined by (6.15) and $P = 1$. Here $\vec{n}$ is a unit vector, $\vec{S}$ is the operator defined by (6.12), and θ is arbitrary.

6.5 Calculate the fluctuations in Stokes operators $\langle S_i S_j \rangle - \langle S_i \rangle \langle S_j \rangle$ for unpolarized state defined by (6.32).

6.6 Verify the normalization (6.43) of the LG modes using the following property of the associated Laguerre polynomials

$$\int_0^{\infty} \mathrm{d}x\, x^k \mathrm{L}_n^k(x) \mathrm{L}_m^k(x) \mathrm{e}^{-x} = \frac{(n+k)!}{n!} \delta_{nm}.$$

6.7 Consider a mask with a transmission function given by Eq. (6.47). Let us launch a Gaussian beam $\exp\{-(x^2+y^2)/w^2\}$. Then calculate the far-field pattern which is proportional to the Fourier transform

$$\iint \left| e^{2\pi i \frac{x'}{\lambda}} + e^{i\theta'} \right|^2 e^{-\frac{x'^2+y'^2}{w^2}} e^{-iq_1 x' - iq_2 y'} dx' dy', \qquad \theta' = \tan^{-1}(y'/x').$$

Here q_1 and q_2 specify the directions of the point in the far field $q_1 = x/r, q_2 = y/r$ and $r = \sqrt{x^2+y^2+z^2}$. Show in particular that the above result represents a combination of vortices v_{01} and v_{10} and the Gaussian beam u_{00}. For a number of interesting experiments using this method of generating optical vortices, see [36].

6.8 Find the coefficients in the expansion of LG modes in terms of the HG modes $v_{nm}(\rho,\theta) = \sum_{p,q} d_{nmpq} u_{pq}(x,y)$. Using this expansion, justify the lower part of Figure 6.6.

References

[1] M. Born and E. Wolf, *Principles of Optics*, 7th ed. (Cambridge: Cambridge University Press, 1999).

[2] D. N. Klyshko, *Sov. Phys.-JETP* **84**, 1065 (1997); A. V. Burlakov and D. N. Klyshko, *JETP Lett.* **69**, 839 (1999).

[3] A. B. Klimov, G. Björk, J. Söderholm *et al.*, *Phys. Rev. Lett.* **105**, 153602 (2010).

[4] E. Wolf, *Introduction to the Theory of Coherence and Polarization of Light* (New York: Cambridge University Press, 2007).

[5] G. S. Agarwal and R. Simon, *Phys. Rev. A* **42**, 6924 (1990).

[6] G. S. Agarwal, J. Lehner, and H. Paul, *Opt. Commun.* **129**, 369 (1996).

[7] G. S. Agarwal, *Nuovo Cim. Lett.* **2**, 53 (1971).

[8] H. Prakash and N. Chandra, *Phys. Rev. A* **4**, 796 (1971).

[9] J. Lehner, U. Leonhardt, and H. Paul, *Phys. Rev. A* **53**, 2727 (1996).

[10] A. Luis, *Phys. Rev. A* **75**, 053806 (2007).

[11] T. Tsegaye, J. Söderholm, M. Atatüre *et al.*, *Phys. Rev. Lett.* **85**, 5013 (2000).

[12] G. S. Agarwal and S. Chaturvedi, *J. Mod. Opt.* **50**, 711 (2003).

[13] N. Korolkova, G. Leuchs, R. Loudon, T. C. Ralph, and C. Silberhorn, *Phys. Rev. A* **65**, 052306 (2002).

[14] L. L. Sánchez-Soto and A. Luis, *Opt. Commun.* **105**, 84 (1994); A. Luis and L. L. Sánchez-Soto, in *Progress in Optics*, edited by E. Wolf (Amsterdam: Elsevier, 2000), **41**, p. 421.

[15] N. Mukunda and T. F. Jordan, *J. Math. Phys.* **7**, 849 (1966).

[16] R. Simon and N. Mukunda, *Phys. Lett. A* **138**, 474 (1989).

[17] U. Schilling, J. von Zanthier, and G. S. Agarwal, *Phys. Rev. A* **81**, 013826 (2010).

[18] D. F. V. James, P. G. Kwiat, W. J. Munro, and A. G. White, *Phys. Rev. A* **64**, 052312 (2001).

[19] R. A. Beth, *Phys. Rev.* **50**, 115 (1936).

[20] J. D. Jackson, *Classical Electrodynamics*, 3rd ed. (New York: Wiley, 1999), Exercise 6.11.

[21] L. Allen, M. W. Beijersbergen, R. J. C. Spreeuw, and J. P. Woerdman, *Phys. Rev. A* **45**, 8185 (1992).

[22] L. Allen, S. Barnett, and M. Padgett, *Optical Angular Momentum* (Bristol, UK: Institute of Physics Publishing, 2003); L. Allen, M. Padgett, and M. Babiker, The orbital angular momentum of light, in *Progress in Optics*, edited by E. Wolf (Amsterdam: Elsevier, 1999), **39**, p. 291.

[23] M. T. L. Hsu, W. P. Bowen, and P. K. Lam, *Phys. Rev. A* **79**, 043825 (2009).

[24] S. Danakas and P. K. Aravind, *Phys. Rev. A* **45**, 1973 (1992).

[25] M. W. Beijersbergen, L. Allen, H. E. L. O. van der Veen, and J. P. Woerdman, *Opt. Commun*. **96**, 123 (1993).

[26] Z. S. Sacks, D. Rozas, and G. A. Swartzlander, Jr., *J. Opt. Soc. Am. B* **15**, 2226 (1998).

[27] H. He, N. R. Heckenberg, and H. Rubinsztein-Dunlop, *J. Mod. Opt*. **42**, 217 (1995).

[28] G. S. Agarwal, *J. Opt. Soc. Am. A* **16**, 2914 (1999).

[29] M. Lassen, G. Leuchs, and U. L. Andersen, *Phys. Rev. Lett*. **102**, 163602 (2009).

[30] M. J. Padgett and J. Courtial, *Opt. Lett*. **24**, 430 (1999).

[31] A. K. Jha, G. S. Agarwal, and R. W. Boyd, *Phys. Rev. A* **83**, 053829 (2011).

[32] H. Di Lorenzo Pires, H. C. B. Florijn, and M. P. van Exter, *Phys. Rev. Lett*. **104**, 020505 (2010).

[33] L. Marrucci, C. Manzo, and D. Paparo, *Phys. Rev. Lett*. **96**, 163905 (2006).

[34] E. Nagali, F. Sciarrino, F. De Martini *et al.*, *Phys. Rev. Lett*. **103**, 013601 (2009).

[35] D. Haefner, S. Sukhov, and A. Dogariu, *Phys. Rev. Lett*. **102**, 123903 (2009).

[36] A. V. Carpentier, H. Michinel, J. R. Salgueiro, and D. Olivieri, *Am. J. Phys*. **76**, 916 (2008).

7 Absorption, emission, and scattering of radiation

So far we have concentrated on the properties of electromagnetic fields. Now we study how electromagnetic fields interact with matter and how the outcome of the radiation–matter interaction can depend on the nonclassical properties of electromagnetic fields. Furthermore, in order to appreciate Glauber's quantum theory of optical coherence [1, 2] in the next chapter we have to understand how a photo detector responds to incoming nonclassical light. In this chapter we treat the radiation–matter interaction perturbatively; the nonperturbative results will be given in later chapters.

7.1 The interaction of radiation and matter in the electric dipole approximation

We keep the analysis of the interaction of radiation and matter as simple as possible. For the purpose of this book it is adequate to treat the interaction using the electric dipole approximation.

We label the states of an atom as $|i\rangle$ with energy E_i. The unperturbed Hamiltonian for the atom is written as

$$H_A = \sum E_i |i\rangle\langle i|. \tag{7.1}$$

The transition frequency between the levels $|i\rangle$ and $|j\rangle$ will be denoted as ω_{ij}

$$\omega_{ij} = \frac{1}{\hbar}(E_i - E_j). \tag{7.2}$$

Let $\vec{p}$ be the electric dipole moment operator

$$\vec{p} = e\vec{r}, \tag{7.3}$$

where $\vec{r}$ is the coordinate of the electron and e its charge. If the atom is located at the position $\vec{R}$, then $\vec{r}$ is the position of the electron relative to the position of the nucleus. For simplicity we assume that the atomic states have no permanent dipole moment. The two states for which the matrix element

$$\vec{p}_{ij} = \langle i|e\vec{r}|j\rangle \neq 0, \tag{7.4}$$

are connected by a dipole transition.

The interaction between an atom located at $\vec{R}$ and the electromagnetic field in the dipole approximation is

$$H_I = -\vec{p} \cdot \vec{E}(\vec{R}), \tag{7.5}$$

where $\vec{E}(\vec{R})$ is the electric field operator at the location $\vec{R}$ of the atom. The form of the electric field operator is given by Eq. (1.22). We treat the electromagnetic field quantum mechanically. The semiclassical limit, if desired, can be taken at the end of the calculation. The advantage of treating the field quantum mechanically is that (i) spontaneous emission comes out naturally, and (ii) fields with low photon numbers can be handled. In Chapter 1, we gave the unperturbed Hamiltonian H_F of the field. For the convenience of the reader we collect all these together and write the net Hamiltonian for radiation–matter interaction

$$\begin{aligned}
H &= H_A + H_F + H_I, \\
H_A &= \sum E_i |i\rangle\langle i|, \\
H_F &= \sum \hbar\omega_k a^{\dagger}_{\vec{k}s} a_{\vec{k}s}, \\
H_I &= -\vec{p} \cdot \vec{E}(\vec{R}), \\
\vec{E}(\vec{R}) &= \sum_{\vec{k}s} \mathrm{i}\sqrt{\frac{2\pi\hbar\omega_k}{V}} \vec{\epsilon}_{\vec{k}s} a_{\vec{k}s} \mathrm{e}^{\mathrm{i}\vec{k}\cdot\vec{R}} + H.c..
\end{aligned} \tag{7.6}$$

The radiation–matter interaction is typically determined by the fine structure parameter $\mathrm{e}^2/(\hbar c) \approx 1/137$ unless the electromagnetic fields get very large. Thus for most purposes we are allowed to treat such an interaction perturbatively.

7.2 Rates for the absorption and emission of radiation

We can now study specifically the question of absorption and emission of radiation by an atom. Let the atom be in the state $|i\rangle$ at $t = 0$. Then by absorption or emission of a photon it makes a transition to the state $|f\rangle$. We specifically want to calculate the rate $\Gamma_{i\to f}$ for the absorption or emission of radiation. Clearly $E_f > E_i$ ($E_f < E_i$) if absorption (emission) of radiation occurs. For simplicity we ignore the motion of the center of mass of the atom. For the calculation of transition probabilities it is convenient to work in the interaction picture in which the interaction Hamiltonian is given by

$$\begin{aligned}
H_I(t) &= -\vec{p}(t) \cdot \vec{E}(\vec{R}, t), \\
\vec{p}(t) &= \sum_{jl} \vec{p}_{jl} |j\rangle\langle l| \mathrm{e}^{\mathrm{i}\omega_{jl}t}, \qquad \vec{p}_{jl} = \langle j|\vec{p}|l\rangle, \\
\vec{E}(\vec{R}, t) &= \sum_{\vec{k}s} \mathrm{i}\sqrt{\frac{2\pi\hbar\omega_k}{V}} \vec{\epsilon}_{\vec{k}s} a_{\vec{k}s} \mathrm{e}^{\mathrm{i}\vec{k}\cdot\vec{R} - \mathrm{i}\omega_k t} + H.c. \\
&= \vec{E}^{(+)}(\vec{R}, t) + \vec{E}^{(-)}(\vec{R}, t),
\end{aligned} \tag{7.7}$$

where $\vec{E}^{(+)}$ ($\vec{E}^{(-)}$) is the term involving annihilation (creation) operators only. Let $|\Phi_i\rangle$ and $|\Phi_f\rangle$ be the initial and final states of the field. The combined state of the atom–field system at $t = 0$ is $|i\rangle|\Phi_i\rangle$. According to quantum mechanics the evolution of the wavefunction $|\Psi_I\rangle$ in the interaction picture is

$$\mathrm{i}\hbar\frac{\partial|\Psi_I\rangle}{\partial t} = H_I(t)|\Psi_I\rangle, \qquad |\Psi_I\rangle = \mathrm{e}^{\mathrm{i}(H_A+H_F)t/\hbar}|\Psi\rangle, \tag{7.8}$$

where $|\Psi\rangle$ is the wavefunction in the Schrödinger picture. From Eq. (7.8) the first-order correction to the wavefunction is found to be

$$\begin{aligned}|\Psi_I(t)\rangle^{(1)} &= \frac{1}{\mathrm{i}\hbar}\int_0^t \mathrm{d}t'\, H_I(t')|\Psi_I(0)\rangle \\ &= \frac{1}{\mathrm{i}\hbar}\int_0^t \mathrm{d}t'\, H_I(t')|\Psi(0)\rangle \\ &= -\frac{1}{\mathrm{i}\hbar}\int_0^t \mathrm{d}t'\, \vec{p}(t')\cdot\vec{E}(\vec{R},t')|i\rangle|\Phi_i\rangle.\end{aligned} \tag{7.9}$$

The probability amplitude of finding the system in the state $|f\rangle|\Phi_f\rangle$ after time t would be

$$\begin{aligned}\langle f|\langle\Phi_f|\Psi_I(t)\rangle &= -\frac{1}{\mathrm{i}\hbar}\int_0^t \mathrm{d}t'\, \langle f|\vec{p}(t')|i\rangle\cdot\langle\Phi_f|\vec{E}(\vec{R},t')|\Phi_i\rangle \\ &= -\frac{1}{\mathrm{i}\hbar}\int_0^t \mathrm{d}t'\, \vec{p}_{fi}\mathrm{e}^{\mathrm{i}\omega_{fi}t'}\cdot\langle\Phi_f|\vec{E}(\vec{R},t')|\Phi_i\rangle.\end{aligned} \tag{7.10}$$

Further simplification depends on whether we are considering absorption or emission processes. We therefore discuss each case separately.

7.2.1 Absorption processes

For absorption we have $E_f > E_i$, $\omega_{fi} > 0$. We also know from our discussion in Chapter 1 that the annihilation operator acting on a Fock state reduces the excitation in the Fock state by unity. Hence the part $E^{(+)}$ in (7.7) corresponds to the absorption of photons. Therefore for absorption of a photon, Eq. (7.10) reduces to

$$\langle f|\langle\Phi_f|\Psi_I(t)\rangle = -\frac{1}{\mathrm{i}\hbar}\int_0^t \mathrm{d}t'\vec{p}_{fi}\mathrm{e}^{\mathrm{i}\omega_{fi}t'}\cdot\langle\Phi_f|\vec{E}^{(+)}(\vec{R},t')|\Phi_i\rangle. \tag{7.11}$$

Let $\mathcal{P}_{if}$ be the probability of finding the atom in the state $|f\rangle$ if it has been initially in state $|i\rangle$. Such a probability has no information on the final state of the field. Thus we are thinking of an experiment where the final state of the atom is measured and the final state of the field is not measured. This probability $\mathcal{P}_{if}$ can be obtained from (7.11) by taking its absolute square and by summing over all the final states of the field:

$$\begin{aligned}\mathcal{P}_{if} &= \sum_{\Phi_f} \frac{1}{\hbar^2} \iint_0^t \mathrm{d}t_1 \mathrm{d}t_2\, \vec{p}_{fi} \cdot \langle \Phi_f | \vec{E}^{(+)}(\vec{R}, t_1) | \Phi_i \rangle \langle \Phi_i | \vec{E}^{(-)}(\vec{R}, t_2) | \Phi_f \rangle \cdot \vec{p}^{\,*}_{fi} \\ &\quad \times \mathrm{e}^{\mathrm{i}\omega_{fi}(t_1 - t_2)} \\ &= \frac{1}{\hbar^2} \iint_0^t \mathrm{d}t_1 \mathrm{d}t_2\, \mathrm{e}^{\mathrm{i}\omega_{fi}(t_1-t_2)} \langle \Phi_i | [\vec{p}^{\,*}_{fi} \cdot \vec{E}^{(-)}(\vec{R}, t_2)][\vec{p}_{fi} \cdot \vec{E}^{(+)}(\vec{R}, t_1)] | \Phi_i \rangle,\end{aligned} \tag{7.12}$$

where we used the completeness of the states $|\Phi_f\rangle$ of the field. Furthermore, the initial state of the field need not be a pure state. It could be a mixed state ρ_F. We thus find that the absorption probability is related to the normally ordered correlation function of the field defined by

$$\begin{aligned}\mathcal{E}^{(N)}_{\alpha\beta}(\vec{r}_2, t_2, \vec{r}_1, t_1) &= \mathrm{Tr}\big[\rho_F E^{(-)}_{\alpha}(\vec{r}_2, t_2) E^{(+)}_{\beta}(\vec{r}_1, t_1)\big] \\ &= \langle E^{(-)}_{\alpha}(\vec{r}_2, t_2) E^{(+)}_{\beta}(\vec{r}_1, t_1) \rangle,\end{aligned} \tag{7.13}$$

which in steady state is a function of $(t_1 - t_2)$ only. Thus (7.12) reduces to

$$\mathcal{P}_{if} = \frac{1}{\hbar^2} \iint_0^t \mathrm{d}t_1 \mathrm{d}t_2\, \mathrm{e}^{\mathrm{i}\omega_{fi}(t_1-t_2)} \vec{p}^{\,*}_{fi} \cdot \overset{\Rightarrow(N)}{\mathcal{E}}(\vec{R}, 0, \vec{R}, t_1 - t_2) \cdot \vec{p}_{fi}, \tag{7.14}$$

where we have used the dyadic notation for the correlation functions (7.13). In order to obtain the rate Γ_{if}, we differentiate $\mathcal{P}_{if}$ and take the limit $t \to \infty$. A simple algebraic calculation yields

$$\Gamma^{\mathrm{abs}}_{if} = \frac{1}{\hbar^2} \int_{-\infty}^{+\infty} \mathrm{d}\tau\, \mathrm{e}^{\mathrm{i}\omega_{fi}\tau} \vec{p}^{\,*}_{fi} \cdot \overset{\Rightarrow(N)}{\mathcal{E}}(\vec{R}, 0, \vec{R}, \tau) \cdot \vec{p}_{fi}, \tag{7.15}$$

where the superscript indicates that this is the rate for absorption. We thus find that the rate of absorption is related to the Fourier transform of the normally ordered correlation (7.13). This is a fundamental result which is central to Glauber's quantum theory of optical coherence. The Fourier component at the transition frequency ω_{fi} directly gives the transition rate. This is very interesting as one can use this relation to find the quantum properties of an unknown field by studying the transition probabilities. The result (7.15) is quite general and can be applied to a variety of fields, both classical and quantum. We will discuss a number of special cases in the next section. It may be noted that if the field is initially in the vacuum, then $\overset{\Rightarrow(N)}{\mathcal{E}} = 0$. Clearly the field has to be present if it is to be absorbed.

7.2.2 Emission processes

For the case of emission we have $E_f < E_i$, $\omega_{fi} < 0$. Furthermore, we use the property of the creation operator, namely that it increases the excitation of a Fock state by unity. Therefore in place of (7.11) we would have

$$\langle f | \langle \Phi_f | \Psi_I(t) \rangle = -\frac{1}{\mathrm{i}\hbar} \int_0^t \mathrm{d}t'\, \vec{p}_{fi} \mathrm{e}^{\mathrm{i}\omega_{fi}t'} \cdot \langle \Phi_f | \vec{E}^{(-)}(\vec{R}, t') | \Phi_i \rangle. \tag{7.16}$$

We can now follow the analysis which led to (7.15). We keep in mind that the roles of $E^{(-)}$ and $E^{(+)}$ are reversed. Therefore in place of (7.13) we need

$$\mathcal{E}^{(A)}_{\alpha\beta}(\vec{r}_2, t_2, \vec{r}_1, t_1) = \langle E^{(+)}_\alpha(\vec{r}_2, t_2) E^{(-)}_\beta(\vec{r}_1, t_1)\rangle, \tag{7.17}$$

which is called the antinormally ordered correlation function of the field. The rate of emission is then calculated to

$$\Gamma^{\text{em}}_{if} = \frac{1}{\hbar^2}\int_{-\infty}^{+\infty} d\tau\, e^{i\omega_{fi}\tau}\, \vec{p}^{\,*}_{fi} \cdot \overset{\Rightarrow(A)}{\mathcal{E}}(\vec{R}, 0, \vec{R}, \tau) \cdot \vec{p}_{fi}. \tag{7.18}$$

7.2.3 Spontaneous emission

We note that if there is no field present initially, then the probability of emission is nonzero. This is because the creation operator acting on the vacuum produces a one-photon state. In view of this we have $\mathcal{E}^{(A)} \neq 0$ even if the field is in the vacuum state, i.e. if

$$\rho_F = |\{0\}\rangle\langle\{0\}|. \tag{7.19}$$

This is the spontaneous emission of radiation. The rate for spontaneous emission was first calculated by Einstein using considerations of thermodynamic equilibrium and Planck's law. It is called the Einstein A coefficient. We show that spontaneous emission naturally follows here from considerations of the second quantization of the electromagnetic field, i.e. we show that $\Gamma_{if} = A$ if the field is in state (7.19). It should also be noted that spontaneous emission can take place in any mode of the field. We can now find the explicit form of the Einstein A coefficient by evaluating $\overset{\Rightarrow(A)}{\mathcal{E}}(\vec{R}, 0, \vec{R}, \tau)$. We first note that $\langle a_{\vec{k}_1 s_1} a^\dagger_{\vec{k}_2 s_2}\rangle = \delta_{\vec{k}_1\vec{k}_2}\delta_{s_1 s_2}$ and therefore

$$\begin{aligned}\mathcal{E}^{(A)}_{\alpha\beta}(\vec{R}, 0, \vec{R}, \tau) &= \sum_{\vec{k}_1 s_1 \vec{k}_2 s_2} \sqrt{\frac{2\pi\hbar\omega_{k_1}}{V}}\sqrt{\frac{2\pi\hbar\omega_{k_2}}{V}} (\epsilon_{\vec{k}_1 s_1})_\alpha (\epsilon^*_{\vec{k}_2 s_2})_\beta \\ &\quad \times e^{i\vec{k}_1\cdot\vec{R} - i\vec{k}_2\cdot\vec{R}} \langle a_{\vec{k}_1 s_1} a^\dagger_{\vec{k}_2 s_2}\rangle e^{i\omega_{k_2}\tau} \\ &= \sum_{\vec{k}s}\left(\frac{2\pi\hbar\omega_k}{V}\right) e^{i\omega_k\tau} (\epsilon_{\vec{k}s})_\alpha (\epsilon^*_{\vec{k}s})_\beta,\end{aligned} \tag{7.20}$$

which in the limit $V \to \infty$ becomes

$$\begin{aligned}\mathcal{E}^{(A)}_{\alpha\beta}(\vec{R}, 0, \vec{R}, \tau) &= \frac{V}{(2\pi)^3}\int d^3k \left(\frac{2\pi\hbar\omega_k}{V}\right) e^{ikc\tau}\left(\delta_{\alpha\beta} - \frac{k_\alpha k_\beta}{k^2}\right) \\ &= \frac{2}{3}\delta_{\alpha\beta}\frac{c\hbar}{\pi}\int k^3 dk\, e^{ikc\tau},\end{aligned} \tag{7.21}$$

which on substituting in (7.18) leads to

$$A = \frac{2c}{3\hbar\pi}\int k^3 \mathrm{d}k \int_{-\infty}^{+\infty} \mathrm{d}\tau\, \mathrm{e}^{\mathrm{i}kc\tau - \mathrm{i}|\omega_{fi}|\tau}|p_{fi}|^2$$
$$= \frac{2c}{3\hbar\pi}|p_{fi}|^2 \int k^3 \mathrm{d}k\, 2\pi\,\delta(kc - |\omega_{fi}|)$$
$$= \frac{4|p_{fi}|^2|\omega_{fi}|^3}{3\hbar c^3}. \tag{7.22}$$

This is the well-known expression for the Einstein A coefficient. An estimate for (7.22) can be made by assuming $|p_{fi}| \approx ea_0$, where a_0 is the Bohr radius, and say for ω_{fi} corresponding to 6000 Å, i.e. $\omega_{fi}/c = 2\pi/(6\times 10^{-5})\ \mathrm{cm}^{-1}$. This yields $A \approx 9.37\times 10^6\ \mathrm{s}^{-1}$. The actual value depends on the dipole matrix element and ω_{fi}, and is equal to $6.14\times 10^7\ \mathrm{s}^{-1}$ for the D_1 line of Na. A good website for alkali D line data is reference [3].

7.2.4 Stimulated emission

It is evident from the discussion in Section 7.2.3 that the difference $(\Gamma_{if}^{em} - A)$ depends on the presence of the input field. This part is then the stimulated emission. We next show that the rate of stimulated emission is equal to the rate of absorption under the same conditions from which (7.15) and (7.18) were derived. It must be added that one can set up systems such that these two rates are unequal and for such systems one has found that lasing can occur under no population inversion [4–6].

In order to show the equality of (7.15) and the probability for stimulated emission, we rewrite (7.17) as

$$\mathcal{E}_{\alpha\beta}^{(A)}(\vec{R},0,\vec{R},\tau) = \langle[E_\alpha^{(+)}(\vec{R},0), E_\beta^{(-)}(\vec{R},\tau)]\rangle + \langle E_\beta^{(-)}(\vec{R},\tau)E_\alpha^{(+)}(\vec{R},0)\rangle. \tag{7.23}$$

For fields in the vacuum, the second term in (7.23) vanishes since this is a normally ordered correlation. The commutator in (7.23) is a number and does not depend on the state of the field. Therefore (7.18) can be written as

$$\Gamma_{if}^{\mathrm{em}} = A + \Gamma_{if}^{\mathrm{st}}, \tag{7.24}$$

where $\Gamma_{if}^{\mathrm{st}}$ is the rate of stimulated emission

$$\Gamma_{if}^{\mathrm{st}} = \frac{1}{\hbar^2}\int_{-\infty}^{+\infty} \mathrm{d}\tau\, \mathrm{e}^{-\mathrm{i}|\omega_{fi}|\tau}\langle(\vec{E}^{(-)}(\vec{R},\tau)\cdot\vec{p}_{fi})(\vec{E}^{(+)}(\vec{R},0)\cdot\vec{p}_{fi}^{\,*})\rangle \tag{7.25}$$
$$= \frac{1}{\hbar^2}\int_{-\infty}^{+\infty} \mathrm{d}\tau\, \mathrm{e}^{\mathrm{i}|\omega_{fi}|\tau}\langle\vec{E}^{(-)}(\vec{R},0)\cdot\vec{p}_{if}^{\,*}\vec{E}^{(+)}(\vec{R},\tau)\cdot\vec{p}_{if}\rangle. \tag{7.26}$$

To obtain (7.26) we changed τ to $-\tau$ and used the fact that $\overset{\Rightarrow}{\mathcal{E}}^{(N)}$ depends only on the difference of two arguments, we also used $\vec{p}_{if} = \vec{p}_{fi}^{\,*}$. On comparison with (7.15) we then obtain

$$\Gamma_{if}^{\mathrm{st}} = \Gamma_{fi}^{\mathrm{abs}}. \tag{7.27}$$

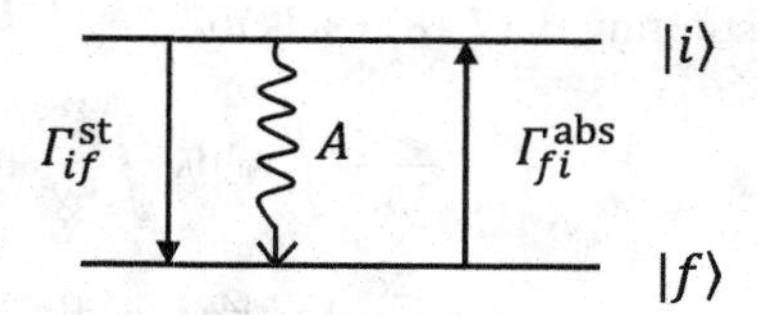

Fig. 7.1 Different radiative processes between the states $|i\rangle$ and $|f\rangle$.

The situation is shown in Figure 7.1. It should be borne in mind that, unlike spontaneous emission, the stimulated emission occurs in the mode in which the field is present.

Let us now consider radiative processes in a thermal (black-body) field. We will show how equilibrium is established. For absorption and emission in thermal fields we have

$$\rho_F = \prod_{\vec{k}s} \big[\exp\big(- \beta\hbar\omega_k a^\dagger_{\vec{k}s} a_{\vec{k}s}\big) / \mathrm{Tr}\big(\exp\big(- \beta\hbar\omega_k a^\dagger_{\vec{k}s} a_{\vec{k}s}\big)\big)\big], \tag{7.28}$$

for which

$$\langle a^\dagger_{\vec{k}_1 s_1} a_{\vec{k}_2 s_2} \rangle = \delta_{\vec{k}_1 \vec{k}_2} \delta_{s_1 s_2} n_{\vec{k}_1 s_1}. \tag{7.29}$$

A derivation similar to that of A leads to

$$\Gamma^{\mathrm{st}}_{if} = \Gamma^{\mathrm{abs}}_{fi} = n(|\omega_{fi}|)A. \tag{7.30}$$

Note that Figure 7.1 suggests that the atomic populations p_i, p_f in the states $|i\rangle$ and $|f\rangle$ obey the equations

$$\begin{aligned} \frac{\partial p_i}{\partial t} &= \Gamma^{\mathrm{abs}}_{fi} p_f - \big(\Gamma^{\mathrm{st}}_{if} + A\big) p_i, \\ \frac{\partial p_f}{\partial t} &= -\Gamma^{\mathrm{abs}}_{fi} p_f + \big(\Gamma^{\mathrm{st}}_{if} + A\big) p_i. \end{aligned} \tag{7.31}$$

Therefore in equilibrium $\dot{p}_i = \dot{p}_f = 0$, we obtain for a black-body field

$$p_f/p_i = \frac{\Gamma^{\mathrm{st}}_{if} + A}{\Gamma^{\mathrm{abs}}_{fi}} \to \frac{1 + n(|\omega_{fi}|)}{n(|\omega_{fi}|)} = \mathrm{e}^{\hbar\beta|\omega_{if}|}, \tag{7.32}$$

which is consistent with the Boltzmann distribution.

If initially no field is present $\Gamma^{\mathrm{abs}} = \Gamma^{\mathrm{st}} = 0$, then the solution of (7.31) under the initial condition $p_i = 1$, $p_f = 0$ is

$$p_i(t) = \mathrm{e}^{-At}, \qquad p_f(t) = 1 - \mathrm{e}^{-At}. \tag{7.33}$$

This is the well-known radiative decay of the excited states first derived by Weisskopf and Wigner [7]. We will give a derivation of this from first principles in Section 7.6.

7.3 Single-mode limit – Einstein's B coefficient and the absorption coefficient $\alpha(\omega)$

We examine (7.15) specifically in the limit of a single-mode monochromatic electromagnetic field and then derive an expression for the absorption coefficient. For a single mode of frequency ω, the normally ordered correlation function is

$$\mathcal{E}^{(N)}_{\alpha\beta}(\vec{R}, 0, \vec{R}, \tau) = \epsilon^*_\alpha \epsilon_\beta \mathrm{e}^{-\mathrm{i}\omega\tau} \left(\frac{2\pi\hbar\omega}{V}\right) \langle a^\dagger a \rangle, \tag{7.34}$$

where $\vec{\epsilon}$ is the polarization vector for the single-mode field. On substituting (7.34) in (7.15) we easily derive

$$\Gamma^{\mathrm{abs}}_{if} = \frac{4\pi^2}{\hbar V} |\vec{\epsilon}_\beta \cdot \vec{p}_{fi}|^2 \omega \delta(\omega - \omega_{fi}) \langle a^\dagger a \rangle. \tag{7.35}$$

Generally the excited states of atoms have finite width, thus we need to sum (7.35) over all the final states. Let $S_A(\omega_{fi})$ be the function which gives the density of the final states of the atom and for simplicity we choose it to be Lorentzian

$$S_A(\omega_{fi}) = \frac{\gamma/\pi}{\gamma^2 + (\omega_{fi} - \bar{\omega}_{fi})^2}. \tag{7.36}$$

Here $\bar{\omega}_{fi} = (\bar{E}_f - E_i)/\hbar$, where $\bar{E}_f$ is the mean energy of the state $|f\rangle$. Then (7.35) goes to

$$\Gamma^{\mathrm{abs}}_{if} = \frac{4\pi}{\hbar V} |\vec{\epsilon}_\beta \cdot \vec{p}_{fi}|^2 \omega \langle a^\dagger a \rangle \frac{\gamma}{\gamma^2 + (\omega - \bar{\omega}_{fi})^2}. \tag{7.37}$$

If there are N atoms in the volume V, then the mean photon number would decrease due to absorption by all the atoms in the volume. This decrease would be described by the rate equation

$$\begin{aligned} \frac{\mathrm{d}\langle a^\dagger a \rangle}{\mathrm{d}t} &= -N\Gamma^{\mathrm{abs}}_{if} \\ &= -\frac{4\pi n |\vec{\epsilon}_\beta \cdot \vec{p}_{fi}|^2 \omega\gamma}{\hbar[\gamma^2 + (\omega - \bar{\omega}_{fi})^2]} \langle a^\dagger a \rangle, \end{aligned} \tag{7.38}$$

where $n = N/V$ is the density of the atomic system. One introduces an absorption parameter α, which is the rate of change of energy per unit length. From (7.38) we have the explicit expression for α

$$\alpha(\omega) = \frac{4\pi n\omega\gamma}{\hbar c} |\vec{\epsilon}_\beta \cdot \vec{p}_{fi}|^2 / [\gamma^2 + (\omega - \bar{\omega}_{fi})^2], \tag{7.39}$$

which is Lorentzian with a halfwidth at half maximum equal to γ. The absorption at the line center $\alpha_0 = \alpha(\bar{\omega}_{fi})$ is

$$\alpha_0 = \frac{4\pi n\omega}{\hbar c\gamma} |\vec{\epsilon}_\beta \cdot \vec{p}_{fi}|^2, \tag{7.40}$$

and therefore

$$\alpha(\omega) = \alpha_0 \gamma^2 / [\gamma^2 + (\omega - \bar{\omega}_{fi})^2]. \tag{7.41}$$

The relation (7.38) also yields the classical Beer's law of absorption for the intensity as a function of the propagation distance

$$\frac{\mathrm{d}I}{\mathrm{d}z} = -\alpha I. \tag{7.42}$$

This is because the intensity is proportional to the mean photon number.

We next introduce the well-known Einstein B coefficient. Let us consider absorption from a classical field $\vec{E} = \vec{\epsilon}\, E_0 \mathrm{e}^{-\mathrm{i}\omega t + \mathrm{i}\vec{k}\cdot\vec{R}} + c.c.$ with amplitude E_0. We can identify $|E_0|^2 = 2\pi\hbar\omega\langle a^\dagger a\rangle/V$. The flux, the modulus S of the Poynting vector, of the incident field is given by $S = c|E_0|^2/(2\pi) = \hbar\omega c\langle a^\dagger a\rangle/V$. Using these definitions in (7.37) we get

$$\Gamma_{if}^{\mathrm{abs}} = \frac{4\pi}{\hbar^2 c}|\vec{\epsilon}_\beta \cdot \vec{p}_{fi}|^2 S \frac{\gamma}{\gamma^2 + (\omega - \bar{\omega}_{fi})^2}. \tag{7.43}$$

The expression (7.43) gives the absorption of a component ω of the incident field. More generally, if the incident field is not monochromatic, then (7.43) is to be replaced by

$$\Gamma_{if}^{\mathrm{abs}} = \int \frac{4\pi}{\hbar^2 c}|\vec{\epsilon}_\beta \cdot \vec{p}_{fi}|^2 S(\omega) \frac{\gamma}{\gamma^2 + (\omega - \bar{\omega}_{fi})^2}\mathrm{d}\omega, \tag{7.44}$$

where $S(\omega)$ is now the spectral distribution of the field. We next introduce the B coefficient via

$$\Gamma_{if}^{\mathrm{abs}} = \int \mathrm{d}\omega B(\omega)\frac{S(\omega)}{c}, \tag{7.45}$$

where

$$B(\omega) = \frac{4\pi^2}{\hbar^2}|\vec{\epsilon}_\beta \cdot \vec{p}_{fi}|^2 \frac{\gamma/\pi}{\gamma^2 + (\omega - \bar{\omega}_{fi})^2}. \tag{7.46}$$

Note that we have introduced a form for $\Gamma_{if}^{\mathrm{abs}}$, which takes into account both the spectral distribution of the field as well as the width of the final states of the atom.

7.4 Scattering of radiation

Let us consider the fundamental process where a photon travelling in the direction $\vec{k}_1$ with polarization $\vec{\epsilon}_1$ is scattered into a photon in the direction $\vec{k}_2$ with polarization $\vec{\epsilon}_2$. The atom goes from the initial state $|i\rangle$ to the final state $|f\rangle$. The conservation of energy requires

$$E_f - E_i = \hbar\omega_1 - \hbar\omega_2. \tag{7.47}$$

When the final state is identical to the initial state, then we have elastic scattering. Otherwise we have inelastic scattering. For inelastic scattering we have two possibilities: (i) Stokes scattering $E_f > E_i$, $\omega_2 < \omega_1$; and (ii) anti-Stokes scattering $E_f < E_i$, $\omega_2 > \omega_1$. These are the Raman processes and are displayed in Figure 7.2.

In order to obtain the rate of scattering we need to do a calculation of the wavefunction to second order in the radiation matter interaction. This calculation is more involved. In order to keep the analysis simple we use Fock states of the radiation field. We use the interaction

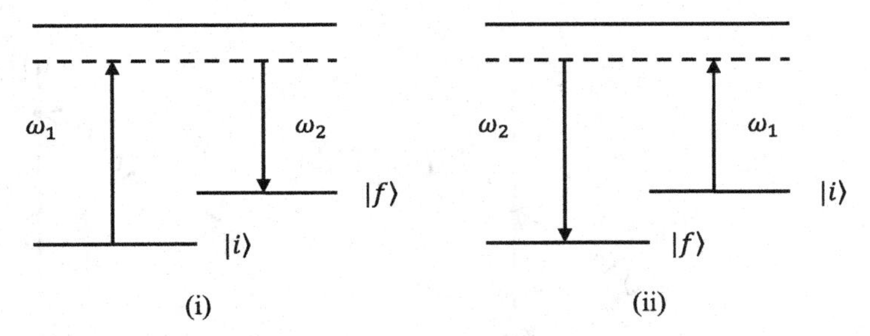

Fig. 7.2 Inelastic scattering: (i) Stokes scattering; (ii) anti-Stokes scattering.

picture (Eq. (7.8)) and the interaction Hamiltonian (7.7). We iterate (7.8) twice to obtain the second-order correction to the wavefunction

$$|\Psi_I(t)\rangle^{(2)} = \left(-\frac{1}{i\hbar}\right)^2 \int_0^t dt_1 \int_0^{t_1} dt_2\, \vec{p}(t_1)\cdot\vec{E}(\vec{R},t_1)\,\vec{p}(t_2)\cdot\vec{E}(\vec{R},t_2)|i\rangle|\Phi_i\rangle. \tag{7.48}$$

The initial state $|\Phi_i\rangle$ for the field is $|n_{\vec{k}_1 s_1}, n_{\vec{k}_2 s_2}\rangle$, i.e. we are assuming that only the incident and scattered modes are occupied, the rest of the modes are empty. The final state of the field is

$$|\Phi_f\rangle = |n_{\vec{k}_1 s_1} - 1, n_{\vec{k}_2 s_2} + 1\rangle. \tag{7.49}$$

Note that $n_{\vec{k}_2 s_2}$ can be zero. This corresponds to spontaneous scattering. The terms proportional to $n_{\vec{k}_2 s_2}$ are called stimulated scattering in analogy to our discussion of spontaneous and stimulated emission. Note also the following result

$$a^\dagger_{\vec{k}_2 s_2} a_{\vec{k}_1 s_1} |n_{\vec{k}_1 s_1}, n_{\vec{k}_2 s_2}\rangle = \sqrt{n_{\vec{k}_1 s_1}(n_{\vec{k}_2 s_2}+1)}\,|n_{\vec{k}_1 s_1} - 1, n_{\vec{k}_2 s_2} + 1\rangle, \tag{7.50}$$

which shows the dependence on the number of photons in the scattering mode. The transition amplitude is $\mathcal{A}_{if} = \langle\Phi_f|\langle f|\Psi_I(t)\rangle^{(2)}$. Since the electric field operator in (7.48) is the sum over all the modes, a nonzero transition amplitude can arise in two possible ways as shown in Figure 7.3: (I) at times t_2 a photon in the mode $(\vec{k}_1 s_1)$ is absorbed and a photon in the mode $(\vec{k}_2 s_2)$ is emitted at t_1; and (II) a photon in the mode $(\vec{k}_2 s_2)$ is emitted at time t_2 followed by the absorption of a photon from the mode $(\vec{k}_1 s_1)$ at time t_1. These two distinct events in time contribute to the second-order transition amplitude; however, the event (II) is highly nonresonant, and its contribution is much smaller than the event (I). Therefore it can be dropped from further considerations. The transition amplitude is then equal to

$$\mathcal{A}_{if} \simeq \frac{1}{\hbar^2}\int_0^t dt_1 \int_0^{t_1} dt_2\, \langle f|\vec{p}(t_1)\cdot\vec{\epsilon}^{\,*}_{\vec{k}_2 s_2}\,\vec{p}(t_2)\cdot\vec{\epsilon}_{\vec{k}_1 s_1}|i\rangle\sqrt{n_{\vec{k}_1 s_1}(n_{\vec{k}_2 s_2}+1)}$$
$$\times\sqrt{\frac{2\pi\hbar\omega_{k_1}}{V}}\sqrt{\frac{2\pi\hbar\omega_{k_2}}{V}}\,e^{i\omega_{k_2}t_1 - i\omega_{k_1}t_2}. \tag{7.51}$$

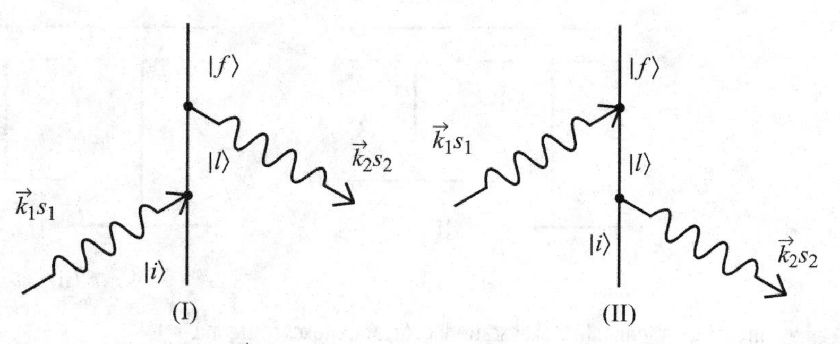

Fig. 7.3 (I) At times t_2 a photon in the mode $(\vec{k}_1 s_1)$ is absorbed and a photon in the mode $(\vec{k}_2 s_2)$ is emitted at t_1; (II) a photon in the mode $(\vec{k}_2 s_2)$ is emitted at time t_2 followed by the absorption of a photon from the mode $(\vec{k}_1 s_1)$ at time t_1.

The matrix element of the dipole moment operators can be simplified by using the completeness of the states and replacing the indices $\vec{k}_i s_i$ by i

$$\begin{aligned}\langle f|\vec{p}(t_1)\cdot\vec{\epsilon}_2^{\,*}\,\vec{p}(t_2)\cdot\vec{\epsilon}_1|i\rangle &= \sum_l \langle f|\vec{p}(t_1)\cdot\vec{\epsilon}_2^{\,*}|l\rangle\langle l|\vec{p}(t_2)\cdot\vec{\epsilon}_1|i\rangle \\ &= \sum_l \langle f|\vec{p}\cdot\vec{\epsilon}_2^{\,*}|l\rangle\langle l|\vec{p}\cdot\vec{\epsilon}_1|i\rangle e^{i\omega_{fl}t_1 + i\omega_{li}t_2}. \end{aligned} \tag{7.52}$$

On using (7.52) and on doing time integrations (7.51) reduces to

$$\begin{aligned}\mathcal{A}_{if} = {}& \frac{1}{\hbar^2}\sqrt{\frac{4\pi^2\hbar^2\omega_1\omega_2}{V^2}}\sqrt{n_1(n_2+1)}\sum_l \langle f|\vec{p}\cdot\vec{\epsilon}_2^{\,*}|l\rangle\langle l|\vec{p}\cdot\vec{\epsilon}_1|i\rangle \\ &\times\left[\frac{e^{i(\omega_2-\omega_1+\omega_{fi})t}-1}{i(\omega_{li}-\omega_1)i(\omega_2-\omega_1+\omega_{fi})} - \frac{e^{i(\omega_2+\omega_{fl})t}-1}{i(\omega_{li}-\omega_1)i(\omega_2+\omega_{fl})}\right].\end{aligned} \tag{7.53}$$

The first term in (7.53) is the only one that can fulfill the energy conservation given by (7.47). The scattering rate defined by

$$\Gamma_{if}^{(sc)} = \lim_{t\to\infty}\frac{d}{dt}|\mathcal{A}_{if}|^2, \tag{7.54}$$

can be obtained from (7.53) using

$$\lim_{t\to\infty}\frac{\sin\Delta t}{\Delta} = \pi\delta(\Delta), \tag{7.55}$$

with the result

$$\Gamma_{if}^{(sc)} = \frac{2\pi}{\hbar^4}\left(\frac{4\pi^2\hbar^2\omega_1\omega_2}{V^2}\right)|m_{fi}|^2\delta(\omega_{fi}-\omega_1+\omega_2)n_1(n_2+1), \tag{7.56}$$

$$m_{fi} = \sum_l \frac{\langle f|\vec{p}\cdot\vec{\epsilon}_2^{\,*}|l\rangle\langle l|\vec{p}\cdot\vec{\epsilon}_1|i\rangle}{\omega_{li}-\omega_1}.$$

The matrix element m_{fi} involves summation over a large number of states and it grows as the frequency ω_1 becomes close to one of the transition frequencies ω_{li}. In this case the scattering is called resonant scattering. The scattering is also sensitive to the polarization of the fields. Equation (7.56) is the key formula for the scattering rate.

7.4.1 Spontaneous Raman scattering

If $n_2 = 0$, then Eq. (7.56) gives the rate for spontaneous scattering. For semiclassical fields $|E_1|^2 = 2\pi\hbar\omega_1 n_1/V$ and on summing over all final states (as spontaneous scattering can take place in any mode) we get

$$\Gamma_{if}^{\text{spon}} = \frac{2\pi}{\hbar^4}|E_1|^2 \int \frac{2\pi\hbar\omega_2}{V} \cdot \frac{V}{(2\pi)^3}\text{d}^3k_2 \sum_{\epsilon_2} \delta(\omega_{fi} - \omega_1 + \omega_2)|m_{fi}|^2. \tag{7.57}$$

We have to remember that the vector $\vec{\epsilon}_2$ depends on $\vec{k}_2$ and that

$$\sum_{\epsilon_2} (\epsilon_2^*)_\alpha (\epsilon_2)_\beta = \delta_{\alpha\beta} - \frac{k_{2\alpha}k_{2\beta}}{k_2^2}, \tag{7.58}$$

which would give the angular distribution of the scattering. The details of the angular distribution depend on the nature of the states involved in the transitions.

7.4.2 Raman gain – stimulated Raman scattering

In stimulated scattering the energy from the field 1 is transferred to the field 2 so that the intensity of the field 2 grows. In this case (7.56) can be used to define the Raman gain g. We proceed as follows: (1) use $|E_1|^2 = 2\pi\hbar\omega_1 n_1/V$; (2) sum over the atomic states $\delta(\omega_{fi} - \omega_1 + \omega_2) \to S_A(\omega_{fi} - \omega_1 + \omega_2)$; (3) multiply (7.56) by N, the number of atoms. Then on ignoring the spontaneous emission term the rate of change of the number of photons at the frequency ω_2 due to stimulated Raman scattering will be

$$\frac{\partial n_2}{\partial t} = |E_1|^2 \left(\frac{N}{V}\right) \cdot \left(\frac{4\pi^2\omega_2}{\hbar^3}\right) S_A(\omega_{fi} - \omega_1 + \omega_2)|m_{fi}|^2 n_2. \tag{7.59}$$

The Raman gain is defined as the increase per unit length and is therefore equal to

$$g = |E_1|^2 \left(\frac{N}{V}\right) \cdot \left(\frac{4\pi^2\omega_2}{\hbar^3 c}\right) S_A(\omega_{fi} - \omega_1 + \omega_2)|m_{fi}|^2. \tag{7.60}$$

The equation (7.59) implies that the intensity of the field at the frequency ω_2 grows as

$$\frac{\partial I_2}{\partial z} = gI_2, \tag{7.61}$$

where g is proportional to the intensity I_1 ($\propto |E_1|^2$) of the field at the frequency ω_1.

7.5 Quantum interferences in scattering

An important feature of the rate for second-order processes is the matrix element m_{fi}, which involves a sum over all the intermediate states with denominators depending on

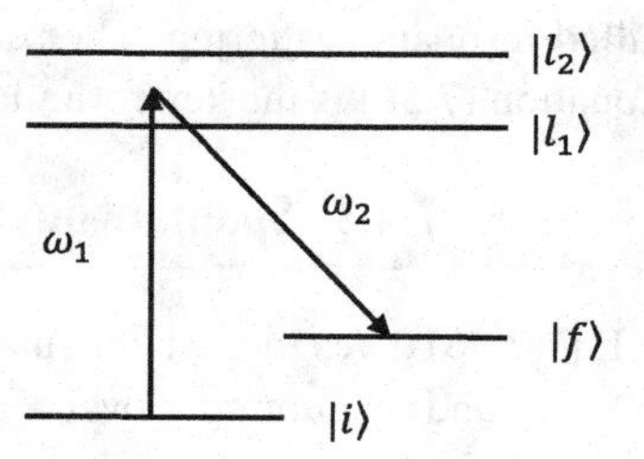

Fig. 7.4 Scattering via two intermediate states, with $\omega_1 - \omega_2 = \omega_{fi}$.

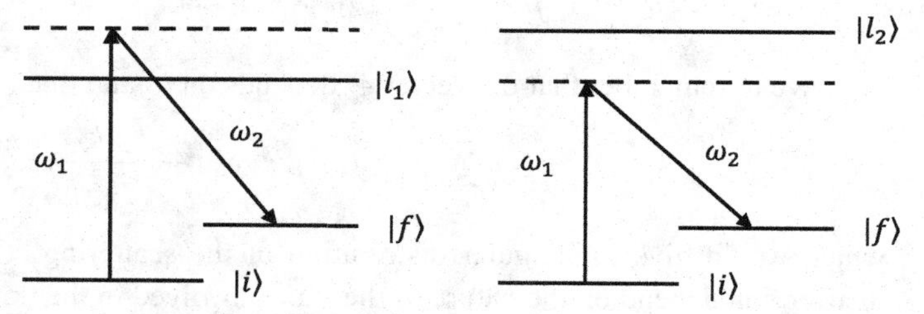

Fig. 7.5 Two different pathways that lead to scattering from $|i\rangle$ to $|f\rangle$.

external parameters, such as the frequency ω_1 of the field. The frequency ω_1 in principle is tunable. Furthermore, there is a dependence on the polarization of the fields ω_1 and ω_2. Depending on the polarizations, the Wigner–Eckert theorem enables one to pick up only specific intermediate states in the summation. The matrix element m_{fi} can become zero for certain values of ω_1. Consider the situation shown in Figure 7.4. The denominators $(\omega_{l_1 i} - \omega_1)^{-1}$ and $(\omega_{l_2 i} - \omega_1)^{-1}$ differ in sign. In this case there are two different pathways that lead to the transition $|i\rangle \rightarrow |f\rangle$, as shown in Figure 7.5. These two pathways can destructively interfere leading to a zero in the transition probability. This is referred to as the quantum interference of the transition amplitudes. The phenomenon of quantum interference is quite ubiquitous to second-order processes [8] and in fact to all higher-order processes. This is because the matrix element m_{fi} involves a coherent sum over different intermediate states. Quantum interferences have been extensively studied and we will return to these in subsequent chapters.

7.6 Radiative decay of states – Weisskopf–Wigner theory

In Section 7.2.3 we have seen how the excited state can decay to a lower state via the spontaneous emission of radiation. This is an intrinsic property of the radiation–matter interaction at the level of quantized fields. We also argued phenomenologically that as a result of spontaneous emission the excited state population decays exponentially. Weisskopf–Wigner proved this from first principles using the fundamental radiation–matter interaction. The idea here is to solve the Schrödinger equation exactly rather than use perturbation theory.

The radiation–matter interaction is given by (7.6). Initially let the atom be in the state $|i\rangle$ and no field be present. Thus the initial state of the combined atom–field system is

$$|\Psi_i\rangle = |i\rangle|\{0_{\vec{k}s}\}\rangle. \tag{7.62}$$

Let us assume that the atom decays to a single final state $|f\rangle$ with the emission of a photon. Let us write $\hbar\omega_0$ as the energy separation of the state $|i\rangle$ relative to $|f\rangle$. The photon can be emitted in any of the modes $\vec{k}s$. The final state can be written as

$$|\Psi_f\rangle = |f\rangle|1_{\vec{k}s}\rangle, \tag{7.63}$$

where $|1_{\vec{k}s}\rangle$ is the state of the field containing only one photon in the mode $\vec{k}s$. It should be borne in mind that we have many possible final states as the mode $\vec{k}s$ is arbitrary. The wavefunction of the combined atom–field system can be written as

$$|\Psi(t)\rangle = c_i(t)|i\rangle|\{0_{\vec{k}s}\}\rangle + \sum_{\vec{k}s} c_{\vec{k}s}(t)|f\rangle|1_{\vec{k}s}\rangle,\ c_i(0) = 1,\ c_{\vec{k}s}(0) = 0. \tag{7.64}$$

Using the Schrödinger equation and Eqs. (7.6) and (7.64) we find equations for the unknown amplitudes $c_i(t)$ and $c_{\vec{k}s}(t)$:

$$\begin{aligned} \dot{c}_i &= -\mathrm{i}\omega_0 c_i - \mathrm{i}\sum_{\vec{k}s} g_{\vec{k}s} c_{\vec{k}s}, \\ \dot{c}_{\vec{k}s} &= -\mathrm{i}\omega_k c_{\vec{k}s} - \mathrm{i}g^*_{\vec{k}s} c_i, \end{aligned} \tag{7.65}$$

where

$$g_{\vec{k}s} = -\mathrm{i}\sqrt{\frac{2\pi\omega_k}{\hbar V}}\mathrm{e}^{\mathrm{i}\vec{k}\cdot\vec{R}}\vec{p}_{if}\cdot\vec{\epsilon}_{\vec{k}s}. \tag{7.66}$$

These coupled equations are to be solved subject to the initial conditions in (7.64). These equations are solved by the Laplace transform defined by

$$\hat{c}(p) = \int_0^\infty c(t)\,\mathrm{e}^{-pt}\mathrm{d}t; \qquad \mathrm{Re}\{p\} > 0. \tag{7.67}$$

The requirement $\mathrm{Re}\{p\} > 0$ must be kept in mind for all calculations. The Laplace transform of (7.65) and (7.66) using $c_{\vec{k}s}(0) = 0$ and $c_i(0) = 1$, gives

$$\hat{c}_{\vec{k}s}(p) = -\mathrm{i}(p + \mathrm{i}\omega_k)^{-1} g^*_{\vec{k}s}\hat{c}_i(p), \tag{7.68}$$

$$p\hat{c}_i(p) - 1 = -\mathrm{i}\omega_0\hat{c}_i(p) - \mathrm{i}\sum_{\vec{k}s} g_{ks}\hat{c}_{\vec{k}s}(p). \tag{7.69}$$

Using (7.68) and (7.69) we obtain the amplitude for the initial state

$$\hat{c}_i(p) = \left[p + \mathrm{i}\omega_0 + \sum_{\vec{k}s}|g_{\vec{k}s}|^2(p + \mathrm{i}\omega_k)^{-1}\right]^{-1}. \tag{7.70}$$

We now have a complete solution to the problem of an atom interacting with the vacuum of the electromagnetic field. So far no approximation has been made. The effect of the interaction is contained in the last term of the denominator in (7.70). In general, the behavior of the function (7.70) is complicated due to the presence of the Laplace variable in the term $(p+i\omega_k)^{-1}$. Typically the interaction is weak (except in the context of cavity QED), so that a reasonable approximation would be to replace p in $(p+i\omega_k)^{-1}$ by $(-i\omega_0+\epsilon)$, where ϵ is an infinitesimal to be set to zero at the end of the calculation. This is because in the absence of interaction the pole of $\hat{c}_i(p)$ occurs at $-i\omega_0$. Thus an approximate expression for $\hat{c}_i(p)$ is

$$\hat{c}_i(p) = (p + i\omega_0 + i\Sigma)^{-1}, \tag{7.71}$$

$$\text{where } \Sigma = \sum_{\vec{k}s} |g_{\vec{k}s}|^2 (\omega_0 - \omega_k + i\epsilon)^{-1}. \tag{7.72}$$

The quantity Σ is called the self-energy term and is typically a complex quantity. The real part of Σ gives the energy shift (Lamb shift) and the imaginary part of Σ gives the energy levels a finite width. In the simple model $\mathrm{Re}\,\Sigma$ diverges and a correct treatment of $\mathrm{Re}\,\Sigma$ requires renormalization methods from QED as well as the inclusion of other atomic levels in the calculation. We will drop $\mathrm{Re}\,\Sigma$ from further considerations, assuming that the frequency ω_0 already includes the effects of $\mathrm{Re}\,\Sigma$. The imaginary part of Σ is related to the Einstein A coefficient as expected. To see this we proceed as follows

$$\begin{aligned} \mathrm{Im}\,\Sigma &= -\sum_{\vec{k}s} |g_{\vec{k}s}|^2 \frac{\epsilon}{\epsilon^2 + (\omega_0 - \omega_k)^2}, \\ &\Rightarrow -\sum_{\vec{k}s} |g_{\vec{k}s}|^2 \pi \delta(\omega_0 - \omega_k) \qquad \text{as} \qquad \epsilon \to 0, \end{aligned} \tag{7.73}$$

which on using (7.66) and the limit $V \to \infty$ reduces to

$$\mathrm{Im}\,\Sigma = -\int \frac{2\pi\omega_k}{\hbar(2\pi)^3} \mathrm{d}^3k \sum_s |\vec{p}_{if} \cdot \vec{\epsilon}_{\vec{k}s}|^2 \pi \delta(\omega_0 - \omega_k). \tag{7.74}$$

On summing over polarizations, this reduces to

$$\mathrm{Im}\,\Sigma = -\frac{A}{2}, \qquad A = \frac{4|p_{fi}|^2\omega_0^3}{3\hbar c^3}. \tag{7.75}$$

Therefore (7.70) reduces to

$$\hat{c}_i(p) = \left[p + i\left(\omega_0 - i\frac{A}{2}\right)\right]^{-1}. \tag{7.76}$$

On inverting the Laplace transform, the probability amplitude for the initial state becomes

$$c_i(t) = \exp\left(-i\omega_0 t - \frac{A}{2}t\right). \tag{7.77}$$

This expression for the decay of the initial state has been derived from first principles. The advantage of the Weisskopf–Wigner method is that it also enables us to calculate the spectral distribution of the emitted field. On combining (7.76) and (7.68) and on inverting

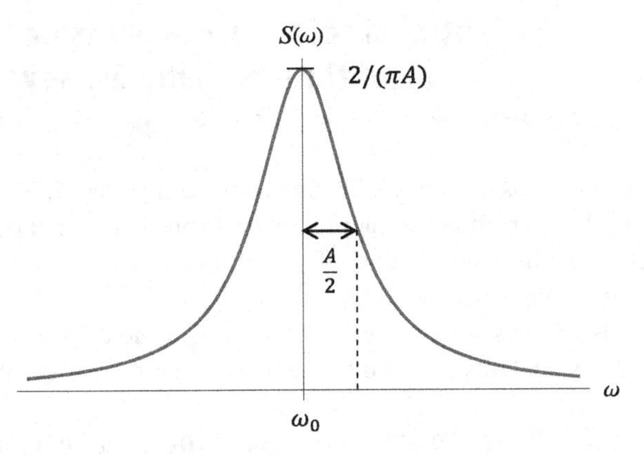

Fig. 7.6 The spectral distribution $S(\omega)$ of the radiation.

the Laplace transform we find

$$c_{\vec{k}s}(t) \Rightarrow -\,\mathrm{i}g^*_{\vec{k}s}\,\mathrm{e}^{-\mathrm{i}\omega_k t}\left[-\mathrm{i}\omega_k + \mathrm{i}\left(\omega_0 - \mathrm{i}\frac{A}{2}\right)\right]^{-1}$$
$$-\,\mathrm{i}g^*_{\vec{k}s}\,\mathrm{e}^{-\mathrm{i}\omega_0 t - \frac{A}{2}t}\left[\mathrm{i}\omega_k - \mathrm{i}\left(\omega_0 - \mathrm{i}\frac{A}{2}\right)\right]^{-1}, \tag{7.78}$$

and hence in the long time limit $t \to \infty$ we obtain

$$|c_{\vec{k}s}|^2 = |g_{\vec{k}s}|^2 \Big/ \left[(\omega_0 - \omega_k)^2 + \frac{A^2}{4}\right]. \tag{7.79}$$

This is the probability of finding a photon in the mode $\vec{k}s$ in the long time limit. The directional dependence of the emitted radiation is contained in the factor $|g_{\vec{k}s}|^2$. On summing over all the directions and polarizations, we obtain the spectral distribution $S(\omega)$ of the radiation

$$\sum_{\vec{k}s} |c_{\vec{k}s}|^2 = 1 = \int_0^{+\infty} S(\omega)\mathrm{d}\omega \approx \int_{-\infty}^{+\infty} S(\omega)\mathrm{d}\omega, \tag{7.80}$$

$$S(\omega) = \frac{A/2\pi}{(\omega_0 - \omega)^2 + (A/2)^2}. \tag{7.81}$$

The spectral distribution of the emitted radiation is Lorentzian centered at the atomic transition frequency (except for the Lamb shift) with a full width at half maximum equal to A, as shown in Figure 7.6. The peak height is $2/(\pi A)$. Thus the basic characteristics of the spectral distribution are governed by the Einstein A coefficient.

The Weisskopf–Wigner wavefunction (7.64) has the property that, for any finite t ($t \neq 0$), it does not factorize in terms of the wavefunctions for the atom and field, indicating the existence of atom–photon entanglement in the transient domain. The importance of this atom–photon entanglement has been realized and can produce atom–atom entanglement [9–13] using appropriate experimental setup. This is treated in detail in Chapter 15.

7.7 Control of spontaneous emission through the design of the electromagnetic vacuum

In the previous section we obtained the exact result (7.70) for the amplitude of the excited state. We then found an approximate expression for the emission in free space. In this section we discuss the case of a structured vacuum, which is the case for emission in a cavity or emission by an atom embedded in a photonic crystal. There are a number of models of the structured vacuum for which the $c_i(p)$ defined by (7.70) can be evaluated explicitly, and this can lead to methods for control of spontaneous emission.

7.7.1 Purcell enhancement of spontaneous emission by an atom in a cavity

From (7.70) it is clear that $c_i(t)$ would have the form

$$c_i(t) = \mathrm{e}^{-\mathrm{i}\omega_0 t}\tilde{c}(t), \tag{7.82}$$

where the Laplace transform of $\tilde{c}$ will be

$$\hat{\tilde{c}}(p) = \left[p + \sum_{\vec{k}s} |g_{\vec{k}s}|^2 (p + \mathrm{i}\omega_k - \mathrm{i}\omega_0)^{-1}\right]^{-1}. \tag{7.83}$$

In a single-mode cavity with frequency ω_c the polarization index is irrelevant. Similarly $\vec{k}$ becomes one dimensional. The summation in the above denominator can be converted to an integral

$$\begin{aligned} &\sum_{\vec{k}s} |g_{\vec{k}s}|^2 (p + \mathrm{i}\omega_k - \mathrm{i}\omega_0)^{-1} \\ &\to \int \mathrm{d}\omega \frac{\kappa/\pi}{(\omega - \omega_c)^2 + \kappa^2} \times \frac{|g|^2}{p + \mathrm{i}\omega - \mathrm{i}\omega_0} \\ &= \frac{|g|^2}{p + \kappa - \mathrm{i}\delta}, \qquad \delta = \omega_0 - \omega_c, \quad |g|^2 = \frac{2\pi\omega}{\hbar V}|p|^2. \end{aligned} \tag{7.84}$$

The factor $\kappa/\{\pi[(\omega-\omega_c)^2+\kappa^2]\}$ gives the mode spectrum in the cavity. For an ideal cavity $\kappa \to 0$ and the mode spectrum reduces to $\delta(\omega - \omega_c)$ as it should. The parameter 2κ is the leakage rate of photons from the cavity. The parameter g has the dimensions of frequency. On substituting (7.84) in (7.83) we get

$$\hat{\tilde{c}}(p) = \left(p + \frac{|g|^2}{p + \kappa - \mathrm{i}\delta}\right)^{-1} = \frac{p + \kappa - \mathrm{i}\delta}{p(p + \kappa - \mathrm{i}\delta) + |g|^2}. \tag{7.85}$$

The full consequences of (7.85) for arbitrary values of $|g|$, κ, δ will be discussed in Chapter 12. Here we discuss a simpler case which yields the celebrated result of Purcell [14]. The poles of (7.85) are given by

$$p = -\frac{\kappa - \mathrm{i}\delta}{2} \pm \frac{1}{2}\sqrt{(\kappa - \mathrm{i}\delta)^2 - 4|g|^2}, \tag{7.86}$$

which on resonance $\delta = 0$ reduce to

$$p = -\frac{\kappa}{2} \pm \frac{1}{2}\sqrt{\kappa^2 - 4|g|^2}, \\ \to -\kappa; -|g|^2/\kappa, \qquad \text{for } \kappa^2 \gg 4|g|^2. \tag{7.87}$$

On substituting (7.87) in (7.85) we obtain for the case $\kappa^2 \gg 4|g|^2$

$$\hat{\tilde{c}}(p) \approx \frac{1}{p + |g|^2/\kappa} \Rightarrow \tilde{c}(t) = \exp\left(-\frac{|g|^2}{\kappa}t\right). \tag{7.88}$$

The excited state of the atom in the cavity decays at the rate $\Gamma_c = 2|g|^2/\kappa$, which is the classical result of Purcell. Note that the decay rate in the cavity is enhanced over that in free space. To see this we calculate the ratio Γ_c/A by using (7.84) and the definition of A (Eq. (7.22))

$$\Gamma_c/A = \frac{3\lambda^3 Q}{4\pi^2 V}, \tag{7.89}$$

where Q is called the quality factor of the cavity given by $Q = \omega_c/2\kappa$. For an experimental verification of Purcell's result see [15]. Goy *etal.* [15] observed the decay of the Rydberg state $23S$ to $22P$ in Na in a cavity with a quality factor given by $\omega_c/Q = 2.8 \times 10^6\ \mathrm{s}^{-1}$. They found the decay rate in the cavity to be $\Gamma_c = 8 \times 10^4\ \mathrm{s}^{-1}$, whereas $A = 150\ \mathrm{s}^{-1}$, and thus a cavity enhancement factor of about five hundred.

The limit $\kappa \gg 2|g|$ is known as the bad cavity limit – this means that the emitted photon leaks out of the cavity and does not get reabsorbed by the atom. A similar analysis in the limit of a detuned bad cavity yields

$$\tilde{c}(t) = \exp\left(-\frac{|g|^2 t}{\kappa - i\delta}\right). \tag{7.90}$$

The decay rate in the detuned cavity would be $2|g|^2\kappa/(\kappa^2 + \delta^2)$, leading to an inhibition of decay as the detuning increases [16, 17]. Furthermore, in a detuned cavity the effective frequency changes – there is a frequency shift of $\delta|g|^2/(\kappa^2 + \delta^2)$. Such a frequency shift has many important consequences and will be discussed in Chapter 12.

7.7.2 Quantum memory effects in spontaneous emission in a photonic crystal

Another case where a structured continuum is very important and where the population can be trapped in excited states arises in the context of a photonic crystal environment. This has been extensively studied by John and collaborators [18, 19]. The situation is shown in Figure 7.7. The photonic bath has a dispersion relation of the form [19]

$$\omega_{\vec{k}} = \omega_c + A(\vec{k} - \vec{k}_0)^2, \tag{7.91}$$

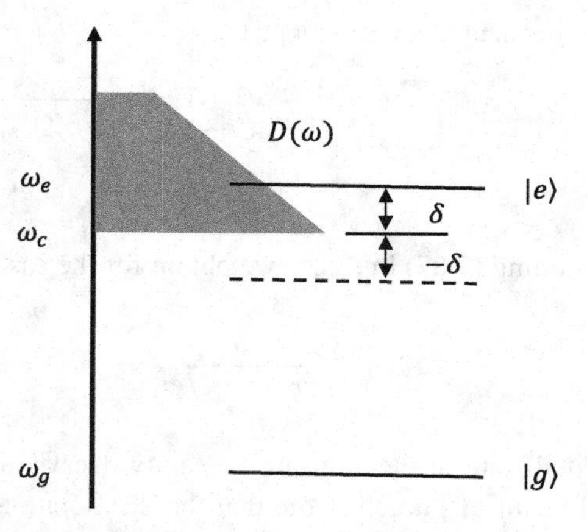

Fig. 7.7 A two-level atom embedded in an ideal photonic bandgap structure whose density of states $D(\omega)$ is shown by the shaded region. The excited level can be positively/negatively detuned δ from the band edge given by ω_c whereby the atomic transition can lie in the bandgap or outside it.

where ω_c is the upper-band edge frequency and $\vec{k}_0$ is related to the point group symmetry of the crystal. In this case one finds that [20–22]

$$\tilde{c}(p) = \frac{1}{p + \alpha e^{i\pi/4}\sqrt{p - i\delta}}, \qquad \delta = \omega_0 - \omega_c, \tag{7.92}$$

$$\alpha \approx \frac{|p_{if}|^2 \omega_0^2}{3\hbar\omega_c A^{3/2}}. \tag{7.93}$$

The Laplace transform (7.92) can be inverted in terms of the error functions with the result

$$\tilde{c}(t) = e^{i\delta t}\frac{1}{a_- - a_+}\left[-a_+ e^{a_+^2 t}\operatorname{erfc}(a_+\sqrt{t}) + a_- e^{a_-^2 t}\operatorname{erfc}(a_-\sqrt{t})\right],$$
$$a_\pm = \frac{1}{2}e^{i\pi/4}(\alpha \mp \sqrt{\alpha^2 - 4\delta}), \tag{7.94}$$

where the function erfc(z) is defined by

$$\operatorname{erfc}(z) = \frac{2}{\sqrt{\pi}}\int_z^\infty e^{-y^2}dy. \tag{7.95}$$

We show in Figure 7.8 the behavior of the population in the excited state $|\tilde{c}(t)|^2$ as a function of $(\alpha^2 t)$ for the two cases when the transition frequency of the atom lies inside the bandgap and outside the bandgap. We find the possibility of population trapping [18] in the exited state if the atomic transition frequency lies inside the bandgap. This is expected as the atom has no density of modes into which it can decay. Leistikow *etal*. [23] report inhibited spontaneous emission by quantum dots in the photonic band gap crystal. They report an inhibition factor of ten when the dot's emission frequency is inside the gap.

From the above two models it follows that the structured continuum has important implications for spontaneous emission. The structured continuum can be designed and thus one can control the spontaneous emission.

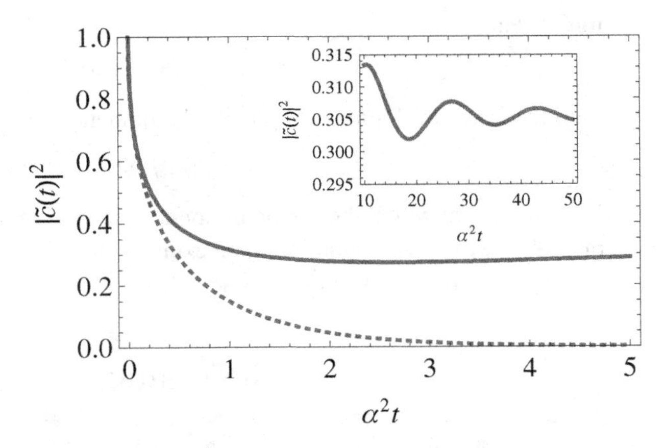

Fig. 7.8 The population in the excited state $|\tilde{c}(t)|^2$ as a function of $\alpha^2 t$ for $\delta = \alpha^2$ (dotted) and $\delta = -\alpha^2$ (solid). The inset shows $|\tilde{c}(t)|^2$ against t from 10 to 50 for $\delta = -\alpha^2$.

Exercises

7.1 Using Eq. (7.14) show that

$$\lim_{t\to\infty} \frac{\mathrm{d}}{\mathrm{d}t} \mathcal{P}_{if}(t) = \Gamma_{if}^{\mathrm{abs}},$$

where $\Gamma_{if}^{\mathrm{abs}}$ is given by Eq. (7.15).

7.2 Prove the relation (7.30).

7.3 Show that the commutator $[E_\alpha^{(+)}(\vec{r}_1, t_1), E_\beta^{(-)}(\vec{r}_2, 0)]$, where the operators $E^{(-)}, E^{(+)}$ are defined by Eq. (7.7), is a number. Find explicitly the value of the Fourier transform of the commutator

$$\int_{-\infty}^{+\infty} \mathrm{d}t_1\, \mathrm{e}^{\mathrm{i}\omega t_1} \left[E_\alpha^{(+)}(\vec{r}_1, t_1), E_\beta^{(-)}(\vec{r}_2, 0)\right].$$

7.4 Estimate the value (7.40) of α_0 for D_1 line of Na given the vapor density of Na as 10^{10} atoms/cm^3.

7.5 Show that in spontaneous emission the mean value of the atomic dipole moment remains zero. Show this using the wave function (7.64), i.e. show that $\langle\Psi(t)|\vec{p}|\Psi(t)\rangle = 0$, where $\vec{p}$ is the dipole moment operator. In Chapter 17, we will show how spontaneous emission can produce atomic coherence in a multilevel system.

7.6 Using the wave function (7.64) find the reduced density operator for the atom defined by $\rho_A(t) = \mathrm{Tr}_F|\Psi(t)\rangle\langle\Psi(t)|$, where Tr_F is the trace over all the field states. Show that $\rho_A(t)$ is a mixed state, i.e. $\mathrm{Tr}\rho_A^2(t) < 1$ for all $t \neq 0$ and ∞.

7.7 Redo the calculation of Section 7.6 for an atom with initial atomic coherence nonzero, i.e. the initial dipole moment is nonzero. Thus instead of (7.62) consider the initial state

$$(\alpha|i\rangle + \beta|f\rangle)|\{0_{\vec{k}s}\}\rangle.$$

Find the time development of the dipole moment

$$\langle\Psi(t)|\vec{p}|\Psi(t)\rangle.$$

Find the rate at which the dipole moment decays. How is this rate different from the rate of decay of the population in an excited state?

7.8 Prove the result (7.92) (for details see [22]).

References

[1] R. J. Glauber, *Phys. Rev.* **130**, 2529 (1963).

[2] R. J. Glauber, in *Quantum Optics and Electronics*, edited by C. deWitt, A. Blanden, and C. Cohen-Tannoudji (New York: Gordon and Breach, 1965), p. 65.

[3] http://steck.us/alkalidata.

[4] S. E. Harris, *Phys. Rev. Lett.* **62**, 1033 (1989).

[5] V. G. Arkhipkin and Yu. I. Heller, *Phys. Lett.* **98**A, 12 (1983).

[6] O. A. Kocharovskaya and Ya. I. Khanin, *Pis'ma Zh. Eksp. Teor. Fiz.* **48**, 581 (1988).

[7] V. Weisskopf and E. Wigner, *Z. Phys.* **63**, 54 (1930); *ibid.* **65**, 18 (1930).

[8] J. E. Bjorkholm and P. F. Liao, *Phys. Rev. Lett.* **33**, 128 (1974).

[9] M. V. Fedorov, M. A. Efremov, A. E. Kazakov, K. W. Chan, C. K. Law, and J. H. Eberly, *Phys. Rev. A* **72**, 032110 (2005).

[10] C. Thiel, J. von Zanthier, T. Bastin, E. Solano, and G. S. Agarwal, *Phys. Rev. Lett.* **99**, 193602 (2007).

[11] W. Wieczorek, N. Kiesel, C. Schmid, W. Laskowski, M. Żukowski, and H. Weinfurter, *IEEE J. Sel. Topics Quantum Electron.* **15**, 1704 (2009).

[12] R. Prevedel, G. Cronenberg, M. S. Tame *et al., Phys. Rev. Lett.* **103**, 020503 (2009).

[13] W. Wieczorek, R. Krischek, N. Kiesel, P. Michelberger, G. Toth, and H. Weinfurter, *Phys. Rev. Lett.* **103**, 020504 (2009).

[14] E. M. Purcell, *Phys. Rev.* **69**, 681 (1946).

[15] P. Goy, J. M. Raimond, M. Gross, and S. Haroche, *Phys. Rev. Lett.* **50**, 1903 (1983).

[16] D. Kleppner, *Phys. Rev. Lett.* **47**, 233 (1981).

[17] R. G. Hulet, E. S. Hilfer, and D. Kleppner, *Phys. Rev. Lett.* **55**, 2137 (1985).

[18] S. John and T. Quang, *Phys. Rev. A* **50**, 1764 (1994).

[19] S. John and J. Wang, *Phys. Rev. Lett.* **64**, 2418 (1990).

[20] F.-Q. Wang, Z.-M. Zhang, and R.-S. Liang, *Phys. Rev. A* **78**, 042320 (2008).

[21] B. Bellomo, R. Lo Franco, S. Maniscalco, and G. Compagno, *Phys. Rev. A* **78**, 060302(R) (2008).

[22] M. Woldeyohannes and S. John, *Phys. Rev. A* **60**, 5046 (1999).

[23] M. D. Leistikow, A. P. Mosk, E. Yeganegi, S. R. Huisman, A. Lagendijk, and W. L. Vos, *Phys. Rev. Lett.* **107**, 193903 (2011).

8 Partial coherence in multimode quantum fields

We have so far mostly concentrated on one or two modes of electromagnetic fields. We have discussed a variety of field states, some of which have unusual quantum properties. In general, a radiating source will produce a multimode field. We have already seen two examples of physical systems where the produced fields are genuinely multimode. These examples are (i) parametric downconverters (Chapter 3) and (ii) spontaneous emission by an atom (Chapter 7). Besides knowing the state of the multimode field, we would also like to know the kind of observables that one can introduce to characterize a multimode field. In addition, we would like to find the experimental schemes to measure these observables.

8.1 Correlation functions for electromagnetic fields

In Chapter 7 we saw that the rate of absorption of a photon from an arbitrary field is (Eq. (7.15))

$$\Gamma_{if}^{(\mathrm{abs})} = \frac{1}{\hbar^2}\int_{-\infty}^{+\infty} \mathrm{d}\tau\, \mathrm{e}^{\mathrm{i}\omega_{fi}\tau}\, \vec{p}_{fi}^{\,*} \cdot \overset{\Rightarrow}{\mathcal{E}}^{(N)}(\vec{R},0,\vec{R},\tau)\cdot \vec{p}_{fi}, \tag{8.1}$$

where $\overset{\Rightarrow}{\mathcal{E}}^{(N)}$ is the correlation function for the electromagnetic field defined by (7.13). In the light of (8.1), we can consider an atom to be a probe of the electromagnetic field, as absorption studies would directly yield the normally ordered correlation function of the electromagnetic field. Clearly $\overset{\Rightarrow}{\mathcal{E}}^{(N)}(\vec{R},0,\vec{R},\tau)$ becomes an observable of the electromagnetic field. Since photodetectors work by absorption of light, they would measure the normally ordered correlations of a quantum field.

In order to characterize the field fully we need, besides the second-order correlation function $\overset{\Rightarrow}{\mathcal{E}}^{(N)}(\vec{r}_1,t_1,\vec{r}_2,t_2)$, all higher-order correlations. A program to fully characterize classical fields in terms of correlation functions of all orders was largely developed by Wolf [1, 2]. Glauber [3–5] developed a characterization of quantum fields in terms of the normally ordered correlation functions of all orders. The normally ordered correlation functions have a special place in quantum optics as all measurements are done with photodetectors. A complete specification of a quantum field would require knowledge of all the correlations defined by

$$\mathcal{E}_{\alpha\beta}(\vec{r}_1, t_1, \vec{r}_2, t_2) = \mathrm{Tr}\left[\rho_F E_\alpha^{(-)}(\vec{r}_1, t_1) E_\beta^{(+)}(\vec{r}_2, t_2)\right], \tag{8.2}$$

$$\mathcal{E}_{\alpha\beta\gamma\delta}(\vec{r}_1, t_1, \vec{r}_2, t_2, \vec{r}_3, t_3, \vec{r}_4, t_4) = \mathrm{Tr}[\rho_F E_\alpha^{(-)}(\vec{r}_1, t_1) E_\beta^{(-)}(\vec{r}_2, t_2) E_\gamma^{(+)}(\vec{r}_3, t_3) \times E_\delta^{(+)}(\vec{r}_4, t_4)], \tag{8.3}$$

$$\vdots$$

$$A_{\alpha\beta}(\vec{r}_1, t_1, \vec{r}_2, t_2) = \mathrm{Tr}\left[\rho_F E_\alpha^{(-)}(\vec{r}_1, t_1) E_\beta^{(-)}(\vec{r}_2, t_2)\right], \tag{8.4}$$

$$\vdots$$

etc.,

where for brevity we have dropped the superscript N. Note that a correlation like $A_{\alpha\beta}$ contains information on the phase characteristics of the field and is very relevant to squeezing of the quantum field. The intensity $I(\vec{r}, t)$ of the field is related to (8.2) via

$$I(\vec{r}, t) = \mathrm{Tr}\left[\rho_F \vec{E}^{(-)}(\vec{r}, t) \cdot \vec{E}^{(+)}(\vec{r}, t)\right] = \sum_\alpha \mathcal{E}_{\alpha\alpha}(\vec{r}, t, \vec{r}, t). \tag{8.5}$$

As discussed earlier, the correlation function $\mathcal{E}_{\alpha\beta}(\vec{r}_1, t_1, \vec{r}_1, t_2)$ is relevant to the absorption of the field by an atom. The information on the spatial coherence of the field is contained in the function $\mathcal{E}_{\alpha\beta}(\vec{r}_1, t_1, \vec{r}_2, t_1)$ for $\vec{r}_1 \neq \vec{r}_2$. In the next section we will see how $\mathcal{E}_{\alpha\beta}(\vec{r}_1, t_1, \vec{r}_2, t_2)$ determines the interference pattern in the Young's interferometer. The correlation function (8.3) is most important in the study of the nonclassical properties of fields. For classical fields, a special form of (8.3) is relevant to Hanbury-Brown–Twiss interferometry (Section 8.3).

For coherent fields, i.e. fields in coherent states,

$$\rho_F = |\Psi\rangle\langle\Psi|, \qquad |\Psi\rangle = \prod_{\vec{k}s} |\alpha_{\vec{k}s}\rangle \equiv |\{\alpha_{\vec{k}s}\}\rangle, \tag{8.6}$$

all normally ordered correlation functions factorize

$$\begin{aligned}
\mathcal{E}_{\alpha\beta}(\vec{r}_1, t_1, \vec{r}_2, t_2) &= E_\alpha^*(\vec{r}_1, t_1) E_\beta(\vec{r}_2, t_2), \\
\mathcal{E}_{\alpha\beta\gamma\delta}(\vec{r}_1, t_1, \vec{r}_2, t_2, \vec{r}_3, t_3, \vec{r}_4, t_4) &= E_\alpha^*(\vec{r}_1, t_1) E_\beta^*(\vec{r}_2, t_2) E_\gamma(\vec{r}_3, t_3) E_\delta(\vec{r}_4, t_4), \\
A_{\alpha\beta}(\vec{r}_1, t_1, \vec{r}_2, t_2) &= E_\alpha^*(\vec{r}_1, t_1) E_\beta^*(\vec{r}_2, t_2),
\end{aligned} \tag{8.7}$$

where

$$\vec{E}_\alpha(\vec{r}, t) = \mathrm{i} \sum_{\vec{k}s} \sqrt{\frac{2\pi\hbar\omega_k}{V}} \vec{\epsilon}_{\vec{k}s} \alpha_{\vec{k}s} \mathrm{e}^{\mathrm{i}\vec{k}\cdot\vec{r} - \mathrm{i}\omega_k t}. \tag{8.8}$$

More generally if we write the density operator in terms of the Glauber–Sudarshan P function [3, 6]

$$\rho_F = \int P(\{\alpha_{\vec{k}s}\}) |\{\alpha_{\vec{k}s}\}\rangle\langle\{\alpha_{\vec{k}s}\}| \mathrm{d}^2\{\alpha_{\vec{k}s}\}, \tag{8.9}$$

then correlations can be written in terms of the fields (8.8) and the averages over the distribution $P(\{\alpha_{\vec{k}s}\})$, which can be singular and need not exist. For example

$$\begin{aligned}&\mathcal{E}_{\alpha\beta\gamma\delta}(\vec{r}_1,t_1,\vec{r}_2,t_2,\vec{r}_3,t_3,\vec{r}_4,t_4)\\&\quad=\int P(\{\alpha_{\vec{k}s}\})E_\alpha^*(\vec{r}_1,t_1)E_\beta^*(\vec{r}_2,t_2)E_\gamma(\vec{r}_3,t_3)E_\delta(\vec{r}_4,t_4)\mathrm{d}^2\{\alpha_{\vec{k}s}\}.\end{aligned}\tag{8.10}$$

In classical coherence theory the correlations are calculated by a formula similar to (8.10) but with $P(\{\alpha_{\vec{k}s}\})$ replaced by a genuine probability distribution. Hence in cases where $P(\{\alpha_{\vec{k}s}\})$ is a genuine probability distribution, then the distinction between classical and quantum descriptions of coherence vanishes. This is the content of Sudarshan's optical equivalence theorem [6]. This, for example, is the case for black-body radiation or more generally for thermal fields characterized by the density matrix

$$\begin{aligned}\rho_F=\prod_{\vec{k}s}\rho_{\vec{k}s},\qquad \rho_{\vec{k}s}&=\sum_{n_{\vec{k}s}}p(n_{\vec{k}s})|n_{\vec{k}s}\rangle\langle n_{\vec{k}s}|,\\ p(n_{\vec{k}s})&=(\bar{n}_{\vec{k}s})^{n_{\vec{k}s}}/(1+\bar{n}_{\vec{k}s})^{n_{\vec{k}s}+1},\end{aligned}\tag{8.11}$$

where for the black-body field we have

$$\bar{n}_{\vec{k}s}=1/[\exp(\hbar\omega_k/K_BT)-1].\tag{8.12}$$

In this case, $P(\{\alpha_{\vec{k}s}\})$ is a genuine probability distribution

$$P(\{\alpha_{\vec{k}s}\})=\prod_{\vec{k}s}\frac{1}{\pi\bar{n}_{\vec{k}s}}\exp\left(-\frac{|\alpha_{\vec{k}s}|^2}{\bar{n}_{\vec{k}s}}\right).\tag{8.13}$$

In view of (8.13) all phase-dependent averages vanish, e.g.

$$A_{\alpha\beta}(\vec{r}_1,t_1,\vec{r}_2,t_2)=0,\qquad \langle E_\alpha^{(+)}(\vec{r}_1,t_1)\rangle=0.\tag{8.14}$$

Thermal fields have the remarkable property, known as the moment theorem, which says that all the higher-order correlations can be expressed in terms of second-order correlations. In particular, for thermal fields (8.3) becomes

$$\begin{aligned}&\mathcal{E}_{\alpha\beta\gamma\delta}(\vec{r}_1,t_1,\vec{r}_2,t_2,\vec{r}_3,t_3,\vec{r}_4,t_4)\\&\quad=\mathcal{E}_{\alpha\gamma}(\vec{r}_1,t_1,\vec{r}_3,t_3)\mathcal{E}_{\beta\delta}(\vec{r}_2,t_2,\vec{r}_4,t_4)+\mathcal{E}_{\alpha\delta}(\vec{r}_1,t_1,\vec{r}_4,t_4)\mathcal{E}_{\beta\gamma}(\vec{r}_2,t_2,\vec{r}_3,t_3).\end{aligned}\tag{8.15}$$

This is a very important property for thermal fields and leads to the phenomenon of photon bunching in thermal fields.

In the next few sections, we would discuss the importance of the correlation functions (8.2)–(8.4) in a variety of measurements.

8.2 Young's interferometer and spatial coherence of the field

Let us consider the Young's interferometer as shown in Figure 8.1. Here light from the source S falls on an opaque screen with two slits Q_1 and Q_2 located at the points $\vec{r}_1$ and $\vec{r}_2$

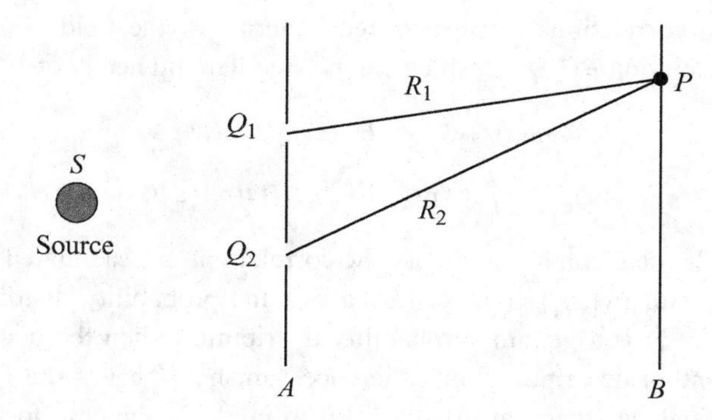

Fig. 8.1 Young's interferometer.

respectively. Let us place a detector on the screen B and measure the intensity of light at a point $\vec{r}(P)$ across B. Let us assume that the source produces quasi-monochromatic light, i.e. the field $E^{(+)}(\vec{r}, t)$ produced by S is centered around some frequency ω. We further assume that the field is stationary in a statistical sense. In this case correlations such as $\mathcal{E}_{\alpha\beta}(\vec{r}_1, t_1, \vec{r}_2, t_2)$ are independent of the origin of time, i.e.

$$\mathcal{E}_{\alpha\beta}(\vec{r}_1, t_1 + \tau, \vec{r}_2, t_2 + \tau) = \mathcal{E}_{\alpha\beta}(\vec{r}_1, t_1, \vec{r}_2, t_2) \qquad \forall\, \tau. \tag{8.16}$$

The intensity of the field, defined by $I(\vec{r}) = \sum_\alpha \langle E_\alpha^{(-)}(\vec{r}, t) E_\alpha^{(+)}(\vec{r}, t)\rangle$, would be independent of time t. For simplicity, let us assume a scalar description of light. We follow the treatment given in standard books [1, 2, 5] except that we replace classical fields by the corresponding operators. Using Huygens–Fresnel theory the field at the point P can be written in terms of the fields at the slits

$$E^{(+)}(\vec{r}, t) = K_1 E^{(+)}(\vec{r}_1, t - t_1) + K_2 E^{(+)}(\vec{r}_2, t - t_2), \tag{8.17}$$

where

$$t_1 = \frac{R_1}{c}, \qquad t_2 = \frac{R_2}{c}, \qquad K_j \approx -\frac{\mathrm{i}}{\lambda R_j} \mathrm{d}\mathcal{A}_j. \tag{8.18}$$

Here $\mathrm{d}\mathcal{A}_j$ is the area of the jth pinhole (slit) and it is assumed that the angles of incidence and diffraction at the pinhole are fairly small. The intensity at the point P would then be

$$\begin{aligned} I(\vec{r}, t) = {} & |K_1|^2 I(\vec{r}_1) + |K_2|^2 I(\vec{r}_2) \\ & + \left[K_1^* K_2 \langle E^{(-)}(\vec{r}_1, t - t_1) E^{(+)}(\vec{r}_2, t - t_2)\rangle + c.c.\right]. \end{aligned} \tag{8.19}$$

Let us introduce the mutual coherence function defined by

$$\Gamma(\vec{r}_1, \vec{r}_2, \tau) = \langle E^{(-)}(\vec{r}_1, t) E^{(+)}(\vec{r}_2, t + \tau)\rangle. \tag{8.20}$$

Then (8.19) can be written as

$$\begin{aligned} I(\vec{r}, t) &= |K_1|^2 I(\vec{r}_1) + |K_2|^2 I(\vec{r}_2) + 2|K_1||K_2| \mathrm{Re}[\Gamma(\vec{r}_1, \vec{r}_2, \tau)], \\ \tau &= t_1 - t_2. \end{aligned} \tag{8.21}$$

This is the basic interference law and shows how the interference depends on the coherence properties of the field. Note that $|K_1|^2 I(\vec{r}_1)$ $[|K_2|^2 I(\vec{r}_2)]$ is the intensity at the screen $I^{(1)}(P)$ $[I^{(2)}(P)]$ if pinhole Q_2 $[Q_1]$ were closed. The last term in (8.21) is the interference term. We can rewrite (8.21) in a more transparent form as

$$I(P) = I^{(1)}(P) + I^{(2)}(P) + 2\mathrm{Re}\left[\sqrt{I^{(1)}(P)I^{(2)}(P)} \cdot \gamma(\vec{r}_1, \vec{r}_2, \tau)\right], \tag{8.22}$$

where γ is the normalized coherence function

$$\gamma(\vec{r}_1, \vec{r}_2, \tau) = \frac{\Gamma(\vec{r}_1, \vec{r}_2, \tau)}{\sqrt{I(\vec{r}_1)I(\vec{r}_2)}}, \tag{8.23}$$

which, in general, is complex. Using the Cauchy–Schwarz inequality it is easily shown that (see Exercise 8.1)

$$0 \leq |\gamma| \leq 1. \tag{8.24}$$

For $|\gamma| = 0$, there is no interference. The fields at $\vec{r}_1$ and $\vec{r}_2$ are incoherent. For $|\gamma| = 1$, the fields are coherent. This follows from (8.7), which gives the form of (8.20) for a field in a coherent state. One can introduce two different measures of coherence depending on the nonzero values of the coherence function γ. We can define a coherence time τ_c as the time over which the function $\gamma(\vec{r}, \vec{r}, \tau)$ is reasonably different from zero. Furthermore, one can define a spatial coherence parameter, i.e. a coherence length l_c, which is the length over which the function $\gamma(\vec{r}, \vec{r}+\vec{l}, 0)$ is different from zero. It is more difficult to characterize the behavior of the full space time function (8.23). It should be borne in mind that the coherence functions propagate in free space according to the Maxwell equations and that the spatial and temporal properties become interlinked as a result of propagation.

If fields were strictly monochromatic, then $\Gamma \propto \mathrm{e}^{-\mathrm{i}\omega(t_1-t_2)}$. For quasi-monochromatic fields, γ would have the form

$$\gamma(\vec{r}_1, \vec{r}_2, \tau) = |\gamma(\vec{r}_1, \vec{r}_2, \tau)| \exp\left[\mathrm{i}\Phi(\vec{r}_1, \vec{r}_2, \tau) - \mathrm{i}\omega\tau\right], \tag{8.25}$$

where $|\gamma|$ and Φ are slowly varying functions of τ. Using (8.25) in (8.22) we obtain

$$I(P) = I^{(1)}(P) + I^{(2)}(P) + 2\left|\gamma(\vec{r}_1, \vec{r}_2, \tau)\right|\sqrt{I^{(1)}(P)I^{(2)}(P)}\cos(\Phi - \omega\tau). \tag{8.26}$$

Clearly as τ changes, i.e. as the detector moves along the screen B, the intensity would exhibit maxima and minima. To understand maxima and minima, let us assume that the region over which the point P moves is such that τ is smaller than the coherence time of the field. The average intensities $I^{(i)}(P)$ and $|\gamma|$, Φ are expected to vary slowly. Thus the intensity pattern (8.26) as a function of τ is approximately periodic with maxima and minima given by

$$\begin{aligned} I_{\max} &= I^{(1)} + I^{(2)} + 2|\gamma|\sqrt{I^{(1)}I^{(2)}}, \\ I_{\min} &= I^{(1)} + I^{(2)} - 2|\gamma|\sqrt{I^{(1)}I^{(2)}}, \end{aligned} \tag{8.27}$$

and hence the visibility of the interference fringes is

$$v = \frac{I_{\max} - I_{\min}}{I_{\max} + I_{\min}} = \frac{2\sqrt{I^{(1)}I^{(2)}}}{I^{(1)} + I^{(2)}} \cdot |\gamma(\vec{r}_1, \vec{r}_2, \tau)|. \tag{8.28}$$

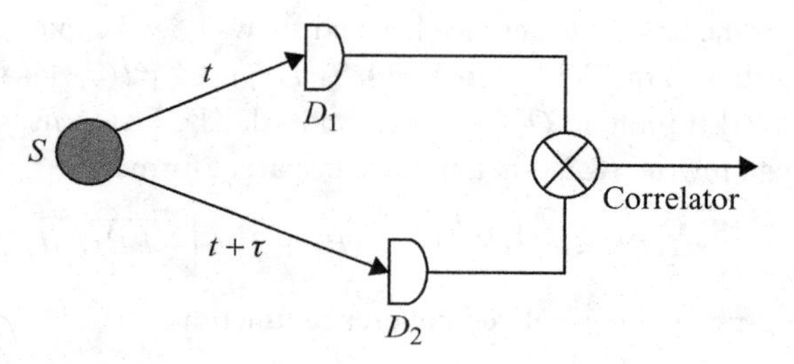

Fig. 8.2 Schematic illustration of an intensity interferometer.

Since the intensities are slow functions, the prefactor in (8.28) is approximately unity. Hence the visibility of the fringes is directly related to the degree of coherence. The Young's interferometer can thus be used to derive information on the coherence properties of the light field. Unlike Eq. (8.1), which gives only the temporal coherence function $\Gamma(\vec{r}, \vec{r}, \tau)$, the Young's interferometer also yields the spatial coherence function of the electromagnetic field.

8.3 Photon–photon correlations – intensity interferometry

One of the outstanding developments in quantum optics has been the development of intensity interferometry or the interferometry where photon–photon correlations are measured. This was the result of borrowing ideas originally introduced in the context of radio astronomy by Hanbury-Brown and Twiss [7] who showed how the measurements of intensity–intensity correlations could provide much more accurate measurements of the diameter of stars. Hanbury-Brown and Twiss were clearly dealing with light sources of thermal origin [8]. In quantum optics we deal with a wide range of sources and the intensity–intensity correlations provide us with a method of investigating the nonclassical nature of the source.

Consider an experimental arrangement schematically shown in Figure 8.2. Here the light from the source falls on two detectors. The signals from the two detectors are electronically correlated and the net current is measured. Let τ be the time delay between the signals at the detectors D_1 and D_2. The arrangement of Figure 8.2 measures the quantity

$$G^{(2)}(\vec{r}_1, \vec{r}_2, \tau) = \sum_{\alpha\beta} \langle E_\alpha^{(-)}(\vec{r}_1, t) E_\beta^{(-)}(\vec{r}_2, t+\tau) E_\beta^{(+)}(\vec{r}_2, t+\tau) E_\alpha^{(+)}(\vec{r}_1, t)\rangle. \tag{8.29}$$

If the fields were classical, then the signal (8.29) would be $\langle I(\vec{r}_1, t) I(\vec{r}_2, t+\tau)\rangle$. The expression (8.29) is the intensity–intensity correlation for quantum fields.

For fields of thermal origin we use the moment theorem and the relation $\langle E_\alpha^{(+)}\rangle = 0$ to simplify (8.29) to

$$\begin{aligned} G^{(2)}(\vec{r}_1, \vec{r}_2, \tau) &= I(\vec{r}_1, t) I(\vec{r}_2, t+\tau) + \sum_{\alpha\beta} \langle E_\alpha^{(-)}(\vec{r}_1, t) E_\beta^{(+)}(\vec{r}_2, t+\tau)\rangle \\ &\quad \times \langle E_\beta^{(-)}(\vec{r}_2, t+\tau) E_\alpha^{(+)}(\vec{r}_1, t)\rangle, \end{aligned} \tag{8.30}$$

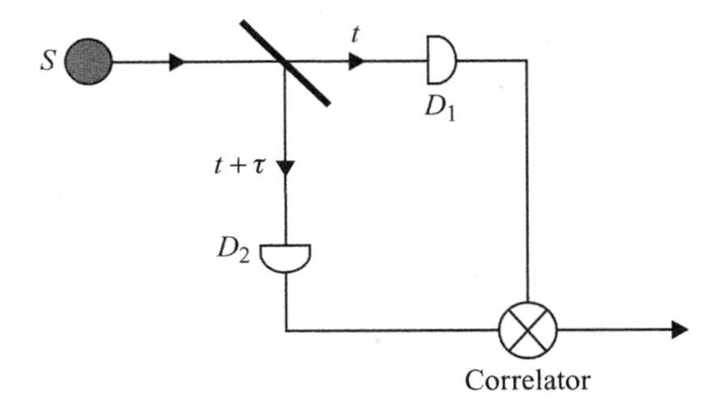

Fig. 8.3 Measurement scheme to test the nonclassicality of the field.

where I is defined by (8.5). The two correlation functions in the second term in (8.30) are complex conjugates of each other and hence

$$G^{(2)}(\vec{r}_1, \vec{r}_2, \tau) = I(\vec{r}_1, t)I(\vec{r}_2, t+\tau) + \sum_{\alpha\beta} \left|\langle E_\alpha^{(-)}(\vec{r}_1, t)E_\beta^{(+)}(\vec{r}_2, t+\tau)\rangle\right|^2. \tag{8.31}$$

It is interesting to note that $G^{(2)}$ involves the same correlation function (8.20) that appears in the Young's interferometer. However, $G^{(2)}$ does not depend on the phase of the correlation function and this was the great advantage of using intensity interferometry. Note further that if the fields at $\vec{r}_1$ and $\vec{r}_2$ were coherent, then

$$G^{(2)}(\vec{r}_1, \vec{r}_2, \tau) = I(\vec{r}_1, t)I(\vec{r}_2, t+\tau). \tag{8.32}$$

Therefore $G^{(2)}$ for thermal fields is in excess over the value for coherent fields of the same average intensity. This excess has been referred to as the phenomenon of photon bunching in thermal fields. The amount of bunching depends on the spatial and temporal coherence of the field.

Let us next consider the single-mode version of $G^{(2)}$, which would be especially relevant for nonclassical fields. Consider the arrangement shown in Figure 8.3. A dynamically evolving field falls on a 50-50 beam splitter which sends light to two detectors with a relative delay τ. The outputs from D_1 and D_2 are correlated and the coincidence signal is measured. The measured signal in this simpler case will be

$$G^{(2)}(\tau) = \langle a^\dagger(t)a^\dagger(t+\tau)a(t+\tau)a(t)\rangle. \tag{8.33}$$

The signal will be a function of τ only if the field is stationary. At the classical level we have the Cauchy–Schwarz inequality for the intensity–intensity correlation function

$$\langle I(t)I(t+\tau)\rangle \leq \sqrt{\langle I^2(t)\rangle\langle I^2(t+\tau)\rangle}. \tag{8.34}$$

Note that for the single mode case $\langle I(t)\rangle = \langle a^\dagger(t)a(t)\rangle$. For stationary fields (8.34) implies

$$\langle I(t)I(t+\tau)\rangle \leq \langle I^2(t)\rangle. \tag{8.35}$$

Thus for classical fields $G^{(2)}$ must satisfy

$$G^{(2)}(\tau) \leq G^{(2)}(0). \tag{8.36}$$

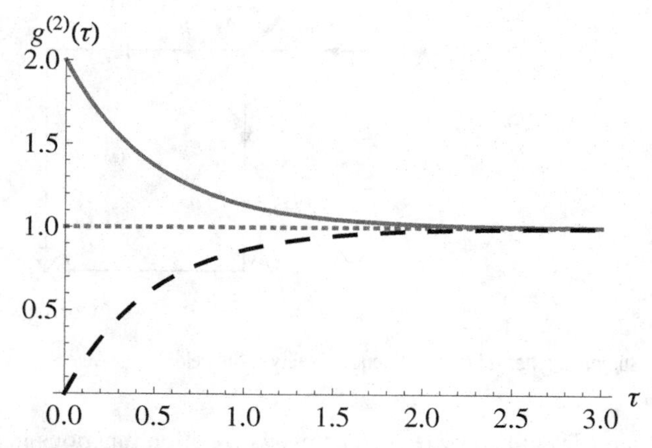

Fig. 8.4 A typical form of the normalized second-order correlation $g^{(2)}(\tau)$.

For quantum fields, the inequality (8.36) can be violated. For example, for a single-photon Fock state we have

$$G^{(2)}(0) = \langle a^{\dagger 2} a^2 \rangle = \langle 1 | a^{\dagger 2} a^2 | 1 \rangle = 0,$$

and thus (8.36) can not hold for the dynamically evolving system for all τ. Note further that

$$G^{(2)}(\tau) \to \langle a^\dagger a \rangle \langle a^\dagger a \rangle \text{ for large } \tau. \tag{8.37}$$

Fields for which (8.36) is violated, i.e. when $G^{(2)}(\tau) \geq G^{(2)}(0)$, are said to exhibit antibunching of photons. We will discuss this extensively in Chapter 13 in connection with antibunching in resonance fluorescence.

For single-mode thermal fields

$$G^{(2)}(\tau) = \langle a^\dagger a \rangle^2 + |\langle a^\dagger(t) a(t+\tau) \rangle|^2, \tag{8.38}$$

which for exponentially correlated fields has the form

$$G^{(2)}(\tau) = \langle a^\dagger a \rangle^2 (1 + \mathrm{e}^{-2\gamma\tau}), \tag{8.39}$$

for $\tau = 0$, $G^{(2)}(0) = 2\langle a^\dagger a \rangle^2$. This is the bunching property of thermal fields. The behavior of the function $g^{(2)}(\tau) = G^{(2)}(\tau)/\langle a^\dagger a \rangle^2$, can typically be of the form shown in Figure 8.4. If $g^{(2)}(\tau)$ falls below unity, then the underlying field must be nonclassical. This is shown as a dashed line in Figure 8.4. Hence a measurement of $g^{(2)}(\tau)$ can be used to determine the nonclassicality of the field. It should be borne in mind that if $g^{(2)}(\tau) > 1$, then no conclusion can be drawn on the nonclassical nature of the field. For thermal fields it follows from (8.39) that $g^{(2)}(\tau) \geq 1$.

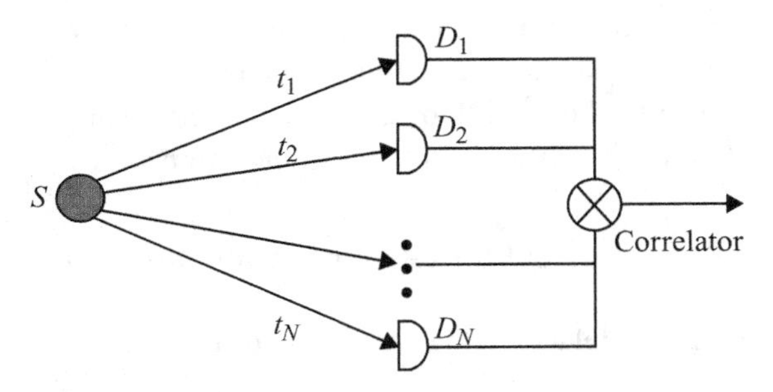

Fig. 8.5 Intensity interferometry with N detectors.

8.4 Higher-order correlation functions of the field

We have seen in the previous section how measurements of $G^{(2)}$ can provide information on the nonclassical properties of light. More can be learnt on the nonclassical properties by studying higher-order correlations. We can think of the setup like Figure 8.2 with two detectors replaced by N detectors, as shown in Figure 8.5. The measured signal in this case will be

$$G^{(N)}(\{\vec{r}_i\}, \{t_i\}) = \langle I(\vec{r}_1, t_1) \cdots I(\vec{r}_N, t_N)\rangle. \tag{8.40}$$

For quantum fields a special ordering is to be used in (8.40): all the creation operators should appear to the left of all the annihilation operators. Furthermore, all the annihilation operators must be arranged in a time-ordered fashion, whereas all the creation operators must be in anti-time-ordered fashion. For a single-mode field, the correlation $G^{(3)}$, for example, would have the structure $\langle a^\dagger(t_1)a^\dagger(t_2)a^\dagger(t_3)a(t_3)a(t_2)a(t_1)\rangle$ where $t_3 > t_2 > t_1$. This order is dictated by the quantum-mechanical perturbation theory [9], and gives the order in which photons are detected.

Another possibility which has been extensively used in studies of statistics of the laser fields is to examine the probability of detecting n photo electrons in a counting interval T. This is given by Mandel's formula [10–12]

$$p_e(n, T) = \left\langle \left\{ \exp\left[-\eta \int_0^T I(\tau)\mathrm{d}\tau\right] \frac{\left[\eta \int_0^T I(\tau)\mathrm{d}\tau\right]^n}{n!} \right\} \right\rangle, \tag{8.41}$$

where η is the quantum efficiency of the detector. For quantum fields the expression in the curly bracket in (8.41) is subject to the ordering as discussed following Eq. (8.40) [5,9]. The expression (8.41) contains information about the dynamics of the field and hence dynamical effects can be probed by studies of the photoelectron distributions. For fields with a very long coherence time compared to the counting interval, (8.41) reduces to

$$p_e(n, T) = \left\langle : \exp(-\eta T I)\frac{\eta^n T^n I^n}{n!} : \right\rangle, \tag{8.42}$$

where : : indicate that we need to arrange the operators in normal order. The derivation of the formula (8.41) is based on the assumption that the bandwidth of the detector is much bigger than the bandwidth of the field (see the result of Exercise 8.3). For a single-mode field (8.42) can be written in terms of the P-function as

$$p_e(n, T) = \int P(\alpha) \exp(-\eta T|\alpha|^2) \frac{(\eta T)^n |\alpha|^{2n}}{n!} \mathrm{d}^2\alpha. \tag{8.43}$$

This is closely related to the probability of finding n photons in the field

$$p(n) = \int P(\alpha) \exp(-|\alpha|^2) \frac{|\alpha|^{2n}}{n!} \mathrm{d}^2\alpha. \tag{8.44}$$

An arrangement like that of Figure 8.5 is very popular in multiphoton interferometry and in applications to quantum information science.

8.5 Interferometry in the spectral domain

The properties of light beams can be studied in either the time domain or the frequency domain. Studies in the frequency domain are especially important when one probes matter using light beams. This is because information on the quantum mechanical states in many cases is more directly accessible in the frequency domain. We first define the Fourier transform of the field via

$$a(t) = \frac{1}{2\pi} \int_{-\infty}^{+\infty} \mathrm{d}\omega\, \mathrm{e}^{-\mathrm{i}\omega t} a(\omega), \tag{8.45}$$

$$a(\omega) = \int_{-\infty}^{+\infty} \mathrm{d}t\, \mathrm{e}^{\mathrm{i}\omega t} a(t). \tag{8.46}$$

Then the adjoint of $a(\omega)$ is

$$a^\dagger(\omega) = \int_{-\infty}^{+\infty} \mathrm{d}t\, \mathrm{e}^{-\mathrm{i}\omega t} a^\dagger(t). \tag{8.47}$$

Note that adjoint of $a(\omega)$ is different from the Fourier transform of $a^\dagger(t)$. Using (8.46) and (8.47) we calculate the correlation function in the frequency domain

$$\langle a^\dagger(\omega_1) a(\omega_2)\rangle = \int_{-\infty}^{+\infty}\int_{-\infty}^{+\infty} \mathrm{d}t_1\, \mathrm{d}t_2\, \mathrm{e}^{-\mathrm{i}\omega_1 t_1 + \mathrm{i}\omega_2 t_2} \langle a^\dagger(t_1) a(t_2)\rangle. \tag{8.48}$$

We assume that the field is stationary so that $\langle a^\dagger(t_1)a(t_2)\rangle$ is a function of $(t_2 - t_1)$ only. Writing $t_2 = t_1 + \tau$ and carrying out t_1 integration we get

$$\langle a^\dagger(\omega_1) a(\omega_2)\rangle = 2\pi\delta(\omega_1 - \omega_2) \int_{-\infty}^{+\infty} \mathrm{d}\tau\, \mathrm{e}^{\mathrm{i}\omega_2 \tau} \langle a^\dagger(t_1) a(t_1 + \tau)\rangle. \tag{8.49}$$

We thus find that for stationary fields the Fourier components ω_1 and ω_2 are uncorrelated. Let $S(\omega)$ be the spectrum of the field defined by

$$S(\omega) = \frac{1}{2\pi}\int_{-\infty}^{+\infty} \mathrm{d}\tau\, \mathrm{e}^{\mathrm{i}\omega\tau}\langle a^\dagger(t_1)a(t_1+\tau)\rangle, \tag{8.50}$$

$$\int_{-\infty}^{+\infty} S(\omega)\mathrm{d}\omega = \langle a^\dagger a\rangle. \tag{8.51}$$

In terms of the spectrum (8.49) becomes

$$\langle a^\dagger(\omega_1)a(\omega_2)\rangle = (2\pi)^2\delta(\omega_1-\omega_2)S(\omega_2). \tag{8.52}$$

The relation (8.52) is essentially the content of the Wiener–Khintchine theorem [13] well known for stochastic processes. It can be proved that $S(\omega) \geq 0$ for all ω. For exponential decay of the correlation function

$$\langle a^\dagger(t_1)a(t_1+\tau)\rangle = \mathrm{e}^{-\mathrm{i}\omega_0\tau-\gamma|\tau|}\langle a^\dagger a\rangle, \tag{8.53}$$

the spectrum is Lorentzian

$$S(\omega) = \langle a^\dagger a\rangle \frac{\gamma/\pi}{(\omega-\omega_0)^2+\gamma^2}. \tag{8.54}$$

The width of the correlation function in the time domain is $1/\gamma$, the half width in the frequency domain is γ. More generally, the widths in the time and frequency domains obey the Fourier relation

$$\Delta\tau\Delta\omega \sim 1. \tag{8.55}$$

Let us now consider the Mach–Zehnder interferometer with an input which is dynamically evolving as shown in Figure 8.6. Let t and $t+\tau$ be the times taken by the light to travel along the paths I and II respectively. A calculation similar to (5.49) would lead to the operator at the detector D_1 given by

$$a_{\text{out}}(D_1) = \frac{1}{\sqrt{2}}\left[-\frac{a(t+\tau)+\mathrm{i}b}{\sqrt{2}} - \mathrm{i}\mathrm{e}^{-\mathrm{i}\varphi}\frac{b+\mathrm{i}a(t)}{\sqrt{2}}\right], \tag{8.56}$$

and hence the signal at D_1, since b is in vacuum, will be

$$\langle a^\dagger_{\text{out}}a_{\text{out}}\rangle = \frac{1}{2}\langle a^\dagger a\rangle - \frac{1}{4}(\langle a^\dagger(t)a(t+\tau)\rangle\mathrm{e}^{\mathrm{i}\varphi} + c.c.). \tag{8.57}$$

The existence of the interference depends on the value of τ relative to the coherence time τ_c of the field. If $\tau \gg \tau_c$, then the interference pattern disappears as $\langle a^\dagger(t)a(t+\tau)\rangle \to 0$. It was realized that even in this case the interferences can be seen if we carry out a different kind of measurement at the detector D_1. Let us consider a measurement of the spectrum at D_1. The Fourier transform of a_{out} is

$$a_{\text{out}}(\omega) = -\frac{1}{2}a(\omega)\mathrm{e}^{-\mathrm{i}\omega\tau} + \frac{1}{2}\mathrm{e}^{-\mathrm{i}\varphi}a(\omega) + b\,\text{terms}. \tag{8.58}$$

Clearly the spectrum of the field at the detector D_1 is

$$S_{\text{out}}(\omega) = \frac{S(\omega)}{4}|\mathrm{e}^{-\mathrm{i}\varphi} - \mathrm{e}^{-\mathrm{i}\omega\tau}|^2 = S(\omega)\sin^2\left(\frac{\varphi-\omega\tau}{2}\right), \tag{8.59}$$

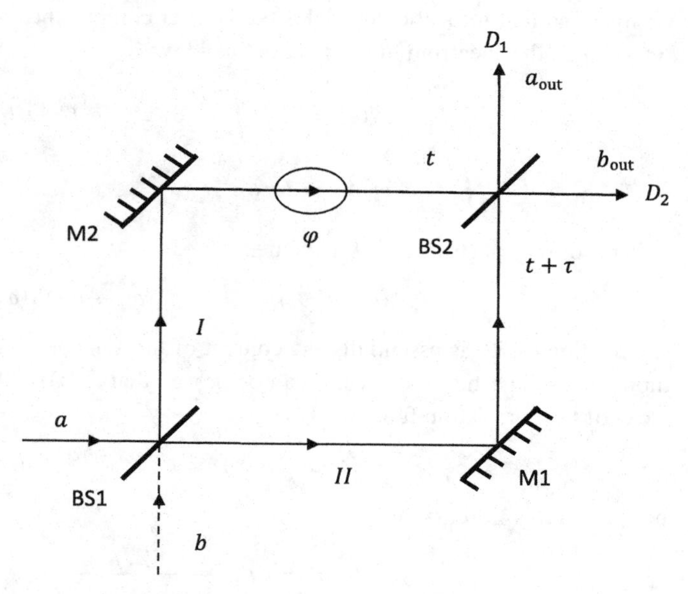

Fig. 8.6 The Mach–Zehnder interferometer with an input which is dynamically evolving.

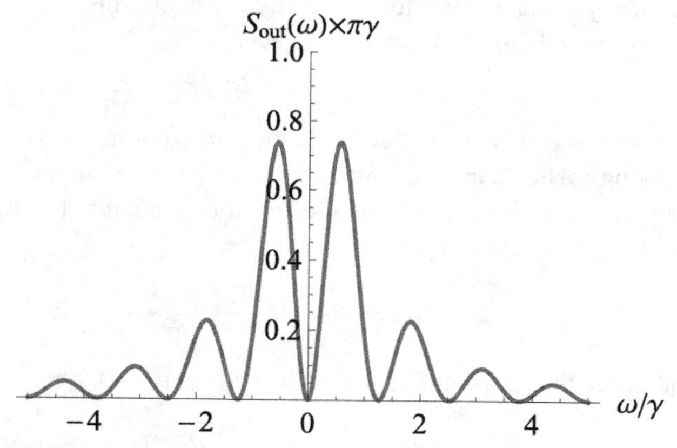

Fig. 8.7 The scaled spectrum $S_{out}(\omega) \times \pi\gamma$ of the field at the detector D_1 as a function of ω/γ for $\gamma\tau = 5$.

where $S(\omega)$ is the spectrum of the input field a. The spectrum would display modulations [14–17] as a function of ω for fixed delay τ even if there is no object ($\varphi = 0$) in one of the arms of the interferometer. The interference is displayed in Figure 8.7 for the Lorentzian spectrum (coherence time γ^{-1}) of the input field when (8.59) becomes

$$S_{out}(\omega) = \frac{\gamma/\pi}{\gamma^2 + (\omega - \omega_0)^2} \sin^2\left(\frac{\omega\tau}{2}\right) \equiv \frac{1/(\pi\gamma)}{1 + x^2} \sin^2\left(\frac{x\gamma\tau}{2}\right), \quad x = \frac{\omega - \omega_0}{\gamma}. \tag{8.60}$$

We choose $\gamma\tau = 5$, i.e. we choose a delay which is five times the coherence time and assume that $\omega_0\tau$ is an integral multiple of 2π. Clearly for $\gamma\tau = 5$, a measurement in the time domain (8.57) would not yield any interference. The presence of the object $\varphi \neq 0$

would lead to a shift of the interference fringes in the spectral domain. Zou *et al.* [18] observed interference in the spectral domain using signal photons produced by two down converters. In their experiments the optical path difference was much bigger than the coherence time thereby ruling out any interference in the time domain.

The interference in the spectral domain is an example of the interference in complementary spaces that we discussed in Chapter 4 (see Section 4.1) because time and frequency are complementary variables. Finally, note that just like the function $\Gamma(\vec{r}_1, \vec{r}_2, \tau)$ (Eq. (8.20)) determines the visibility of the interference fringes in the Young's interferometer, its Fourier transform $\Gamma(\vec{r}_1, \vec{r}_2, \omega)$

$$\Gamma(\vec{r}_1, \vec{r}_2, \omega) = \frac{1}{2\pi} \int_{-\infty}^{+\infty} \mathrm{d}\tau \mathrm{e}^{\mathrm{i}\omega\tau} \Gamma(\vec{r}_1, \vec{r}_2, \tau) \tag{8.61}$$

would determine the result of a spectral measurement at the point P. This can be shown by using the relation (8.17) in the Fourier domain. Many practical applications of interference in the spectral domain can be found in [2].

8.6 Squeezing spectrum and spectral homodyne measurement

So far we have concentrated on measurements of intensity–intensity correlations. In Chapters 2 and 3 we have shown how the studies of the quadratures of the fields provide us with a method to study the phase-dependent nonclassical properties of the field. For dynamically evolving systems, this brings us naturally to the spectrum of squeezing. To study this, we define time dependent quadrature operators in analogy to (1.37) – we remove fast optical frequencies by working with slowly varying operators $a(t) = a_0(t)\mathrm{e}^{-\mathrm{i}\omega t}$, and define the two quadratures as

$$X(t) = \frac{a_0(t) + a_0^\dagger(t)}{\sqrt{2}}, \qquad Y(t) = \frac{a_0(t) - a_0^\dagger(t)}{\sqrt{2}\mathrm{i}}, \tag{8.62}$$

and their Fourier transforms

$$X(\omega) = \frac{a_0(\omega) + a_0^\dagger(-\omega)}{\sqrt{2}}, \qquad Y(\omega) = \frac{a_0(\omega) - a_0^\dagger(-\omega)}{\sqrt{2}\mathrm{i}}, \tag{8.63}$$

where $a_0^\dagger(\omega)$ is the adjoint of $a_0(\omega)$. We now need a relation analogous to (8.52) for $\langle a_0(\omega_1)a_0(\omega_2)\rangle$. We note that for a monochromatic field $a_0(t) = a_0$ and hence

$$a_0(\omega) = 2\pi a_0 \delta(\omega), \tag{8.64}$$

and therefore

$$\langle a_0(\omega_1)a_0(\omega_2)\rangle = (2\pi)^2 \langle a_0^2\rangle \delta(\omega_1)\delta(\omega_2). \tag{8.65}$$

For quasi-monochromatic fields, we assume that $\langle a_0(t)a_0(t+\tau)\rangle$ is a function of τ only. Then an argument similar to the one that led to (8.49) yields

$$\langle a_0(\omega_1)a_0(\omega_2)\rangle = 2\pi\delta(\omega_1 + \omega_2) \int_{-\infty}^{+\infty} \mathrm{d}\tau\, \mathrm{e}^{\mathrm{i}\omega_2\tau} \langle a_0(0)a_0(\tau)\rangle. \tag{8.66}$$

Using (8.66) and (8.49) we can calculate the spectrum of the quadratures (8.63). For instance, the spectrum of X-quadrature would be

$$\langle : X(\omega_1)X(\omega_2) : \rangle = 2\pi\delta(\omega_1+\omega_2)\mathrm{Re}\left[\int_{-\infty}^{+\infty} \mathrm{d}\tau\, \mathrm{e}^{\mathrm{i}\omega_2\tau}\left(\langle a_0(0)a_0(\tau)\rangle + \langle a_0^\dagger(0)a_0(\tau)\rangle\right)\right]. \tag{8.67}$$

In Section 5.18 we showed how the quadratures can be measured by the balanced homodyne setup. Our equation (5.87) already shows that the time-dependent input to the spectrum analyzer is

$$\begin{aligned} S(t) &= \mathrm{i}\varepsilon_0 a_0^\dagger(t) - \mathrm{i}\varepsilon_0^* a_0(t) = \sqrt{2}|\varepsilon_0|[X(t)\sin\varphi_L + Y(t)\cos\varphi_L], \\ a_0(t) &= \frac{X(t)+\mathrm{i}Y(t)}{\sqrt{2}}, \qquad \varepsilon_L = \varepsilon_0 \mathrm{e}^{-\mathrm{i}\omega_0 t}, \qquad a = a_0(t)\mathrm{e}^{-\mathrm{i}\omega_0 t}, \end{aligned} \tag{8.68}$$

where φ_L is the phase of ε_0. The spectrum analyzer in Figure 5.17 would thus directly measure the spectrum of the quadratures X or Y depending on the phase φ_L of the local oscillator.

8.7 Coherence effects in two-photon absorption

We next discuss how the efficiency of the optical processes can depend on the coherence and nonclassical properties of the field used to excite the system. We have already seen in Chapter 7 that the absorption of radiation by an atom depends on the temporal coherence characteristics of the exciting light (Eq. (7.15)). We then expect that higher order correlations of the light field would be relevant to multiphoton processes [19–22]. We know, for example, that in monochromatic deterministic fields, the probability of two-photon absorption is proportional to the square of the intensity. Thus more generally the two-photon absorption probability is expected to be determined by correlations like (8.3). Thus nonclassical effects are especially significant for two-photon absorption.

Before giving a general discussion, consider the absorption of two photons from a single-mode field. The two-photon absorption probability $p^{(2)}$ would be proportional to

$$p^{(2)} \propto \langle a^{\dagger 2} a^2 \rangle. \tag{8.69}$$

A very useful technique to measure the quantity $\langle a^{\dagger 2} a^2 \rangle$ was developed by Boitier *et al.* [23] using two-photon processes in semiconductors. The normally ordered correlation enters in (8.69) because we have two absorptions. In fact, the rate is given by a formula very similar to (7.56)

$$\begin{aligned} p^{(2)} &= \frac{2\pi}{\hbar^4}\left(\frac{4\pi^2\hbar^2\omega^2}{V^2}\right)|m_{fi}|^2\delta(\omega_{fi}-2\omega)\langle a^{\dagger 2}a^2\rangle, \\ m_{fi} &= \sum_l \frac{\langle f|\vec{p}\cdot\vec{\epsilon}|l\rangle\langle l|\vec{p}\cdot\vec{\epsilon}|i\rangle}{\omega_{li}-\omega}. \end{aligned} \tag{8.70}$$

Equation (8.70) is the Fermi Golden rule for a two-photon absorption process.

The matrix element for two-photon absorption is little different from that for the scattering problem as here both photons are absorbed. Thus $p^{(2)}$ can be written in terms of Mandel's $Q_{\rm M}$ parameter as

$$p^{(2)} \propto \langle a^\dagger a\rangle^2 + Q_{\rm M}\langle a^\dagger a\rangle. \tag{8.71}$$

For coherent fields $p_{\rm c}^{(2)} \propto \langle a^\dagger a\rangle^2$ as $Q_{\rm M} = 0$. Thus depending on the quantum nature of the field $p^{(2)} > p_{\rm c}^{(2)}$ or $p^{(2)} < p_{\rm c}^{(2)}$. Thermal fields lead to enhancement of $p^{(2)}$ as $\langle a^{\dagger 2}a^2\rangle = 2\langle a^\dagger a\rangle^2$, i.e. $p_{\rm th}^{(2)}/p_{\rm c}^{(2)} = 2$, provided the average photon number is same in both thermal and coherent fields. For squeezed fields (Eq. (2.20)), $p^{(2)} \approx 3p_{\rm c}^{(2)}$. On the other hand, for fields with sub-Poissonian statistics $Q_{\rm M} < 0$, we would have an inhibition of two-photon process as compared to that in coherent fields.

For absorption of photons from two different modes a and b of the field, we would have

$$p^{(2)} \propto \langle a^\dagger b^\dagger ab\rangle. \tag{8.72}$$

As discussed in Chapter 3, the nonclassical properties of the two-mode field are contained in $\langle a^\dagger b^\dagger ab\rangle$ and hence the two-photon transition would be especially sensitive to such nonclassical properties. In particular for the photon pairs produced in the down-conversion process with wavefunction

$$|\Psi\rangle = \alpha|0, 0\rangle + \beta|1, 1\rangle, \tag{8.73}$$

we obtain

$$p^{(2)} = |\beta|^2. \tag{8.74}$$

Note that the mean $\langle a^\dagger a\rangle$ is also $|\beta|^2$. Hence it has been argued [19,20] that the two-photon absorption in such a field is proportional to the intensity rather than the square of the intensity even though the probability of two-photon absorption would be negligibly small. However, one can use the nonclassical light produced by a doubly resonant OPO to have sufficient flux for two-photon absorption. This was done by Georgiades *et al.* [21] who studied the two-photon transition $6S_{1/2} \rightarrow 6P_{3/2} \rightarrow 6D_{5/2}$ in trapped Cs vapor in a MOT. They observed that the two-photon transition rate depends on the intensity as $d_1 I + d_2 I^2$ and thus for low flux varies linearly in intensity, in agreement with the predictions [19,20].

A formula for the two photon absorption rate in terms of the correlation functions of the field can be obtained from (7.48) and by following arguments similar to those that led to (7.14). Let us assume that the field is centered around some frequency $\bar{\omega}$ and that its coherence time is large so that the two-photon matrix element in (8.70) is a slow function of ω. In such a case (8.70) generalizes to [22]

$$p^{(2)} = \frac{2\pi}{\hbar^4}|m_{fi}(\bar{\omega})|^2 \int_{-\infty}^{+\infty} {\rm d}t\, {\rm e}^{{\rm i}\omega_{fi}t-\gamma|t|}\langle E^{(-)}(0)E^{(-)}(0)E^{(+)}(t)E^{(+)}(t)\rangle, \tag{8.75}$$

where γ is the half width of the final state and where we have assumed that the field is polarized with polarization $\vec{\epsilon}$. Note that the correlation function in (8.75) is different from the intensity–intensity correlation function. The form in (8.75) may be anticipated from the fact that two photons are absorbed and energy conservation requires the resonance condition $\omega_{fi} = 2\bar{\omega}$. Therefore one would expect the two annihilation operators at the same time. If $\gamma\tau_c \gg 1$, then (8.75) is reduced to (8.70). Here τ_c is the correlation time of the correlation function in (8.75). Given the fourth-order correlation function, $p^{(2)}$ can be evaluated. For thermal fields, the correlation in (8.75) can be obtained from the moment theorem. Such a correlation function is also known for fields produced by an OPA [24].

In Section 7.7, we had discussed the possibility of interference in the scattering process. Similar interferences do occur in two-photon absorption. This should be clear from the structure of the matrix element in two-photon absorption from two different beams of light with frequencies ω_1 and ω_2. In this case (8.70) is replaced by

$$p^{(2)} = \frac{2\pi}{\hbar^4}\left(\frac{4\pi^2\hbar^2\omega_1\omega_2}{V^2}\right)|m_{fi}|^2\delta(\omega_{fi}-\omega_1-\omega_2)\langle a^\dagger b^\dagger ab\rangle, \tag{8.76}$$

where m_{fi} is obtained from (8.70) by replacing ω by ω_1. Interferences in two-photon absorption were first observed in sodium using the transition $3S_{1/2} \to 3P_{1/2} \to 4D_{3/2}$, $3S_{1/2} \to 3P_{3/2} \to 4D_{3/2}$ [25]. The two-photon transition proceeds via two intermediate states $3P_{1/2}$ and $3P_{3/2}$. Thus, depending on matrix elements, if ω_1 is tuned between the levels $3P_{1/2}$ and $3P_{3/2}$, then an interference minimum is observed in two-photon absorption. Thus two-photon absorption has many unusual features arising from the structure of the two-photon matrix element m_{fi} as well as the dependence of $p^{(2)}$ on the fourth-order correlation function $\langle E^{(-)}(0)E^{(-)}(0)E^{(+)}(t)E^{(+)}(t)\rangle$. The entanglement of photons is also important for two-photon absorption [19,20,26]. We have presented the simplest nonlinear optical process where photon statistics and even phase fluctuations can affect the efficiency of the process. Other nonlinear processes such as multi-wave mixing and multi-photon absorption also depend on photon statistics and extensive discussion of these can be found in [27–29].

8.8 Two-photon imaging – ghost imaging using $G^{(2)}$

As a further application of intensity–intensity correlations, we consider entangled two-photon imaging which has attracted considerable attention. The basic idea is to image an object by scanning the light which never interacted with the object. Here one uses the entanglement between the idler and signal photons. The signal photon goes through the object (Figure 8.8). The idler photon is detected in coincidence with the signal photon after it has passed through the object. The information on the object is obtained by scanning the position of the detector D_2. The technique was introduced by Klyshko [30] and was implemented by several groups [31–33]. The technique is also called ghost imaging as the light that is scanned never passes through the object [32, 34].

After the original experiments on ghost imaging using the light produced by a down-converter, it was found that one can also use thermal light to perform ghost imaging quite efficiently [34–37]. Let us now present an argument to explain how the result of coincidence measurements is related to the $G^{(2)}$ of the input fields. For simplicity we present the argument for a thermal source using the scheme of Figure 8.9. The field reaching the

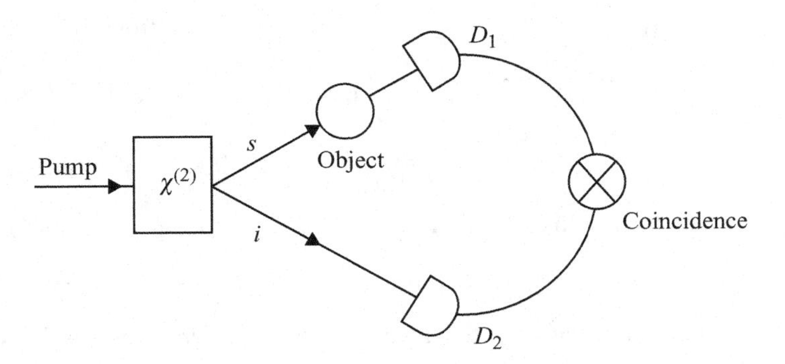

Fig. 8.8 Schematics of the entangled two photon or ghost imaging. Each arm in addition will consist of optical elements.

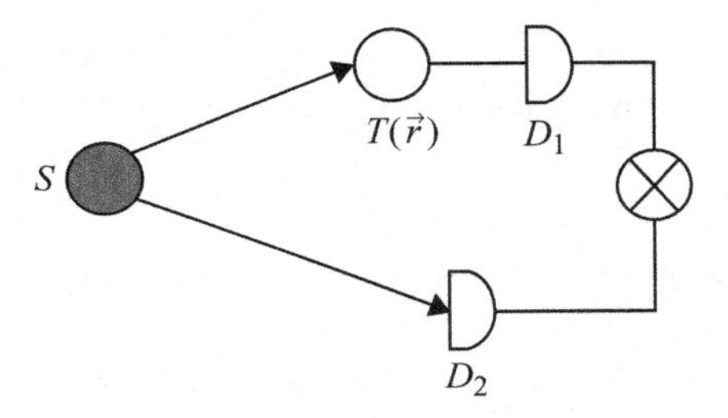

Fig. 8.9 Ghost imaging with an arbitrary source.

detectors D_1 and D_2 can be written in terms of the transmission function $T(\vec{r})$ of the object

$$E^{(+)}(\vec{r}_1) = \int K(\vec{r}_1, \vec{r}_s)E^{(+)}(\vec{r}_s)\mathrm{d}^3r_s, \tag{8.77}$$

$$E^{(+)}(\vec{r}_2) = \int h(\vec{r}_2, \vec{r}_s)E^{(+)}(\vec{r}_s)\mathrm{d}^3r_s. \tag{8.78}$$

The function $h(\vec{r}_2, \vec{r}_1)$ represents the propagation of the field from the source to the detector D_2. The function K is linearly related to the function $T(\vec{r})$ and the propagation of the field from the source to the object and from the object to the detector D_1. The measured intensity–intensity correlation

$$G^{(2)}(\vec{r}_1, \vec{r}_2) = \langle E^{(-)}(\vec{r}_1)E^{(-)}(\vec{r}_2)E^{(+)}(\vec{r}_2)E^{(+)}(\vec{r}_1)\rangle \tag{8.79}$$

can be simplified since the detected fields would also have thermal statistics. This is because the output fields (8.77) and (8.78) are linear functionals of the input fields. Thus for thermal fields $G^{(2)}(\vec{r}_1, \vec{r}_2)$ becomes

$$G^{(2)}(\vec{r}_1, \vec{r}_2) = I(\vec{r}_1)I(\vec{r}_2) + |\langle E^{(-)}(\vec{r}_1)E^{(+)}(\vec{r}_2)\rangle|^2. \tag{8.80}$$

The correlation $\langle E^{(-)}(\vec{r}_1)E^{(+)}(\vec{r}_2)\rangle$ can be obtained from (8.77) and (8.78)

$$\begin{aligned}\langle E^{(-)}(\vec{r}_1)E^{(+)}(\vec{r}_2)\rangle &= \iint K^*(\vec{r}_1,\vec{r}_s{}')h(\vec{r}_2,\vec{r}_s{}'')\langle E^{(-)}(\vec{r}_s{}')E^{(+)}(\vec{r}_s{}'')\rangle \mathrm{d}r_s \\ &= \int K^*(\vec{r}_1,\vec{r}_s)h(\vec{r}_2,\vec{r}_s)I(\vec{r}_s)\mathrm{d}r_s,\end{aligned} \tag{8.81}$$

where the last line follows from the incoherent nature of the thermal fields

$$\langle E^{(-)}(\vec{r}_s{}')E^{(-)}(\vec{r}_s{}'')\rangle \equiv I(\vec{r}_s{}')\delta(\vec{r}_s{}' - \vec{r}_s{}'').$$

Thus the fluctuation in the intensity–intensity correlation has the form

$$\begin{aligned}\Delta G^{(2)}(\vec{r}_1,\vec{r}_2) &= G^{(2)}(\vec{r}_1,\vec{r}_2) - I(r_1)I(r_2) \\ &= \left|\int K^*(\vec{r}_1,\vec{r}_s)h(\vec{r}_2,\vec{r}_s)I(\vec{r}_s)\mathrm{d}r_s\right|^2.\end{aligned} \tag{8.82}$$

Note that the information on the object is contained in the function $K(\vec{r}_1,\vec{r}_s)$.

For two-photon fields with initial state $|\chi\rangle$,

$$E^{(+)}(\vec{r}_s{}')E^{(+)}(\vec{r}_s{}'')|\chi\rangle = \Psi(\vec{r}_s{}',\vec{r}_s{}'')|0\rangle, \tag{8.83}$$

where Ψ is called the two-photon wavefunction [34, 37], the correlation $G^{(2)}$ becomes

$$G^{(2)}(\vec{r}_1,\vec{r}_2) = \left|\iint K(\vec{r}_1,\vec{r}_s{}')h(\vec{r}_2,\vec{r}_s{}'')\Psi(\vec{r}_s{}',\vec{r}_s{}'')\mathrm{d}r_s'\mathrm{d}r_s''\right|^2. \tag{8.84}$$

If the coherence length of the two-photon wavefunction is much smaller than the typical scale associated with the features of the object, then we can approximate $\Psi(\vec{r}_s{}',\vec{r}_s{}'')$ as $|\Psi(\vec{r}_s{}')|^2\delta(\vec{r}_s{}' - \vec{r}_s{}'')$. Under this assumption (8.84) reduces to

$$G^{(2)}(\vec{r}_1,\vec{r}_2) = \left|\int K(\vec{r}_1,\vec{r}_s)h(\vec{r}_2,\vec{r}_s)|\Psi(\vec{r}_s)|^2\mathrm{d}r_s\right|^2. \tag{8.85}$$

Note the similarity of (8.85) to (8.82), although there are some differences which arise because of the appearance of K in (8.85) and K^* in (8.82). Ghost imaging with both thermal light and entangled light has been extensively studied (for reviews see [34, 37]). Here we present a very simple example using thermal light to clearly demonstrate how it works. For simplicity, let us consider a one-dimensional version ($\vec{r} \to x$) of (8.82) for imaging a one-dimensional object. The point x_1 is fixed say at $x = 0$ and x_2 is scanned. The response function $h(x_2, x_s)$ can be approximated by

$$h(x_2,x_s) \approx \frac{\mathrm{e}^{\mathrm{i}kd_2}}{\sqrt{\mathrm{i}\lambda d_2}}\exp\left[\frac{\mathrm{i}\pi}{\lambda d_2}(x_2 - x_s)^2\right], \tag{8.86}$$

where d_2 is the distance between the source plane and the detector plane. The function K is more complicated as we need the propagation from the source plane to the object plane and then from the object plane to the detector

$$K(0,x) \approx \int \mathrm{d}x' \frac{\mathrm{e}^{\mathrm{i}kd_1}}{\sqrt{\mathrm{i}\lambda d_1}}\exp\left[\frac{\mathrm{i}\pi}{\lambda d_1}(x - x')^2\right]T(x')\frac{\mathrm{e}^{\mathrm{i}kd_1'}}{\sqrt{\mathrm{i}\lambda d_1'}}\exp\left(\frac{\mathrm{i}\pi}{\lambda d_1'}x'^2\right), \tag{8.87}$$

where d_1 (d_1') is the distance from source (object) to object (detector) and $T(x)$ is the transmission function of the object. Equation (8.82) can now be simplified by assuming: (i) a uniform source $I(\vec{r}_s) \to I_0$, (ii) large dimensions of the source so that the integration is over all transverse space, and (iii) by fixing the distances such that $d_1 - d_2 = -d_1'$. Under these assumptions a very simple result is obtained [35]

$$\Delta G^{(2)}(0, x) = \frac{I_0^2}{\lambda^2 d_1'^2} \left| \hat{t}\left(\frac{2\pi x}{\lambda d_1'}\right) \right|^2, \tag{8.88}$$

where $\hat{t}$ is the Fourier transform of $T(x)$. This type of ghost imaging does not require any lenses. Thus scanning the intensity intensity correlation as a function of the position of the detector in the arm which does not contain the object, yields the Fourier transform of the transmission function of the object. Note that the method above recovers only the modulus of $|\hat{t}|$. Methods have been developed, and implemented experimentally, to obtain the phase information on the object as well [34, 36].

Exercises

8.1 Prove the inequality (8.24) for the normalized coherence function. You can start with the property $\mathrm{Tr}\{\rho f^\dagger f\} \geq 0$, $f = \alpha E^{(+)}(\vec{r}_1, t_1) + \beta E^{(+)}(\vec{r}_2, t_2)\ \forall\ \alpha, \beta$.

8.2 Show that the correlation function (8.20) satisfies the following two-wave equations

$$\nabla_i^2 \Gamma(\vec{r}_1, \vec{r}_2, \tau) = \frac{1}{c^2}\frac{\partial^2}{\partial \tau^2}\Gamma(\vec{r}_1, \vec{r}_2, \tau), \qquad i = 1, 2,$$

where ∇_1^2, for example, is the Laplacian with respect to the position $\vec{r}_1$ (for details see [2]).

8.3 Show that (8.1) in the limit of very large bandwidth of the final states goes over to

$$\Gamma_{if}^{(A)} = \frac{2}{\gamma \hbar^2}\vec{p}_{fi}^{\,*} \cdot \overset{\Rightarrow(N)}{\mathcal{E}}(\vec{R}, 0, \vec{R}, 0) \cdot \vec{p}_{fi}.$$

This can be proved by averaging (8.1) over ω_{fi} with a distribution $\frac{\gamma/\pi}{\gamma^2 + (\omega_{fi} - \bar{\omega}_{fi})^2}$ and by taking the limit $\gamma \tau_c \gg 1$, where τ_c is the correlation time of the fields correlation function.

8.4 Prove the inequality (8.34) by starting from the fact that for any classical probability distribution

$$\langle [aI(t) + \beta I(t + \tau)]^2 \rangle$$

is always positive for arbitrary α, β, and τ.

8.5 Evaluate the counting distribution (8.43) for thermal light, i.e. for

$$P(\alpha) = \frac{1}{\pi \bar{n}} \mathrm{e}^{-|\alpha|^2/\bar{n}}.$$

8.6 Define $P(\alpha) = P(I)\mathrm{d}I\mathrm{d}\varphi$, then show that (8.43) can be inverted to obtain $P(I)$ from the measurements of $p_e(n, T)$. It is useful to introduce a generating function $\sum_{n=0}^{\infty} \beta^n p_e(n, T)$ in order to obtain $P(I)$ (for details see [38]).

8.7 Consider a field with Gaussian correlation

$$\langle a^\dagger(t)a(t+\tau)\rangle = \mathrm{e}^{-\mathrm{i}\omega_0\tau - \gamma^2\tau^2}\langle a^\dagger a\rangle.$$

Find the spectrum (8.50) and its half width at half maximum.

8.8 Calculate the correlation function (7.13), i.e.

$$\mathcal{E}_{\alpha\beta}(\vec{r}_1, t_1, \vec{r}_2, t_2) = \mathrm{Tr}\left[\rho_F E_\alpha^{(-)}(\vec{r}_1, t_1)E_\beta^{(+)}(\vec{r}_2, t_2)\right]$$

for thermal radiation

$$\rho_F = \prod_{\vec{k}s} \rho_{\vec{k}s}, \qquad \rho_{\vec{k}s} = \exp(-\beta\hbar\omega_k a_{\vec{k}s}^\dagger a_{\vec{k}s})/\mathrm{Tr}[\exp(-\beta\hbar\omega_k a_{\vec{k}s}^\dagger a_{\vec{k}s})].$$

Find the Fourier transform

$$\int_{-\infty}^{+\infty} \mathcal{E}_{\alpha\beta}(\vec{r}_2, t_2, \vec{r}_1, 0)\mathrm{e}^{-\mathrm{i}\omega t_2}\mathrm{d}t_2.$$

(For details see [39,40].)

8.9 Verify the result (8.88) using (8.86) and (8.87) and the simplifying assumptions listed after Eq. (8.87).

References

[1] M. Born and E. Wolf, *Principles of Optics*, 7th ed. (Cambridge: Cambridge University Press, 1999), Chap. 10.
[2] E. Wolf, *Introduction to the Theory of Coherence and Polarization of Light* (Cambridge: Cambridge University Press, 2007).
[3] R. J. Glauber, *Phys. Rev. Lett.* **10**, 84 (1963).
[4] R. J. Glauber, *Phys. Rev.* **130**, 2529 (1963).
[5] R. J. Glauber, in *Quantum Optics and Electronics, Les Houches Lectures 1964*, edited by C. deWitt, A. Blandin, and C. Cohen-Tannoudji (New York: Gordon and Breach, 1965).
[6] E. C. G. Sudarshan, *Phys. Rev. Lett.* **10**, 277 (1963).
[7] A comprehensive account of intensity interferometry, both with radiowaves and with light, is given in R. Hanbury Brown, *The Intensity Interferometer* (London: Taylor and Francis, 1974).
[8] R. Hanbury-Brown and R. Q. Twiss, *Nature (London)* **177**, 27 (1956).
[9] P. L. Kelley and W. H. Kleiner, *Phys. Rev.* **136**, A316 (1964).
[10] L. Mandel, *Proc. Phys. Soc. (London)* **72**, 1037 (1958); *ibid.* **74**, 223 (1959).
[11] L. Mandel, E. C. G. Sudarshan, and E. Wolf, *Proc. Phys. Soc. (London)* **84**, 435 (1964).

[12] L. Mandel, *Progress in Optics* **2**, edited by E. Wolf (Amsterdam: North-Holland, 1963).
[13] See Ref. [2], p. 25, Section 2.5.
[14] L. Mandel, *J. Opt. Soc. Am.* **52**, 1335 (1962).
[15] D. F. V. James and E. Wolf. *Opt. Commun.* **81**, 150 (1991).
[16] G. S. Agarwal and D. F. V. James, *J. Mod. Opt.* **40**, 1431 (1993).
[17] N. Kumar and D. N. Rao, *J. Mod. Opt.* **48**, 1455 (2001).
[18] X. Y. Zou, T. P. Grayson, and L. Mandel, *Phys. Rev. Lett.* **69**, 3041 (1992).
[19] J. Gea-Banacloche, *Phys. Rev. Lett.* **62**, 1603 (1989).
[20] J. Javanainen and P. L. Gould, *Phys. Rev. A* **41**, 5088 (1990).
[21] N. Ph. Georgiades, E. S. Polzik, K. Edamatsu, H. J. Kimble, and A. S. Parkins, *Phys. Rev. Lett.* **75**, 3426 (1995).
[22] B. R. Mollow, *Phys. Rev.* **175**, 1555 (1968).
[23] F. Boitier, A. Godard, N. Dubreuil, P. Delaye, C. Fabre, and E. Rosencher, *Nature Commun.* **2**, 425 (2011).
[24] M. J. Collett and C. W. Gardiner, *Phys. Rev. A* **30**, 1386 (1984).
[25] J. E. Bjorkholm and P. F. Liao, *Phys. Rev. Lett.* **33**, 128 (1974).
[26] A. Muthukrishnan, G. S. Agarwal, and M. O. Scully, *Phys. Rev. Lett.* **93**, 093002 (2004).
[27] G. S. Agarwal, *Phys. Rev. A* **1**, 1445 (1970).
[28] M. Tsang, *Phys. Rev. Lett.* **101**, 033602 (2008).
[29] O. Roslyak, C. A. Marx, and S. Mukamel, *Phys. Rev. A* **79**, 033832 (2009).
[30] D. Klyshko, *Sov. Phys. JETP* **67**, 1131 (1988).
[31] P. H. S. Ribeiro, S. Pádua, J. C. Machado da Silva, and G. A. Barbosa, *Phys. Rev. A* **49**, 4176 (1994).
[32] D. V. Strekalov, A. V. Sergienko, D. N. Klyshko, and Y. H. Shih, *Phys. Rev. Lett.* **74**, 3600 (1995).
[33] T. B. Pittman, Y. H. Shih, D. V. Strekalov, and A. V. Sergienko, *Phys. Rev. A* **52**, R3429 (1995).
[34] A. Gatti, E. Brambilla, and L. Lugiato, in *Progress in Optics*, edited by E. Wolf (Amsterdam: Elsevier, 2008), **51**, pp. 251–348.
[35] J. Cheng and S. S. Han, *Phys. Rev. Lett.* **92**, 093903 (2004).
[36] E. Baleine, A. Dogariu, and G. S. Agarwal, *Opt. Lett.* **31**, 2124 (2006).
[37] Y. Shih, in *International Conference on Quantum Information* (Optical Society of America, 2008), paper QTuB1.
[38] E. Wolf and C. L. Mehta, *Phys. Rev. Lett.* **13**, 705 (1964).
[39] C. L. Mehta and E. Wolf, *Phys. Rev. A* **134**, 1143 (1964).
[40] C. L. Mehta and E. Wolf, *Phys. Rev. A* **134**, 1149 (1964).

9 Open quantum systems

This chapter is devoted to the dynamical evolution of open quantum system [1–3]. An open quantum system is one where it interacts with the environment. A system undergoing relaxation is an example of an open quantum system. We have already come across an example of open quantum system in Chapter 7 where we have discussed spontaneous emission from a two-level system. The two-level system interacts with the vacuum of the electromagnetic field. The vacuum consists of infinite number of modes and is a large system. The vacuum in this case is the environment. The population in the excited state decays. A photon is emitted and the emitted photon leaves the vicinity of the atom, i.e. the emitted photon is not reabsorbed by the atom. Another example of an open system is the case of atoms colliding with the atoms of a buffer gas. Here the buffer gas is the environment. Other examples of open systems are the fields confined in the cavities. The case of ideal cavities, i.e. cavities bounded by mirrors with 100% reflectivity, is uninteresting. We need the photons from the cavity to leak out in order to learn about the photons in the cavity. Thus we need to have mirrors with nonzero transmission. In this case, the electromagnetic field inside the cavity couples to the vacuum modes outside the cavity; thus the vacuum outside is the environment.

Since the environment is usually an infinitely large system, it is almost impossible to solve for the quantum dynamics of a system coupled to the environment. However, by imposing physically relevant conditions, one can study the dynamics of the system alone, while approximately including the effects of the environment. This is the strategy one adopts for the case of open systems. In the next few sections we will describe methods used to study the dynamics of open systems.

9.1 Master equation description of open systems

Let us consider the interaction of a system S with a bath B (environment) (Figure 9.1). We would assume that this interaction is weak. We write the total Hamiltonian H as

$$H = H_S + H_B + H_{SB}, \tag{9.1}$$

where H_S is the Hamiltonian of the system S and H_B is the Hamiltonian of the bath B. The interaction between S and B is represented by H_{SB}. For simplicity we assume that H_S, H_B, and H_{SB} are time independent. For the case of spontaneous emission discussed in

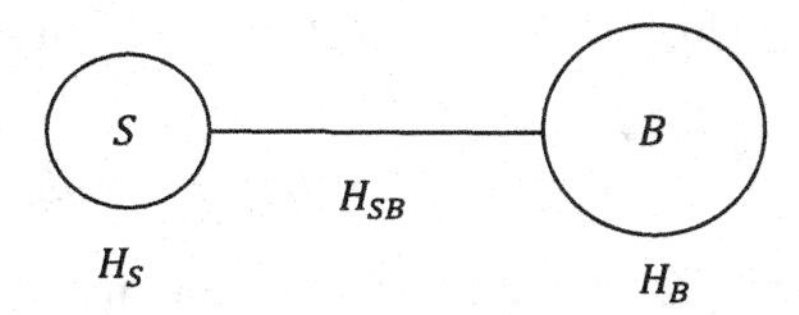

Fig. 9.1 A small system S interacts with a bath B.

Section 7.6, H is given by (7.6), more specifically in the notation of this section

$$H_S = E_i|i\rangle\langle i| + E_f|f\rangle\langle f|, \qquad H_B = \sum \hbar\omega_k a^\dagger_{\vec{k}s} a_{\vec{k}s},$$
$$H_{SB} = -(\vec{p}_{if}|i\rangle\langle f| + \vec{p}_{fi}|f\rangle\langle i|) \times \left(\sum_{\vec{k}s} \mathrm{i}\sqrt{\frac{2\pi\hbar\omega_k}{V}} \vec{\epsilon}_{\vec{k}s} a_{\vec{k}s} e^{\mathrm{i}\vec{k}\cdot\vec{R}} + H.c. \right). \tag{9.2}$$

Let ρ be the density matrix for the combined system $S + B$. The reduced density matrix ρ_S for the system S alone is related to ρ by

$$\rho_S = \mathrm{Tr}_B \rho, \tag{9.3}$$

i.e. it is obtained by tracing out the degrees of freedom associated with the bath. Let ρ_B be the density matrix associated with the bath in the equilibrium state. The bath is a large system and we assume that the changes in the state of the bath would be insignificant. Typically the bath would be in a state of thermal equilibrium at temperature T, then

$$\rho_B = \exp(-H_B/K_BT)/\mathrm{Tr}\exp(-H_B/K_BT). \tag{9.4}$$

The bath could be at zero temperature as in the case of spontaneous emission, then

$$\rho_B = |\{0_{\vec{k}s}\}\rangle\langle\{0_{\vec{k}s}\}|. \tag{9.5}$$

In order to obtain the dynamical equation for ρ_S, we need to specify the initial state of $S + B$. We take it to be uncorrelated state

$$\rho_{S+B}(0) = \rho_S(0)\rho_B, \tag{9.6}$$

and we leave $\rho_S(0)$ to be arbitrary. The density matrix ρ obeys the Liouville equation

$$\frac{\partial \rho}{\partial t} = -\frac{\mathrm{i}}{\hbar}[H, \rho] = -\mathrm{i}\mathcal{L}\rho, \tag{9.7}$$

where $\mathcal{L}$ is the Liouvillian defined by $\mathcal{L} = \frac{1}{\hbar}[H, \;]$. For the derivation of the master equation, it is convenient to write H_{SB} in the form

$$H_{SB} = \sum_\alpha S_\alpha B_\alpha, \tag{9.8}$$

where S_α (B_α) is a system (bath) operator. Since H_{SB} is Hermitian, it follows that $S^\dagger_\alpha = S_\alpha$, $B^\dagger_\alpha = B_\alpha$. A master equation can be derived provided there is a clear separation of time scales over which system and bath evolve. We assume that the time scale over which system evolves is much bigger than the scale over which B evolves. For simplicity we also assume that

$$\langle B_\alpha \rangle = \mathrm{Tr}\rho_B B_\alpha = 0. \tag{9.9}$$

Several methods can be used to derive the master equation. Here we follow the method based on projection operators [1, 2]. Let $\mathcal{P}$ be the projection operator defined by

$$\mathcal{P} \cdot\cdot = \rho_B \mathrm{Tr}_B \cdot\cdot, \tag{9.10}$$

where $\cdot\cdot$ stand for any operator on which $\mathcal{P}$ acts. Then we clearly have

$$\mathcal{P}\rho = \rho_B \rho_S, \qquad \mathcal{P}\rho(0) = \rho_B \rho_S(0) = \rho(0). \tag{9.11}$$

We now take the Laplace transform of (9.7) and repeatedly use the relations $\mathcal{P}+(1-\mathcal{P}) = 1$, $\mathcal{P}^2 = \mathcal{P}$, $\mathcal{P}(1-\mathcal{P}) = 0$, $(1-\mathcal{P})^2 = 1-\mathcal{P}$, $\mathcal{P}\rho(0) = \rho(0)$, $(1-\mathcal{P})\rho(0) = 0$, to obtain

$$\mathcal{P}p\hat{\rho}(p) - \rho(0) = -\mathrm{i}\mathcal{P}\mathcal{L}\mathcal{P}\hat{\rho} - \mathrm{i}\mathcal{P}\mathcal{L}(1-\mathcal{P})\hat{\rho}, \tag{9.12}$$

$$(1-\mathcal{P})p\hat{\rho}(p) = -\mathrm{i}(1-\mathcal{P})\mathcal{L}(1-\mathcal{P})\hat{\rho} - \mathrm{i}(1-\mathcal{P})\mathcal{L}\mathcal{P}\hat{\rho}. \tag{9.13}$$

The quantity of interest is $\mathcal{P}\hat{\rho}$. Hence we formally solve (9.13) for $(1-\mathcal{P})\hat{\rho}$ and use it in (9.12). According to (9.1), $\mathcal{L}$ would be a sum of three Liouvillians $\mathcal{L}_\alpha = \frac{1}{\hbar}[H_\alpha, \ ]$, $\alpha = S, B, SB$, with properties

$$\mathcal{P}\mathcal{L}_S\mathcal{P} = \mathcal{L}_S\mathcal{P}^2 = \mathcal{L}_S\mathcal{P}, \quad \mathcal{P}\mathcal{L}_B \cdot\cdot = 0, \quad \mathcal{L}_B\mathcal{P}\cdot\cdot = \frac{1}{\hbar}[H_B, \rho_B\mathrm{Tr}_B \cdot\cdot] = 0, \tag{9.14}$$

where the latter property follows from (9.4). Furthermore, $\mathcal{P}\mathcal{L}_{BS}\mathcal{P} = 0$ because of the assumption (9.9). The second term in (9.12) can be simplified as

$$\begin{aligned}
\mathcal{P}\mathcal{L}_S(1-\mathcal{P})\hat{\rho} &= \mathcal{L}_S\mathcal{P}(1-\mathcal{P})\hat{\rho} = 0, \\
\mathcal{P}\mathcal{L}_B(1-\mathcal{P})\hat{\rho} &= \frac{1}{\hbar}\rho_B\mathrm{Tr}_B[H_B, (1-\mathcal{P})\hat{\rho}] \\
&= \frac{1}{\hbar}\rho_B\mathrm{Tr}_B\{H_B(1-\mathcal{P})\hat{\rho} - (1-\mathcal{P})\hat{\rho}H_B\} \\
&= \frac{1}{\hbar}\rho_B\mathrm{Tr}_B\{H_B(1-\mathcal{P})\hat{\rho} - H_B(1-\mathcal{P})\hat{\rho}\} \\
&= 0.
\end{aligned} \tag{9.15}$$

Thus using (9.14) and (9.15), Eq. (9.12) reduces to

$$\rho_B[p\hat{\rho}_S(p) - \rho_S(0)] = -\mathrm{i}\rho_B\mathcal{L}_S\hat{\rho}_S - \mathrm{i}\mathcal{P}\mathcal{L}_{SB}(1-\mathcal{P})\hat{\rho}. \tag{9.16}$$

We can solve (9.13) formally

$$(1-\mathcal{P})\hat{\rho} = [p + \mathrm{i}(1-\mathcal{P})\mathcal{L}(1-\mathcal{P})]^{-1}(-\mathrm{i})(1-\mathcal{P})\mathcal{L}\mathcal{P}\hat{\rho}. \tag{9.17}$$

We can simplify the last part $(1-\mathcal{P})\mathcal{L}\mathcal{P}\hat{\rho}$ as follows

$$\begin{aligned}
(1-\mathcal{P})\mathcal{L}_S\mathcal{P}\hat{\rho} &= \mathcal{L}_S(1-\mathcal{P})\mathcal{P}\hat{\rho} = 0, \\
(1-\mathcal{P})\mathcal{L}_B\mathcal{P}\hat{\rho} &= (1-\mathcal{P})\frac{1}{\hbar}[H_B, \rho_B\hat{\rho}_S] = (1-\mathcal{P})\frac{1}{\hbar}\hat{\rho}_S[H_B, \rho_B] = 0,
\end{aligned}$$

and hence (9.17) reduces to

$$(1-\mathcal{P})\hat{\rho} = [p + \mathrm{i}(1-\mathcal{P})\mathcal{L}(1-\mathcal{P})]^{-1}(-\mathrm{i})(1-\mathcal{P})\mathcal{L}_{SB}\mathcal{P}\hat{\rho}. \tag{9.18}$$

On substituting (9.18) in (9.16) and removing the common factor ρ_B, we obtain

$$p\hat{\rho}_S - \rho_S(0) = -\mathrm{i}\mathcal{L}_S\hat{\rho}_S - \mathrm{i}\tilde{\mathcal{L}}(p)\hat{\rho}_S, \tag{9.19}$$

$$-\mathrm{i}\tilde{\mathcal{L}}(p) = -\mathrm{Tr}_B\{\mathcal{L}_{SB}[p + \mathrm{i}(1-\mathcal{P})\mathcal{L}(1-\mathcal{P})]^{-1}(1-\mathcal{P})\mathcal{L}_{SB}\rho_B\}. \tag{9.20}$$

This is the final master equation for the density matrix ρ_S of the system S alone. The effect of the bath is contained in the second term, $-\mathrm{i}\tilde{\mathcal{L}}\hat{\rho}_S$ on the right-hand side of (9.19). The equation (9.19) is exact and thus valid to all orders in the interaction with the bath. However, for most purposes the dissipative effects are adequately described by considering lowest-order terms in (9.19). In Section 9.7 we will treat some exactly soluble models. To second order in H_{SB}, we approximate (9.20) by replacing $\mathcal{L}$ in the denominator by the unperturbed $\mathcal{L} = \mathcal{L}_S + \mathcal{L}_B$

$$\begin{aligned} -\mathrm{i}\tilde{\mathcal{L}} &\approx -\mathrm{Tr}_B\{\mathcal{L}_{SB}[p + \mathrm{i}(1-\mathcal{P})(\mathcal{L}_S + \mathcal{L}_B)(1-\mathcal{P})]^{-1}(1-\mathcal{P})\mathcal{L}_{SB}\rho_B\} \\ &= -\mathrm{Tr}_B\{\mathcal{L}_{SB}[p + \mathrm{i}(\mathcal{L}_S + \mathcal{L}_B)]^{-1}(1-\mathcal{P})\mathcal{L}_{SB}\rho_B\} \\ &= -\mathrm{Tr}_B\{\mathcal{L}_{SB}[p + \mathrm{i}(\mathcal{L}_S + \mathcal{L}_B)]^{-1}\mathcal{L}_{SB}\rho_B\}, \end{aligned} \tag{9.21}$$

as $\mathcal{P}\mathcal{L}_{SB}\rho_B = 0$. On substituting (9.21) in (9.19) and on taking the Laplace transform we obtain

$$\frac{\partial \rho_S}{\partial t} = -\mathrm{i}\mathcal{L}_S\rho_S - \int_0^t \mathrm{d}\tau\, \mathrm{Tr}_B[\mathcal{L}_{SB}U(\tau)\mathcal{L}_{SB}\rho_B\rho_S(t-\tau)], \tag{9.22}$$

where $U(\tau)$ is the free-time evolution in Liouville space

$$U(\tau) = \exp[-\mathrm{i}(\mathcal{L}_S + \mathcal{L}_B)\tau]. \tag{9.23}$$

The master equation (9.22) for the system has the integro-differential form. This can be reduced to an equation local in time if there is a clear separation of time scales. In order to see this we would first write the integral term in terms of the equilibrium correlation functions of the bath. We denote the integrand in (9.22) by $I(\tau, t-\tau)$ and rewrite it in terms of the Hamiltonian

$$\begin{aligned} I(\tau, t-\tau) &= \frac{1}{\hbar^2}\mathrm{Tr}_B\left[H_{SB}, U_0(\tau)[H_{SB}, \rho_B\rho_S(t-\tau)]U_0^\dagger(\tau)\right], \\ U_0(\tau) &= \exp\left[-\frac{\mathrm{i}}{\hbar}(H_S + H_B)\tau\right], \end{aligned} \tag{9.24}$$

which can be rewritten in a more transparent form as

$$\begin{aligned} I(\tau, t-\tau) &= \frac{1}{\hbar^2}\mathrm{Tr}_B\left[H_{SB}, [H_{SB}(-\tau), \rho_B\tilde{\rho}_S]\right], \\ H_{SB}(-\tau) &= U_0(\tau)H_{SB}U_0^\dagger(\tau), \\ \tilde{\rho}_S &= U_0(\tau)\rho_S(t-\tau)U_0^\dagger(\tau), \end{aligned} \tag{9.25}$$

Note that if there were no interaction with bath, then $\tilde{\rho}_S$ would be equal to $\rho_S(t)$. Thus any difference between $\tilde{\rho}_S$ and $\rho_S(t)$ is due to the interaction with the bath over a time interval τ. This time interval is determined by the correlation functions of the bath. If there is a clear separation of time scales, i.e. the correlation time of the bath is short compared to the scale over which system evolves, then we use the approximation

$$\tilde{\rho}_S \cong \rho_S(t), \tag{9.26}$$

which is called the Markov approximation. On combining (9.22), (9.25), and (9.26) we obtain the final master equation which is valid in Born and Markov approximations

$$\frac{\partial \rho_S}{\partial t} = -\frac{\mathrm{i}}{\hbar}[H_S, \rho_S] - \frac{1}{\hbar^2}\int_0^\infty \mathrm{d}\tau \, \mathrm{Tr}_B[H_{SB}, [H_{SB}(-\tau), \rho_B\rho_S(t)]]. \tag{9.27}$$

We have extended the limit on integration to infinity because of the short correlation time of the bath. By expanding the double commutator in (9.27) and by using (9.8) it is straightforward to simplify (9.27) to

$$\frac{\partial \rho_S}{\partial t} = -\frac{\mathrm{i}}{\hbar}[H_S, \rho_S] - \left\{ \int_0^\infty \sum_{\alpha,\beta} C_{\alpha\beta}(-\tau)[S_\alpha S_\beta(-\tau)\rho_S - S_\beta(-\tau)\rho_S S_\alpha]\mathrm{d}\tau + c.c. \right\}, \tag{9.28}$$

where $C_{\alpha\beta}(-\tau)$ is the bath correlation defined by

$$C_{\alpha\beta}(-\tau) = \frac{1}{\hbar^2}\langle B_\alpha B_\beta(-\tau)\rangle = \frac{1}{\hbar^2}\mathrm{Tr}[\rho_B B_\alpha B_\beta(-\tau)], \tag{9.29}$$

and the operators have time dependence determined by

$$\begin{aligned} S_\beta(-\tau) &= \exp\left(-\frac{\mathrm{i}H_S\tau}{\hbar}\right) S_\beta \exp\left(\frac{\mathrm{i}H_S\tau}{\hbar}\right), \\ B_\beta(-\tau) &= \exp\left(-\frac{\mathrm{i}H_B\tau}{\hbar}\right) B_\beta \exp\left(\frac{\mathrm{i}H_B\tau}{\hbar}\right). \end{aligned} \tag{9.30}$$

Note that if Ω is a typical frequency associated with the time development of the system operator, then the effect of the bath is determined by the Fourier component of the correlation function (9.29) at the frequency Ω. We next consider several dissipative systems used in quantum optics.

9.2 Dissipative dynamics of harmonic oscillators

As a first example we consider the interaction of a harmonic oscillator with a bath – we model the bath to be made of harmonic oscillators. Thus we write

$$\begin{aligned} H_S &= \hbar\omega_0 a^\dagger a, \qquad H_B = \sum_j \hbar\omega_j a_j^\dagger a_j, \\ H_{SB} &= \hbar\sum_j \mathrm{i}g_j(a + a^\dagger)(a_j - a_j^\dagger). \end{aligned} \tag{9.31}$$

We have specifically used this form of H_{SB} keeping in view the form of dipolar interaction (9.2). The equilibrium density matrix of the bath is given by (9.4) with H_B given by (9.31).

The system operators are a and $a^\dagger$ and their time dependence as defined by (9.30) will be

$$a(-\tau) = e^{i\omega_0\tau}a, \qquad a_j(-\tau) = e^{i\omega_j\tau}a_j. \tag{9.32}$$

In the notation of the previous section

$$\begin{aligned} S(-\tau) &= e^{i\omega_0\tau}a + e^{-i\omega_0\tau}a^\dagger, \\ B(-\tau) &= \hbar\sum_j ig_j(a_j e^{i\omega_j\tau} - a_j^\dagger e^{-i\omega_j\tau}). \end{aligned} \tag{9.33}$$

The correlation function $C(-\tau) = \frac{1}{\hbar^2}\langle BB(-\tau)\rangle$ can be evaluated by using (9.4) and the properties

$$\begin{aligned} &\langle a_j a_l\rangle = \langle a_j^\dagger a_l^\dagger\rangle = 0, \quad \langle a_j^\dagger a_l\rangle = \delta_{jl}n(\omega_j), \quad \langle a_j a_l^\dagger\rangle = \delta_{jl}[1+n(\omega_j)], \\ &n(\omega_j) = 1/[\exp(\beta\hbar\omega_j)-1], \quad \beta = \frac{1}{K_B T}, \end{aligned} \tag{9.34}$$

with the result

$$C(-\tau) = \sum_j g_j^2\left\{n(\omega_j)e^{i\omega_j\tau} + [n(\omega_j)+1]e^{-i\omega_j\tau}\right\}. \tag{9.35}$$

The substitution of (9.35) and (9.33) in (9.28) would result in integrals which are well known

$$\int_0^\infty d\tau\, e^{i(\omega_0\mp\omega_j)\tau} = \pi\delta(\omega_0\mp\omega_j) + iP\frac{1}{\omega_0\mp\omega_j}, \tag{9.36}$$

where P stands for the principal value part. Note that $\delta(\omega_0+\omega_j) = 0$ as all the frequencies ω_0 and ω_j are positive by definition. Clearly the substitution of (9.35) and (9.33) in (9.28) would also result in terms like a^2 and $a^{\dagger 2}$ besides $a^\dagger a$ in the density matrix equation. The terms like a^2 and $a^{\dagger 2}$ oscillate very fast and can be dropped. This is known as the rotating wave approximation. All the principal value terms have the effect of changing $H_S \to \hbar\tilde{\omega}_0 a^\dagger a$, where $\tilde{\omega}_0$ differs slightly from ω_0. In further consideration we drop such frequency shifts, i.e. we set $\tilde{\omega}_0 \approx \omega_0$. It is now a matter of straightforward algebra to derive the final master equation for the oscillator

$$\begin{aligned} \frac{\partial\rho_S}{\partial t} &= -i\omega_0[a^\dagger a, \rho_S] - \kappa(1+n_0)(a^\dagger a\rho_S - 2a\rho_S a^\dagger + \rho_S a^\dagger a) \\ &\quad -\kappa n_0(aa^\dagger\rho_S - 2a^\dagger\rho_S a + \rho_S aa^\dagger), \end{aligned} \tag{9.37}$$

where

$$\kappa = \pi\sum_j g_j^2\delta(\omega_j - \omega_0), \tag{9.38}$$

and n_0 is given by (9.34) with $\omega_j \to \omega_0$. The equation (9.37) describes the dissipative dynamics of the harmonic oscillator – the parameters κ and n_0 are dependent on the bath. We will discuss detailed solutions of (9.37) in Chapter 10 when we treat the attenuation and amplification of the radiation. We note that (9.37) implies

$$\begin{aligned} \langle\dot{a}\rangle &= -i\omega_0\langle a\rangle - \kappa\langle a\rangle, \\ \langle\dot{\overline{a^\dagger a}}\rangle &= -2\kappa\langle a^\dagger a\rangle + 2\kappa n_0, \end{aligned} \tag{9.39}$$

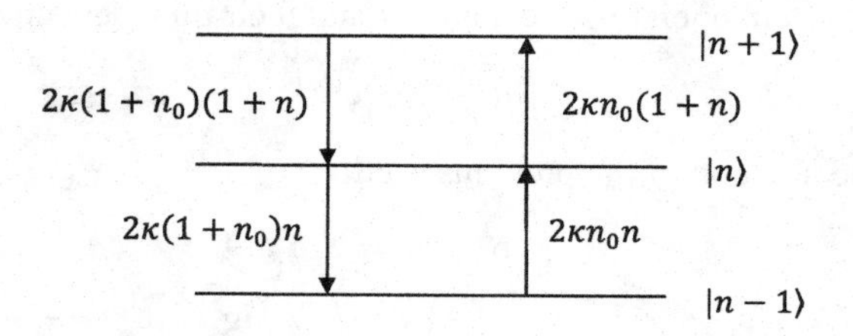

Fig. 9.2 Detailed balance among the reservoir created transitions in harmonic oscillator energy levels.

The amplitude of the oscillator decays at the rate κ. The mean excitation of the oscillator has a time dependence

$$\langle a^\dagger a\rangle_t = e^{-2\kappa t}\langle a^\dagger a\rangle_0 + n_0(1 - e^{-2\kappa t}). \tag{9.40}$$

In the long time limit (9.37) leads to

$$\rho_S \xrightarrow{t\to\infty} e^{-\beta\hbar\omega_0}/\mathrm{Tr}(e^{-\beta\hbar\omega_0}), \tag{9.41}$$

and hence the oscillator acquires the temperature of the bath. From (9.37) we find that the distribution $p_n = \langle n|\rho_S|n\rangle$ of populations in various states of the oscillator obeys the equation

$$\frac{\partial p_n}{\partial t} = 2\kappa(1+n_0)(n+1)p_{n+1} - 2\kappa[n(1+2n_0)+n_0]p_n + 2\kappa n_0 n p_{n-1}. \tag{9.42}$$

This equation has a simple interpretation in terms of rates of upward and downward transitions as shown in Figure 9.2. The steady-state populations satisfy the principle of detailed balance

$$p_{n+1}2\kappa(1+n_0)(1+n) = n_0 2\kappa p_n(n+1). \tag{9.43}$$

The solution of (9.43) agrees with (9.41).

9.3 Dissipative dynamics of a two-level system

Our next example is of a two-level system interacting with a bath of oscillators. A two-level system can be represented by spin $\frac{1}{2}$ angular momentum operators $S_\pm$, S_z, defined via

$$S_+ = |e\rangle\langle g|, \qquad S_- = |g\rangle\langle e|, \qquad S_z = \frac{1}{2}(|e\rangle\langle e| - |g\rangle\langle g|), \tag{9.44}$$

which are related to the Pauli matrices $\sigma^\pm$, σ^z via $S_+ = \frac{1}{2}\sigma^+$, $S_- = \frac{1}{2}\sigma^-$, $S_z = \frac{1}{2}\sigma^z$. The interaction Hamiltonian H_{SB} for this system is given by (9.31) with $a \to S_-$, $a^\dagger \to S_+$. Furthermore, $H_S = \hbar\omega_0 S_z$, so that the excited (ground) state has the energy $\hbar\omega_0/2$ $(-\hbar\omega_0/2)$. The steps in the derivation of the master equation are identical to those in Section 9.2 if we keep in mind that in the derivation (9.37) we did not use the commutation relation between the system operators a and $a^\dagger$. Therefore the master equation for the two-level system interacting with a bath with coupling (9.31) with $a \to S_-$, $a^\dagger \to S_+$, will be

$$\frac{\partial \rho_S}{\partial t} = -\mathrm{i}\omega_0[S_z, \rho_S] - \gamma(1+n_0)(S_+S_-\rho_S - 2S_-\rho_S S_+ + \rho_S S_+S_-) \\ - \gamma n_0(S_-S_+\rho_S - 2S_+\rho_S S_- + \rho_S S_-S_+). \tag{9.45}$$

From (9.45) we can derive the equation for the mean values of $S_\pm$, S_z:

$$\frac{\mathrm{d}}{\mathrm{d}t}\langle S_\pm\rangle = -\gamma(1+2n_0)\langle S_\pm\rangle \pm \mathrm{i}\omega_0\langle S_\pm\rangle, \tag{9.46}$$

$$\frac{\mathrm{d}}{\mathrm{d}t}\langle S_z\rangle = -2\gamma(1+2n_0)(\langle S_z\rangle - \eta), \tag{9.47}$$

where η is the value of $\langle S_z\rangle$ in equilibrium

$$\eta = -\frac{1}{2(1+2n_0)} = -\frac{1}{2}\tanh\left(\frac{1}{2}\beta\hbar\omega_0\right). \tag{9.48}$$

Note that $\langle S_z\rangle$ is equal to half of the population inversion. The dipole moment of the two-level atom, which is related to the density matrix element between $|e\rangle$ and $|g\rangle$, can be obtained from $\langle S_+\rangle$. Note that the rate of decay of the dipole moment is half the rate at which the inversion decays. In magnetic resonance one introduces two relaxation times: the longitudinal relaxation time T_1 and the transverse relaxation time T_2. For our system these are given by

$$\frac{1}{T_1} = 2\gamma(1+2n_0), \qquad \frac{1}{T_2} = \gamma(1+2n_0), \qquad \frac{1}{T_1} = \frac{2}{T_2}. \tag{9.49}$$

Generally $\frac{1}{T_1} \neq \frac{2}{T_2}$ if dephasing interactions are also included. The dephasing interaction can be described in a simple manner by a very simple stochastic Hamiltonian

$$H_{SB} = \hbar S_z f(t), \tag{9.50}$$

where $f(t)$ is a Gaussian stochastic field. Clearly the variable S_z does not evolve due to this H_{SB}. The dipole moment operators evolve according to the Heisenberg equation

$$\dot{S}_+ = \mathrm{i}[\omega_0 + f(t)]S_+. \tag{9.51}$$

In the light of this the dephasing interaction would change only the decay of the dipole moment by an amount Γ, i.e. T_2 becomes

$$\frac{1}{T_2} = \gamma(1+2n_0) + \Gamma. \tag{9.52}$$

This can be seen by integrating (9.51) and using the property of a Gaussian stochastic process

$$\langle S_+(t)\rangle = \mathrm{e}^{\mathrm{i}\omega_0 t}\mathrm{e}^{-\frac{1}{2}\int\int\langle f(t_1)f(t_2)\rangle \mathrm{d}t_1\mathrm{d}t_2}\langle S_+(0)\rangle. \tag{9.53}$$

If we further assume that $f(t)$ is delta correlated

$$\langle f(t_1)f(t_2)\rangle = 2\Gamma\delta(t_1 - t_2), \tag{9.54}$$

then (9.53) reduces to

$$\langle S_+(t)\rangle = \mathrm{e}^{\mathrm{i}\omega_0 t - \Gamma t}\langle S_+(0)\rangle. \tag{9.55}$$

The master equation (9.45) acquires an extra term and can be written in the final form as

$$\begin{aligned}\frac{\partial \rho_S}{\partial t} &= -\mathrm{i}\omega_0[S_z, \rho_S] - \Gamma(S_z S_z \rho_S - 2S_z\rho_S S_z + \rho_S S_z S_z)\\ &\quad - \gamma(1+n_0)(S_+S_-\rho_S - 2S_-\rho_S S_+ + \rho_S S_+ S_-)\\ &\quad - \gamma n_0(S_-S_+\rho_S - 2S_+\rho_S S_- + \rho_S S_- S_+),\end{aligned} \tag{9.56}$$

instead of (9.46) and (9.47) we now have

$$\begin{aligned}\frac{\mathrm{d}}{\mathrm{d}t}\langle S_\pm\rangle &= \left(\pm\mathrm{i}\omega_0 - \frac{1}{T_2}\right)\langle S_\pm\rangle,\\ \frac{\mathrm{d}}{\mathrm{d}t}\langle S_z\rangle &= -\frac{1}{T_1}(\langle S_z\rangle - \eta),\\ \frac{1}{T_2} &= \frac{1}{2T_1} + \Gamma.\end{aligned} \tag{9.57}$$

There are many systems, particularly solid-state systems, where $T_2 \ll T_1$; for example, in a homogeneously broadened system like Ruby, $T_1 \approx 5$ ms, $T_2 =$ femtoseconds. In such systems the decay of $\langle S_+\rangle$ is much faster than the decay of the populations given by $\langle S_z\rangle$. Thus any initial coherence between two levels which is described by the mean value $\langle S_+\rangle$ will disappear much faster. The disappearance of coherence is nowadays referred to as decoherence [4].

9.4 Dissipative dynamics of a multilevel system

As a final example we consider a multilevel system with states labeled as $|i\rangle$ with energy E_i. Let A_{ij} be the operator which takes the system from the state $|j\rangle$ to $|i\rangle$

$$A_{ij} = |i\rangle\langle j|, \qquad A_{ij}^\dagger = |j\rangle\langle i| = A_{ji}. \tag{9.58}$$

The free evolution of the operator A_{ij} is given by

$$A_{ij}(t) = \mathrm{e}^{\mathrm{i}\omega_{ij}t}A_{ij}(0), \qquad \omega_{ij} = \frac{1}{\hbar}(E_i - E_j). \tag{9.59}$$

Let us write the interaction (9.8) with the bath as

$$H_{SB} = \sum_{ij} B_{ij}A_{ij}, \qquad B_{ij}^\dagger = B_{ji}, \qquad H_S = \sum E_i A_{ii}. \tag{9.60}$$

Substituting of (9.60) in (9.28) yields the equation

$$\frac{\partial \rho_S}{\partial t} = -\frac{\mathrm{i}}{\hbar}\left[\sum_i E_i A_{ii}, \rho_S\right] + \sum_{ijkl}[(A_{ij}\rho_S A_{kl} - A_{kj}\rho_S\delta_{li})\gamma^{+}_{klij} + (A_{kl}\rho_S A_{ij} - \rho_S A_{il}\delta_{jk})\gamma^{-}_{ijkl}], \tag{9.61}$$

where $\gamma^{\pm}$ are related to the bath correlations

$$\gamma^{+}_{klij} = \int_0^{\infty} \mathrm{d}t\, \mathrm{e}^{-\mathrm{i}\omega_{ij}t}\langle B_{kl}(t)B_{ij}(0)\rangle, \tag{9.62}$$

$$\gamma^{-}_{ijkl} = \int_0^{\infty} \mathrm{d}t\, \mathrm{e}^{-\mathrm{i}\omega_{ij}t}\langle B_{ij}(0)B_{kl}(t)\rangle, \tag{9.63}$$

$$= (\gamma^{+}_{lkji})^{*}. \tag{9.64}$$

where the last property follows from $B^{\dagger}_{ij} = B_{ji}$. We next make the rotating wave approximation, i.e. drop the rapidly oscillating terms from (9.61). To see what these are, we can transform (9.61) to the interaction picture and use (9.59). The terms like $A_{ij}\rho A_{kl}$ would have a time dependence $\mathrm{e}^{\mathrm{i}(\omega_{ij}+\omega_{kl})t}$. Therefore in (9.61) we retain only those terms for which

$$\omega_{ij} + \omega_{kl} = 0. \tag{9.65}$$

We assume that the energy levels are nondegenerate and unevenly spaced. Then (9.65) can be satisfied if

$$(a)\ k = j,\ l = i, \qquad (b)\ i = j,\ k = l. \tag{9.66}$$

The latter case implies terms in the interaction with the bath which are diagonal, i.e. of the form $B_{ii}A_{ii}$. Such terms would correspond to dephasing interactions. Using (9.61) and (9.66) we can write the equations for the matrix elements of ρ as

$$\begin{aligned}\frac{\partial \rho_{ii}}{\partial t} &= \sum_{k\neq i}(\gamma_{ik}\rho_{kk} - \gamma_{ki}\rho_{ii}),\\ \frac{\partial \rho_{ij}}{\partial t} &= -\mathrm{i}\omega_{ij}\rho_{ij} - \left[\Gamma_{ij} + \frac{1}{2}\sum_k(\gamma_{ki} + \gamma_{kj})\right]\rho_{ij}, \qquad i \neq j,\end{aligned} \tag{9.67}$$

where γ_{ik} is the rate of transition from the states $|k\rangle$ and $|i\rangle$, and Γ_{ij} is the dephasing rate. These are obtained from (9.62) and (9.63)

$$\gamma_{ik} = \int_{-\infty}^{+\infty}\langle B_{ki}(t)B_{ik}(0)\rangle \mathrm{e}^{-\mathrm{i}\omega_{ik}t}\mathrm{d}t, \tag{9.68}$$

$$\Gamma_{ij} = \pi\sum_{\alpha}|\langle\alpha|B_{ii} - B_{jj}|\alpha\rangle|^2\rho(\alpha), \tag{9.69}$$

where $|\alpha\rangle$ are the eigenstates of the bath Hamiltonian H_B. The diagonal elements of ρ satisfy the Pauli master equation [5] and the rates of transition are shown schematically in Figure 9.3. The decay of the off-diagonal element ρ_{ij} is the sum of the dephasing rate and a term that is half the net decay out of both the levels $|i\rangle$ and $|j\rangle$. The equations

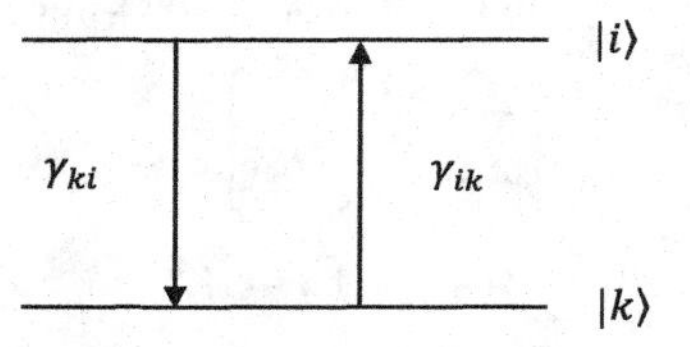

Fig. 9.3 Population changes are governed by the Pauli master equation with γ'^s representing the transfer of populations among different levels.

(9.67)–(9.69) are the basic equations for the dynamics of a multilevel system interacting with the environment [6].

It should be borne in mind that γ'^s and Γ'^s are to be obtained from microscopic considerations (Eqs. (9.68) and (9.69)) and should not be taken from phenomenological considerations as one has to insure the positivity of the density matrix. The positivity question has been investigated in detail for both two-level and multilevel systems [7–9]. One can show [7, 8] that the master equation

$$\frac{\partial \rho_s}{\partial t} = -\mathrm{i}[H, \rho] + \frac{1}{2}\sum_{i,j=1}^{N^2-1} c_{ij}\left([F_i, \rho F_j^\dagger] + [F_i\rho, F_j^\dagger]\right), \tag{9.70}$$

with the operators satisfying the conditions

$$H = H^\dagger, \quad \mathrm{Tr}H = 0, \quad \mathrm{Tr}F_i = 0, \quad \mathrm{Tr}(F_i^\dagger F_j) = \delta_{ij}, \tag{9.71}$$

will maintain the positivity of ρ_s if the coefficients c_{ij} form a positive definite matrix. The derivations in references [7] and [8] depend on the considerations of dynamical subgroups, whereas our derivation of (9.67) is from microscopic considerations. References [9, 10] consider explicitly the question of the positivity of ρ for three- and four-level systems. A simple case where phenomenological considerations can lead to difficulties is that of pure dephasing in three-level systems. The condition for the positivity of the density matrix for a three-level system undergoing only dephasing (see Exercise 9.7) is

$$\Gamma_{ab} + \Gamma_{bc} + \Gamma_{ca} \leq 2\sqrt{\Gamma_{ab}\Gamma_{bc} + \Gamma_{ac}\Gamma_{bc} + \Gamma_{ab}\Gamma_{ac}}. \tag{9.72}$$

Note that if one were to put phenomenologically one of the decay constants, say Γ_{ac}, equal to zero, then the above condition would yield $(\Gamma_{ab} - \Gamma_{bc})^2 \leq 0$, which can not be satisfied if either of Γ_{ab}, Γ_{bc} is different from zero. Hence in phenomenological theory numerical values of Γ'^s have to be used with great caution.

9.5 Time correlation functions for multilevel systems

The master equations derived in the previous section give the dynamics of the system and in particular enable us to calculate the time dependence of the expectation values. In

Chapter 8 we saw that complete physical specification of the system requires not only the time dependence of the expectation values but also time-dependent correlation functions. For example, we have seen the importance of intensity–intensity correlations (Eq. (8.31)) in interferometry. Furthermore, the spectral properties of the field given by (8.50) again require the evaluation of two-time correlation functions. Thus we need methods to calculate two-time and multitime correlation functions from the master equation for the dissipative system. Lax [11] discovered the quantum regression theorem which enables one to calculate the multitime correlation functions. The theorem relies on the Markovian dynamics of the system. The theorem states that if the mean values at time t are written in terms of the expectation values at an earlier time t' as

$$\langle G(t)\rangle = \sum_{\alpha} C_\alpha(t,t')\langle G_\alpha(t')\rangle, \tag{9.73}$$

then the multitime correlation can be computed as

$$\langle\cdots D(t'')C(t')G(t)A(t')B(t'')\cdots\rangle = \sum_{\alpha} C_\alpha(t,t')\langle\cdots D(t'')C(t')G_\alpha(t')A(t')B(t'')\cdots\rangle. \tag{9.74}$$

The functions $C_\alpha(t,t')$ are already known from the solution of the master equation. The theorem tells how to obtain the n-time correlation function in terms of $(n-1)$-time correlation functions. Thus the repeated use of (9.74) would enable one to calculate multitime correlation functions.

Let us illustrate this for the case of a harmonic oscillator by calculating $\langle a^\dagger(t+\tau)a(t)\rangle$. We write the solution of (9.39) as

$$\langle a^\dagger(t+\tau)\rangle = \mathrm{e}^{\mathrm{i}\omega_0\tau-\kappa\tau}\langle a^\dagger(t)\rangle, \tag{9.75}$$

and hence according to (9.74)

$$\langle a^\dagger(t+\tau)a(t)\rangle = \mathrm{e}^{(\mathrm{i}\omega_0-\kappa)\tau}\langle a^\dagger(t)a(t)\rangle. \tag{9.76}$$

The mean $\langle a^\dagger(t)a(t)\rangle$ can be obtained from the second equation in (9.39)

$$\langle a^\dagger(t)a(t)\rangle = \mathrm{e}^{-2\kappa t}\langle a^\dagger a\rangle_0 + n_0(1-\mathrm{e}^{-2\kappa t}). \tag{9.77}$$

As a further example we evaluate the intensity–intensity correlation

$$\langle a^\dagger(t)a^\dagger(t+\tau)a(t+\tau)a(t)\rangle,$$

which can be evaluated by writing (9.77) as

$$\langle a^\dagger(t+\tau)a(t+\tau)\rangle = \mathrm{e}^{-2\kappa\tau}\langle a^\dagger(t)a(t)\rangle + n_0(1-\mathrm{e}^{-2\kappa\tau}),$$

and hence

$$\langle a^\dagger(t)a^\dagger(t+\tau)a(t+\tau)a(t)\rangle = \mathrm{e}^{-2\kappa\tau}\langle[a^\dagger(t)]^2[a(t)]^2\rangle + n_0(1-\mathrm{e}^{-2\kappa\tau})\langle a^\dagger(t)a(t)\rangle. \tag{9.78}$$

The one-time expectation value $\langle a^{\dagger 2}a^2\rangle$ can be obtained from (9.37). As a special case in the steady state $t\to\infty$, $\langle a^{\dagger 2}a^2\rangle = 2n_0^2$, $\langle a^\dagger a\rangle = n_0$, and then (9.78) reduces to

$$\langle a^\dagger(t)a^\dagger(t+\tau)a(t+\tau)a(t)\rangle \longrightarrow n_0^2 + n_0^2\mathrm{e}^{-2\kappa\tau}. \tag{9.79}$$

This is something we used earlier (Eq. (8.39)) in connection with our discussion of the bunching of photons. The calculation of the two-time correlation function for a dissipative two-level system is left as an exercise – Exercise 9.8.

9.6 Quantum Langevin equations

In quantum mechanics we have two pictures – the Schrödinger picture and the Heisenberg picture. In the Heisenberg picture the operators evolve with time such that the equal time commutation relations are obeyed. For dissipative systems the analog of the Heisenberg equations are the quantum Langevin equations. These equations can be derived from the master equation (9.27). The Langevin equations contain force terms which arise from the interaction with the bath. These force terms are absolutely necessary to preserve the commutation relation. We write the Langevin equation for the system operator as

$$\frac{\mathrm{d}S_\alpha}{\mathrm{d}t} = A_\alpha + F_\alpha, \tag{9.80}$$

where $F_\alpha(t)$ is the operator Langevin force with zero mean value $\langle F_\alpha(t)\rangle = 0$, so that the mean value of S_α obeys the equation

$$\frac{\mathrm{d}\langle S_\alpha\rangle}{\mathrm{d}t} = \langle A_\alpha\rangle. \tag{9.81}$$

Clearly A_α can be obtained from the master equation (9.27). The master equation can also be used to obtain

$$\frac{\mathrm{d}\langle S_\alpha S_\beta\rangle}{\mathrm{d}t} = \langle A_{\alpha\beta}\rangle, \qquad \frac{\mathrm{d}\langle S_\beta S_\alpha\rangle}{\mathrm{d}t} = \langle A_{\beta\alpha}\rangle, \tag{9.82}$$

where $A_{\alpha\beta} \neq A_{\beta\alpha}$ as S_α and S_β do not commute. The correlation function of $F_\alpha(t)$ is a delta function under the assumed Markovian dynamics of the system

$$\langle F_\alpha(t)F_\beta(t')\rangle = 2\langle D_{\alpha\beta}(t)\rangle\delta(t-t'). \tag{9.83}$$

The diffusion term $2\langle D_{\alpha\beta}(t)\rangle$ is related to A'^s as follows

$$2\langle D_{\alpha\beta}(t)\rangle = \langle A_{\alpha\beta}\rangle - \langle A_\alpha S_\beta\rangle - \langle S_\alpha A_\beta\rangle. \tag{9.84}$$

Thus $\langle D_{\alpha\beta}(t)\rangle$ can be obtained from the master equation (9.27). In the present book we will mostly work with density matrix equations and the quantum regression theorem, although in Chapter 20 we mostly work with Langevin equations.

As an example of quantum Langevin equations we write the Langevin equations for the dissipative dynamics (9.37) of the harmonic oscillator

$$\dot{a} = -\mathrm{i}\omega_0 a - \kappa a + F(t), \tag{9.85}$$

where

$$\begin{aligned} \langle F^\dagger(t)F(t')\rangle &= 2\kappa n_0\delta(t-t'), \\ \langle F(t)F^\dagger(t')\rangle &= 2\kappa(n_0+1)\delta(t-t'). \end{aligned} \tag{9.86}$$

The quantum Langevin equation (9.85) can be used directly to calculate the time-dependent correlation functions. We demonstrate this calculation for the steady-state case. For this case it is sufficient to consider the contribution of the force term in the integration of (9.85)

$$a(t) = \int_0^t dt_1\, e^{-(\kappa+i\omega_0)t_1} F(t-t_1), \tag{9.87}$$

and therefore

$$\begin{aligned}
&\langle a^\dagger(t+\tau)a(t)\rangle \\
&= \lim_{t\to\infty}\int_0^{t+\tau} dt_2 \int_0^t dt_1\, e^{-(\kappa+i\omega_0)(t-t_1)} e^{-(\kappa-i\omega_0)(t+\tau-t_2)} \langle F^\dagger(t_2)F(t_1)\rangle, \\
&= \lim_{t\to\infty}\int_0^{t+\tau} dt_2 \int_0^t dt_1\, e^{-(\kappa+i\omega_0)(t-t_1)} e^{-(\kappa-i\omega_0)(t+\tau-t_2)} 2\kappa n_0 \delta(t_1-t_2).
\end{aligned} \tag{9.88}$$

The integrals in (9.88) are tricky. The best way is to write the t_2 integral as $\int_0^{t+\tau} dt_2 \cdots = \int_0^t dt_2 \cdots + \int_t^{t+\tau} dt_2 \cdots$ and show that the second integral would be zero as the t_1 and t_2 intervals do not overlap for this integral. The remaining integrals are straightforward and lead to

$$\langle a^\dagger(t+\tau)a(t)\rangle = n_0 e^{i\omega_0\tau - \kappa\tau}, \tag{9.89}$$

which is identical to (9.76) in the limit $t \to \infty$.

9.7 Exactly soluble models for the dissipative dynamics of the oscillator

In this section we discuss models where the effect of the bath can be treated exactly [12, 13]. This is advantageous when either the Markov approximation or the Born approximation is not a good approximation.

We first consider the case of a harmonic oscillator interacting with a zero temperature bath. We use a simplified form of the Hamiltonian

$$H = \hbar\omega_0 a^\dagger a + \sum_j \hbar\omega_j a_j^\dagger a_j + \sum_j (g_j a^\dagger a_j + H.c.). \tag{9.90}$$

We work in the Heisenberg picture. The equations of motion for a and a_j are

$$\dot{a} = -i\omega_0 a - i\sum_j g_j a_j, \tag{9.91}$$

$$\dot{a}_j = -i\omega_j a_j - i\sum_j g_j^* a. \tag{9.92}$$

We use Laplace transforms as we did in Section 7.6. We solve (9.92) and substitute it in (9.91) to obtain the Laplace transform $\hat{a}(p)$ of the system operator a

$$\hat{a}(p) = \left[a(0) - \mathrm{i}\sum_j g_j (p + \mathrm{i}\omega_j)^{-1} a_j \right] \Big/ \left[p + \mathrm{i}\omega_0 + \sum_j |g_j|^2 (p + \mathrm{i}\omega_j)^{-1} \right]. \quad (9.93)$$

This is in fact the exact solution. The bath operators are contained in (9.93) and this very term is like the Langevin force term required to maintain the commutation relation for the Heisenberg operators. Further simplification depends on the model of the bath, i.e. on the form of the coupling constants g_j. In the time domain we can write (9.93) as

$$a(t) = v(t)a(0) + f(t), \quad (9.94)$$

where $v(t)$ is the Laplace transform of the denominator in (9.93). Note further that since the bath is at zero temperature and since f contains the annihilation operators a_j, the normally ordered correlations of f would be zero

$$\langle f^\dagger(t) f(t') \rangle = 0, \qquad \langle f(t) \rangle = 0. \quad (9.95)$$

The antinormally ordered correlation can be obtained by the requirement $[a(t), a^\dagger(t)] = 1$, i.e.

$$\langle f(t) f^\dagger(t) \rangle = 1 - |v|^2. \quad (9.96)$$

Clearly from (9.94) we have

$$\langle a^\dagger(t) a(t) \rangle = |v(t)|^2 \langle a^\dagger(0) a(0) \rangle. \quad (9.97)$$

$$\langle a^\dagger(\tau) a(0) \rangle = v^*(\tau) \langle a^\dagger a \rangle. \quad (9.98)$$

Thus many important characteristics of the harmonic oscillator can be obtained from the functional form of $v(\tau)$. In order to evaluate $v(t)$ we consider a model of a bath consisting of a Lorentzian [12] distribution of oscillators, then

$$\sum_j |g_j|^2 (p + \mathrm{i}\omega_j)^{-1} \longrightarrow \int |g(\nu)|^2 (\mathcal{Z} + \mathrm{i}\nu - \mathrm{i}\omega_0)^{-1} \mathrm{d}\nu$$
$$\longrightarrow \frac{\kappa\Gamma}{\mathcal{Z} + \Gamma}, \qquad \mathcal{Z} = p + \mathrm{i}\omega_0, \quad (9.99)$$

for

$$|g(\nu)|^2 = \kappa\Gamma \frac{\Gamma/\pi}{\Gamma^2 + \nu^2}. \quad (9.100)$$

Therefore $v(t)$ is

$$v(t) = \mathrm{e}^{-\mathrm{i}\omega_0 t} \frac{1}{2\pi\mathrm{i}} \int \mathrm{d}z \left(p + \frac{\kappa\Gamma}{p + \Gamma} \right)^{-1} \mathrm{e}^{pt}$$
$$\rightarrow \mathrm{e}^{-\mathrm{i}\omega_0 t - \kappa t} \qquad \text{for large } \Gamma, \quad (9.101)$$

where Γ^{-1} is a measure of the memory time of the bath. The Laplace transform in (9.101) can be evaluated with the result

$$v(t) = \mathrm{e}^{-\mathrm{i}\omega t} [(\Gamma + \Lambda_1)\mathrm{e}^{\Lambda_1 t} - (\Gamma + \Lambda_2)\mathrm{e}^{\Lambda_2 t}]/(\Lambda_1 - \Lambda_2), \quad (9.102)$$

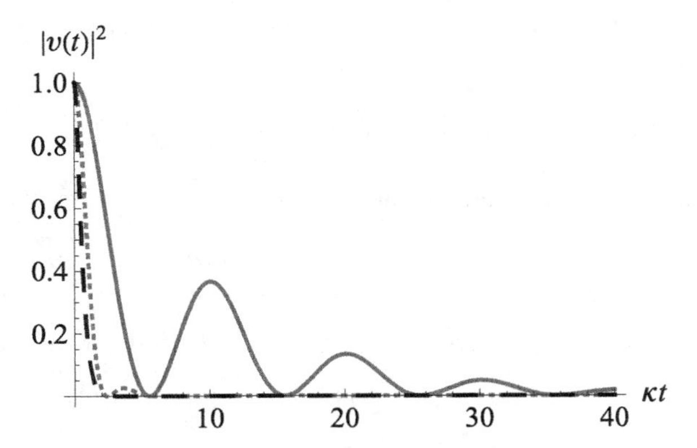

Fig. 9.4 The function $|v(t)|^2$ as a function of κt for different values of the bath correlation time. $\Gamma/\kappa = 0.1$ (solid), 1 (dotted), and 5 (dashed).

where Λ_1 and Λ_2 are the roots of the equation

$$p(p+\Gamma)+\kappa\Gamma = 0, \qquad \Lambda_1\Lambda_2 = \kappa\Gamma, \qquad \Lambda_1+\Lambda_2 = -\Gamma. \tag{9.103}$$

The finite memory time of the bath yields two different time scales for the evolution of $v(t)$. We show in Figure 9.4 the behavior of the function $|v(t)|^2$ as a function of κt for different values of the bath correlation time. From this figure, it is clear that for non-Markovian baths the decoherence is much slower than for Markovian baths.

9.8 Exact dissipative dynamics of a two-level system under dephasing

We now consider another exactly soluble model of dissipative dynamics: the dephasing of the atomic coherence. On a microscopic scale the dephasing can be considered to arise from the interaction of a two-level system with a bath of oscillators, i.e. from the Hamiltonian

$$H = \hbar\sum_i \omega_i a_i^\dagger a_i + \hbar S_z \sum_i g_i(a_i + a_i^\dagger). \tag{9.104}$$

The bath is taken to have a broad spectrum. In particular, for an Ohmic bath [14] we take the spectrum of the bath as

$$J \to \sum_i |g_i|^2\delta(\omega-\omega_i) = 2\alpha\omega\Theta(\omega_D-\omega), \tag{9.105}$$

or

$$J \to 2\alpha\omega e^{-\omega/\omega_D}, \tag{9.106}$$

where ω_D is the cut-off frequency. It essentially determines the correlation time of the bath. Such a bath leads to dephasing, i.e. the spin polarization decays at the rate T_2. We need to calculate the dynamical evolution of the off-diagonal element of the density matrix for the two-level system. We work in the interaction picture; hence the Hamiltonian (9.104) becomes

$$H = \hbar S_z \sum_i g_i (a_i e^{-i\omega_i t} + a_i^\dagger e^{i\omega_i t}) = \hbar S_z B(t), \tag{9.107}$$

where $B(t)$ is the bath operator given by

$$B(t) = \sum_i g_i (a_i e^{-i\omega_i t} + a_i^\dagger e^{i\omega_i t}). \tag{9.108}$$

It is easy to see that the off-diagonal element of the density matrix ρ is

$$\rho_{eg}(t) = \mathrm{Tr}_B \langle e|U(t)\rho_B \rho(0) U^\dagger(t)|g\rangle, \tag{9.109}$$

where Tr_B is over the initial bath density matrix ρ_B and where

$$U(t) = T \exp\left[-i \int_0^t S_z B(\tau) d\tau\right]. \tag{9.110}$$

This can be simplified to

$$\rho_{eg}(t) = \rho_{eg}(0) \mathrm{Tr}_B V(t) \rho_B V(t) = \rho_{eg}(0) \langle V^2(t)\rangle, \tag{9.111}$$

where

$$V(t) = T \exp\left[-\frac{i}{2} \int_0^t B(\tau) d\tau\right]. \tag{9.112}$$

Thus we can write

$$\rho_{eg}(t) = \rho_{eg}(0) \zeta(t), \tag{9.113}$$

$$\zeta(t) = \langle V^2(t)\rangle = \langle W(t)\rangle, \qquad W(t) = T \exp\left[-i \int_0^t B(\tau) d\tau\right]. \tag{9.114}$$

So far no approximation has been made.

We now examine the calculation of the function $W(t)$. We note that the bath operator $B(\tau)$ is such that the commutator $[B(\tau_1), B(\tau_2)]$ is a c-number. In such a case it has been shown by Glauber [15] that the time ordering can be simplified. It can be shown that

$$W = \exp\left[-i \int_0^t B(\tau) d\tau\right] \exp\left\{-\frac{1}{2} \int_0^t d\tau_1 \int_0^{\tau_1} d\tau_2 [B(\tau_1), B(\tau_2)]\right\}. \tag{9.115}$$

Since B is a Hermitian operator, the last exponential is just a c-number phase factor $\Phi(t)$ and hence

$$W = \exp[i\Phi(t)] \exp\left[-i \int_0^t B(\tau) d\tau\right], \tag{9.116}$$

$$= \exp[i\Phi(t)] \prod_j \exp(i f_j a_j + i f_j^* a_j^\dagger), \tag{9.117}$$

where

$$f_j = -g_j \int_0^t e^{-i\omega_j \tau} d\tau. \tag{9.118}$$

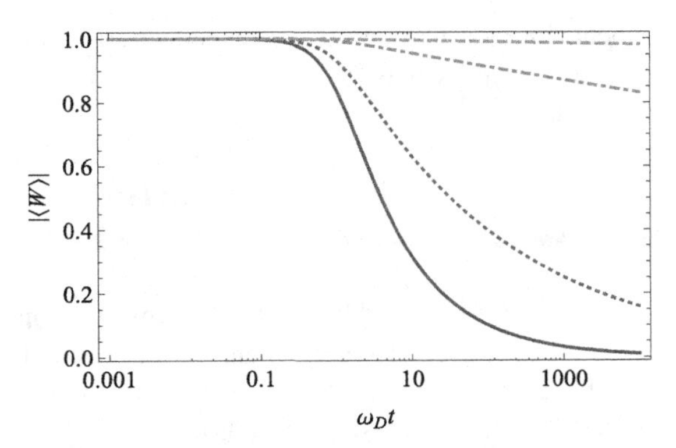

Fig. 9.5 The behavior of $|\langle W\rangle|$ as a function of $\omega_D t$ for a zero temperature bath and for $\alpha = 0.25$ (solid), 0.1 (dotted), 0.01 (dotdashed), and 0.001 (dashed).

On using the Baker–Hausdorff identity, (9.117) can be further simplified to

$$W = \exp[\mathrm{i}\Phi(t)]\prod_j \exp(\mathrm{i}f_j^* a_j^\dagger)\exp(\mathrm{i}f_j a_j)\exp\left(-\frac{1}{2}|f_j|^2\right), \tag{9.119}$$

The thermal expectation value of W can be obtained using, for example, the P-representation of the thermal density matrix [16, 17]

$$\rho_{thj} = \frac{1}{\pi n_j}\int \exp\left(-\frac{|\alpha|^2}{n_j}\right)|\alpha\rangle\langle\alpha|\mathrm{d}^2\alpha, \qquad n_j = \frac{1}{\mathrm{e}^{\beta\hbar\omega_j}-1}. \tag{9.120}$$

Using (9.119) in (9.120) we get

$$\langle W\rangle = \exp[\mathrm{i}\Phi(t)] \times \prod_j \exp\left(-\frac{1}{2}|f_j|^2\right) \times \frac{1}{\pi n_j}\int \exp\left(-\frac{|\alpha|^2}{n_j}\right)\exp(\mathrm{i}f_j^*\alpha^* + \mathrm{i}f_j\alpha)\mathrm{d}^2\alpha, \tag{9.121}$$

which on simplification reduces to

$$\langle W\rangle = \exp[\mathrm{i}\Phi(t)]\prod_j \exp\left[-\left(n_j+\frac{1}{2}\right)|f_j|^2\right]. \tag{9.122}$$

On using the form of f_j and on introducing the spectral density of the bath oscillators, the expression (9.122) becomes [14, 18]

$$\langle W\rangle = \exp[\mathrm{i}\Phi(t)]\exp\left\{-\int \mathrm{d}(\omega)J(\omega)\left[n(\omega)+\frac{1}{2}\right]|f(\omega)|^2\right\}, \tag{9.123}$$

where now

$$f(\omega) = -\int_0^t \mathrm{e}^{-\mathrm{i}\omega\tau}\mathrm{d}\tau. \tag{9.124}$$

For Ohmic baths at zero temperature the integral in (9.123) can be calculated with the result

$$|\langle W\rangle| = (1+\omega_D^2 t^2)^{-\alpha}, \tag{9.125}$$

and its behavior as a function of $\omega_D t$ is shown in Figure 9.5. It is important to note that the time-dependent decay is far from exponential decay. This is due to the non-Markovian behavior of the system.

Exercises

9.1 Prove that $\mathcal{P}$ as defined by (9.10) is a projection operator.

9.2 Prove that $\mathrm{Tr}_B[H_B, A] = 0$, where A is any operator dependent on both the system and the bath variables and H_B depends only on the bath operators.

9.3 Prove the relations (9.34) and (9.35).

9.4 The equation (9.42) is a recursion relation for p_n. Find explicitly $p_n(t)$ in terms of $p_m(0)$ for the special case of a zero temperature bath $n_0 = 0$. Show that

$$p_n(t) = \sum_{m \geq n} \binom{m}{n} \mathrm{e}^{-2\kappa t n}(1 - \mathrm{e}^{-2\kappa t})^{m-n} p_m(0).$$

Show that if $p_m(0) = \delta_{Nm}$, then $p_n(t)$ is a binomial distribution. What is the solution for $p_n(t)$ in the limit $t \to \infty$?

9.5 Solve directly (9.42) by setting $\frac{\partial p_n}{\partial t} = 0$ and show that $p_n = n_0^n/(1+n_0)^{n+1}$.

9.6 Using (9.44) prove that $S_\pm$, S_z are related to the Pauli matrices as stated in the line following (9.44), i.e. show that $S_\pm$, S_z satisfy all the algebraic properties of the Pauli matrices.

9.7 Consider the relaxation of a three-level atom with levels $|a\rangle$, $|b\rangle$, and $|c\rangle$ only via a dephasing mechanism. Then the diagonal elements of ρ do not change with time. The off-diagonal elements such as ρ_{ab} decay as $\exp(-\Gamma_{ab} t)$. Show that the three-level master equation for this case has the form (9.70) with

$$F_1 = (|a\rangle\langle a| - |b\rangle\langle b|)/\sqrt{2}, \qquad F_2 = \sqrt{\frac{3}{2}}\left(\frac{1}{3} - |c\rangle\langle c|\right),$$

$$c_{11} = \Gamma_{ab}, \qquad c_{22} = -\frac{\Gamma_{ab}}{3} + \frac{2}{3}(\Gamma_{ac} + \Gamma_{bc}),$$

$$c_{12} = \frac{1}{\sqrt{3}}(\Gamma_{ac} - \Gamma_{bc}) = c_{21}, \qquad H = 0.$$

The requirement that the matrix c be positive yields the condition

$$\Gamma_{ab} + \Gamma_{bc} + \Gamma_{ca} \leq 2\sqrt{\Gamma_{ab}\Gamma_{bc} + \Gamma_{ac}\Gamma_{bc} + \Gamma_{ab}\Gamma_{ac}},$$

which indeed would be the condition for the positivity of the density matrix for all times.

9.8 Using the master equation (9.56) and the quantum regression theorem show that the two-time correlation functions are given by

$$\lim_{t\to\infty} \langle S_+(t+\tau)S_-(t)\rangle = \frac{n_0}{1+2n_0}\mathrm{e}^{[i\omega_0 - \gamma(1+2n_0) - \Gamma]\tau},$$

$$\lim_{t\to\infty} \langle S_+(t)S_+(t+\tau)S_-(t+\tau)S_-(t)\rangle = \left(\frac{n_0}{1+2n_0}\right)^2 [1 - \mathrm{e}^{-2\gamma(1+2n_0)\tau}].$$

Confirm the Markov property of the last correlation function, i.e. in the limit $\tau \to \infty$, it goes to $\langle S_+ S_- \rangle^2$.

9.9 Using (9.94) show that all the normally ordered moments are given by

$$\langle a^{\dagger m}(t) a^n(t) \rangle = v^{*m}(t) v^n(t) \langle a^{\dagger m}(0) a^n(0) \rangle.$$

9.10 Using the result of Exercise 9.9 calculate the behavior of the Mandel Q_M parameter for an oscillator initially in a single-photon state $|1\rangle$. Plot $Q_M(t)$ as a function of κt for a range of values of Γ/κ.

References

[1] G. S. Agarwal, in *Progress in Optics* **XI**, edited by E. Wolf (Amsterdam: North-Holland, 1973).
[2] F. Haake, *Statistical Treatment of Open Systems by Generalized Master Equations*, 3rd ed. (Berlin: Springer-Verlag, 1973).
[3] H. J. Carmichael, *An Open Systems Approach to Quantum Optics* (Berlin: Springer-Verlag, 1993).
[4] W. H. Zurek, *Rev. Mod. Phys.* **75**, 715 (2003).
[5] W. Pauli, 1928, in *Probleme der Modernen Physik*, Arnold Sommerfeld zum 60, pp. 30–45; reprinted in *Collected Scientific Papers*, edited by P. Kronig and V. F. Weisskopf (New York: Interscience Publishers, 1964), p. 549.
[6] W. H. Louisell, in *Quantum Optics*, edited by R. J. Glauber (New York: Academic Press, 1969), p. 680.
[7] V. Gorini, A. Kossakowski, and E. C. G. Sudarshan, *J. Math. Phys.* **17**, 821 (1976).
[8] G. Lindblad, *Rep. Math. Phys.* **10**, 393 (1976).
[9] S. G. Schirmer and A. I. Solomon, *Phys. Rev. A* **70**, 022107 (2004).
[10] P. R. Berman and R. C. O'Connell, *Phys. Rev. A* **71**, 022501 (2005).
[11] M. Lax, *Phys. Rev.* **172**, 350 (1968).
[12] F. Haake and R. Reibold, *Phys. Rev. A* **32**, 2462 (1985).
[13] R. R. Puri, *Mathematical Methods of Quantum Optics* (Berlin: Springer-Verlag, 2001), Section 8.5.
[14] L. Viola and S. Lloyd, *Phys. Rev. A* **58**, 2733 (1998).
[15] R. J. Glauber, *Quantum Optics and Electronics*, edited by C. deWitt, A. Blanden, and C. Cohen-Tannoudji (New York: Gordon and Breach, 1965), p. 132.
[16] R. J. Glauber, *Phys. Rev. Lett.* **10**, 84 (1963).
[17] E. C. G. Sudarshan, *Phys. Rev. Lett.* **10**, 277 (1963).
[18] G. S. Uhrig, *Phys. Rev. Lett.* **98**, 100504 (2007).

10 Amplification and attenuation of quantum fields

As an application of the theory of open quantum systems, we consider the theory of optical amplifiers and absorbers. We would like to understand how an input field, particularly a nonclassical field, is amplified or attenuated. We specifically discuss changes in the quantum character of the field. This is especially important in connection with the amplification of fields with very low photon numbers since the amplifiers are known to add quantum noise [1–5]. Thus even if the input signal is shot noise limited, the output is not. We will also discuss how the entanglement is significantly affected by both amplifiers and attenuators.

10.1 Quantum theory of optical amplification

Consider first a classical field propagating through an optical amplifier, as shown in Figure 10.1. The output field is related to the input field via

$$\mathcal{E}_{\text{out}} = \exp\left(\frac{Gt}{2}\right)\mathcal{E}_{\text{in}}\text{e}^{-\text{i}\theta}, \tag{10.1}$$

where θ is the phase acquired in propagation and G is the gain of the amplifier. The index t is related to the length of the amplifier. In quantum theory one might anticipate that the Heisenberg operators for the output and input fields would be related by

$$a_{\text{out}} = \exp\left(-\text{i}\theta + \frac{Gt}{2}\right)a_{\text{in}}. \tag{10.2}$$

However, (10.2) leads inconsistency with quantum mechanics as

$$[a_{\text{out}}(t), a_{\text{out}}^{\dagger}(t)] = \exp(Gt) \neq 1. \tag{10.3}$$

In order to maintain consistency with quantum mechanics we have to supplement (10.2) with a quantum force term $f(t)$ such that

$$a_{\text{out}} = \exp\left(-\text{i}\theta + \frac{Gt}{2}\right)a_{\text{in}} + f(t). \tag{10.4}$$

Clearly the mean value of $f(t)$ should be zero. Furthermore, the requirement $[a_{\text{out}}(t), a_{\text{out}}^{\dagger}(t)] = 1$ yields

$$[f(t), f^{\dagger}(t)] = 1 - \exp(Gt). \tag{10.5}$$

We need to know the detailed properties of the force $f(t)$ besides the commutator, however. For example, we need to know $\langle f^{\dagger}(t)f(t)\rangle$ in order to obtain the mean number of photons

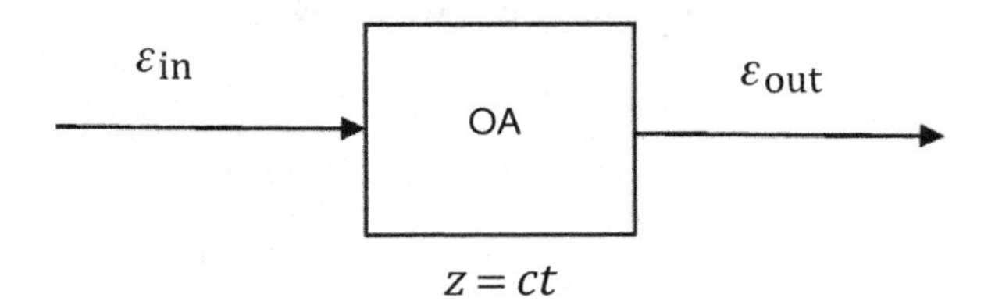

Fig. 10.1 A classical field propagates through an optical amplifier.

$\langle a^\dagger_{out} a_{out}\rangle$ in the output. The properties of $f(t)$ can be obtained from a microscopic model of the amplifier. It may be noted that Eq. (10.4) has the structure of the solution for the quantum Langevin equation that we derived in Section 9.6.

We can treat the amplifier as a bath of two-level atoms with N atoms such that we have N_1 atoms (a large number) in the excited state and $N_2 < N_1$ atoms in the ground state. In principle, N_2 could be zero. In order to achieve this we have to pump the atoms by external means. A state of thermal equilibrium would lead to $N_1 < N_2$. We can effectively consider the amplifier problem as a case of equilibrium but with an effective negative temperature. The derivation of the master equation is almost parallel with that in Section 9.2. The bath is now a set of two-level atoms which can be represented by spin operators (Eq. (9.44)). The Hamiltonian would be (9.31) with $a_j \to S_j^-$, $a_j^\dagger \to S_j^+$. Furthermore, instead of (9.34) we have

$$\begin{aligned} &\sum_j \langle S_j^+ S_j^-\rangle = N_1, \qquad \sum_j \langle S_j^- S_j^+\rangle = N_2, \\ &\langle S_j^+ S_l^-\rangle = 0, \qquad \langle S_j^+ S_l^+\rangle = 0, \qquad j \neq l. \end{aligned} \tag{10.6}$$

This yields the master equation (cf. Eq. (9.37))

$$\begin{aligned} \frac{\partial \rho}{\partial t} &= -\mathrm{i}\omega[a^\dagger a, \rho] - \kappa n_2 (a^\dagger a \rho - 2a\rho a^\dagger + \rho a^\dagger a) \\ &\quad - \kappa n_1 (a a^\dagger \rho - 2a^\dagger \rho a + \rho a a^\dagger), \end{aligned} \tag{10.7}$$

where $n_i = \frac{N_i}{N_1+N_2}$, $i = 1, 2$, and $N = N_1 + N_2$. The parameter κ is defined by (9.38). As before the master equation would be valid in the limit of weak coupling and the short correlation time for the bath. Following the procedure of Section 9.6, we can write the quantum Langevin equation

$$\dot{a} = -\mathrm{i}\omega a - \kappa(n_2 - n_1)a + F(t), \tag{10.8}$$

where the Langevin force $F(t)$ is Gaussian with the properties

$$\begin{aligned} &\langle F(t)\rangle = 0, \qquad \langle F^\dagger(t) F(t')\rangle = 2\kappa\delta(t-t')n_1, \\ &\langle F(t) F^\dagger(t')\rangle = 2\kappa\delta(t-t')n_2. \end{aligned} \tag{10.9}$$

The Gaussian property of the Langevin force enables one to compute all the quantum statistical properties of the fields. We can write the solution of (10.8) as

$$\begin{aligned} a(t) &= g(t)a(0) + f(t), \\ f(t) &= \int_0^t \mathrm{d}\tau\, g(\tau)F(t-\tau), \\ g(\tau) &= \exp[-\mathrm{i}\omega\tau + \kappa(n_1 - n_2)\tau]. \end{aligned} \tag{10.10}$$

We thus obtain the structure (10.4) as required by elementary quantum-mechanical consistency considerations. However, detailed correlation properties of $F(t)$ can now be obtained from Eq. (10.9). From (10.9) and (10.10) it is easily shown that

$$\langle a(t)\rangle = g(t)\langle a(0)\rangle, \tag{10.11}$$

$$\langle a^\dagger(t)a(t)\rangle = |g(t)|^2\langle a^\dagger(0)a(0)\rangle + \frac{n_1}{n_1 - n_2}(|g(t)|^2 - 1). \tag{10.12}$$

Here is the key aspect of the optical amplification: it adds noise photons given by the second term in (10.12). An ideal amplifier would be one for which $N_2 = 0$, $N_1 = N$. For an ideal amplifier all antinormally ordered correlations like $\langle F(t)F^\dagger(t')\rangle$ are zero. This when combined with (10.10) yields the following very interesting result for the antinormally ordered moments of a

$$\langle a^m(t)a^{\dagger n}(t)\rangle = g^m(t)g^{*n}(t)\langle a^m(0)a^{\dagger n}(0)\rangle. \tag{10.13}$$

10.1.1 Amplification of a coherent field – the noise figure of an ideal amplifier

Let us consider the amplification of a quantum field in a coherent state. We calculate the signal-to-noise ratio as a function of the gain of the amplifier. For simplicity, we consider only the case of an ideal amplifier. We also derive an expression for the noise figure of the amplifier [6]. The signal S is $\langle a^\dagger(t)a(t)\rangle$, which for a coherent state $|\alpha\rangle$ can be obtained from (10.12)

$$S = |g(t)|^2|\alpha|^2 + |g(t)|^2 - 1, \qquad n_2 = 0. \tag{10.14}$$

The noise ΔS can be calculated from

$$(\Delta S)^2 = \langle [a^\dagger(t)a(t)]^2\rangle - \langle a^\dagger(t)a(t)\rangle^2, \tag{10.15}$$

which in turn can be calculated using the relation (10.13) for a perfect amplifier. Using the commutation relation $[a, a^\dagger] = 1$, we can write (10.15) as

$$\begin{aligned} (\Delta S)^2 &= \langle a^2(t)a^{\dagger 2}(t)\rangle - \langle a(t)a^\dagger(t)\rangle^2 - \langle a(t)a^\dagger(t)\rangle \\ &= |g(t)|^4(\langle a^2a^{\dagger 2}\rangle - \langle aa^\dagger\rangle^2) - |g(t)|^2\langle aa^\dagger\rangle \\ &= |g(t)|^4[\langle (a^\dagger a)^2\rangle - \langle a^\dagger a\rangle^2 + \langle aa^\dagger\rangle] - |g(t)|^2\langle aa^\dagger\rangle \\ &= |g(t)|^4(\langle a^{\dagger 2}a^2\rangle - \langle a^\dagger a\rangle^2 + \langle a^\dagger a\rangle + \langle aa^\dagger\rangle) - |g(t)|^2\langle aa^\dagger\rangle \\ &= |g(t)|^4(1 + 2|\alpha|^2) - |g(t)|^2(1 + |\alpha|^2), \end{aligned} \tag{10.16}$$

which in the limit of large gain and large input field goes over to

$$(\Delta S)^2 \to 2|\alpha|^2|g(t)|^4. \tag{10.17}$$

In this limit, the signal $S \to |g(t)|^2|\alpha|^2$. Hence the signal to noise ratio for the output field is

$$S/(\Delta S) = |g(t)|^2|\alpha|^2/\sqrt{2|\alpha|^2|g(t)|^4} = \sqrt{|\alpha|^2/2}. \tag{10.18}$$

For a coherent field, the initial value of the ratio is $\sqrt{|\alpha|^2}$. Thus the noise figure $\mathcal{F}$ of an ideal amplifier is equal to

$$\mathcal{F} = (S/\Delta S)^2_{\text{in}}/(S/\Delta S)^2_{\text{out}} \cong 2. \tag{10.19}$$

The situation is different for a very weak coherent source $|\alpha|^2 \ll 1$, $S \sim |g(t)|^2$, $(\Delta S)^2 \sim |g(t)|^4$ and hence $(S/\Delta S)_{\text{out}} \approx 1$.

We can also compute the noise figure if a homodyne detection of signal is made [7]. Let us detect the quadrature $x(t) = [\mathrm{e}^{\mathrm{i}\omega t}a(t) + a^\dagger(t)\mathrm{e}^{-\mathrm{i}\omega t}]/\sqrt{2}$, which from (10.10) is

$$x(t) = |g(t)|x(0) + \frac{\mathrm{e}^{\mathrm{i}\omega t}f(t) + \mathrm{e}^{-\mathrm{i}\omega t}f^\dagger(t)}{\sqrt{2}}. \tag{10.20}$$

The mean value of $x(t)$ is $|g(t)|\langle x(0)\rangle$. The fluctuation in $x(t)$ is

$$\begin{aligned}\langle x^2(t)\rangle &= |g(t)|^2\langle x^2(0)\rangle + \frac{1}{2}\langle ff^\dagger + f^\dagger f + \mathrm{e}^{2\mathrm{i}\omega t}f^2 + \mathrm{e}^{-2\mathrm{i}\omega t}f^{\dagger 2}\rangle \\ &= |g(t)|^2\langle x^2(0)\rangle + \frac{1}{2}(|g(t)|^2 - 1),\end{aligned} \tag{10.21}$$

and hence if the input signal is in a coherent state, then

$$\begin{aligned}\Delta x^2(t) &= \langle x^2(t)\rangle - \langle x(t)\rangle^2 \\ &= |g(t)|^2(\langle x^2(0)\rangle - \langle x(0)\rangle^2) + \frac{1}{2}(|g(t)|^2 - 1) \\ &= \frac{1}{2}|g(t)|^2 + \frac{1}{2}(|g(t)|^2 - 1) \\ &= |g(t)|^2 - \frac{1}{2} \\ &\to |g(t)|^2 \qquad \text{for large gain.}\end{aligned} \tag{10.22}$$

Defining the noise figure $\mathcal{F}_x$ for quadrature detection as

$$\begin{aligned}\mathcal{F}_x &= \frac{(\langle x\rangle^2/\Delta x^2)_{\text{in}}}{\langle x(t)\rangle^2/[\Delta x(t)]^2} \\ &\cong 2 \qquad \text{for large gain.}\end{aligned} \tag{10.23}$$

Hence for a phase-insensitive amplifier the noise figure for the input coherent fields is two for both direct detection (photon number detection) and homodyne detection.

10.2 Loss of nonclassicality in the amplification process

Since the process of amplification adds noise to the input field, we expect that the amplifier would lead to the loss of nonclassicality of the input fields. We also expect that for sufficiently large gain of the amplifier, the nonclassical states would become classical. As discussed in Chapter 2, a complete description of nonclassicality is in terms of the P-function of the field. We would therefore find the P-function of the output field in terms of the P-function of the input field.

10.2.1 Time evolution of the phase-space distributions of quantum fields

Starting with the basic density matrix equation (10.7), we derive equations of motion for the phase-space distributions $P(\alpha)$, $Q(\alpha)$, and $W(\alpha)$ introduced in Chapter 1. In the literature one can find well-formulated procedures to transform the density matrix equations into equations for the phase-space distributions. However, the structure of Eq. (10.7) is rather simple and therefore one can use a simpler procedure. The following properties can be proved by expanding the coherent states $|\alpha\rangle$ in Fock states

$$\begin{gathered} a|\alpha\rangle\langle\alpha| = \alpha|\alpha\rangle\langle\alpha|, \qquad |\alpha\rangle\langle\alpha|a^\dagger = \alpha^*|\alpha\rangle\langle\alpha|, \\ a^\dagger|\alpha\rangle\langle\alpha| = \left(\frac{\partial}{\partial\alpha} + \alpha^*\right)|\alpha\rangle\langle\alpha|, \; |\alpha\rangle\langle\alpha|a = \left(\frac{\partial}{\partial\alpha^*} + \alpha\right)|\alpha\rangle\langle\alpha|, \\ a|\alpha\rangle\langle\alpha|a^\dagger = |\alpha|^2|\alpha\rangle\langle\alpha|, \\ a^\dagger|\alpha\rangle\langle\alpha|a = \left(\frac{\partial}{\partial\alpha} + \alpha^*\right)\left(\frac{\partial}{\partial\alpha^*} + \alpha\right)|\alpha\rangle\langle\alpha|. \end{gathered} \tag{10.24}$$

Consider a term like $a^\dagger a\rho$, which on using the P-function and (10.24) can be written as

$$\begin{aligned} a^\dagger a\rho &= \int P(\alpha)\mathrm{d}^2\alpha\, a^\dagger a|\alpha\rangle\langle\alpha| \\ &= \int \mathrm{d}^2\alpha\, P(\alpha)\alpha\left(\frac{\partial}{\partial\alpha} + \alpha^*\right)|\alpha\rangle\langle\alpha|, \end{aligned} \tag{10.25}$$

which on integration by parts reduces to

$$a^\dagger a\rho = \int \mathrm{d}^2\alpha\, |\alpha\rangle\langle\alpha|\left(\alpha^* - \frac{\partial}{\partial\alpha}\right)\alpha P(\alpha). \tag{10.26}$$

Thus all the terms in (10.7) can be written in the form (10.26), which then leads to a differential equation for the P-function

$$\frac{\partial P}{\partial t} = [\mathrm{i}\omega - \kappa(n_1 - n_2)]\frac{\partial}{\partial\alpha}(\alpha P) + \kappa n_1\frac{\partial^2 P}{\partial\alpha\partial\alpha^*} + c.c.. \tag{10.27}$$

The P-function satisfies a second-order differential equation. Note that since the Q-function is defined by $\frac{1}{\pi}\langle\alpha|\rho|\alpha\rangle = \frac{1}{\pi}\mathrm{Tr}\{\rho|\alpha\rangle\langle\alpha|\}$, we can use (10.24) to derive the differential equation for $Q(\alpha)$

$$\frac{\partial Q}{\partial t} = [\mathrm{i}\omega - \kappa(n_1 - n_2)]\frac{\partial}{\partial\alpha}(\alpha Q) + \kappa n_2\frac{\partial^2 Q}{\partial\alpha\partial\alpha^*} + c.c.. \tag{10.28}$$

The derivation of the equation for the Wigner function is more involved. We state the procedure without proof – we need to transform the product of operators like $G\rho$, ρG into differential operators acting on $W(\alpha)$. For this purpose we first write G in a form which is symmetric in a and $a^\dagger$. For example, $a^\dagger a \to \frac{a^\dagger a + aa^\dagger}{2} - \frac{1}{2}$. Then we replace a by α and

$a^\dagger$ by α^*. Let us denote the function thus obtained from the operator G by $\chi(\alpha)$, then the transformation rules are [8,9]

$$G\rho \to \chi(\alpha)\exp\left[\frac{1}{2}\left(\frac{\overleftarrow{\partial}}{\partial\alpha}\frac{\overrightarrow{\partial}}{\partial\alpha^*} - \frac{\overleftarrow{\partial}}{\partial\alpha^*}\frac{\overrightarrow{\partial}}{\partial\alpha}\right)\right]W(\alpha),$$

$$\rho G \to \chi(\alpha)\exp\left[-\frac{1}{2}\left(\frac{\overleftarrow{\partial}}{\partial\alpha}\frac{\overrightarrow{\partial}}{\partial\alpha^*} - \frac{\overleftarrow{\partial}}{\partial\alpha^*}\frac{\overrightarrow{\partial}}{\partial\alpha}\right)\right]W(\alpha). \tag{10.29}$$

The function $\chi(\alpha)$ is the Wigner function representation of the operator G. We can use these repeatedly to transform terms like $G\rho H$. For example,

$$a\rho \to \alpha W(\alpha) + \frac{1}{2}\frac{\partial W(\alpha)}{\partial\alpha^*},$$

$$\rho a \to \alpha W(\alpha) - \frac{1}{2}\frac{\partial W(\alpha)}{\partial\alpha^*}. \tag{10.30}$$

This procedure transforms the master equation (10.7) into the equation for the Wigner function

$$\frac{\partial W}{\partial t} = [\mathrm{i}\omega - \kappa(n_1 - n_2)]\frac{\partial}{\partial\alpha}(\alpha W) + \frac{1}{2}\kappa\frac{\partial^2 W}{\partial\alpha\partial\alpha^*} + c.c.. \tag{10.31}$$

The differential equations (10.27), (10.28), and (10.31) for the phase-space distributions have a structure similar to the Fokker–Planck equations for classical stochastic processes [10]. In particular, the similarity to the Ornstein–Uhlenbeck process should be noted. The techniques for solving the Fokker–Planck equations are well known and these can be adopted for obtaining the solutions of (10.27), (10.28), and (10.31) subject to arbitrary initial conditions. We summarize the key results of the linearized Fokker–Planck equations in Table 10.1. Using Eqs. (7) and (8) of the table we find that the time-dependent solutions for any of these functions have the form

$$\Phi(\alpha, t) = \int \mathrm{d}^2\alpha_0\, \Phi(\alpha_0, 0)\frac{1}{\pi\eta}\exp\left(-\frac{|\alpha - \alpha_0 g(t)|^2}{\eta}\right), \tag{10.32}$$

where Φ stands for any of the functions P, Q, and W, and η depends on the phase-space distribution

$$\begin{aligned}\eta &= \frac{n_1}{n_1 - n_2}(|g(t)|^2 - 1), && \text{for the } P\text{-function,}\\ &= \frac{n_2}{n_1 - n_2}(|g(t)|^2 - 1), && \text{for the } Q\text{-function,}\\ &= \frac{1/2}{n_1 - n_2}(|g(t)|^2 - 1), && \text{for the } W\text{-function.}\end{aligned} \tag{10.33}$$

Clearly the process of amplification leads to the broadening of phase-space distributions as $\eta \neq 0$. An exception occurs for a perfect amplifier, $n_2 \to 0$, when the Q-function has a very simple evolution

$$Q(\alpha, t) = \int \mathrm{d}^2\alpha_0\, Q(\alpha_0, 0)\delta^{(2)}(\alpha - g(t)\alpha_0), \tag{10.34}$$

$$= \frac{1}{|g(t)|^2}Q(g^{-1}(t)\alpha, 0). \tag{10.35}$$

Table 10.1 Solutions of the linearized Fokker–Planck equation.

Fokker–Plank equation for N real variables x_i

$$\frac{\partial P}{\partial t} = \sum_{ij} \gamma_{ij} \frac{\partial}{\partial x_i}(x_j P) + \sum_{ij} D_{ij} \frac{\partial^2 P}{\partial x_i \partial x_j}, \quad D_{ij} = D_{ji}. \tag{1}$$

Solution of (1) in terms of Green's function

$$P(\{x\}, t) = \int G(\{x\}, t|\{x'\}, 0) P(\{x'\}, 0) \mathrm{d}\{x'\}. \tag{6}$$

Initial condition for obtaining Green's function $G(\{x\}, t|\{x'\}, 0)$

$$G(\{x\}, 0|\{x'\}, 0) = \delta(\{x\} - \{x'\}). \tag{2}$$

Result

$$G(\{x\}, t|\{x'\}, 0) = (2\pi)^{-N/2}[\det \sigma(t)]^{-1/2} \times \exp\left\{-\tfrac{1}{2}[x - x'(t)]^T \sigma^{-1}(t)[x - x'(t)]\right\}. \tag{3}$$

x [x'] column vector with components $x_1, \ldots, x_N$ [$x'_1, \ldots, x'_N$].

$x'(t)$ column vector$= \mathrm{e}^{-\gamma t} x'$. (4)

The matrix γ has elements γ_{ij}.

The matrix σ is the solution of

$$\dot{\sigma} = -\gamma\sigma - \sigma\gamma^{\mathrm{T}} + 2D. \tag{5}$$

For diagonal γ and D matrices $\gamma_{ii} = \gamma_i$, $D_{ii} = D_i$,

$$G = \prod_i \frac{1}{\sqrt{2\pi\sigma_i(t)}} \exp\left[-\frac{1}{2}(x_i - \mathrm{e}^{-\gamma_i t} x'_i)^2/\sigma_i(t)\right], \tag{7}$$

$$\sigma_i(t) = D_i(1 - \mathrm{e}^{-\gamma_i t})/\gamma_i. \tag{8}$$

Furthermore, the structure of the differential equations (10.27), (10.28), and (10.31) implies that if initially any of the distributions P, Q, and W are Gaussian, then these would remain Gaussian for all times, and the parameters of the Gaussian would depend on time. Assuming for simplicity that $\langle a(0)\rangle = 0$, then we can write any of the functions P, Q, and W in the form (2.77) for any time t

$$\Phi(\alpha) = \frac{1}{\pi^2\sqrt{\tau^2 - 4|\mu|^2}} \exp\left(-\frac{\mu\alpha^2 + \mu^*\alpha^{*2} + \tau|\alpha|^2}{\tau^2 - 4|\mu|^2}\right), \tag{10.36}$$

where

$$2\mu^* = -\langle a^2\rangle = -g^2(t)\langle a^2\rangle_0, \tag{10.37}$$

and τ depends on the chosen phase-space function

$$\tau_W = \frac{1}{2} + \langle a^\dagger a \rangle = |g(t)|^2 \langle a^\dagger a \rangle_0 + \frac{1}{2} + \frac{n_1}{n_1 - n_2}(|g(t)|^2 - 1), \tag{10.38}$$

$$\tau_P = \tau_W - \frac{1}{2} = \tau_Q - 1. \tag{10.39}$$

10.2.2 Complete loss of nonclassicality

The basic result (10.32) on the transformation of phase-space distributions can be used to understand the loss of nonclassicality [11, 12] in the amplification process. Let us write (10.32) for the P-function

$$P(\alpha, t) = \int \mathrm{d}^2\alpha_0\, P(\alpha_0, 0) \frac{1}{\pi\eta} \exp\left[-\frac{|g(t)|^2}{\eta} |\alpha_0 - \alpha g^{-1}(t)|^2 \right], \tag{10.40}$$

where, in the context of the P-function, η is given by the first line of (10.33). For a gain given by

$$\frac{\eta}{|g(t)|^2} = \frac{n_1}{n_1 - n_2}\left(1 - \frac{1}{|g(t)|^2}\right) = 1, \tag{10.41}$$

we find that the P-function at time t is equal to the scaled Q-function at $t = 0$ [11]

$$P(\alpha, t) = \frac{1}{\pi |g(t)|^2} \int \mathrm{d}^2\alpha_0\, P(\alpha_0, 0) \exp\left[-|\alpha_0 - \alpha g^{-1}(t)|^2 \right] \tag{10.42}$$

$$= \frac{1}{|g(t)|^2} Q(\alpha g^{-1}(t), 0), \tag{10.43}$$

where we have used the relation (1.95) between the Q-function and the P-function. Noting that the Q-function always exists, we find that for a gain given by (10.41), the field becomes classical, i.e. the initial field has lost all its nonclassical properties. Further amplification will make the field more and more classical. Thus for all gains given by

$$\frac{\eta}{|g(t)|^2} \geq 1, \tag{10.44}$$

the field would become completely classical [12]. However, for special cases of input fields, the nonclassical fields can become classical for even lower gains given by $\eta/|g(t)|^2 < 1$. We give an important example. Using (10.40) and the relation (1.103) between the P-function and the Wigner function, we find that the P-function at time t is equal to the Wigner function at $t = 0$

$$P(\alpha, t) = \frac{1}{|g(t)|^2} W(\alpha g^{-1}(t), 0), \tag{10.45}$$

$$\frac{\eta}{|g(t)|^2} = \frac{1}{2}. \tag{10.46}$$

Consider the input field which is in the squeezed vacuum state, then the Wigner function is positive, see Eq. (2.39), although the P-function doesn't exist. Hence under the condition (10.46), which is a weaker condition than (10.44), the squeezed vacuum state on amplification loses its nonclassical character completely.

The result (10.45) has an important consequence – any field for which the P-function is highly singular will evolve into a field whose P-function would exist if the gain satisfies the condition (10.46), although it can be negative. The last property follows from the fact that, in general, the Wigner function can be negative.

10.2.3 Partial loss of nonclassicality

In the previous section we derived the conditions under which nonclassicality is completely lost. However, it is possible that some of the observable nonclassical characteristics such as squeezing and sub-Poissonian statistics are lost while the P-function still remains nonclassical. For simplicity we consider the case of the perfect amplifier, $n_2 \to 0$. Let us first consider the degradation in squeezing as the amplifier gain increases. The relevant mean values are

$$\begin{aligned} \langle a(t)\rangle &= g(t)\langle a(0)\rangle, \qquad \langle a^2(t)\rangle = g^2(t)\langle a^2(0)\rangle, \\ \langle a^\dagger(t)a(t)\rangle &= |g(t)|^2\langle a^\dagger(0)a(0)\rangle + |g(t)|^2 - 1. \end{aligned} \tag{10.47}$$

We define the time-dependent quadratures (cf. Eq. (1.37))

$$X(t) = \frac{a\mathrm{e}^{\mathrm{i}\omega t} + a^\dagger \mathrm{e}^{-\mathrm{i}\omega t}}{\sqrt{2}}, \qquad Y(t) = \frac{a\mathrm{e}^{\mathrm{i}\omega t} - a^\dagger \mathrm{e}^{-\mathrm{i}\omega t}}{\sqrt{2}\,\mathrm{i}}. \tag{10.48}$$

Using (10.47) and (10.48) we find

$$\begin{aligned} \langle X(t)\rangle &= |g(t)|\langle X(0)\rangle, \\ \langle X^2(t)\rangle &= |g(t)|^2\langle X^2(0)\rangle + \frac{|g(t)|^2 - 1}{2}. \end{aligned} \tag{10.49}$$

Therefore the squeezing parameter S_0 defined by (2.7) is

$$S_0(t) = \langle X^2(t)\rangle - \langle X(t)\rangle^2 - \frac{1}{2} = |g(t)|^2 S_0(0) + |g(t)|^2 - 1. \tag{10.50}$$

If initially the field is squeezed, i.e. $-\frac{1}{2} \le S_0 < 0$, then $S_0(t)$ would become nonnegative if

$$|g(t)|^2 \ge \frac{1}{1 + S_0(0)}. \tag{10.51}$$

Since the smallest value of $S_0(0)$ is $-\frac{1}{2}$, we find that under the condition

$$|g(t)|^2 \ge 2, \tag{10.52}$$

the squeezing character of the field would be definitely lost [12] no matter what the initial value of S_0 is.

We next consider the degradation in the sub-Poissonian nature of the field. For this purpose we use the result (10.16)

$$\langle (a^\dagger a)_t^2\rangle - \langle (a^\dagger a)_t\rangle^2 = |g(t)|^4[\langle (a^\dagger a)^2\rangle - \langle a^\dagger a\rangle^2 + \langle a^\dagger a\rangle + 1] - |g(t)|^2(\langle a^\dagger a\rangle + 1). \tag{10.53}$$

From this the Mandel $Q_{\rm M}$ parameter is

$$Q_{\rm M}(t) = \frac{\langle (a^\dagger a)_t^2 \rangle - \langle (a^\dagger a)_t \rangle^2 - \langle (a^\dagger a)_t \rangle}{\langle (a^\dagger a)_t \rangle}$$
$$= \frac{|g(t)|^4 [Q_{\rm M}(0)\langle a^\dagger a\rangle + 2\langle a^\dagger a\rangle + 1] - 2|g(t)|^2(1 + \langle a^\dagger a\rangle) + 1}{|g(t)|^2 \langle a^\dagger a\rangle + |g(t)|^2 - 1}. \quad (10.54)$$

Noting that the smallest value of $Q_{\rm M}(0)$ is -1, it is clear that $Q_{\rm M} > 0$ if

$$|g|^2 \geq 2. \quad (10.55)$$

Thus for a perfect amplifier all squeezing and sub-Poissonian characteristics are lost if $|g|^2 \geq 2$ [12]. For the case of $n_2 \neq 0$, a similar treatment leads to the condition

$$|g|^2 \geq \frac{2n_1}{n_1 + n_2} = 2n_1. \quad (10.56)$$

10.3 Amplification of single-photon states

The states of a field with a fixed number of photons are mostly nonclassical and hence we specifically consider amplification of such states. For the case of the vacuum state, $P(\alpha_0, 0) = \delta^{(2)}(\alpha_0)$ and then

$$P(\alpha, t) = \frac{1}{\pi\eta} \exp\left(-\frac{|\alpha|^2}{\eta}\right). \quad (10.57)$$

Thus the vacuum states evolves into a thermal state with mean number equal to η, where η is given by the first line of (10.33). We next consider the consequence of the result (10.45) obtained under the condition (10.46). The Wigner function for a Fock state $|n\rangle$ is oscillatory as shown by Eq. (1.111). Hence for a Fock state $|n\rangle$ the highly singular P-function evolves into an ordinary function which is oscillatory around the origin when $\eta/|g(t)|^2 \geq \frac{1}{2}$. Thus the number state continues to exhibit nonclassical properties since $P(\alpha)$ can be negative. However, for $\eta/|g(t)|^2 \geq 1$, all oscillations disappear (Eq. (10.43)) and the P-function becomes classical. Let us consider specifically the amplification of a single-photon Fock state $|1\rangle$ for which

$$P(\alpha_0, 0) = {\rm e}^{|\alpha_0|^2} \frac{\partial^2}{\partial\alpha_0 \partial\alpha_0^*} \delta^2(\alpha_0), \quad (10.58)$$

and hence the P-function at time t is

$$P(\alpha, t) = \frac{1}{\pi\eta} \frac{\partial^2}{\partial\alpha_0 \partial\alpha_0^*} \exp\left[|\alpha_0|^2 - \frac{|\alpha - \alpha_0 g|^2}{\eta}\right]_{\alpha_0 = 0}$$
$$= \frac{|g|^2}{\pi\eta^2}\left[\frac{|\alpha|^2}{\eta} - \left(1 - \frac{\eta}{|g|^2}\right)\right] \exp\left(-\frac{|\alpha|^2}{\eta}\right). \quad (10.59)$$

The negativity of the P-function is especially evident in the region $\frac{\eta}{|g|^2} \leq 1$.

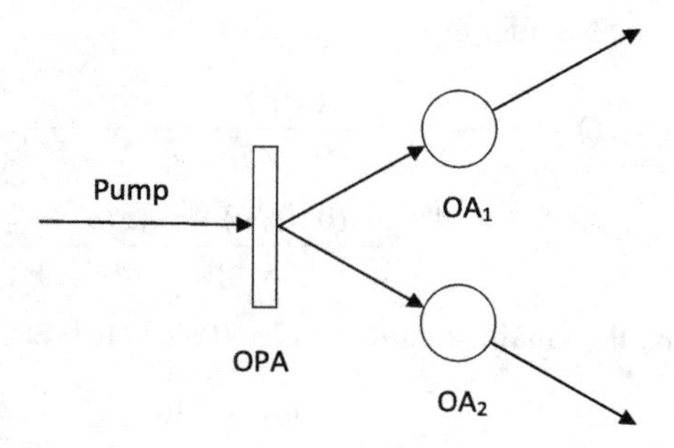

Fig. 10.2 Schematic diagram for the amplification of a two-mode entangled Gaussian state by a phase-insensitive amplifier. The optical parametric amplifier (OPA) produces the two-mode squeezed vacuum state of a and b. In the symmetric case, both the optical amplifiers (OA) are present. In the asymmetric case the OA_2 from the b arm is removed.

10.4 Amplification of entangled fields

We now consider changes in the entanglement of the fields if the fields are subjected to amplification [13]. Consider the scheme in Figure 10.2. The downconverter produces the two-mode squeezed vacuum, defined by (3.3), i.e. by

$$|\xi\rangle = \exp\left(\xi a^\dagger b^\dagger - \xi^* ab\right)|0,0\rangle. \tag{10.60}$$

Each mode is now subjected to amplification. Since the Wigner function for the state (10.60) is Gaussian, we can use the theory of entanglement from Section 3.13. We assume that OA_1 and OA_2 are two independent amplifiers, each described by the density matrix equation (10.7). The covariance matrix σ defined by Eq. (3.97) requires knowledge of all second-order moments like $\langle a^2\rangle$, $\langle b^2\rangle$, $\langle ab\rangle$, $\langle a^\dagger a\rangle$, $\langle b^\dagger b\rangle$, $\langle a^\dagger b\rangle$, etc. For the initial state (10.60), these are given in Section 3.1. The time dependence of the covariance matrix can be obtained from results like (10.11) and (10.12). For instance

$$\langle a(t)b(t)\rangle = g^2(t)\langle a(0)b(0)\rangle, \qquad \langle a^2(t)\rangle = g^2(t)\langle a^2(0)\rangle = 0. \tag{10.61}$$

The time dependence of the covariance matrix defined by Eq. (3.97) can be obtained by using the values of α, β, and γ, obtained from a procedure like the one that led to (10.61)

$$\begin{aligned} \alpha = \beta &= \frac{|g|^2\cosh 2r + (1+2\tilde{\eta})(|g|^2-1)}{2}\begin{pmatrix}1 & 0\\ 0 & 1\end{pmatrix}, \\ \gamma &= \frac{1}{2}|g|^2\sinh 2r\begin{pmatrix}\cos\theta & \sin\theta\\ \sin\theta & -\cos\theta\end{pmatrix}, \end{aligned} \tag{10.62}$$

where $\tilde{\eta} = N_2/(N_1 - N_2)$. Using (10.62) in Eqs. (3.97) and (3.98), we find the lowest symplectic eigenvalue

$$\tilde{\nu}_< = \{|g|^2[e^{-2r} + (1+2\tilde{\eta})] - (1+2\tilde{\eta})\}/2. \tag{10.63}$$

Requiring that the output state remains an entangled state, i.e. the condition $\tilde{\nu}_< < 1/2$ then translates into the following condition for the gain

$$|g|^2 < \frac{2 + 2\tilde{\eta}}{1 + 2\tilde{\eta} + e^{-2r}}. \tag{10.64}$$

In the special case of a fully inverted amplifier, $\tilde{\eta} \to 0$, and we have

$$|g|^2 < \frac{2}{1 + e^{-2r}} = \frac{2}{1 + e^{-E_N}}, \tag{10.65}$$

where E_N is the logarithmic negativity (Section 3.7) of the input state. We find the condition for the loss of entanglement for an ideal amplifier to be similar to the condition we had found for the loss of squeezing of a single-mode field.

In the above we considered the case when both the modes a and b were subjected to identical amplification. It would be interesting to examine the case when only one mode, say "a," is amplified. In this case the elements of the covariance matrix (3.97) are found to be

$$\begin{aligned}
\alpha &= \frac{|g|^2 \cosh 2r + (1 + 2\tilde{\eta})(|g|^2 - 1)}{2} \begin{pmatrix} 1 & 0 \\ 0 & 1 \end{pmatrix}, \\
\beta &= \frac{\cosh 2r}{2} \begin{pmatrix} 1 & 0 \\ 0 & 1 \end{pmatrix}, \\
\gamma &= \frac{|g| \sinh 2r}{2} \begin{pmatrix} \cos\theta & \sin\theta \\ \sin\theta & -\cos\theta \end{pmatrix}.
\end{aligned} \tag{10.66}$$

The smallest simplectic eigenvalue is now equal to

$$\begin{aligned}
\tilde{\nu}_< = \frac{1}{4}\Big[&(|g^2| + 1)\cosh 2r + (1 + 2\tilde{\eta})(|g|^2 - 1) \\
&- \sqrt{(|g|^2 - 1)^2(\cosh 2r + 1 + 2\tilde{\eta})^2 + 4|g|^2 \sinh^2 2r}\Big].
\end{aligned} \tag{10.67}$$

As before, requiring $\tilde{\nu}_< < 1/2$, so that the output state remains an entangled state, we obtain the following condition on the gain

$$|g|^2 < 1 + \frac{1}{\tilde{\eta}}. \tag{10.68}$$

In contrast to the symmetric case, here one finds $\tilde{\nu}_<$ is always less than $1/2$ for $\tilde{\eta} = 0$ and therefore the entanglement survives no matter how large the gain is. The variation of $\tilde{\nu}_<$ in the two cases as a function of the gain is shown in Figure 10.3. The situation is different, however, if $\tilde{\eta}$ is nonzero. In this case there is a threshold value of gain beyond which the entanglement of the initial state is lost. The threshold value of the gain depends on the value of $\tilde{\eta}$. Clearly for larger $\tilde{\eta}$ the entanglement degrades faster. We show this behavior in Figure 10.4. The difference from (10.62) comes from the fact that now

$$\langle a(t)b(t)\rangle = g(t)\langle a(0)b(0)\rangle, \qquad \langle b^\dagger b\rangle_t = \langle b^\dagger b\rangle_0. \tag{10.69}$$

Pooser *et al.* [14] have demonstrated the survival of entanglement when only one mode is amplified over a significant region of gain values. Finally, we note that some results on the amplification of non-Gaussian entangled states are available [15, 16].

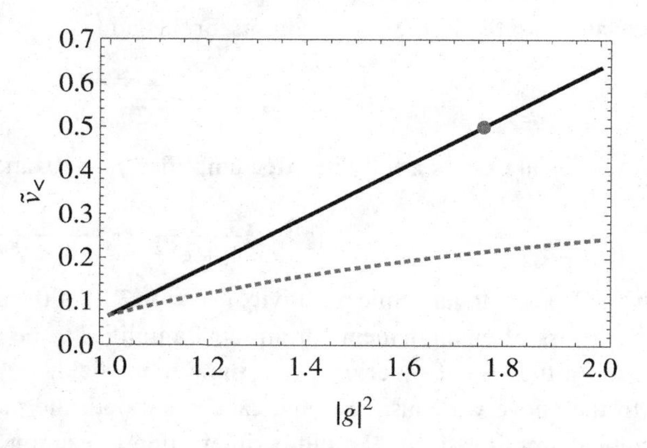

Fig. 10.3 Variation of the entanglement measure $\tilde{\nu}_<$ as a function of the gain $|g|^2$ for the symmetric case (solid line) and the asymmetric case (dashed line) for $r = 1; \tilde{\eta} = 0$. The dot marks the critical value of the gain in the symmetric case beyond which the entanglement in the output state vanishes, after [13].

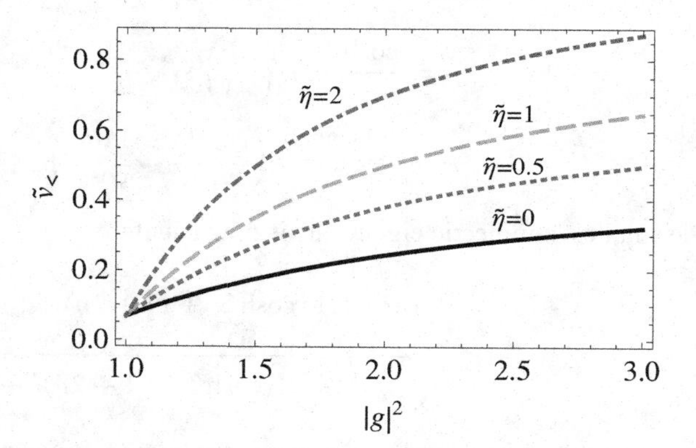

Fig. 10.4 Variation of the entanglement measure $\tilde{\nu}_<$ as a function of the gain $|g|^2$ for different $\tilde{\eta}$ and the squeezing parameter $r = 1$. $\tilde{\eta} = 0$ (solid line), $\tilde{\eta} = 0.5$ (dotted line), $\tilde{\eta} = 1$ (dashed line), $\tilde{\eta} = 2$ (dot-dashed line), after [13].

10.5 Realising a phase-insensitive amplifier from a phase-sensitive amplifier

In Section 3.9 we briefly considered the parametric amplification of signals. In particular, we considered the process of down-conversion with the mode a in a coherent state and the mode b in the vacuum state. We showed that the mean number of photons in the mode "a" grows as $e^{2r}|\alpha|^2/4$, where r is the squeezing parameter associated with the downconverter. We will now establish the following: the full quantum dynamics of the mode "a" is identical to the dynamics under the amplification process as long as no information on the mode "b" is gained. In Section 3.8 we showed that in the down-conversion process the Heisenberg operators evolve as (Eq. (3.61))

$$a(t) = a \cosh gt + b^{\dagger} e^{i\varphi} \sinh gt, \tag{10.70}$$

where $gt = r$, r being the squeezing parameter of Section 3.1. The Heisenberg solution (10.70) is especially useful as we can consider any initial states. The mode "a" is the one whose amplification we are considering. We take the mode b to be in the vacuum state. The solution (10.70) has the form (10.4) or (10.10), i.e.

$$\begin{gathered} a_{\text{out}} = Va + F, \qquad V = \cosh gt, \\ \langle FF^{\dagger} \rangle = 0, \qquad \langle F^{\dagger}F \rangle = \sinh^2 gt = V^2 - 1, \\ V \to e^{gt}/2 = e^r/2. \end{gathered} \tag{10.71}$$

There is a slight difference in the form of the coefficient of a but this is of no significance. Thus all the results obtained in Sections 10.2 to 10.4 would apply to the amplifier described by (10.71). We have to use the replacement $|g(t)| \to V$. This is an interesting way to realize a phase-insensitive amplifier using a phase-sensitive one and in fact has been implemented experimentally by Pooser *et al.* [14]. Furthermore, we get an ideal amplifier as $\langle FF^{\dagger} \rangle = 0$. This is true as long as the mode b is in the vacuum state.

10.6 Degradation of nonclassicality and entanglement due to the absorption of quantum fields

An amplifier adds noise to the optical signals. However, an ideal absorber does not add noise to the signals. Thus one might expect that an absorber would affect the nonclassical properties of fields in a much less severe way than an amplifier. For an ideal absorber, the basic master equation is (10.7) with $N_1 = 0$, $N_2 = N$, i.e.

$$\frac{\partial \rho}{\partial t} = -i\omega[a^{\dagger}a, \rho] - \kappa(a^{\dagger}a\rho - 2a\rho a^{\dagger} + \rho a^{\dagger}a). \tag{10.72}$$

The corresponding Langevin equation is (cf. (9.85))

$$\begin{gathered} \dot{a} = -i\omega a - \kappa a + F(t), \qquad \langle F \rangle = \langle F^{\dagger}(t)F(t') \rangle = 0, \\ \langle F(t)F^{\dagger}(t') \rangle = 2\kappa\delta(t - t'). \end{gathered} \tag{10.73}$$

We write the solution to (10.73) as

$$\begin{gathered} a(t) = T(t)a(0) + f(t), \qquad T(t) = e^{-i\omega t - \kappa t}, \\ \langle f(t) \rangle = 0, \qquad \langle f^{\dagger}(t)f(t) \rangle = 0, \qquad \langle f(t)f^{\dagger}(t) \rangle = 1 - |T(t)|^2. \end{gathered} \tag{10.74}$$

We have noticed earlier that all the normally ordered moments of $a(t)$ are related to their value at $t = 0$

$$\langle [a^{\dagger}(t)]^m [a(t)]^n \rangle = T^{*m}(t)T^n(t)\langle a^{\dagger m}(0)a^n(0) \rangle. \tag{10.75}$$

We can use (10.75) to examine the nonclassical properties at the output in terms of the nonclassical properties at the input [4]

$$\langle a^\dagger(t)a(t)\rangle = |T(t)|^2\langle a^\dagger a\rangle, \qquad \langle a^{\dagger 2}(t)a^2(t)\rangle = |T(t)|^4\langle a^{\dagger 2}a^2\rangle, \tag{10.76}$$

and hence Mandel's Q_M parameter at the output would be

$$Q_M(t) = \frac{\langle a^{\dagger 2}a^2\rangle - \langle a^\dagger a\rangle^2}{\langle a^\dagger a\rangle} = |T(t)|^2 Q_M(0). \tag{10.77}$$

Thus the field always remains sub-Poissonian. This is in contrast to the case of an amplifier where the sub-Poissonian property is completely lost if the gain exceeds a certain value (Eq. (10.52)). Using (10.75), it is straightforward to show that the squeezing at the output is given by

$$S_0(t) = |T(t)|^2 S_0(0), \tag{10.78}$$

which exhibits a simple decay behavior; however, the initial squeezing is not completely lost.

We next describe the effect of the absorber [17–19] on the entanglement of two modes initially in the squeezed vacuum state (10.60). We assume that each mode decays according to (10.72) with the same decay constant κ. For simplicity we assume that the free evolution ($e^{-i\omega t}$) has been removed by working in the interaction picture. According to the theory of Gaussian entanglement (Section 3.13), we need to know the covariance matrix σ defined by Eq. (3.97) in terms of the matrices α, β, γ. Once we know the time dependent σ, then we can calculate the lowest simplectic eigenvalue and the log-negativity parameter E_N. For the state (10.60), using (10.75), the matrices α, β, and γ are found to be

$$\begin{aligned} \alpha = \beta &= \left(e^{-2\kappa t}\sinh^2 r + \frac{1}{2}\right)\begin{pmatrix}1 & 0\\ 0 & 1\end{pmatrix}, \\ \gamma &= \frac{1}{2}e^{-2\kappa t}\sinh 2r\begin{pmatrix}-1 & 0\\ 0 & 1\end{pmatrix}. \end{aligned} \tag{10.79}$$

Hence on using Eq. (3.102), the lowest simplectic eigenvalue is

$$\tilde{\nu}_<(t) = \frac{1}{2}e^{-2\kappa t}(e^{-2r} + e^{2\kappa t} - 1). \tag{10.80}$$

The eigenvalue $\tilde{\nu}_<(t)$ is always less than $\frac{1}{2}$, as shown in Figure 10.5 and hence the entanglement between the modes survives as the entangled light propagates through an absorber. The entanglement disappears only for $t \to \infty$.

10.6.1 The phase-space distribution after propagation through an absorber

The time-dependent phase-space distributions after propagation through an ideal absorber can be obtained from (10.32) by letting $N_1 \to 0$, $N_2 \to N$

$$\Phi(\alpha, t) = \int d^2\alpha_0 \Phi(\alpha_0, 0)\frac{1}{\pi\eta}\exp\left[-\frac{|\alpha - \alpha_0 T(t)|^2}{\eta}\right], \tag{10.81}$$

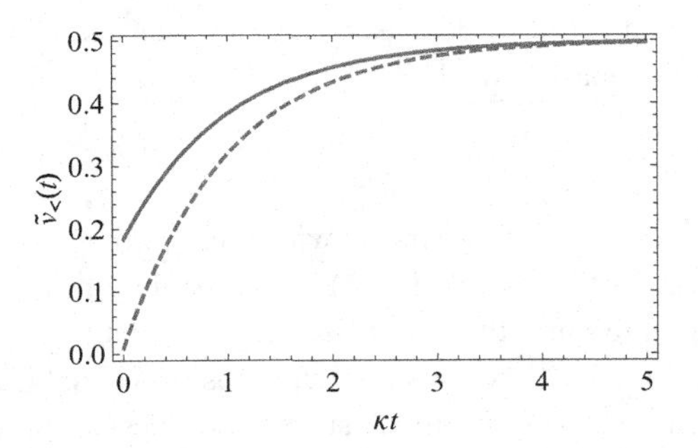

Fig. 10.5 The eigenvalue $\tilde{\nu}_<(t)$ as a function of the loss κt for different values of the squeezing parameter. $r = 0.5$ (solid line), $r = 2$ (dashed line).

where $T(t)$ is given by (10.74). The fluctuation parameter η is now equal to zero for the P-function, $\eta = 1 - |T(t)|^2$ for the Q-function, and equal to $[1 - |T(t)|^2]/2$ for the Wigner function. Thus the time-dependent solution for the P-function is rather simple

$$P(\alpha, t) = \frac{1}{|T(t)|^2} P[\alpha / |T(t)|]. \tag{10.82}$$

This implies that a field in a coherent state $|\beta\rangle$ remains coherent

$$|\beta\rangle \rightarrow |\beta T(t)\rangle. \tag{10.83}$$

The absorber does not change the character of the coherent state.

The Q-function and the Wigner function broaden as for these functions the parameter η is nonzero. Furthermore, if any of these functions are Gaussian at $t = 0$, these would remain Gaussian, i.e. the result (10.36) applies with

$$\begin{gathered}
2\mu^* = -\langle a^2\rangle = -T^2(t)\langle a^2\rangle_0, \\
\tau_W = \frac{1}{2} + \langle a^\dagger a\rangle = |T(t)|^2 \langle a^\dagger a\rangle_0 + \frac{1}{2}, \qquad \tau_W(t=0) = \frac{1}{2} + \langle a^\dagger a\rangle_0, \\
\tau_P = \tau_W - \frac{1}{2}, \qquad \tau_Q = \tau_W + \frac{1}{2}.
\end{gathered} \tag{10.84}$$

10.7 Loss of coherence on interaction with the environment

The interaction with the environment is usually modeled in terms of system–bath interaction. We saw in the previous chapter in connection with the relaxation of a two-level atom that the coherence usually decays much faster than the population. For the radiation field, the exact analog of dephasing would be the master equation with a structure similar to the Γ terms in Eq. (9.56)

$$\frac{\partial \rho}{\partial t} = -\Gamma[(a^\dagger a)^2 \rho - 2a^\dagger a \rho a^\dagger a + \rho (a^\dagger a)^2]. \tag{10.85}$$

Clearly as a result of dephasing the diagonal elements of ρ do not change with time, but the off-diagonal elements decay as

$$\rho_{mn}(t) = \mathrm{e}^{-\Gamma(m+n)t}\rho_{mn}(0), \tag{10.86}$$

and hence in the long time ρ will become diagonal. Thus the coherence is completely lost [20]. For the model (10.72), both populations and coherences decay and as a result in the long time limit, the field is in a vacuum state.

In Section 4.1 we studied the superposition (Eq. (4.2)) of two coherent states and showed that such a superposition has strong nonclassical properties which arise due to the interference of the two coherent states. We investigate how sensitive such a superposition is to environmental perturbations [21]. The density matrix associated with (4.2) is

$$\rho = \mathcal{N}^{-2}(|\alpha_0\rangle + \mathrm{e}^{\mathrm{i}\varphi}|-\alpha_0\rangle)(\langle\alpha_0| + \mathrm{e}^{-\mathrm{i}\varphi}\langle-\alpha_0|). \tag{10.87}$$

The time evolution of diagonal terms in (10.87) is very simple (Eq. (10.83))

$$|\alpha_0\rangle\langle\alpha_0| \to |\alpha_0\mathrm{e}^{-\kappa t}\rangle\langle\alpha_0\mathrm{e}^{-\kappa t}|. \tag{10.88}$$

The off-diagonal terms are more complicated. To obtain the full time evolution of the density matrix, consider the characteristic function (Section 2.5.3) defined by

$$C_P(\beta, t) = \mathrm{Tr}\{\rho(t)\mathrm{e}^{\beta a^\dagger}\mathrm{e}^{-\beta^* a}\}. \tag{10.89}$$

This is the Fourier transform of the P-function. Its time dependence is easily obtained since in the expansion of (10.89) we only have normally ordered quantities. We can then use (10.75) and resum the series to obtain the characteristic function at time t in terms of its value at $t = 0$

$$\begin{gathered} C_P(\beta, t) = C_P(\beta\mathrm{e}^{-\kappa t}, 0), \\ C_P(\beta, 0) = \mathcal{N}^{-2}\left(\exp(\beta\alpha_0^* - \beta^*\alpha_0) + \mathrm{e}^{\mathrm{i}\varphi}\exp(\beta\alpha_0^* + \beta^*\alpha_0)\mathrm{e}^{-2|\alpha_0|^2} + c.c.\right), \end{gathered} \tag{10.90}$$

where we used (10.87) and the property $\langle\alpha|\mathrm{e}^{\beta a^\dagger}\mathrm{e}^{-\beta^* a}|\gamma\rangle = \langle\alpha|\gamma\rangle\exp(\beta\alpha^* - \beta^*\gamma)$. By using (10.90), the density matrix at time t is clearly equal to

$$\begin{aligned} \rho(t) = \mathcal{N}^{-2}(&|\alpha_0\mathrm{e}^{-\kappa t}\rangle\langle\alpha_0\mathrm{e}^{-\kappa t}| + |-\alpha_0\mathrm{e}^{-\kappa t}\rangle\langle-\alpha_0\mathrm{e}^{-\kappa t}| \\ &+ \{\mathrm{e}^{\mathrm{i}\varphi}|-\alpha_0\mathrm{e}^{-\kappa t}\rangle\langle\alpha_0\mathrm{e}^{-\kappa t}| + \mathrm{e}^{-\mathrm{i}\varphi}|\alpha_0\mathrm{e}^{-\kappa t}\rangle\langle-\alpha_0\mathrm{e}^{-\kappa t}|\} \\ &\times \exp[-2|\alpha_0|^2(1 - \mathrm{e}^{-2\kappa t})]). \end{aligned} \tag{10.91}$$

The interference term (curly bracket) is scaled down by a factor $\mathrm{e}^{-2|\alpha_0|^2(1-\mathrm{e}^{-2\kappa t})} \approx \mathrm{e}^{-4|\alpha_0|^2\kappa t}$ for $\kappa t \ll 1$. This leads to a decoherence time τ_d

$$\tau_d \sim \frac{1}{4|\alpha_0|^2\kappa}, \tag{10.92}$$

which is inversely proportional to $|\alpha_0|^2$, i.e. to the size of the coherent state [20]. As $|\alpha_0|$ increases, the decoherence time goes down. The Wigner function for the state (10.91) can

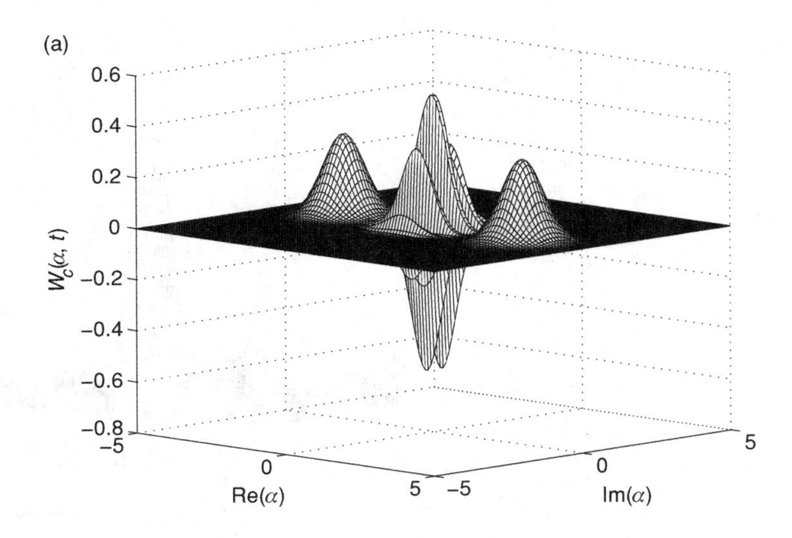

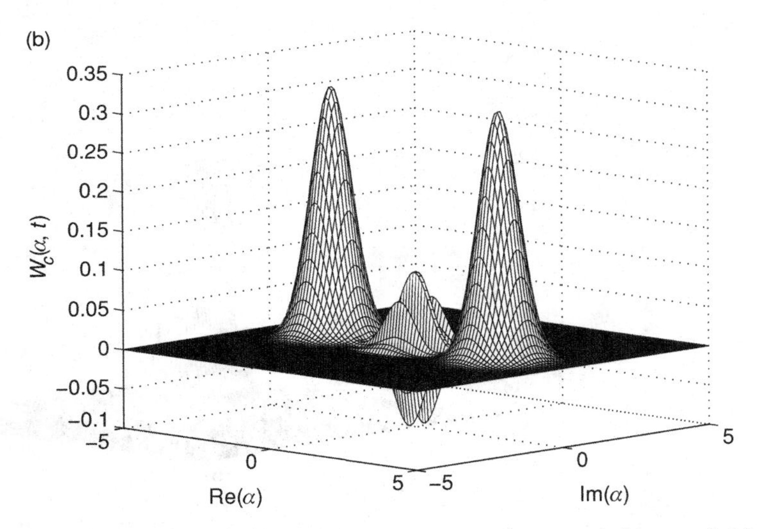

Fig. 10.6 The Wigner function $W_c(\alpha, t)$ for the state (10.91) for $\alpha_0 = 3, \varphi = 0$. (a) $t = 0$, (b) $\kappa t = 0.05$, (c) $\kappa t = 0.25$, (d) $\kappa t = 2$.

be written down by inspection as we already know the Wigner function (4.15) for (10.87)

$$\begin{aligned} W_c(\alpha, t) =& \frac{2\mathcal{N}^{-2}}{\pi}\{\exp(-2|\alpha - \alpha_0 e^{-\kappa t}|^2) + \exp(-2|\alpha + \alpha_0 e^{-\kappa t}|^2) \\ &+ 2e^{-2|\alpha|^2}\exp[-2|\alpha_0|^2(1 - e^{-2\kappa t})]\cos[\varphi + 4\mathrm{Im}\{\alpha_0^*\alpha e^{-\kappa t}\}]\}. \end{aligned} \tag{10.93}$$

We note that as a result of interaction with the environment the two Gaussians move towards each other eventually merging into one Gaussian. The amplitude of the oscillatory term goes down by a factor $\exp[-2|\alpha_0|^2(1 - e^{-2\kappa t})]$ and the period of oscillation increases by

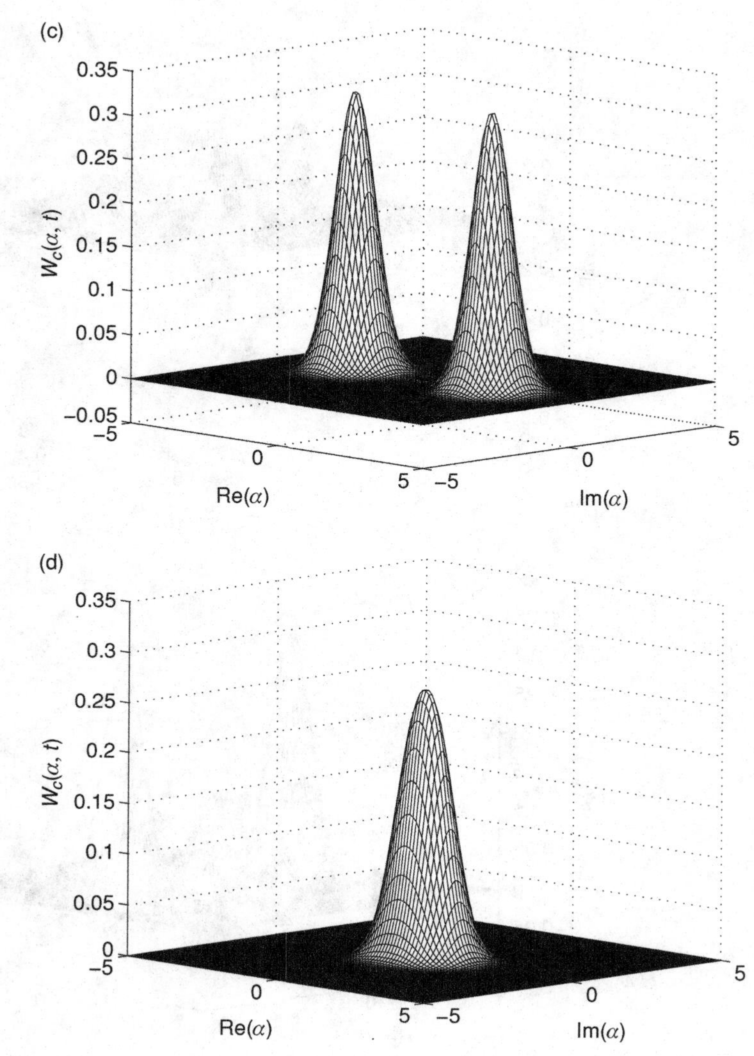

Fig. 10.6 (*continued*).

$e^{\kappa t}$. We show this evolution in Figure 10.6 for different values of κt. The nature of transition from (a) to (c) is especially noteworthy as it shows how the nonclassical oscillations in the Wigner function disappear on interaction with the environment.

Exercises

10.1 Prove the relations (10.24).

10.2 Prove the result (10.56).

10.3 Using (10.32) show that the P-function of the amplified Fock state $|n\rangle$ is

$$P(\alpha, t) = \frac{e^{-|\alpha|^2/\eta}}{\pi\eta} \mathrm{L}_n \left[-\frac{|\alpha|^2}{\eta\left(\frac{\eta}{|g|^2} - 1\right)} \right] \left(1 - \frac{g^2}{\eta}\right)^n,$$

where L_n is a Laguerre polynomial of order n [11].

10.4 Write the differential equations (10.27), (10.28), and (10.31) in terms of the quadrature variables x and y defined by $\alpha = \frac{x+\mathrm{i}y}{\sqrt{2}}$, $\alpha^* = \frac{x-\mathrm{i}y}{\sqrt{2}}$. Show that the second-order differential operator is separable in x and y.

10.5 Consider a field initially in a coherent state $|\beta\rangle$. Find the P-function at any time for the amplified field using (10.32), and using this P-function show that the photon number distribution in the amplified field is given by

$$p(n) = \frac{\eta^n}{(1+\eta)^{n+1}} \exp\left(-\frac{|g\beta|^2}{\eta+1}\right) \mathrm{L}_n \left[-\frac{|g\beta|^2}{\eta(1+\eta)} \right].$$

10.6 For an ideal amplifier, using (10.35) find the photon number distribution of the amplified field if initially the field is in a single-photon state $|1\rangle$. The following series expansion

$$Q(\alpha) = \sum_{m,n} \rho_{mn} \frac{\alpha^{*m}\alpha^n}{\sqrt{m!n!}} \mathrm{e}^{-|\alpha|^2}/\pi$$

will be useful.

10.7 Consider the single-mode one-photon-subtracted squeezed vacuum state (4.69)

$$\begin{aligned} |\xi\rangle^{(s)} &= \frac{a}{\sinh r} \frac{1}{\sqrt{\cosh r}} \sum_{n=0}^{\infty} \mathrm{e}^{\mathrm{i}n\varphi} (\tanh r)^n \frac{\sqrt{(2n)!}}{n! 2^n} |2n\rangle \\ &= \sum_{n=0}^{\infty} c_{2n+1} |2n+1\rangle, \end{aligned}$$

where

$$c_{2n+1} = \frac{\mathrm{e}^{\mathrm{i}(n+1)\varphi} (\tanh r)^n \sqrt{(2n+1)!}}{(\cosh r)^{3/2} n! 2^n}.$$

Consider the dephasing interaction with the environment, use (10.85) to obtain the density matrix in the limit $t \to \infty$. Check if this still has nonclassical properties. Prove that the Wigner function of the density matrix in the limit $t \to \infty$ is given by

$$\begin{aligned} W(\alpha) = {} & \frac{2}{\pi} \exp[-2|\alpha|^2 \cosh(2r)] \{ [4|\alpha|^2 \cosh(2r) - 1] \\ & \times I_0(2|\alpha|^2 \sinh(2r)) - 4|\alpha|^2 \sinh(2r) I_1(2|\alpha|^2 \sinh(2r)) \}, \end{aligned}$$

where $I_n(z) = \frac{1}{\pi} \int_0^{\pi} \mathrm{e}^{z\cos\theta} \cos^n\theta \, \mathrm{d}\theta$. Furthermore, show that $W(\alpha)$ can be negative [22].

10.8 Consider a combination of a phase-insensitive amplifier and an absorber. Assume that both are operating in the ideal limit. Let us denote the output fields as shown in the figure below. Clearly we have the relations (Eqs. (10.10) and (10.75))

$$
\begin{aligned}
b &= ga + f_1, \qquad \langle f_1^\dagger f_1\rangle = |g|^2 - 1, \qquad \langle f_1 f_1^\dagger\rangle = 0,\\
c &= Tb + f_2, \qquad \langle f_2^\dagger f_2\rangle = 0, \qquad \langle f_2 f_2^\dagger\rangle = 1 - |T|^2,\\
&\qquad \langle f_1^\dagger f_2\rangle = 0, \qquad \langle f_1\rangle = \langle f_2\rangle = 0.
\end{aligned}
$$

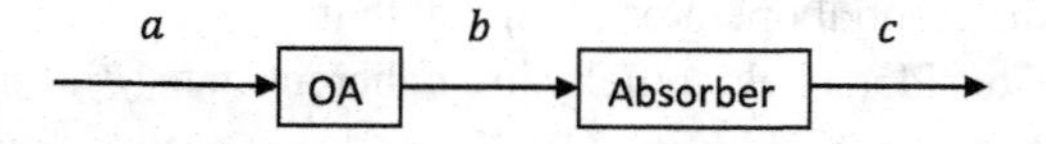

Calculate the Mandel Q_M and squeezing parameters of the field c in terms of the characteristics of the input field. What happens in the limit $|g|^2|T|^2 = 1$? Are the results different if the order of the amplifier and absorber are interchanged? (For details see [6, 23].)

References

[1] K. Shimoda, H. Takahasi, and C. H. Townes, *J. Phys. Soc. Japan* **12**, 686 (1957).
[2] H. A. Haus and J. A. Mullen, *Phys. Rev.* **128**, 2407 (1962).
[3] C. M. Caves, *Phys. Rev. D* **26**, 1817 (1982).
[4] R. W. Boyd, G. S. Agarwal, K. W. C. Chan, A. K. Jha, and M. N. O' Sullivan, *Opt. Comm.* **281**, 3732 (2008).
[5] P. Diament and M. C. Teich, *IEEE J. Quantum Electron.* **28**, 1325 (1992).
[6] H. P. Yuen, *Opt. Lett.* **17**, 73 (1992).
[7] H. P. Yuen and V. W. S. Chan, *Opt. Lett.* **8**, 177 (1983).
[8] G. S. Agarwal and E. Wolf, *Phys. Rev. D* **2**, 2187 (1970).
[9] K. E. Cahill and R. J. Glauber, *Phys. Rev.* **177**, 1857 (1969); *ibid* **177**, 1882 (1969).
[10] H. Risken, *The Fokker-Plank Equation: Methods of Solution and Applications* (Berlin: Springer-Verlag, 1984).
[11] G. S. Agarwal and K. Tara, *Phys. Rev. A* **47**, 3160 (1993).
[12] C. K. Hong, S. R. Friberg, and L. Mandel, *J. Opt. Soc. Am. B* **2**, 494 (1985).
[13] G. S. Agarwal and S. Chaturvedi, *Opt. Commun.* **283**, 839 (2010).
[14] R. C. Pooser, A. M. Marino, V. Boyer, K. M. Jones, and P. D. Lett, *Phys. Rev. Lett.* **103**, 010501 (2009).
[15] C. Vitelli, N. Spagnolo, F. Sciarrino, and F. De Martini, *J. Opt. Soc. Am. B* **26**, 892 (2009).
[16] G. S. Agarwal, S. Chaturvedi, and A. Rai, *Phys. Rev. A* **81**, 043843 (2010).
[17] L.-M. Duan, G. Giedke, J. I. Cirac, and P. Zoller, *Phys. Rev. Lett.* **84**, 2722 (2000).
[18] S. Scheel and D.-G. Welsch, *Phys. Rev. A* **64**, 063811 (2001).
[19] M. Hillery and M. S. Zubairy, *Phys. Rev. A* **74**, 032333 (2006).

[20] W. H. Zurek, *Rev. Mod. Phys*. **75**, 715 (2003).
[21] V. Buzek and P. L. Knight, in *Progress in Optics*, edited by E. Wolf (Amsterdam: North-Holland, 1995), **34**, p. 1.
[22] A. Biswas and G. S. Agarwal, *Phys. Rev. A* **75**, 032104 (2007).
[23] C. J. McKinstrie, M. Karlsson, and Z. Tong, *Opt. Express* **18**, 19792 (2010).

11 Quantum coherence, interference and squeezing in two-level systems

In the earlier chapters we have discussed extensively concepts of coherence, interference, squeezing, and entanglement for radiation fields. Many of these concepts are equally applicable to other quantum systems and specifically to any kind of two-level system. For example, once we introduce the concept of coherence in two-level atoms, then we can discuss many types of atomic interferometers. Similarly, for a collection of atoms we can define atomic coherent states, squeezed states, and introduce analogs of various types of phase-space distributions for the radiation fields. This chapter is devoted to the study of coherence and interference in two-level systems and some applications of these ideas.

11.1 Two-level approximation: atomic dynamics in a monochromatic field

The interaction Hamiltonian between a system with states $|\psi_j\rangle$ with energies E_j and the electromagnetic field $\vec{E}(\vec{R},t)$ can be written in dipole approximation as

$$H_1 \cong -\vec{p}\cdot\vec{E}(\vec{R},t), \tag{11.1}$$

where $\vec{R}$ denotes the position of the atom and $\vec{p}$ is the dipole matrix element, which can be expanded as

$$\vec{p} = \sum_{ij} \vec{p}_{ij}|\psi_i\rangle\langle\psi_j|. \tag{11.2}$$

The unperturbed Hamiltonian can then be written as

$$H_0 = \sum_j E_j|\psi_j\rangle\langle\psi_j|. \tag{11.3}$$

In resonant physics very often the two-level approximation [1] for the atom is adequate. This is so if the frequency ω_l of the external field is tuned close to the transition frequency ω_0 between two levels designated as $|e\rangle$ and $|g\rangle$, which we would call the excited and ground levels. The two-level approximation works well as long as the width of the external fields is small compared to the energy separation between $|e'\rangle$ and $|e\rangle$, where $|e'\rangle$ could be a neighboring state to which the atom can also get excited. If the electromagnetic field is a plane wave of frequency moving in the direction $\vec{k}$

$$\vec{E}(\vec{R},t) = \vec{\varepsilon}_l\, \mathrm{e}^{\mathrm{i}\vec{k}\cdot\vec{R}-\mathrm{i}\omega_l t} + c.c., \tag{11.4}$$

and if the two-level approximation is made, then the interaction (11.1) can be expressed as

$$H_1 = -\hbar(g|e\rangle\langle g|e^{-i\omega_l t} + H.c.) - \hbar(g'|e\rangle\langle g|e^{i\omega_l t} + H.c.), \tag{11.5}$$

where

$$g = \frac{\vec{p}_{eg}\cdot\vec{\varepsilon}_l\, e^{i\vec{k}\cdot\vec{R}}}{\hbar}, \qquad g' = \frac{\vec{p}_{eg}\cdot\vec{\varepsilon}_l^{\,*} e^{-i\vec{k}\cdot\vec{R}}}{\hbar}. \tag{11.6}$$

On choosing the zero of the energy halfway between two levels, the Schrödinger equation for the two-level system will be

$$\frac{\partial|\psi\rangle}{\partial t} = -i\frac{\omega_0}{2}(|e\rangle\langle e| - |g\rangle\langle g|)|\psi\rangle - \frac{i}{\hbar}H_1|\psi\rangle. \tag{11.7}$$

In dealing with intense fields it is useful to write (11.7) in a frame rotating with frequency ω_l of the external fields. This is done by defining the wave function $|\phi\rangle$ as

$$|\phi\rangle = \exp\{i\omega_l S_z t\}|\psi\rangle, \tag{11.8}$$

then the equation for $|\phi\rangle$ is

$$\frac{\partial|\phi\rangle}{\partial t} = -i\frac{H_{\text{eff}}}{\hbar}|\phi\rangle, \tag{11.9}$$

where

$$H_{\text{eff}} = \hbar(\omega_0 - \omega_l)S_z - \hbar(gS_+ + H.c.) - \hbar(g'S_+e^{2i\omega_l t} + H.c.). \tag{11.10}$$

Here we have introduced the operators defined by

$$S_+ = |e\rangle\langle g|, \qquad S_- = |g\rangle\langle e|, \qquad S_z = \frac{1}{2}(|e\rangle\langle e| - |g\rangle\langle g|). \tag{11.11}$$

It can be shown that these operators satisfy spin $-\frac{1}{2}$ angular momentum algebra

$$\begin{aligned} &[S_+, S_-] = 2S_z, \qquad [S_z, S_+] = S_+, \qquad [S_z, S_-] = -S_-, \\ &S_+S_- = \frac{1}{2} + S_z, \qquad S_+^2 = 0, \qquad S_zS_z = \frac{1}{4}. \end{aligned} \tag{11.12}$$

Note that the Hamiltonian (11.10) contains terms oscillating at twice the optical frequency. Such terms lead to negligible contribution as long as $|g| \ll \omega_0$. This is indeed the case for typical optical fields used in resonant experiments. Hence in what follows we ignore these counter-rotating terms. Thus the interaction (11.10) is approximated by

$$H_{\text{eff}} \simeq \hbar\Delta S_z - \hbar(gS_+ + g^*S_-), \qquad \Delta = \omega_0 - \omega_l. \tag{11.13}$$

This approximation is known as the rotating wave approximation. We also notice that the effective Hamiltonian (11.13) can be written as

$$\begin{aligned} &H_{\text{eff}} = \hbar(\vec{S}\cdot\vec{\Omega}), \qquad |\vec{\Omega}| = \sqrt{\Delta^2 + 4|g|^2} = \Omega, \\ &\Omega_x = -(g + g^*), \qquad \Omega_y = -ig + ig^*, \qquad \Omega_z = \Delta. \end{aligned} \tag{11.14}$$

One thus finds that the problem of a spin in a magnetic field and the problem of a two-level atom interacting with an electromagnetic field are isomorphic. This was first shown by

Feynman *et al.* [2]. Note that the detuning factor Δ is like the static magnetic field which is used to define the quantization axis of the spin.

We next discuss the dynamical behavior in the rotating frame. The time evolution operator $U(t)$ is easily computed

$$U(t) = \exp\left\{-\frac{\mathrm{i}}{\hbar}H_{\mathrm{eff}}t\right\} = \exp\{-\mathrm{i}\vec{S}\cdot\vec{\Omega}t\}$$

$$= \cos\frac{\Omega}{2}t - \frac{\mathrm{i}(\vec{S}\cdot\vec{\Omega})}{(\Omega/2)}\sin\frac{\Omega t}{2} \tag{11.15}$$

$$= \cos\frac{\Omega t}{2} - \mathrm{i}\frac{H_{\mathrm{eff}}}{\hbar}\sin\frac{\Omega t}{2}\bigg/\frac{\Omega}{2}. \tag{11.16}$$

The wave function at time t can be obtained from (11.16) and (11.8) assuming that $|\phi(0)\rangle = |g\rangle$:

$$|\phi(t)\rangle = \left(\cos\frac{\Omega t}{2} + \frac{\mathrm{i}\Delta}{\Omega}\sin\frac{\Omega t}{2}\right)|g\rangle + \frac{2\mathrm{i}g}{\Omega}\sin\frac{\Omega t}{2}|e\rangle. \tag{11.17}$$

The probability $p_e(t)$ of finding the atom in the excited state is

$$p_e(t) = \frac{4|g|^2}{\Omega^2}\sin^2\left(\frac{\Omega t}{2}\right) = \frac{2|g|^2}{\Omega^2}(1-\cos\Omega t), \tag{11.18}$$

$$\sim |g|^2t^2 \qquad \text{if } \Omega t \ll 1. \tag{11.19}$$

Note that for short times p_e is proportional to the square of time rather than proportional to t. This is because we are dealing with discrete levels and the external field is assumed to have no width. For arbitrary values of the detuning Δ and the field strength g, the excitation probability exhibits oscillatory behavior. The oscillation frequency is Ω, which is called the generalized Rabi frequency. For the field on resonance, $\Omega = 2|g|$, which is called the Rabi frequency. Note that the maximum value of the population in the excited state depends on the detuning

$$p_{e,\max} = \frac{4|g|^2}{\Delta^2 + 4|g|^2} \to 1 \qquad \text{if } \Delta = 0. \tag{11.20}$$

Furthermore, for $\frac{\Omega t}{2} = \pi$, Eq. (11.16) gives

$$U(t) = -1. \tag{11.21}$$

Thus after a complete cycle

$$U(\Omega t = 2\pi) = -1, \tag{11.22}$$

the wave function acquires a phase shift of π, which is useful in many different applications, particularly in the context of the control of decoherence (Chapter 16) and realizations of Berry's phase in spin systems. A pulse is called a $q\pi$-pulse if $\Omega t = q\pi$. Under a π-pulse and for a field at resonance ($\Delta = 0$), Eq. (11.16) yields

$$U(\Omega t = \pi) = \frac{\mathrm{i}}{|g|}(gS_+ + g^*S_-), \tag{11.23a}$$

and hence, under a π-pulse, the state $|g\rangle$ and $|e\rangle$ get interchanged

$$|e\rangle \rightarrow \frac{\mathrm{i}g^*}{|g|}|g\rangle, \qquad |g\rangle \rightarrow \frac{\mathrm{i}g}{|g|}|e\rangle. \tag{11.23b}$$

For a $\frac{\pi}{2}$-pulse with $\Delta = 0$, we have from Eq. (11.16)

$$U(\Omega t = \pi/2) = \frac{1}{\sqrt{2}} + \frac{\mathrm{i}}{\sqrt{2}|g|}(gS_+ + g^*S_-), \tag{11.24a}$$

and hence

$$|g\rangle \rightarrow \frac{1}{\sqrt{2}}\left(|g\rangle + \frac{\mathrm{i}g}{|g|}|e\rangle\right) \rightarrow \frac{1}{\sqrt{2}}(|g\rangle + |e\rangle), \quad \text{if } g = -\mathrm{i}|g|, \tag{11.24b}$$

$$|e\rangle \rightarrow \frac{1}{\sqrt{2}}\left(|e\rangle + \frac{\mathrm{i}g^*}{|g|}|g\rangle\right) \rightarrow \frac{1}{\sqrt{2}}(|e\rangle - |g\rangle), \quad \text{if } g = -\mathrm{i}|g|. \tag{11.24c}$$

The time evolution of any initial state can be similarly obtained. The $\frac{\pi}{2}$-pulse produces a coherent superposition of excited and ground states with a relative phase which depends on the phase of g. In the context of quantum information science the transformations (11.24b) and (11.24c) are related to the Hadamard transformations $|g\rangle \rightarrow (|g\rangle + |e\rangle)/\sqrt{2}$, $|e\rangle \rightarrow (|g\rangle - |e\rangle)/\sqrt{2}$.

We next examine the behavior of the Bloch vector $\langle \vec{S}(t)\rangle = \langle\psi(t)|\vec{S}|\psi(t)\rangle$. From (11.14) we can write the Heisenberg equation of motion for $\vec{S}(t)$

$$\frac{\mathrm{d}\vec{S}}{\mathrm{d}t} = \frac{\mathrm{i}}{\hbar}[H_{\text{eff}}, \vec{S}] = -\mathrm{i}[\vec{S}, \vec{S}\cdot\vec{\Omega}] = \vec{\Omega}\times\vec{S}, \tag{11.25}$$

and hence

$$\frac{\mathrm{d}\langle\vec{S}\rangle}{\mathrm{d}t} = \vec{\Omega}\times\langle\vec{S}\rangle. \tag{11.26}$$

From (11.26), the conservation law is obvious

$$\langle\vec{S}\rangle\cdot\langle\vec{S}\rangle = \text{constant}. \tag{11.27}$$

The vector $\langle\vec{S}\rangle$ precesses about the direction of $\vec{\Omega}$. Such a precession equation (11.26) has the well-known solution [3]

$$\langle\vec{S}(t)\rangle = \vec{S}_0\cos\Omega t + 2\hat{n}(\vec{S}_0\cdot\hat{n})\sin^2\frac{\Omega t}{2} + \hat{n}\times\vec{S}_0\sin\Omega t, \tag{11.28}$$

where $\hat{n}$ is the unit vector in the direction $\vec{\Omega}$ and $\vec{S}_0$ is the value of $\langle\vec{S}(t)\rangle$ at $t = 0$. It is traditional to represent the dynamical behavior by a sphere as $\langle\vec{S}(t)\rangle$ for each time t would correspond to a point on the sphere with polar angles θ and φ

$$\langle S_z(t)\rangle = -\frac{1}{2}\cos\theta(t), \qquad \langle S_\pm(t)\rangle = \frac{1}{2}\sin\theta(t)\mathrm{e}^{\mathrm{i}\varphi(t)}. \tag{11.29}$$

The explicit values of θ and φ, although complicated, can be obtained from (11.28). In the special case of a field at resonance $\Delta = 0$ and the atom initially in ground state $\vec{S}_0 = (0, 0, -\frac{1}{2})$, (11.28) leads to

$$\langle S_x(t)\rangle = -\frac{n_y}{2}\sin\Omega t, \qquad \langle S_y(t)\rangle = \frac{n_x}{2}\sin\Omega t, \qquad \langle S_z(t)\rangle = -\frac{1}{2}\cos\Omega t. \tag{11.30}$$

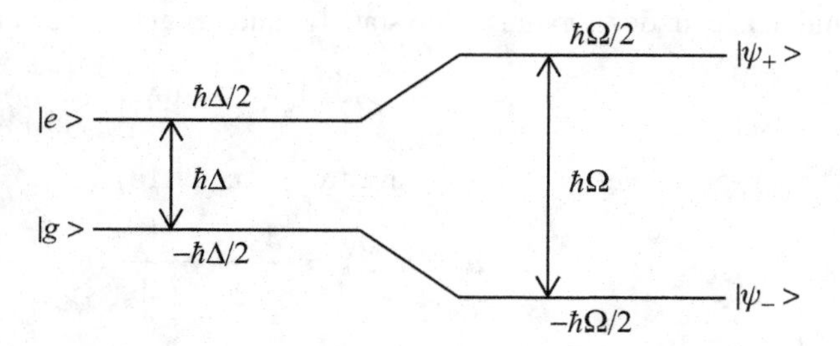

Fig. 11.1 Energy levels of dressed states.

A $\pi/2$-pulse ($\Omega t = \pi/2$) would therefore produce a maximum dipole moment or maximum coherence as is also evident from (11.24b). A state with a maximum dipole moment gives rise to coherent radiation in a way analogous to dipole radiation in classical physics. Clearly states like (11.24b) and (11.24c) can be considered to be analogs of coherent states for the atomic systems and we discuss these in detail in Section 11.3.

11.1.1 Semiclassical dressed states

We have seen that in a rotating frame the effective Hamiltonian, describing the interaction of a two-level atom with external electromagnetic fields, is given by (11.13). The eigenstates $|\psi_\pm\rangle$ of (11.13) are called the dressed states. For $\Delta > 0$, these states and their energies are given by

$$E_\pm = \pm\frac{\hbar\Omega}{2}, \qquad |\psi_\pm\rangle = \left(|e\rangle + \frac{1}{2g}(\Delta \mp \Omega)|g\rangle\right)\left(1 + \frac{(\Delta \mp \Omega)^2}{4|g|^2}\right)^{-\frac{1}{2}}\begin{pmatrix}1\\ g/|g|\end{pmatrix}. \tag{11.31}$$

These are schematically shown in Figure 11.1. The phase coefficient $\begin{pmatrix}1\\ g/|g|\end{pmatrix}$ is needed so that $|\psi_+\rangle \to |e\rangle$, $|\psi_-\rangle \to |g\rangle$ if $g \to 0$. Clearly if a system is prepared in one of these dressed states, then no further evolution takes place unless decay effects are included. Bai *et al.* [4] showed experimentally how the system can be prepared in dressed states. Let us consider for simplicity the case $\Delta = 0$, $g = |g|e^{i\chi}$, then

$$E_\pm = \pm\hbar|g|, \qquad |\psi_\pm\rangle = \frac{1}{\sqrt{2}}(|e\rangle \mp e^{-i\chi}|g\rangle)\begin{pmatrix}1\\ e^{i\chi}\end{pmatrix}. \tag{11.32}$$

Let the $\pi/2$-pulse interact with an atom in the ground state, let χ' be the phase of the field, $g = |g|e^{i\chi'}$. Then from (11.24b) the state at time t is

$$|\varphi(t)\rangle = \frac{1}{\sqrt{2}}(|g\rangle + ie^{i\chi'}|e\rangle). \tag{11.33}$$

On comparison with (11.32) we see that $|\varphi(t)\rangle$ is the eigenstate of H, where the phase of the field χ is $\chi' + \pi/2$, i.e.

$$H(\chi = \chi' + \pi/2)|\varphi(t)\rangle = -\hbar|g|(S_+e^{i\chi} + S_-e^{-i\chi})|\varphi(t)\rangle = -\hbar|g|\,|\varphi(t)\rangle. \tag{11.34}$$

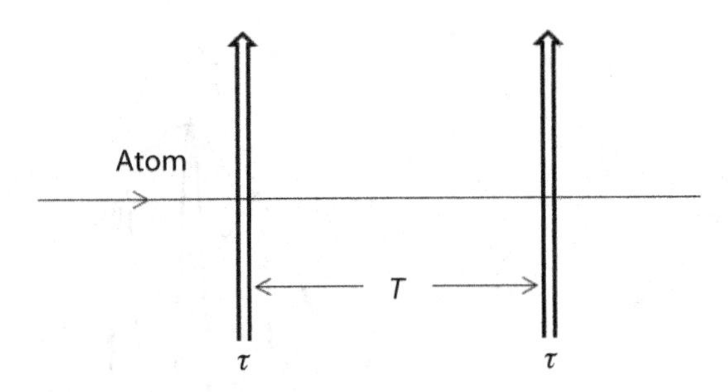

Fig. 11.2 Schematic illustration of the two-field Ramsey interferometer.

This suggests that, at the end of the $\pi/2$-pulse, the phase of the field χ' is to be changed by $\pi/2$ to $\chi' + \pi/2$ in order that the state $|\varphi(t)\rangle$ would become a dressed state $|\psi_-\rangle$ of the Hamiltonian $-\hbar|g|(S_+e^{i\chi} + S_-e^{-i\chi}) = -\hbar|g|i(S_+e^{i\chi'} - S_-e^{-i\chi'})$.

11.2 Application of atomic coherence – Ramsey interferometry

The pulsed excitation of a two-level atom leads us to a high-resolution atomic interferometer by which the frequency of transition can be measured very accurately. This method was introduced by Ramsey in the context of microwaves and has been adopted in the optical domain [5–8]. Consider first the excitation of the atom by a short pulse of duration τ, such that $|g|^2 \ll \Delta^2$, then the excitation probability is given by (11.18), i.e.

$$\begin{aligned} p_e(\tau) &\approx \frac{4|g|^2\tau^2}{\Delta^2\tau^2}\sin^2\frac{\Delta\tau}{2} \\ &= |g|^2\tau^2\text{sinc}^2X, \qquad X = \frac{\Delta\tau}{2}, \qquad \text{sinc}^2X = \frac{\sin^2 X}{X^2}. \end{aligned} \tag{11.35}$$

The sinc function is maximum at $X = 0$ and becomes zero at $X = \pi$. Thus the scan of the probability as a function of the frequency of the field or equivalently as a function of $\Delta\tau$ would have a half width of the order $\Delta\tau \sim 2\pi$. The width of the resonance is determined by the duration of the pulse.

Ramsey introduced a novel scheme where the atom was coherently excited in two separate zones separated by the time interval $T \gg \tau$, as shown in Figure 11.2. The time interval T has to be much smaller than the lifetime of the atom. Let the atom be initially in the ground state. We calculate the probability of finding the atom in the excited state after passing through two zones. We calculate p_e to second order in the coupling constant g. The Hamiltonian (11.13) in the interaction picture is

$$H = -\hbar(gS_+e^{i\Delta t} + g^*S_-e^{-i\Delta t}). \tag{11.36}$$

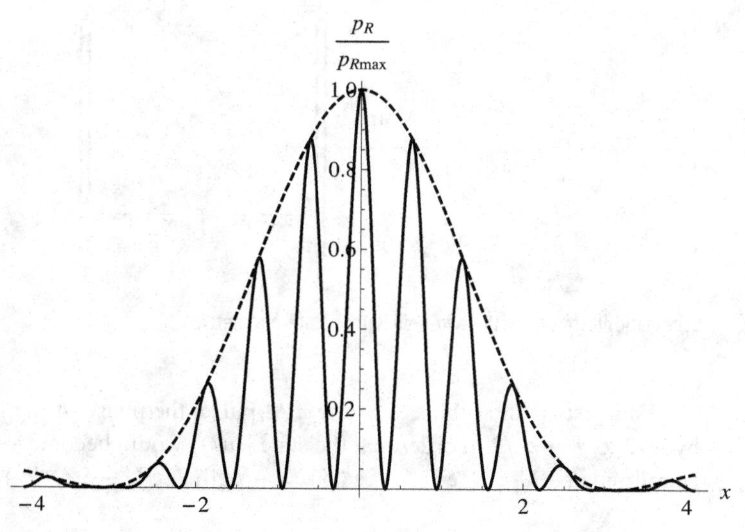

Fig. 11.3 The fringes for two-zone atomic resonance frequency measurement for $(T/\tau) = 5$ (solid curve), compared with the single-zone measurement (dashed curve).

The wave function of the atom to first order in the coupling constant g is

$$\begin{aligned} |\psi(\tau)\rangle &\cong |g\rangle + \mathrm{i}\int_0^\tau \mathrm{d}t\,(gS_+\mathrm{e}^{\mathrm{i}\Delta t} + g^*S_-\mathrm{e}^{-\mathrm{i}\Delta t})|g\rangle \\ &\approx |g\rangle + \frac{g}{\Delta}(\mathrm{e}^{\mathrm{i}\Delta\tau} - 1)|e\rangle = |g\rangle + \alpha|e\rangle. \end{aligned} \tag{11.37}$$

In the domain $\tau \le t \le \tau + T$, there is no field. In the second region only the ground state part will evolve as the excited part is already of first order in the coupling. The state $|g\rangle$ to first order would evolve into

$$\begin{aligned} |g\rangle &\to |g\rangle + \mathrm{i}\int_{T+\tau}^{T+2\tau} \mathrm{d}t\,(gS_+\mathrm{e}^{\mathrm{i}\Delta t} + g^*S_-\mathrm{e}^{-\mathrm{i}\Delta t})|g\rangle \\ &\equiv |g\rangle + \mathrm{e}^{\mathrm{i}\Delta(T+\tau)}\alpha|e\rangle. \end{aligned} \tag{11.38}$$

Hence the probability of finding the atom in the excited state is

$$\begin{aligned} p_R(T+2\tau) &= |\alpha|^2|1 + \mathrm{e}^{\mathrm{i}\Delta(T+\tau)}|^2 \\ &= |g|^2\tau^2\mathrm{sinc}^2(X)\cdot 4\cos^2\left(\frac{T}{\tau}X\right), \quad T \gg \tau. \end{aligned} \tag{11.39}$$

The addition of the second zone improves the resolution for the measurement of atomic resonance frequency by a factor (T/τ), as shown in Figure 11.3. We show the behavior of (11.39) $p_R/p_{R\max}$ as a function of X, say for $(T/\tau) = 5$. This is compared with the excitation probability (11.35). The Ramsey method has been extensively used and is now a standard tool for the measurement of atomic phases. Let us assume that between the two

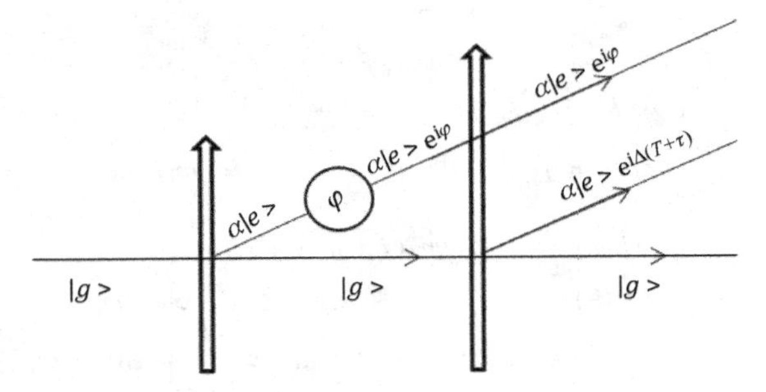

Fig. 11.4 Optical Ramsey excitation of an atomic beam by separated laser fields interpreted as a matter wave interferometer.

zones there is some disturbance which shifts the phase of the excited state by φ, then the result (11.39) would be modified to

$$\cos^2\left(\frac{\Delta T}{2}\right) \rightarrow \cos^2\left(\frac{\Delta T-\varphi}{2}\right). \tag{11.40}$$

This is because the wave function $|\psi(\tau)\rangle$ just before the second zone would be

$$|g\rangle + \frac{g}{\Delta}(\mathrm{e}^{\mathrm{i}\Delta\tau}-1)\mathrm{e}^{\mathrm{i}\varphi}|e\rangle, \tag{11.41}$$

therefore φ can be determined from the measurement of p_R.

We now explain why this setup is an interferometer. We show pictorially (Figure 11.4) what happens to the atomic wave function. The two amplitudes have to be summed up coherently. The scheme shown as before is very similar to the optical interferometers. One can think of splitting the amplitudes as the excited state is produced by the absorption of a laser photon and thus the excited state carries extra momentum.

11.3 Atomic coherent states

Clearly for a two-level system, we can define the coherent states $|\theta,\varphi\rangle$ by

$$|\theta,\varphi\rangle = \cos\frac{\theta}{2}|g\rangle + \sin\frac{\theta}{2}\mathrm{e}^{-\mathrm{i}\varphi}|e\rangle, \quad 0\le\theta\le\pi,\ 0\le\varphi\le 2\pi. \tag{11.42}$$

In the coherent state, the mean value of the dipole moment which is proportional to $\langle S_+\rangle$ is nonzero

$$\langle\theta,\varphi|S_+|\theta,\varphi\rangle = \frac{1}{2}\sin\theta\,\mathrm{e}^{\mathrm{i}\varphi}. \tag{11.43}$$

The mean inversion is

$$\langle\theta,\varphi|S_z|\theta,\varphi\rangle = -\frac{1}{2}\cos\theta. \tag{11.44}$$

The states with nonzero $\langle S_+\rangle$ are said to possess atomic coherence. We have already seen in Section 11.2 how interferometers can be built using nonzero coherence. If many two-level

Table 11.1 Angular momentum algebra.

$[J_\alpha, J_\beta] = \mathrm{i}\epsilon_{\alpha\beta\gamma}J_\gamma, \qquad [J_\alpha, J^2] = 0.$	(1)
$J^2\|j,m\rangle = j(j+1)\|j,m\rangle, \quad J_z\|j,m\rangle = m\|j,m\rangle.$ $J_\pm\|j,m\rangle = \sqrt{(j\mp m)(j\pm m+1)}\|j,m\pm 1\rangle,$	(2)
$D(\xi) = \exp(\xi J_+ - \xi^* J_-), \quad \xi = \frac{\theta}{2}\mathrm{e}^{-\mathrm{i}\varphi}$ $= \exp(\zeta J_+)\exp\{J_z \ln(1+\|\zeta\|^2)\}\exp(-\zeta^* J_-), \quad \zeta = \tan\frac{\theta}{2}\mathrm{e}^{-\mathrm{i}\varphi}.$	(3)
$\exp(\mathrm{i}\theta\vec{n}\cdot\vec{J})\vec{A}\exp(-\mathrm{i}\theta\vec{n}\cdot\vec{J}) = \vec{n}(\vec{n}\cdot\vec{A}) - \vec{n}\times(\vec{n}\times\vec{A})\cos\theta + (\vec{n}\times\vec{A})\sin\theta.$	(4)

atoms interact with an external field, then each two-level atom gets prepared in the state (11.42). If we further assume that the parameters (θ, φ) are same for each two-level atom, then the state can be written as

$$|\theta,\varphi\rangle = \prod_{j=1}^{N} |\theta,\varphi\rangle_j, \tag{11.45}$$

where $|\theta,\varphi\rangle_j$ is the coherent state for the jth atom. The state $|\theta,\varphi\rangle$ is called the atomic coherent state for a collection of two-level atoms.

A collection of two-level systems can be characterized by the angular momentum algebra corresponding to $j = N/2$ provided we choose to work in the completely symmetric representation. For convenience, many important properties of angular momentum algebra are summarized in Table 11.1. Let us write the collective angular momentum operators as

$$\vec{J} = \sum_{j=1}^{N} \vec{S}^{(j)}; \qquad J^2|j,m\rangle = j(j+1)|j,m\rangle; \qquad J_z|j,m\rangle = m|j,m\rangle;$$

$$J_\pm = J_x \pm \mathrm{i}J_y; \quad J_\pm|j,m\rangle = \sqrt{(j\mp m)(j\pm m+1)}|j,m\pm 1\rangle; \quad -j \le m \le j. \tag{11.46}$$

Note that two-level (spin) operators are denoted by the symbol $\vec{S}$ whereas the collective operators are denoted by $\vec{J}$. The atomic coherent states $|\theta,\varphi\rangle$ are then given by [9]

$$|\theta,\varphi\rangle = D(\xi)|j,-j\rangle, \tag{11.47}$$

where the unitary operator D is

$$D(\xi) = \exp\{\xi J_+ - \xi^* J_-\} = \exp\{\mathrm{i}\theta\vec{n}\cdot\vec{J}\}, \quad \xi = \frac{\theta}{2}\mathrm{e}^{-\mathrm{i}\varphi}, \quad \vec{n} = (-\sin\varphi, \cos\varphi, 0). \tag{11.48}$$

The generator $D(\xi)$ of the coherent states can be separated in a normally ordered form by using the BCH identity for $SU(2)$ operators [10]

$$D(\xi) = \exp\{\zeta J_+\}\exp\{J_z\ln(1+|\zeta|^2)\}\exp\{-\zeta^* J_-\}, \quad \zeta = \tan\frac{\theta}{2}\mathrm{e}^{-\mathrm{i}\varphi}. \tag{11.49}$$

The relation (11.47) can be simplified by noting that

$$J_-|j,-j\rangle = 0, \qquad J_z|j,-j\rangle = (-j)|j,-j\rangle,$$
$$D(\xi)|j,-j\rangle = (1+|\zeta|^2)^{-j}\exp\{\zeta J_+\}|j,-j\rangle, \tag{11.50}$$

and hence

$$|\theta,\varphi\rangle = \sum_{m=-j}^{+j}\binom{2j}{j+m}^{\frac{1}{2}}\sin^{j+m}\frac{\theta}{2}\cos^{j-m}\frac{\theta}{2}\,\mathrm{e}^{-\mathrm{i}(j+m)\varphi}|j,m\rangle. \tag{11.51}$$

This is the final expression for the atomic coherent states in terms of the basis states of the collective angular momentum operators J^2 and J_z.

The atomic coherent states form a complete set

$$\frac{2j+1}{4\pi}\int|\theta,\varphi\rangle\langle\theta,\varphi|\sin\theta\mathrm{d}\theta\mathrm{d}\varphi = 1, \tag{11.52}$$

and are nonorthogonal

$$\langle\theta,\varphi|\theta',\varphi'\rangle = \left[\sin\frac{\theta}{2}\sin\frac{\theta'}{2} + \mathrm{e}^{i(\varphi-\varphi')}\cos\frac{\theta}{2}\cos\frac{\theta'}{2}\right]^{2j},$$
$$|\langle\theta,\varphi|\theta',\varphi'\rangle|^2 = \left(\cos^2\frac{\Theta}{2}\right)^{2j}, \tag{11.53}$$

where Θ is the angle between the directions (θ,φ) and (θ',φ') defined by $\cos\Theta = \cos\theta\cos\theta' + \sin\theta\sin\theta'\cos(\varphi-\varphi')$. The relations (11.52) and (11.53) are easily verified by using (11.51). Thus the states lying in opposite directions on the Bloch sphere $\Theta=\pi$ are orthogonal.

We recall that the field coherent states are the eigenstates of the annihilation operator. We can then ask: what is the operator for which $|\theta,\varphi\rangle$ is the eigenstate? To consider this we observe that

$$J_z|j,-j\rangle = -j|j,-j\rangle, \tag{11.54}$$

and hence

$$J_zD^\dagger(\xi)|\theta,\varphi\rangle = -jD^\dagger(\xi)|\theta,\varphi\rangle, \tag{11.55}$$

i.e.

$$\tilde{J}_z|\theta,\varphi\rangle = -j|\theta,\varphi\rangle. \tag{11.56}$$

Thus the atomic coherent is the eigenstate of the operator

$$\tilde{J}_z = D(\xi)J_zD^\dagger(\xi) \tag{11.57}$$

which on using the result of Exercise 11.5 reduces to

$$\tilde{J}_z = J_z\cos\theta - (\sin\varphi J_y + \cos\varphi J_x)\sin\theta. \tag{11.58}$$

Unlike the coherent states for the field for which all the normally ordered correlations have the property $\langle\alpha|a^{\dagger m}a^n|\alpha\rangle = \alpha^{*m}\alpha^n$, the atomic coherent states do not have such a property, e.g. $\langle J_+J_-\rangle \neq \langle J_+\rangle\langle J_-\rangle$.

11.4 Minimum uncertainty states for two-level systems – spin squeezing

The coherent and squeezed states of harmonic oscillators are known to be minimum uncertainty states (see Exercise 2.7). Now we can check if the atomic coherent states satisfy any type of minimum uncertainty relation. Furthermore, we enquire if we can define squeezed states for two-level systems. The derivation of the uncertainty relation for two Hermitian operators A and B satisfying

$$[A, B] = \mathrm{i}C, \tag{11.59}$$

shows that

$$\Delta A\Delta B = \left|\frac{\langle C\rangle}{2}\right|, \quad \Delta A^2 = \langle(A - \langle A\rangle)^2\rangle, \quad \Delta B^2 = \langle(B - \langle B\rangle)^2\rangle, \tag{11.60}$$

if

$$\Delta A|\psi\rangle = -\mathrm{i}\lambda\Delta B|\psi\rangle, \qquad \langle(\Delta A)^2\rangle = \lambda^2\langle(\Delta B)^2\rangle, \qquad \lambda = \text{real}. \tag{11.61}$$

For $\lambda = \pm1$, the variances in A and B are equal. The solution of (11.61) for $\lambda = \pm1$ defines coherent states, whereas the solution for $|\lambda| \neq 1$ defines squeezed states [11]. Let us first consider the eigenvalue problem for $\lambda = -1$

$$(A - \mathrm{i}B)|\psi\rangle = (\langle A\rangle - \mathrm{i}\langle B\rangle)|\psi\rangle. \tag{11.62}$$

Clearly if we choose $A = J_x$, $B = J_y$, then the only solution for $|\psi\rangle$ is $|\psi\rangle = |j, -j\rangle$. We will now show how the whole class of minimum uncertainty states can be obtained for $\lambda = -1$. We consider the operators obtained by rotating J_x and J_y by $D(\xi)$, with $D(\xi)$ defined by (11.48)

$$\tilde{J}_\alpha = D(\xi)J_\alpha D^\dagger(\xi), \qquad \alpha = x, y, z, \tag{11.63}$$

such that $[\tilde{J}_x, \tilde{J}_y] = \mathrm{i}\tilde{J}_z$. Hence the eigenvalue problem (11.62) with $A = \tilde{J}_x$, $B = \tilde{J}_y$ will have a solution $|\psi\rangle = D(\xi)|j, -j\rangle$, which is the atomic coherent state (11.47). This is because we can rewrite the eigenvalue equation $J_-|j, -j\rangle = 0$ as $D(\xi)J_-D^\dagger(\xi)D(\xi)|j, -j\rangle = 0$, i.e. $\tilde{J}_-|\psi\rangle = 0$. Thus we can generate a whole class of coherent states which are now parameterized by (θ, φ).

For $|\lambda| \neq 1$, we choose $A = J_x$, $B = J_y$,

$$(J_x - \mathrm{i}\lambda J_y)|\psi\rangle_s = (\langle J_x\rangle - \mathrm{i}\lambda\langle J_y\rangle)|\psi\rangle_s. \tag{11.64}$$

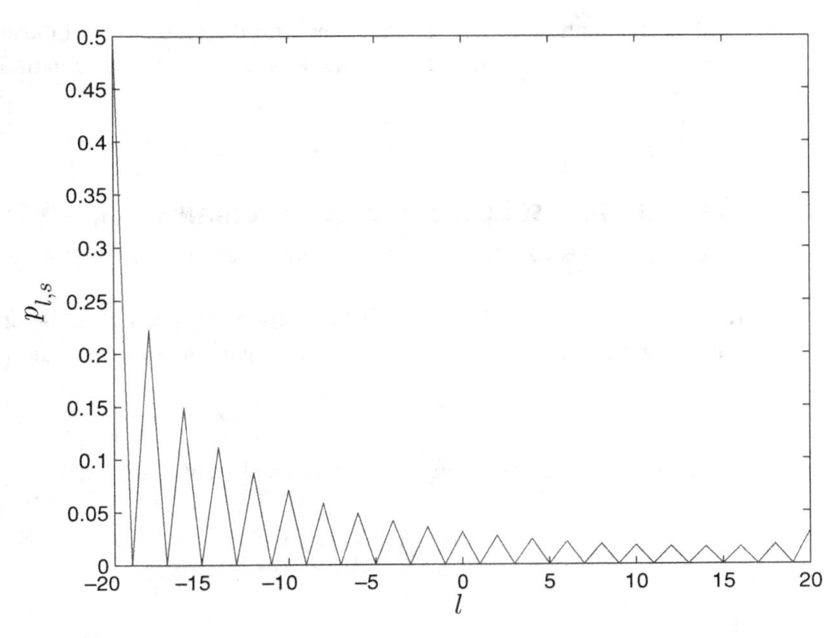

Fig. 11.5 The probability $p_{l,s}$ of occupation of collective atomic states $|j, l\rangle$ as a function of l for $j = 20$ and $\alpha = 0.0345$.

To solve this eigenvalue problem, note that $(J_x - \mathrm{i}\lambda J_y)$ can be written as

$$\mathrm{e}^{-\alpha J_z}\mathrm{e}^{-\mathrm{i}\frac{\pi}{2}J_y}J_z\mathrm{e}^{+\mathrm{i}\frac{\pi}{2}J_y}\mathrm{e}^{+\alpha J_z} = \mathrm{e}^{-\alpha J_z}J_x\mathrm{e}^{+\alpha J_z} = \cosh\alpha J_x - \mathrm{i}\sinh\alpha J_y, \tag{11.65}$$

and hence (11.64) can be solved in terms of the eigenstates $|j, m\rangle$ of J_z

$$|\psi\rangle_s = \mathrm{e}^{-\alpha J_z}\mathrm{e}^{-\mathrm{i}\frac{\pi}{2}J_y}|j, m\rangle \equiv |j, m\rangle_s, \qquad \lambda = \tanh\alpha. \tag{11.66}$$

The states $|\psi\rangle_s$ satisfy the equality sign in the minimum uncertainty relation. Furthermore

$$\begin{gathered}\Delta J_x^2 = (\tanh^2\alpha)\Delta J_y^2 < \Delta J_y^2,\\ \Delta J_x\Delta J_y = \frac{\langle J_z\rangle}{2}, \qquad \Delta J_x^2 = \left|\frac{\langle J_z\rangle}{2}\right|\tanh\alpha,\end{gathered} \tag{11.67}$$

and hence the x component of the angular momentum is squeezed. The solution to the eigenvalue problem (11.64) was obtained in [12, 13].

In view of the squeezing property (11.67), we call the state (11.66) the atomic squeezed state. These are generated from the state $|j, m\rangle$ by a nonunitary transformation as α is real. For integral values of j, i.e. for even values of the number of atoms, m can be zero. The state $|j, 0\rangle_s$ has some properties similar to those of the squeezed vacuum for the radiation field. Let $p_{l,s}$ be the probability of finding the collective system in the state $|j, l\rangle$ for integer values of j, then up to a normalization factor

$$p_{l,s} = |\langle j, l|j, 0\rangle_s|^2. \tag{11.68}$$

As shown in Figure 11.5, $p_{l,s}$ exhibits oscillatory behavior and vanishes for odd values of $l + j$. This should be compared with Figure 2.2 for the squeezed vacuum for the radiation field. Agarwal and Puri [14] give a method for the production of the atomic squeezed state $|j, 0\rangle_s$ by transforming irreversibly the squeezing from the radiation field to the atomic system.

11.5 Atomic squeezed states by nonlinear unitary transformations

Kitagawa and Ueda [15] realized that the atomic squeezing can be produced by a unitary transformation involving a nonlinear Hamiltonian, the simplest example of which is

$$H = -\hbar\chi J_z^2. \tag{11.69}$$

Let us consider the initial state as an atomic coherent state $|\theta, \varphi\rangle$ with $\theta = \pi/2$, $\varphi = 0$,

$$\left|\frac{\pi}{2}, 0\right\rangle = \left(\frac{1}{2}\right)^j \sum_m \binom{2j}{j+m}^{\frac{1}{2}} |j, m\rangle, \tag{11.70}$$

then

$$\begin{aligned}\rho_{m,m'}(t) &= \langle j, m|\rho|j, m'\rangle \\ &= \langle j, m|\mathrm{e}^{\mathrm{i}\chi J_z^2 t}\rho(0)\mathrm{e}^{-\mathrm{i}\chi J_z^2 t}|j, m'\rangle \\ &= \exp\{\mathrm{i}\chi t(m^2 - m'^2)\}\left(\frac{1}{2}\right)^{2j}\binom{2j}{j+m}^{\frac{1}{2}}\binom{2j}{j+m'}^{\frac{1}{2}}.\end{aligned} \tag{11.71}$$

As in the case of radiation fields we want to find the operators whose uncertainties are minimum. Therefore we introduce the rotated operators

$$\begin{aligned}\tilde{\vec{J}} = \mathrm{e}^{\mathrm{i}\nu J_x}\vec{J}\mathrm{e}^{-\mathrm{i}\nu J_x}, \qquad & \tilde{J}_x = J_x, \\ \tilde{J}_y = \cos\nu J_y - \sin\nu J_z, \qquad & \tilde{J}_z = \sin\nu J_y + \cos\nu J_z.\end{aligned} \tag{11.72}$$

The parameter ν is fixed by the condition that the variance in either $\tilde{J}_y$ or $\tilde{J}_z$ is minimum. All the uncertainties can be calculated by the repeated use of (11.71). The calculations are straightforward. Note that we need expectation values involving at most quadratic forms of angular momentum operators, thus $m^2 - m'^2 = (m - m')(m + m')$ and m' can be at most $m \pm 2$, $m \pm 1$, m, and hence we would have time-dependent exponentials that are linear in m. The calculations of variances would involve summing over binomial coefficients and these can be done by noting

$$\begin{aligned}\left(\frac{1}{2}\right)^{2j}\sum_m \mathrm{e}^{im\varphi}\binom{2j}{j+m} &= (\cos\varphi/2)^{2j}, \\ \left(\frac{1}{2}\right)^{2j}\sum_m m^p \mathrm{e}^{im\varphi}\binom{2j}{j+m} &= \frac{\partial^p}{\partial(\mathrm{i}\varphi)^p}(\cos\varphi/2)^{2j}.\end{aligned} \tag{11.73}$$

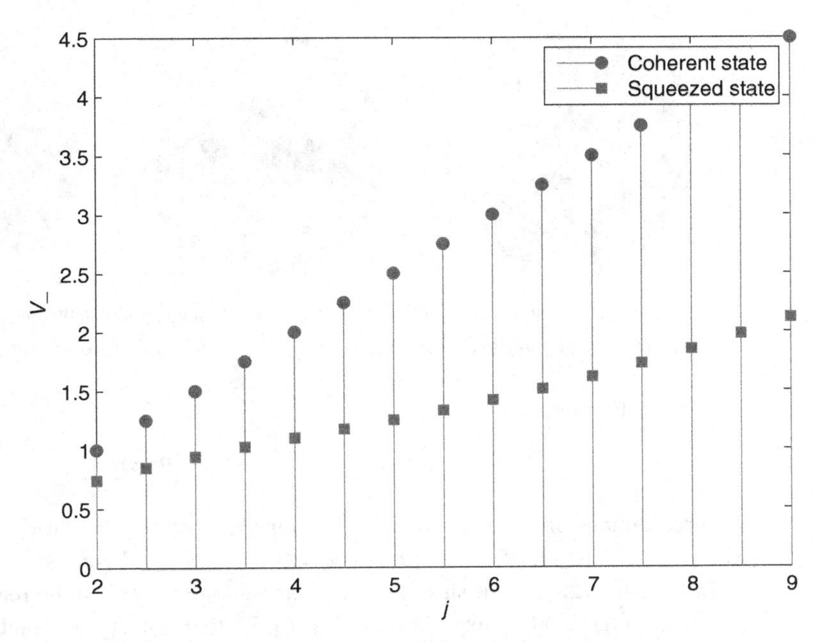

Fig. 11.6 The variances of the spin components of the coherent state and squeezed state.

This procedure yields

$$\langle \tilde{J}_x \rangle = j \cos^{2j-1} \frac{\mu}{2}, \qquad \langle \tilde{J}_y \rangle = \langle \tilde{J}_z \rangle = 0, \tag{11.74}$$

$$\begin{aligned} \Delta \tilde{J}_x^{\,2} &= \frac{j}{2} \left[2j \left(1 - \cos^{4j-2} \frac{\mu}{2} \right) - \left(j - \frac{1}{2} \right) A \right], \\ \Delta \tilde{J}_{\substack{y \\ z}}^{\,2} &= \frac{j}{2} \left[1 + \frac{1}{2} \left(j - \frac{1}{2} \right) \left[A \pm \sqrt{A^2 + B^2} \cos(2\nu + 2\delta) \right] \right], \\ A &= 1 - \cos^{2j-2} \mu, \qquad B = 4 \sin \frac{\mu}{2} \cos^{2j-2} \frac{\mu}{2}, \\ \delta &= \frac{1}{2} \tan^{-1} \left(\frac{B}{A} \right), \qquad \mu = 2\chi t. \end{aligned} \tag{11.75}$$

The relation (11.75) show that in order to obtain minimum variance V_- in either $\tilde{J}_y$ or $\tilde{J}_z$, we set $2\nu + 2\delta$ appropriately so that V_- is given by

$$V_- = \frac{j}{2} \left[1 + \frac{1}{2} \left(j - \frac{1}{2} \right) (A - \sqrt{A^2 + B^2}) \right]. \tag{11.76}$$

The behavior of V_- as a function of j is shown in Figure 11.6, which also shows for comparison the result $j/2$ for the atomic coherent state obtained by setting $\mu = 0$ in (11.76) ($j \neq 1$). We can find the smallest value of V_- as a function of μ. The minimum value of V_- is about $\frac{1}{2}(\frac{j}{3})^{\frac{1}{3}}$, which occurs at $|\mu| \approx 24^{\frac{1}{6}} j^{-\frac{2}{3}}$ [15].

For the case of radiation fields, we have represented squeezing in phase space as an ellipse. A similar thing can be done for atomic squeezing. Let us introduce the analog of

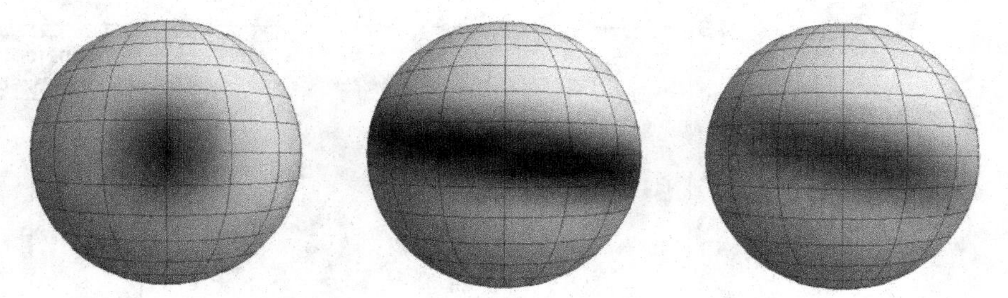

Fig. 11.7 State evolutions by twisting in terms of the quasiprobability distribution (QDP) on the sphere for $j = 20$: (a) the initial atomic coherent state; (b) optimally squeezed at $\mu = 0.199$; and (c) excessively twisted at $\mu = 0.399$.

the Q-function via

$$Q(\theta, \varphi) = \frac{1}{4\pi}\langle\theta, \varphi|\rho|\theta, \varphi\rangle. \tag{11.77}$$

A three-dimensional polar plot of $\frac{Q(\theta,\varphi)}{Q_{\max}}$ for the state (11.71) for different values of μ is shown in Figure 11.7 for $j = 20$. For $\mu = 0$, we have a coherent state.

In Chapter 12, we will show how the interaction (11.69) can be realized by atoms contained in a high-quality dispersive cavity [16]. Furthermore, the interaction (11.69) naturally arises in other systems, such as two-component Bose condensates [17, 18]. The interaction in each condensate and the interactions between two condensates gives effectively the interaction (11.69). This has been considered as a way to realize the squeezed states of Bose condensates. Spin squeezing has also been realized in collective atomic ensembles via continuous quantum nondemolition measurements [19–21].

11.6 Atomic squeezed states produced by supersensitivity of Ramsey interferometers

In this section we show an important application of atomic squeezed states in producing supersensitive measurements of phase in Ramsey interferometry. Consider the setup shown in Figure 11.8. Consider a set of atoms prepared in an initial state $|\psi\rangle_{\text{in}}$ and sent through the Ramsey interferometer. Traditionally $|\psi\rangle_{\text{in}}$ is chosen as $|g, g, \ldots, g\rangle$, i.e. all the atoms are in the lower state. We can calculate the sensitivity of the interferometer when $|\psi\rangle_{\text{in}}$ is a squeezed state (11.71) and compare it with the sensitivity when $|\psi\rangle_{\text{in}} = |g, g, \ldots, g\rangle$. We would describe the above setup in terms of the collective operators $\vec{J}$. We assume that pulses are so short that we can drop the effect of atomic detuning during the time the pulse is on. In this case we can relate $|\psi\rangle_{\text{out}}$ to $|\psi\rangle_{\text{in}}$ via a product of unitary transformations

$$\begin{aligned}
|\psi\rangle_{\text{out}} &= \exp\left\{-\mathrm{i}\frac{\pi}{2}J_x\right\}\exp\{-\mathrm{i}\varphi J_z\}\exp\left\{-\mathrm{i}\frac{\pi}{2}J_x\right\}|\psi\rangle_{\text{in}} \\
&= \exp\{-\mathrm{i}\pi J_x\}\exp\left\{\mathrm{i}\frac{\pi}{2}J_x\right\}\exp\{-\mathrm{i}\varphi J_z\}\exp\left\{-\mathrm{i}\frac{\pi}{2}J_x\right\}|\psi\rangle_{\text{in}} \\
&= \exp\{-\mathrm{i}\pi J_x\}\exp\left\{-\mathrm{i}\varphi J_y\right\}|\psi\rangle_{\text{in}}.
\end{aligned} \tag{11.78}$$

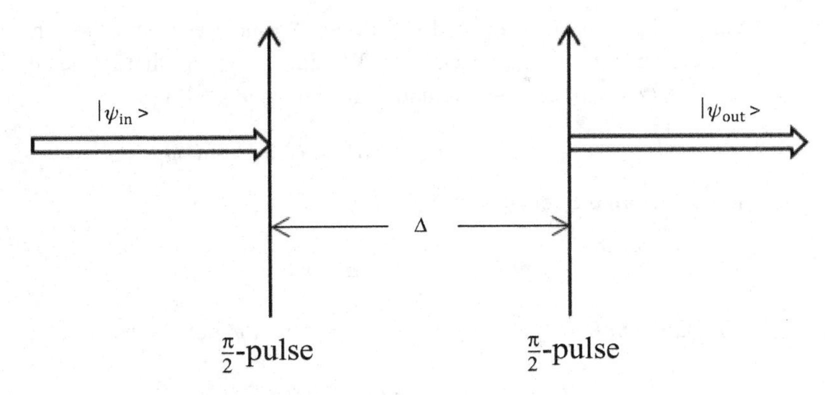

Fig. 11.8 Setup of Ramsey interferometry with supersensitive measurement of phase.

At the output we measure the population in the excited state which is related to $\langle J_z \rangle$. Thus the signal and the noise in the signal would be given by

$$\begin{aligned} I &= \langle J_z \rangle = {}_{\text{out}}\langle \psi | J_z | \psi \rangle_{\text{out}} = {}_{\text{in}}\langle \psi | \tilde{J}_z | \psi \rangle_{\text{in}}, \\ \Delta I^2 &= \langle J_z^2 \rangle - \langle J_z \rangle^2 = {}_{\text{in}}\langle \psi | \tilde{J}_z^2 | \psi \rangle_{\text{in}} - {}_{\text{in}}\langle \psi | \tilde{J}_z | \psi \rangle_{\text{in}}^2, \end{aligned} \tag{11.79}$$

where

$$\tilde{J}_z = \exp\left\{ \mathrm{i}\varphi J_y \right\} \exp\left\{ \mathrm{i}\pi J_x \right\} J_z \exp\left\{ -\mathrm{i}\pi J_x \right\} \exp\left\{ -\mathrm{i}\varphi J_y \right\}. \tag{11.80}$$

We define the sensitivity $\Delta\varphi$ via

$$(\Delta\varphi)^2 = \frac{\Delta I^2}{|\frac{\partial I}{\partial \varphi}|^2}. \tag{11.81}$$

Using the angular momentum algebra, (11.80) can be simplified to

$$\tilde{J}_z = -J_z \cos\varphi + J_x \sin\varphi. \tag{11.82}$$

Using (11.79) and (11.82), the mean signal is

$$\begin{aligned} I &= -\langle J_z \rangle \cos\varphi + \langle J_x \rangle \sin\varphi, \\ \frac{\partial I}{\partial \varphi} &= \langle J_z \rangle \sin\varphi + \langle J_x \rangle \cos\varphi. \end{aligned} \tag{11.83}$$

For the state $|g, g, \ldots, g\rangle = |\frac{N}{2}, -\frac{N}{2}\rangle$, $\langle J_x \rangle = 0$,

$$\begin{aligned} I &= -\frac{N}{2} \cos\varphi, \\ \Delta I^2 &= \sin^2\varphi \langle J_x^2 \rangle = \sin^2\varphi \left\langle \frac{N}{2}, -\frac{N}{2} \middle| J_x^2 \middle| \frac{N}{2}, -\frac{N}{2} \right\rangle = \frac{N \sin^2\varphi}{4}, \end{aligned} \tag{11.84}$$

and therefore the sensitivity $\Delta\varphi$ (Eq. (11.81)) is

$$(\Delta\varphi)^2 = \frac{N \sin^2\varphi / 4}{(\frac{N}{2})^2 \sin^2\varphi} = \frac{1}{N}. \tag{11.85}$$

This is the shot noise limited sensitivity. We next estimate the sensitivity for the case of atoms prepared in a squeezed state. We start by noting that in the neighborhood of $\varphi \sim \frac{\pi}{2}$, $|\frac{\partial I}{\partial \varphi}|^2 \sim \langle J_z \rangle^2 \sin^2 \varphi$. The fluctuation for the squeezed state is

$$\Delta I^2 \approx (\Delta J_x)^2 \sin^2 \varphi, \tag{11.86}$$

and therefore the sensitivity $\Delta\varphi_s$ for the squeezed state is

$$(\Delta\varphi_s)^2 = \frac{(\Delta J_x)^2}{\langle J_z \rangle^2}. \tag{11.87}$$

Wineland *et al.* [22] introduced the parameter ξ_R, defined by

$$\xi_R = \frac{\Delta\varphi_s}{\Delta\varphi} = \frac{\sqrt{N}(\Delta J_x)}{|\langle J_z \rangle|}, \tag{11.88}$$

as a measure of the improvement in the sensitivity of the interferometer if the atomic squeezed state is used as the state of the incoming atoms. Clearly the input state should be such that $\xi_R < 1$. The parameter ξ_R is also known as the spectroscopic squeezing parameter. Note that the squeezing based on Heisenberg uncertainty relations implies

$$(\Delta J_x)^2 < \left| \frac{\langle J_z \rangle}{2} \right|. \tag{11.89}$$

The condition $\xi_R < 1$ is much stronger than the condition (11.89) as $\left| 2\frac{\langle J_z \rangle}{N} \right| < 1$.

11.7 Phase-space representation for a collection of two-level systems

Just as for radiation fields we can use a variety of c-number functions to visualize the quantum states of two-level systems. We have already used the function $Q(\theta, \varphi)$ in Section 11.5 to discuss the squeezing in a system of two-level systems. The analog of the Glauber–Sudarshan representation for the atomic system is

$$\begin{aligned} \rho &= \int P(\theta, \varphi)|\theta, \varphi\rangle\langle\theta, \varphi| \sin\theta \, \mathrm{d}\theta \mathrm{d}\varphi, \\ &\int P(\theta, \varphi) \sin\theta \, \mathrm{d}\theta \mathrm{d}\varphi = 1. \end{aligned} \tag{11.90}$$

In the theory of angular momentum, the multipole operators T_{KQ} are basic as these form a complete orthogonal set of operators [23]. Thus these can be utilized to develop phase-space distributions for angular momentum systems [24]. The T_{KQ}'s enable us to prove (11.90) and also suggest a definition of the Wigner function for spins or two-level atoms. The state-multipole operators are defined by

$$T_{KQ} = \sum_{mm'} (-1)^{j-m} (2K+1)^{1/2} \begin{pmatrix} j & K & j \\ -m & Q & m' \end{pmatrix} |jm\rangle\langle jm'|, \tag{11.91}$$

where $\begin{pmatrix} j & K & j \\ -m & Q & m' \end{pmatrix}$ is the Wigner $3j$ symbol and $T_{00} = 1/\sqrt{2j+1}$. Note that K is an integer taking values $0, 1, 2, \ldots, 2j$ and $-K \leq Q \leq +K$. The state multipoles have the following important orthogonality property

$$\mathrm{Tr}(T^{\dagger}_{K_1Q_1} T_{K_2Q_2}) = \delta_{K_1K_2}\delta_{Q_1Q_2}, \qquad T^{\dagger}_{KQ} = (-1)^{Q} T_{K,-Q}, \tag{11.92}$$

and hence any function of angular momentum operators can be expanded in terms of T_{KQ}. In particular, for the density matrix, we get the expansion

$$\rho = \sum_{KQ} \rho_{KQ} T_{KQ} = \sum_{KQ} \langle T^{\dagger}_{KQ}\rangle T_{KQ}. \tag{11.93}$$

Thus the density matrix is completely characterized by the state multipoles $\langle T^{\dagger}_{KQ}\rangle$. These expectation values are closely related to the moments of the angular momentum operators

$$\begin{aligned}
\langle T^{\dagger}_{10}\rangle &= \left(\frac{3}{j(2j+1)(j+1)}\right)^{\frac{1}{2}} \langle J_z\rangle, \\
\langle T^{\dagger}_{11}\rangle &= -\left(\frac{3}{2j(j+1)(2j+1)}\right)^{\frac{1}{2}} \langle J_-\rangle, \\
\langle T^{\dagger}_{20}\rangle &= \left(\frac{5}{(2j+3)(j+1)(2j+1)j(2j-1)}\right)^{\frac{1}{2}} [3\langle J_z^2\rangle - j(j+1)].
\end{aligned} \tag{11.94}$$

The tensor $\langle T^{\dagger}_{2Q}\rangle$ is also known as the alignment tensor. It may also be noticed that the properties of the atomic coherences will be reflected in the properties of $\langle T^{\dagger}_{KQ}\rangle$ with $Q \neq 0$.

Using the expansion (11.93) we can define the Wigner function $W(\theta, \varphi)$ for the angular momentum systems via

$$W(\theta, \varphi) = \sqrt{\frac{2j+1}{4\pi}} \sum_{KQ} \langle T^{\dagger}_{KQ}\rangle Y_{KQ}(\theta, \varphi), \qquad W^* = W, \tag{11.95}$$

where $Y_{KQ}(\theta, \varphi)$ is the spherical harmonic. The Wigner function $W(\theta, \varphi)$ is normalized

$$\int W(\theta, \varphi) \sin\theta \, \mathrm{d}\theta \mathrm{d}\varphi = 1. \tag{11.96}$$

Note that the expectation values can be evaluated using the Wigner function by writing the operator and their Wigner representations as

$$G = \sum_{KQ} G_{KQ} T_{KQ}, \qquad G_W(\theta, \varphi) = \sum_{KQ} G_{KQ} Y_{KQ}, \tag{11.97}$$

then

$$\begin{aligned}\langle G\rangle &= \mathrm{Tr}\{\rho G\}\\ &= \sum_{K_1Q_1K_2Q_2} \rho_{K_1Q_1} G_{K_2Q_2} \mathrm{Tr}\{T_{K_1Q_1} T_{K_2Q_2}\}\\ &= \sum_{K_1Q_1} \rho_{K_1Q_1} G_{K_1-Q_1} (-1)^{Q_1}\\ &= \sqrt{\frac{4\pi}{2j+1}} \int W(\theta,\varphi) G_W(\theta,\varphi) \sin\theta \, \mathrm{d}\theta \mathrm{d}\varphi. \end{aligned} \tag{11.98}$$

To derive (11.98) we need to use the orthogonality of spherical harmonics and $Y^*_{KQ}(\theta,\varphi) = (-1)^Q Y_{K-Q}(\theta,\varphi)$. Arecchi *et al.* [9] gave a very useful relation

$$\langle\theta,\varphi|T^{\dagger}_{KQ}|\theta,\varphi\rangle = Y^*_{KQ}(\theta,\varphi)\frac{(-1)^{K-Q}\sqrt{4\pi}\,(2j)!}{\sqrt{(2j-K)!(2j+K+1)!}}. \tag{11.99}$$

Using (11.99) we can write the phase-space function $Q(\theta,\varphi)$ as

$$\begin{aligned}Q(\theta,\varphi) &= \frac{2j+1}{4\pi}\langle\theta,\varphi|\rho|\theta,\varphi\rangle\\ &= \frac{2j+1}{4\pi}\sum_{KQ}\rho_{KQ}Y_{KQ}(\theta,\varphi)\frac{(-1)^{K-Q}\sqrt{4\pi}\,(2j)!}{\sqrt{(2j-K)!(2j+K+1)!}}.\end{aligned} \tag{11.100}$$

The $P(\theta,\varphi)$ can also be written in the form (11.100)

$$P(\theta,\varphi) = \sum_{KQ}\rho_{KQ}Y_{KQ}(\theta,\varphi)(-1)^{K-Q}\frac{\sqrt{(2j-K)!(2j+K+1)!}}{\sqrt{4\pi}\,(2j)!}. \tag{11.101}$$

The results (11.95), (11.100), and (11.101) show the relations between different phase-space distributions for angular momentum systems. For convenience we collect these results in Table 11.2.

The Wigner function for a system in an atomic coherent state $|\theta_0,\varphi_0\rangle$ can be obtained by using (11.99) in (11.95)

$$W(\theta,\varphi) = \sum_{KQ} Y^*_{KQ}(\theta_0,\varphi_0)Y_{KQ}(\theta,\varphi)(-1)^{K-Q}\sqrt{\frac{(2j)!(2j+1)!}{(2j-K)!(2j+K+1)!}}. \tag{11.102}$$

The Wigner function for an atomic coherent state $|\pi/2, 0\rangle$ is shown in Figure 11.9 [25]. It is similar to that for a coherent state for the radiation field. For a cat state $\frac{1}{\sqrt{2}}(|3,2\rangle + |3,-2\rangle)$, the function as shown in Figure 11.10 displays regions of negativity. Such a cat state is experimentally realized in [26] by the excitation of the hyperfine state corresponding to the angular momentum $F = 3$ of the Cs atom.

For a spin-1/2 system, the generalized density matrix can be written as

$$\rho = \frac{1}{2} + 2\langle\vec{J}\rangle\cdot\vec{J}, \tag{11.103}$$

Table 11.2 Angular momentum: states and phase-space distributions.

Coherent states

$$|\theta,\varphi\rangle = D(\xi)|j,-j\rangle$$
$$= \sum_{m=-j}^{+j} \binom{2j}{j+m}^{\frac{1}{2}} \sin^{j+m}\frac{\theta}{2}\cos^{j-m}\frac{\theta}{2}\mathrm{e}^{-(j+m)\varphi}|j,m\rangle. \quad (1)$$

Squeezed states defined via:

(a) nonunitary transformation $|\psi\rangle_s = \mathrm{e}^{-\alpha J_z}\mathrm{e}^{-\mathrm{i}\frac{\pi}{2}J_y}|j,m\rangle,\ \alpha = \text{Real},$ (2)

(b) unitary transformation $|\psi\rangle_u = \mathrm{e}^{-\mathrm{i}\mu J_z^2}|\theta,\varphi\rangle.$ (3)

Phase-space functions [24]

$$\Phi(\theta,\varphi) = \sum_{KQ} \rho_{KQ} Y_{KQ} f_{KQ}, \quad (4)$$

$$f_{KQ} = \sqrt{\frac{2j+1}{4\pi}} \quad \text{for Wigner function} \quad (5a)$$

$$= \frac{\frac{2j+1}{4\pi}(-1)^{K-Q}\sqrt{4\pi}(2j)!}{\sqrt{(2j-K)!(2j+K+1)!}} \quad \text{for Q-function} \quad (5b)$$

$$= \frac{(-1)^{K-Q}\sqrt{(2j-K)!(2j+K+1)!}}{\sqrt{4\pi}(2j)!} \quad \text{for P-function.} \quad (5c)$$

which can be rewritten in terms of the multipole operators T_{KQ} by using

$$J_z = \left(\frac{j(2j+1)(j+1)}{3}\right)^{\frac{1}{2}} T_{10},$$
$$J_{\mp} = \pm\left(\frac{2j(2j+1)(j+1)}{3}\right)^{\frac{1}{2}} T_{1,\mp 1}. \quad (11.104)$$

Let $\vec{n}$ be the unit vector in the direction (θ,ϕ), which we write in terms of spherical harmonics

$$n_x = \sin\theta\cos\phi = -\sqrt{\frac{2\pi}{3}}(Y_{11} - Y_{1,-1}),$$
$$n_y = \sin\theta\sin\phi = \sqrt{\frac{2\pi}{3}}\mathrm{i}(Y_{11} + Y_{1,-1}), \quad (11.105)$$
$$n_z = \cos\theta = \sqrt{\frac{4\pi}{3}}Y_{10}.$$

Using (11.103)–(11.105) in (11.95) and (11.101), we obtain the Wigner function and the P-function for the state (11.103)

$$W(\theta,\phi) = \frac{1}{4\pi}\left[1 + 2\sqrt{3}\vec{n}\cdot\langle\vec{J}\rangle\right], \quad (11.106)$$

$$P(\theta,\phi) = \frac{1}{4\pi}\left[1 + 6\vec{n}\cdot\langle\vec{J}\rangle\right]. \quad (11.107)$$

Noting that $|2\langle\vec{J}\rangle| \leq 1$, we find that both the P- and W-functions could be negative, which is a reflection of the quantum nature of the two-level system. Furthermore, as expected, $P(\theta,\varphi)$ has a higher level of negativity than the Wigner function.

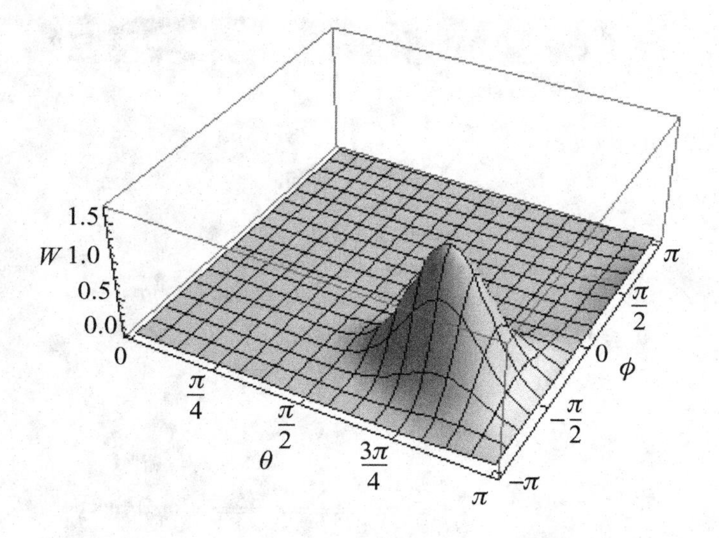

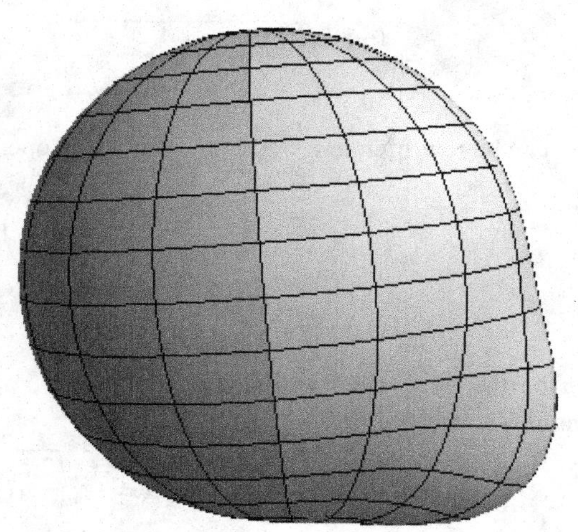

Fig. 11.9 The Wigner function of a coherent state $|\frac{\pi}{2}, 0\rangle$: top – planar representation; bottom – representation on a Bloch sphere.

11.8 Phase-space description of EPR correlations of spin systems

Consider a system of two spin-1/2 particles in the state

$$|\psi\rangle = \frac{1}{\sqrt{2}}(|\uparrow\downarrow\rangle - |\downarrow\uparrow\rangle). \tag{11.108}$$

For this state it is easily proved that

$$\langle \vec{J}^{(1)} \cdot \vec{a} \rangle = \langle \vec{J}^{(2)} \cdot \vec{b} \rangle = 0, \tag{11.109}$$

$$\langle \vec{J}^{(1)} \cdot \vec{a} \vec{J}^{(2)} \cdot \vec{b} \rangle = -\frac{1}{4}(\vec{a} \cdot \vec{b}) = \frac{1}{4} P(\vec{a}, \vec{b}). \tag{11.110}$$

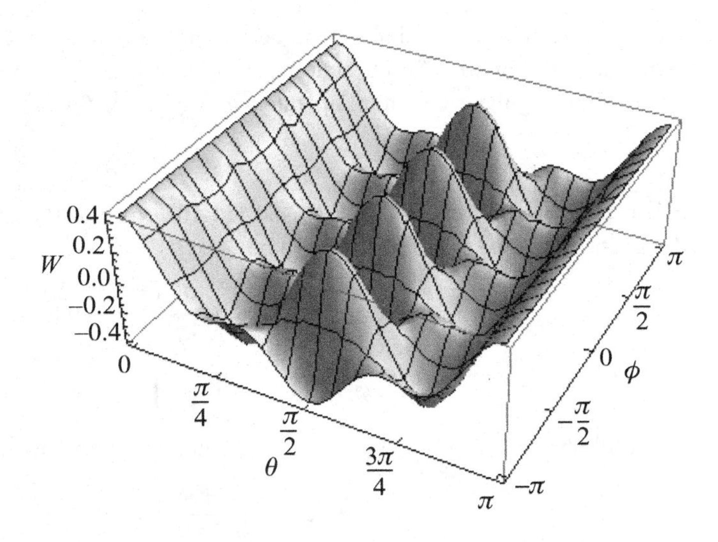

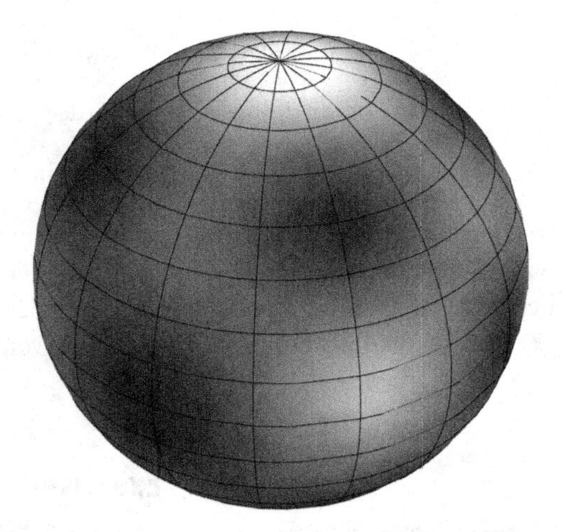

Fig. 11.10 The Wigner function of the cat state $\frac{1}{\sqrt{2}}(|3, 2\rangle + |3, -2\rangle)$. The regions of interference can be clearly seen.

The state $|\psi\rangle$ is the well-known singlet state $|0, 0\rangle$ of two spins. It is an example of an entangled state. The correlations in the state $|\psi\rangle$ cannot be explained in terms of the local hidden variable theories which led to the well-known Bell inequalities [27, 28]

$$|P(\vec{a}, \vec{b}) - P(\vec{a}, \vec{b}')| + |P(\vec{a}', \vec{b}) + P(\vec{a}', \vec{b}')| \leq 2, \tag{11.111}$$

whereas the quantum-mechanical correlations satisfy the relation

$$|P(\vec{a}, \vec{b}) - P(\vec{a}, \vec{b}')| + |P(\vec{a}', \vec{b}) + P(\vec{a}', \vec{b}')| \leq 2\sqrt{2}. \tag{11.112}$$

The difference between (11.111) and (11.112) can be traced back to the negativity of the underlying phase space distribution that determines the quantum correlations. Using the results of the previous section we can determine the phase-space distributions, such as the P-, Q- and W-functions for the state (11.108). The density matrix for the state (11.108) in terms of the angular momentum operators $\vec{J}^{(1)}$ and $\vec{J}^{(2)}$ for the two spins can be written as

$$\rho = \frac{1}{4} - \vec{J}^{(1)} \cdot \vec{J}^{(2)}. \tag{11.113}$$

It is clear from the results (11.106) and (11.107) for the state (11.103), that the P- and W-functions for the state (11.113) will be

$$P(\theta_1, \varphi_1, \theta_2, \varphi_2) = \left(\frac{1}{4\pi}\right)^2 [1 - 9\vec{n}_1 \cdot \vec{n}_2], \tag{11.114}$$

$$W(\theta_1, \varphi_1, \theta_2, \varphi_2) = \left(\frac{1}{4\pi}\right)^2 [1 - 3\vec{n}_1 \cdot \vec{n}_2], \tag{11.115}$$

and would be negative in certain directions whereas the Q-function is always positive,

$$Q(\theta_1, \varphi_1, \theta_2, \varphi_2) = \left(\frac{1}{4\pi}\right)^2 [1 - \vec{n}_1 \cdot \vec{n}_2] > 0. \tag{11.116}$$

The mean value of $\langle \vec{J}^{(1)}\vec{J}^{(2)} \rangle$ in terms of the P-function is

$$\begin{aligned}\langle \vec{J}^{(1)}\vec{J}^{(2)} \rangle &= \int P(\theta_1, \varphi_1, \theta_2, \varphi_2) \langle \theta_1, \varphi_1 | \vec{J}^{(1)} | \theta_1, \varphi_1 \rangle \langle \theta_2, \varphi_2 | \vec{J}^{(2)} | \theta_2, \varphi_2 \rangle \\ &\quad \times \sin\theta_1 \mathrm{d}\theta_1 \mathrm{d}\varphi_1 \sin\theta_2 \mathrm{d}\theta_2 \mathrm{d}\varphi_2 \\ &= \frac{1}{4} \int P(\theta_1, \varphi_1, \theta_2, \varphi_2) \vec{n}_1 \vec{n}_2 \sin\theta_1 \mathrm{d}\theta_1 \mathrm{d}\varphi_1 \sin\theta_2 \mathrm{d}\theta_2 \mathrm{d}\varphi_2. \end{aligned} \tag{11.117}$$

Thus the quantum nature of the EPR correlations in the state (11.108) is due to the strong negativity of the underlying P-function given by (11.114) [29–31]. It should be noted that the calculation of $\langle \vec{J}^{(1)}\vec{J}^{(2)} \rangle$ using hidden variables would be done using a relation like (11.117), except that the underlying distribution would be positive over the entire range of the integration variables.

Exercises

11.1 Verify the solution (11.28) for the precession equation (11.26).

11.2 Prove the relation (11.58) by using

$$\mathrm{e}^B A\, \mathrm{e}^{-B} = A + [B, A] + \frac{1}{2!}[B, [B, A]] + \cdots .$$

11.3 Prove the BCH identity (11.49) for the SU(2) algebra.

11.4 Prove that the product of two generators of angular momentum coherent states is

$$D(\xi_1)D(\xi_2) = D(\xi_3)\exp\{\mathrm{i}\Phi(\xi_1, \xi_2)J_z\},$$

$$\Phi(\xi_1, \xi_2) = -\mathrm{i}\ln\left(\frac{1 - \zeta_1\zeta_2^*}{1 - \zeta_1^*\zeta_2}\right), \qquad \zeta_3 = \frac{\zeta_1 + \zeta_2}{1 - \zeta_1^*\zeta_2}.$$

11.5 Find the explicit form of $\tilde{J}_\alpha$ defined by (11.63) using the general result for the rotation operators

$$\exp(\mathrm{i}\theta\vec{n}\cdot\vec{J})\vec{A}\exp(-\mathrm{i}\theta\vec{n}\cdot\vec{J}) = \vec{n}(\vec{n}\cdot\vec{A}) - \vec{n}\times(\vec{n}\times\vec{A})\cos\theta + (\vec{n}\times\vec{A})\sin\theta.$$

Here $\vec{n}$ is the unit vector. Choose $\vec{A}$ to be J_z and take the z component of the right-hand side to obtain (11.58)

11.6 Using (11.66) and the properties of the rotation operators, evaluate (11.68) and show explicitly that $p_{ls} = 0$ for odd values of $l + j$.

11.7 Prove the result given by (11.75) using (11.73).

11.8 Prove the relation (11.113) for the singlet state (11.108).

11.9 Consider two spins $\vec{S}$ and $\vec{F}$ with arbitrary values, interacting via the Hamiltonian $H = \alpha S_z F_z$. Clearly both S_z and F_z are conserved quantities. Obtain the operators $\vec{S}(t)$ and $\vec{F}(t)$ in terms of the conserved quantities. Find fluctuations in F_y i.e. $\langle F_y^2(t)\rangle - \langle F_y(t)\rangle^2$ if both spins $\vec{S}$ and $\vec{F}$ are initially prepared in coherent states. This H $(= \alpha S_z F_z)$ is called the quantum nondemolition Hamiltonian, as we can use the conjugate variable F_y to obtain information on the variable S_z without disturbing it. Such Hamiltonians have been used in realizing spin squeezing of atomic ensembles [19–21], where the spin F stands for the polarization or Stokes operator of the light (Eq. (6.12)).

11.10 Landau–Zener problem: Consider a two-level atom interacting with a chirped field, so that the coupling constant g (Eq. (11.6)) is replaced by $ge^{\mathrm{i}\varphi^2 t^2}$. Take the field on resonance $\omega_l = \omega_0$. Assume that the atom is in the ground state at time $t = -\infty$. Show that the probability of finding it in the excited state at $t \to +\infty$ is $1 - \exp\left(-\frac{\pi|g|^2}{\varphi^2}\right)$. The limit $\varphi \to \infty$ corresponds to the adiabatic limit. The problem can be done by finding a second-order differential equation for the excited state amplitudes. The second-order differential equation can be solved in terms of the confluent hypergeometric functions [32].

References

[1] L. Allen and J. H. Eberly, *Optical Resonance and Two-level Atoms* (New York: Wiley, 1975).
[2] R. P. Feynman, F. L. Vernon, Jr., and R. W. Hellwarth, *J. Appl. Phys.* **28**, 49 (1957).
[3] E. Merzbacher, *Quantum Mechanics*, 2nd ed. (New York: J. Wiley & Sons, 1970), p. 396.
[4] Y. S. Bai, A. G. Yodh, and T. W. Mossberg, *Phys. Rev. Lett.* **55**, 1277 (1985).
[5] M. M. Salour and C. Cohen-Tannoudji, *Phys. Rev. Lett.* **38**, 757 (1977).
[6] J. C. Bergquist, S. A. Lee, and J. L. Hall, *Phys. Rev. Lett.* **38**, 159 (1977).
[7] M. M. Salour, *Rev. Mod. Phys.* **50**, 667 (1978)
[8] Ye. V. Baklanov, B. Ya. Dubetsky, and V. P. Chebotayev, *Appl. Phys.* **9**, 171 (1976).
[9] F. T. Arecchi, E. Courtens, R. Gilmore, and H. Thomas, *Phys. Rev. A* **6**, 2211 (1972).

[10] S. M. Barnett and P. M. Radmore, *Methods in Theoretical Quantum Optics* (New York: Oxford University, 2003),
[11] R. R. Puri, *Phys. Rev. A* **49**, 2178 (1994).
[12] C. Aragone, G. Guerri, S. Salamo, and J. L. Tani, *J. Phys. A* **7**, L149 (1974).
[13] M. A. Rashid, *J. Math. Phys.* **19**, 1391 (1978).
[14] G. S. Agarwal and R. R. Puri, *Phys. Rev. A* **41**, 3782 (1990).
[15] M. Kitagawa and M. Ueda, *Phys. Rev. A* **47**, 5138 (1993).
[16] G. S. Agarwal, R. R. Puri, and R. P. Singh, *Phys. Rev. A* **56**, 2249 (1997).
[17] J. Esteve, C. Gross, A. Weller, S. Giovanazzi, and M. K. Oberthaler, *Nature (London)* **455**, 1216 (2008).
[18] A. Sørensen, L.-M. Duan, J. I. Cirac, and P. Zoller, *Nature (London)* **409**, 63 (2001).
[19] A. Kuzmich, L. Mandel, and N. P. Bigelow, *Phys. Rev. Lett.* **85**, 1594 (2000).
[20] T. Takano, M. Fuyama, R. Namiki, and Y. Takahashi, *Phys. Rev. Lett.* **102**, 033601 (2009).
[21] M. H. Schleier-Smith, I. D. Leroux, and V. Vuletić, *Phys. Rev. Lett.* **104**, 073604 (2010).
[22] D. J. Wineland, J. J. Bollinger, W. M. Itano, F. L. Moore, and D. J. Heinzen, *Phys. Rev. A* **46**, 6797 (1992).
[23] D. M. Brink and G. R. Satchler, *Angular Momentum* (Oxford: Oxford University Press, 1994).
[24] G. S. Agarwal, *Phys. Rev. A* **24**, 2889 (1981).
[25] J. P. Dowling, G. S. Agarwal, and W. P. Schleich, *Phys. Rev. A* **49**, 4101 (1994).
[26] S. Chaudhury, S. Merkel, T. Herr, A. Silberfarb, I. H. Deutsch, and P. S. Jessen, *Phys. Rev. Lett.* **99**, 163002 (2007).
[27] J. S. Bell, *Physics* **1**, 195 (1964).
[28] J. S. Bell, *Speakable and Unspeakable in Quantum Mechanics* (Cambridge: Cambridge University Press, 1987).
[29] K. Wodkiewicz, *Phys. Lett. A* **129**, 1 (1988).
[30] M. O. Scully and K. Wodkiewicz, in *Coherence and Quantum Optics* VI, edited by J. H. Eberly, L. Mandel, and E. Wolf (New York: Plenum, 1990).
[31] G. S. Agarwal, D. Home, and W. Schleich, *Phys. Lett. A* **170**, 359 (1992).
[32] P. Horwitz, *Appl. Phys. Lett.* **26**, 306 (1975).

12 Cavity quantum electrodynamics

In Chapter 7 we considered the absorption and emission of electromagnetic radiation when the interaction with the electromagnetic field was weak. However, it is possible to design cavities such that the interaction of atoms inside cavities with the cavity field would be strong. The interaction of an atom with the field in the cavity is essentially given by Eq. (7.6). We now write the quantized field as

$$\vec{E}(\vec{r}) = \sum_{\alpha} \mathrm{i}\sqrt{\frac{2\pi\hbar\omega_\alpha}{V}}\vec{\epsilon}_\alpha a_\alpha u_\alpha(\vec{r}) + H.c.. \tag{12.1}$$

Here the sum is over all the cavity modes ω_α, $u_\alpha(\vec{r})$ is the cavity mode function and V is the volume of the cavity. The cavity mode function is normalized over the volume V. For a two-level atom, we can restrict the field (12.1) to a single mode of the cavity field, if the atomic transition frequency ω_0 is close to the frequency ω_c of the mode. In this case, we reduce (12.1) to

$$\vec{E}(\vec{r}) = \mathrm{i}\sqrt{\frac{2\pi\hbar\omega_c}{V}}\vec{\epsilon}\, a u(\vec{r}) + H.c.. \tag{12.2}$$

We write the dipole moment operator in the form

$$\begin{aligned} \vec{p} &= \vec{p}_{eg}|e\rangle\langle g| + \vec{p}_{ge}|g\rangle\langle e| \\ &= \vec{p}_{eg}S_+ + \vec{p}_{ge}S_-, \end{aligned} \tag{12.3}$$

where the operators $S_\pm$ satisfy the algebra of spin-1/2 operators (Eq. (9.44)). The interaction Hamiltonian H_1 in the rotating wave approximation is

$$H_1 = \hbar g S_+ a + \hbar g^* S_- a^\dagger, \tag{12.4}$$

where the coupling constant is

$$g = -\sqrt{\frac{2\pi\omega_c}{V\hbar}}\mathrm{i}\vec{\epsilon}\cdot\vec{p}_{eg}u(\vec{r}). \tag{12.5}$$

Adding the unperturbed Hamiltonians, we have the full Hamiltonian

$$H = \hbar\omega_0 S_z + \hbar\omega_c a^\dagger a + \hbar g S_+ a + \hbar g^* S_- a^\dagger. \tag{12.6}$$

This is the Jaynes–Cummings model for the cavity QED [1]. In a cavity, the coupling constant g could be made quite large. Several situations involving both Rydberg atoms and optical transitions have been extensively studied. In order to distinguish atom coupling effects g should be large compared to the incoherent or dissipative processes in the cavity. The important dissipative processes are (i) the leakage of photons through the cavity mirrors at the rate 2κ, and (ii) the spontaneous emission of the atoms to the side modes at the rate 2γ. The strong coupling condition means

$$g \gg \kappa, \qquad g \gg \gamma. \tag{12.7}$$

The condition $g \gg \kappa$ means that a photon emitted in the cavity mode can interact repeatedly with the atom before it leaks out. The condition $g \gg \gamma$ implies that the emission is primarily in the cavity mode rather than the side modes.

12.1 Exact solution of the Jaynes–Cummings model: dressed states

The Jaynes–Cummings model (JCM) has been the workhorse of quantum optics. Its predictions have been extensively investigated and in more recent years many other systems particularly in the context of superconducting qubits and quantum dots have been treated after the original JCM. The model is exactly soluble and this provides great insight into the new regime of quantum electrodynamics where the radiation–matter interaction is strong. For the unperturbed Hamiltonian we have two sets of basis states $|n, g\rangle$ and $|n, e\rangle$, with $n = 0, 1, 2, \ldots, \infty$. The interaction Hamiltonian (12.4) has the important feature that

$$[H_c, H_1] = 0, \qquad H_c = \hbar\omega_c(a^\dagger a + S_z). \tag{12.8}$$

Thus H_c is a constant of motion. Hence it is convenient to write (12.6) as

$$H = H_c + H_I, \qquad [H_c, H_I] = 0, \tag{12.9}$$

where

$$H_I = \hbar\Delta S_z + \hbar g S_+ a + \hbar g^* S_- a^\dagger, \tag{12.10}$$

and where $\Delta = \omega_0 - \omega_c$ is the detuning between the atomic transition frequency and the cavity frequency. We will now treat H_c as the equivalent of the unperturbed Hamiltonian. The advantage of this is that the basic states $|g, n\rangle$ and $|e, n-1\rangle$ have identical unperturbed energies.

The interaction Hamiltonian H_I has the property that it connects only two states for each value of n except for $n = 0$

$$\begin{aligned} H_I|e, n-1\rangle &= \hbar g^* \sqrt{n}|g, n\rangle + \frac{\hbar\Delta}{2}|e, n-1\rangle, \\ H_I|g, n\rangle &= \hbar g\sqrt{n}|e, n-1\rangle - \frac{\hbar\Delta}{2}|g, n\rangle, \\ H_I|g, 0\rangle &= -\frac{\hbar\Delta}{2}|g, 0\rangle. \end{aligned} \tag{12.11}$$

We illustrate these couplings in Figure 12.1.

Clearly H_I has the block diagonal structure in terms of the states $|g, n\rangle$ and $|e, n-1\rangle$ and the Hamiltonian can be diagonalized in terms of these by using the diagonalization of

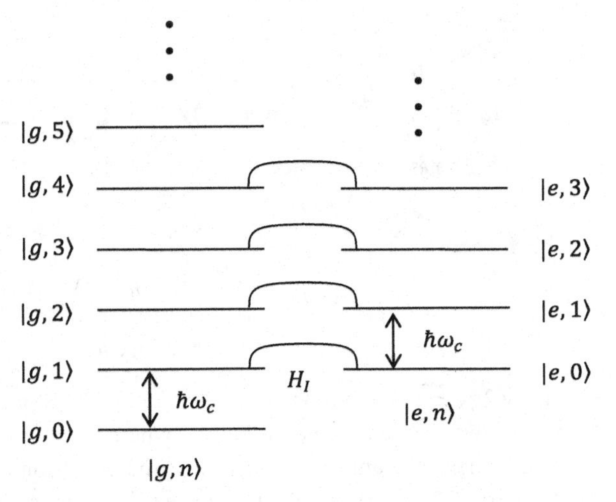

Fig. 12.1 The couplings in the Jaynes–Cummings model.

2×2 matrices of the form

$$\begin{pmatrix} \frac{\hbar\Delta}{2} & \hbar g^* \sqrt{n} \\ \hbar g \sqrt{n} & -\frac{\hbar\Delta}{2} \end{pmatrix}. \tag{12.12}$$

Such a diagonalization is straightforward. In order to avoid the unnecessary phase factors associated with g, we assume g to be real and positive. Any phase of g can be absorbed in the definition of the angular momentum operators. Because of the structure (12.12) of H_I, all eigenstates would be linear combinations of $|g, n+1\rangle$ and $|e, n\rangle$. Let us denote these by $|\Psi_n^{\pm}\rangle$ and are given by [2]

$$\begin{aligned} H|g, 0\rangle &= -\frac{\hbar\omega_0}{2}|g, 0\rangle, \\ H|\Psi_n^{\pm}\rangle &= \hbar\left[\omega_c\left(n+\frac{1}{2}\right) \pm \frac{\Omega_{n\Delta}}{2}\right]|\Psi_n^{\pm}\rangle = \hbar\omega_n^{\pm}|\Psi_n^{\pm}\rangle, \\ \Omega_{n\Delta}^2 &= 4g^2(n+1) + \Delta^2, \qquad \Delta = \omega_0 - \omega_c, \\ |\Psi_n^{\pm}\rangle &= \begin{pmatrix} \cos\theta_n \\ -\sin\theta_n \end{pmatrix}|g, n+1\rangle + \begin{pmatrix} \sin\theta_n \\ \cos\theta_n \end{pmatrix}|e, n\rangle, \\ n &= 0, 1, 2, \ldots, \infty, \qquad \tan\theta_n = 2g\sqrt{n+1}/(\Omega_{n\Delta} - \Delta). \end{aligned} \tag{12.13}$$

The states $|\Psi_n^{\pm}\rangle$ are called the dressed states – these are the linear combinations of the atomic and field states as shown in Figure 12.2. The form depends on the coupling parameter g. Furthermore, the eigen energies of the dressed states depend on the coupling parameter. The parameter θ is the mixing parameter and depends on the coupling g and the detuning. For a cavity tuned to atomic resonance $\Delta = 0$, $\theta_n = \pi/4$, and then the mixing of states is

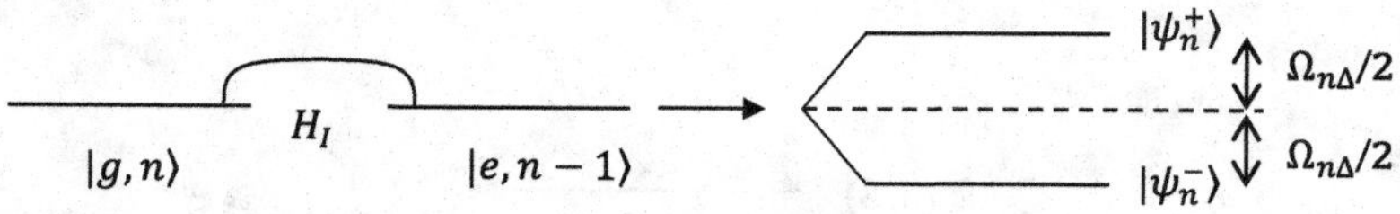

Fig. 12.2 Formation of dressed states from bare states.

maximum

$$|\Psi_n^{\pm}\rangle = \frac{1}{\sqrt{2}}(\pm|g, n+1\rangle + |e, n\rangle), \\ \Omega_{n\Delta} = \pm 2g\sqrt{n+1}. \tag{12.14}$$

The quantity $2g\sqrt{n+1}$ and more generally $\Omega_{n\Delta}$ is the n-photon Rabi frequency. In order to see this consider an initial value problem – where the atom field system has been prepared in the state $|e, n\rangle$. We want to calculate the time evolution of such an initial state. Because of the property (12.11), the state at any time t is a linear combination of the states $|e, n\rangle$ and $|g, n+1\rangle$

$$|\Psi(t)\rangle = c(t)|e, n\rangle + d(t)|g, n+1\rangle, \tag{12.15}$$

where the coefficients $c(t)$ and $d(t)$ are given by

$$\dot{c} = -\mathrm{i}\omega_c\left(n+\frac{1}{2}\right)c - \mathrm{i}\frac{\Delta}{2}c - \mathrm{i}g\sqrt{n+1}d, \\ \dot{d} = -\mathrm{i}\omega_c\left(n+\frac{1}{2}\right)d + \mathrm{i}\frac{\Delta}{2}d - \mathrm{i}g\sqrt{n+1}c. \tag{12.16}$$

These are to be solved subject to the initial condition $c(0) = 1$, $d(0) = 0$. It is straightforward to show that the probability of finding the atom–field system in the state $|g, n+1\rangle$ is

$$p_{en\to gn+1}(t) = \frac{4g^2(n+1)}{\Omega_{n\Delta}^2}\sin^2\frac{\Omega_{n\Delta}t}{2}. \tag{12.17}$$

Hence we have periodic transfer from the state $|e, n\rangle$ to $|g, n+1\rangle$ and then back to $|e, n\rangle$. The period of oscillation is given by $2\pi/\Omega_{n\Delta}$. This period or the frequency of oscillation $\Omega_{n\Delta}$ depends on the number of photons. We call $\Omega_{n\Delta}$, the n-photon Rabi frequency. The oscillation continues to occur even if there are no photons in the cavity. This is called the vacuum field Rabi oscillation, which has been experimentally studied by a number of groups [3] – the atom emits a photon into the cavity mode, which is then reabsorbed by the atom

$$|e, 0\rangle \to |g, 1\rangle \to |e, 0\rangle. \tag{12.18}$$

The parameter $2g$ is the on-resonance vacuum Rabi frequency. The n-photon Rabi oscillation can be understood in terms of the dressed states. The state $|e, n\rangle$ can be written as a linear combination of the dressed states $|\Psi_n^+\rangle$ and $|\Psi_n^-\rangle$, which evolve at the frequencies ω_n^+ and ω_n^- respectively. The relative frequency of evolution is $\omega_n^+ - \omega_n^-$, which is equal to $\Omega_{n\Delta}$. A very clear demonstration of n-photon Rabi oscillation for n up to 2 was given by Varcoe *et al.* [4].

In the case of resonant interaction $\omega_c = \omega_0$, we find from (12.17) that the probability of finding the atom in the excited state is

$$p_{en\to en}(t) = \cos^2 gt\sqrt{n+1}. \tag{12.19}$$

Using (12.19), Filipowicz *et al.* [5] predicted the existence of trapping states in the cavity field. Consider sending an atom in an excited state through the cavity, then under the condition

$$g\sqrt{n+1}t = q\pi, \qquad q, n \text{ integers}, \tag{12.20}$$

the atom comes out in an excited state, which was the state with which it went in. Hence if successive atoms under fixed interaction time come out in the excited state, then the cavity field must be a mixture of states $|n\rangle$ with values of n that satisfy (12.20). These are called as the trapping states and show up in the dynamics of micromasers at very low temperatures [6]. The trapping states have been used to generate single photons on demand [7].

12.2 Collapse and revival phenomena in JCM

If the field in the cavity is in a state which is a superposition of different Fock states $\sum_n \alpha_n|n\rangle$, then the probability for an atom to make a transition from the state $|e\rangle$ to the state $|g\rangle$ would be obtained by summing (12.17) over the distribution of $|n\rangle$

$$p_{e\to g}(t) = \sum_{n=0}^{\infty} p_n \frac{4g^2(n+1)}{\Omega_{n\Delta}^2} \sin^2 \frac{t\Omega_{n\Delta}}{2}, \quad p_n = |\alpha_n|^2, \tag{12.21}$$

$$\overset{\Delta\to 0}{\longrightarrow} \sum_{n=0}^{\infty} p_n \sin^2(gt\sqrt{n+1}) = \frac{1}{2} - \frac{1}{2}\mathrm{Re}\left(\sum_{n=0}^{\infty} p_n \mathrm{e}^{2\mathrm{i}gt\sqrt{n+1}}\right). \tag{12.22}$$

A plot of $p_{e\to g}(t)$ as a function of gt for a field with a low number of photons is shown in Figure 12.3. The field is assumed to be a coherent state, i.e. $p_n = |\alpha|^{2n}\mathrm{e}^{-|\alpha|^2}/n!$ with $|\alpha|^2 = 9$. The transition probability (12.22) has a very rich structure primarily due to the occurrence of the factor $\sqrt{n+1}$ in the argument of the sine function. For a small enough number of photons in the field, many terms in (12.22) contribute and each such term has a different period, leading to the collapse and revival of the Rabi oscillations [8]. Collapse and revival phenomena in nonlinear quantum dynamics have been extensively studied not only for the JCM but also for other systems such as H-atoms, square well potentials, and nonlinear oscillators [9, 10]. An understanding of the behavior can be obtained if the distribution p_n is sharply peaked around $\bar{n}$, which allows us to expand $\sqrt{n+1}$ as

$$gt\sqrt{n+1} \approx gt\sqrt{\bar{n}}\left[1 + \frac{n-\bar{n}}{2\bar{n}} - \frac{(n-\bar{n})^2}{8\bar{n}^2} + \cdots\right]. \tag{12.23}$$

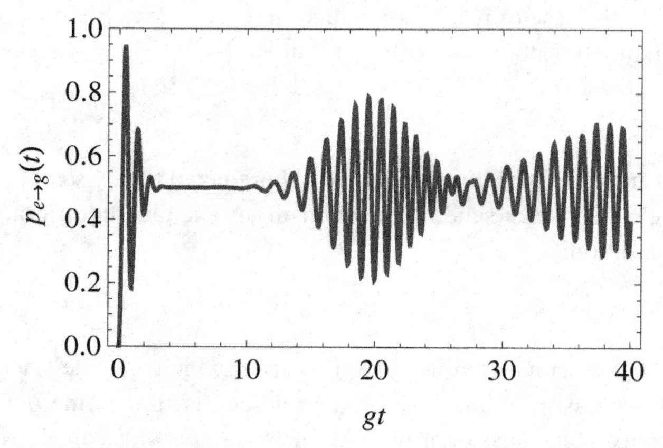

Fig. 12.3 The transition probability $p_{e\to g}(t)$ as a function of gt for a coherent field with $p_n = |\alpha|^{2n}e^{-|\alpha|^2}/n!$, $|\alpha|^2 = 9$.

The dynamics is then governed by different time scales. The Rabi flopping time is $2\pi/(2g\sqrt{\bar{n}})$. The other time scales are

$$T_{\rm R} = 2\pi\left(\frac{g}{\sqrt{\bar{n}}}\right)^{-1}, \qquad T_{\rm FR} = 2\pi\left(\frac{g}{4\bar{n}^{3/2}}\right)^{-1}. \tag{12.24}$$

Here $T_{\rm R}$ is the revival time and $T_{\rm FR}$ is the time for fractional revivals. The cubic terms in the expansion (12.23) lead to superrevivals at a time scale $T_{\rm SR}$. Thus the very early stage of time evolution of (12.22) is determined by the Rabi oscillation, and the next stage is determined by $T_{\rm R}$. The dynamics at longer time scales is determined by $T_{\rm FR}$ and $T_{\rm SR}$, etc. In the early stages there is dephasing of the Rabi oscillations due to the $(n - \bar{n})$ term in (12.23) until t becomes of the order of $T_{\rm R}$, whereupon different terms in the exponent start interfering constructively, leading to revival of the Rabi oscillation. This can be seen much more clearly if we evaluate (12.22) for a field with sub-Poissonian statistics [10]

$$p_n = \frac{1}{\sqrt{2\pi}\,\Delta n}\exp\left[-\frac{(n-\bar{n})^2}{2(\Delta n)^2}\right], \qquad \Delta n < \sqrt{\bar{n}}. \tag{12.25}$$

We show the result for the time dependent transition probability in Figures 12.4 and 12.5 for $t \sim T_{\rm R}$ and $t \sim T_{\rm FR}$. Figure 12.4 shows the collapse of the Rabi oscillation followed by a revival for a time $t \sim T_{\rm R}$. Figure 12.5 shows the fractional revivals in the region $t \sim \frac{1}{3}T_{\rm FR}$. The first experimental observation of collapse and revival is due to Rempe *et al.* [11]. Subsequently several refined observations were made [12, 13]. Raimond *et al.* [12] clearly identified several n-photon Rabi frequencies in an experiment done with a coherent field containing only 0.85 photons. This is one case where approximate analysis based on (12.23) does not apply as the dispersion $\Delta n = \sqrt{\bar{n}} \sim \bar{n}$.

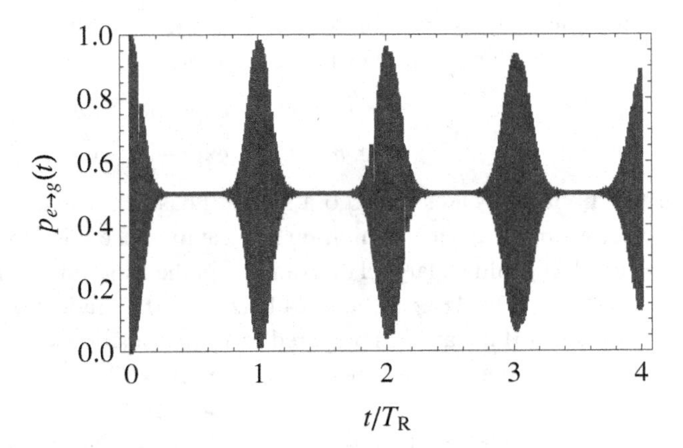

Fig. 12.4 The transition probability $p_{e\to g}(t)$ as a function of t/T_R, where T_R is the revival time. The parameters used are $\bar{n} = 50, \Delta n = 2$ (redrawn after [10]).

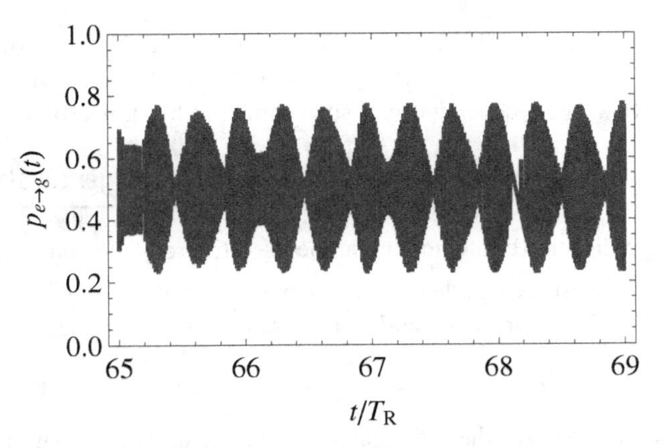

Fig. 12.5 The transition probability $p_{e\to g}(t)$ as a function of t/T_R for times in the range where fractional revivals can occur. The parameters used are $\bar{n} = 50, \Delta n = 2$ (redrawn after [10]).

12.3 Dispersive limit of the JCM

A very useful limit of the JCM is the dispersive limit, where Δ is large compared to the coupling constant g and the leakage rate κ of the photons. The Rabi frequency $\Omega_{n\Delta}$ can be approximated by

$$\Omega_{n\Delta} \approx |\Delta| + \frac{2g^2(n+1)}{|\Delta|}. \tag{12.26}$$

The dressed state energies then become

$$\omega_n^{\pm} = \omega_c\left(n + \frac{1}{2}\right) \pm \frac{|\Delta|}{2} \pm \frac{g^2(n+1)}{|\Delta|}. \tag{12.27}$$

The parameter $\theta_n \to \pi/2$ for $\Delta > 0$. Thus the state $|\Psi_n^+\rangle \to |e, n\rangle$, $|\Psi_n^-\rangle \to -|g, n+1\rangle$. The time evolution of the states is very simple

$$\begin{aligned} |e, n\rangle &\to \exp(-\mathrm{i}\omega_n^+ t)|e, n\rangle, \\ |g, n+1\rangle &\to \exp(-\mathrm{i}\omega_n^- t)|g, n+1\rangle. \end{aligned} \tag{12.28}$$

In this limit, the states undergo a simple phase shift which is dependent on the number of photons in the cavity. In high-quality cavities, the n independent part of the phase shift $\varphi = g^2 t/|\Delta|$ could be large. For example, in the experiment of Brune *et al.* [14] $t \sim 19\ \mu$s, $|\Delta| \sim 2\pi \times 100$ kHz, $g \sim 2\pi \times 24$ kHz, $\varphi \sim 0.69$ radian. An interesting consequence of (12.28) is that if the atom is prepared in a superposition state $(c|e\rangle + d|g\rangle)$ and the field in the Fock state, then the output state will be $(\Delta > 0)$

$$\begin{aligned} (c|e\rangle + d|g\rangle)|n\rangle \to &\left\{ c \exp\left[-\mathrm{i}\frac{\Delta}{2}t - \mathrm{i}\varphi(n+1)\right] |e, n\rangle \right. \\ &\left. + d \exp\left(\mathrm{i}\frac{\Delta}{2}t + \mathrm{i}\varphi n\right) |g, n\rangle \right\} \mathrm{e}^{-\mathrm{i}\omega_c t(n+\frac{1}{2})}. \end{aligned} \tag{12.29}$$

We observe from (12.29) that the states $|e, n\rangle$ and $|g, n\rangle$ undergo frequency shifts $\varphi(n+1)$ and φn respectively. The excited state has an extra contribution due to the vacuum of the cavity. The n dependence of such shifts has been measured by Brune *et al.* [15].

12.3.1 Generation of the Schrödinger cat states for the field

The dispersive interaction can be used to produce entangled states of the atom and field. Let us assume that the field is prepared in a coherent state $|\alpha\rangle = \sum_{n=0}^{\infty} \frac{\alpha^n \mathrm{e}^{-\frac{1}{2}|\alpha|^2}}{\sqrt{n!}}|n\rangle$, then we use (12.29) for each n and sum over all n values to obtain

$$(c|e\rangle + d|g\rangle)|\alpha\rangle \to \mathrm{e}^{-\frac{\mathrm{i}\omega_c t}{2}} \left[c\, \mathrm{e}^{-\mathrm{i}\frac{\Delta}{2}t - \mathrm{i}\varphi}|e\rangle|\alpha \mathrm{e}^{-\mathrm{i}\varphi - \mathrm{i}\omega_c t}\rangle + d\, \mathrm{e}^{\mathrm{i}\frac{\Delta}{2}t}|g\rangle|\alpha \mathrm{e}^{\mathrm{i}\varphi - \mathrm{i}\omega_c t}\rangle\right], \tag{12.30}$$

which displays the atom–photon entanglement. The state (12.30) involves two coherent states which are phase shifted in opposite directions from the initial coherent state. The state (12.30) has the structure of the original Schrödinger cat state. To see it more explicitly, let us calculate the probability of finding the atom in the state $|\chi\rangle = c'|e\rangle + d'|g\rangle$, which is a superposition of the states of the atom. This probability can be obtained from the projection of (12.30) onto the state $|\chi\rangle$

$$p_\chi = \left|\left| cc'^* \mathrm{e}^{-\mathrm{i}\frac{\Delta}{2}t - \mathrm{i}\varphi}|\alpha \mathrm{e}^{-\mathrm{i}\omega_c t - \mathrm{i}\varphi}\rangle + dd'^* \mathrm{e}^{\mathrm{i}\frac{\Delta}{2}t}|\alpha \mathrm{e}^{-\mathrm{i}\omega_c t + \mathrm{i}\varphi}\rangle \right|\right|^2. \tag{12.31}$$

The coefficients c, c', d, and d' depend on the initial preparation and the measured basis. The interferences in the probability p_χ are determined by the overlap of the coherent states $|\alpha \mathrm{e}^{-\mathrm{i}\omega_c t - \mathrm{i}\varphi}\rangle$ and $|\alpha \mathrm{e}^{-\mathrm{i}\omega_c t + \mathrm{i}\varphi}\rangle$, which is equal to

$$\langle \alpha \mathrm{e}^{-\mathrm{i}\omega_c t + \mathrm{i}\varphi}|\alpha \mathrm{e}^{-\mathrm{i}\omega_c t - \mathrm{i}\varphi}\rangle = \exp\left[-|\alpha|^2(1 - \mathrm{e}^{-2\mathrm{i}\varphi})\right]. \tag{12.32}$$

The modulus of $\exp\left[-|\alpha|^2(1 - \mathrm{e}^{-2\mathrm{i}\varphi})\right] = \exp(-2|\alpha|^2 \sin^2\varphi)$ determines the visibility of the interference pattern. For small φ, the visibility remains close to 1, as φ increases the overlap between two coherent states decreases and the visibility goes down. The size of the

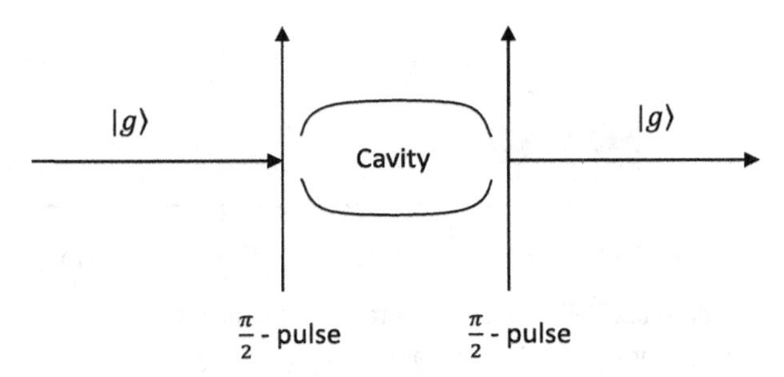

Fig. 12.6 Generation of cat states.

coherent state also determines how fast the visibility goes down. This is due to presence of the factor $|\alpha|^2$ in the exponent in (12.32). The first successful demonstration of cat states in CQED was given by Brune *et al.* [14]. In their experiment, a Rydberg atom was prepared in a coherent superposition state by a $\pi/2$-microwave pulse. Then it was sent through a dispersive cavity containing a coherent field with $|\alpha| = 3.1$. The outgoing Rydberg atom was exposed to another microwave field ($\pi/2$-pulse) (Figure 12.6). These $\pi/2$-pulses determine the coefficients c, c', d, and d' in (12.31). Note that in the absence of the cavity, the setup is a Ramsey interferometer [16] and the probability p_g would show interference fringes as a function of the frequency difference between the microwave field and the atomic transition frequency (Eq. (11.39)). The cavity would modify the interference fringes in the manner discussed above. Brune *et al.* measured the interference pattern for different values of the detuning between the cavity frequency and the atomic frequency.

The dispersive interaction can be used to produce a variety of superposition states. Zurek [17] introduced the idea of a compass state defined by

$$|\Psi\rangle \sim (|\alpha\rangle + |\mathrm{i}\alpha\rangle + |-\alpha\rangle + |-\mathrm{i}\alpha\rangle). \tag{12.33}$$

Zurek especially noticed that such states in phase space can have sub-Planck structure, i.e. the area over which the two quadratures vary can be smaller than $\hbar$. The state (12.33) can be realized by using the dispersive interactions [18]. We need to send two atoms in succession and measure the joint probabilities for the two atoms. We already showed in (12.30) that the passage of a single atom produces a cat state. The passage of a second atom splits each coherent state in the superposition in (12.31) into two coherent states, leading to the generation of the compass state.

12.4 Dissipative processes in cavity QED – the master equation

In order to examine realistic systems in cavity QED, we have to include the dissipative processes in the cavity. The basic dissipative processes are (i) the leakage of photons from the cavity at the rate 2κ (Eq. (9.38)) and (ii) the spontaneous emission of atom into all the other modes. The rate at which the population decays from the atomic excited state is $2\gamma = A$. We have already discussed in Chapter 9 the basic master equations for such

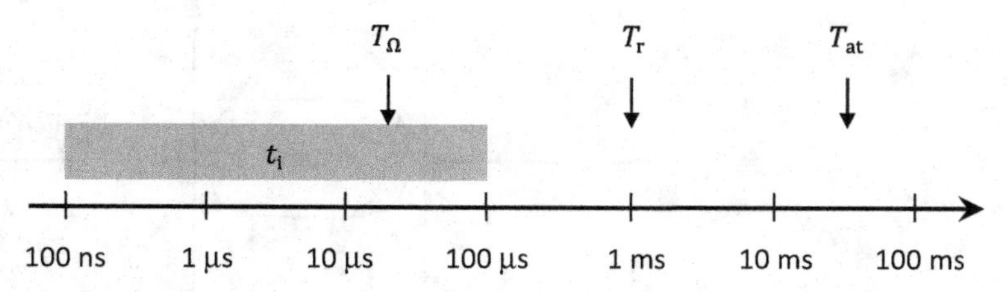

Fig. 12.7 Relevant times of a CQED experiment, plotted on logarithmic scale. T_{at}, T_r, T_Ω, and t_i are defined in the text. t_i can be varied in the range depicted by the gray bar [12].

dissipative processes. We can now put everything together to obtain the key master equation for cavity QED

$$\begin{aligned}\frac{\partial \rho}{\partial t} &= -\frac{\mathrm{i}}{\hbar}[H,\rho] - \gamma(1+\bar{n})(S_+S_-\rho - 2S_-\rho S_+ + \rho S_+S_-) \\ &\quad - \gamma\bar{n}(S_-S_+\rho - 2S_+\rho S_- + \rho S_-S_+) \\ &\quad - \kappa(1+\bar{n})(a^\dagger a\rho - 2a\rho a^\dagger + \rho a^\dagger a) \\ &\quad - \kappa\bar{n}(aa^\dagger\rho - 2a^\dagger\rho a + \rho aa^\dagger),\end{aligned} \tag{12.34}$$

where

$$H = \hbar\omega_c(a^\dagger a + S_z) + \hbar\Delta S_z + \hbar g(S_+a + S_-a^\dagger), \tag{12.35}$$

and where $\bar{n}$ is the average number of thermal photons.

For trapped atoms in optical cavities, we can set $\bar{n}$-zero. Typical values of γ and κ [19,20] are of the order of: $\gamma/(2\pi) \sim 2.6$ MHz, $\kappa/(2\pi) = 4.1$ MHz, and the vacuum Rabi coupling $g/(2\pi) \sim 34$ MHz. These are extremely high finesse cavities, $F \sim 4\times10^5$, with very small volume $L \sim 42$ μm, so as to make g large. The value of γ depends on the atomic transition – the quoted value is for the D_2 line of Cs. It must be borne in mind that the CQED community continues to make unprecedented progress with the fabrication of better and better cavities, so as to make $g \gg \kappa, \gamma$.

The experiments with Rydberg atoms have been performed by sending atoms through open cavities [12] or almost closed cavities [21]. The majority of experiments, as done by Haroche *et al.* [13], correspond to time scales as shown in Figure 12.7. Here the atomic relaxation time $T_{at} = (2\gamma)^{-1}$, Rabi flipping time $T_\Omega = (2g)^{-1}$, cavity relaxation time $T_r = (2\kappa)^{-1}$, t_i = interaction time. For Rydberg atoms, the thermal photons could be important unless one is working at very low temperatures.

The master equation (12.34) is to be supplemented by adding the coherent term H_ε representing the coupling of the cavity to the external field. The coherent term can be written as

$$H_\varepsilon = \mathrm{i}\hbar\varepsilon(a^\dagger \mathrm{e}^{-\mathrm{i}\omega_l t} - a\mathrm{e}^{\mathrm{i}\omega_l t}), \tag{12.36}$$

where ω_l is the frequency of the driving field and ε is related to the power $\wp$ of the driving field via

$$\varepsilon = \sqrt{\frac{2\kappa\wp}{\hbar\omega_l}}. \tag{12.37}$$

Fig. 12.8 The input and output relations.

Note that ε has the dimension of frequency. In order to obtain the output fields we need the input–output relations, i.e. we need to relate the output fields to the cavity fields. Consider the scheme shown in Figure 12.8 [22]. Let the leakage rates from each mirror be $2\kappa_1$ and $2\kappa_2$ so that $\kappa = \kappa_1 + \kappa_2$. We would take the incoming quantum fields as satisfying the commutation relations

$$[a_{l\text{in}}(t), a^\dagger_{l\text{in}}(t')] = \delta(t - t') \qquad \text{etc.,} \tag{12.38}$$

whereas the field inside the cavity satisfies $[a, a^\dagger] = 1$. Then the input–output relations are

$$\begin{aligned} a_{r\text{out}}(t) + a_{r\text{in}}(t) &= \sqrt{2\kappa_2}a(t), \\ a_{l\text{out}}(t) + a_{l\text{in}}(t) &= \sqrt{2\kappa_1}a(t), \end{aligned} \tag{12.39}$$

and we would generally take $\kappa_1 = \kappa_2$. The parameter κ can be related to the finesse of the cavity defined by the reflectivity R of each mirror by

$$F \cong \pi R^{1/2}/(1 - R) \cong \frac{\pi c}{\kappa L} \qquad \text{for } R \to 1,\ \kappa = \frac{c}{L}(1 - R), \tag{12.40}$$

where L is the length of the cavity.

12.5 Spectroscopy of the ladder of dressed states

In Section 12.1 we obtained the exact eigenstates of a system consisting of a strongly interacting two-level system and a single mode of the cavity field. A natural question is now how one can probe the exact eigenstates. The two lowest states $|\psi_0^+\rangle$ and $|\psi_0^-\rangle$ (Eq. (12.13)) can be studied by monitoring the vacuum Rabi oscillation (Eq. (12.17) for $n = 0$). If the field in the cavity can be prepared in the Fock state $|n\rangle$, then the states $|\psi_n^\pm\rangle$ can be studied by monitoring the n-photon Rabi oscillation as was done by Varcoe *et al.* [4]. Furthermore, by considering a small coherent field in the cavity and by examining the population of the atom, say, in the excited state, some of the low lying states $|\psi^\pm\rangle$ can be studied.

12.5.1 Vacuum Rabi splittings in the absorption/transmission of a weak field

In spectroscopy a standard tool to probe the spectrum or energy levels of a system is the absorption spectroscopy. This is especially useful in the optical domain. One can consider

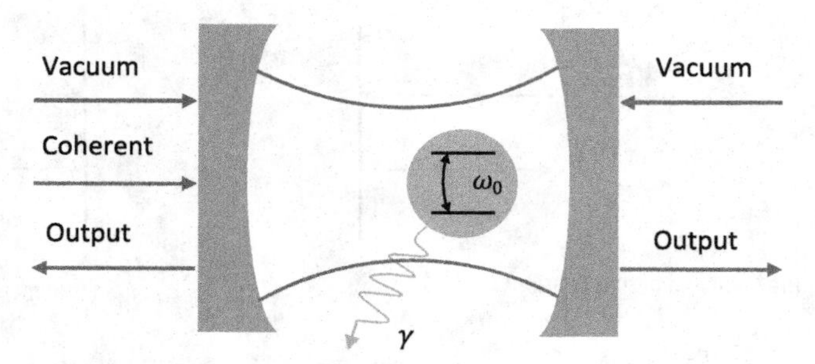

Fig. 12.9 A two-level atom of transition frequency ω_0 is trapped in a high-quality cavity. The cavity has a photon decay rate 2κ, while the atom has a spontaneous emission rate 2γ.

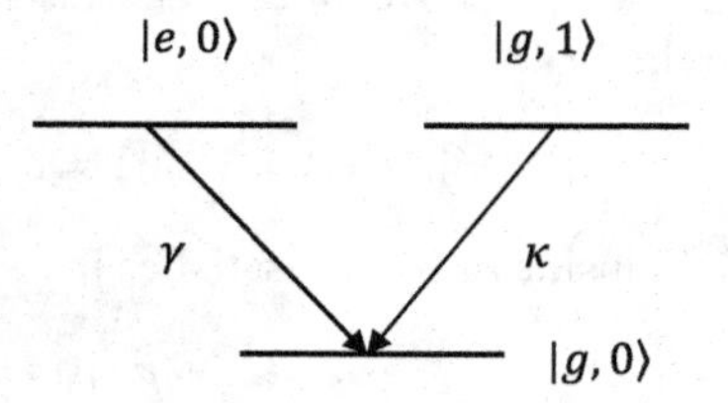

Fig. 12.10 The dissipative process would couple the states $|e, 0\rangle$ and $|g, 1\rangle$ to the state $|g, 0\rangle$.

probing the dressed states by examining the absorption of a very weak probe field [23]. Consider an atom which is trapped [19] in a high-quality cavity (Figure 12.9). Let us assume that initially the cavity is empty and the atom is in its ground state. Thus the initial state of the atom–field system is $|g, 0\rangle$, which is an eigenstate of the Hamiltonian (12.6). Consider next the interaction of the atom with a very weak external field of frequency ω_l. The atom–field system now makes transitions under the influence of the external field $|g, 0\rangle \rightarrow |\psi_0^+\rangle$, $|g, 0\rangle \rightarrow |\psi_0^-\rangle$. Clearly the absorption must show resonances whenever

$$\omega_l = \frac{\omega_0}{2} + \omega_0^{\pm} = \omega_0 - \frac{\Delta}{2} \pm \frac{1}{2}(\Delta^2 + 4g^2)^{1/2} \rightarrow \omega_0 \pm g, \quad \text{if } \Delta = 0. \tag{12.41}$$

This is to be contrasted with the free-space result where absorption will exhibit a single resonance at $\omega_l = \omega_0$. Thus the interaction of the atom with the vacuum field in the cavity leads to a doublet that is resolvable if g is large. One refers to the doublet (12.41) as the vacuum field Rabi splitting [24] of the degenerate levels $|e, 0\rangle$, $|g, 1\rangle$, to $|\psi_0^{\pm}\rangle$. The width of the doublet depends on the losses of the cavity and the spontaneous emissions into other modes; the doublet is resolvable when the width of each peak in the doublet is much smaller than the separation $2g$.

To estimate these widths we examine the states involved in vacuum Rabi splittings. The states $|e, 0\rangle$ and $|g, 1\rangle$ are coupled via the constant g. Furthermore, the dissipative process would couple these to the state $|g, 0\rangle$. Thus the processes involved are shown in Figure 12.10. Note that both the state $|e, 0\rangle$ and $|g, 1\rangle$ decay to states outside $|e, 0\rangle$ and $|g, 1\rangle$. Thus the dynamical behavior in the space $|e, 0\rangle$ and $|g, 1\rangle$ can be described in terms of the wave function. The equations (12.16) are now modified to

$$\dot{\Phi} = -\mathrm{i}M\Phi - \mathrm{i}\frac{\omega_c}{2}\Phi, \qquad \Phi = \begin{pmatrix} c \\ d \end{pmatrix}, \tag{12.42}$$

where the matrix M is

$$M = \begin{pmatrix} \frac{\Delta}{2} - \mathrm{i}\gamma & g \\ g & -\frac{\Delta}{2} - \mathrm{i}\kappa \end{pmatrix}. \tag{12.43}$$

The eigenvalues of M are given by

$$\begin{aligned} \lambda_\pm &= -\mathrm{i}\frac{\kappa+\gamma}{2} \pm \frac{1}{2}\sqrt{-(\kappa+\gamma)^2 + 4g^2 + 4\left(\frac{\Delta}{2} - \mathrm{i}\gamma\right)\left(\frac{\Delta}{2} + \mathrm{i}\kappa\right)} \\ &\xrightarrow{g\to 0} \left(-\frac{\Delta}{2} - \mathrm{i}\kappa\right), \left(\frac{\Delta}{2} - \mathrm{i}\gamma\right) \\ &\xrightarrow{\Delta\to 0} -\mathrm{i}\frac{\kappa+\gamma}{2} \pm \sqrt{g^2 - \frac{(\kappa-\gamma)^2}{4}}. \end{aligned} \tag{12.44}$$

Thus for $\Delta = 0$ and in the limit of strong coupling $g \gg \kappa, \gamma$, the two eigenvalues are

$$\lambda_\pm = \pm g - \mathrm{i}\frac{\kappa+\gamma}{2}. \tag{12.45}$$

Each line in the absorption has a half width at half maximum equal to $(\frac{\kappa+\gamma}{2})$. We show the behavior of the complex roots $\lambda_\pm$ in Figure 12.11 for two values of the ratio of the decay parameter γ/κ. For nonzero delta we show the roots in Figure 12.12. The imaginary part of the roots are shown as widths of the lines. The dressed state energies in Figure 12.12 are reminescent of the polariton dispersion curves in macroscopic dielectrics [25]. It should be borne in mind that we are considering a microscopic system of a single-mode cavity and a single two-level atom. The Purcell result for the atomic decay, which we derived in Section 7.7.1, also follows from (12.44) (Exercise 12.3). The vacuum Rabi splittings have been most extensively studied first in atomic systems [26,27] and now in the framework of circuit QED [28,29]. For details of how superconducting qubits work and how these can be used to realize JCM, we refer to [30]. The circuit QED systems are generally more flexible with regard to the change of parameters.

12.5.2 Dressed states anharmonicity and interferences in two-photon processes

Now we consider nonlinearities in cavity QED. We have already seen in Section 12.5.1 that in the regime of strong coupling the absorption from a probe field in a cavity, even for a single atom, is large. We thus expect that even single atom nonlinearities will be large. Let us consider the process of two-photon absorption. Let us first consider two-photon absorption from two different beams with frequencies ω_1 and ω_2. The rate of absorption clearly depends on how the fields are tuned. We show the possible transitions in Figure 12.13. For simplicity, we assume that the cavity is on resonance with the atomic frequency. We have the two-photon resonance condition

$$\omega_1 + \omega_2 = 2\omega_c \pm \sqrt{2}g. \tag{12.46}$$

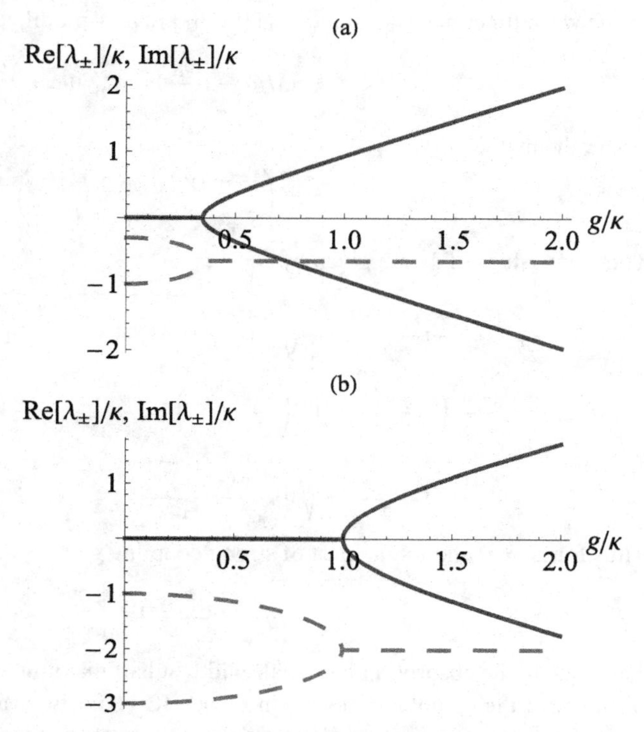

Fig. 12.11 The real parts (solid) and the imaginary parts (dashed) of the scaled complex roots $\lambda_\pm/\kappa$ as a function of g/κ for different values of γ/κ at $\Delta = 0$. (a) $\gamma/\kappa = 0.3$, (b) $\gamma/\kappa = 3$.

With two beams we have the possibility of tuning ω_1 say to one of the intermediate states

$$\omega_1 = \omega_c \pm g, \tag{12.47}$$

and thus two-photon absorption as a function of ω_2 will exhibit resonances, provided the dressed state matrix element is nonzero, at

$$\begin{aligned} \omega_2 &= \omega_c + (\pm\sqrt{2} - 1)g, \qquad |\psi_0^+\rangle \to |\psi_1^\pm\rangle \\ &= \omega_c + (\pm\sqrt{2} + 1)g, \qquad |\psi_0^-\rangle \to |\psi_1^\pm\rangle. \end{aligned} \tag{12.48}$$

This would be doubly resonant two-photon absorption, say, for $\omega_1 = \omega_c + g$, $\omega_2 = \omega_c + (\pm\sqrt{2} - 1)g$. Note that the anharmonic nature of the Jaynes–Cummings Hamiltonian determines all such transition frequencies. The evidence of the anharmonic spectrum can thus be observed by doing two-beam two-photon absorption. Fink *et al.* [31] reported the resonances (12.48) using circuit QED experiments. Kubanek *et al.* [32] reported the excitation of the states $|\psi_1^\pm\rangle$ (Eq. (12.13)) by measuring the $g^{(2)}$ function of the transmitted field.

In addition, the cavity QED provides the possibility of observing the interferences of the type discussed in Section 8.7. Let us consider the case when the beam ω_1 is tuned to the cavity frequency and ω_2 is tuned to $\omega_c + \sqrt{2}g$. We now have two channels as shown in

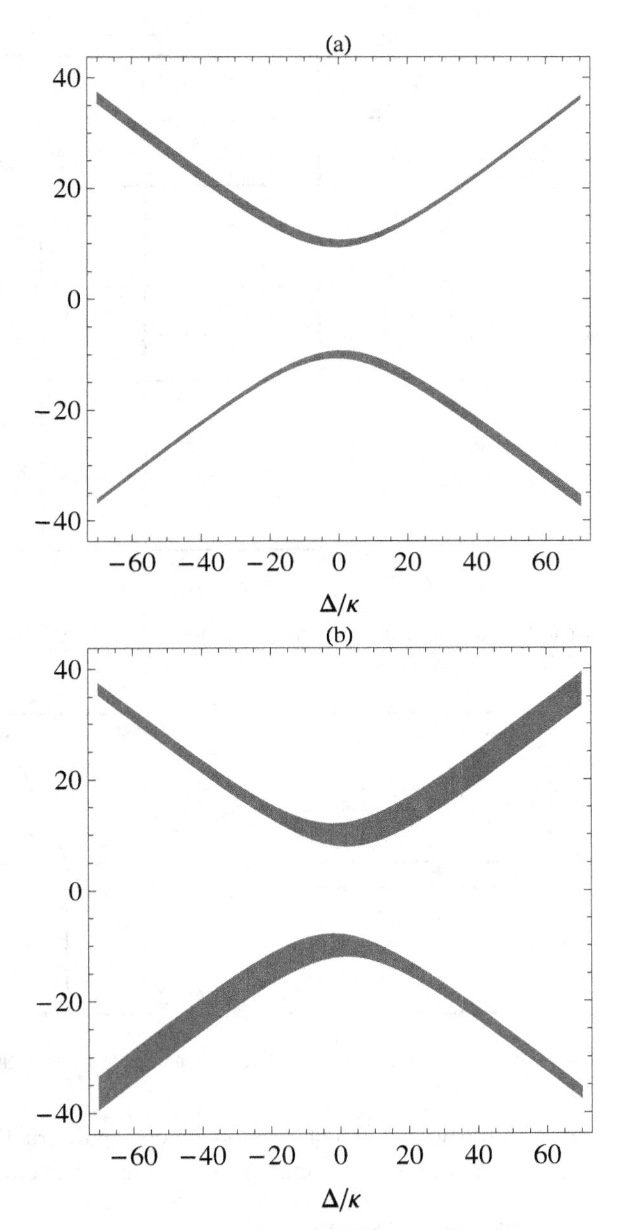

Fig. 12.12 The scaled complex roots $\lambda_\pm/\kappa$ as a function of Δ/κ for different values of γ/κ and $g/\kappa = 10$. The positions of the curves represent the real parts of the roots $\lambda_\pm/\kappa$, and the widths of the curves represent the imaginary parts of the roots $\lambda_\pm/\kappa$. (a) $\gamma/\kappa = 0.3$, (b) $\gamma/\kappa = 3$.

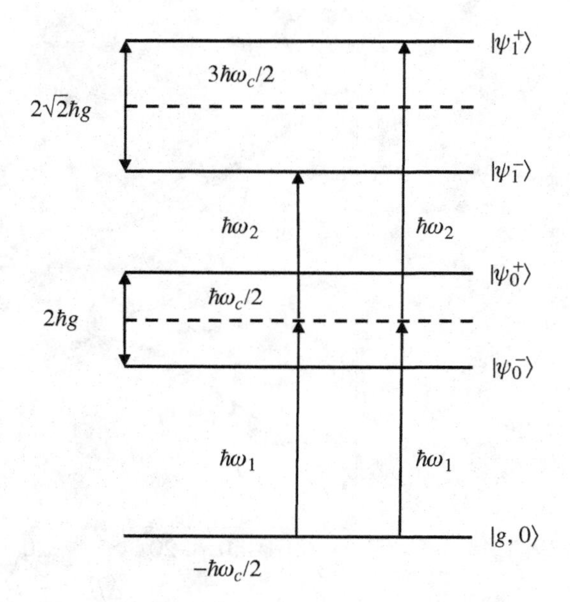

Fig. 12.13 Two-photon transitions among dressed states; there are similar transitions with ω_1 and ω_2 interchanged.

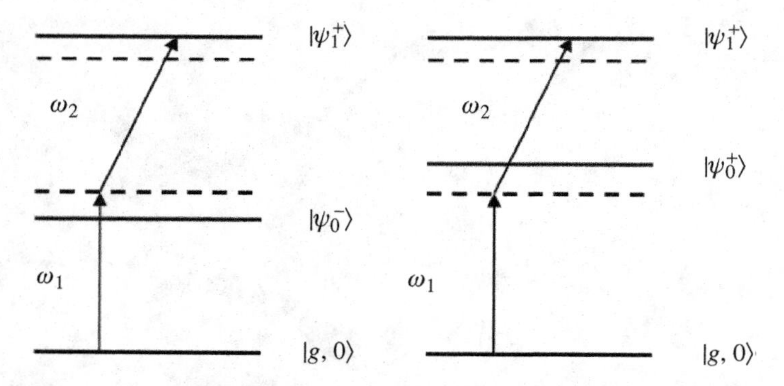

Fig. 12.14 Two channels for two-photon absorption leading to the possibility of interference in cavity QED.

Figure 12.14 for two-photon absorption which can destructively interfere with each other leading to an interference minimum in two-beam two-photon absorption.

12.5.3 Photon blockade in two-photon absorption from a single beam

An important consequence of the dressed states of a strongly coupled atom–cavity system is the inhibition of resonant absorption of a second photon if one photon is already resonantly absorbed. This possibility was called the photon blockade effect by Birnbaum *et al.* [20]. This is displayed in Figure 12.15. If the single frequency field is tuned to the dressed state $|\psi_0^-\rangle$, then the two-photon transition is detuned from both the levels $|\psi_1^\pm\rangle$. In this case the two-photon transition would be nonresonant and the probability for two-photon transition

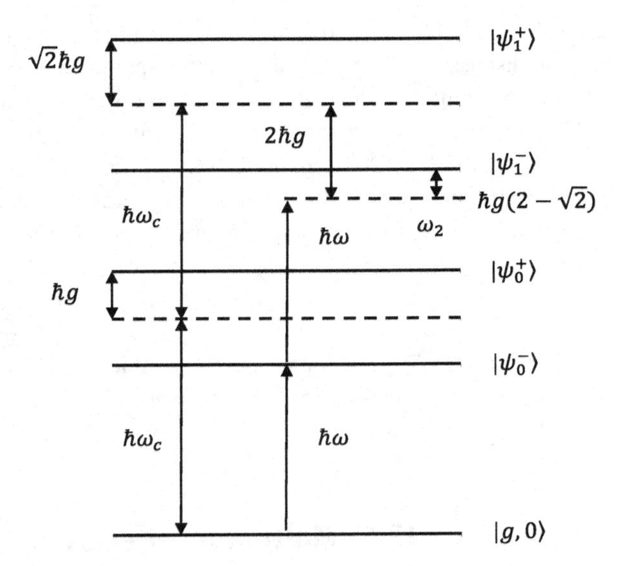

Fig. 12.15 The photon blockade effect.

would be small. Clearly we can not make both single-photon absorption and two-photon absorption resonant at the same time. The transmission of a second photon by the cavity is blocked if the first photon excites one of the dressed states $|\psi_0^{\pm}\rangle$. This leads to sub-Poissonian statistics of the transmitted light as predicted by Brecha *et al.* [33] and observed by Birnbaum *et al.* [20].

12.5.4 Probe of the full spectrum of dressed states

In the previous sections we have seen how single-photon and two-photon absorptions can probe the low lying dressed states. Clearly in order to probe other dressed states we have to drive the system hard. The driving field could be either a coherent field or even an incoherent field. All the information on the dynamical behavior of the system can be obtained by solving the fundamental master equation of Section 12.4. A quantity of interest is the spontaneous emission by the atom, which is usually calculated [34] in terms of the spectrum defined as the Fourier transform of the dipole–dipole correlation function

$$S(\omega) = \mathrm{Re}\int_0^\infty \mathrm{d}\tau\, \mathrm{e}^{\mathrm{i}\omega\tau} \lim_{t\to\infty} \langle S_+(t)S_-(t+\tau)\rangle. \tag{12.49}$$

For the JCM, $S(\omega)$ has resonances at frequencies given by

$$\begin{aligned}\omega-\omega_c &= \mp\frac{1}{2}(\Omega_{n-1,\Delta}-\Omega_{n,\Delta}) \quad \text{for } |\psi_n^{\pm}\rangle \to |\psi_{n-1}^{\pm}\rangle \\ &= \pm\frac{1}{2}(\Omega_{n-1,\Delta}+\Omega_{n,\Delta}) \quad \text{for } |\psi_n^{\pm}\rangle \to |\psi_{n-1}^{\mp}\rangle.\end{aligned} \tag{12.50}$$

For $\Delta = 0$, $\Omega_{n\Delta} = 2g\sqrt{n+1}$ and hence we have resonances at

$$\omega-\omega_c = \pm g,\ \pm g(\sqrt{n+1}+\sqrt{n}),\ \pm g(\sqrt{n+1}-\sqrt{n}). \tag{12.51}$$

The resolution of the peaks depends on γ, κ, and $\bar{n}$. Sanchez-Mondragon *et al.* [24] were the first to discuss the resonances at $\pm g$ in the spontaneous emission spectrum by an atom in a cavity with infinite Q. The existence of other resonances in the limit $Q \to \infty$ is discussed in [35]. This paper also presents calculations in a realistic cavity showing the appearance of the peaks at $\omega_c \pm (\sqrt{2}-1)g$, $\omega_c \pm (\sqrt{2}+1)g$.

Furthermore, the absorption from a weak probe field can be obtained by using the well-known linear response theory [36] and is directly proportional to

$$A(\omega) = \mathrm{Re} \int_0^\infty \mathrm{d}\tau\, \mathrm{e}^{\mathrm{i}\omega\tau} \lim_{t\to\infty} \langle [S_-(t+\tau), S_+(t)] \rangle, \tag{12.52}$$

which is related to $S(\omega)/\bar{n}$ for no pumping field. Thus in a cavity containing a thermal field ($\bar{n} \leq 1$), dressed state resonances such as $\pm g(\sqrt{2} \pm 1)$ besides the ones at $\pm g$ can be seen [35].

12.6 Multi-atom effects in cavity QED

We now study the effect of many atoms interacting with a single mode in a cavity. The regime of strong coupling is of special interest. For simplicity we consider identical two-level atoms. The interaction (12.4) is modified to

$$H_1 = \hbar \sum (g_j S_+^{(j)} a + H.c.), \tag{12.53}$$

where g_j is the coupling constant for the jth atom and $S_+^{(j)}$ is the raising operator for the jth atom. The variation of g with j makes it difficult to calculate the eigenstates and eigenvalues and generally one has to resort to numerical methods. However, for certain cases g_j can be approximated by a number independent of j. These include (a) Rydberg atoms where the distance between atoms is smaller than a wavelength, and (b) if the atoms can be located at the antinodes of the cavity. Then we can introduce a collective description for the atoms. We define the collective operators

$$J_\pm = \sum_j S_\pm^{(j)}, \qquad J_z = \sum_j S_z^{(j)}, \tag{12.54}$$

and then H_1 becomes

$$H_1 = \hbar g(J_+ a + J_- a^\dagger). \tag{12.55}$$

The multi-atom model (12.55) is known as the Tavis–Cummings model [37]. The collective description was first introduced by Dicke [38]. The collective operators satisfy angular momentum algebra. The eigenstates can be obtained from the addition of angular momenta. We level the eigenstates as $|j, m, \eta\rangle$, where

$$J_z|j, m, \eta\rangle = m|j, m, \eta\rangle, \qquad -j \leq m \leq +j, \tag{12.56}$$

$$J^2|j, m, \eta\rangle = j(j+1)|j, m, \eta\rangle, \qquad J^2 = \vec{J}\cdot\vec{J}. \tag{12.57}$$

The parameter η is a degeneracy parameter. The maximum value of j is $\frac{N}{2}$. The states $|\frac{N}{2}, m\rangle$ have no degeneracy and are symmetric with respect to all the atoms. If all the atoms are initially prepared in a symmetric state, then the state would remain symmetric under (12.55). In what follows we consider only the symmetric situations.

12.6.1 Some low lying dressed states of the multi-atom JCM

In order to keep the analysis as simple as possible, we consider the special case of resonance, i.e. $\omega_0 = \omega_c$. The total Hamiltonian is then

$$H = \hbar\omega_c(J_z + a^\dagger a) + \hbar g(J_+ a + J_- a^\dagger). \tag{12.58}$$

We consider the eigenstates with $j = \frac{N}{2}$. Note that $(J_z + a^\dagger a)$ is a constant of motion. We will use $(J_z + a^\dagger a)$ to label the eigenstates. The smallest eigenvalue of J_z is $-\frac{N}{2}$. The lowest state of (12.58) will be

$$|\psi_g\rangle = \left|-\frac{N}{2}, 0\right\rangle, \qquad E = -\hbar\omega_c \frac{N}{2}. \tag{12.59}$$

We next consider states where $(J_z + a^\dagger a)$ has the eigenvalue $(-\frac{N}{2} + 1)$. Such states clearly would be combinations of $|-\frac{N}{2}, 1\rangle$ and $|-\frac{N}{2} + 1, 0\rangle$. Thus we have to diagonalize the Hamiltonian (12.58) in 2×2 space consisting of $|\phi_1\rangle = |-\frac{N}{2} + 1, 0\rangle$ and $|\phi_2\rangle = |-\frac{N}{2}, 1\rangle$. The corresponding matrix that needs to be diagonalized is

$$\begin{pmatrix} H_{11} & H_{12} \\ H_{21} & H_{22} \end{pmatrix},$$

$$H_{11} = H_{22} = \hbar\omega_c\left(-\frac{N}{2} + 1\right), \qquad H_{12} = H_{21} = \hbar g\sqrt{N}. \tag{12.60}$$

The corresponding eigenstates and eigenvalues are

$$|\psi_0^\pm\rangle = \frac{1}{\sqrt{2}}\left(\left|-\frac{N}{2} + 1, 0\right\rangle \pm \left|-\frac{N}{2}, 1\right\rangle\right), \qquad \hbar\omega_c\left(-\frac{N}{2} + 1\right) \pm \hbar g\sqrt{N}. \tag{12.61}$$

We can now continue this process: we can write the Hamiltonian in a space with fixed eigenvalue of $(J_z + a^\dagger a)$ say $\left(-\frac{N}{2} + p\right)$. For a given value of p, there would be $p + 1$ basis states $|-\frac{N}{2} + q, p - q\rangle$, $q = 0, 1, \ldots, p$. Therefore we have to diagonalize the $(p + 1) \times (p + 1)$ matrix. The matrix elements of H_1 in terms of these are

$$\begin{aligned}&\left\langle -\frac{N}{2} + q, p - q\right| (J_+ a + J_- a^\dagger) \left| -\frac{N}{2} + q + q', p - q - q'\right\rangle \\ &= \delta_{q',-1}\sqrt{p - q + 1}\sqrt{(N - q + 1)q} + \delta_{q',1}\sqrt{p - q}\sqrt{(q + 1)(N - q)}.\end{aligned} \tag{12.62}$$

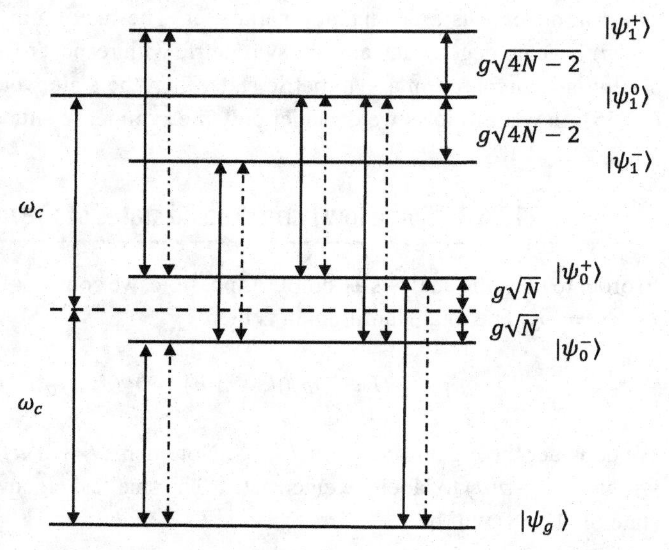

Fig. 12.16 Schematic diagram of the relevant dressed states of a system of a large number of atoms interacting with the cavity field. For large N, dot-dashed (solid) lines give the transitions due to the cavity leakage and spontaneous emission (external field). The allowed transitions depend on the number of atoms [39].

For $p = 2$, we find the eigenstates and eigenvalues as [39]

$$\begin{pmatrix} |\psi_1^+\rangle \\ |\psi_1^0\rangle \\ |\psi_1^-\rangle \end{pmatrix} = \begin{pmatrix} \left(\frac{N-1}{4N-2}\right)^{1/2} & \frac{1}{2^{1/2}} & \left(\frac{N}{4N-2}\right)^{1/2} \\ -\left(\frac{N}{2N-1}\right)^{1/2} & 0 & \left(\frac{N-1}{2N-1}\right)^{1/2} \\ \left(\frac{N-1}{4N-2}\right)^{1/2} & -\frac{1}{2^{1/2}} & \left(\frac{N}{4N-2}\right)^{1/2} \end{pmatrix} \begin{pmatrix} |-\frac{N}{2}+2, 0\rangle \\ |-\frac{N}{2}+1, 1\rangle \\ |-\frac{N}{2}, 2\rangle \end{pmatrix},$$

$$E_1^{\pm} = \left(-\frac{N}{2}+2\right)\omega_c \pm g(4N-2)^{1/2},$$

$$E_1^0 = \left(-\frac{N}{2}+2\right)\omega_c. \tag{12.63}$$

The energy levels are displayed in Figure 12.16.

12.6.2 Atomic number dependence of vacuum Rabi splittings

Let us now consider the process of single-photon absorption in a cavity on resonance with the atomic transition frequency. Clearly there are two channels of absorption $|\psi_g\rangle \rightarrow |\psi_0^{\pm}\rangle$. Thus, one will find resonances at

$$\omega = \omega_0 \pm gN^{1/2}. \tag{12.64}$$

The usual absorption resonance at ω_0 splits into two resonances. The resolution will depend on $gN^{1/2}$ and the width of each resonance. This result is true down to $N = 1$.

To make an estimate of the width of each peak in vacuum Rabi splittings we proceed as in Section 12.5.1. We need to examine the decay of the state $|\frac{N}{2}, -\frac{N}{2}+1\rangle$ to the state $|\frac{N}{2}, -\frac{N}{2}\rangle$. In terms of the states of each atom we have

$$\left|\frac{N}{2}, -\frac{N}{2}\right\rangle = |g, \ldots, g\rangle,$$
$$\left|\frac{N}{2}, -\frac{N}{2}+1\right\rangle = \frac{1}{\sqrt{N}} \sum_j |g, \ldots, e_j, \ldots, g\rangle. \tag{12.65}$$

Thus the spontaneous decay of $|\frac{N}{2}, -\frac{N}{2}+1\rangle$ is due to the decay of $|e_j\rangle$. Therefore the dynamical behavior in the space $|-\frac{N}{2}+1, 0\rangle$, $|-\frac{N}{2}, 1\rangle$ is described by the matrix M (Eq. (12.43)) with $g \to g\sqrt{N}$. Hence in the limit of large $g\sqrt{N}$, the half width of each doublet will be $(\kappa+\gamma)/2$.

Bernardot *et al.* [40] studied the absorption spectra by using the transition $^{39}S_{1/2}F = 3 \to {}^{39}P_{3/2}$ in Rb atoms crossing a cylindrical superconducting Nb cavity at 1.7 K, with $Q \sim 10^8$. They were able to resolve the vacuum field Rabi splittings down to effectively five atoms on average. By increasing the atomic flux they also verified the $N^{1/2}$ dependence of the vacuum field Rabi splitting as predicted by (12.64). It is remarkable to note that the line splitting is produced by coupling to the vacuum mode of the cavity and thus the response of the atoms is renormalized [23, 39] by the vacuum field. The equation (12.64) also suggests that the lines can be resolved by increasing the number of atoms. Fink *et al.* [41] have used circuit QED to make very precise measurements of the N dependence of vacuum Rabi splittings. They report results up to three qubits.

12.6.3 Atomic number dependent two-photon absorption

We now consider the atomic number dependence of the two-photon absorption spectra. Here the sum of the frequencies $\omega_1+\omega_2$ of two photons must match the transition frequency between any of the levels $|\psi_1^\pm\rangle$, $|\psi_1^0\rangle$ and the level $|\psi_g\rangle$. For example, if we tune, say, one of the external fields of frequency ω_1 on the transition $|\psi_g\rangle \to |\psi_0^+\rangle$, then the two-photon absorption spectrum will display a resonance at

$$\begin{aligned} \omega_2 &= \omega_c + [g(4N-2)^{1/2} - gN^{1/2}] && \text{for } |\psi_g\rangle \to |\psi_0^+\rangle \\ &\to \omega_c + gN^{1/2} && \text{for large } N \\ &\to \omega_c + g(2^{1/2}-1) && \text{for } N = 1. \end{aligned} \tag{12.66}$$

For $N=2$, the resonance will occur at $\omega_c + 2^{1/2}g(3^{1/2}-1)$. Note also that for two-photon absorption from the same field $\omega_1 = \omega_2 = \omega$, the resonances should occur at

$$\begin{aligned} \omega &= \omega_c \pm g(N-1/2)^{1/2} && \\ &\to \omega_c \pm gN^{1/2} && \text{for large } N \\ &\to \omega_c \pm \frac{g}{2^{1/2}} && \text{for } N = 1. \end{aligned} \tag{12.67}$$

Thus for $N = 1$ one will have new resonances at $\omega = \omega_c \pm (g/2^{1/2})$. However, for large N the two-photon absorption can be very large as it has a double resonance due to the single-photon resonance condition $\omega = \omega_c \pm g\sqrt{N}$ being satisfied at the same time.

The strength of the two-photon absorption is determined by the dipole matrix elements and hence we next consider the N dependence of the transition matrix elements. From Eqs. (12.61) and (12.63), we have for the matrix elements of the collective dipole moment operator $\tilde{d} = d(J_+ + J_-)$

$$d_{\pm g} \equiv \frac{d}{2^{1/2}} N^{1/2}, \qquad d_{1\pm} \equiv \frac{d}{2}\left(\frac{2N-2}{[2(2N-1)]^{1/2}} \pm N^{1/2}\right), \tag{12.68}$$

where $d_{\pm g} = \langle \psi_0^{\pm} | \tilde{d} | \psi_g \rangle$ and $d_{1\pm} = \langle \psi_1^{+} | \tilde{d} | \psi_0^{\pm} \rangle$. The transition matrix elements are strongly dependent on N. In the limit of large N, $d_{1-} \to 0$, $d_{1+} \to dN^{1/2}$. The N dependence of the matrix elements results in very large two-photon absorption cross-sections. For example, the transition $|\psi_g\rangle \to |\psi_0^{+}\rangle \to |\psi_1^{+}\rangle$ will exhibit N^2 dependence, i.e. superradiant character, if ω_1 and ω_2 are chosen to be resonant with the corresponding transition frequencies. Moreover, the possibility of two-photon quantum interference will disappear as the pathway $|0\rangle \to |-\rangle \to |1\rangle$, which is allowed for $N = 1$, is forbidden for large N. We expect that circuit QED would be excellent system to probe size dependent two-photon transitions.

12.7 Effective dipole–dipole interaction in a dispersive cavity from Lamb shift of the vacuum

In Section 12.3, we have seen the great utility of the dispersive JCM. If many atoms interact with a dispersive cavity, then the interaction with a common mode of the cavity can produce an effective interaction between noninteracting atoms. This effective interaction turns out to have a number of important applications for quantum logic operations and for the production of multiparticle entangled states [42–45].

We now calculate the effective interaction. As discussed in Section 12.6, the Hamiltonian describing the interaction of many identical two-level atoms with the single mode of the cavity field is given by

$$H = \hbar\omega_0 J_z + \hbar\omega_c a^\dagger a + \hbar g(J_+ a + J_- a^\dagger). \tag{12.69}$$

In a frame rotating with frequency ω_c, the interaction Hamiltonian becomes

$$H = \hbar\Delta J_z + \hbar g(J_+ a + J_- a^\dagger). \tag{12.70}$$

The basis states for the atoms are the Dicke states $|\frac{N}{2}, m\rangle$ and Fock states for the field. Let us then consider an initial state $|n, \frac{N}{2}, m\rangle$ and examine the shift of this state to second order in the coupling constant g. This is justified as we are in the dispersive limit. According to the perturbation theory, the correction to the energy will be

$$\Delta E_{n,\frac{N}{2},m} = \sum_j \frac{|\langle n, \frac{N}{2}, m|\hbar g(J_+ a + J_- a^\dagger)|\Psi_j\rangle|^2}{E_{n,\frac{N}{2},m} - E_j}, \qquad E_{n,\frac{N}{2},m} = \hbar\Delta m, \tag{12.71}$$

where $|\Psi_j\rangle$ are the states connected to the initial state $|n, \frac{N}{2}, m\rangle$ via the perturbation. Clearly, we have only two states $|\Psi_j\rangle = |n-1, \frac{N}{2}, m+1\rangle$; $|n+1, \frac{N}{2}, m-1\rangle$. Then (12.71) on simplification yields

$$\Delta E_{n,\frac{N}{2},m} = \frac{\hbar g^2}{\Delta}\left[2nm + \left(\frac{N}{2}\right)^2 - m^2 + \frac{N}{2} + m\right]. \tag{12.72}$$

This energy shift is equivalent to an effective interaction

$$V_{\text{eff}} = \frac{\hbar g^2}{\Delta}\left[\frac{N}{2}\left(\frac{N}{2}+1\right) - (J_z)^2 + J_z + 2a^\dagger a J_z\right]. \tag{12.73}$$

If the cavity is initially in the vacuum state, then

$$V_{\text{eff}} = \frac{\hbar g^2}{\Delta} J_+ J_-. \tag{12.74}$$

Combining (12.70) and (12.74), we get

$$H = \hbar\Delta J_z + \frac{\hbar g^2}{\Delta} J_+ J_-, \tag{12.75}$$

which on transforming to a frame rotating with frequency ω_0, becomes

$$H = \frac{\hbar g^2}{\Delta} J_+ J_-. \tag{12.76}$$

For two two-level atoms, (12.76) gives

$$H = \frac{\hbar g^2}{\Delta}(S_+^{(1)} S_-^{(1)} + S_+^{(2)} S_-^{(2)} + S_+^{(1)} S_-^{(2)} + S_+^{(2)} S_-^{(1)}). \tag{12.77}$$

The last two terms in (12.77) are the interaction terms between two atoms. We thus conclude that the dispersive interaction with a common mode of the field produces an interaction like the dipole–dipole interaction. Note that for this effective interaction to be significant, the atoms need not be within a distance much smaller than a wavelength. The interaction (12.77) arises with no photons in the cavity. Thus (12.77) can be seen to arise from the Lamb shift of the cavity vacuum.

The effective dipole–dipole interaction can be used in a number of applications. One can use it to produce quantum logic gates. It is well established [43, 44] that dipole–dipole interaction can be used to implement swap and CNOT gates. Thus cavity QED interactions can be used to realize CNOT gates [45].

12.8 Atomic cat states using multi-atom dispersive JCM

In this section, we show that the unitary evolution generated by the Hamiltonian (12.76) transforms an atomic coherent state into a superposition of distinct atomic coherent states [42]. The superposition of atomic coherent states is an atomic Schrödinger cat state in the same sense that the superposition of electromagnetic field coherent states is a Schrödinger cat state.

Consider an atomic system prepared in the atomic coherent state defined by (Eq. (11.51))

$$\begin{aligned} |\Psi(0)\rangle &\equiv |\theta, \phi\rangle \\ &= \sum_{k=0}^{2j} \binom{2j}{k}^{\frac{1}{2}} \sin^k \frac{\theta}{2} \cos^{2j-k} \frac{\theta}{2} \, \mathrm{e}^{-\mathrm{i}k\phi} |j, k-j\rangle. \end{aligned} \tag{12.78}$$

Here $|j, k-j\rangle$ is the Dicke state with cooperation number j and total z component of angular momentum equal to $k-j$. We would take $j = N/2$. The state (12.78) is an eigenstate of the component of the angular momentum in the (θ, ϕ) direction with eigenvalue $N/2$. The atoms in the state (12.78) are uncorrelated as (12.78) can be written as a product of states for individual atoms

$$|\theta, \phi\rangle = \prod_{j=1}^{N} \left(\sin \frac{\theta}{2} \mathrm{e}^{-\mathrm{i}\phi} \, |e\rangle_j + \cos \frac{\theta}{2} \, |g\rangle_j \right). \tag{12.79}$$

In particular, one will have

$$\langle S_+^{(i)} S_-^{(j)} \rangle \equiv \langle S_+^{(i)} \rangle \langle S_-^{(j)} \rangle, \qquad i \neq j. \tag{12.80}$$

An atomic coherent state, like a harmonic oscillator coherent state, is most classical in the sense that it is a minimum uncertainty state of a pair of operators for two mutually orthogonal spin components orthogonal to the average direction of spin in the atomic coherent state. An evolution generated by a linear combination $(\vec{\alpha} \cdot \vec{J})$ of the atomic operators transforms an atomic coherent state to another one – this is equivalent to motion on the Bloch sphere. The atomic coherent state (12.78), under the action of the nonlinear unitary evolution generated by Eq. (12.76), transforms to

$$\begin{aligned} |\Psi(t)\rangle &\equiv \exp\left(-\mathrm{i}\frac{H}{\hbar} t \right) \Big| \theta, \phi \Big\rangle \\ &= \sum_{k=0}^{N} \binom{2j}{k}^{\frac{1}{2}} \mathrm{e}^{-\mathrm{i}k\phi} \sin^k \left(\frac{\theta}{2} \right) \cos^{2j-k} \left(\frac{\theta}{2} \right) \exp\left\{ -\mathrm{i}\tau [k(2j+1) - k^2] \right\} |j, k-j\rangle, \\ \tau &= \frac{g^2 t}{\Delta}. \end{aligned} \tag{12.81}$$

From now on we consider Eq. (12.81) at special times $\tau = \pi/m$, where m is an integer,

$$|\Psi(t)\rangle = \sum_{k=0}^{N} \binom{2j}{k}^{\frac{1}{2}} \mathrm{e}^{-\mathrm{i}k\phi'} \sin^k \left(\frac{\theta}{2} \right) \cos^{2j-k} \left(\frac{\theta}{2} \right) \exp\left[\mathrm{i}\frac{\pi}{m} k(k+1) \right] |j, k-j\rangle, \tag{12.82}$$

where

$$\phi' = \phi + \frac{\pi}{m}(2j+2). \tag{12.83}$$

To simplify (12.82), we would use the relations

$$\begin{aligned}
\exp\left[\frac{\mathrm{i}\pi}{m}k(k+1)\right] &= \sum_{q=0}^{m-1} f_q^{(o)} \exp\left(\frac{2\pi \mathrm{i}q}{m}k\right), \\
\exp\left(\frac{\mathrm{i}\pi}{m}k^2\right) &= \sum_{q=0}^{m-1} f_q^{(e)} \exp\left(\frac{2\pi \mathrm{i}q}{m}k\right),
\end{aligned} \tag{12.84}$$

for m odd and even, respectively, along with the inverse relations

$$\begin{aligned}
f_q^{(o)} &= \frac{1}{m}\sum_{k=0}^{m-1} \exp\left(-\frac{2\pi \mathrm{i}q}{m}k\right) \exp\left[\frac{\mathrm{i}\pi}{m}k(k+1)\right], \\
f_q^{(e)} &= \frac{1}{m}\sum_{k=0}^{m-1} \exp\left(-\frac{2\pi \mathrm{i}q}{m}k\right) \exp\left(\frac{\mathrm{i}\pi}{m}k^2\right).
\end{aligned} \tag{12.85}$$

These relations are based on the periodicity of the left-hand side of Eq. (12.84). The importance of Eq. (12.84) lies in the fact that an exponentially quadratic form has been converted into sums of exponentials linear in k. On combining Eqs. (12.82) and (12.84) we obtain, for odd m,

$$\exp\left(-\mathrm{i}\frac{H}{\hbar}t\right)\left|\theta,\phi\right\rangle = \sum_{q=0}^{m-1} f_q^{(o)} \left|\theta, \phi + \pi\frac{2j+2-2q}{m}\right\rangle, \qquad t = \frac{\pi\Delta}{mg^2}, \tag{12.86}$$

and for even m,

$$\exp\left(-\mathrm{i}\frac{H}{\hbar}t\right)\left|\theta,\phi\right\rangle = \sum_{q=0}^{m-1} f_q^{(e)} \left|\theta, \phi + \pi\frac{2j+1-2q}{m}\right\rangle, \qquad t = \frac{\pi\Delta}{mg^2}. \tag{12.87}$$

The expression (12.86) and (12.87) show that, at the particular time $t = \pi\Delta/(mg^2)$, an atomic coherent state evolves into a superposition of the atomic coherent states, i.e. it becomes an atomic cat state. The states in the superposition differ in phase. In particular, for $m = 2$ and $g^2t/\Delta = \pi/2$ it follows that

$$\begin{aligned}
\exp\left(-\mathrm{i}\frac{H}{\hbar}t\right)\left|\theta,\phi\right\rangle = \frac{1}{\sqrt{2}}\Bigg[&\exp\left(\frac{\mathrm{i}\pi}{4}\right)\left|\theta,\phi+\pi\frac{N+1}{2}\right\rangle \\
&+\exp\left(-\frac{\mathrm{i}\pi}{4}\right)\left|\theta,\phi+\pi\frac{N-1}{2}\right\rangle\Bigg].
\end{aligned} \tag{12.88}$$

Note that the value $g^2t/\Delta = \pi/2$ should be feasible with better cavity Q values. One can similarly obtain expressions for the Schrödinger cat states for other values of m. For a detailed discussion and diagrammatic representation of the atomic cat states, see [42].

In the special case of $\theta = \pi/2$, $\phi = -\pi/2$, the state (12.79) is the product state

$$\left|\frac{\pi}{2}, -\frac{\pi}{2}\right\rangle = \prod_{j=1}^{N} \frac{1}{\sqrt{2}} \left(|g\rangle_j + \mathrm{i}\,|e\rangle_j\right), \tag{12.89}$$

and then (12.88) goes over to

$$\mathrm{e}^{-\mathrm{i}\frac{H}{\hbar}t} \left|\frac{\pi}{2}, -\frac{\pi}{2}\right\rangle = \frac{1}{\sqrt{2}} \left\{ \prod_j \frac{1}{\sqrt{2}} \left[|g\rangle_j + (-\mathrm{i})^N |e\rangle_j\right] \mathrm{e}^{\mathrm{i}\pi/4} + \prod_j \frac{1}{\sqrt{2}} \left[|g\rangle_j - (-\mathrm{i})^N |e\rangle_j\right] \mathrm{e}^{-\mathrm{i}\pi/4} \right\}. \tag{12.90}$$

It should be noted that the two parts in the superposition state (12.90) are orthogonal

$$\left({}_j\langle g| + (\mathrm{i})^N {}_j\langle e|\right)\left(|g\rangle_j - (-\mathrm{i})^N |e\rangle_j\right) = 0. \tag{12.91}$$

The cat state (12.90) has the same structure as the multi-particle Greenberger–Horne–Zeilinger (GHZ) states [46]

$$|\Psi\rangle_{\mathrm{GHZ}} = \frac{1}{\sqrt{2}} \left(|g, g, \ldots, g\rangle + \mathrm{e}^{\mathrm{i}\theta} |e, e, \ldots, e\rangle\right). \tag{12.92}$$

The only difference is that (12.90) is like a GHZ state but in a rotated basis. A rotation of the spin-1/2 state by $\pi/2$ about the y-axis yields

$$\begin{aligned} \mathrm{e}^{-\mathrm{i}\frac{\pi}{2}S_y} |e\rangle &= \frac{1}{\sqrt{2}} \left(|e\rangle + |g\rangle\right), \\ \mathrm{e}^{-\mathrm{i}\frac{\pi}{2}S_y} |g\rangle &= \frac{1}{\sqrt{2}} \left(|g\rangle - |e\rangle\right). \end{aligned} \tag{12.93}$$

To see this, we can write S_y in terms of the Pauli matrix and then expand the exponential

$$\begin{aligned} \mathrm{e}^{-\mathrm{i}\frac{\pi}{2}S_y} = \mathrm{e}^{-\mathrm{i}\frac{\pi}{4}\sigma_y} &= \frac{1}{\sqrt{2}}(1 - \mathrm{i}\sigma_y) \\ &= \frac{1}{\sqrt{2}}(1 - 2\mathrm{i}S_y) = \frac{1}{\sqrt{2}}[1 - (S_+ - S_-)], \end{aligned} \tag{12.94}$$

and hence

$$\frac{1}{\sqrt{2}}[1 - (S_+ - S_-)]\begin{pmatrix} |e\rangle \\ |g\rangle \end{pmatrix} = \frac{1}{\sqrt{2}}\begin{pmatrix} (|e\rangle + |g\rangle) \\ (|g\rangle - |e\rangle) \end{pmatrix}. \tag{12.95}$$

On combining (12.90) and (12.93), we see that

$$\mathrm{e}^{-\mathrm{i}\frac{\pi}{2}J_y} \mathrm{e}^{-\mathrm{i}\frac{H}{\hbar}t} \left|\frac{\pi}{2}, -\frac{\pi}{2}\right\rangle = \frac{1}{\sqrt{2}} \left(\mathrm{e}^{\mathrm{i}\alpha} |g, g, \ldots, g\rangle + \mathrm{e}^{-\mathrm{i}\alpha} |e, e, \ldots, e\rangle\right), \tag{12.96}$$

where α is $\pi/4$ $(-\pi/4)$ for even (odd) values of $N/2$. The last state is exactly the GHZ state. Clearly the dispersive interaction (12.76) can be used to generate a whole class of multiparticle entangled states.

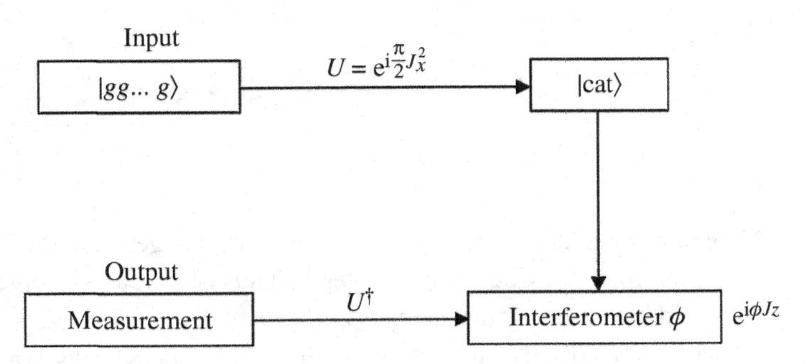

Fig. 12.17 Atomic interferometer with cat states as input.

12.9 Application of atomic cat states in Heisenberg limited measurements

We consider measurement of the phase φ by an atomic interferometer. The quantum mechanical uncertainty relation between the number N and phase is

$$\Delta N \Delta\varphi \sim 1. \tag{12.97}$$

For coherent beams $\Delta N \sim \sqrt{N}$ and hence

$$\Delta\varphi \sim \frac{1}{\sqrt{N}}. \tag{12.98}$$

This is known as the standard quantum limit, and gives the maximum achievable accuracy in phase with coherent beams. However, the maximum fluctuation in the number could be equal to the number of particles and thus

$$\Delta\varphi \geq \frac{1}{N}. \tag{12.99}$$

Consider now the interferometric scheme depicted in Figure 12.17. The working of this interferometric setup can be modeled as a successive operation of the collective spin operators J_x and J_z on the input state. Hence the output state generated from the setup can be written as

$$|\Psi\rangle_{\text{out}} = \mathrm{e}^{-\mathrm{i}\frac{\pi}{2}J_x^2}\mathrm{e}^{\mathrm{i}\phi J_z}\mathrm{e}^{\mathrm{i}\frac{\pi}{2}J_x^2}|\Psi\rangle_{\text{in}}. \tag{12.100}$$

Note that if $\phi = 0$, i.e. in the absence of the interferometer, we get

$$|\Psi\rangle_{\text{out}} = U^{\dagger}U|\Psi\rangle_{\text{in}} = |\Psi\rangle_{\text{in}}. \tag{12.101}$$

We next calculate the output signal and the sensitivity of the phase measurements using the setup of Figure 12.17. In order to perform the calculation of the output state, we need to calculate the action of the operator $U = \exp\left[\mathrm{i}\frac{\pi}{2}(J_x)^2\right] = \exp\left[\mathrm{i}\frac{\pi}{2}(\sum_i S_x^{(i)})^2\right]$ on the input state. Following the method of the previous section, we prove the following relations for even values of N

$$U|gg\cdots g\rangle = e^{i\frac{\pi}{4}} \frac{1}{\sqrt{2}} \left[|gg\cdots g\rangle + (i)^{N-1} |ee\cdots e\rangle \right], \tag{12.102}$$

$$U|ee\cdots e\rangle = e^{i\frac{\pi}{4}} \frac{1}{\sqrt{2}} \left[|ee\cdots e\rangle + (i)^{N-1} |gg\cdots g\rangle \right]. \tag{12.103}$$

Note that, for even N, the eigenvalues of J_z are integers and therefore the eigenvalues of J_x^2 are of the form k^2, where k is an integer. Thus in the basis of eigenstates of J_x, the operator U would have the eigenvalue $\exp\left(i\frac{\pi}{2}k^2\right)$.

Let us consider the quantity $\exp\left(i\frac{\pi}{m}k^2\right)$. This is periodic under $k \to k + 2m$, then we have

$$\exp\left(i\frac{\pi}{m}k^2\right) = \exp\left(i\frac{\pi}{m}k^2\right) \exp[4i\pi(m+k)], \qquad k \text{ and } m \text{ are integers.} \tag{12.104}$$

Therefore we can write a Fourier series for even values of m

$$\exp\left(i\frac{\pi}{m}k^2\right) = \sum_{q=0}^{m-1} f_q^{(e)} \exp\left(\frac{2i\pi q}{m}k\right), \tag{12.105}$$

where

$$f_q^{(e)} = \frac{1}{m} \sum_{k=0}^{m-1} \exp\left(i\frac{\pi}{m}k^2\right) \exp\left(-\frac{2i\pi q}{m}k\right). \tag{12.106}$$

Since J_x^2 has eigenvalues k^2, we immediately get from Eq. (12.105)

$$\exp\left(i\frac{\pi}{m}J_x^2\right) = \sum_{q=0}^{m-1} f_q^{(e)} \exp\left(\frac{2i\pi q}{m}J_x\right). \tag{12.107}$$

Now for $m = 2$, we get

$$\exp\left(i\frac{\pi}{2}J_x^2\right) = f_0 + f_1 \exp(i\pi J_x), \tag{12.108}$$

and $f_0 = \frac{1}{2}(1+i) = \frac{1}{\sqrt{2}}\exp\left(i\frac{\pi}{4}\right)$; $f_1 = \frac{1}{2}(1-i) = \frac{1}{\sqrt{2}}\exp\left(-i\frac{\pi}{4}\right)$. Thus for any state $|\Psi\rangle$ we obtain

$$\exp\left(i\frac{\pi}{2}J_x^2\right)|\Psi\rangle = \frac{1}{\sqrt{2}}\left[\exp\left(i\frac{\pi}{4}\right)|\Psi\rangle + \exp\left(-i\frac{\pi}{4}\right)\exp(i\pi J_x)|\Psi\rangle\right]. \tag{12.109}$$

The last term in (12.109) can be simplified by using the properties of spin-1/2 operators S_x^i

$$e^{i\pi J_x} = \prod_j \exp\left(i\frac{\pi}{2}\sigma_x^j\right) = \prod_j (i\sigma_x^j), \tag{12.110}$$

and therefore

$$e^{i\pi J_x}|gg\cdots g\rangle = i^N |ee\cdots e\rangle. \tag{12.111}$$

Hence on combining Eqs. (12.109) and (12.111) we get

$$U|gg\cdots g\rangle = \exp\left(i\frac{\pi}{4}\right)\frac{1}{\sqrt{2}}\left[|gg\cdots g\rangle + i^N \exp\left(-i\frac{\pi}{2}\right)|ee\cdots e\rangle\right]. \tag{12.112}$$

A similar procedure can be used to prove (12.103). Using Eqs. (12.100), (12.102), and (12.103), we can calculate the output state $|\Psi\rangle_{\text{out}}$

$$\begin{aligned}|\Psi\rangle_{\text{out}} &= U^{\dagger}\mathrm{e}^{\mathrm{i}\varphi J_z}U|gg\cdots g\rangle \\ &= U^{\dagger}\mathrm{e}^{\mathrm{i}\varphi J_z}\frac{\mathrm{e}^{\mathrm{i}\frac{\pi}{4}}}{\sqrt{2}}(|gg\cdots g\rangle + \mathrm{i}^{N-1}|ee\cdots e\rangle) \\ &= U^{\dagger}\frac{\mathrm{e}^{\mathrm{i}\frac{\pi}{4}}}{\sqrt{2}}(\mathrm{e}^{-\frac{N}{2}\mathrm{i}\varphi}|gg\cdots g\rangle + \mathrm{i}^{N-1}\mathrm{e}^{\frac{N}{2}\mathrm{i}\varphi}|ee\cdots e\rangle),\end{aligned} \tag{12.113}$$

which, on using (12.102), and (12.103) again (with $\mathrm{i} \to (-\mathrm{i})$), reduces to

$$\begin{aligned}|\Psi\rangle_{\text{out}} = &\frac{\mathrm{e}^{\mathrm{i}\frac{\pi}{4}}}{\sqrt{2}}\mathrm{e}^{-\frac{N}{2}\mathrm{i}\varphi}\frac{\mathrm{e}^{-\mathrm{i}\frac{\pi}{4}}}{\sqrt{2}}(|gg\cdots g\rangle + (-\mathrm{i})^{N-1}|ee\cdots e\rangle) \\ &+ \frac{\mathrm{e}^{\mathrm{i}\frac{\pi}{4}}}{\sqrt{2}}\mathrm{e}^{\frac{N}{2}\mathrm{i}\varphi}(\mathrm{i})^{N-1}\frac{\mathrm{e}^{-\mathrm{i}\frac{\pi}{4}}}{\sqrt{2}}(|ee\cdots e\rangle + (-\mathrm{i})^{N-1}|gg\cdots g\rangle) \\ = &\cos\left(\frac{N}{2}\varphi\right)|gg\cdots g\rangle + (\mathrm{i})^{N}\sin\left(\frac{N}{2}\varphi\right)|ee\cdots e\rangle.\end{aligned} \tag{12.114}$$

Note that the output state depends on N times the phase φ of the interferometer. This leads to both superresolution and supersensitivity of the atomic interferometer. Let the mean inversion $\langle J_z\rangle$ be the signal. Clearly it is given by

$$\langle J_z\rangle = -\frac{N}{2}\cos N\varphi, \tag{12.115}$$

which shows that the superresolution of the signal as the number of fringes in a given interval is N times greater than what would be obtained if one had not used a cat state. Now the noise in the signal is given by

$$\begin{aligned}\Delta J_z &= \sqrt{\langle J_z^2\rangle - \langle J_z\rangle^2} \\ &= \sqrt{\frac{N^2}{4}(1-\cos^2 N\varphi)} \\ &= \frac{N}{2}\sin N\varphi,\end{aligned} \tag{12.116}$$

and hence the signal to noise ratio is

$$\frac{\langle J_z\rangle}{\Delta J_z} = -\cot N\varphi. \tag{12.117}$$

The sensitivity of the measurement can be defined by

$$\Delta\varphi \sim \left[\frac{(\Delta J_z)^2}{(\partial\langle J_z\rangle/\partial\varphi)^2}\right]^{\frac{1}{2}} = \frac{1}{N}. \tag{12.118}$$

We have a precision of measurement at the Heisenberg limit. Leibfried *et al.* [47] first demonstrated the superresolution and supersensitivity of an atomic interferometer by preparing GHZ states of three trapped ions. They obtained a fringe pattern with sinusoidal oscillation that was three times faster and showed that the sensitivity of their measurement

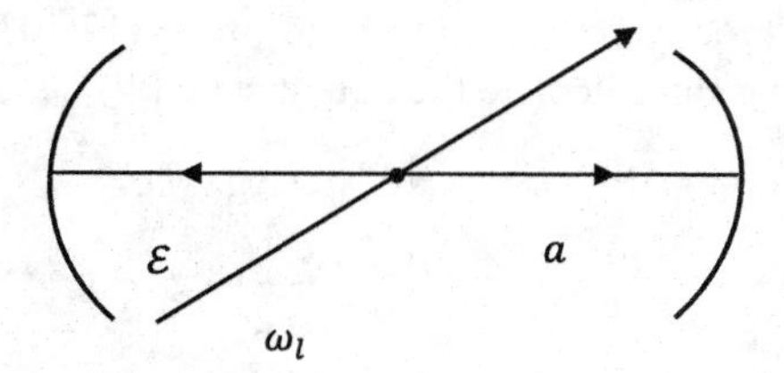

Fig. 12.18 Engineering different interactions via pumping of the atom from the side and by tuning external parameters such as field strength ε and frequency ω_l.

was about 1.45 times better (ideally $\sqrt{3}$ times) than that for separable states. A later experiment [48] demonstrated superresolution and supersensitivity with up to six ion cat states.

12.10 Engineering anti-Jaynes–Cummings interaction

We now consider the case of an atom in a single-mode cavity where the atom is driven strongly by a coherent field. We will show how a class of interesting model Hamiltonians [49] can be tailored this way. Such models enable one to generate a variety of states with atom–photon entanglement.

Consider a field on resonance with the atomic transition frequency, which is applied from the side, as shown in Figure 12.18. The Hamiltonian in a frame rotating with the frequency of the field is

$$H = \hbar\delta a^{\dagger}a + \hbar\frac{\Omega}{2}(S_+ + S_-) + \hbar g(S_+a + S_-a^{\dagger}), \qquad \delta = \omega_c - \omega_l, \tag{12.119}$$

where Ω is the Rabi frequency of the applied field. We now define the interaction picture by using the unperturbed Hamiltonian

$$H_0 = \hbar\delta a^{\dagger}a + \hbar\frac{\Omega}{2}(S_+ + S_-), \tag{12.120}$$

which includes the interaction with the external field. The interaction Hamiltonian in the interaction picture is then

$$\begin{aligned} H_I &= \exp\left(\mathrm{i}\frac{H_0}{\hbar}t\right)\hbar g(S_+a + S_-a^{\dagger})\exp\left(-\mathrm{i}\frac{H_0}{\hbar}t\right) \\ &= \hbar g \exp(\mathrm{i}\Omega S_x t)\, S_+ \exp(-\mathrm{i}\Omega S_x t)\, a e^{-\mathrm{i}\delta t} + H.c. \\ &= \hbar g\left(\cos\frac{\Omega t}{2} + 2\mathrm{i}S_x\sin\frac{\Omega t}{2}\right) S_+ \left(\cos\frac{\Omega t}{2} - 2\mathrm{i}S_x\sin\frac{\Omega t}{2}\right) a e^{-\mathrm{i}\delta t} + H.c. \\ &= \hbar g\left[S_+\cos^2\frac{\Omega t}{2} + S_-\sin^2\frac{\Omega t}{2} - \mathrm{i}S_z\sin(\Omega t)\right] a e^{-\mathrm{i}\delta t} + H.c.. \end{aligned} \tag{12.121}$$

Let us assume that the Rabi frequency is large. Then we can tune the cavity in such a way that δ is either Ω or $-\Omega$. In this case, the interaction Hamiltonian has terms which are oscillating at frequencies like $(\Omega + \delta)$, etc. and these terms can be ignored. The approximation would be good as long as $|\Omega + \delta| \gg g$. Thus we get the following approximate Hamiltonians:

I. $\delta = \Omega$

$$H_I \approx \hbar g a \left[\frac{1}{4}(S_+ - S_-) - \frac{1}{2} S_z \right] + H.c.. \tag{12.122}$$

II. $\delta = -\Omega$

$$H_I \approx \hbar g a \left[\frac{1}{4}(S_+ - S_-) + \frac{1}{2} S_z \right] + H.c.. \tag{12.123}$$

On introducing the eigenstates $|\pm\rangle = \frac{1}{\sqrt{2}}(|g\rangle \pm |e\rangle)$ of the operator S_x, we can write (12.122) and (12.123) as

$$H_{\mathrm{JC}} \sim \frac{\hbar g}{2}(a|+\rangle\langle-| + a^\dagger|-\rangle\langle+|), \tag{12.124}$$

$$H_{\mathrm{AJC}} \sim -\frac{\hbar g}{2}(a|-\rangle\langle+| + a^\dagger|+\rangle\langle-|). \tag{12.125}$$

The Hamiltonian (12.124) is the standard Jaynes–Cummings (JC) interaction where the atomic raising and lowering operators are defined in terms of the basis using the eigenstates of the S_x operator. The Hamiltonian (12.125) is clearly the anti-Jaynes–Cummings interaction (AJC), which one would traditionally throw away. However, now the interaction (12.125) is slowly varying as we are already in the interaction picture defined by the unperturbed Hamiltonian given by Eq. (12.120).

Another interesting model Hamiltonian can arise if the cavity's detuning is nearly zero, and the Rabi frequency is large, then (12.121) can be approximated by

$$H_I = \frac{\hbar g}{2}\left[(S_+ + S_-)a + (S_+ + S_-)a^\dagger\right] = \hbar g S_x (a + a^\dagger). \tag{12.126}$$

In this case, we get JC interaction with both counter-rotating and rotating terms. However, the counter-rotating terms can be as important as the rotating ones as these terms are no longer rapidly varying. The Hamiltonian (12.126) is rather simple and can be solved in terms of the eigenstates of S_x and the properties of the displacement operator for the field. For example, consider the action of $\exp\{\frac{i}{\hbar}H_I\theta\}$ on the state $|+\rangle$ of S_x and the Fock state $|n\rangle$ of the photon. In terms of the displacement operator D, we get

$$\exp\left\{\frac{i}{\hbar}H_I\theta\right\}|+, n\rangle = \exp\left\{\frac{ig\theta}{2}(a + a^\dagger)\right\}|+, n\rangle = D\left(\frac{ig\theta}{2}\right)|+, n\rangle, \tag{12.127}$$

which can be simplified by using Exercise 1.7. Hamiltonians like (12.126) provide us with a cavity QED model for quantum random walks [50–52]. The quantum random walk can also be realized by using optical elements and we discuss it in detail in Section 19.5.

Finally, we provide solution to the anti-JC interaction. Consider an initial state, say $|+, n\rangle$, then the action of H_{AJC} produces the state $|-, n-1\rangle$. The action of H_{AJC} on $|-, n-1\rangle$

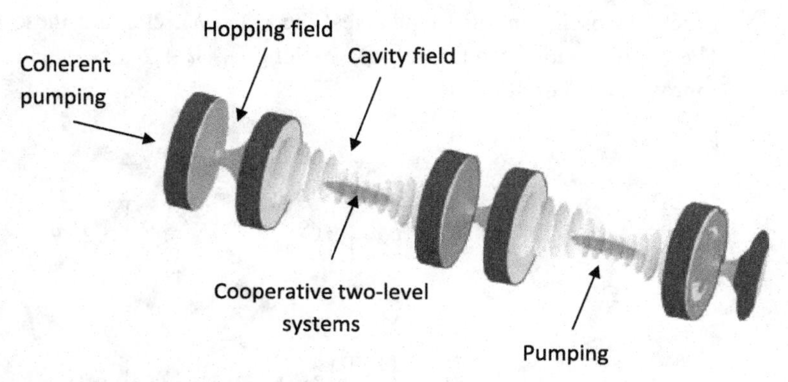

Fig. 12.19 An array of coupled cavities containing atoms [56].

produces the state $|+, n\rangle$. Hence in the space of two states $|+, n\rangle$ and $|-, n-1\rangle$, the H_{AJC} has the structure

$$H_{\mathrm{AJC}} \rightarrow -\frac{\hbar g}{2}\begin{pmatrix} 0 & \sqrt{n} \\ \sqrt{n} & 0 \end{pmatrix}, \tag{12.128}$$

and therefore

$$\exp\left(-\mathrm{i}\frac{H_{\mathrm{AJC}}}{\hbar}t\right) = \cos\frac{gt\sqrt{n}}{2} + \mathrm{i}\sin\frac{gt\sqrt{n}}{2}\begin{pmatrix} 0 & 1 \\ 1 & 0 \end{pmatrix}. \tag{12.129}$$

The evolution operator (12.129) has to act on the column vector $\binom{|-,n-1\rangle}{|+,n\rangle}$. It should be borne in mind that (12.129) is in the interaction picture and the state $|\pm\rangle$ are the eigenstates of S_x. In order to obtain physical results, one finally has to transform to the original picture by using Eq. (12.120).

12.11 QED in coupled cavity arrays – single-photon switch

We have discussed how different types of cavities can be used to produce new regimes of atom–photon interaction. Arrays of cavities containing two-level atoms or multilevel atoms have been proposed to simulate strongly correlated many body systems. Consider the scheme shown in Figure 12.19. Each cavity can contain a single atom or many atoms strongly coupled to photons in the cavity. Different cavities are coupled by a hopping like interaction. Furthermore, the atoms can be pumped externally or the cavity field may be pumped externally. The Hamiltonian of the whole system can be written as

$$H = \sum_l \hbar\omega_c a_l^\dagger a_l + \hbar\omega_0 \sum_l J_z^l + \hbar\sum_{l=1}^{M} \beta_l (a_l a_{l+1}^\dagger + a_l^\dagger a_{l+1}) + \sum_l \hbar g_l (a_l^\dagger J_+^l + a_l J_-^l) + \text{pumping terms}, \tag{12.130}$$

where β_l is the hopping interaction and the sum over l is over all the cavities. For each cavity $J_{\pm}^l, J_z^l$ are the collective operators for the two-level atoms in that cavity. The Hamiltonian dynamics is to be supplemented by the effects of atomic spontaneous emission and cavity leakage. The above Hamiltonian has been extensively investigated mostly numerically [53–55] except in the semiclassical limit which yields the solitonic behavior of the field in the cavity [56]. The numerical simulations even for one or two atoms in each cavity clearly established the existence of Mott like phase transition in the excitations of the coupled array as the relative magnitudes of the Jaynes–Cummings coupling and the hopping between the cavities are varied [57]. Recent reviews of the subject are given in [58, 59].

12.11.1 Single-photon switch

The coupled array of resonators allow the possibility of realizing a single-photon switch [60]. Let us imagine a single two-level atom placed in the pth resonator and imagine a single photon as an input to the entire system. The Hamiltonian (12.130) then becomes

$$H = \hbar\omega_c \sum_l a_l^\dagger a_l + \hbar\omega_0 S_z + \hbar \sum_l \beta(a_l a_{l+1}^\dagger + a_l^\dagger a_{l+1}) + \hbar g_p(a_p^\dagger S_- + a_p S_+). \tag{12.131}$$

We only assume nearest neighbor coupling β. A single photon can now be shared between different cavities and the two-level atom in pth cavity. Thus in the space of a single photon the allowed states are $|\{1_l\}\rangle|g\rangle$ and $|\{0\}\rangle|e\rangle$. Here $\{1_l\}$ means one photon in the lth cavity. We now look for eigensolutions of (12.131) in the form

$$\begin{aligned} H|\Psi\rangle &= \hbar\left(\Omega - \frac{\omega_0}{2}\right)|\Psi\rangle, \\ |\Psi\rangle &= \sum_l c_l|\{1_l\}\rangle|g\rangle + c_e|\{0\}\rangle|e\rangle, \end{aligned} \tag{12.132}$$

which leads to coupled equations

$$\begin{aligned} (\Omega - \omega_0)c_e &= g_p c_p, \\ (\Omega - \omega_c)c_l &= g_p c_e \delta_{pl} + \beta(c_{l-1} + c_{l+1}). \end{aligned} \tag{12.133}$$

On eliminating c_e we obtain

$$(\Omega - \omega_c)c_l - \beta(c_{l-1} + c_{l+1}) = \frac{g_p^2 c_p \delta_{pl}}{\Omega - \omega_0}. \tag{12.134}$$

Note that if $g_p = 0$, then (12.134) admits a solution of the form $c_l \sim \mathrm{e}^{ikl}$,

$$\Omega_k = \omega_c + 2\beta\cos k. \tag{12.135}$$

Thus, in the absence of the atom, the eigenvalues are characterized by the index k. We now find the solution of (12.134) for scattering of a photon with energy given by (12.135). For $p \neq l$, we clearly can write

$$c_l = \begin{cases} \mathrm{e}^{ikl} + r\mathrm{e}^{-ikl}, & l < p, \\ t\mathrm{e}^{ikl}, & l > p, \end{cases} \tag{12.136}$$

where r and t are the reflection and transmission amplitudes. We extend these solutions into the resonator $l = p$ by demanding continuity

$$e^{ikp} + re^{-ikp} = te^{ikp}, \tag{12.137}$$

and then using (12.134)

$$(\Omega_k - \omega_c)c_p - \beta\left(c_{p-1} + c_{p+1}\right) = \frac{g_p^2 c_p}{\Omega_k - \omega_0}, \tag{12.138}$$

to obtain the reflection amplitude

$$r = -g_p^2 e^{2ikp}\{g_p^2 + 2i\beta(\Omega_k - \omega_0)\sin k(p+1)e^{ikp}\}^{-1}. \tag{12.139}$$

This result for reflection amplitude leads to the possibility of a single-photon switch. In the absence of the atom $g_p = 0$, $\beta \neq 0$, $r = 0$. In the presence of the atom, $|r|^2 = 1$ if the energy of the single photon $\hbar\Omega_k$ is equal to the atomic excitation energy $\hbar\omega_0$. Thus a single two-level atom can reflect completely a resonant photon. From (12.139), on setting $p = 0$ for simplicity, we get

$$|r|^2 = \frac{1}{1 + \frac{4(\Omega_k - \omega_0)^2}{g_p^2} \cdot \frac{\beta^2}{g_p^2} \cdot \sin^2 k}. \tag{12.140}$$

It should be borne in mind that the line shape (12.140) is rather complicated [60] as Ω depends on k via $\sin^2 k = 1 - [(\Omega_k - \omega_c)/2\beta]^2$, although $|r| = 1$ for $\Omega_k = \omega_0$. We have thus shown the possibility of a single-photon switch. It should be borne in mind that a full theory must include the input–output coupling of the photon to the field outside the structure. The single-photon switch can also be realized by using other schemes, such as plasmonics [61] and electromagnetically induced transparency [62–64].

Exercises

12.1 Investigate the dependence of $p_{e\to g}(t)$ (defined by (12.17)) when the radiation field is in a single-mode squeezed vacuum state with the same photon number as used in Figure 12.3. Compare your result with that shown in Figure 12.3.

12.2 For the compass state (12.33) examine the regimes in phase space with sub-Planck structure. This can be done by studying the Wigner function for the state (12.33). (For details see [17].)

12.3 Use the eigenvalues of (12.43) to derive the Purcell result for the decay of an atom in a cavity in the limit $\gamma \to 0$ and when the decay constant $\kappa \gg g$. This can be done by showing that the complex root relevant for atomic decay is approximately equal to

$$\frac{\Delta}{2} + \frac{g^2(\Delta - i\kappa)}{\Delta^2 + \kappa^2} \to \frac{g^2}{\kappa}, \text{ when } \Delta \to 0.$$

12.4 Diagonalize the matrix defined by (12.62) to obtain the eigenstates (12.63).

12.5 Prove the Fourier relations (12.84) and (12.85).

12.6 Use the Hamiltonian (12.126) to obtain the generation of cat states of the cavity field. Assume that the atom at $t = 0$ is in the state $|g\rangle$. Find then the joint state of the atom field system. Find the conditional state of the field subject to the conditions, (I) an atom is detected in the state $|g\rangle$ at time t, (II) an atom is detected in the state $|e\rangle$ at time t.

12.7 Consider a field in a coherent state $|\alpha\rangle$ propagating through a Kerr medium with an effective Hamiltonian $\hbar\chi a^{\dagger 2}a^2$. Find the state of the field at time t. Calculate the state at a time $\chi t = \pi/m$, where m is a nonzero integer. Show that the state is a superposition of coherent states lying on a ring. Generalize the calculation to the case of a field in a single mode squeezed coherent state Eq. (2.40). Hint: use relations like (12.84) and (12.85). (For details see [65].)

12.8 An exactly soluble model describing off-resonant Raman transitions between degenerate levels is given by $H = \hbar g a^\dagger a(S_+ + S_-)$ [66]. Calculate the time evolution operator explicitly. Assuming that the atom is in the state $|g\rangle$ and the field is in a coherent state $|\alpha\rangle$, find the probability of finding the atom in the excited state. Plot the result for $|\alpha|^2 = 10$. For a whole class of exactly soluble models of cavity QED as applicable to three-level systems see [67]. A two-mode Hamiltonian

$$\hbar\delta|e\rangle\langle e| + \hbar g(|g\rangle\langle g|a^\dagger a + |e\rangle\langle e|b^\dagger b) + \hbar g(ab^\dagger S_+ + a^\dagger bS_-)$$

for Raman transitions is more useful for quantum gate operations [68].

References

[1] E. T. Jaynes and F. W. Cummings, *Proc. IEEE* **51**, 89 (1963).
[2] G. S. Agarwal, *J. Opt. Soc. Am. B* **2**, 480 (1985).
[3] M. Brune, F. Schmidt-Kaler, A. Maali *et al.*, *Phys. Rev. Lett.* **76**, 1800 (1996).
[4] B. T. H. Varcoe, S. Brattke, M. Weidinger, and H. Walther, *Nature (London)* **403**, 743 (2000).
[5] P. Filipowicz, J. Javanainen, and P. Meystre, *J. Opt. Soc. Am. B* **3**, 906 (1986).
[6] P. Meystre, G. Rempe, and H. Walther, *Opt. Lett.* **13**, 1078 (1988).
[7] H. Walther, *Ann. Phys. (Leipzig)* **14**, 7 (2005).
[8] J. H. Eberly, N. B. Narozhny, and J. J. Sanchez-Mondragon, *Phys. Rev. Lett.* **44**, 1323 (1980).
[9] J. A. Yeazell, M. Mallalieu, J. Parker, and C. R. Stroud, Jr., *Phys. Rev. A* **40**, 5040 (1989).
[10] I. Sh. Averbukh, *Phys. Rev. A* **46**, R2205 (1992).
[11] G. Rempe, H. Walther, and N. Klein, *Phys. Rev. Lett.* **58**, 353 (1987).
[12] J. M. Raimond, M. Brune, and S. Haroche, *Rev. Mod. Phys.* **73**, 565 (2001).
[13] S. Haroche and J.-M. Raimond, *Exploring the Quantum: Atoms, Cavities, and Photons* (New York: Oxford University Press, 2006).
[14] M. Brune, E. Hagley, J. Dreyer *et al.*, *Phys. Rev. Lett.* **77**, 4887 (1996).
[15] M. Brune, P. Nussenzveig, F. Schmidt-Kaler *et al.*, *Phys. Rev. Lett.* **72**, 3339 (1994).

[16] M. M. Salour, *Rev. Mod. Phys.* **50**, 667 (1978).
[17] W. H. Zurek, *Nature (London)* **412**, 712 (2001).
[18] G. S. Agarwal and P. K. Pathak, *Phys. Rev. A* **70**, 053813 (2004).
[19] A. Boca, R. Miller, K. M. Birnbaum, A. D. Boozer, J. McKeever, and H. J. Kimble, *Phys. Rev. Lett.* **93**, 233603 (2004).
[20] K. M. Birnbaum, A. Boca, R. Miller, A. D. Boozer, T. E. Northup, and H. J. Kimble, *Nature (London)* **436**, 87 (2005).
[21] W. Becker, F. Grasbon, R. Kopold, D. B. Milosevic, G. G. Paulus, and H. Walther, *Adv. Atom. Molec. Opt. Phys.* **48**, 35 (2002).
[22] D. F. Walls and G. J. Milburn, *Quantum Optics* (Berlin: Springer-Verlag, 1994).
[23] G. S. Agarwal, *Phys. Rev. Lett.* **53**, 1732 (1984).
[24] J. J. Sanchez-Mondragon, N. B. Narozhny, and J. H. Eberly, *Phys. Rev. Lett.* **51**, 550 (1983).
[25] R. S. Knox, *Theory of Excitons* (New York: Academic Press, 1963).
[26] P. Maunz, T. Puppe, I. Schuster, N. Syassen, P. W. H. Pinkse, and G. Rempe, *Phys. Rev. Lett.* **94**, 033002 (2005).
[27] M. G. Raizen, R. J. Thompson, R. J. Brecha, H. J. Kimble, and H. J. Carmichael, *Phys. Rev. Lett.* **63**, 240 (1989).
[28] A. Wallraff, D. I. Schuster, A. Blais *et al.*, *Nature (London)* **431**, 162 (2004).
[29] J. Johansson, S. Saito, T. Meno *et al.*, *Phys. Rev. Lett.* **96**, 127006 (2006).
[30] J. Koch, T. M. Yu, J. Gambetta *et al.*, *Phys. Rev. A* **76**, 042319 (2007).
[31] J. M. Fink, M. Göppl, M. Baur *et al.*, *Nature (London)* **454**, 315 (2008).
[32] A. Kubanek, A. Ourjoumtsev, I. Schuster *et al.*, *Phys. Rev. Lett.* **101**, 203602 (2008).
[33] R. J. Brecha, P. R. Rice, and M. Xiao, *Phys. Rev. A* **59**, 2392 (1999).
[34] B. R. Mollow, *Phys. Rev.* **188**, 1969 (1969).
[35] G. S. Agarwal, R. K. Bullough, and N. Nayak, *Opt. Commun.* **85**, 202 (1991).
[36] B. R. Mollow, *Phys. Rev. A* **5**, 2217 (1972).
[37] M. Tavis and F. W. Cummings, *Phys. Rev.* **170**, 379 (1968).
[38] R. H. Dicke, *Phys. Rev.* **93**, 99 (1954).
[39] G. V. Varada, M. S. Kumar, and G. S. Agarwal, *Opt. Commun.* **62**, 328 (1987).
[40] F. Bernardot, P. Nussenzveig, M. Brune, J. M. Raimond, and S. Haroche, *Europhys. Lett.* **17**, 33 (1992).
[41] J. M. Fink, R. Bianchetti, M. Baur *et al.*, *Phys. Rev. Lett.* **103**, 083601 (2009).
[42] G. S. Agarwal, R. R. Puri, and R. P. Singh, *Phys. Rev. A* **56**, 2249 (1997).
[43] A. Barenco, D. Deutsch, A. Ekert, and R. Jozsa, *Phys. Rev. Lett.* **74**, 4083 (1995).
[44] D. Loss and D. P. DiVincenzo, *Phys. Rev. A* **57**, 120 (1998).
[45] A. Gábris and G. S. Agarwal, *Phys. Rev. A* **71**, 052316 (2005).
[46] D. M. Greenberger, M. A. Horne, A. Shimony, and A. Zeilinger, *Am. J. Phys.* **58**, 1131 (1990).
[47] D. Leibfried, M. D. Barrett, T. Schaetz *et al.*, *Science* **304**, 1476 (2004).
[48] D. Leibfried, E. Knill, S. Seidelin *et al.*, *Nature (London)* **438**, 639 (2005).
[49] E. Solano, G. S. Agarwal, and H. Walther, *Phys. Rev. Lett.* **90**, 027903 (2003).
[50] P. K. Pathak and G. S. Agarwal, *Phys. Rev. A* **75**, 032351 (2007).
[51] P. L. Knight, E. Roldán, and J. E. Sipe, *Phys. Rev. A* **68**, 020301(R) (2003).

[52] Y. Aharonov, L. Davidovich, and N. Zagury, *Phys. Rev. A* **48**, 1687 (1993).
[53] M. J. Hartmann, F. G. S. L. Brandao, and M. B. Plenio, *Nat. Phys.* **2**, 849 (2006).
[54] A. D. Greentree, C. Tahan, J. H. Cole, and L. C. L. Hollenberg, *Nat. Phys.* **2**, 856 (2006).
[55] D. G. Angelakis, M. F. Santos, and S. Bose, *Phys. Rev. A* **76**, 031805(R) (2007).
[56] M. Paternostro, G. S. Agarwal, and M. S. Kim, *New J. Phys.* **11**, 013059 (2009).
[57] D. Rossini and R. Fazio, *Phys. Rev. Lett.* **99**, 186401 (2007).
[58] A. Tomadin and R. Fazio, *J. Opt. Soc. Am. B* **27**, A130 (2010).
[59] M. J. Hartmann, F. G. S. L. Brandao, and M. P. Plenio, *Laser Photonics Rev.* **2**, 527 (2008).
[60] L. Zhou, Z. R. Gong, Y. X. Liu, C. P. Sun, and F. Nori, *Phys. Rev. Lett.* **101**, 100501 (2008).
[61] D. E. Chang, A. S. Sørensen, E. A. Demler, and M. D. Lukin, *Nature Physics* **3**, 807 (2007).
[62] S. E. Harris and Y. Yamamoto, *Phys. Rev. Lett.* **81**, 3611 (1998).
[63] I. C. Hoi, C. M. Wilson, G. Johansson, T. Palomaki, B. Peropadre, and P. Delsing, *Phys. Rev. Lett.* **107**, 073601 (2011).
[64] G. S. Agarwal and S. Huang, *Phys. Rev. A* **85**, 021801 (2012).
[65] K. Tara, G. S. Agarwal, and S. Chaturvedi, *Phys. Rev. A* **47**, 5024 (1993).
[66] S. J. D. Phoenix and P. L. Knight, *J. Opt. Soc. Am. B* **7**, 116 (1990).
[67] H. I. Yoo and J. H. Eberly, *Phys. Rep.* **118**, 239 (1985).
[68] A. Biswas and G. S. Agarwal, *Phys. Rev. A* **69**, 062306 (2004).

13 Absorption, emission, and scattering from two-level atoms

It is well known that electromagnetic fields are important probes of the properties of matter. We can learn about atomic molecular energy levels by studying the absorption, emission, and scattering of electromagnetic waves. For example, the rate at which a system absorbs energy and its dependence on the frequency of the electromagnetic field gives information on the allowed transitions. From such studies one can determine the energy levels and their lifetimes. Similarly, scattering processes provide a wealth of information. The traditional probing of matter is restricted to weak fields; however, in Chapter 11 we saw how strong fields dress the energy levels of a system. Strong fields also modify the transition rates. In order to study the characteristics of such modifications we need to probe a coherently driven system using a probe field. In this chapter we study the absorption, emission, and scattering processes in strongly driven systems. A novel characteristic of radiation from strongly driven systems is its nonclassical nature.

13.1 Effects of relaxation: optical Bloch equations

So far we have considered only interactions with external electromagnetic fields. In reality, one has to account for various sources of decay of the atomic population and coherences. For example, an atom can decay radiatively by emitting a photon. The resulting collisions change the populations and coherences. In Chapter 9 we discussed in detail how various relaxation processes can be included from first principles in the master equation framework. In Chapter 11, we obtained an effective Hamiltonian for a two-level atom in a rotating frame. On combining the master equation (9.56) and the effective Hamiltonian (11.13), we can obtain the basic master equation to describe the dynamics of a coherently driven two-level atom under the influence of various relaxation processes. The density matrix for the atom evolves according to the equation

$$\begin{aligned}\frac{\partial\rho}{\partial t} &= -\mathrm{i}[\Delta S_z - (gS_+ + g^*S_-), \rho] - \Gamma(S_zS_z\rho - 2S_z\rho S_z + \rho S_zS_z)\\ &\quad - \gamma(1+n_0)(S_+S_-\rho - 2S_-\rho S_+ + \rho S_+S_-)\\ &\quad - \gamma n_0(S_-S_+\rho - 2S_+\rho S_- + \rho S_-S_+),\end{aligned} \tag{13.1}$$

$$\Delta = \omega_0 - \omega_l, \qquad g = \frac{\vec{p}_{eg}\cdot\vec{\varepsilon}_l \mathrm{e}^{\mathrm{i}\vec{k}_l\cdot\vec{R}}}{\hbar}.$$

To recap: Δ is the detuning, $2g$ is the Rabi frequency, 2γ is the Einstein A coefficient, Γ gives the rate of elastic collisions, and n_0 is the number of thermal photons at the frequency ω_0. For optical transitions and the laboratory temperatures, n_0 can be set equal to zero. Using (13.1) we can obtain the equations for the mean values of the dipole moment operator $\langle S_\pm \rangle$ and the inversion $\langle S_z \rangle$, which we write in compact form as

$$\frac{\mathrm{d}\psi}{\mathrm{d}t} = M\psi + f, \tag{13.2}$$

where the matrices M, ψ, and f are given by

$$\psi = \begin{pmatrix} \langle S_+ \rangle \\ \langle S_- \rangle \\ \langle S_z \rangle \end{pmatrix}, \qquad f = \begin{pmatrix} 0 \\ 0 \\ \eta/T_1 \end{pmatrix},$$
$$M = \begin{pmatrix} \mathrm{i}\Delta - \frac{1}{T_2} & 0 & 2\mathrm{i}g^* \\ 0 & -\mathrm{i}\Delta - \frac{1}{T_2} & -2\mathrm{i}g \\ \mathrm{i}g & -\mathrm{i}g^* & -\frac{1}{T_1} \end{pmatrix}. \tag{13.3}$$

Here η (Eq. (9.48)) is the equilibrium value of $\langle S_z \rangle$, i.e. the value of the population inversion in the absence of an external field. The variables T_1 and T_2 are the longitudinal and transverse relaxation times defined by (9.49) and (9.57). For radiative relaxation

$$\frac{1}{T_1} = 2\gamma = \frac{2}{T_2}, \qquad \eta = -\frac{1}{2}, \tag{13.4}$$

where 2γ is the Einstein A coefficient. Elastic collisions are included if we take

$$\frac{1}{T_1} = 2\gamma, \qquad \frac{1}{T_2} = \gamma + \Gamma, \qquad \eta = -\frac{1}{2}, \tag{13.5}$$

where Γ is the collisional line width. The equations (13.3) are known as the optical Bloch equations and describe a variety of optical resonance phenomena [1]. Let us make an estimate of the Rabi frequency $2|g|$ for the $3s \leftrightarrow 3p$ transition in sodium: if we use $|p| = 5.5 \times 10^{-18}$ esu, $I = \frac{c}{2\pi}|E|^2 = 1\,\mu\mathrm{W/cm^2}$, then the Rabi frequency $2|g|$ is about $5 \times 10^5\,\mathrm{s^{-1}}$.

13.1.1 Steady-state susceptibilities and saturation behavior

We first discuss the characteristics of the steady-state solution of (13.2). This is obtained in the limit $t \to \infty$, when all time derivatives $\mathrm{d}\psi/\mathrm{d}t$ can be set zero. This immediately yields the solution

$$\langle S_z \rangle = \frac{(1 + (\Delta T_2)^2)\eta}{1 + (\Delta T_2)^2 + 4|g|^2 T_1 T_2}, \tag{13.6}$$

$$\langle S_- \rangle = \frac{-2\mathrm{i}gT_2(-\mathrm{i}\Delta T_2 + 1)\eta}{1 + (\Delta T_2)^2 + 4|g|^2 T_1 T_2}. \tag{13.7}$$

For high fields $4|g|^2 T_1 T_2/(\Delta T_2)^2 \gg 1$, $\langle S_z \rangle \to 0$ and hence high fields equalize the population in the ground and excited states. In the special case where $\eta = -1/2$, the

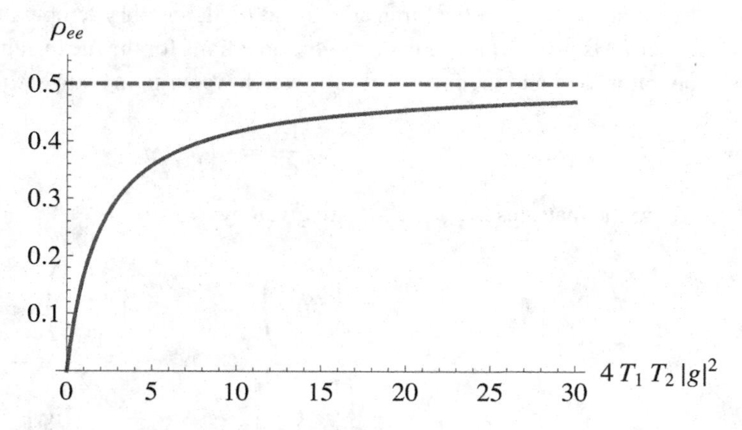

Fig. 13.1 Saturation behavior of the two-level atom for a fixed detuning $\Delta T_2 = 1$.

excited state population in steady state is

$$\rho_{ee} = \frac{1}{2} + \langle S_z \rangle = \frac{2|g|^2 T_1 T_2}{1 + (\Delta T_2)^2 + 4|g|^2 T_1 T_2} \tag{13.8}$$

$$\rightarrow \frac{2|g|^2 T_1 T_2}{1 + (\Delta T_2)^2} \quad \text{in the limit of small fields} \tag{13.9}$$

$$\rightarrow \frac{1}{2} \quad \text{in the limit of large fields.} \tag{13.10}$$

For small fields an increase of the field leads to an increase in the population of the excited state. However, for large fields nonlinear effects set in and the excited state population does not indefinitely increase but saturates as shown in Figure 13.1. The dimensionless parameter $4|g|^2 T_1 T_2$ is the on-resonance saturation parameter S_0 for the two-level atom

$$S_0 = 4|g|^2 T_1 T_2. \tag{13.11}$$

Again for the sodium $3s \rightarrow 3p$ transition, if $T_1 = 16$ ns, $T_2 = 32$ ns, then the saturation intensity I, corresponding to $S_0 = 1$, is $8\,\text{mW/cm}^2$.

The excited state population as a function of the detuning exhibits Lorentzian behavior with a peak height $S_0/[2(1 + S_0)]$ and a full width at half maximum $2\sqrt{1 + S_0}$. Both peak height and width depend on the intensity of the field. The dependence of the width on the power of the field is called power broadening. The steady state fluorescence is a direct measure of the excited state population. We notice an important feature now – the fields can produce a finite dipole moment in the steady state $\langle S_- \rangle \neq 0$ (Eq. (13.7)). This is in contrast to the interaction with a thermal reservoir, which induces no dipole moment. Note further that the dipole moment depends on the strength of the field, which could be arbitrary. The dipole moment results in nonzero polarization which can be used to obtain nonlinear optical susceptibilities of two-level systems.

We write the polarization $\vec{p}$ as $\vec{p}^{\,*}_{eg}\langle S_-\rangle + c.c.$. From (13.7) and from the definition (11.6) of g, we can write the induced polarization as

$$p_\alpha = \sum_\beta \chi_{\alpha\beta}(\omega_l, E) E_\beta e^{i\vec{k}_l\cdot\vec{R} - i\omega_l t} + c.c., \tag{13.12}$$

$$\chi_{\alpha\beta}(\omega_l, \varepsilon_l) = \frac{(-2\eta T_2)(\vec{p}^{\,*}_{eg})_\alpha (\vec{p}_{eg})_\beta (\Delta T_2 + i)}{[1 + (\Delta T_2)^2 + 4|g|^2 T_1 T_2]\hbar} n, \tag{13.13}$$

where n is the density of atoms. Note that this χ depends on all powers of E because of the $|g|^2 (\propto |E|^2)$ term in the denominator. The usual linear susceptibility is obtained by setting $g = 0$ in the denominator,

$$\chi_{\alpha\beta}(\omega_l) = \frac{-2\eta T_2 (\vec{p}^{\,*}_{eg})_\alpha (\vec{p}_{eg})_\beta \, n}{\hbar(\Delta T_2 - i)}. \tag{13.14}$$

If $\vec{p}$ is randomly oriented, then we get an isotropic susceptibility

$$\chi(\omega_l) = \frac{n\chi_0(\Delta T_2 + i)}{(\Delta T_2)^2 + 1}, \qquad \chi_0 = -\frac{2\eta T_2}{\hbar}|p|^2,$$

$$\mathrm{Re}\ \chi(\omega_l) = \frac{n\chi_0 T_2(\omega_0 - \omega_l)}{1 + (T_2(\omega_0 - \omega_l))^2}, \tag{13.15}$$

$$\mathrm{Im}\ \chi(\omega_l) = \frac{n\chi_0}{1 + (T_2(\omega_0 - \omega_l))^2}. \tag{13.16}$$

Let us make an estimate of χ_0: on taking $n = 10^{10}$ atoms cm^{-3}, $\eta = -1/2$, $T_2 = 32$ ns, $|p| \approx 5.5 \times 10^{-18}$ esu ($3s \leftrightarrow 3p$ transition in sodium), we obtain $n\chi_0 = 9 \times 10^{-6}$.

For $\eta < 0$, Im $\chi(\omega_l) > 0$. Such a medium is an absorber. For $\eta > 0$, we have the case of an inverted two-level medium and the medium is an amplifier. The parameter χ_0 is the absorption at resonance. The behavior of the real and imaginary parts of χ are shown in Figure 13.2. These display the dispersive and absorptive behavior as a function of the frequency ω_l of the field. The region of absorption is the region of anomalous dispersion, i.e. Re $\chi(\omega_l)$ decreases as ω_l increases. Far away Re $\chi(\omega_l)$ exhibits normal dispersion, i.e. Re $\chi(\omega_l)$ increases with ω_l. Note that the refractive index $n(\omega_l) = \sqrt{1 + 4\pi\chi(\omega_l)}$ of a medium such as a gas of two-level atoms is approximately equal to $1 + 2\pi\chi(\omega_l)$, since $|\chi| \ll 1$.

Note that if a weak plane wave $e^{ik_l z}$ propagates through a medium of two-level atoms, then the field at the output face $z = l$ will be $e^{ik_l l n(\omega_l)}$. Because the refractive index $n(\omega_l)$ is complex, the field suffers attenuation, and the intensity gets attenuated to $\exp\{-2k_l l\ \mathrm{Im}\ n(\omega_l)\}$. Furthermore, there is a phase shift given by $k_l l\ \mathrm{Re}\ n(\omega_l)$. A measurement of the phase shifts produced by a single trapped atom was reported by Aljunid *et al.* [2]. Their experiment was done using a Mach–Zehnder interferometer, one arm of which contained a single trapped atom. The phase shift showed the dispersive character (see Exercise 13.1) in conformity with the behavior of Re $\chi(\omega_l)$, and the maximum value of the phase shift was about 1°.

In the limit of large ΔT_2, (13.13) can be used to define Kerr media for which the susceptibility has a term linearly proportional to the intensity of the field

$$\chi_{\alpha\beta}(\omega_l, \varepsilon_l) \cong \frac{-2\eta T_2 (\vec{p}^{\,*}_{eg})_\alpha (\vec{p}_{eg})_\beta n}{\hbar \Delta T_2} \left[1 - \frac{4|g|^2 T_1 T_2}{(\Delta T_2)^2}\right]. \tag{13.17}$$

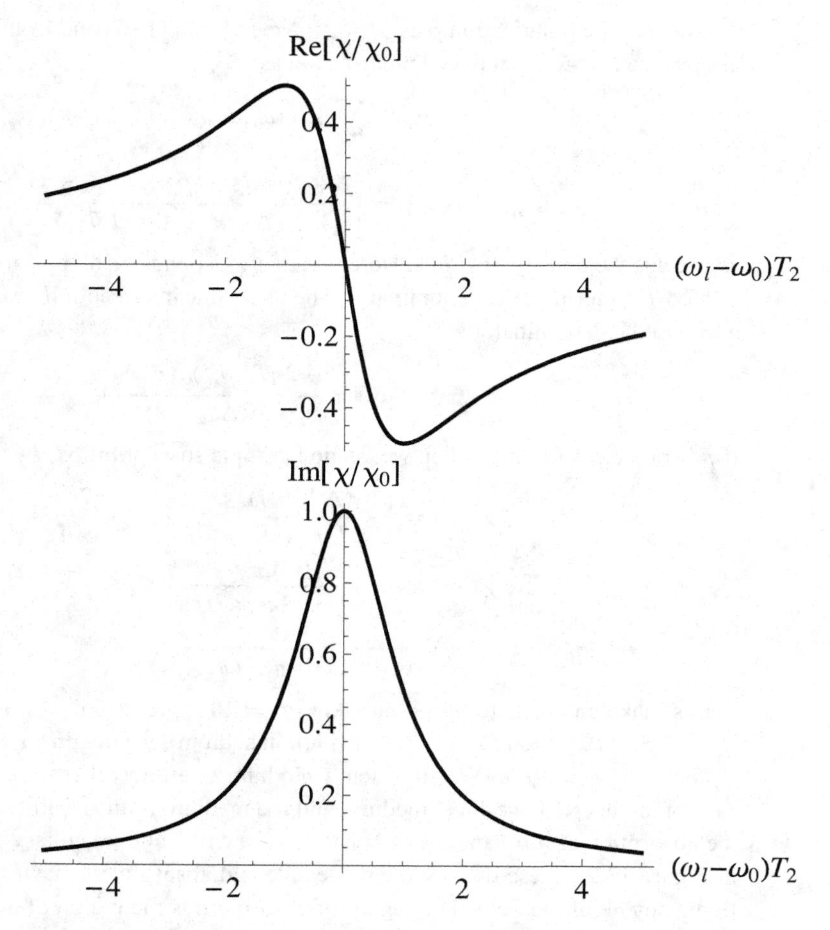

Fig. 13.2 The dispersion and absorption behavior of a two-level system.

The second term in (13.17) can be used to define intensity dependent refractive index usually denoted by n_2 [3].

13.1.2 Rate equations

In certain situations the optical Bloch equations lead to a rate equation for the inversion. This is the case if the detuning is large or if T_2 is small so that the polarization follows the population adiabatically. In this limit, we can set $\frac{\mathrm{d}}{\mathrm{d}t}\langle S_-(t)\rangle \approx 0$, which leads to the following relation between the atomic polarization and the inversion

$$\langle S_-(t)\rangle \approx \frac{-2\mathrm{i}g\langle S_z(t)\rangle T_2}{(1+\mathrm{i}\Delta T_2)}. \tag{13.18}$$

Now by substituting (13.18) in (13.2) we get the rate equation for the inversion

$$\frac{\mathrm{d}}{\mathrm{d}t}\langle S_z(t)\rangle = -\frac{1}{T_1}\langle S_z\rangle\left[1+\frac{4|g|^2T_1T_2}{1+\Delta^2T_2^2}\right]+\frac{\eta}{T_1}. \tag{13.19}$$

The rate equation for $\langle S_z(t)\rangle$ can be easily solved for the time dependence of inversion. Even cases of time-dependent field envelopes, i.e. when g depends on t, can be handled. The second term in the square bracket in (13.19) is the stimulated contribution to the decay of inversion.

13.1.3 Time-dependent solutions of the Bloch equations

The transient behavior of the solutions of the Bloch equations can be obtained by using Laplace transform techniques. The eigenvalues of the matrix M determine the dynamical behavior. The following polynomial is the key to the dynamical behavior

$$D(p) = \det(p - M) = \left(p + \frac{1}{T_1}\right)\left[\Delta^2 + \left(p + \frac{1}{T_2}\right)^2\right] + 4|g|^2\left(p + \frac{1}{T_2}\right), \tag{13.20}$$

where p is the Laplace variable. In the limit of $T_1, T_2 \to \infty$, the roots are

$$p = 0, \qquad \pm \mathrm{i}\sqrt{4|g|^2 + \Delta^2}. \tag{13.21}$$

These roots give Rabi oscillations. For finite T_1 and T_2, the zeroes of the cubic polynomial (13.20) can be obtained analytically in the limits

$$\text{(a) } T_1 = T_2, \qquad p = -\frac{1}{T_2}, \ -\frac{1}{T_2} \pm \mathrm{i}\sqrt{4|g|^2 + \Delta^2}, \tag{13.22}$$

$$\text{(b) } \Delta = 0, \qquad p = -\frac{1}{T_2}, \ -\frac{1}{2}\left(\frac{1}{T_1} + \frac{1}{T_2}\right) \pm \frac{1}{2}\sqrt{\left(\frac{1}{T_1} - \frac{1}{T_2}\right)^2 - 16|g|^2}. \tag{13.23}$$

In other cases the roots can be obtained numerically. In the case (b), we have either all three real roots or one real root and two complex roots. In the limit $4|g| > |\frac{1}{T_1} - \frac{1}{T_2}|$, the roots are approximately $-\frac{1}{T_2}$, $\pm 2\mathrm{i}|g| - \frac{1}{2}(\frac{1}{T_1} + \frac{1}{T_2})$. The roots corresponding to Rabi oscillations have a decay rate determined by both T_1 and T_2. We will see in Section 13.4.2 that the roots of (13.20) determine the behavior of the spectrum of resonance fluorescence.

13.2 Absorption and amplification of radiation by a strongly pumped two-level system

We next consider an important topic: how to probe the dynamical properties of a system dressed by a strong pump field. The strongly pumped system has a dynamical behavior determined by the dressed states (11.31) and the solutions of the polynomial (13.20). To probe a strongly pumped system, we can imagine a probe field of frequency ω_p acting on the two-level system shown in Figure 13.3 [4]. We take the probe field to be weak and thus consider the linear response of the atomic system while treating the pump field to all orders.

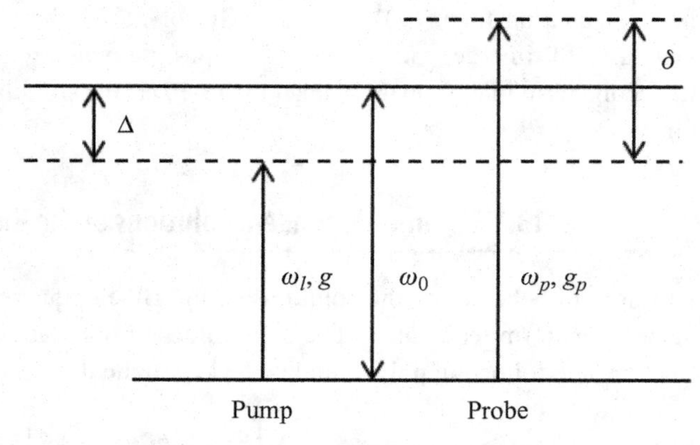

Fig. 13.3 A probe field acting on the two-level system driven strongly by another coherent field.

In the rotating frame the interaction with the probe field ε_p can be written as

$$H_p = -(g_p S_+ \mathrm{e}^{-\mathrm{i}\delta t} + H.c.),$$
$$g_p = \frac{\vec{p}_{eg} \cdot \vec{\varepsilon}_p}{\hbar} \mathrm{e}^{\mathrm{i}\vec{k}_p \cdot \vec{R}}, \qquad \delta = \omega_p - \omega_l, \tag{13.24}$$

where $\vec{k}_p$ is the direction of the propagation of the probe field. The optical Bloch equations (13.2) are modified to

$$\frac{\mathrm{d}\psi}{\mathrm{d}t} = M\psi + f + (M_+ \mathrm{e}^{-\mathrm{i}\delta t} + M_- \mathrm{e}^{\mathrm{i}\delta t})\psi,$$
$$(M_+)_{23} = -2\mathrm{i}g_p, \qquad (M_+)_{31} = \mathrm{i}g_p,$$
$$(M_-)_{32} = -\mathrm{i}g_p^*, \qquad (M_-)_{13} = 2\mathrm{i}g_p^*, \tag{13.25}$$

where only nonvanishing elements of the matrices $M_\pm$ are given. It should be borne in mind that the matrix M depends on the pump field. We now have equations with periodic time dependence. These can be solved by a variety of techniques such as the Floquet method or by continued fractions [5]. For the present, we need to find a solution to the lowest order in g_p. Then the solution of (13.25) in the long time limit can be expressed as

$$\psi = \psi^{(0)} + \psi^{(+)} \mathrm{e}^{-\mathrm{i}\delta t} + \psi^{(-)} \mathrm{e}^{\mathrm{i}\delta t},$$
$$\psi^{\pm} = [\mp \mathrm{i}\delta - M]^{-1} M_\pm \psi^{(0)}, \qquad \psi^{(0)} = -M^{-1} f. \tag{13.26}$$

It should be borne in mind that the solution (13.26) is in the rotating frame. We next examine the induced dipole moment in the original frame, which will be

$$\vec{p}(t) = \vec{p}_{eg}^{\,*} \langle S_-(t) \rangle + c.c. = \vec{p}_{eg}^{\,*} \mathrm{e}^{-\mathrm{i}\omega_l t} \psi_2(t) + c.c.$$
$$= \vec{p}_{eg}^{\,*} \psi_2^{(0)} \mathrm{e}^{-\mathrm{i}\omega_l t} + \vec{p}_{eg}^{\,*} \psi_2^{(+)} \mathrm{e}^{-\mathrm{i}\omega_p t} + \vec{p}_{eg}^{\,*} \psi_2^{(-)} \mathrm{e}^{-\mathrm{i}(2\omega_l - \omega_p)t} + c.c.. \tag{13.27}$$

The induced polarization at the probe frequency ω_p is given by $\vec{p}_{eg}^{\,*} \psi_2^{(+)}$. The term at the frequency $(2\omega_l - \omega_p)$ corresponds to the four-wave mixing in a system of two-level atoms. Thus, as shown in Figure 13.4, the system of two-level atoms can generate newer frequencies by the process of nonlinear mixing. Note also that the solution of (13.25) to higher orders

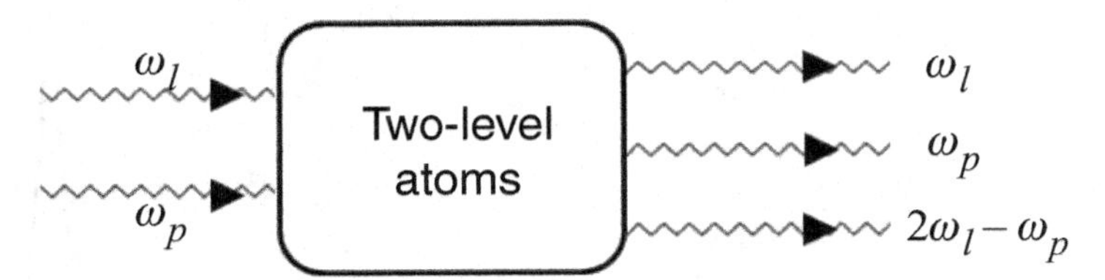

Fig. 13.4 Four-wave mixing in a system of two-level atoms.

of the probe field yields many new frequencies such as $\omega_l \pm q\delta = \omega_l \pm q(\omega_p - \omega_l)$, where q is an integer [5].

We concentrate on the response at the probe frequency. We write the induced polarization at ω_p as

$$\begin{aligned}\vec{p}(t) &= \vec{p}_{eg}^{\,*}\psi_2^{(+)}\mathrm{e}^{-\mathrm{i}\omega_p t} + c.c.\\ &= \chi(\omega_p)\mathrm{e}^{-\mathrm{i}\omega_p t}\vec{\varepsilon}_p\mathrm{e}^{-\mathrm{i}\vec{k}_p\cdot\vec{r}} + c.c.,\end{aligned} \tag{13.28}$$

where $\chi(\omega_p)$ is the single atom susceptibility which depends on all powers of the pump field. In writing (13.28) we assumed that $\vec{p}$ is randomly distributed so that χ is isotropic. This can be evaluated using (13.26) with the result

$$\begin{aligned}\chi = &-\chi_0\frac{1+\Delta^2T_2^2}{1+\Delta^2T_2^2+4|g|^2T_1T_2}\times\frac{1}{\delta T_2-\Delta T_2+\mathrm{i}}\\ &\times\left[1+\frac{2|g|^2T_2^2(\Delta T_2+\mathrm{i})^{-1}(\delta T_2+2\mathrm{i})(\delta T_2+\Delta T_2+\mathrm{i})}{(\delta T_2+\mathrm{i}T_2/T_1)(\delta T_2-\Delta T_2+\mathrm{i})(\delta T_2+\Delta T_2+\mathrm{i})-4|g|^2T_2^2(\delta T_2+\mathrm{i})}\right].\end{aligned} \tag{13.29}$$

The strong pump modifies the susceptibility in two ways: (i) it changes the overall numerical factor which is the new inversion (13.6); (ii) it leads to extra resonances in the susceptibility (second term in the square bracket). Let us now relate the susceptibility (13.29) to the rate at which the energy is absorbed from the probe field. According to classical electrodynamics the rate at which the fields do work is given by

$$\begin{aligned}\frac{\mathrm{d}\vec{p}}{\mathrm{d}t}\cdot\vec{E}_p &= (-\mathrm{i}\omega_p\vec{p}\,\mathrm{e}^{-\mathrm{i}\omega_p t}+c.c.)\cdot(\vec{\varepsilon}_p\mathrm{e}^{-\mathrm{i}\omega_p t}+c.c.)\\ &\approx 2\omega_p\mathrm{Im}(\chi(\omega_p)|\varepsilon_p|^2) = \hbar\omega_p S_A(\omega_p),\end{aligned} \tag{13.30}$$

where we have retained only the dc terms and used (13.28). The quantity $S_A(\omega_p)$ is the rate of absorption and has the dimensions of frequency. It is related to (13.29) via

$$S_A(\omega_p) = \frac{2|\varepsilon_p|^2}{\hbar}\mathrm{Im}(\chi(\omega_p)). \tag{13.31}$$

The absorption spectrum (13.31) was first derived by Mollow [4] and has been studied experimentally [6–8].

We first note that in the limit of no pump field $|g| \to 0$, (13.31) reduces to

$$S_A(\omega_p) = \frac{-4T_2\eta}{1+T_2^2(\omega_0-\omega_p)^2}\frac{|\vec{p}|^2|\varepsilon_p|^2}{\hbar^2}. \tag{13.32}$$

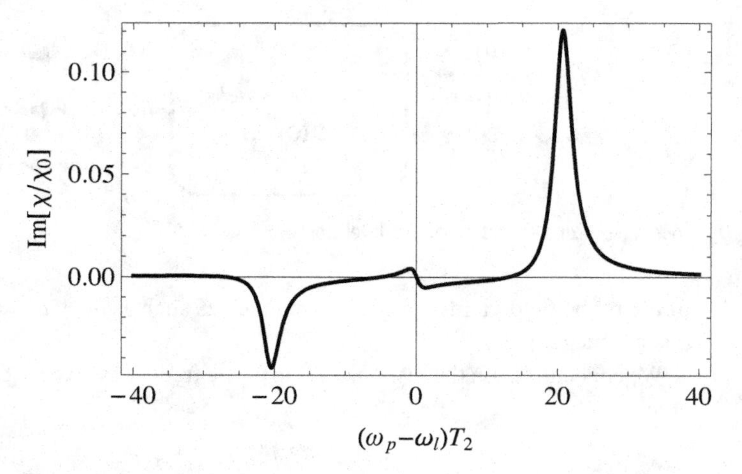

Fig. 13.5 The probe absorption spectrum as modified by the intense pump-laser field for $T_2/T_1 = 2, \Delta T_2 = -5$, and $g = 10$.

This expression (13.32) is equivalent to the rate of absorption $\Gamma_{if}^{(\text{abs})}$ derived in Chapter 7. For atoms in ground state $\eta = -1/2$ and $S_A(\omega_p) > 0$, the peak absorption for $|g| \to 0$ is $-4|g_p|^2 T_2 \eta$, where g_p is defined by (13.24). We show the behavior of (13.31) normalized to $-4|g_p|^2 T_2 \eta$ in Figure 13.5 for a strongly driven two-level system, which is subject to radiative relaxation only, $T_2/T_1 = 2$.

We first note that the absorption $S_A(\omega_p)$ exhibits new features at frequencies different from the atomic transition frequency. Furthermore, it has the unusual property that it becomes negative, which implies that the probe field gets amplified in the region given by

$$\omega_p = \omega_l - \sqrt{4|g|^2 + \Delta^2}, \tag{13.33}$$

whereas in the region

$$\omega_p = \omega_l + \sqrt{4|g|^2 + \Delta^2}, \tag{13.34}$$

the probe is absorbed. The existence of the gain region can be understood in terms of the dressed states (Section 11.1.1) of a two-level system. Let us level the states of the pump and probe as $|n, p\rangle$, which consist of n pump photons and p probe photons. Consider the four-wave mixing process shown in Figure 13.6. Clearly the excitation of the atom would be a resonant process if

$$2\omega_l - \omega_p = \omega_0, \qquad \text{i.e. } \delta = -\Delta. \tag{13.35}$$

Such a three-photon process leads to the gain of the probe. Two quanta of energy taken out from the pump field are converted into a probe photon and a photon of frequency $2\omega_l - \omega_p$. For intense pump fields, the bare states of the atom are to be replaced by the dressed states whence the condition (13.35) is replaced by the condition (13.33). This is also seen from the denominator in (13.29). A direct absorption of the probe photon $|g, n, p\rangle \to |e, n, p-1\rangle$ yields a resonance at $\omega_p = \omega_0$ or $\delta = \Delta$, which in the presence of a strong pump is modified to (13.34). In addition, a driven two-level atom produces a dispersive structure at $\delta = 0$, i.e. at $\omega_p = \omega_l$. This structure is associated with the population oscillations. To

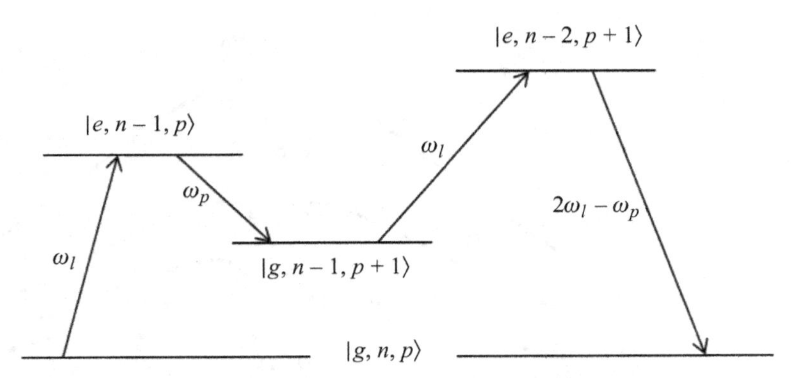

Fig. 13.6 Four-wave mixing process with n pump photons and p probe photons leading to the generation of a photon of frequency $2\omega_l - \omega_p$.

understand all this more clearly, we give a discussion in terms of the semiclassical dressed states introduced in Section 11.1.1. We first note that if initially the atom is in the state $|g\rangle$, then both the dressed states $|\psi_\pm\rangle$ are populated. The interaction with the probe field is given by (13.24). It is easily seen that the probe interaction has nonzero matrix elements like

$$\langle\psi_\pm|H_p(t)|\psi_\pm\rangle \neq 0. \tag{13.36}$$

These nonzero matrix elements would lead to resonances at $\delta = 0$, since in the rotating frame a probe photon (effective frequency 0) is exchanged without a change in the dressed state. The nonzero matrix elements like

$$\langle\psi_\pm|H_p(t)|\psi_\mp\rangle \neq 0, \tag{13.37}$$

would lead to resonances at

$$\pm\,\delta = \Omega = \sqrt{4|g|^2 + \Delta^2}. \tag{13.38}$$

Here we have an exchange of probe photon effective frequency δ with a change in the dressed state. It should be borne in mind that the two dressed states are separated by the Rabi frequency Ω.

13.2.1 Holes in the absorption spectrum

We next consider strongly homogeneously broadened systems where $T_2 \ll T_1$. The Ruby crystal is a well-known example of such a system. We will show that in such systems the absorption lines exhibit a well-defined hole at the frequency $\delta = 0$. Such holes are caused by coherent population oscillations and are useful, for example, in the production of slow light (Chapter 18).

To demonstrate the existence of holes, we take the value of T_2^{-1} to be large compared to the pump field Rabi frequency g and to the detuning of the probe field δ. In this limit, the susceptibility (13.29) of the driven two-level system reduces to

$$\chi(\delta) = \mathrm{i}\frac{|p|^2 T_2}{\hbar}\left[\frac{1}{1 + 4|g|^2 T_1 T_2} - 4|g|^2\frac{T_2}{T_1}\frac{1 + \mathrm{i}\delta/\beta}{\delta^2 + \beta^2}\right], \tag{13.39}$$

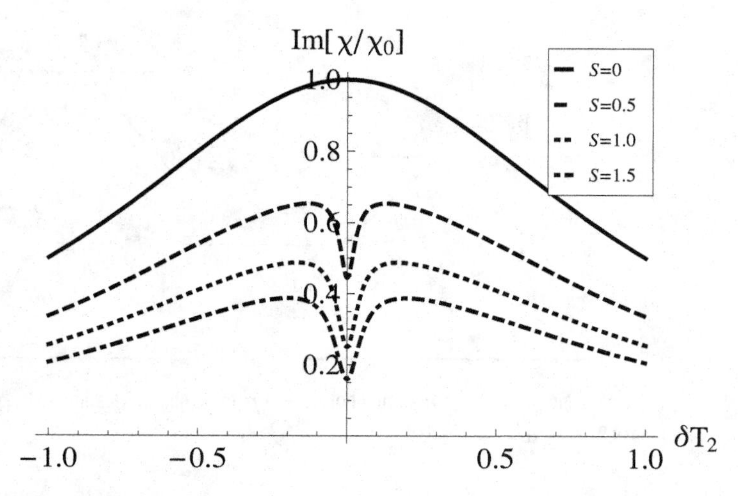

Fig. 13.7 The frequency variation of the imaginary part of the susceptibility for different strengths of the pump field (e.g. (13.11)).

where the pump field has been set in resonance with the atomic system, i.e. $\omega_l = \omega_0$. Figure 13.7 shows that a spectral hole is created in the probe response due to coherent population oscillations [9, 10]. The half width at half maximum of the spectral hole is given by

$$\beta = \frac{1 + 4g^2 T_1 T_2}{T_1}. \tag{13.40}$$

It is clear from the above expression that β has two contributions. The first contribution comes from the longitudinal relaxation and originates from the oscillations induced by the pump and probe fields acting simultaneously. For a weak pump field, the dip has a half width of $1/(2\pi T_1)$ Hz. The second contribution arises from the intensity of the pump field, which causes the power broadening of the system. The hole diminishes as the pump field intensity goes to zero. The width of the hole increases with increasing intensity of the pump field.

In the foregoing we have concentrated on the absorption spectrum. The real part of (13.29) would be important in connection with slow light (Chapter 18).

13.3 Resonance fluorescence from a coherently driven two-level atom

In the previous section, we have seen how the absorption of radiation from a weak probe is affected by a coherent pump driving the two-level atom. We now study the emission of radiation by a coherently driven atom and derive the Mollow spectrum [11]. We will see that the dressed states of the two-level atom determine the characteristics of the emitted radiation. Let the atom with dipole matrix element $\vec{p}$ be located at the point $\vec{R} = 0$ and

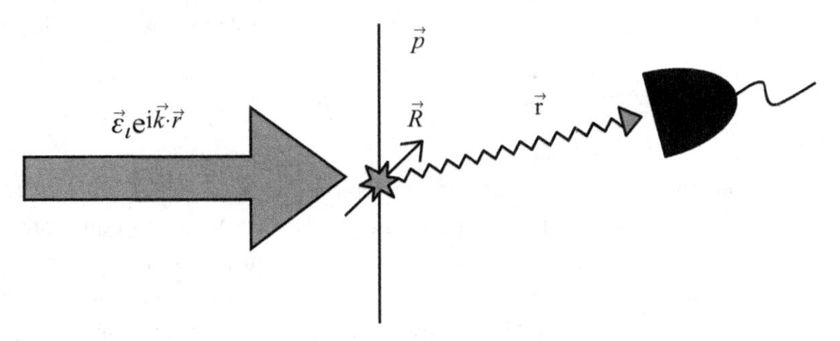

Fig. 13.8 Schematic illustration of the radiation from a coherently driven atom.

the spectrum be measured at the point $\vec{r}$ which is in the far zone $\frac{\omega}{c}r \gg 1$, as shown in Figure 13.8. The coherent pump is denoted by $\vec{\varepsilon}_l e^{i\vec{k}_l\cdot\vec{r}}$. Let $\vec{E}^{(+)}(\vec{r},t)$ be the positive frequency part of the field at the detector. The steady-state spectrum $S(\vec{r},\omega)$ is then defined by Eq. (8.50) with $a \to E^{(+)}$

$$S(\vec{r},\omega) = \frac{1}{2\pi}\int_{-\infty}^{+\infty} d\tau e^{i\omega\tau}\langle \vec{E}^{(-)}(\vec{r},t)\cdot\vec{E}^{(+)}(\vec{r},t+\tau)\rangle, \tag{13.41}$$

so that

$$\langle \vec{E}^{(-)}(\vec{r},t)\cdot\vec{E}^{(+)}(\vec{r},t)\rangle = \int_{-\infty}^{+\infty} S(\vec{r},\omega)d\omega. \tag{13.42}$$

The two time correlation function in (13.41) is the steady-state correlation, i.e. it stands for $\lim_{t\to\infty}\langle \vec{E}^{(-)}(\vec{r},t)\cdot\vec{E}^{(+)}(\vec{r},t+\tau)\rangle$. In order to obtain the spectrum we need to relate the two time correlation function for the field to the dipole moment operators of the atom.

13.3.1 Relation of the field operator $\vec{E}^{(+)}(\vec{r},t)$ to the dipole moment operator

In classical electrodynamics, the radiated field is related to the oscillating dipole moment. If the dipole moment is oscillating at the frequency ω_l, i.e.

$$\vec{p}(t) = \vec{p}e^{-i\omega_l t} + \vec{p}^{\,*}e^{i\omega_l t}, \tag{13.43}$$

then the field in the far zone is [12]

$$\vec{E}(\vec{r},t) = -\frac{\omega_l^2}{c^2}\vec{n}\times(\vec{n}\times\vec{p})\frac{e^{ik_l r - i\omega_l t}}{r}e^{-ik_l\vec{n}\cdot\vec{R}} + c.c.,$$
$$k_l = \frac{\omega_l}{c}, \qquad \vec{n} = \frac{\vec{r}}{r}. \tag{13.44}$$

From (13.44) we have the positive frequency part of the field

$$\vec{E}^{(+)}(\vec{r},t) = -\frac{\omega_l^2}{c^2}\vec{n}\times(\vec{n}\times\vec{p})\frac{e^{ik_l r - i\omega_l t}}{r}e^{-ik_l\vec{n}\cdot\vec{R}}. \tag{13.45}$$

Note also the way we have defined $\vec{p}$ in (13.43), which is the positive frequency part of the dipole moment. We also note that if $\vec{p}(t)$ were a slowly varying function of time compared to the optical oscillations, then in place of (13.45) we would have

$$\vec{E}^{(+)}(\vec{r},t) = -\frac{\omega_l^2}{c^2}\vec{n}\times\left[\vec{n}\times\vec{p}\left(t-\frac{|\vec{r}-\vec{R}|}{c}\right)\right]\frac{e^{ik_l r - i\omega_l t}}{r}e^{-ik_l\vec{n}\cdot\vec{R}}. \tag{13.46}$$

Note that the appearance of the retarded time $t - |\vec{r}-\vec{R}|/c$ in (13.46) is based on the classical description of sources and fields. The Maxwell equations in full quantum theory can be considered as equations for the Heisenberg operators for the electromagnetic fields and the sources are to be replaced by the corresponding polarization or current operators. Thus in place of (13.43) we would have, for the two-level atom,

$$\vec{p}(t) = \vec{p}_{ge}^{\,*}S_+(t)e^{+i\omega_l t} + \vec{p}_{ge}S_-(t)e^{-i\omega_l t}. \tag{13.47}$$

Here we have used the frequency ω_l of the driving field rather than the atomic frequency ω_{eg}. It is advantageous to work in the rotating frame. The correct expression for the electric field operator would be obtained by using (13.46) and by including the contribution $\vec{E}_0^{(+)}$ of the vacuum

$$\vec{E}^{(+)}(\vec{r},t) = \vec{E}_0^{(+)} - \frac{\omega_l^2}{c^2}\vec{n}\times(\vec{n}\times\vec{p})\frac{e^{ik_l r - i\omega_l t}}{r}e^{-ik_l\vec{n}\cdot\vec{R}}S_-\left(t-\frac{|\vec{r}-\vec{R}|}{c}\right), \tag{13.48}$$

where for brevity we use $\vec{p}$ for $\vec{p}_{eg}$. A complete quantum-mechanical derivation of (13.48) is given in Section 7 of [13].

We can now use (13.48) in (13.41) to obtain the spectrum of resonance fluorescence. The free field part $\vec{E}_0^{(+)}$ would not contribute to the normally ordered correlation function. On dropping the retardation terms, we can express the spectrum of the radiated field as

$$S(\vec{r},\omega) = \frac{\omega_l^4}{c^4}|\vec{n}\times\vec{p}_{ge}|^2\frac{1}{r^2}S(\omega), \tag{13.49}$$

$$S(\omega) = \frac{1}{2\pi}\int_{-\infty}^{+\infty} d\tau e^{i\delta\tau}\langle S_+(t)S_-(t+\tau)\rangle, \qquad \delta = \omega - \omega_l, \tag{13.50}$$

where $S(\omega)$ is the spectrum of the correlation function of the atomic dipole moment operators. This is the quantity which is of main interest. Other factors in (13.49) are essentially geometrical factors. Using the fact that in steady state the correlation depends on τ and that $S_- = S_+^\dagger$, we can write $S(\omega)$ in terms of the Laplace transforms

$$S(\omega) = \frac{1}{\pi}\mathrm{Re}\left[\int_0^\infty \langle S_+(t)S_-(t+\tau)\rangle e^{-p\tau}d\tau\right]_{p=-i\delta}. \tag{13.51}$$

The total radiation emitted is

$$\begin{aligned} I &= \int S(\omega)d\omega = \langle S_+S_-\rangle = \frac{1}{2} + \langle S_z\rangle \\ &= \frac{1}{2} + \frac{\eta[1+(\Delta T_2)^2]}{1+(\Delta T_2)^2 + 4|g|^2T_1T_2} \end{aligned} \tag{13.52}$$

where (13.6) has been used.

13.4 Quantum dynamics of the two-level atom and spectrum of fluorescence

The two time correlation $\langle S_+(t)S_-(t+\tau)\rangle$ can be computed from the solution of optical Bloch equations (13.2) and the quantum regression theorem (Section 9.5). We first note that we can write the solution of optical Bloch equations as

$$\langle S_-(t+\tau)\rangle = f(\tau)\langle S_+(t)\rangle + g(\tau)\langle S_-(t)\rangle + h(\tau)\langle S_z(t)\rangle + l(\tau), \tag{13.53}$$

and hence on using the quantum regression theorem

$$\begin{aligned}\langle S_+(t)S_-(t+\tau)\rangle &= f(\tau)\langle S_+(t)S_+(t)\rangle + g(\tau)\langle S_+(t)S_-(t)\rangle \\ &\quad + h(\tau)\langle S_+(t)S_z(t)\rangle + l(\tau)\langle S_+(t)\rangle,\end{aligned} \tag{13.54}$$

which on using the properties of two-level operators reduces to

$$\langle S_+(t)S_-(t+\tau)\rangle = g(\tau)\left(\frac{1}{2} + \langle S_z(t)\rangle\right) - \frac{1}{2}h(\tau)\langle S_+(t)\rangle + l(\tau)\langle S_+(t)\rangle. \tag{13.55}$$

The last line (13.55) requires the steady-state solutions given by (13.6) and (13.7). On combining (13.55) and (13.51), we get the spectrum

$$S(\omega) = \frac{1}{\pi}\mathrm{Re}\left[\hat{g}(-\mathrm{i}\delta)\left(\frac{1}{2} + \langle S_z(t)\rangle\right) - \langle S_+(t)\rangle\left(\hat{l}(-\mathrm{i}\delta) - \frac{1}{2}\hat{h}(-\mathrm{i}\delta)\right)\right]. \tag{13.56}$$

From (13.2) it is seen that

$$\hat{g} = [(p-M)^{-1}]_{22}, \qquad \hat{h} = [(p-M)^{-1}]_{23}, \qquad \hat{l} = [(p-M)^{-1}]_{23}p^{-1}f, \tag{13.57}$$

and therefore the explicit form of the spectrum can be computed. In next few subsections we will analyze different aspects of the spectrum (13.56).

13.4.1 Coherent part of the spectrum

In classical electrodynamics, a coherently excited dipole at the frequency ω radiates at the frequency ω. Thus we expect a similar result for the coherently driven two-level atom. This implies that $S(\omega)$ must have a term like $\delta(\omega-\omega_l)$ with the coefficient proportional to the square of the dipole moment. In order to see this we note that no eigenvalue of M is zero (Section 13.1.3). The contribution from the pole at $p=0$ arises from $\hat{l}$, which we can write as

$$\hat{l} = [(p-M)^{-1}M^{-1} - p^{-1}M^{-1}]_{23}f. \tag{13.58}$$

Hence we identify the coherent part of the spectrum by the contribution from the second term in (13.58). On using (13.58) in (13.56) and noticing from (13.3) that the steady-state value of $\langle S_-\rangle$ is $-(M^{-1})_{23}f$, we get the coherent part of the spectrum

$$\begin{aligned}S_c(\omega) &= |\langle S_-\rangle|^2\frac{1}{\pi}\mathrm{Re}\left[\left(\frac{1}{p}\right)_{-\mathrm{i}\delta}\right] \\ &= |\langle S_-\rangle|^2\delta(\omega-\omega_l).\end{aligned} \tag{13.59}$$

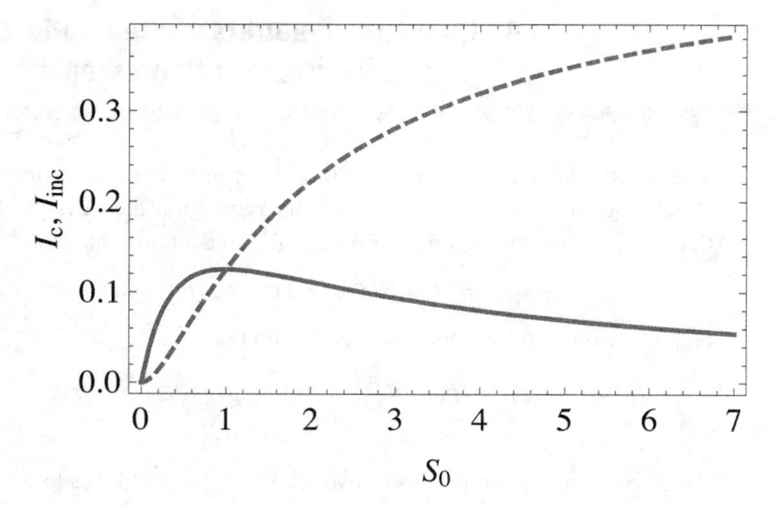

Fig. 13.9 The intensities I_c (solid) and I_{inc} (dashed) as a function of the saturation parameter S_0 for $\Delta = 0$, and for radiative relaxation $\eta = -\frac{1}{2}$, and $\frac{1}{T_1} = \frac{2}{T_2}$.

Thus we always have a coherent contribution to spectrum as $\langle S_- \rangle$ (Eq. (13.7)) is nonzero. It is interesting to note that such a coherent contribution is present in spite of the T_1 and T_2 relaxation processes. Furthermore, (13.59) holds to all powers in the strength of field. Gibbs and Venkatesan [14] measured the coherent part of the spectrum for a weakly driven two-level atom. Höffges *et al.* [15] made very precise measurements of the coherent part of the spectrum by using heterodyning, i.e. by beating the coherently scattered light with a known signal. They showed that the width of the scattered spectrum was less than 1 Hz, thus confirming that the coherent part has a spectrum that is indeed a delta function. These measurements were not limited by the laser linewidth as the local oscillator was derived from the same laser that was used to excite the atom. The intensity I_c of the coherent part from (13.7) and (13.59) is

$$I_c = \frac{4|g|^2 T_2^2 (1 + \Delta^2 T_2^2)\eta^2}{[1 + (\Delta T_2)^2 + 4|g|^2 T_1 T_2]^2}. \tag{13.60}$$

Note that $I_c \neq I$ with I given by (13.52). Hence $I_{inc} = I - I_c$ is the net incoherent part of the radiation. We show in Figure 13.9 the behavior of I_c and $I - I_c = I_{inc}$ as a function of the saturation parameter S_0 (13.11) for $\Delta = 0$ and $\eta = -\frac{1}{2}$, $\frac{1}{T_1} = \frac{2}{T_2}$. We note that for small S_0, the coherent part dominates whereas for large S_0, the incoherent part I_{inc} dominates.

13.4.2 Incoherent part of the spectrum

In view of (13.58) the incoherent part of the spectrum $S_{inc}(\omega)$ is

$$S_{inc}(\omega) = \frac{1}{\pi}\mathrm{Re}\left\{\hat{g}(-i\delta)\left(\frac{1}{2} + \langle S_z(t)\rangle\right) - \langle S_+(t)\rangle\left[((-i\delta - M)^{-1}M^{-1})_{23} f - \frac{1}{2}\hat{h}(-i\delta)\right]\right\}. \tag{13.61}$$

We first examine (13.61) in the limit of the weak driving fields i.e. we evaluate $S_{\text{inc}}(\omega)$ to order g^2. A calculation leads to

$$S_{\text{inc}}(\omega) = \frac{1}{\pi}\text{Re}\left\{\frac{g^2T_2^2\left(\frac{2T_1}{T_2}-1\right)}{1+(\Delta T_2)^2}\cdot\frac{1}{-\text{i}\delta+\text{i}\Delta+\frac{1}{T_2}}\right\}$$

$$= \frac{g^2T_2^2}{1+(\Delta T_2)^2}\cdot\left(\frac{2T_1}{T_2}-1\right)\cdot\frac{1}{\pi}\cdot\frac{\frac{1}{T_2}}{\left(\frac{1}{T_2}\right)^2+(\omega-\omega_0)^2}. \tag{13.62}$$

The incoherent part of the spectrum to the lowest order in the driving field has a peak at the atomic frequency provided that $2T_1/T_2 \neq 1$. For radiative relaxation $S_{\text{inc}}(\omega) = 0$ as $1/T_1 = 2/T_2$. Thus the spectral peak at the atomic frequency is produced by elastic collisions as then $1/T_1 \neq 2/T_2$. Collision-induced spectral features in nonlinear wave mixing and fluorescence are quite ubiquitous and have been extensively discussed in the literature [16–20]. The nonvanishing of $S_{\text{inc}}(\omega)$ for $1/T_1 \neq 2/T_2$ is also known by the name spectral redistribution of radiation [21].

The expression for $S_{\text{inc}}(\omega)$ to all orders in g was first derived by Mollow [11] and is rather complicated. Using (13.61) it can be written as

$$S_{\text{inc}}(\omega) = \frac{g^2T_2^2}{\pi}\text{Re}\left\{\frac{N_1(p)(T_2-2T_1)+N_2(p)}{D^2(0)D(p)}\right\}_{p=-\text{i}\delta}, \tag{13.63}$$

$$\begin{aligned}
D(p) &= (1+pT_1)[\Delta^2T_2^2+(1+pT_2)^2]+4g^2T_1T_2(1+pT_2),\\
N_1(p) &= \text{i}\Delta T_2(1+pT_1)(1+\Delta^2T_2^2+4g^2T_1T_2)\\
&\quad -(1+\Delta^2T_2^2)[(1+pT_1)(1+pT_2)+4g^2T_1T_2],\\
N_2(p) &= 4g^2T_1T_2(1+pT_1)(2pT_1T_2+T_2+2T_1)+(4g^2T_1T_2)^2T_1.
\end{aligned}$$

Note that the spectral features of the incoherent part are determined by the roots of $D(p) = 0$. This is the same polynomial (13.20) which determines the transient solution of the Bloch equations. For an external field on resonance ($\Delta = 0$) and for large Rabi frequency the roots are (13.23) and thus the incoherent spectrum would be a three-peaked spectrum with peaks located at $\delta = 0$, $\delta = \pm 2|g|$, and with widths (at half-peak height) $\frac{2}{T_2}$, $(\frac{1}{T_1}+\frac{1}{T_2})$. The nature of the Mollow spectrum at high driving fields is quite different than in weak fields (13.62). In case of radiative relaxation $T_2 = 2T_1$, the result (13.63) simplifies considerably ($\delta = \omega - \omega_l$)

$$S_{\text{inc}}(\omega) = \frac{g^4T_2^5}{\pi(1+\Delta^2T_2^2+2g^2T_2^2)^2}$$
$$\times\,\text{Re}\left\{\frac{2(2+pT_2)^2+4g^2T_2^2}{(2+pT_2)[(1+pT_2)^2+\Delta^2T_2^2]+4g^2T_2^2(1+pT_2)}\right\}_{p=-\text{i}\delta}, \tag{13.64}$$

which for the driving field on resonance with the atomic transition, i.e. $\Delta = 0$, further reduces to

$$S_{\text{inc}}(\omega) = \frac{2g^4}{\pi(\frac{1}{T_2^2}+2g^2)^2}\text{Re}\left\{\frac{(\frac{2}{T_2}+p)^2+2g^2}{(\frac{2}{T_2}+p)(\frac{1}{T_2}+p)^2+4g^2(\frac{1}{T_2}+p)}\right\}_{p=-\text{i}\delta}. \tag{13.65}$$

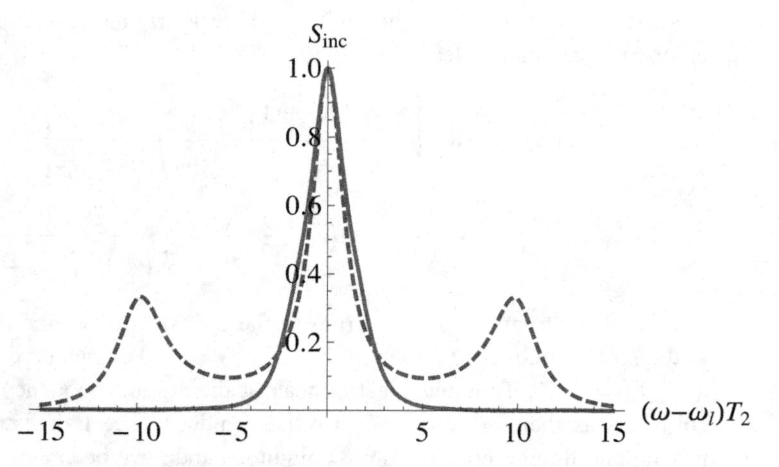

Fig. 13.10 The normalized spectrum $S_{inc}(\omega)$ as a function of detuning δ for different values of g : $g = 1$ (solid), $g = 5$ (dashed).

We show a typical behavior of the incoherent part of the spectrum in Figure 13.10. The peak at the atomic frequency can arise only at high intensities. Note that (13.65) is of the order of g^4 at low powers. For large driving fields, $2|g|T_2 \gg 1$, (13.65) simplifies to

$$S_{inc}(\omega) = \frac{1}{4\pi}\left\{\frac{1}{(\omega-\omega_0)^2T_2^2+1}+\frac{3/4}{(\omega-\omega_0+2g)^2T_2^2+(\frac{3}{2})^2}\right.$$
$$\left.+\frac{3/4}{(\omega-\omega_0-2g)^2T_2^2+(\frac{3}{2})^2}\right\}. \tag{13.66}$$

We have a very interesting feature of the incoherent spectrum: the height of the central peak is three times the height of the side peaks, whereas the side peaks are one and half times broader than the central peak. A detailed experimental confirmation of Mollow's predictions is given in [22–25] by using low-density, well-collimated atomic beams with well-defined velocities. A clear physical explanation of the Mollow spectrum was given by Cohen-Tannoudji *et al.* using the dressed states of the two-level atom interacting with a quantized field [26]. The dressed states were discussed in Chapter 12 (Eq. (12.13)). Let us consider two neighboring set of doublets (cf. Figure 12.2), as shown in Figure 13.11. The wavy arrows give the transitions among dressed states occurring due to spontaneous emission. The frequency of the spontaneously emitted photon can be obtained for each transition with $\Omega_{n\Delta}$ defined by (12.13)

$$\begin{aligned}
|\Psi_{n+1}^+\rangle \rightarrow |\Psi_n^+\rangle &: \omega_l + \frac{\Omega_{n+1\,\Delta}}{2} - \frac{\Omega_{n\,\Delta}}{2} \approx \omega_l,\\
|\Psi_{n+1}^+\rangle \rightarrow |\Psi_n^-\rangle &: \omega_l + \frac{\Omega_{n+1\,\Delta}}{2} + \frac{\Omega_{n\,\Delta}}{2} \approx \omega_l + \Omega_0,\\
|\Psi_{n+1}^-\rangle \rightarrow |\Psi_n^+\rangle &: \omega_l - \frac{\Omega_{n+1\,\Delta}}{2} - \frac{\Omega_{n\,\Delta}}{2} \approx \omega_l - \Omega_0,\\
|\Psi_{n+1}^-\rangle \rightarrow |\Psi_n^-\rangle &: \omega_l - \frac{\Omega_{n+1\,\Delta}}{2} + \frac{\Omega_{n\,\Delta}}{2} \approx \omega_l,
\end{aligned} \tag{13.67}$$

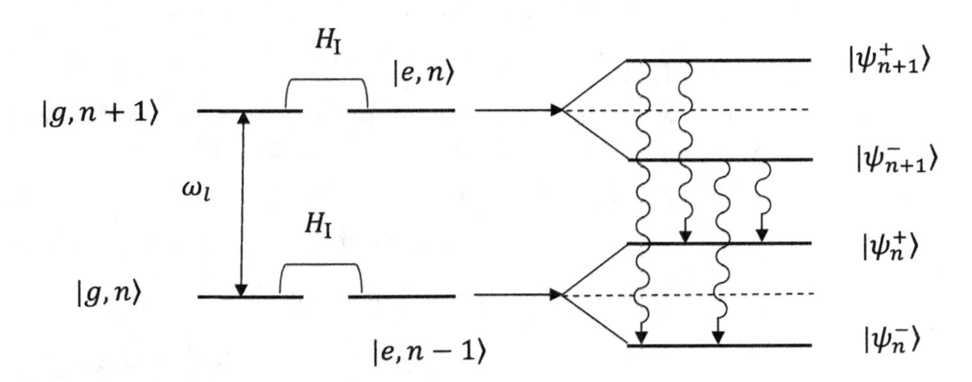

Fig. 13.11 Two neighboring set of doublets. H_I is the interaction with the strong coherent drive.

where the last line is for large values of n, i.e. for large Rabi frequency of the driving field, and where $\Omega_0 \approx \sqrt{\Delta^2 + 4g^2}$. It should be borne in mind that the coupling constant g in Section 12.1 was the coupling with the vacuum of the field. The Rabi frequency ($2g$ in the notation of Chapter 13) of the driving field will be $\sqrt{n}$ times the coupling to the vacuum. The widths and heights of the peaks can also be obtained by calculating the transition rates among the dressed states. A simpler approach is to use the semiclassical dressed states of Section 11.1.1 (Exercise 13.5).

13.4.3 Photon–photon correlation functions of radiation from a coherently driven two-level atom – antibunching

In Section 8.3 we saw how the study of photon–photon correlations enables us to deduce the nonclassical character of the radiation field. In particular, we showed the possibility of antibunching, which is a nonclassical feature. The radiation from the two-level atom posseses such a feature as first demonstrated by Carmichael and Walls [27] and experimentally verified by Kimble *et al.* [28, 29].

Consider the measurement of the nth order correlation function (8.40). Since, according to Eq. (13.48), the field operator $E^{(+)}$ is related to the atomic dipole moment operator S_-, the intensity–intensity correlation function (8.40)

$$\sum_{\alpha\beta} \langle E_\alpha^{(-)}(\vec{r},t) E_\beta^{(-)}(\vec{r},t+\tau) E_\beta^{(+)}(\vec{r},t+\tau) E_\beta^{(+)}(\vec{r},t) \rangle$$

becomes proportional to the one involving atomic dipole moment operators for $n = 2$

$$\Gamma_{nn} \equiv \left\langle S_+(t) S_+(t+\tau_1) S_+(t+\tau_1+\tau_2) \cdots S_+\left(t + \sum_{i=1}^{n-1} \tau_i\right) S_-\left(t + \sum_{i=1}^{n-1} \tau_i\right) \right.$$
$$\left. \cdots S_-(t+\tau_1) S_-(t) \right\rangle \ (\tau_i > 0). \tag{13.68}$$

The correlation function can be evaluated by the repeated use of the quantum regression theorem (Section 9.5) and the spin-$\frac{1}{2}$ property of the dipole operators. From the optical

Bloch equation (13.2) it is clear that $\langle S_z(t)\rangle$ can be written as

$$\langle S_z(t+\tau)\rangle = A(\tau)\langle S_+(t)\rangle + A^*(\tau)\langle S_-(t)\rangle + B(\tau)\langle S_z(t)\rangle + C(\tau), \tag{13.69}$$

where A, B, and C are easily obtained from (13.2). We now make use of the quantum regression theorem to obtain for the two time correlation functions

$$\begin{aligned}\langle S_+(t)S_z(t+\tau)S_-(t)\rangle &= A(\tau)\langle S_+(t)S_+(t)S_-(t)\rangle + A^*(\tau)\langle S_+(t)S_-(t)S_-(t)\rangle \\ &\quad + B(\tau)\langle S_+(t)S_z(t)S_-(t)\rangle + C(\tau)\langle S_+(t)S_-(t)\rangle,\end{aligned} \tag{13.70}$$

which on using the algebra of spin-$\frac{1}{2}$ operators simplifies to

$$\langle S_+(t)S_z(t+\tau)S_-(t)\rangle = \left[C(\tau) - \frac{1}{2}B(\tau)\right]\langle S_+(t)S_-(t)\rangle, \tag{13.71}$$

and hence

$$\begin{aligned}\Gamma(t,t+\tau) &\equiv \langle S_+(t)S_+(t+\tau)S_-(t+\tau)S_-(t)\rangle \\ &= \frac{1}{2}\langle S_+(t)S_-(t)\rangle + \langle S_+(t)S_z(t+\tau)S_-(t)\rangle \\ &= \langle S_+(t)S_-(t)\rangle\left[\frac{1}{2} + C(\tau) - \frac{1}{2}B(\tau)\right] \\ &= f(t)g(\tau),\end{aligned} \tag{13.72}$$

with

$$\begin{aligned} f(t) &= \langle S_+(t)S_-(t)\rangle, \\ g(\tau) &= \frac{1}{2} + C(\tau) - \frac{1}{2}B(\tau).\end{aligned} \tag{13.73}$$

We have thus shown an interesting factorization property of the second-order intensity correlation using the operator algebra and the atomic dynamics. It is clear that the function $g(\tau)$ does not depend on the initial state of the atomic system (i.e. it depends only on dynamics), whereas $f(t)$ does. The function $f(t)$ obviously gives the probability of finding the atom in the excited state if at time $t = 0$ it was in some initial state $\rho(0)$. Let us now examine the meaning of the function $g(\tau)$ by relating to $f(t)$. From (13.69) one finds that

$$\begin{aligned} f(t) \equiv \langle S_+(t)S_-(t)\rangle &= \frac{1}{2} + \langle S_z(t)\rangle \\ &= A(t)\langle S_+(0)\rangle + A^*(t)\langle S_-(0)\rangle + B(t)[\langle S_z(0)\rangle + 0.5] + g(t),\end{aligned} \tag{13.74}$$

and thus if at time $t = 0$ the atom is in the ground state $\langle S_+(0)\rangle = \langle S_-(0)\rangle = 0$, $\langle S_z(0)\rangle = -1/2$, then

$$\langle S_+(t)S_-(t)\rangle = g(t). \tag{13.75}$$

Hence $g(t)$ gives the probability of finding the atom in the excited state given that it was in the ground state at time $t = 0$, and thus $g(0) = 0$. It is also interesting to note that

$$\lim_{t\to\infty} f(t) = \lim_{t\to\infty} g(t), \tag{13.76}$$

which follows from the act that the steady state should be independent of the initial condition.

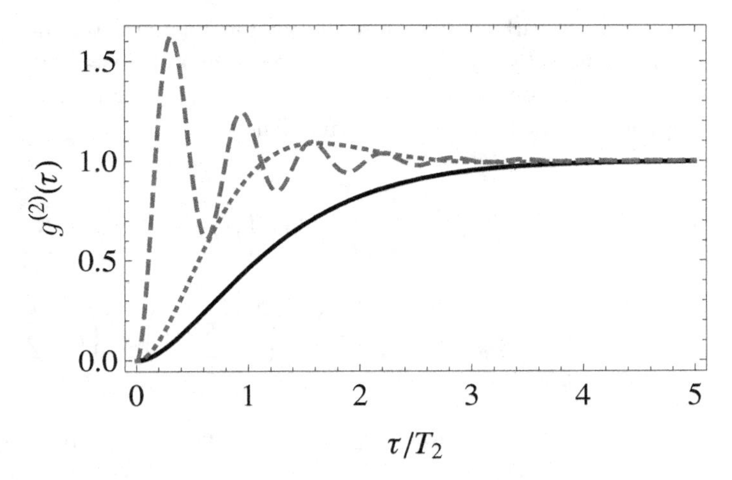

Fig. 13.12 The behavior of $g^{(2)}(\tau)$ as a function of the delay time τ for different Rabi frequencies: $gT_2 = 0.3$ (solid), 1 (dotted), 5 (dashed). Other parameters are $\Delta = 0, \eta = -\frac{1}{2}, \frac{1}{T_1} = \frac{2}{T_2}$.

Clearly from (13.72) and the property $g(0) = 0$, we have $\Gamma(t, t) = 0$, which reflects the fact that once a photon is emitted the atom is in the ground state and it needs time to get excited in order to emit a second photon. It also then follows since $g(\tau) > 0$ that

$$\Gamma(t, t + \tau) > \Gamma(t, t), \tag{13.77}$$

which is a violation of the condition (8.36) for classical fields and hence the radiation from a single two-level atom exhibits a nonclassical feature like antibunching. Another nonclassical feature of the radiation from a two-level atom is squeezing (which is discussed in Exercise 13.6).

From (13.72), the normalized value of the intensity–intensity correlation in the steady state would be

$$g^{(2)}(\tau) = \lim_{t\to\infty} \frac{\Gamma(t, t + \tau)}{\langle S_+(t)S_-(t)\rangle^2} = \frac{g(\tau)}{g(\infty)}. \tag{13.78}$$

We show the behavior of $g^{(2)}(\tau)$ in Figure 13.12 for $\Delta = 0$, and for the radiative relaxation $\eta = -\frac{1}{2}, \frac{1}{T_1} = \frac{2}{T_2}$. The behavior depends on the Rabi frequency of the driving fields. Experiments on antibunching with single trapped ions are described in [30]. This experiment also reports how the data is deteriorated with increase in the number of trapped ions from one to three. Much more refined data on antibunching with a single trapped ion under minimal micromotion is presented by Höffges *et al.* [15]. They also observed Rabi oscillations in $g^{(2)}(\tau)$ at higher driving powers as expected from theory (Figure 13.12).

We can prove a result similar to (13.72) for the higher order correlations like (13.68). All the higher-order correlations satisfy the factorization theorem [31]

$$\Gamma_n(t, \{\tau_i\}) = f(t) \sum_{i=1}^{n-1} g(\tau_i). \tag{13.79}$$

It is quite remarkable that all order photon–photon correlations for the radiation from a single coherently driven atom can be computed in closed form. The proof for the higher-order correlation functions is similar to the one leading to (13.72). We use the operator algebra to write the nth-order correlation as

$$\Gamma_n\left(t, t+\tau_1, \ldots, t+\sum_{i=1}^{n-1}\tau_i\right) = \frac{1}{2}\Gamma_n\left(t, t+\tau_1, \ldots, t+\sum_{i=1}^{n-2}\tau_i\right)$$

$$+\left\langle S_+(t)\cdots S_+\left(t+\sum_{i=1}^{n-2}\tau_i\right) S_z\left(t+\sum_{i=1}^{n-1}\tau_i\right) S_-\left(t+\sum_{i=1}^{n-2}\tau_i\right)\cdots S_-(t)\right\rangle, \quad (13.80)$$

which by using the quantum regression theorem and the operator algebra reduces to

$$\Gamma_n\left(t, t+\tau_1, \ldots, t+\sum_{i=1}^{n-1}\tau_i\right) = g(\tau_n)\Gamma_{n-1}\left(t, t+\tau_1, \ldots, t+\sum_{i=1}^{n-2}\tau_i\right). \quad (13.81)$$

The Eq. (13.81) relates the nth-order intensity correlation to the $(n-1)$th-order intensity correlation; hence by the repeated use of (13.81) we arrive at the result

$$\Gamma_n\left(t, t+\tau_1, \ldots, t+\sum_{i=1}^{n-1}\tau_i\right) = f(t)\sum_{i=1}^{n-1} g(\tau_i). \quad (13.82)$$

Such higher-order correlations can be used in calculations of the counting distributions, i.e. in the formula (8.41).

A very different kind of measurement of photon correlations was done by Aspect *et al.* [32]. Here the correlations between the photons leading to two side bands in the Mollow spectrum were measured. One assumes that the detunings and driving strengths are such that the side bands are well separated and thus one can use frequency filters to separate each side band. In the experiment a highly detuned pump was used to excite the atom. In this case the side bands occur at ω_A and ω_B given by

$$\begin{aligned} \omega_A &= \omega_l + \sqrt{\Delta^2 + 4g^2} \approx 2\omega_l - \omega_0, \quad \omega_l > \omega_0 \\ \omega_B &= \omega_l - \sqrt{\Delta^2 + 4g^2} \approx \omega_0, \end{aligned} \quad (13.83)$$

so that $\omega_A + \omega_B = 2\omega_l$. The occurrence of the side band can be understood as a perturbative process as shown in Figure 13.13. This is because $|\Delta| \gg 2g$. As Figure 13.13 shows there is a definite order in the emission of the two photons – ω_A is emitted first, followed by the emission of ω_B. Furthermore, one has the possibility of resonant excitation of the atom by the three photon process $\omega_l - \omega_A + \omega_l \cong \omega_0$. Therefore the scale of the correlation function of the intensities of the A and B photons is determined by the life time of the excited state as demonstrated by the experiment [32].

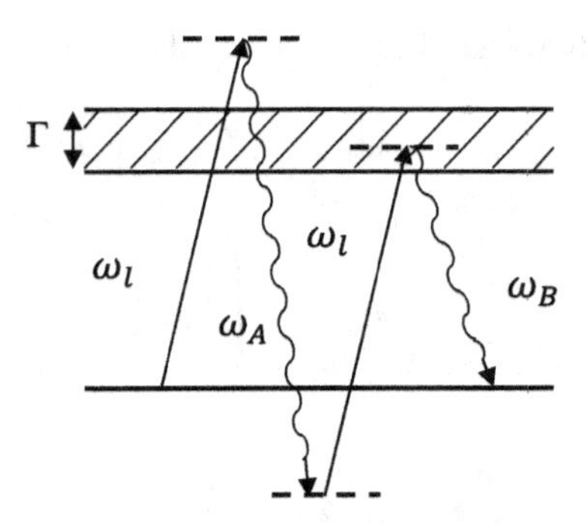

Fig. 13.13 Second-order nonlinear scattering process giving rise to a definite time order in which the side bands of the fluorescence triplet are emitted [32].

Exercises

13.1 Use the results (13.12) and (13.14) for the dipole moment induced by the weak Gaussian probe field $\vec{E}$ to obtain the scattered field by an atom in the direction $\vec{n}$

$$\vec{E}_{\rm sc} \sim -\frac{\omega_l^2}{c^2}\frac{{\rm e}^{{\rm i}k_l r}}{r}\vec{n}\times(\vec{n}\times\vec{p}).$$

Assuming that the probe field is traveling in the direction z and with polarization vector $\hat{\epsilon}$ in the xy plane. Let the direction of the dipole matrix element $\vec{p}_{eg}$ be given by the vector $\hat{\epsilon}_0$. This would be the case for transition between levels with definite azimuthal quantum number. Find then $\hat{\epsilon}^*\cdot(\vec{E}+\vec{E}_{sc})$ and relate the phase shift of the probe beam to detuning Δ.

13.2 Using (13.17) and assuming that the dipole is randomly oriented, obtain the expression for n_2 using intensity I defined by $I=\frac{c}{2\pi}|E|^2$ and $n=n_0+n_2 I\approx 1+2\pi\chi(\omega_l)$.

13.3 Using (13.26) and (13.27) calculate the explicit form of the induced polarization for four-wave mixing, i.e. for the radiation generated at the frequency $2\omega_l-\omega_p$.

13.4 Simplify (13.29) in the limit of large Δ up to order g^2. Show that the term proportional to g^2 has a resonant character when $\delta=0$ and that this has a half width $1/T_1$. Show further that this resonance disappears if the system is subject to only radiative relaxation. Discuss the regions of δ when the probe is amplified. For experiments see [33].

13.5 Consider the dressed state transformation of the optical Bloch equations (13.2) for $\eta=-\frac{1}{2}$ and $\Delta=0$. Let M_1 be the matrix obtained from M by setting $\frac{1}{T_1}=\frac{1}{T_2}=0$. Diagonalize the matrix M_1 such that $S^{-1}M_1S=\Lambda$, and show that

$$\Lambda = 2ig\begin{pmatrix}0&0&0\\0&1&0\\0&0&-1\end{pmatrix},\quad S=\frac{1}{\sqrt{3}}\begin{pmatrix}\sqrt{\frac{3}{2}}&1&-1\\\sqrt{\frac{3}{2}}&-1&1\\0&1&1\end{pmatrix},$$

$$S^{-1}=\frac{\sqrt{3}}{4}\begin{pmatrix}2\sqrt{\frac{2}{3}}&2\sqrt{\frac{2}{3}}&0\\1&-1&2\\-1&1&2\end{pmatrix}.$$

Then show that $\Phi = S^{-1}\Psi$ satisfies

$$\dot{\Phi} = (B_0 + B_1 + \Lambda)\Phi + S^{-1}f,$$

$$B_0 = -\begin{pmatrix} \frac{1}{T_2} & 0 & 0 \\ 0 & \frac{1}{2}\left(\frac{1}{T_1}+\frac{1}{T_2}\right) & 0 \\ 0 & 0 & \frac{1}{2}\left(\frac{1}{T_1}+\frac{1}{T_2}\right) \end{pmatrix},$$

$$B_1 = -\frac{1}{2}\left(\frac{1}{T_1} - \frac{1}{T_2}\right)\begin{pmatrix} 0 & 0 & 0 \\ 0 & 0 & 1 \\ 0 & 1 & 0 \end{pmatrix}$$

Note that Φ_2 is coupled to Φ_3 and vice versa as the matrix B_1 is off-diagonal. However Φ_2 (Φ_3) also oscillates at the frequency $2g$ ($-2g$). For large driving fields, show that this coupling can be ignored leading to

$$\dot{\Phi}_1 = -\frac{\Phi_1}{T_2} + (S^{-1}f)_1,$$

$$\dot{\Phi}_2 = \left[-\frac{1}{2}\left(\frac{1}{T_1} + \frac{1}{T_2}\right) + 2ig\right]\Phi_2 + (S^{-1}f)_2,$$

$$\dot{\Phi}_3 = \left[-\frac{1}{2}\left(\frac{1}{T_1} + \frac{1}{T_2}\right) - 2ig\right]\Phi_3 + (S^{-1}f)_3.$$

These are the optical Bloch equations in the dressed state picture and under the assumption of large driving fields. These equations have great utility in optical resonance and in understanding the Mollow spectrum. The dipole moment or the Bloch vector is related to the dressed state components by $\Psi = S\Phi$, i.e.

$$\begin{pmatrix} \Psi_1 \\ \Psi_2 \\ \Psi_3 \end{pmatrix} = \frac{1}{\sqrt{3}}\begin{pmatrix} \sqrt{\frac{3}{2}} & 1 & -1 \\ \sqrt{\frac{3}{2}} & -1 & 1 \\ 0 & 1 & 1 \end{pmatrix}\begin{pmatrix} \Phi_1 \\ \Phi_2 \\ \Phi_3 \end{pmatrix}.$$

13.6 The uncertainty relation $[S_x, S_y] = iS_z$ leads to $\Delta S_x \Delta S_y \geq |\langle S_z\rangle|/2$. Using optical Bloch equations, fluctuations like ΔS_x can be calculated by using

$$\Delta S_x^2 = \langle S_x^2\rangle - \langle S_x\rangle^2 = \frac{1}{4} - \langle S_x\rangle^2.$$

Show using (13.6) and (13.7) that the condition for squeezing $\Delta S_\alpha^2 < \frac{1}{2}|\langle S_z\rangle|$ for $\alpha = x, y$ can be satisfied for radiative relaxation $\frac{1}{T_1} = \frac{2}{T_2}, \eta = -\frac{1}{2}$ if $(\Delta T_2)^2 > (1+4g^2 T_1 T_2)$ for $\alpha = x$ or $(\Delta T_2)^2 + 4g^2 T_1 T_2 < 1$ for $\alpha = y$. Therefore one can find squeezing in either the x or y component of the atomic dipole moment operator, for details see [34].

References

[1] L. Allen and J. H. Eberly, *Optical Resonance and Two-Level Atoms* (New York: Wiley, New York, 1975).
[2] S. A. Aljunid, M. K. Tey, B. Chng *et al., Phys. Rev. Lett.* **103**, 153601 (2009).
[3] R. W. Boyd, *Nonlinear Optics*, 3rd edition (London: Academic Press, 2008), p.11.
[4] B. R. Mollow, *Phys. Rev. A* **5**, 2217 (1972).
[5] G. S. Agarwal and N. Nayak, *J. Opt. Soc. Am. B* **1**, 164 (1984).
[6] F. Y. Wu, S. Ezekiel, M. Ducloy, and B. R. Mollow, *Phys. Rev. Lett.* **38**, 1077 (1977).
[7] M. T. Gruneisen, K. R. MacDonald, and R. W. Boyd, *J. Opt. Soc. Am. B* **5**, 123 (1988).
[8] W. V. Davis, A. L. Gaeta, R. W. Boyd, and G. S. Agarwal, *Phys. Rev. A* **53**, 3625 (1996).
[9] M. Sargent III, *Phys. Rep. C* **43**, 223 (1978).
[10] L. W. Hillman, R. W. Boyd, J. Krasinski, and C. R. Stroud Jr., *Opt. Commun.* **45**, 416 (1983).
[11] B. R. Mollow, *Phys. Rev.* **188**, 1969 (1969).
[12] J. D. Jackson, *Classical Electrodynamics*, 3rd edition (New York: Wiley, 1999), p. 411.
[13] G. S. Agarwal, *Quantum Optics* (Berlin: Springer-Verlag, 1974).
[14] H. M. Gibbs and T. N. C. Venkatesan, *Optics Commun.* **17**, 87 (1976).
[15] J. T. Höffges, H. W. Baldauf, W. Lange, and H. Walther, *J. Mod. Opt.* **44**, 1999 (1997).
[16] N. Bloembergen, Y. H. Zou, and L. J. Rothberg, *Phys. Rev. Lett.* **54**, 186 (1985).
[17] Y. Prior, A. R. Bogdan, M. Dagenais, and N. Bloembergen, *Phys. Rev. Lett.* **46**, 111 (1981).
[18] G. Grynberg, *Opt. Commun.* **38**, 439 (1981).
[19] W. Lange, *Opt. Commun.* **59**, 243 (1986).
[20] G. S. Agarwal, *Opt. Commun.* **57**, 129 (1986).
[21] K. Burnett, J. Cooper, R. J. Ballagh, and E. W. Smith, *Phys. Rev. A* **22**, 2005 (1980).
[22] F. Schuda, C. R. Stroud. Jr., and M. Hercher, *J. Phys. B* **1**, L198 (1974).
[23] F. Y. Wu, R. E. Grove, and S. Ezekiel, *Phys. Rev. Lett.* **35**, 1426 (1975).
[24] R. E. Grove, F. Y. Wu, and S. Ezekiel, *Phys. Rev. A* **15**, 227 (1977).
[25] W. Hartig, W. Rasmussen, R. Schieder, and H. Walther, Z. *Phys. A* **278**, 205 (1976).
[26] C. Cohen-Tannoudji, J. Dupont-Roc, and G. Grynberg, *Atom-Photon-Interactions: Basic Processes and Applications* (New York: Wiley-Interscience, 1998).
[27] H. J. Carmichael and D. F. Walls, *J. Phys. B* **9**, L43 (1976).
[28] H. J. Kimble, M. Dagenais, and L. Mandel, *Phys. Rev. Lett.* **39**, 691 (1977).
[29] H. J. Kimble and L. Mandel, *Phys. Rev. A* **13**, 2123 (1976).
[30] F. Diedrich and H. Walther, *Phys. Rev. Lett.* **58**, 203 (1987).
[31] G. S. Agarwal, *Phys. Rev. A* **15**, 814 (1977).
[32] A. Aspect, G. Roger, S. Reynaud, J. Dalibard, and C. Cohen-Tannoudji, *Phys. Rev. Lett.* **45**, 617 (1980).
[33] D. Grandclément, G. Grynberg, and M. Pinard, *Phys. Rev. Lett.* **59**, 40 (1987).
[34] D. F. Walls and P. Zoller, *Phys. Rev. Lett.* **47**, 709 (1981).

14 Quantum interference and entanglement in radiating systems

This chapter and the next deal with many aspects of quantum interference and entanglement in a system of radiating atoms. We treat Young's interference and the Hanbury-Brown–Twiss effect using the radiation produced from microscopic sources, which could be neutral atoms, trapped ions, quantum dots, etc. Many of the ideas from Chapter 5 could be applied to radiating single atoms and in fact the interference between different quantum pathways enables us to understand basic phenomena, such as superradiance. Furthermore, the detection of photons can yield heralded entanglement among atoms. This is because atom–photon entanglement is intrinsic to any radiating system. Even the possibility of nonlocal entanglement among remote systems emerges.

14.1 Young's interference with microscopic slits – atoms as slits

In view of the importance of Young's interference (Section 8.2) in optics, it is natural to ask if such an interference would occur if the slits are replaced by atoms [1–4]. Consider the arrangement shown in Figure 14.1. Here two identical atoms located at $\vec{R}_A$ and $\vec{R}_B$ are coherently driven by a common laser field $\varepsilon e^{i\vec{k}_l\cdot\vec{r}-i\omega_l t}$. We take each atom to be a two-level atom as in Chapters 11 and 13. We further assume that the atoms are far apart so that the interaction between them can be ignored. Each coherently driven atom would emit radiation as discussed in Section 13.3. We now examine whether the measurement of intensity in the far zone could display any fringes as the detector is moved along the screen. Using (13.48), the total electric field operator can be written in the form

$$\begin{aligned} \vec{E}^{(+)}(\vec{r},t) &= \vec{E}_0^{(+)}(\vec{r},t) + \vec{e}(\mathcal{E}_A^{(+)}(\vec{r},t) + \mathcal{E}_B^{(+)}(\vec{r},t)), \\ \mathcal{E}_j^{(+)}(\vec{r},t) &= S_-^{(j)}(t)e^{-ik_l\vec{n}\cdot\vec{R}_j}; \\ \vec{e} &= -\frac{\omega_l^2}{c^2}\vec{n}\times(\vec{n}\times\vec{p})\frac{e^{ik_l r-i\omega_l t}}{r}. \end{aligned} \tag{14.1}$$

We recall that the coupling constant g, as defined in (13.1), $g = \frac{\vec{p}^*\cdot\vec{\varepsilon}}{\hbar}e^{i\vec{k}_l\cdot\vec{R}}$, depends on the location of the atom and the direction of propagation of the laser beam. For a system with many atoms it is convenient to work with a single coupling constant $g = \frac{\vec{p}\cdot\vec{\varepsilon}}{\hbar}$. However, we need to change the factor $e^{-ik_l\vec{n}\cdot\vec{R}_j}$ in (14.1) to $e^{i(\vec{k}_l-k_l\vec{n})\cdot\vec{R}_j}$. The expression for the intensity can be obtained from (14.1) as

$$I(\vec{r},t) = I_0\sum_{i,j}\left\langle S_+^{(i)}(t)S_-^{(j)}(t)\right\rangle e^{-i(\vec{R}_i-\vec{R}_j)\cdot(\vec{k}_l-k_l\vec{n})}, \qquad I_0 = |\vec{e}\,|^2. \tag{14.2}$$

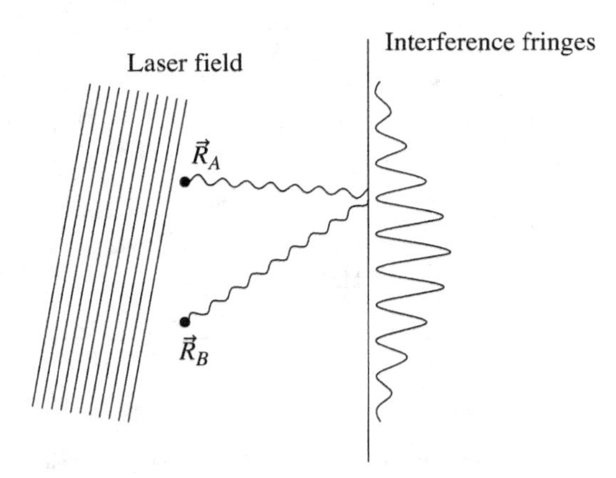

Fig. 14.1 Two atoms in a coherent laser field show interference fringes in the first-order correlation function, i.e. in the intensity, after [4].

Note that we have taken the atoms to be radiating independently and hence $\langle S_+^{(i)}(t)S_-^{(j)}(t)\rangle = \langle S_+^{(i)}(t)\rangle\langle S_-^{(j)}(t)\rangle = |\langle S_-\rangle|^2$ $(i \neq j)$, $\langle S_+^{(i)}(t)S_-^{(i)}(t)\rangle = 1/2 + \langle S_z(t)\rangle$. Furthermore, the intensity of emission from a single atom will be $|\vec{e}|^2\langle S_+^{(i)}S_-^{(i)}\rangle = I_0(1/2 + \langle S_z(t)\rangle)$. The steady-state values of $\langle S_-\rangle$ and $\langle S_z\rangle$ are given by (13.6) and (13.7) and hence (14.2) becomes

$$I(\vec{r},t) = 2I_0\left(\frac{1}{2} + \langle S_z\rangle\right)(1 + \upsilon\cos\delta),$$
$$\delta = \vec{R}\cdot(\vec{k}_l - k_l\vec{n}), \qquad \upsilon = \frac{|\langle S_-\rangle|^2}{\left(\frac{1}{2} + \langle S_z\rangle\right)}, \qquad \vec{R} = \vec{R}_A - \vec{R}_B. \tag{14.3}$$

Clearly we get fringes for the microscopic system of two atoms which are coherently driven. The visibility of the fringes depends on the strength of the driving field and the detuning Δ. For radiative relaxation $\frac{1}{T_1} = \frac{2}{T_2}$, $\eta = -\frac{1}{2}$, and for a coherent field on resonance, we show the behavior of υ in Figure 14.2.

At low driving powers, $4|g|^2T_1T_2 \ll \Delta^2T_2^2$, and for radiative relaxation, the visibility becomes 50%, as from (13.6) and (13.7) we find

$$|\langle S_-\rangle|^2 = |g|^2T_2^2/[1 + (\Delta T_2)^2], \qquad \frac{1}{2} + \langle S_z\rangle \approx \frac{2|g|^2T_1T_2}{1 + (\Delta T_2)^2}. \tag{14.4}$$

Eichmann *et al.* [3] observed such an interference by using an off-resonant excitation $6s^2S_{1/2} \to 6p^2P_{1/2}$ of two $^{198}Hg^+$ ions in a linear trap and by observing π-polarized fluorescence. They showed the dependence of the periodicity of fringes on the separation $|\vec{R}|$ of two ions. The visibility of the fringes was affected by the thermal motion of the ions in the trap.

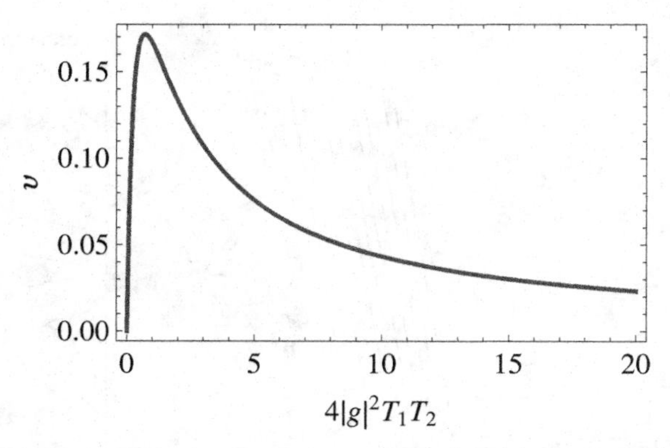

Fig. 14.2 The behavior of υ as a function of $4|g|^2 T_1 T_2$ for radiative relaxation $\frac{1}{T_1} = \frac{2}{T_2}$, $\eta = -\frac{1}{2}$, and for a coherent field on resonance $\Delta = 0$.

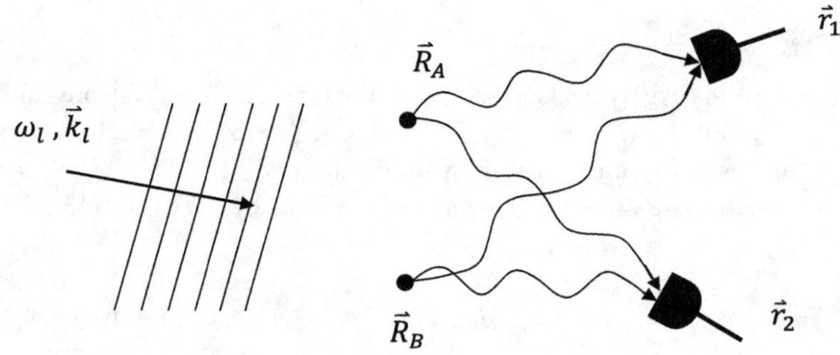

Fig. 14.3 Measurement of photon–photon correlation from a two-atom system considered: a plane wave with wave vector $\vec{k}_l$ is impinging on two atoms fixed at positions $\vec{R}_A$ and $\vec{R}_B$. The light scattered by the two atoms is registered in the far field by two detectors positioned at $\vec{r}_1$ and $\vec{r}_2$.

14.2 Spatial bunching and antibunching of photons

Having established that the radiation from two coherently driven independent atoms can produce interference fringes, we now examine the nature of the photon–photon correlation function for radiation from a coherently driven two-atom system. The scheme is as shown in Figure 14.3. Let us consider the intensity–intensity correlation defined by

$$G^{(2)}(\vec{r}_1, t, \vec{r}_2, t) = \sum_{\alpha,\beta} \langle \vec{E}_\alpha^{(-)}(\vec{r}_1, t) \vec{E}_\beta^{(-)}(\vec{r}_2, t) \vec{E}_\beta^{(+)}(\vec{r}_2, t) \vec{E}_\alpha^{(+)}(\vec{r}_1, t) \rangle, \tag{14.5}$$

which gives us the probability of detecting one photon by the detector 1 at time t and another photon by the detector 2 at the same time t. We will calculate all the quantities in the steady state and hence t is redundant. Using Eq. (14.1), we can reduce (14.5) to

$$G^{(2)}(1, 2) = I_0^2 \bar{G}^{(2)}(1, 2), \tag{14.6}$$

where we have used the compact notation (α) for the space point $(\vec{r}_\alpha)$ and where

$$\bar{G}^{(2)}(1,2) = \langle [\mathcal{E}_A^{(-)}(1) + \mathcal{E}_B^{(-)}(1)][\mathcal{E}_A^{(-)}(2) + \mathcal{E}_B^{(-)}(2)] \\ \times [\mathcal{E}_A^{(+)}(2) + \mathcal{E}_B^{(+)}(2)][\mathcal{E}_A^{(+)}(1) + \mathcal{E}_B^{(+)}(1)]\rangle. \quad (14.7)$$

Note that the vacuum terms do not contribute to normally ordered correlations. Since the single atom operators satisfy the property of the spin-$\frac{1}{2}$ algebra, the terms like $\mathcal{E}_A^{(-)}(1)\mathcal{E}_A^{(-)}(2)$ do not contribute. Hence Eq. (14.7) reduces further to

$$\bar{G}^{(2)}(1,2) = \langle [\mathcal{E}_A^{(-)}(1)\mathcal{E}_B^{(-)}(2) + \mathcal{E}_B^{(-)}(1)\mathcal{E}_A^{(-)}(2)] \\ \times [\mathcal{E}_B^{(+)}(2)\mathcal{E}_A^{(+)}(1) + \mathcal{E}_A^{(+)}(2)\mathcal{E}_B^{(+)}(1)]\rangle. \quad (14.8)$$

We next make use of the uncorrelated nature of the atoms A and B to simplify Eq. (14.8) in the following manner

$$\bar{G}^{(2)}(1,2) = \langle \mathcal{E}_A^{(-)}(1)\mathcal{E}_A^{(+)}(1)\rangle\langle \mathcal{E}_B^{(-)}(2)\mathcal{E}_B^{(+)}(2)\rangle + \langle \mathcal{E}_A^{(-)}(2)\mathcal{E}_A^{(+)}(2)\rangle\langle \mathcal{E}_B^{(-)}(1)\mathcal{E}_B^{(+)}(1)\rangle \\ + (\langle \mathcal{E}_A^{(-)}(1)\mathcal{E}_A^{(+)}(2)\rangle\langle \mathcal{E}_B^{(-)}(2)\mathcal{E}_B^{(+)}(1)\rangle + c.c.). \quad (14.9)$$

Clearly, the existence of interference terms in $\bar{G}^{(2)}(1,2)$ depends on the nonvanishing of the amplitude correlation function

$$G_A^{(1)}(1,2) = \langle \mathcal{E}_A^{(-)}(1)\mathcal{E}_A^{(+)}(2)\rangle. \quad (14.10)$$

Note that $G_A^{(1)}(1,2)$, which is a measure of spatial coherence, is not necessarily zero even if $\langle \mathcal{E}_A^{(-)}(1)\rangle$ is zero. We can rewrite Eq. (14.9) in the form

$$\bar{G}^{(2)}(1,2) = [G_A^{(1)}(1,1)G_B^{(1)}(2,2) + G_A^{(1)}(2,2)G_B^{(1)}(1,1)] \cdot [1 + \Gamma^{(2)}(1,2)], \quad (14.11)$$

where

$$\Gamma^{(2)}(1,2) = \frac{\left(G_A^{(1)}(1,2)G_B^{(1)}(2,1) + c.c.\right)}{\left(G_A^{(1)}(1,1)G_B^{(1)}(2,2) + G_A^{(1)}(2,2)G_B^{(1)}(1,1)\right)}. \quad (14.12)$$

Note that Eq. (14.11) has resemblance to the well-known result for thermal light. However, it should be borne in mind that for radiation produced by coherently driven single atoms $\Gamma^{(2)}(1,2)$ can also be negative, whereas for thermal light this is always positive.

Using now the relation $\mathcal{E}_j^{(+)}(\vec{r},t) = S_-^{(j)}(t)e^{i(\vec{k}_l - k_l\vec{n})\cdot\vec{R}_j}$, we obtain from (14.11)

$$\bar{G}^{(2)}(1,2) = 2\langle S_+S_-\rangle^2[1 + \cos k_l(\vec{n}_1 - \vec{n}_2)\cdot\vec{R}], \qquad \vec{r}_i = n_i r. \quad (14.13)$$

The prefactor in (14.13) depends on the driving field. The square bracket gives interferences in $\bar{G}^{(2)}$ as the detectors are moved around. Note that the fringes in $\bar{G}^{(2)}$ would have 100% visibility obtained for example by fixing one detector and by moving the other detector around [5]. Note further that the interference fringes in (14.13) always survive even in the limit of large driving field when $\langle S_+S_-\rangle \to \frac{1}{2}$, whereas in this limit the fringes in mean intensity (Eq. (14.3)) disappear. We also note the remarkable antibunching property of $\bar{G}^{(2)}$ which occurs when $(\vec{n}_1 - \vec{n}_2)\cdot\vec{R} = \pi$. This is the spatial antibunching property [4] as we are varying the locations of the detectors. Similarly, for $(\vec{n}_1 - \vec{n}_2)\cdot\vec{R} = 0$ or 2π, $\bar{G}^{(2)} \to 4\langle S_+S_-\rangle^2$, and one thus gets spatial bunching [4].

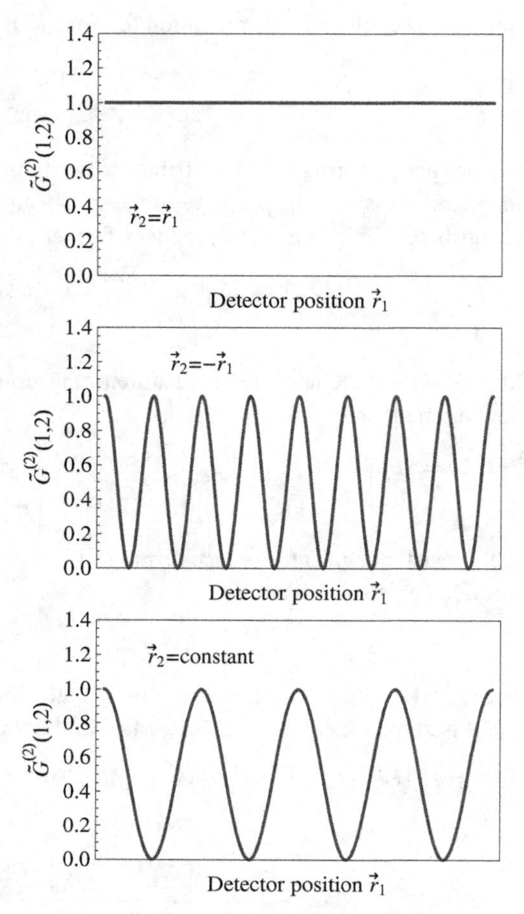

Fig. 14.4 Second-order correlation function $\tilde{G}^{(2)}(1,2)$ as a function of detector position $\vec{r}_1$.

Figure 14.4 shows the character of $\tilde{G}^{(2)}(1,2) = \bar{G}^{(2)}(1,2)/4\langle S_+S_-\rangle^2$ depending on how the detectors are scanned. Note the very interesting possibility obtained by moving the two detectors simultaneously in opposite directions. In this case, one gets twice as many fringes, and this clearly has implications in the consideration of superresolution of atomic positions [6].

For quantitative characterization of bunching and antibunching, it is useful to introduce the normalized $g^{(2)}$ given by (14.3) by dividing $G^{(2)}$ by the product of the mean intensities

$$g^{(2)}(1,2) = \frac{G^{(2)}(1,2)}{I(1)I(2)}, \tag{14.14}$$

where I is given by (14.3) and hence

$$g^{(2)}(1,2) = \frac{1+\cos(\delta_1-\delta_2)}{2(1+\upsilon\cos\delta_1)(1+\upsilon\cos\delta_2)}, \tag{14.15}$$

where δ_j is given as in (14.3) with $\vec{n} \rightarrow \vec{n}_j$.

For unequal times we can relate $G^{(2)}$ to the quantities that we introduced in Section 13.4 in connection with single-atom resonance fluorescence. Many additional terms now appear in (14.8). A closed form expression can be found in Skornia *et al.* [4]. A measurement of $g^{(2)}(1, 2, \tau)$ was reported by Gerber *et al.* [7] using two independent ion traps. In their experiment a relative random phase was introduced between the fields collected from the two traps. Under these conditions $G^{(2)}$ is given by (14.8) with two additional single-atom contributions of the form $\langle \mathcal{E}_i^{(-)}(1)\mathcal{E}_i^{(-)}(2)\mathcal{E}_i^{(+)}(2)\mathcal{E}_i^{(+)}(1)\rangle$ $(i = A, B)$. The indices 1 and 2 now stand for $(\vec{r}_1, t)$ and $(\vec{r}_2, t + \tau)$ respectively.

14.3 Interference in radiation from two incoherently excited atoms

Let us examine the possibility of Young's interference with two incoherently excited independent atoms. This would be the case if the atoms were excited by a π-pulse. Then $\langle S_+^{(i)}\rangle = 0$ and hence there are no interferences according to (14.3) as $v \to 0$. This is exactly as in the classical double slit experiment (Eq. (8.22)). The next question is then: can there be interference in the intensity–intensity correlation (14.5) and, if yes, then what is the physical mechanism leading to such an interference? The derivation of (14.13) remains valid with $\langle S_+^{(i)} S_-^{(i)}\rangle \to 1$ leading to

$$\bar{G}^{(2)}(1, 2) = 2[1 + \cos k_l(\vec{n}_1 - \vec{n}_2) \cdot \vec{R}]. \tag{14.16}$$

Thus we conclude that the radiation from two incoherently excited independent atoms exhibits interference fringes even though there are no interferences in the corresponding Young's interference experiment [5]. We now explain why one sees interferences in the intensity–intensity correlations. Note that there are two distinct quantum paths leading to the simultaneous clicking of the two detectors as shown in Figure 14.5. The two quantum paths correspond to transitions via different intermediate states as shown in Figure 14.6. The initial and final states of the two atoms are identical and hence, according to the general discussion of Section 7.7, these two paths would lead to interferences in the two photon process of detecting two photons, one at each detector.

The interference in $G^{(2)}$ can also be understood in terms of the entanglement produced by detection. We start with the uncorrelated state $|e_A, e_B\rangle$. The detection of a photon by a detector in the direction $\vec{n}_1$ projects the state of the atomic system to

$$\rho_D = \frac{D^-(\vec{r}_1)\rho D^+(\vec{r}_1)}{\langle D^+(\vec{r}_1)D^-(\vec{r}_1)\rangle}, \tag{14.17}$$

where

$$D^-(\vec{r}_1) \equiv (|g\rangle\langle e|)_A + e^{ik_l\vec{n}_1\cdot\vec{R}}(|g\rangle\langle e|)_B, \qquad \vec{R} = \vec{R}_A - \vec{R}_B. \tag{14.18}$$

If $\rho \equiv |e_A, e_B\rangle\langle e_A, e_B|$, then $\rho_D = |\Psi\rangle_D\,{}_D\langle\Psi|$, where

$$|\Psi\rangle_D = \frac{|g_A, e_B\rangle + e^{ik_l\vec{n}_1\cdot\vec{R}}|e_A, g_B\rangle}{\sqrt{2}}, \tag{14.19}$$

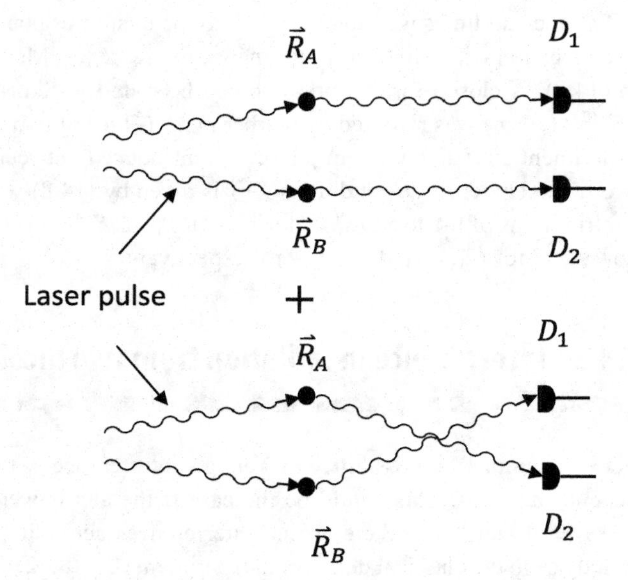

Fig. 14.5 Different paths contributing to intensity–intensity correlations.

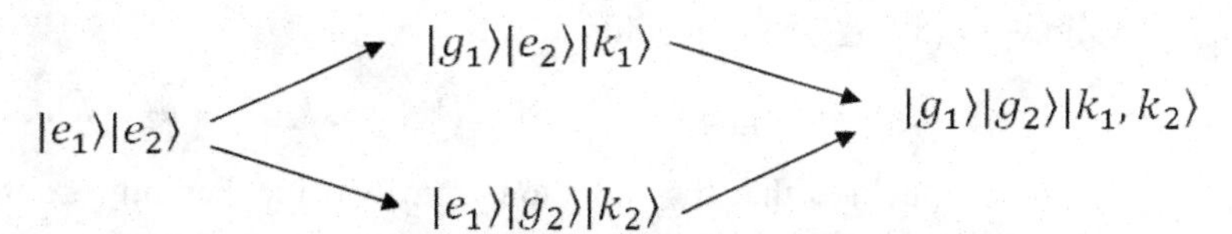

Fig. 14.6 Two different quantum paths in terms of the states lead to detection of one photon at each of the detectors D_1 and D_2.

which is an entangled state (EPR state). The detection of the second photon would now produce interference fringes. This follows from (14.2), which shows that the interference fringes would be present provided that $\langle S_+^{(A)} S_-^{(B)} \rangle \neq 0$. For the state (14.19) we can show that

$$\langle S_+^{(A)} S_-^{(B)} \rangle = \frac{\mathrm{e}^{-ik_l \vec{n}_1 \cdot \vec{R}}}{2}, \tag{14.20}$$

which is nonzero. Thus the interferences fringes in $G^{(2)}$ can be considered as due to the entanglement of the initially uncorrelated atoms produced by the process of detection.

The remarkable property of the interference in $G^{(2)}$ is its visibility which is 100%. If the visibility of the interference pattern is more than 50%, then the underlying field is nonclassical. We will now use a version of the Cauchy–Schwarz inequality (Section 8.3) to establish the nonclassicality. On using (14.16), we can define

$$\begin{aligned} g^{(2)}(1,2) &= \frac{G^{(2)}(1,2)}{\langle I(1)\rangle\langle I(2)\rangle}, \\ &= \frac{1}{2}[1 + \cos k_l(\vec{n}_1 - \vec{n}_2)\cdot\vec{R}]. \end{aligned} \tag{14.21}$$

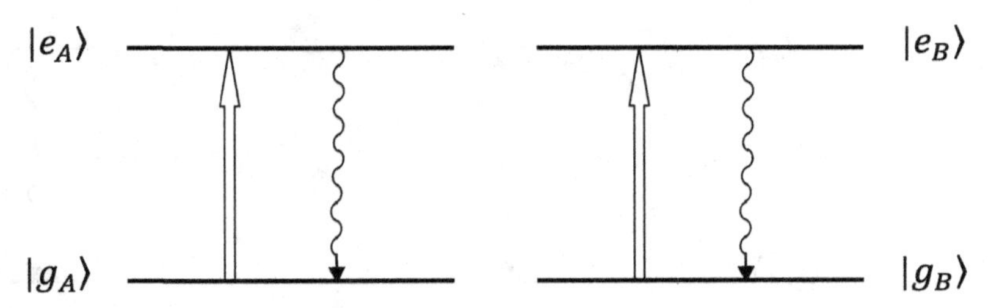

Fig. 14.7 Emission from two coherently excited independent atoms.

Let us consider the derivation of the Cauchy–Schwarz inequality starting from the obvious property

$$\langle(\alpha\delta I(1) + \beta\delta I(2))^2\rangle \geq 0 \qquad \forall\, \alpha, \beta, \tag{14.22}$$

where $\delta I(1) = I(1) - \langle I(1)\rangle$ and where the quantum-mechanical ordering as in (14.5) is implied. If the underlying function $P(\alpha)$ associated with the density matrix of the field has the properties of a classical probability distribution then (14.22) leads to

$$\langle\delta I(1)\delta I(1)\rangle\langle\delta I(2)\delta I(2)\rangle \geq \langle\delta I(1)\delta I(2)\rangle\langle\delta I(2)\delta I(1)\rangle. \tag{14.23}$$

Any violation of (14.23) would imply the nonclassical nature of the field. We can write the inequality (14.23) in terms of $g^{(2)}$ as

$$(g^{(2)}(1,1) - 1)(g^{(2)}(2,2) - 1) \geq (g^{(2)}(1,2) - 1)(g^{(2)}(2,1) - 1). \tag{14.24}$$

Note that for $g^{(2)}$ given by (14.21), $g^{(2)}(1,1) = g^{(2)}(2,2) = 1$ and hence the left-hand side of (14.24) is zero, leading to complete violation of the Cauchy–Schwarz inequality

$$0 \geq \sin^4[k_l(\vec{n}_1 - \vec{n}_2)\cdot\vec{R}/2]. \tag{14.25}$$

Thus we have established the nonclassical character of the radiation produced by two independently excited atoms. Quantum interferences in photon–photon correlations have been observed by Beugnon *et al.* [8] in radiation from two incoherently excited independent atoms, in two-ion systems by Maunz *et al.* [9], and in a system of two quantum dots by Flagg *et al.* [10].

Since we have discussed a variety of situations where interferences are possible it is good to summarize these:

(A) Two coherently driven independent atoms, Figure 14.7.
- Interferences in intensity measurements – an analog of the Young's two slit experiment at a microscopic level.
- Interferences in intensity–intensity correlations; spatial bunching and antibunching for photon–photon correlations measured even with zero delay.

(B) Two incoherently excited atoms, Figure 14.8.
No interferences in mean intensity; however, intensity–intensity correlations exhibit interferences with 100% visibility.

(C) Single excitation in the two-atom system, Figure 14.9.
For an initial excitation to an entangled state $\frac{1}{\sqrt{2}}(|e_A, g_B\rangle + e^{i\varphi}|g_A, e_B\rangle)$, intensity

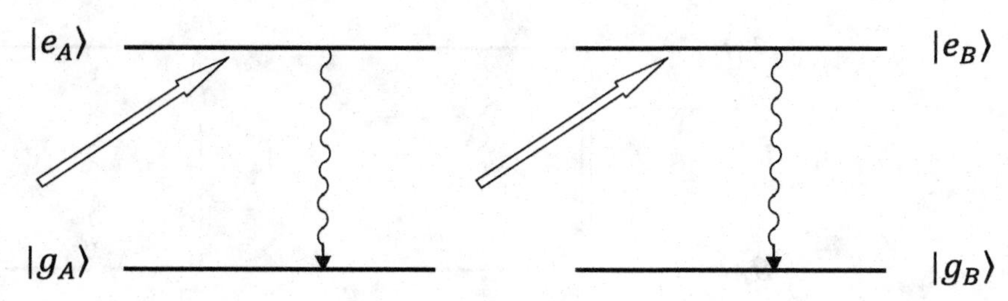

Fig. 14.8 Emission from incoherently excited atoms.

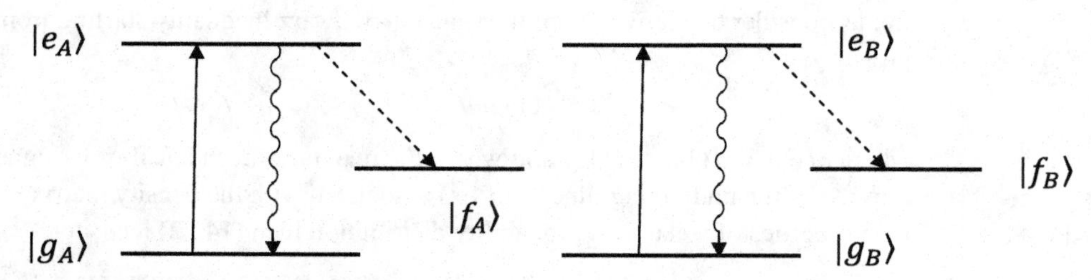

Fig. 14.9 Emission from two atoms in an entangled state containing signle excitation.

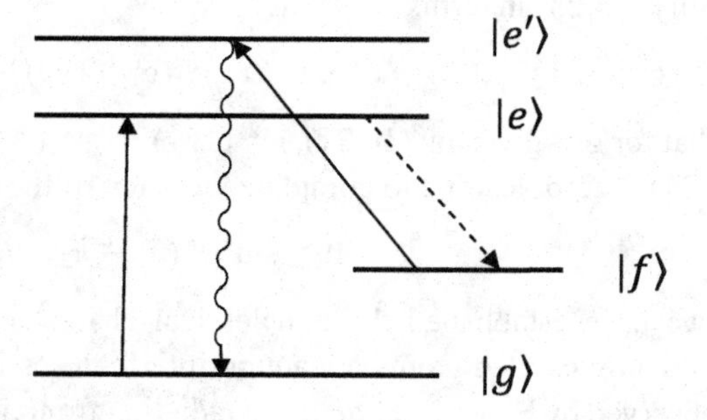

Fig. 14.10 Quantum eraser scheme.

exhibits interferences in radiation on the transition $|e\rangle \rightarrow |g\rangle$ (c.f. our discussion following Eq. (14.19)). However, radiation on the transition $|e\rangle \rightarrow |f\rangle$ exhibits no interferences. This is because for the transition $|e\rangle \rightarrow |f\rangle$ the interference can occur according to (14.2) only if the mean value of the corresponding operator $\langle(|e_A\rangle\langle f_A|)(|f_B\rangle\langle e_B|)\rangle$ is nonzero, which is not the case for the initially prepared entangled state. Scully and Drüh [11] have shown that by using a weak laser pulse on the transition $|f\rangle \rightarrow |e'\rangle$ (Figure 14.10), one can restore interferences in the radiation emitted on the transition $|e\rangle \rightarrow |f\rangle$ if a simultaneous detection of the photons emitted on the transition $|e'\rangle \rightarrow |g\rangle$ is also made. This is the idea of the quantum eraser [11, 12], where path information for photons emitted on the transition $|e\rangle \rightarrow |f\rangle$ is erased by excitation from $|f\rangle$ to $|e'\rangle$ followed by the detection of photons on the transition $|e'\rangle \rightarrow |g\rangle$ (see Exercise 14.4).

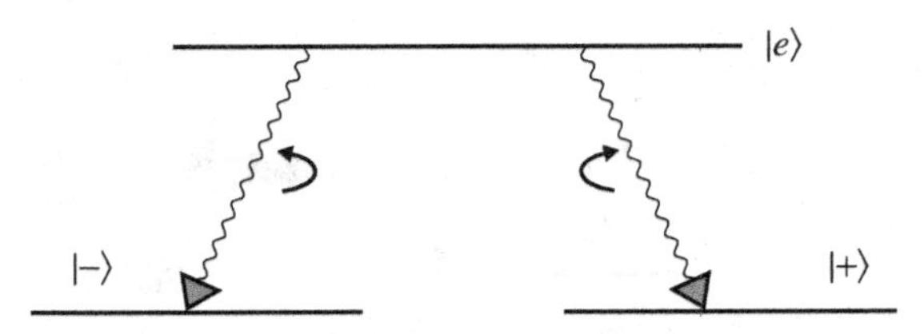

Fig. 14.11 The Λ-type atomic system in which a photon with either polarization is emitted.

14.4 Atom–photon entanglement

In Section 7.6 we have discussed the decay of the excited state of an atom. We found that the wavefunction of a system consisting of an atom and a spontaneously emitted photon can be written in the form (7.64), i.e.

$$|\psi\rangle = c_i(t)|i\rangle|\{0\}\rangle + \sum_{\vec{k}s} c_{\vec{k}s}(t)|f\rangle|1_{\vec{k}s}\rangle. \tag{14.26}$$

The state (14.26) is an entangled state of the atom–photon system and thus in the process of spontaneous emission, entanglement arises naturally [13]. The state of the photon is also an entangled state as the photon can be found in any of the infinitely available modes. More remarkable consequences of photon–atom entanglement appear in spontaneous emission in multilevel systems where one can use the extra degrees of freedom, such as polarization.

We illustrate the important features by considering the Λ-system as shown in Figure 14.11 where the two lower states are the Zeeman states. Thus in terms of the angular momentum of the states this would correspond to the $j = 0$ to $j = 1$ transition. We have a left (right) circularly polarized photon emitted in the transition $|e\rangle \to |-\rangle$ ($|e\rangle \to |+\rangle$). By analogy to (14.26), the wavefunction at time t will have the structure

$$|\psi\rangle = c_e(t)|e\rangle|\{0\}\rangle + \sum_{\vec{k}} c_{-\vec{k}}(t)|-\rangle|\vec{k}, \circlearrowleft\rangle + \sum_{\vec{k}} c_{+\vec{k}}(t)|+\rangle|\vec{k}, \circlearrowright\rangle, \tag{14.27}$$

where $|\vec{k}, \circlearrowleft\rangle$ denotes the state of the left circularly polarized photon in the direction $\vec{k}$. In the long time limit $c_e(t) \to 0$, we have entanglement between the atom and the photon with the important property that the entanglement is long lasting as the states $|\pm\rangle$ are long lived. The amplitudes c can be determined by following the procedure of Section 7.6.

The atom–photon entanglement can be used to produce atomic coherence. Suppose we detect, as shown in Figure 14.12, a photon in the direction $\vec{k}$ but with arbitrary polarization, i.e. we detect the photon in the state $\alpha_+|\vec{k}, \circlearrowleft\rangle + \alpha_-|\vec{k}, \circlearrowright\rangle$, then the wavefunction of the atom gets projected to

$$|\psi\rangle_c = \alpha_+^* c_{-\vec{k}}|-\rangle + \alpha_-^* c_{+\vec{k}}|+\rangle. \tag{14.28}$$

The conditional atomic wavefunction is a coherent superposition of two Zeeman states as long as $|\alpha_+|$, $|\alpha_-|$ are nonzero. An observation of atom–photon entanglement was made by Volz *et al.* [14], who considered the decay of the state $2P_{3/2}$, $F = 0$, $m_F = 0$ of ^{87}Rb

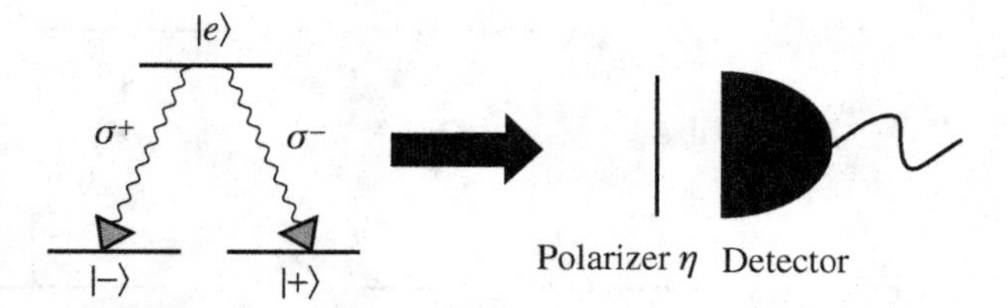

Fig. 14.12 Atom–photon entanglement and projective measurement by choosing the polarization of the emitted radiation.

atom by first exciting it using a laser pulse from the ground state $2S_{1/2}$, $F = 1$, $m_F = 0$. The linear polarization of the detected photon was seen at different angles ($\alpha_\pm \neq 0$). The conditional superpositions like (14.28) were detected by using the stimulated Raman adiabatic passage technique [15]. In another experiment [16] where atoms were continuously pumped, coherence in the ground state was produced by the detection of the first photon (as in (14.28)) and was measured by the detection of a second photon.

14.5 Atom–atom entanglement via detection of spontaneously emitted photons

Consider now spontaneous emission from two identical Lambda systems. We will assume that the atoms have a relative distance which is much bigger than a wavelength so that one can consider the atoms as independent. Initially each atom is in the excited state. Let us label the atoms as A and B. Then the wavefunction for each atom at time t will have the form (14.27) and the wavefunction of the combined system of two atoms will be a product of wavefunctions of the form (14.27). Consider now a detection scheme where the two photons are detected. We thus need a projection of the wavefunction in the space consisting of two photons. After multiplying wavefunctions of the form (14.27), the part containing two photons is

$$|\psi\rangle_{\text{2photons}} \sim \sum_{kq} c_{--}(\vec{k},\vec{q})|-\rangle_A|-\rangle_B|\vec{k},\circlearrowleft\rangle|\vec{q},\circlearrowleft\rangle + c_{++}(\vec{k},\vec{q})|+\rangle_A|+\rangle_B|\vec{k},\circlearrowright\rangle|\vec{q},\circlearrowright\rangle$$
$$+ c_{-+}(\vec{k},\vec{q})|-\rangle_A|+\rangle_B|\vec{k},\circlearrowleft\rangle|\vec{q},\circlearrowright\rangle + c_{+-}(\vec{k},\vec{q})|+\rangle_A|-\rangle_B|\vec{k},\circlearrowright\rangle|\vec{q},\circlearrowleft\rangle. \tag{14.29}$$

Now suppose we detect the left circularly polarized photon in the direction $\vec{k}$ and a right circularly polarized photon in the direction $\vec{q}$, then the conditional state of the two atoms would be

$$|\psi\rangle^{(c)}_{AB} = c_{-+}(\vec{k},\vec{q})|-\rangle_A|+\rangle_B + c_{+-}(\vec{q},\vec{k})|+\rangle_A|-\rangle_B, \tag{14.30}$$

which is an EPR state, i.e. an entangled state. It is clear from the structure of (14.29) that we can produce all four Bell states of the two atoms by selecting appropriately the polarization of the detected photons. It should be noted that such Bell states are stable as both $|\pm\rangle$ states of the atom could be long-lived ground states.

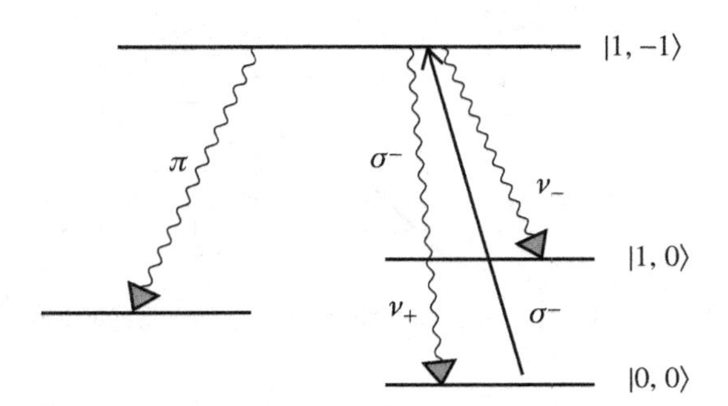

Fig. 14.13 The relevant energy levels in a Yb ion that produce atom–atom entanglement. The ions are excited by a σ^- polarized pulse from the state $|0, 0\rangle$ to the state $|1, -1\rangle$, after [17].

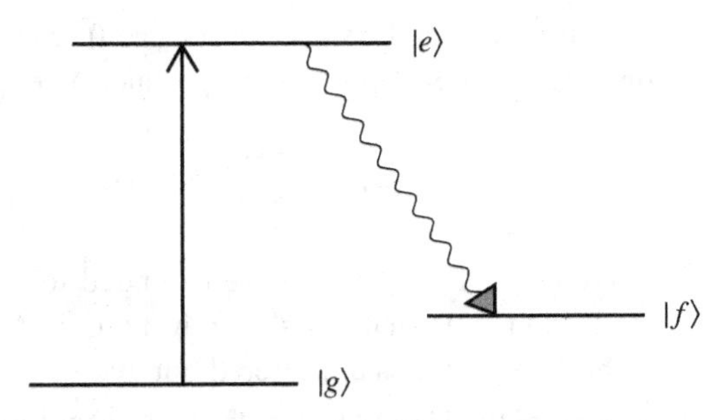

Fig. 14.14 A three-level atom with levels $|g\rangle$, $|e\rangle$, and $|f\rangle$. A weak laser pulse excites the atom from $|g\rangle$ to $|e\rangle$.

This method of atom–atom entanglement does not require any interatomic interactions. The atoms could thus be far apart. In fact, entanglement between remote atoms can only be produced by the detection process. Moehring *et al.* [17] demonstrated entanglement between two ^{171}Yb ions contained in two different traps a meter apart. The principle of their scheme is similar to what we outlined earlier; however, in their scheme (Figure 14.13) the polarizations of the two detected photons are identical but the two photons differ in frequencies, which are denoted by $\nu_\pm$. We briefly discuss their scheme to make it clear that the polarization degrees of photons are not the only ones that can be used. The states are labeled in terms of their F and m_F values. The solid line shows the excitation pulse. The collection of σ^- photons would produce an atom–photon entangled state $(|1, 0\rangle|\nu_-\rangle - |0, 0\rangle|\nu_+\rangle)/\sqrt{2}$, with relative sign determined by the Clebsch–Gordon coefficients. When σ^- photons are detected, one with frequency ν_- and the other with frequency ν_+, then we would generate entangled states of two ions.

The atom–photon entanglement discussed in Section 14.4 can also be used to produce an entangled state of an ensemble of atoms [18]. Let us consider an ensemble of atoms with a level scheme as shown in Figure 14.14. Let us assume that the ensemble is excited by a very weak pulse leading to the scattering of a Raman photon, say, in the forward direction. Thus in the ensemble only one atom at a time would produce a scattered photon. Thus the

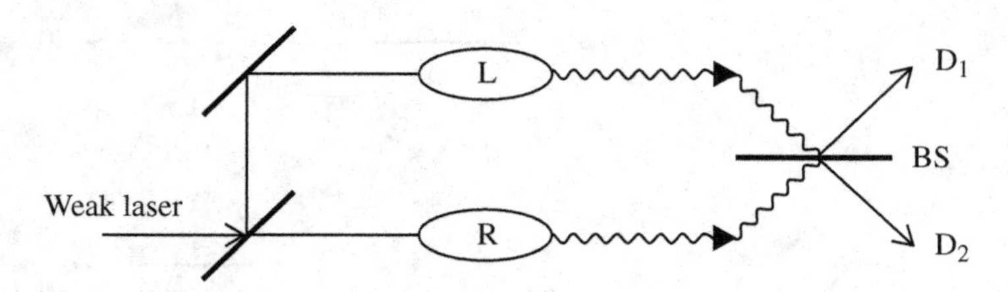

Fig. 14.15 Schematic of generating entanglement between two remote atomic ensembles.

state of the atom–photon system would be

$$|\psi\rangle \sim |g_1 g_2 \cdots g_N\rangle|0\rangle + \sum_{j=1}^{N} c_j |g_1 g_2 \cdots f_j \cdots g_N\rangle|1\rangle, \tag{14.31}$$

where $|1\rangle$ and $|0\rangle$ represent the states with 1 and 0 scattered photons. Hence conditional on the detection of the scattered photon, we produce an entangled state of the ensemble

$$|\phi\rangle \sim \sum_{j=1}^{N} c_j |g_1 g_2 \cdots f_j \cdots g_N\rangle, \tag{14.32}$$

which is the W state. This idea can be extended to produce entanglement between two remote ensembles [18, 19]. This is shown in Figure 14.15, where L and R are two atomic ensembles. The laser photons are filtered out and only the scattered photons are collected by optical fibers and made to interfere at the beam splitter BS. A click in D_1 or D_2 prepares an entangled state of the form $(|\phi\rangle_L \pm |\phi_R\rangle e^{i\varphi})/\sqrt{2}$, with $\pm$ depending on which detector clicks and φ represents the phase difference between the two photons arriving at the BS. Chou *et al.* [20] demonstrated such an entanglement between two remote ensembles with each ensemble containing about 10^5 atoms. The entanglement between two ensembles was detected by using read pulses which produce anti-Stokes photons and by doing tomography on the anti-Stokes photons. These ideas on creation of entanglement by detection have been extended to produce entanglement between two remote Bose–Einstein condensates [21]. Furthermore, Duan *et al.* [18] have extended these ideas to propose long-distance quantum communication based on entanglement.

Finally, we mention that a Rydberg dipole blockade can also be used to entangle two atoms [22]. Consider two atoms each with a lower state $|g\rangle$ and a Rydberg state $|e\rangle$. When both atoms are in the state $|e\rangle$, then there is strong interaction between them leading to a shift of the level $|e, e\rangle$. The amount of shift ΔE depends on the distance between two atoms. The situation is shown in Figure 14.16. For example, for ^{87}Rb, one can choose $|e\rangle = |58d_{3/2}, F = 3, m_F = 3\rangle$ [23]. In this case one has a quasi-degeneracy between $|e, e\rangle$ and the Rydberg state $(60p_{1/2}, 56f_{5/2})$ for two atoms. This leads to the Förster resonance, which results in even larger shifts $\pm\Delta E$. The effect of the shift is that the double excitation becomes off-resonant and thus strongly suppressed. Furthermore, if the atomic locations are perpendicular to the direction in which the excitation field is traveling, then both the atoms see the same coupling to the external fields and the antisymmetric state $|a\rangle$ is not excited. Clearly under these conditions, the two-atom system gets excited to the state $|s\rangle$, which is

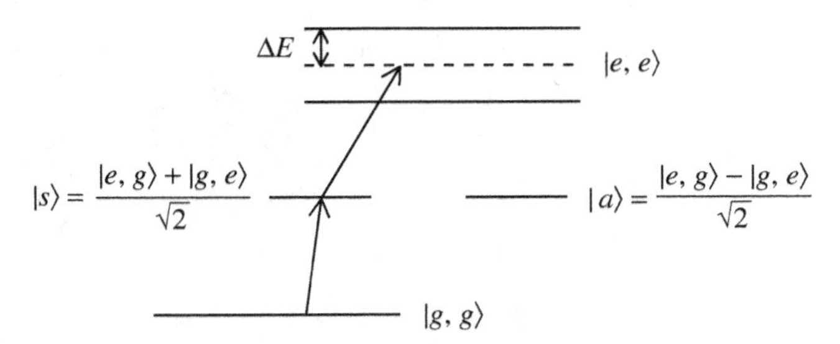

Fig. 14.16 Entangling two atoms by a Rydberg dipole blockade.

a Bell state. The inhibition of the doubly excited state (dipole blockade) was demonstrated in [23, 24].

14.6 Multi-atom entanglement

We have seen in Section 14.5 how the detection of photons can lead to the production of entanglement between two atoms. A natural question is: can these ideas be applied for the production of multi-atom or multi-particle entanglement? The answer is indeed yes. Greenberger *et al.* [25] introduced a new paradigm in the study of nonlocality of quantum mechanics by introducing three particle entangled states of the form

$$|GHZ\rangle = \frac{1}{\sqrt{2}}(|+++\rangle - |---\rangle). \tag{14.33}$$

Using these states, they could demonstrate nonlocal character of quantum mechanics without requiring Bell-like inequalities. Methods to produce such states are of great interest and one has already demonstrated such states using type-II downconverters [26]. Another class of states for three atoms are W states of the form

$$|W\rangle = \frac{1}{\sqrt{3}}(|++-\rangle + |+-+\rangle + |-++\rangle). \tag{14.34}$$

The $|W\rangle$ states are more robust to decoherence. For example, the $|GHZ\rangle$ states completely decohere under the loss of one particle, whereas the $|W\rangle$ state still retains its entangled character.

In this section we discuss the possibility of realizing such multi-particle or multi-atom entangled states. The basic idea is to generalize the procedure of Sections 14.4 and 14.5 to N atoms. Following the discussion of Bastin *et al.* [27], we use the arrangement of Figure 14.17 to produce multiparticle entanglement. As indicated in Section 14.4 (Eq. (14.28)), the detection of a photon by a detector, say the jth, with a polarizer with orientations defined by α_j, β_j [$\alpha = \alpha_+$, $\beta = \alpha_-$] can come by the emission from any of the N atoms, with each detection projecting the atomic state into combinations like (14.28). Thus many different final states of the atomic system can be produced depending on the kind of detections performed. Let us specifically consider the case of three atoms. We show in Figure 14.18

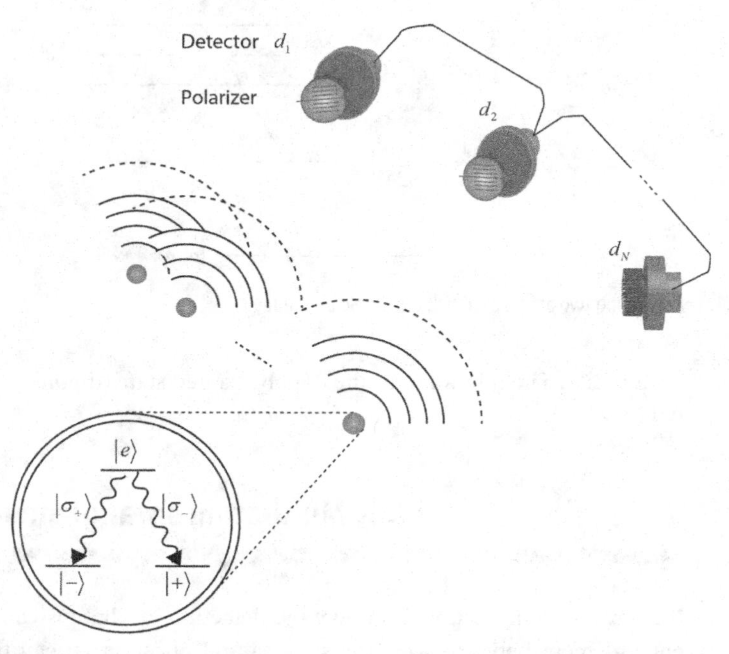

Fig. 14.17 Proposed experimental arrangement. N excited emitters are aligned in a row, each of defining a three-level Λ system. A long-lived entangled state is obtained in the N emitter qubits, defined by $|\pm\rangle$, after detecting the N spontaneously emitted photons with N detectors equipped with polarizers. The final N-qubit state is tuned and determined by the polarizer orientations, after [27].

all possible paths leading to different final states. The final state which is the sum of all the states in the box is arbitrary up to an overall normalization factor. Figure 14.18 shows the interesting result that if αs and βs are chosen such that

$$\begin{aligned} \alpha_1\alpha_2\beta_3 + \alpha_1\beta_2\alpha_3 + \beta_1\alpha_2\alpha_3 &= 0, \\ \alpha_1\beta_2\beta_3 + \beta_1\alpha_2\beta_3 + \beta_1\beta_2\alpha_3 &= 0, \end{aligned} \tag{14.35}$$

then the final state will be the GHZ state up to a normalization factor

$$|\mathrm{GHZ}\rangle = \alpha_1\alpha_2\alpha_3|---\rangle + \beta_1\beta_2\beta_3|+++\rangle. \tag{14.36}$$

The condition (14.35) can be satisfied if we choose $\alpha_j = \frac{e^{-i\theta_j}}{\sqrt{2}}$, $\beta_j = \frac{e^{i\theta_j}}{\sqrt{2}}$, $\theta_1 = \frac{\phi}{6}$, $\theta_2 = \frac{2\pi}{3} + \frac{\phi}{6}$, $\theta_3 = \frac{4\pi}{3} + \frac{\phi}{6}$, i.e. the θ_is are the roots of unity except for the factor $\frac{\phi}{6}$. Using these values, the GHZ state (14.36) reduces to

$$|\mathrm{GHZ}\rangle = \frac{1}{\sqrt{2}}(|+++\rangle + e^{-i\phi}|---\rangle). \tag{14.37}$$

Furthermore, a simple choice $\alpha_1 = 1$, $\alpha_2 = 1$, $\alpha_3 = 0$ (or $\beta_1 = 1$, $\beta_2 = 1$, $\beta_3 = 0$) would produce the W state (14.34) (or the one obtained by the interchange $|\pm\rangle \rightarrow |\mp\rangle$). As explained by Bastin *et al.* [27], the method is applicable in principle to any number of atoms and one can produce certain classes of entangled states.

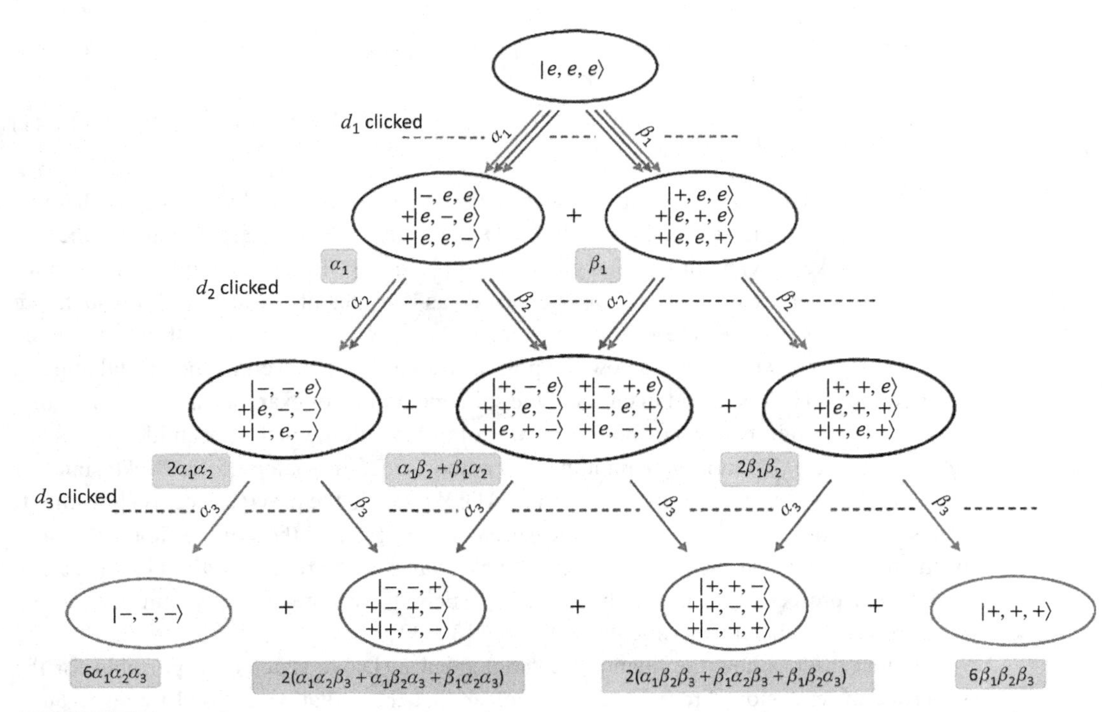

Fig. 14.18 Pyramid of entanglement paths for the case of three emitters initialized in the excited state $|e, e, e\rangle$, after [27].

14.7 Quantum entanglement in Dicke states and superradiance

We introduced Dicke states in Section 12.6. There we considered a collection of N two-level atoms and defined collective angular momentum operators

$$\vec{J} = \sum_{i=1}^{N} \vec{S}^{(i)}, \qquad [J_\alpha, J_\beta] = \mathrm{i}\epsilon_{\alpha\beta\gamma} J_\gamma. \tag{14.38}$$

The eigenstates (12.56) and (12.57) were labeled as $|j, m, \eta\rangle$ with η standing for the degeneracy parameter. The parameter η is zero if $j = N/2$. Let us label the states of a single spin as $|\frac{1}{2}, -\frac{1}{2}, 0\rangle = |g\rangle$, $|\frac{1}{2}, \frac{1}{2}, 0\rangle = |e\rangle$. Then it turns out that many of the Dicke states are entangled states. For example, for $N = 3$, we list the Dicke states that exhibit entanglement

$$\begin{aligned} \left|\frac{3}{2}, \frac{1}{2}, 0\right\rangle &= \frac{1}{\sqrt{3}}(|eeg\rangle + |ege\rangle + |gee\rangle), \\ \left|\frac{3}{2}, -\frac{1}{2}, 0\right\rangle &= \frac{1}{\sqrt{3}}(|egg\rangle + |geg\rangle + |gge\rangle), \end{aligned} \tag{14.39}$$

$$\left|\frac{1}{2},\frac{1}{2},\eta\right\rangle = \frac{1}{\sqrt{6}}(2|gee\rangle - |eeg\rangle - |ege\rangle),\ \frac{1}{\sqrt{2}}|e\rangle(|eg\rangle - |ge\rangle), \tag{14.40}$$

$$\left|\frac{1}{2},-\frac{1}{2},\eta\right\rangle = \frac{1}{\sqrt{6}}(|geg\rangle + |gge\rangle - 2|egg\rangle),\ \frac{1}{\sqrt{2}}|g\rangle(|eg\rangle - |ge\rangle). \tag{14.41}$$

The Dicke states $|\frac{3}{2},\pm\frac{1}{2},0\rangle$ are W states, generally denoted here by $W_{n_e,N-n_e}$, where n_e $(N-n_e)$ atoms are in the state $|e\rangle$ $(|g\rangle)$; the Dicke states $|\frac{1}{2},\pm\frac{1}{2},\eta\rangle$ are doubly degenerate. The only Dicke states with no entanglement are $|\frac{3}{2},\pm\frac{3}{2},0\rangle$. In fact all the Dicke states for $j=N/2$ (except for $m=\pm N/2$) can be written as W states of the form (14.39) such that $n_e(n_g=N-n_e)$ atoms are in the states $|e\rangle$ $(|g\rangle)$ with $m=n_e-N/2$. It may be noted that in Chapter 6 we discussed how the polarization of light as well as the orbital angular momentum can be discussed in terms of the spin operators. For example, a photon traveling in a given direction has two orthogonal states of polarization which we can identify as $|e\rangle$, $|g\rangle$. Thus for a collection of N photons we can construct an analog of the Dicke states or the entangled states of the form (14.39)–(14.41). We saw in the previous section how the W states can be produced by projective measurements. In fact, all the symmetric Dicke states of the form (14.39) as well as asymmetric Dicke states of the form (14.40), (14.41) can be realized by projective measurements [28, 29]. Such Dicke states for a system of four and six photons have been experimentally realized [30–32].

We next discuss how the quantum entanglement in Dicke states is responsible for the superradiant emission. From (14.2), it is clear that the radiation emitted by an N-atom system would be

$$I(t) = I_0 \sum_{ij} \langle S_+^{(i)}(t) S_-^{(j)}(t)\rangle e^{ik_l\vec{n}\cdot(\vec{R}_i-\vec{R}_j)}, \tag{14.42}$$

where

$$I_0 = \frac{\omega_l^4}{c^4}|\vec{n}\times\vec{p}|^2/r^2. \tag{14.43}$$

Let us assume that the atomic system is prepared in an initial state $|\frac{N}{2},m\rangle$ which is completely symmetric. This state has no degeneracy. We can use this symmetry property to evaluate all the expectation values that we need in (14.42). Note that

$$\begin{aligned}
\langle J_z\rangle &= m = \sum_i \langle S_z^{(i)}\rangle = N\langle S_z^{(i)}\rangle,\\
\langle S_+^{(i)}S_-^{(i)}\rangle &= \frac{1}{2} + \langle S_z^{(i)}\rangle = \frac{1}{2} + \frac{m}{N},\\
\langle J_+J_-\rangle &= \left\langle \frac{N}{2},m\middle|J_+J_-\middle|\frac{N}{2},m\right\rangle = \left(\frac{N}{2}+m\right)\left(\frac{N}{2}-m+1\right)\\
&= \sum_{ij}\langle S_+^{(i)}S_-^{(j)}\rangle = \sum_i \langle S_+^{(i)}S_-^{(i)}\rangle + \sum_{i\neq j}\langle S_+^{(i)}S_-^{(j)}\rangle\\
&= \frac{N}{2} + m + N(N-1)\langle S_+^{(i)}S_-^{(j)}\rangle,
\end{aligned} \tag{14.44}$$

and hence we find atom–atom correlation in the state $|\frac{N}{2}, m\rangle$

$$\langle S_+^{(i)} S_-^{(j)} \rangle = \frac{\left(\frac{N}{2}\right)^2 - m^2}{N(N-1)}, \qquad i \neq j, \tag{14.45}$$

which is maximum for $m = 0$. On using (14.44) and (14.45) in (14.42) we find

$$I = I_0 \left\{ \frac{N}{2} + m + \frac{\left(\frac{N}{2}\right)^2 - m^2}{N(N-1)} \cdot \sum_{i \neq j} \mathrm{e}^{\mathrm{i}k_l \vec{n} \cdot (\vec{R}_i - \vec{R}_j)} \right\}. \tag{14.46}$$

Dicke considered a system whose dimensions were much smaller than a single wavelength, then by setting all exponentials to unity, (14.46) reduces to

$$I = I_0 \left(\frac{N}{2} + m \right) \left(\frac{N}{2} - m + 1 \right). \tag{14.47}$$

For $m = 0$, $I \propto N^2$. This is superradiance when half the atoms are in the excited state and half are in the ground state. Clearly this behavior results from the nonvanishing of the atom–atom correlation $\langle S_+^{(i)} S_-^{(j)} \rangle$ (Eq. (14.45)), which takes a maximum value for the Dicke state that exhibits superradiance. The fact that the correlation is nonzero is also the property of the entangled state $|\frac{N}{2}, 0\rangle$. Thus we conclude that the superradiance arises from the entangled nature of the Dicke state $|\frac{N}{2}, 0\rangle$.

14.8 Multi-path quantum interference as the source of Dicke superradiance

In this section we develop a physical picture for Dicke superradiance. We show that the Dicke superradiance can be considered as arising from multipath quantum interference [33].

Let us examine the transition amplitude for each individual photon detection event. The net result would then be obtained by coherently summing over all the paths via which photons reach the detector. In the following we will demonstrate that the interference of various quantum paths gives us a transparent physical picture of the superradiant emission from symmetric W states.

Let us first investigate the different quantum paths of the initially separable state $|S_{2,0}\rangle = |e\,e\rangle$ (cf. Figure 14.19) which lead to a successful photon detection event. For a particular event the detector cannot resolve from which of the two atoms the photon was emitted due to the far-field condition. There are thus two distinct possibilities: either the photon (black arrow) was emitted by the first excited atom (black circle) transferring it into the ground state (white circle), where a phase $\mathrm{e}^{-\mathrm{i}\varphi_1}$ is accumulated by the photon, or the photon was emitted by the second atom resulting in the accumulation of the phase $\mathrm{e}^{-\mathrm{i}\varphi_2}$. Each quantum path leads to a different final state, so in principle they are distinguishable, and we do not expect interference terms to appear. Explicitly, from Figure 14.19, we obtain the intensity distribution

$$\mathrm{I}_{|S_{2,0}\rangle} = \| \mathrm{e}^{-\mathrm{i}\varphi_1} |g\,e\rangle \|^2 + \| \mathrm{e}^{-\mathrm{i}\varphi_2} |e\,g\rangle \|^2 = 2\,, \tag{14.48}$$

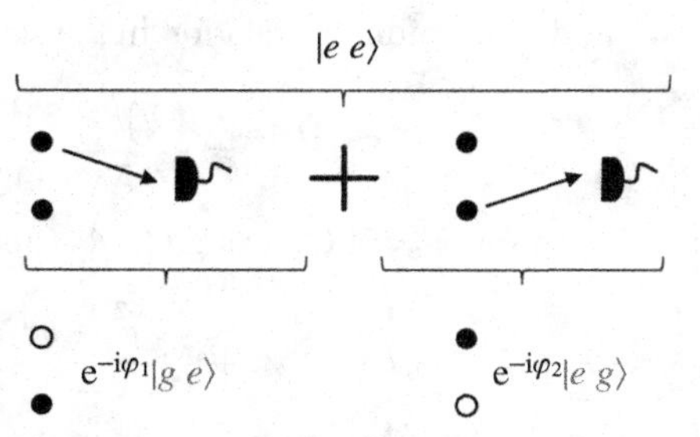

Fig. 14.19 Possible quantum paths of the initially separable state $|S_{2,0}\rangle$. Black circles denote atoms in the excited state and white circles denote atoms in the ground state. The middle row depicts the different quantum paths. The lower row displays the final states of the atoms and the phases accumulated by the photon along the different quantum paths. See text for details, after [33].

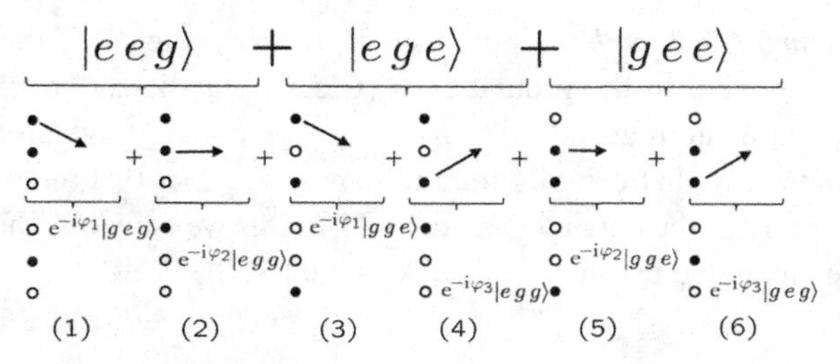

Fig. 14.20 Possible quantum paths leading to the detection of a photon emitted by atoms in the W state $|W_{2,1}\rangle$.

where the norm of the state vector $|\Psi\rangle$ is denoted by $\||\Psi\rangle\|^2 = \langle\Psi|\Psi\rangle$. Let us compare these results to the superposition of quantum paths and the intensity distribution obtained in the case of an initial W state $|W_{2,1}\rangle$ with two excited atoms and one in the ground state. From Eq. (14.34) this state reads

$$|W_{2,1}\rangle = \frac{1}{\sqrt{3}} \left(|e\,e\,g\rangle + |e\,g\,e\rangle + |g\,e\,e\rangle\right) . \tag{14.49}$$

Figure 14.20 depicts the different quantum paths leading to a successful measurement event. Let us exemplify the emerging interference by considering only the first term in the coherent sum of $|W_{2,1}\rangle$. The state $|e\,e\,g\rangle$ basically leads to the same quantum paths as the state $|S_{2,0}\rangle$: either the first atom emits the photon leading to the final state $|g\,e\,g\rangle$ and to an accumulation of the phase $\mathrm{e}^{-\mathrm{i}\varphi_1}$ or the second atom emits the photon, so that the final state is $|e\,g\,g\rangle$ and the accumulated phase corresponds to $\mathrm{e}^{-\mathrm{i}\varphi_2}$. However, different from the separable state $|S_{2,0}\rangle$, we have here a superposition of three different terms in the state $|W_{2,1}\rangle$ leading to six quantum paths in total. These quantum paths can lead to a constant contribution to the intensity as in the case of an initial separable state; however, they are also capable of interference, namely, with another *indistinguishable quantum path*. Taking into account the normalization factor $1/\sqrt{3}$ we find the same constant contribution to the intensity in the case of the W state $|W_{2,1}\rangle$ (namely $6/3$ due to the six quantum paths that do not interfere) and in the case of the separable state $|S_{2,0}\rangle$ (namely 2). However, photons which were emitted from the atomic state $|W_{2,1}\rangle$ can occupy more than one quantum path leading to the same final state (cf., e.g., the far-left and far-right quantum paths in Figure 14.20, both leading to the final state $|geg\rangle$). Furthermore, for all quantum paths the initial states are equal from the detectors' point of view – the detector is unable to identify from which atom the photon was emitted due to the far-field assumption. Thus we obtain interfering quantum paths exclusively for nonseparable states which are leading to interference terms in the intensity distribution.

Let us explicitly calculate the intensity produced by the state $|W_{2,1}\rangle$ to quantitatively investigate the validity of our quantum path interpretation. From Figure 14.20, it is clearly equal to

$$\mathrm{I}_{|W_{2,1}\rangle} = \frac{1}{3}\left\|(\mathrm{e}^{-\mathrm{i}\varphi_1} + \mathrm{e}^{-\mathrm{i}\varphi_2})|gge\rangle\right\|^2 + \frac{1}{3}\left\|(\mathrm{e}^{-\mathrm{i}\varphi_1} + \mathrm{e}^{-\mathrm{i}\varphi_3})|geg\rangle\right\|^2 + \frac{1}{3}\left\|(\mathrm{e}^{-\mathrm{i}\varphi_2} + \mathrm{e}^{-\mathrm{i}\varphi_3})|egg\rangle\right\|^2, \tag{14.50}$$

whose maximum value is four, which is twice that from a separable system of two excited atoms – the extra contribution arises from interfering pathways. In the following we demonstrate that the enhanced maximal emission of radiation by W states can be explained purely by additional constructive interference terms created by indistinguishable quantum paths. To this end we cast the foregoing argument into a formula for the maximum of the intensity from a W state

$$\left[\mathrm{I}_{|W\rangle}\right]^{\mathrm{Max}} = \left[\mathrm{I}_{|S\rangle}\right]^{\mathrm{Max}} + (\mathcal{P}^{\mathrm{pair}}) \times (f) \times (\mathcal{N})\,. \tag{14.51}$$

In Eq. (14.51) $(\mathcal{P}^{\mathrm{pair}})$ abbreviates the number of interfering quantum path *pairs* leading to the same final state, which when multiplied by the number of final states (f) allows us to arrive at the total number of interfering quantum paths pairs, i.e. interference terms, contributing to the intensity maximum of the signal. Together with the squared normalization constant $(\mathcal{N})$ of the corresponding W state the expression $(\mathcal{P}^{\mathrm{pair}}) \cdot (f) \cdot (\mathcal{N})$ equals the constructive contribution of the interference terms to the maximum of intensity.

Now we adopt the foregoing reasoning to an initial generalized symmetric W state $|W_{n_e,N-n_e}\rangle$ with n_e excited atoms and $N - n_e$ atoms in the ground state. The general formula for the maximum intensity of the W state $|W_{n_e,N-n_e}\rangle \equiv |W_\star\rangle$ can be derived using combinatorial considerations and the maximum of the intensity n_e of the separable state $\mathrm{I}_{|S_{n_e,N-n_e}\rangle}$. It reads

$$\begin{aligned}\left[\mathrm{I}_{|W_\star\rangle}\right]^{\max} &= n_e + \left[(\mathcal{P}^{\mathrm{pair}}) \times (f) \times (\mathcal{N})\right]_{|W_\star\rangle} \\ &= n_e + n_g(n_g+1) \times \binom{N}{n_e - 1} \times \binom{N}{n_e}^{-1} \\ &= n_e(n_g+1) = \left(\frac{N}{2} + m\right)\left(\frac{N}{2} - m + 1\right), \quad m = -\frac{N}{2} + n_e.\end{aligned} \tag{14.52}$$

Let us investigate the different terms of Eq. (14.52) in more detail. As stated earlier, $\mathcal{N} = \binom{N}{n_e}^{-1}$ is the squared normalization constant of the generalized symmetric W state. The number of final states (f) can be derived by taking into account that after the detection of a photon there are $n_e - 1$ excited atoms left which are able to occupy N different position in the chain of N atoms, which leads to $\binom{N}{n_e-1}$. The crucial term $\mathcal{P}^{\mathrm{pair}}$ needs more

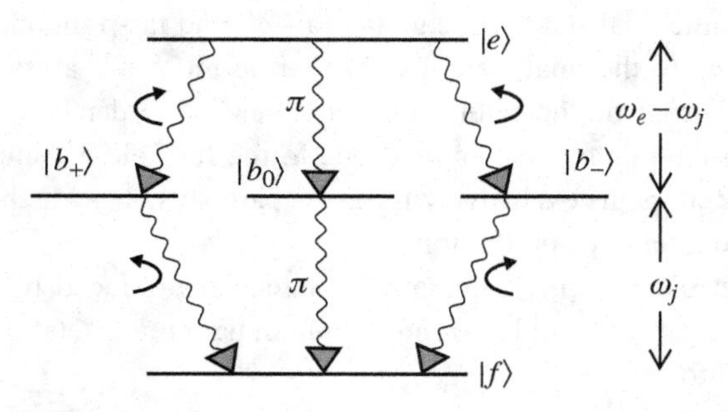

Fig. 14.21 Entanglement of photons emitted from the excited level $|e\rangle$ to the levels $|b_j\rangle$ and then to the level $|f\rangle$.

explanation: n_e different single quantum paths lead to a detection event for every term of the initial W state (cf., e.g., Eq. (14.50)). If we now multiply these single quantum paths by the number of terms of the initial W state (given by $\binom{N}{n_e}$) we arrive at the total number of single quantum paths. The number of single quantum paths leading to the same final states – abbreviated by $\mathcal{P}^{\text{single}}$ – is then obtained by dividing the total number of single quantum paths by the number of final states:

$$\mathcal{P}^{\text{single}} = \frac{n_e \binom{N}{n_e}}{\binom{N}{n_e - 1}} = n_g + 1 . \tag{14.53}$$

These $n_g + 1$ single quantum paths which lead to the same final state now interfere among each other, producing in total $n_g(n_g + 1)$ interfering quantum path pairs [33].

14.9 Entanglement of photons produced in an atomic cascade

Consider the emission of radiation from the excited level $|e\rangle$ to the levels $|b_j\rangle$ and then to the level $|f\rangle$ as shown in Figure 14.21. The levels $|e\rangle$ and $|f\rangle$ can be taken with angular momentum $j = 0$, $m = 0$. The levels $|b_j\rangle$ correspond to $j = 1$, $m = \pm, 0$. The generated photons are entangled in momentum and polarization. As indicated in Figure 14.21, the polarization of photons depends on the atomic transition. The $|e\rangle \rightarrow |f\rangle$ transition with the emission of circularly polarized photons proceeds via different pathways. The two pathways involving the first emission of a ↻ (↺) photon followed by the emission of a ↺ (↻) photon lead to quantum entanglement. The cascade emission is extensively discussed in the literature [34], because of its importance in the first experiments on the nonlocal character of quantum mechanics and tests of the violations of the Bell inequalities [35–37].

We follow the Weisskopf–Wigner approach to evaluate the wavefunction of the generated photons. In analogy to (7.64), the wavefunction for the combined system of atom and field can be written as

$$|\psi\rangle = c_e(t)|e\rangle|\{0_{\vec{k}s}\}\rangle + \sum_{j\vec{k}s} c_{j\vec{k}s}(t)|b_j\rangle|\{1_{\vec{k}s}\}\rangle + \sum_{\vec{k}s\vec{q}\sigma} c_{\vec{k}s\vec{q}\sigma}(t)|f\rangle|\{1_{\vec{k}s}, 1_{\vec{q}\sigma}\}\rangle . \tag{14.54}$$

In analogy to (7.65) we can write the Schrödinger equation for the amplitudes c

$$\begin{aligned}
\dot{c}_e &= -\mathrm{i}\omega_e c_e - \mathrm{i}\sum_{j\vec{k}s} g_{j\vec{k}s} c_{j\vec{k}s}, \\
\dot{c}_{j\vec{k}s} &= -\mathrm{i}(\omega_j + \omega_{ks}) c_{j\vec{k}s} - \mathrm{i}g^*_{j\vec{k}s} c_e - \mathrm{i}\sum_{\vec{q}\sigma} \tilde{g}_{j\vec{q}\sigma} c_{\vec{k}s\vec{q}\sigma}, \\
\dot{c}_{\vec{k}s\vec{q}\sigma} &= -\mathrm{i}(\omega_{ks} + \omega_{q\sigma}) c_{\vec{k}s\vec{q}\sigma} - \mathrm{i}\sum_{j} \tilde{g}^*_{j\vec{q}\sigma} c_{j\vec{k}s}.
\end{aligned} \tag{14.55}$$

The unperturbed frequencies of the atomic levels are defined as in Figure 14.21. The coupling constants have forms like (7.66)

$$\begin{aligned}
g_{j\vec{k}s} &= -\mathrm{i}\sqrt{\frac{2\pi\omega_{ks}}{\hbar V}}\mathrm{e}^{\mathrm{i}\vec{k}\cdot\vec{R}}\vec{p}_{ej}\cdot\vec{\epsilon}_{\vec{k}s}, \\
\tilde{g}_{j\vec{q}\sigma} &= -\mathrm{i}\sqrt{\frac{2\pi\omega_{q\sigma}}{\hbar V}}\mathrm{e}^{\mathrm{i}\vec{q}\cdot\vec{R}}\vec{p}_{jf}\cdot\vec{\epsilon}_{\vec{q}\sigma}.
\end{aligned} \tag{14.56}$$

The initial condition is $c_e(0) = 1$, $c_{j\vec{k}s} = c_{\vec{k}s\vec{q}\sigma} = 0$. As in Section 7.6, Eq. (14.55) is solved by the Laplace transform technique leading to

$$\begin{aligned}
\hat{c}_{\vec{k}s\vec{q}\sigma}(p) &= -\mathrm{i}\sum_{j}\frac{\tilde{g}^*_{j\vec{q}\sigma}\hat{c}_{j\vec{k}s}(p)}{p + \mathrm{i}(\omega_{ks} + \omega_{q\sigma})}, \\
[p + \mathrm{i}(\omega_j + \omega_{ks})]\hat{c}_{j\vec{k}s}(p) &= -\mathrm{i}g^*_{j\vec{k}s}\hat{c}_e(p) - \sum_{j'}\sum_{\vec{q}\sigma}\frac{\tilde{g}_{j\vec{q}\sigma}\tilde{g}^*_{j'\vec{q}\sigma}\hat{c}_{j'\vec{k}s}(p)}{p + \mathrm{i}(\omega_{ks} + \omega_{q\sigma})},
\end{aligned} \tag{14.57}$$

where p is the Laplace variable. It should be borne in mind that, according to Figure 14.13 for each σ, there is a definite nonzero dipole matrix element $\vec{p}_{jf}$, i.e. the value of j is fixed by σ. Hence $j' = j$ in (14.57). Furthermore, the summation in (14.57) can be handled in the same way as in (7.70) by assuming weak coupling to the vacuum of the radiation field. In such a case (14.57) leads to

$$\hat{c}_{j\vec{k}s}(p) = -\mathrm{i}\frac{g^*_{j\vec{k}s}\hat{c}_e(p)}{p + \mathrm{i}(\omega_j + \omega_{ks}) + \frac{1}{2}A_b}, \tag{14.58}$$

where A_b is the Einstein A coefficient for the state $|b_j\rangle$. Now using (14.58) and following a similar procedure for c_e and introducing the A coefficient for the state $|e\rangle$, we get

$$\hat{c}_e(p) = \frac{1}{p + \mathrm{i}\omega_e + \frac{1}{2}A_e}. \tag{14.59}$$

On substituting (14.58) and (14.59) in (14.57), we get the two-photon amplitude

$$\hat{c}_{\vec{k}s\vec{q}\sigma}(p) = -\frac{[p + \mathrm{i}(\omega_{ks} + \omega_{q\sigma})]^{-1}}{p + \mathrm{i}\omega_e + \frac{1}{2}A_e}\sum_{j}\frac{\tilde{g}^*_{j\vec{q}\sigma}g^*_{j\vec{k}s}}{p + \mathrm{i}(\omega_j + \omega_{ks}) + \frac{1}{2}A_b}, \tag{14.60}$$

and hence in the long time limit $A_e t, A_b t \gg 1$, we obtain our final result for the two-photon amplitude

$$c_{\vec{k}s\vec{q}\sigma} \to -\sum_{j}\frac{\exp\{-\mathrm{i}(\omega_{ks} + \omega_{q\sigma})t\}}{\left[\frac{1}{2}A_e + \mathrm{i}(\omega_e - \omega_{ks} - \omega_{q\sigma})\right]}\frac{\tilde{g}^*_{j\vec{q}\sigma}g^*_{j\vec{k}s}}{\left[\frac{1}{2}A_b + \mathrm{i}(\omega_j - \omega_{q\sigma})\right]}. \tag{14.61}$$

The two photons with polarizations ↻ and ↺ are clearly entangled: the wavefunction for two photons traveling in directions $\vec{k}$ and $\vec{\sigma}$ would have the form $(|\circlearrowright, \circlearrowleft\rangle + |\circlearrowleft, \circlearrowright\rangle)/\sqrt{2}$.

Having obtained the form of the two-photon amplitude, we now find the nature of the two-photon correlation function (8.29) produced by cascade emission. The probability of detecting one photon at $(\vec{r}_1, t_1)$ and another at $(\vec{r}_2, t_2)$ is

$$G^{(2)}(\vec{r}_1, t_1, \vec{r}_2, t_2) = \sum_{\alpha\beta} \langle E_\alpha^{(-)}(\vec{r}_1, t_1) E_\beta^{(-)}(\vec{r}_2, t_2) E_\beta^{(+)}(\vec{r}_2, t_2) E_\alpha^{(+)}(\vec{r}_1, t_1)\rangle. \tag{14.62}$$

Now the action of $E^{(+)}E^{(+)}$ on a two-photon wavefunction would yield the vacuum state and therefore

$$G^{(2)}(\vec{r}_1, t_1; \vec{r}_2, t_2) = \sum_{\alpha\beta} |\psi_{\alpha\beta}(\vec{r}_1, t_1; \vec{r}_2, t_2)|^2, \tag{14.63}$$

where

$$\psi_{\alpha\beta}(\vec{r}_1, t_1; \vec{r}_2, t_2) = \langle 00| E_\beta^{(+)}(\vec{r}_2, t_2) E_\alpha^{(+)}(\vec{r}_1, t_1)|\psi\rangle, \tag{14.64}$$

which on using (1.22) becomes

$$\begin{aligned}\psi_{\alpha\beta}(\vec{r}_1, t_1; \vec{r}_2, t_2) = -\sum_{\vec{k}s\vec{q}\sigma} \sqrt{\frac{2\pi\hbar\omega_k \cdot 2\pi\hbar\omega_q}{V^2}}\, \mathrm{e}^{\mathrm{i}\vec{k}\cdot\vec{r}_1 - \mathrm{i}\omega_k t_1 + \mathrm{i}\vec{q}\cdot\vec{r}_2 - \mathrm{i}\omega_q t_2} \\ \times \langle 00| a_{\vec{q}\sigma} a_{\vec{k}s}|\psi\rangle (\vec{\epsilon}_{\vec{k}s})_\alpha (\vec{\epsilon}_{\vec{q}\sigma})_\beta.\end{aligned} \tag{14.65}$$

Note that the dimension of ψ is energy per unit volume. We next have to use the value obtained from (14.61)

$$\langle 00| a_{\vec{q}\sigma} a_{\vec{k}s}|\psi\rangle = c_{\vec{k}s\vec{q}\sigma} + c_{\vec{q}\sigma\vec{k}s} \tag{14.66}$$

in (14.65) and carry out the summations by letting $V \to \infty$. The resulting integrations can be done using methods from complex variables. We cite the result [34]

$$\begin{aligned}&\psi_{\alpha\beta}(\vec{r}_1, t_1; \vec{r}_2, t_2) \\ &= \frac{-\hbar\sqrt{\omega_{ej}\omega_{jf}A_e A_b}}{c|\vec{r}_1 - \vec{R}| \cdot |\vec{r}_2 - \vec{R}|} \exp\left[-\left(\mathrm{i}\omega_e + \frac{A_e}{2}\right)\left(t_1 - \frac{|\vec{r}_1 - \vec{R}|}{c}\right)\right] \\ &\times \Theta\left(t_1 - \frac{|\vec{r}_1 - \vec{R}|}{c}\right) \exp\left\{-\left(\mathrm{i}\omega_j + \frac{A_b}{2}\right)\left[\left(t_2 - \frac{|\vec{r}_2 - \vec{R}|}{c}\right) - \left(t_1 - \frac{|\vec{r}_1 - \vec{R}|}{c}\right)\right]\right\} \\ &\times \Theta\left[\left(t_2 - \frac{|\vec{r}_2 - \vec{R}|}{c}\right) - \left(t_1 - \frac{|\vec{r}_1 - \vec{R}|}{c}\right)\right] + (1 \leftrightarrow 2),\end{aligned} \tag{14.67}$$

where the Θ function is defined by

$$\Theta(\alpha) \equiv \frac{1}{2\pi\mathrm{i}} \int_{-\infty}^{+\infty} \frac{\mathrm{e}^{\mathrm{i}\alpha\tau}\,\mathrm{d}\tau}{\tau - \mathrm{i}\epsilon} = \begin{cases} 1 & \text{if } \alpha > 0, \\ 0 & \text{if } \alpha < 0. \end{cases} \tag{14.68}$$

The Θ functions basically give the arrival times of photons at the detectors. These also reflect the fact that, in a cascade, photons are emitted in a definite time order. As already mentioned, the cascade emission is extremely important as it enables one to obtain definitive tests of Bell inequalities [37].

Exercises

14.1 Generalize the results of Section 14.2 when the source A is an excited single two-level atom whereas the source B is an arbitrary independent source. In (14.1) use $S^{(B)} \to a$, where a is the Bosonic operator. Show that the visibility in terms of the Mandel Q parameter of the source B is $\upsilon = 1/\left[1 + \frac{1}{2}(\bar{n}_1 + Q_{\mathrm{M}})\right]$, $\bar{n}_1 = \langle a^\dagger a\rangle$. A whole class of special cases are discussed in [38].

14.2 Consider Hong–Ou–Mandel interference between a single photon source and another one with arbitrary statistics, described by its density matrix ρ. The operators c and d are related to a and b via (5.3) and (5.7). Calculate the joint probability $\langle c^\dagger(t)d^\dagger(t+\tau)d(t+\tau)c(t)\rangle$ of detecting one photon in each of the ports c and d at times t and $t+\tau$ respectively. Find the maximum and minimum value of the joint probability as a function of the statistical properties of b and for $\langle a^\dagger(t)a(t+\tau)\rangle = \mathrm{e}^{-\Gamma|\tau|}$. Calculate the contrast when the source b is a thermal source, assuming that $\langle b^\dagger(t)b(t+\tau)\rangle \approx \langle b^\dagger b\rangle$, and $\langle b^\dagger b\rangle \leq 1$. For experimental work see [39].

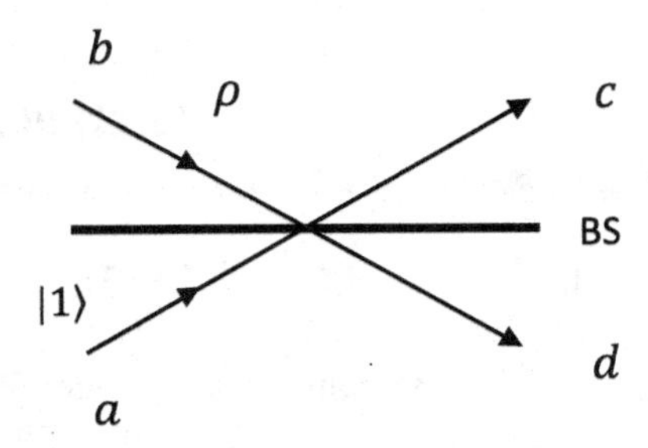

14.3 Prove the inequality (14.23) for the intensity–intensity correlations.

14.4 Calculate for the case of two identical atoms located at $\vec{R}_A$ and $\vec{R}_B$ and with the scheme shown in Figure 14.10, the probability of detecting jointly a photon γ from the transition $|e\rangle \to |f\rangle$ and another one φ from the transition $|e'\rangle \to |g\rangle$, i.e. calculate the correlation function

$$G^{(2)}(\vec{r}, t, \vec{\rho}, \tau) = \langle E_\gamma^{(-)}(\vec{r}, t)E_\varphi^{(-)}(\vec{\rho}, t+\tau)E_\varphi^{(+)}(\vec{\rho}, t+\tau)E_\gamma^{(+)}(\vec{r}, t)\rangle.$$

Show the existence of interference terms in $G^{(2)}$, for details see [11]. Note that each field operator would be given by an expression like (13.48). The corresponding dipole moment operator S_- is $|f\rangle\langle e|$ [$|g\rangle\langle e'|$] for γ [φ] photon.

14.5 Prove (14.67); for details see [34].

14.6 Show that for a system of four two-level atoms, all possible Dicke states for $j = 1$ are given by

$$\frac{1}{2\sqrt{3}}(3|eeeg\rangle - |eege\rangle - |egee\rangle - |geee\rangle),$$

$$\frac{1}{\sqrt{6}}(|eegg\rangle + |egeg\rangle + |geeg\rangle - |egge\rangle - |gege\rangle - |ggee\rangle),$$

$$\frac{1}{2\sqrt{3}}(|eggg\rangle + |gegg\rangle + |ggeg\rangle - 3|ggge\rangle),$$

$$\frac{1}{\sqrt{6}}(2|eege\rangle - |egee\rangle - |geee\rangle),$$

$$\frac{1}{\sqrt{12}}(2|eegg\rangle - |egeg\rangle - |geeg\rangle + |egge\rangle + |gege\rangle - 2|ggee\rangle),$$

$$\frac{1}{\sqrt{6}}(|eggg\rangle + |gegg\rangle - 2|ggeg\rangle),$$

$$\frac{1}{\sqrt{2}}(|egee\rangle - |geee\rangle),$$

$$\frac{1}{2}(|egeg\rangle - |geeg\rangle + |egge\rangle - |gege\rangle),$$

$$\frac{1}{\sqrt{2}}(|eggg\rangle - |gegg\rangle).$$

References

[1] L. Mandel, *Phys. Rev. A* **28**, 929 (1983).

[2] Th. Richter, *Opt. Commun.* **80**, 285 (1991).

[3] U. Eichmann, J. C. Bergquist, J. J. Bollinger, J. M. Gilligan, W. M. Itano, D. J. Wineland, and M. G. Raizen, *Phys. Rev. Lett.* **70**, 2359 (1993).

[4] C. Skornia, J. von Zanthier, G. S. Agarwal, E. Werner, and H. Walther, *Phys. Rev. A* **64**, 063801 (2001).

[5] G. S. Agarwal, J. von Zanthier, C. Skornia, and H. Walther, *Phys. Rev. A* **65**, 053826 (2002).

[6] C. Thiel, T. Bastin, J. Martin, E. Solano, J. von Zanthier, and G. S. Agarwal, *Phys. Rev. Lett.* **99**, 133603 (2007).

[7] S. Gerber, D. Rotter, M. Hennrich *et al.*, *New J. Phys.* **11**, 013032 (2009).

[8] J. Beugnon, M. Jones, J. Dingjan *et al.*, *Nature (London)* **440**, 779 (2006).

[9] P. Maunz, D. L. Moehring, S. Olmschenk, K. C. Younge, D. N. Matsukevich, and C. Monroe, *Nature Physics* **3**, 538 (2007).

[10] E. B. Flagg, A. Muller, S. V. Polyakov, A. Ling, A. Migdall, and G. S. Solomon, *Phys. Rev. Lett.* **104**, 137401 (2010).

[11] M. O. Scully and K. Drühl, *Phys. Rev. A* **25**, 2208 (1982).
[12] Y. Aharonov and M. S. Zubairy, *Science* **307**, 875 (2005).
[13] K. W. Chan, C. K. Law, and J. H. Eberly, *Phys. Rev. Lett.* **88**, 100402 (2002).
[14] J. Volz, M. Weber, D. Schlenk *et al.*, *Phys. Rev. Lett.* **96**, 030404 (2006).
[15] K. Bergmann, H. Theuer, and B. W. Shore, *Rev. Mod. Phys.* **70**, 1003 (1998).
[16] D. G. Norris, L. A. Orozco, P. Barberis-Blostein, and H. J. Carmichael, *Phys. Rev. Lett.* **105**, 123602 (2010).
[17] D. L. Moehring, P. Maunz, S. Olmschenk *et al.*, *Nature (London)* **449**, 68 (2007).
[18] L. M. Duan, M. D. Lukin, J. I. Cirac, and P. Zoller, *Nature (London)* **414**, 413, (2001).
[19] D. Matsukevich and A. Kuzmich, *Science* **306**, 663 (2004).
[20] C. W. Chou, H. de Riedmatten, D. Felinto, S. V. Polyakov, S. J. van Enk, and H. J. Kimble, *Nature (London)* **438**, 828 (2005).
[21] B. Deb and G. S. Agarwal, *Phys. Rev. A* **78**, 013639 (2008).
[22] M. Saffman, T. G. Walker, and K. Molmer, *Rev. Mod. Phys.* **82**, 2313 (2010).
[23] A. Gaetan, Y. Miroshnychenko, T. Wilk *et al.*, *Nature Physics* **5**, 115 (2009).
[24] E. Urban, T. A. Johnson, T. Henage *et al.*, *Nat. Phys.* **5**, 110 (2009).
[25] D. M. Greenberger, M. A. Horne, A. Shimony, and A. Zeilinger, *Am. J. Phys.* **58**, 1131 (1990).
[26] D. Bouwmeester, J. Pan, M. Daniell, H. Weinfurter, and A. Zeilinger, *Phys. Rev. Lett.* **82**, 1345 (1999).
[27] T. Bastin, C. Thiel, J. von Zanthier, L. Lamata, E. Solano, and G. S. Agarwal, *Phys. Rev. Lett.* **102**, 053601 (2009).
[28] C. Thiel, J. von Zanthier, T. Bastin, E. Solano, and G. S. Agarwal, *Phys. Rev. Lett.* **99**, 193602 (2007).
[29] A. Maser, U. Schilling, T. Bastin, E. Solano, C. Thiel, and J. von Zanthier, *Phys. Rev. A* **79**, 033833 (2009).
[30] R. Prevedel, G. Cronenberg, M. S. Tame *et al.*, *Phys. Rev. Lett.* **103**, 020503 (2009).
[31] W. Wieczorek, R. Krischek, N. Kiesel, P. Michelberger, G. Tóth, and H. Weinfurter, *Phys. Rev. Lett.* **103**, 020504 (2009).
[32] N. Kiesel, C. Schmid, G. Tóth, E. Solano, and H. Weinfurter, *Phys. Rev. Lett.* **98**, 063604 (2007).
[33] R. Wiegner, J. von Zanthier, and G. S. Agarwal, *Phys. Rev. A* **84**, 023805 (2011).
[34] M. O. Scully and M. S. Zubairy, *Quantum Optics* (Cambridge: Cambridge University Press, 1997), p.617.
[35] J. F. Clauser, M. A. Horne, A. Shimony, and R. A. Holt, *Phys. Rev. Lett.* **23**, 880 (1969).
[36] J. F. Clauser and M. A. Horne, *Phys. Rev. D* **10**, 526 (1974).
[37] A. Aspect, P. Grangier, and G. Roger, *Phys. Rev. Lett.* **47**, 460 (1981).
[38] R. Wiegner, J. von Zanthier, and G. S. Agarwal, *J. Phys. B: At. Mol. Opt. Phys.* **44**, 055501 (2011).
[39] X. Li, L. Yang, L. Cui, Z. Ou, and D. Yu, *Optics Express* **16**, 12505 (2008).

15 Near field radiative effects

Dipole–dipole interactions between atoms or molecules profoundly affect the light absorption that occurs in matter. The spectral characteristics of light absorption can be strongly modified. For example, the weak field absorption spectrum splits into a doublet [1]. The separation of the doublet depends on the strength of the dipole–dipole interaction. Furthermore, the photon antibunching exhibited by a single-atom fluorescence starts becoming bunching due to a nearby atom [2]. The dipole–dipole interactions also give rise to fascinating applications in quantum information science, such as quantum logic operations in neutral atoms [3]. The dipole–dipole interaction can transfer excitation from one atom to the other and this transfer process produces entanglement between two atoms. For two atoms with the first one in the excited state $|e_A, g_B\rangle$, the excitation would be on the atom B, i.e. $|e_A, g_B\rangle \rightarrow |g_A, e_B\rangle$ after a certain time. Clearly halfway through one would expect the state of the two atom system would be of the form $(|e_A, g_B\rangle + |g_A, e_B\rangle)/\sqrt{2}$, which is a state of maximum entanglement The dipole–dipole interaction is known to aid the process of simultaneous excitation of two atoms leading to the possibility of nanometric resolution of atoms [2,4–6]. There are other types of dipole–dipole interactions, such as van der Waal interaction [7] which involves two atoms each in a state, which could be an excited state or the ground state. The level shifts associated with these interactions can strongly modify the laser excitation of adjacent atoms leading to the dipole blockade effect [8–11]. Thus the first excited atom prevents any further excitation in a confined volume, by shifting the resonance for its nonexcited neighbors. This results in the production of singly excited collective states which naturally exhibit entanglement. The entanglement resulting from dipole–dipole interactions is deterministic and hence quite useful. This is in contrast to the heralded entanglement discussed in Chapter 14.

15.1 Near field radiative effects – coupling between dipoles

It is well known that the electrostatic interaction between two dipoles $\vec{p}_A$ and $\vec{p}_B$ located at $\vec{r}_A$ and $\vec{r}_B$, respectively, is given by

$$\Omega_{AB} = \frac{\vec{p}_A \cdot \vec{p}_B - 3(\vec{n} \cdot \vec{p}_A)(\vec{n} \cdot \vec{p}_B)}{|\vec{r}_A - \vec{r}_B|^3}, \tag{15.1}$$

where $\vec{n}$ is the unit vector in the direction $\vec{r}_A - \vec{r}_B$. The interaction (15.1) can be thought of as arising, say, from the electric field at the position of the dipole A due to the dipole B

$$\vec{E}(\vec{r}_A) = \frac{3\vec{n}(\vec{p}_B \cdot \vec{n}) - \vec{p}_B}{|\vec{r}_A - \vec{r}_B|^3}, \qquad \Omega_{AB} = -\vec{p}_A \cdot \vec{E}(\vec{r}_A). \tag{15.2}$$

Let us now consider the electromagnetic version of (15.1). Consider a dipole which is oscillating at the frequency ω, then from the Maxwell equations, we can write the electric field as

$$\vec{E}_i(\vec{r}_A,\omega) = \frac{\omega^2}{c^2}\sum_j G_{ij}(\vec{r}_A,\vec{r}_B,\omega)\vec{p}_{Bj}, \tag{15.3}$$

where G_{ij} is the free space Green's function (tensor) given by [12]

$$G_{ij}(\vec{r},\vec{r}\,',\omega) = \left(\delta_{ij} + \frac{c^2}{\omega^2}\frac{\partial^2}{\partial r_i \partial r_j}\right)\frac{e^{i\frac{\omega}{c}|\vec{r}-\vec{r}\,'|}}{|\vec{r}-\vec{r}\,'|}. \tag{15.4}$$

We write the real and imaginary parts of $\overset{\Rightarrow}{G}$ as

$$G_{ij}(\vec{r}_A,\vec{r}_B,\omega) = k(-\Omega_{ij}(\vec{r}_A,\vec{r}_B,\omega) + i\gamma_{ij}(\vec{r}_A,\vec{r}_B,\omega)),\ k=\omega/c, \tag{15.5}$$

$$\Omega_{ij} = \delta_{ij}\left[-\frac{\cos x}{x} + \frac{\cos x}{x^3} + \frac{\sin x}{x^2}\right] + n_i n_j\left[\frac{\cos x}{x} - \frac{3\cos x}{x^3} - \frac{3\sin x}{x^2}\right], \tag{15.6}$$

$$\gamma_{ij} = \delta_{ij}\left[\frac{\sin x}{x} - \frac{\sin x}{x^3} + \frac{\cos x}{x^2}\right] + n_i n_j\left[-\frac{3\cos x}{x^2} + \frac{3\sin x}{x^3} - \frac{\sin x}{x}\right], \tag{15.7}$$

$$x = k|\vec{r}_A - \vec{r}_B|.$$

The Green's tensor is symmetric under $\vec{n}\to-\vec{n}$. Note that in the static limit

$$c\to\infty (x\to 0),\quad \Omega_{ij}\to\frac{1}{x^3}(\delta_{ij}-3n_i n_j),\quad \gamma_{ij}\to\frac{2}{3}\delta_{ij}, \tag{15.8}$$

and therefore (15.3) goes over to (15.2), i.e.

$$\vec{E}_i(\vec{r}_A,\omega)\to -\frac{1}{|\vec{r}_A-\vec{r}_B|^3}(\vec{p}_B - 3\vec{n}(\vec{n}\cdot\vec{p}_B)). \tag{15.9}$$

For $c\neq\infty$, the force on the dipole A is given by $e\vec{E}(\vec{r}_A,t)$, and therefore its motion given by

$$\left(\frac{d^2}{dt^2}+\omega_0^2\right)\vec{p}_A = \frac{e^2}{m}\vec{E}(\vec{r}_A,t), \tag{15.10}$$

is determined by the complex Green's function. Note that $\vec{E}(\vec{r}_A,t)$ would have two contributions: one from the dipole B and the other from the dipole A. The contribution from the dipole A (self-contribution) is obtained from (15.3) as

$$\vec{E}_i(\vec{r}_A,\omega) = \lim_{\vec{r}\to\vec{r}_A}\frac{\omega^2}{c^2}\sum G_{ij}(\vec{r},\vec{r}_A,\omega)\vec{p}_{Aj} \tag{15.11}$$

$$= \frac{2}{3}ik^3\vec{p}_A + \text{a divergent contribution}, \tag{15.12}$$

where we have used (15.8). The imaginary part of $\vec{E}(\vec{r}_A,\omega)$ leads to radiative damping and for a detailed discussion see [13] (especially Eqs. (16.3) and (16.71a)). It is well known that the real parts of the self fields diverge and a proper treatment requires renormalization of mass and frequencies. We would drop this divergent contribution by assuming that the dipole's frequency has been properly renormalized. Writing $\vec{p}_A$ and $\vec{p}_B$ in terms of slowly varying quantities $\vec{\alpha}$ and $\vec{\beta}$

$$\vec{p}_A = \vec{\alpha}\mathrm{e}^{-\mathrm{i}\omega_0 t} + c.c., \qquad \vec{p}_B = \vec{\beta}\mathrm{e}^{-\mathrm{i}\omega_0 t} + c.c., \tag{15.13}$$

and dropping $\ddot{\vec{\alpha}}$, $\ddot{\vec{\beta}}$, (15.10) and a similar equation for $\vec{p}_B$ reduce to

$$\begin{aligned} \frac{\partial \vec{\alpha}}{\partial t} &= \frac{\mathrm{e}^2 k_0^3}{2m\omega_0}\mathrm{i}(-\overset{\Rightarrow}{\Omega} + \mathrm{i}\overset{\Rightarrow}{\gamma}) \cdot \vec{\beta} - \frac{\Gamma}{2}\vec{\alpha}, \quad \Gamma = \frac{2\mathrm{e}^2 k_0^3}{3m\omega_0}, \quad k_0 = \frac{\omega_0}{c}, \\ \frac{\partial \vec{\beta}}{\partial t} &= \frac{\mathrm{e}^2 k_0^3}{2m\omega_0}\mathrm{i}(-\overset{\Rightarrow}{\Omega} + \mathrm{i}\overset{\Rightarrow}{\gamma}) \cdot \vec{\alpha} - \frac{\Gamma}{2}\vec{\beta}, \end{aligned} \tag{15.14}$$

The result (15.14) shows how the radiative effects lead to the coupling of two dipoles besides the radiative damping of each dipole. The coupling terms depend on the distance between dipoles. The behavior of some components of $\overset{\Rightarrow}{\Omega}$ and $\overset{\Rightarrow}{\gamma}$ as a function of x is shown in Figures 15.1 and 15.2. Note the oscillatory character of these quantities. Also note the divergent behavior of $\overset{\Rightarrow}{\Omega}$ as $x \to 0$.

The radiative effects couple dipoles and this coupling can lead to the entanglement of dipoles in a fully quantum-mechanical treatment. The above discussion has been classical. We now present a quantum treatment of dipole–dipole coupling. Let us consider the situation shown in Figure 15.3, which shows the transfer of excitation from one two-level atom to the other. This transfer occurs via spontaneous emission from atom A and absorption of the spontaneously emitted photon by the atom B. We calculate the probability of transferring excitation from atom A to B. The interaction with the vacuum of the electromagnetic field in the interaction picture can be written as

$$H_1(t) \equiv -\sum_{\alpha} \vec{p}_\alpha \cdot \vec{E}(\vec{r}_\alpha, t)|e_\alpha\rangle\langle g_\alpha|\mathrm{e}^{\mathrm{i}\omega_0 t} + H.c.\,, \alpha = A, B. \tag{15.15}$$

In (15.15) $\vec{p}_\alpha$ is the dipole matrix element for the dipole moment operator for the αth atom $\vec{p}_\alpha = \langle e_\alpha|\vec{p}|g_\alpha\rangle$. The wave function up to second order in perturbation is

$$|\psi(t)\rangle \equiv -\frac{1}{\hbar^2}\int_{-\infty}^{t} \mathrm{d}t_1 \int_{-\infty}^{t_1} \mathrm{d}t_2 H_1(t_1)H_2(t_2)|e_A, g_B, \{0\}\rangle, \tag{15.16}$$

where $\{0\}$ denotes the vacuum of the electromagnetic field. We need to calculate the transition amplitude: $A = \langle g_A, e_B, \{0\}|\psi(t)\rangle$. A calculation shows that

$$\frac{\partial A}{\partial t} \to \frac{\mathrm{i}\omega_0^2}{\hbar c^2}\vec{p}_B \cdot \overset{\Rightarrow}{G}(\vec{r}_B, \vec{r}_A, \omega_0) \cdot \vec{p}_A{}^*, \tag{15.17}$$

where $\overset{\Rightarrow}{G}$ is the Green's dyadic (15.4). It is remarkable that the result of the quantum-mechanical problem is related to the dipole fields (15.3). To prove (15.17), we simplify (15.16) by substituting the initial and final states and we take time derivative to reduce the

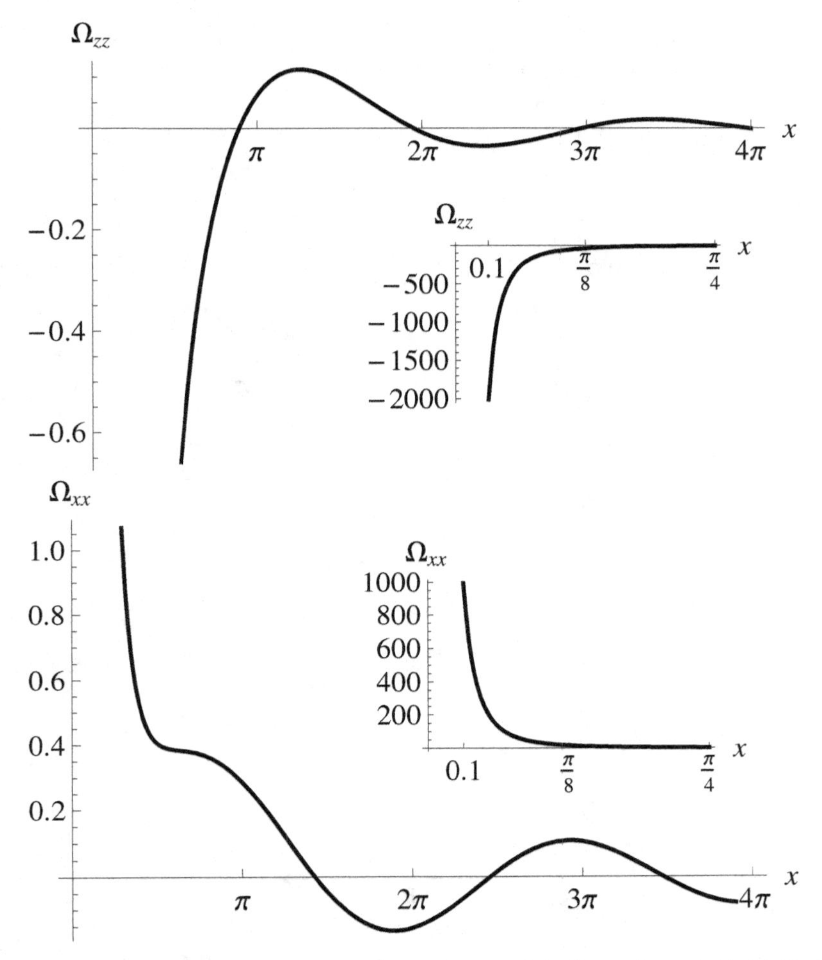

Fig. 15.1 Behavior of the real part of some of the components of G as a function of the distance between dipoles. The inset shows the behavior for much smaller distances. We have chosen a coordinate system such that $\vec{n}$ is along the z axis.

double time integral to single integral

$$\begin{aligned}\frac{\partial A}{\partial t} \to & -\frac{1}{\hbar^2}\int_0^\infty \langle \vec{p}_B \cdot \vec{E}(\vec{r}_B, \tau)\vec{p}_A{}^* \cdot \vec{E}(\vec{r}_A, 0)\rangle e^{i\omega_0\tau}d\tau \\ & -\frac{1}{\hbar^2}\int_0^\infty \langle \vec{p}_A{}^* \cdot \vec{E}(\vec{r}_A, \tau)\vec{p}_B \cdot \vec{E}(\vec{r}_B, 0)\rangle e^{-i\omega_0\tau}d\tau,\end{aligned} \tag{15.18}$$

where the expectation values are with respect to the vacuum of the electromagnetic field. Writing $\vec{E}$ in terms of the mode decomposition Eq. (1.22) and using the properties of the vacuum, (15.18) reduces to (15.17) (calculations are similar to the one in Exercise 1.9). The two terms in (15.18) correspond to two different quantum-mechanical pathways via which an excitation is transformed from the atom A to the atom B – a resonant one $|e_A, g_B\rangle \to |g_A, g_B\rangle \to |g_A, e_B\rangle$, and a nonresonant one $|e_A, g_B\rangle \to |e_A, e_B\rangle \to |g_A, e_B\rangle$.

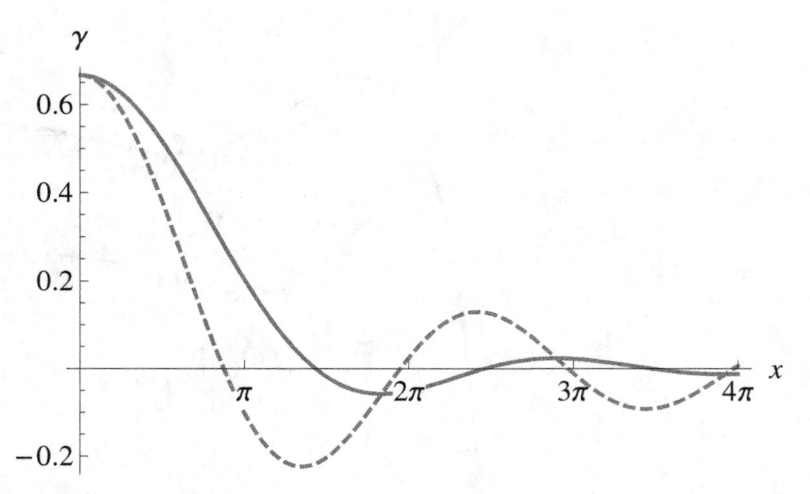

Fig. 15.2 Behavior of γ_{xx} and γ_{zz} as functions of x.

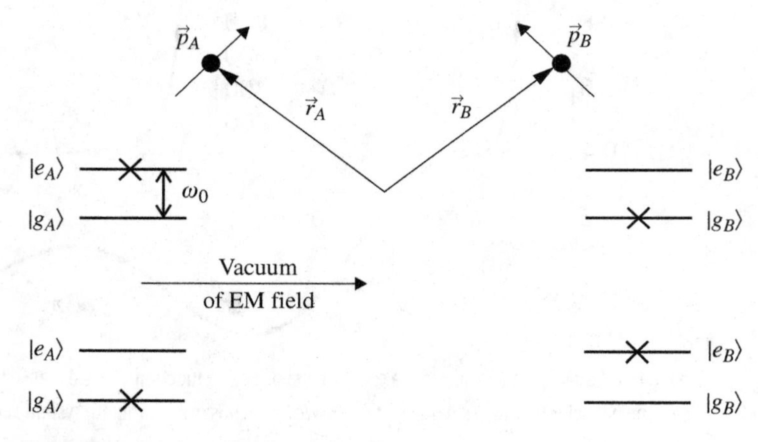

Fig. 15.3 Electromagnetic-vacuum-induced transfer of excitation.

15.2 Radiative coupling between dipoles and dynamics

In Section 15.1, we discussed how the field emitted from one dipole can interact with the other dipole leading to an effective coupling between dipoles. We presented both the classical argument as well as a simple quantum-mechanical perturbation analysis to demonstrate this coupling. Such couplings determine the dynamical evolution of the dipoles. In this section we give the basic equation for the dynamics of two atoms interacting with the vacuum of the electromagnetic field. The master equation can be derived by following the procedure of Section 9.1. In place of the Hamiltonian (9.2), we would have for two atoms

$$H_S = \sum_{j=1}^{2}(E_e|e\rangle\langle e| + E_g|g\rangle\langle g|)_j, \qquad H_B = \sum_{\vec{k}s}\hbar\omega_{ks}a^{\dagger}_{\vec{k}s}a_{\vec{k}s},$$

$$H_{SB} = -\sum_{j=1}^{2}(\vec{p}_{eg}|e\rangle\langle g| + H.c.)_j \cdot \left(\mathrm{i}\sum_{\vec{k}s}\sqrt{\frac{2\pi\hbar\omega_{ks}}{V}}\epsilon_{\vec{k}s}a_{\vec{k}s}\mathrm{e}^{\mathrm{i}\vec{k}\cdot\vec{r}_j} + H.c.\right). \tag{15.19}$$

The initial state of the field is given by (9.5). We need to simplify (9.27). The variable $H_{SB}(\tau)$ appearing in (9.27) is obtained from (15.19) by attaching exponentials to various operators $|e\rangle\langle g| \to |e\rangle\langle g|\mathrm{e}^{\mathrm{i}\omega_0\tau}$, $a_{\vec{k}s} \to a_{\vec{k}s}\mathrm{e}^{-\mathrm{i}\omega_{ks}\tau}$. The algebra is long and is left to the reader as Exercise 15.1. The final result can be written in terms of the basic quantities that we already introduced in Section 15.1 [14–16].

$$\begin{aligned}\frac{\partial\rho}{\partial t} &= -\mathrm{i}\omega_0\sum_{j=1}^{2}[S_z^{(j)},\rho] - \mathrm{i}\sum_{i\neq j}\Omega_{ij}[S_+^{(i)}S_-^{(j)},\rho] \\ &\quad -\sum_{i,j}\gamma_{ij}[S_+^{(i)}S_-^{(j)}\rho - 2S_-^{(j)}\rho S_+^{(i)} + \rho S_+^{(i)}S_-^{(j)}],\end{aligned} \tag{15.20}$$

where

$$(-\Omega_{ij} + \mathrm{i}\gamma_{ij}) = \left(\frac{\omega_0^2}{c^2}\right)\frac{1}{\hbar}\vec{p}\cdot\overset{\Rightarrow}{G}(\vec{r}_i,\vec{r}_j,\omega_0)\cdot\vec{p}, \tag{15.21}$$

$$\begin{aligned}\gamma_{ij} &= \frac{3}{2}\gamma\left\{\left[\frac{\sin x}{x} - \frac{\sin x}{x^3} + \frac{\cos x}{x^2}\right] + \cos^2\theta\left[\frac{3\sin x}{x^3} - \frac{3\cos x}{x^2} - \frac{\sin x}{x}\right]\right\} \\ &\to \gamma \quad \text{for } x \to 0,\end{aligned} \tag{15.22}$$

$$\begin{aligned}\Omega_{ij} &= \frac{3}{2}\gamma\left\{\left[\frac{\cos x}{x^3} + \frac{\sin x}{x^2} - \frac{\cos x}{x}\right] + \cos^2\theta\left[\frac{\cos x}{x} - \frac{3\cos x}{x^3} - \frac{3\sin x}{x^2}\right]\right\} \\ &\to \frac{3}{2}\gamma\left(1 - 3\cos^2\theta\right)/x^3 \quad \text{for } x \to 0, \quad x = \frac{\omega_0}{c}|\vec{r}_i - \vec{r}_j|,\end{aligned} \tag{15.23}$$

$$\cos^2\theta = \left(\frac{\vec{p}\cdot(\vec{r}_i - \vec{r}_j)}{|p|\,|\vec{r}_i - \vec{r}_j|}\right)^2, \qquad \gamma_{ii} = \gamma = \frac{2}{3}\frac{|\vec{p}|^2\omega_0^3}{c^3\hbar}. \tag{15.24}$$

It is assumed that the frequency ω_0 has been renormalized. The vacuum-field-induced dipolar coupling, given by the Ω term in (15.20), modifies the energy levels. The effective unperturbed Hamiltonian is now

$$H = \hbar\omega_0\sum_j S_z^{(j)} + \hbar\Omega(S_+^{(1)}S_-^{(2)} + S_-^{(1)}S_+^{(2)}), \qquad \Omega = \Omega_{12}. \tag{15.25}$$

Its eigenstates are as displayed in Figure 15.4. Here $|s\rangle$ and $|a\rangle$ are the symmetric and antisymmetric states of the two-atom system

$$|s\rangle = (|e,g\rangle + |g,e\rangle)/\sqrt{2}, \qquad |a\rangle = (|e,g\rangle - |g,e\rangle)/\sqrt{2}. \tag{15.26}$$

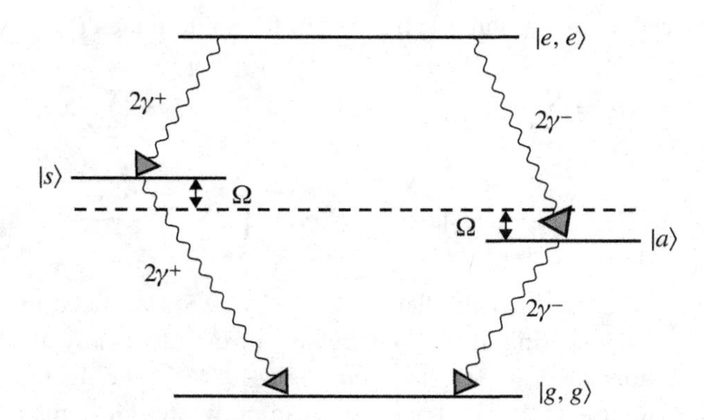

Fig. 15.4 Transitions as implied by the master equation (15.20).

The basic master equation in the basis of states $|s\rangle$, $|a\rangle$, $|e, e\rangle \equiv |e\rangle$, $|g, g\rangle \equiv |g\rangle$ has a very simple form

$$\begin{aligned} \dot{\rho}_{ee} &= -4\gamma\rho_{ee}, & \dot{\rho}_{eg} &= -2(\gamma + \mathrm{i}\omega_0)\rho_{eg}, \\ \dot{\rho}_{as} &= -2(\gamma + \mathrm{i}\Omega)\rho_{as}, & \dot{\rho}_{ss} &= -2\gamma^+(\rho_{ss} - \rho_{ee}), \\ \dot{\rho}_{aa} &= -2\gamma^-(\rho_{aa} - \rho_{ee}), & \gamma^\pm &= \gamma \pm \gamma_{12}. \end{aligned} \tag{15.27}$$

The diagonal elements in the collective basis satisfy rate equations and off-diagonal elements are coupled to off-diagonal ones only. The doubly excited state $|e, e\rangle$ has two channels of decay (Figure 15.4) $|e\rangle \to |s\rangle$ at the rate $2\gamma^+$, $|e\rangle \to |a\rangle$ at the rate $2\gamma^-$. The symmetric (antisymmetric) state decays at the rate $2\gamma^+(2\gamma^-)$ to the ground state. Note also that if the two atoms are so close that $\gamma_{12} \approx \gamma$, then the antisymmetric state hardly decays. However, the effect of Ω is dominant at such distances. The rate equations (15.27) have simple solutions

$$\begin{aligned} \rho_{ee}(t) &= \rho_{ee}(0)\mathrm{e}^{-4\gamma t}, \qquad \rho_{eg}(t) = \mathrm{e}^{-2(\gamma+\mathrm{i}\omega_0)t}\rho_{eg}(0), \\ \rho_{as}(t) &= \mathrm{e}^{-2(\gamma+\mathrm{i}\Omega)t}\rho_{as}(0), \\ \rho_{ss}(t) &= \rho_{ss}(0)\mathrm{e}^{-2\gamma^+ t} + \rho_{ee}(0)\frac{\gamma^+}{\gamma^-}(\mathrm{e}^{-2\gamma^+ t} - \mathrm{e}^{-4\gamma t}), \\ \rho_{aa}(t) &= \rho_{aa}(0)\mathrm{e}^{-2\gamma^- t} + \rho_{ee}(0)\frac{\gamma^-}{\gamma^+}(\mathrm{e}^{-2\gamma^- t} - \mathrm{e}^{-4\gamma t}). \end{aligned} \tag{15.28}$$

These solutions can be used to obtain all the dynamical features of spontaneous emission from a system of two atoms.

15.3 Vacuum-induced deterministic entanglement

We saw in the previous section that the interaction of atoms with the vacuum of the field essentially couples them. This coupling gives rise to atom–atom correlations and deterministic entanglement between atoms. Let us first consider the possibility of the atom–atom correlation $\langle S_+^{(1)} S_-^{(2)} \rangle$ being nonzero. Let us assume asymmetric excitation at $t = 0$, in the state $|e_1, g_2\rangle$, then on expressing $S_+^{(1)} S_-^{(2)} = (|e_1\rangle\langle g_1|)(|g_2\rangle\langle e_2|)$ in terms of the symmetric and antisymmetric states

$$\langle S_+^{(1)} S_-^{(2)} \rangle = \frac{1}{2}(\rho_{ss} - \rho_{aa} + \rho_{sa} - \rho_{as}), \tag{15.29}$$

and using the solutions (15.28), we find

$$\langle S_+^{(1)} S_-^{(2)} \rangle = \frac{1}{4}(\mathrm{e}^{-2\gamma^+ t} - \mathrm{e}^{-2\gamma^- t}) + \frac{1}{2}\mathrm{i}\mathrm{e}^{-2\gamma t}\sin 2\Omega t. \tag{15.30}$$

Clearly the near-field effects have produced atom–atom correlation. If the atoms are far apart, then $\gamma^\pm \to \gamma$ and $\langle S_+^{(1)} S_-^{(2)} \rangle$ becomes small leading to almost no correlation. The entanglement due to near field radiative effects can be characterized in terms of the concurrence C introduced by Wootters [17]. This parameter is especially useful for mixed states. The parameter C is defined by

$$C = \max\big(0, \sqrt{\lambda_1} - \sqrt{\lambda_2} - \sqrt{\lambda_3} - \sqrt{\lambda_4}\big), \tag{15.31}$$

where the λ_is are the eigenvalues of $\rho\tilde{\rho}$, where $\tilde{\rho}$ is defined in terms of the Pauli matrix σ_y

$$\tilde{\rho} = \sigma_y \bigotimes \sigma_y \rho^* \sigma_y \bigotimes \sigma_y. \tag{15.32}$$

If the state is separable, then $C = 0$. The parameter $C = 1$ for a maximally entangled state. Given ρ, we can calculate the concurrence. We take the result given by Tanaś and Ficek [18]. For an initial density matrix with nonzero elements ρ_{gg}, ρ_{ee}, ρ_{ge}, ρ_{eg}, $\rho_{\alpha\alpha}$, $\rho_{\beta\beta}$, $\rho_{\alpha\beta}$, $\rho_{\beta\alpha}$ with $|g\rangle = |g_1, g_2\rangle$, $|e\rangle = |e_1, e_2\rangle$, $|\alpha\rangle = |g_1, e_2\rangle$, $|\beta\rangle = |e_1, g_2\rangle$, the concurrence is (see Exercise 15.4)

$$\begin{gathered} C = \max(0, C_1, C_2), \\ C_1 = 2(|\rho_{ge}| - \sqrt{\rho_{\alpha\alpha}\rho_{\beta\beta}}), \qquad C_2 = 2(|\rho_{\alpha\beta}| - \sqrt{\rho_{gg}\rho_{ee}}). \end{gathered} \tag{15.33}$$

For two atoms starting in the initial state $|e_1, g_2\rangle = |\beta\rangle$, $C = 0$; however, during the course of time development $C \neq 0$. Using (15.28), $\rho_{ee} = 0$, $\rho_{ge} = 0$, $C_1 < 0$, $C_2 = 2|\rho_{\alpha\beta}|$ and hence

$$\begin{aligned} C(t) &= 2|\rho_{\alpha\beta}(t)| = 2|\langle S_+^{(1)} S_-^{(2)} \rangle| \\ &= \mathrm{e}^{-2\gamma t}\sqrt{\sinh^2(2\gamma_{12}t) + \sin^2(2\Omega t)} > 0. \end{aligned} \tag{15.34}$$

In the long time limit $C(t) \to 0$ as expected. The time dependence of the concurrence is shown in Figure 15.5. We have thus shown how the vacuum of the electromagnetic field can produce deterministic entanglement between two separable atoms. This happens because both atoms see the same electromagnetic vacuum. This property is quite generic and applies to many different types of baths. In general for the situation shown in Figure 15.6, we would always produce deterministic entanglement between two systems A and B. Other examples of the situation of Figure 15.6 are known in the literature [19, 20]. The entanglement that we have discussed in this section is different from the one discussed in Section 14.5, for example, which was the result of the measurement process.

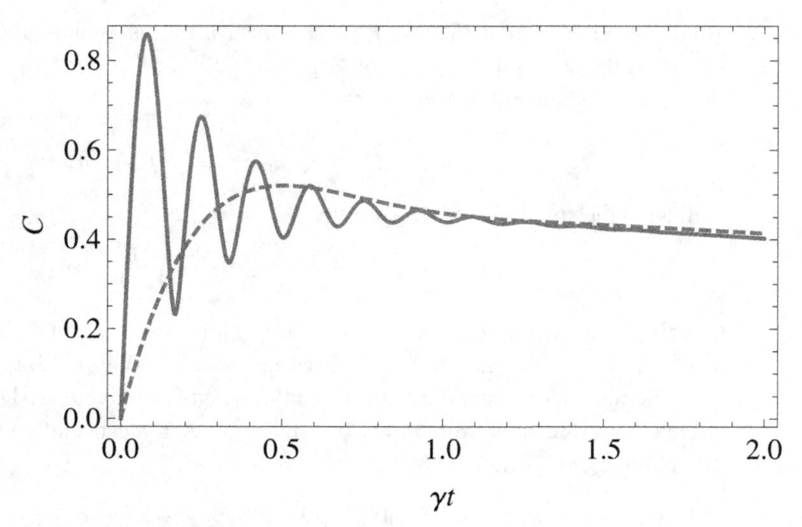

Fig. 15.5 Time dependence of concurrence with one atom in an excited state. The parameters are: $x = \frac{\omega_0}{c}|\vec{r}_1 - \vec{r}_2| = \pi/6$, $\cos\theta = 0$ (solid), and $\cos\theta = 1/\sqrt{2}$ (dashed). For $\theta = \pi/4$, the oscillatory behavior is not seen as $\Omega \approx 1$.

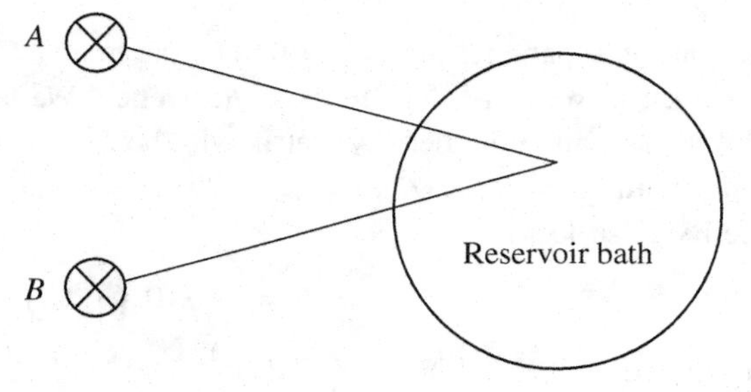

Fig. 15.6 Idea of bath-induced entanglement.

Finally, we note that if the two atoms are coherently driven, then the master equation (15.20) is modified by the presence of an extra term $[\mathrm{i}\sum_j g_j S_+^{(j)} + g_j^* S_-^{(j)}, \rho]$ on the right-hand side. Many publications discuss coherently driven two-level systems [21–23]. Such cases are best studied by numerical methods and therefore we do not discuss these.

15.4 Two-photon resonance induced by near field radiative effects

In this section, we discuss another remarkable effect arising from the near field radiative effects. We show how two-photon absorption in a system of two unidentical atoms becomes allowed due to dipole–dipole interaction effects. Consider the absorption of two photons by a system of two unidentical atoms (Figure 15.7). Let the two atoms be initially in the state $|g_A, g_B\rangle$. The incident field is nonresonant with each of atoms as $\omega_l \neq \omega_1$, $\omega_l \neq \omega_2$.

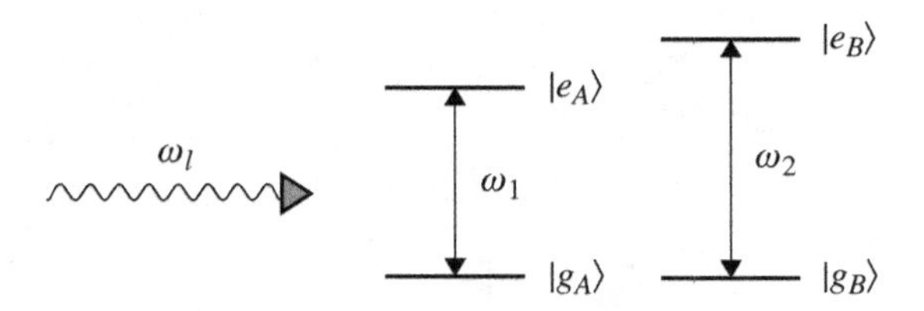

Fig. 15.7 Absorption of two photons by two unidentical atoms.

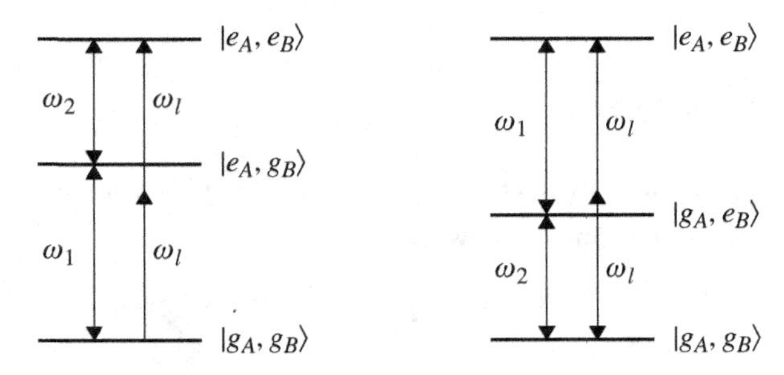

Fig. 15.8 The two pathways for two-photon absorption by a system of two unidentical atoms.

However, the absorption of two photons with the simultaneous excitation of each atom would be a resonant process if $2\omega_l = \omega_1 + \omega_2$. We calculate the probability or rate for two-photon absorption using the second-order Fermi Golden rule (8.70). If the two atoms are noninteracting and if we ignore the near-field radiative effects, then the Hamiltonian for the two-atom system

$$H = \sum_{i=A,B} \hbar\omega_i S_z^{(i)} - \hbar \sum_{i=A,B} (g_i S_+^{(i)} \mathrm{e}^{-\mathrm{i}\omega_l t} + c.c.). \tag{15.35}$$

There are two pathways by which the two atoms get excited (Figure 15.8) and the rate of excitation is obtained by summing over the two contributions coherently

$$R_{ee} = 2\pi \left| \frac{g_B g_A}{\omega_1 - \omega_l} + \frac{g_A g_B}{\omega_2 - \omega_l} \right|^2 \delta(\omega_1 + \omega_2 - 2\omega_l). \tag{15.36}$$

The two terms are according to the two pathways shown in Figure 15.8, and the matrix elements in (8.70). The Eq. (15.36) on simplification yields

$$R_{ee} = \frac{2\pi |g_A g_B|^2}{(\omega_1 - \omega_l)^2 (\omega_2 - \omega_l)^2} (\omega_1 + \omega_2 - 2\omega_l)^2 \delta(\omega_1 + \omega_2 - 2\omega_l) = 0. \tag{15.37}$$

The rate of simultaneous excitation is zero due to destructive interference of the two pathways of Figure 15.8. The near field radiative effects make the interference of the two pathways incomplete leading to the possibility of simultaneous excitation of two atoms [2, 4–6]. The Hamiltonian (15.35) is now modified to (c.f. (15.25))

$$h = H + \hbar\Omega(S_+^{(1)} S_-^{(2)} + S_-^{(1)} S_+^{(2)}), \tag{15.38}$$

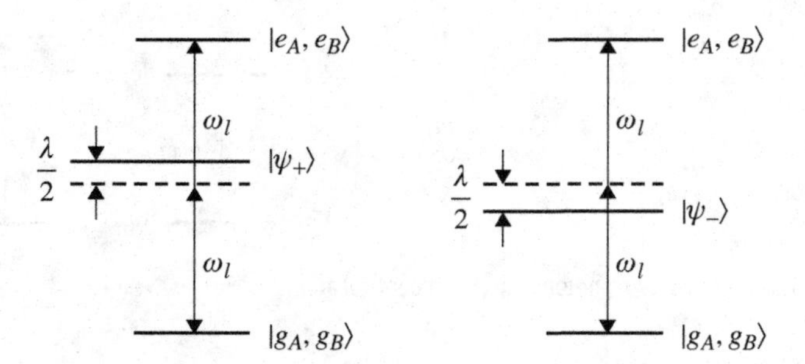

Fig. 15.9 Pathways in the presence of the dipole–dipole interaction for resonant two-photon absorption, $\omega_l = (\omega_1 + \omega_2)/2$.

where we can now evaluate Ω at $(\omega_1 + \omega_2)/2$. The coupling term modifies the bare energy levels. Let us consider now the unperturbed H_0 as

$$H_0 = \sum_i \hbar\omega_i S_z^{(i)} + \hbar\Omega(S_+^{(1)}S_-^{(2)} + S_-^{(1)}S_+^{(2)}). \tag{15.39}$$

The states $|g_A, g_B\rangle$ and $|e_A, e_B\rangle$ are unaffected by the coupling Ω. However, the states $|e_A, g_B\rangle$ and $|g_A, e_B\rangle$ are affected by the coupling. The new states and eigenvalues are obtained by the diagonalization of the 2×2 matrix $\hbar\left(\begin{smallmatrix}\omega_1 & \Omega \\ \Omega & \omega_2\end{smallmatrix}\right)$ in which we have subtracted the energy $-\frac{\hbar}{2}(\omega_1 + \omega_2)$ of the state $|g_A, g_B\rangle$. This has the eigenvalues $\hbar(\omega_1 + \omega_2 \pm \lambda)/2$ and eigenfunctions

$$\begin{aligned} |\psi_+\rangle &= \cos\theta |e_A, g_B\rangle + \sin\theta |g_A, e_B\rangle, \\ |\psi_-\rangle &= \cos\theta |g_A, e_B\rangle - \sin\theta |e_A, g_B\rangle, \end{aligned} \tag{15.40}$$

where

$$\frac{\lambda}{2} = \sqrt{\frac{(\omega_1 - \omega_2)^2}{4} + \Omega^2}, \qquad \tan\theta = \frac{\Omega}{\frac{\omega_1 - \omega_2}{2} + \sqrt{\frac{(\omega_1 - \omega_2)^2}{4} + \Omega^2}}, \tag{15.41}$$

$$\omega_1 > \omega_2.$$

The two paths of Figure 15.8 would now change as shown in Figure 15.9. The transition rate is now equal to

$$R_{ee} = 2\pi \left| \frac{(g_A\cos\theta + g_B\sin\theta)(g_B\cos\theta + g_A\sin\theta)}{\frac{\lambda}{2} + (\frac{\omega_1+\omega_2}{2} - \omega_l)} + \frac{(g_A\cos\theta - g_B\sin\theta)(g_B\cos\theta - g_A\sin\theta)}{-\frac{\lambda}{2} + (\frac{\omega_1+\omega_2}{2} - \omega_l)} \right|^2 \delta(\omega_1 + \omega_2 - 2\omega_l). \tag{15.42}$$

The first (second) term in the two-photon matrix element is via the intermediate state $|\psi_+\rangle$ ($|\psi_-\rangle$). Each energy denominator consists of the energy of the state $|\psi_\pm\rangle$ relative to $|g_A, g_B\rangle$ minus the energy of the absorbed photon. The numerators are the matrix elements $\langle e_A, e_B|H_1|\psi_\pm\rangle\langle\psi_\pm|H_1|g_A, g_B\rangle$, where H_1 is the interaction with the external field. In general, $R_{ee} \neq 0$ as long as $\Omega \neq 0$.

The two-photon matrix element at resonance $\omega_1 + \omega_2 = 2\omega_l$ simplifies to $16(g_A^2 + g_B^2)^2 \sin^2\theta \cos^2\theta/\lambda^2$, which is nonzero as long as $\Omega \neq 0$, i.e. as long as the near-field radiative effects are accounted for. A very detailed discussion of the near-field-induced two-photon resonance for the case of intense fields is given in [6]. Hettich *et al.* [2] experimentally studied the effect of intense fields on such a two-photon resonance. They were able to study two emitters within a distance of the order of $\lambda/50$, thus making Ω very large. They also showed how the fluorescence from such a doubly excited system exhibits photon bunching. The bunching basically arises as in presence of strong near field radiative effects, the probability of simultaneously emitting two photons is very high.

15.5 The dipole blockade

The near field radiative effect that we have so far discussed couples the state $|e_A, g_B\rangle$ to $|g_A, e_B\rangle$. This results in a number of interesting effects like the generation of quantum entanglement. It also helps us in discriminating near by atoms within nanometers [2]. There are other kinds of interatomic interactions that can arise even when the two atoms are in the ground state. These include the well-known van der Waal's forces, which vary as R^{-6} for short distances and as R^{-7} for large distances, where R is the distance between two atoms. These are obtained by using fourth order perturbation theory [7] in terms of the interaction of atoms with the electromagnetic field. One consequence of such van der Waal's interaction is the possibility of dipole blockade between two Rydberg atoms, i.e. the interactions forbid the excitation of a second nearby atom if the first atom is excited. We have already discussed briefly in Section 14.5 how the Rydberg blockade can lead to the entanglement between two atoms. We can build up a simple model to see such a Rydberg blockade effect [11].

We consider two atoms at fixed positions R_1 and R_2 with internal levels $|e\rangle$ and $|g\rangle$, dipolar transition frequency $\omega_0 = 2\pi c/\lambda$. In experiments of [8], the two atoms are located more than 20λ apart. The system is conveniently described in the Dicke basis $|ee\rangle$, $|gg\rangle$, $|s\rangle \equiv (|eg\rangle + |ge\rangle)/\sqrt{2}$ and $|a\rangle \equiv (|eg\rangle - |ge\rangle)/\sqrt{2}$. We consider that the two atoms strongly interact when in state $|ee\rangle$ resulting in a shift $\hbar\delta$ of this doubly excited state. They are driven by a resonant external laser field with wave vector $\vec{k}_l$ and Rabi frequency $2g$. In the rotating-wave approximation, the coherent evolution of the system is described by the interaction Hamiltonian

$$H = \hbar\delta|ee\rangle\langle ee| + \hbar g(\mathrm{e}^{\mathrm{i}\vec{k}_l\cdot\vec{R}_1} S_+^{(1)} + \mathrm{e}^{\mathrm{i}\vec{k}_l\cdot\vec{R}_2} S_+^{(2)} + H.c.), \tag{15.43}$$

where $S_+^{(i)} = (S_-^{(i)})^\dagger$ $(i = 1, 2)$ is the atomic raising operator $|e\rangle_i\langle g|$. The term $\hbar\delta|ee\rangle\langle ee|$ accounts for the shift of the doubly excited state of the system induced by the dipole–dipole interaction. For Rydberg atom experiments, the shift is produced by the Förster interaction as discussed earlier in the context of Figure 14.16. Let $\vec{k}_l$ be perpendicular to the two-atom line and the reference frame is properly chosen so $\vec{k}_l \cdot \vec{R}_1 = \vec{k}_l \cdot \vec{R}_2 = 0$. We further allow for the loss of atomic excitation via incoherent processes at the rate 2Γ, so that the density

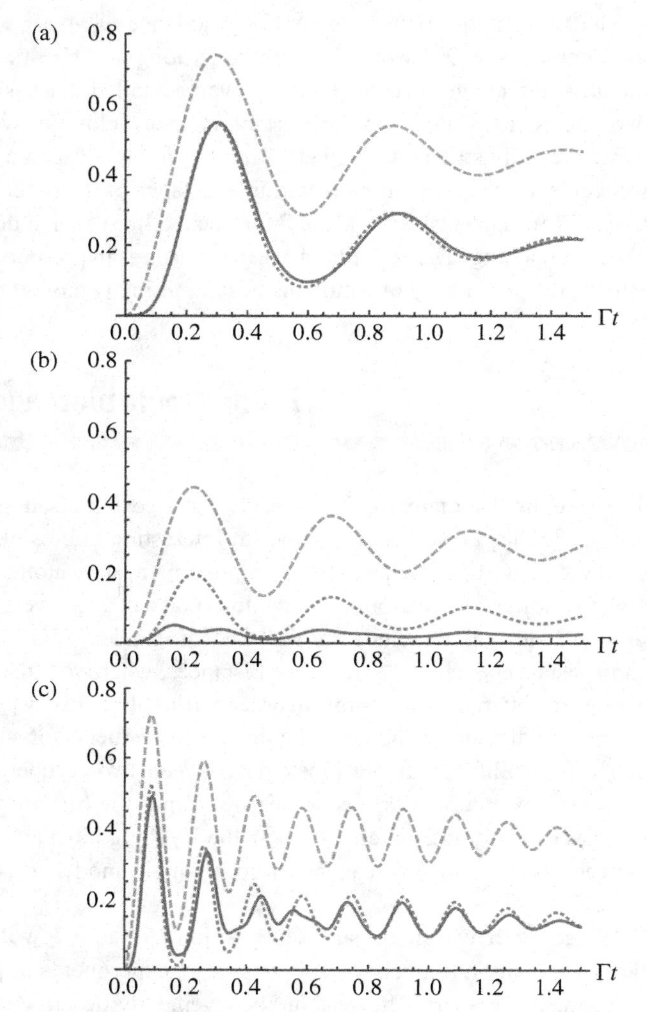

Fig. 15.10 Time evolution of the excitation probability p_e (dashed curve), its square (dotted curve), and the probability p_{ee} of having both atoms excited (solid curve), for (a) $\Omega/\Gamma = 5, \delta/\Gamma = 5$; (b) $\Omega/\Gamma = 5, \delta/\Gamma = 30$; (c) $\Omega/\Gamma = 15, \delta/\Gamma = 30$. The dipole blockade effect is well marked in (b), where $p_{ee} \ll p_e^2$ [11].

matrix of the two-atom system evolves according to

$$\dot{\rho} = -\frac{\mathrm{i}}{\hbar}[H, \rho] - \Gamma \sum_{i=1}^{2} (S_{+}^{(i)} S_{-}^{(i)} \rho + \rho S_{+}^{(i)} S_{-}^{(i)} - 2 S_{-}^{(i)} \rho S_{+}^{(i)}). \tag{15.44}$$

This is the basic equation that we use to establish the phenomenon of dipole blockade. We can obtain bath transient and steady-state solutions.

In the presence of the dipole blockade mechanism, the doubly excited state $|ee\rangle$ is expected to be poorly populated although not totally unpopulated. This is illustrated quantitatively in Figure 15.10 where we compare the time evolution of the square of the probability

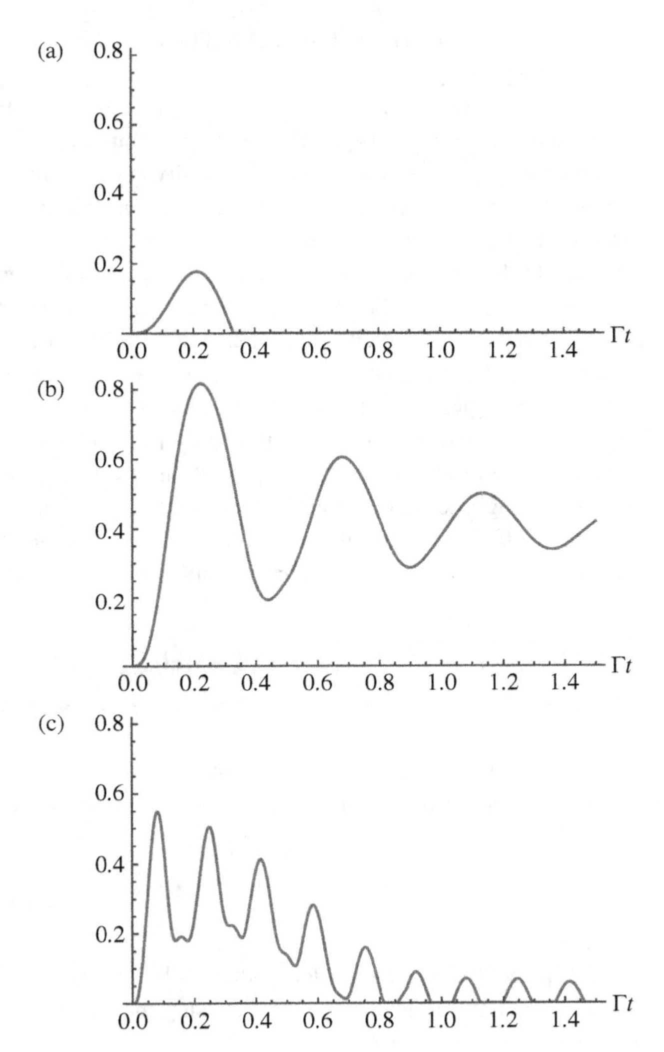

Fig. 15.11 Time evolution of the concurrence C of the two-atom system, for (a) $\Omega/\Gamma = 5, \delta/\Gamma = 5$; (b) $\Omega/\Gamma = 5$, $\delta/\Gamma = 30$; (c) $\Omega/\Gamma = 15, \delta/\Gamma = 30$, after [11].

$p_e = \langle e|\mathrm{Tr}_1\rho|e\rangle = \langle e|\mathrm{Tr}_2\rho|e\rangle$ of having one of the two atoms excited with the probability $p_{ee} = \langle ee|\rho|ee\rangle$ of finding both atoms excited, considering them initially in the ground state. When the dipole–dipole interaction is not strong enough (Figure 15.10(a)) it has negligible effect and the atoms act as independent systems ($p_{ee} \approx p_e^2$). For a greater dipole–dipole interaction (Figure 15.10(b)), the double excitation is blocked and the population of the $|ee\rangle$ state remains at insignificant levels though not zero. More importantly the double-excitation probability p_{ee} is much lower than p_e^2, giving a direct signature of the blockade mechanism. When the laser intensity is increased (Figure 15.10(c)), we observe that p_{ee} is again very similar to p_e^2. The dipole blockade is lifted and the atoms behave again as if they were independent without mutual influence. The dipole blockade effect can thus be circumvented by using strong laser fields. Figure 15.10(b) corresponds to what is observed experimentally [9].

The experimental results reported in [8, 9] clearly imply the entanglement in the two-atom system. We can quantify such an entanglement. From the master equation (15.44) we can obtain the complete time-dependent density matrix, which then can be used to compute the concurrence [17] – a well-known measure of entanglement. For this purpose we use the prescription as given in (15.31) and (15.32). We calculate the time-dependent density matrix from Eq. (15.44) and use these solutions to calculate the eigenvalues and the concurrence. We show the results in Figure 15.11. The concurrence is maximized when the dipole blockade mechanism is itself optimized. In Figure 15.11(a), the dipole–dipole interaction is too weak and the two-atom system behaves as a collection of independent atoms. No significant entanglement is produced. In Figure 15.11(b), the dipole blockade prevents the doubly excited state from being significantly populated and the two-atom system shares a collective single excitation. More population in the entangled $(|eg\rangle + |ge\rangle)/\sqrt{2}$ state is expected and significant amount of entanglement is produced. In Figure 15.11(c), the dipole blockade is lifted and more population in the separable doubly excited state is expected. The concurrence is again less important than in Figure 15.11(b).

Exercises

15.1 Using Maxwell equations obtain an equation for the electric field in the frequency domain assuming a source of polarization and an electric field of the atom

$$\vec{p}(\vec{r}, t) = \vec{p}\delta(\vec{r} - \vec{r}_B)\mathrm{e}^{-i\omega t} + c.c.,$$
$$\vec{E}(\vec{r}, t) = \vec{E}(\vec{r}, \omega)\mathrm{e}^{-i\omega t} + c.c..$$

Then prove Eqs. (15.3) and (15.4).

15.2 Simplify (15.18) so as to derive the result (15.17).

15.3 Derive the master equation (15.20) using the method of Chapter 9 and Eq. (9.27).

15.4 Prove the result (15.33).

15.5 Derive the results (15.40) and (15.41).

15.6 Show the validity of (15.42).

References

[1] S. Das, G. S. Agarwal, and M. O. Scully, *Phys. Rev. Lett.* **101**, 153601 (2008).
[2] C. Hettich, C. Schmitt, J. Zitzmann, S. Kuühn, I. Gerhardt, and V. Sandoghdar, *Science* **298**, 385 (2002).
[3] D. Loss and D. P. DiVincenzo, *Phys. Rev. A* **57**, 120 (1998).
[4] F. Varsanyi and G. H. Dieke, *Phys. Rev. Lett.* **7**, 442 (1961).
[5] D. L. Dexter, *Phys. Rev.* **126**, 1962 (1962).

[6] G. V. Varada and G. S. Agarwal, *Phys. Rev. A* **45**, 6721 (1992).
[7] C. Cohen-Tannoudji, J. Dupont-Roc, and G. Grynberg, *Atom–Photon-Interactions: Basic Processes and Applications* (New York: Wiley-Interscience, 1998), p.124.
[8] E. Urban, T. A. Johnson, T. Henage *et al.*, *Nature Physics* **5**, 110 (2009).
[9] A. Gaëtan, Y. Miroshnychenko, T. Wilk *et al.*, *Nature Physics* **5**, 115 (2009).
[10] T. Pohl and P. R. Berman, *Phys. Rev. Lett.* **102**, 013004 (2009).
[11] J. Gillet, G. S. Agarwal, and T. Bastin, *Phys. Rev. A* **81**, 013837 (2010).
[12] C.-T. Tai, *Dyadic Green's Functions in Electromagnetic Theory* (New York: Intext Educational Publishers, 1971).
[13] J. D. Jackson, *Classical Electrodynamics*, 3rd edn. (New York: John Wiley & Sons, Inc., 1998).
[14] G. S. Agarwal, *Phys. Rev. A* **2**, 2038 (1970).
[15] R. H. Lehmberg, *Phys. Rev. A* **2**, 883 (1970).
[16] R. H. Lehmberg, *Phys. Rev. A* **2**, 889 (1970).
[17] W. K. Wootters, *Phys. Rev. Lett.* **80**, 2245 (1998).
[18] R. Tanaś and Z. Ficek, *J. Opt. B: Quantum Semiclass. Opt.* **6**, S90 (2004).
[19] G. S. Agarwal, *Phys. Rev. Lett.* **57**, 827 (1986).
[20] F. Benatti, R. Floreanini, and M. Piani, *Phys. Rev. Lett.* **91**, 070402 (2003).
[21] G. S. Agarwal, A. C. Brown, L. M. Narducci, and G. Vetri, *Phys. Rev. A* **15**, 1613 (1977).
[22] A. Beige and G. C. Hegerfeldt, *Phys. Rev. A* **59**, 2385 (1999).
[23] J. von Zanthier, T. Bastin, and G. S. Agarwal, *Phys. Rev. A* **74**, 061802(R) (2006).

16 Decoherence and disentanglement in two-level systems

In Chapters 11–15 we discussed how coherence in atomic systems can be produced by using coherent driving fields. Furthermore, we have discussed how quantum entanglement can be produced by near field radiative effects. Clearly the applications of coherence and entanglement require that these characteristics should survive over the scale of time at which applications would be carried out. However, all systems are subject to interaction with the environment (Chapter 9). The interaction is typically determined by two time scales T_1 and T_2. These time scales also determine the scale over which coherence and entanglement would survive.

In this chapter, we discuss both loss of coherence and entanglement due to environmental interaction. We also discuss some of the methods to protect against such deteriorating effects of the environment [1–10].

16.1 Decoherence due to the interaction of a two-level system with the environment

In order to understand how decoherence arises, let us consider the interaction of a two-level system with the environment, where the environment is treated as a bath. We can then use the master equation (9.56) for a two-level system, which in terms of the two states $|e\rangle$ and $|g\rangle$ (for spin $|e\rangle = |\uparrow\rangle$, $|g\rangle = |\downarrow\rangle$) is

$$\begin{aligned} \frac{\partial \rho_{ee}}{\partial t} &= -\frac{1}{T_1}(\rho_{ee} - \rho_{ee}^{(0)}), \\ \frac{\partial \rho_{eg}}{\partial t} &= -\left(\frac{1}{T_2} + \mathrm{i}\omega\right)\rho_{eg}, \end{aligned} \tag{16.1}$$

where $\rho_{ee}^{(0)}$ is the population of the state $|e\rangle$ in equilibrium. Here $1/T_2$ gives the dephasing rate and $1/T_1$ gives the relaxation rate for the population difference between two levels. The solution of (16.1) is

$$\begin{aligned} \rho_{eg}(t) &= \rho_{eg}(0)\mathrm{e}^{-\mathrm{i}\omega t - t/T_2}, \\ \rho_{ee}(t) &= \rho_{ee}^{(0)} + (\rho_{ee}(0) - \rho_{ee}^{(0)})\mathrm{e}^{-t/T_1}. \end{aligned} \tag{16.2}$$

Let us assume that the two-level system is prepared in a coherent superposition of the states $|e\rangle$ and $|g\rangle$

$$|\psi\rangle = \alpha|e\rangle + \beta|g\rangle, \qquad |\alpha|^2 + |\beta|^2 = 1. \tag{16.3}$$

On combining (16.2) and (16.3), we get

$$\begin{aligned}\rho_{eg}(t) &= \alpha\beta^* \mathrm{e}^{-\mathrm{i}\omega t - t/T_2},\\ \rho_{ee}(t) &= \rho_{ee}^{(0)} + (|\alpha|^2 - \rho_{ee}^{(0)})\mathrm{e}^{-t/T_1}.\end{aligned} \tag{16.4}$$

In most solid-state environments, the transverse relaxation is much faster than the longitudinal relaxation. Therefore, the coherence term decays quickly and then ρ becomes an incoherent mixture of $|e\rangle$ and $|g\rangle$ states. The vanishing of the off-diagonal terms is called decoherence [11] and a pure state becomes a mixed state. Hence we need methods which can protect against decoherence. We discuss a particularly useful technique in Section 16.4. For purely radiative processes T_1 and T_2 are comparable and strategies to protect against decoherence are more difficult to realize. For radiative decay $1/T_1 = 2\gamma$, $1/T_2 = \gamma$, $\rho_{ee}^{(0)} = 0$ and (16.4) becomes

$$\rho_{eg} = \alpha\beta^* \mathrm{e}^{-\mathrm{i}\omega t - \gamma t}, \qquad \rho_{ee}(t) = |\alpha|^2 \mathrm{e}^{-2\gamma t}. \tag{16.5}$$

Note that the coefficient $\mathrm{e}^{-2\gamma t}$ represents the decay of the state $|e\rangle$. In the limit $\gamma t \gg 1$, the state $|e\rangle$ completely decays and $\rho \to |g\rangle\langle g|$. For radiative decay the loss of coherence is connected with the decay of the state $|e\rangle$. However, if the environment is such that the state $|e\rangle$ does not decay completely, then complete decoherence does not occur. An example of this is the photonic crystal environment for which we have shown (Section 7.7.2) that the excited state, under the condition that it lies close to the band edge, does not completely decay. More generally, for survival of coherence we need to have a situation where the bath has a nonzero correlation time, i.e. the environment is non-Markovian [12, 13]. A simpler example is discussed in Section 9.7, albeit in the context of oscillator dynamics – see especially Figure 9.4. This figure shows that an initially prepared coherence survives over much longer intervals than in Markovian theory. To see this consider that the oscillator is initially in a coherent state $|\alpha\rangle$, then (9.94) implies $|\langle a(t)\rangle|^2 = |v(t)|^2|\alpha|^2$. Thus survival of $v(t)$ over long times implies the survival of coherence.

16.2 Disentanglement in two-level systems

The relaxation of the diagonal and off-diagonal elements of the density matrix also leads to the loss of entanglement between the two systems. Consider, for example, the entangled state

$$\begin{aligned}|\psi\rangle &= \frac{1}{\sqrt{2}}(|e,g\rangle - |g,e\rangle),\\ \rho &= \frac{1}{2}(|e,g\rangle\langle e,g| + |g,e\rangle\langle g,e| - |e,g\rangle\langle g,e| - |g,e\rangle\langle e,g|).\end{aligned} \tag{16.6}$$

If terms like $|e,g\rangle\langle g,e|$ were absent in (16.6), then the state would be a separable state. Thus if such terms decayed much faster than diagonal terms like $|e,g\rangle\langle e,g|$, then the state would become disentangled. Let us then examine the interaction of the system with the environment. First assume that the two systems are far apart, so that each two-level system decays according to (16.1). Then the off-diagonal terms in (16.6) would decay as

$$\rho_{eg,ge}(t) = \rho_{eg,ge}(0) \exp\left(-\mathrm{i}\omega t - \frac{t}{T_2} + \mathrm{i}\omega t - \frac{t}{T_2}\right) = -\frac{1}{2} \exp\left(-\frac{2t}{T_2}\right). \tag{16.7}$$

Thus the term which is responsible for the entanglement of the state (16.6) decays at the rate $2/T_2$, which is twice as fast as the decay of the superposition (16.3). The state (16.6) is fully entangled. However, on interaction with the environment it becomes a mixed state and we need to calculate the time dependence of both coherence and populations to have an estimate of the entanglement at time t.

16.2.1 Pure dephasing-induced disentanglement

We next discuss a very simple situation where the relaxation arises purely from dephasing, as is the case in most solid-state environments. This amounts to letting $T_1 \to \infty$. In this case

$$\rho_{eg}(t) = \rho_{eg}(0)\mathrm{e}^{-\mathrm{i}\omega t - t/T_2}, \quad \rho_{ee}(t) = \rho_{ee}(0), \quad \rho_{gg}(t) = \rho_{gg}(0). \tag{16.8}$$

Let us consider an initial density matrix for a system of two two-level atoms having nonzero $\rho_{gg}, \rho_{ee}, \rho_{\alpha\alpha}, \rho_{\beta\beta}, \rho_{\alpha\beta}$, where $|g\rangle = |g_1, g_2\rangle$, $|e\rangle = |e_1, e_2\rangle$, $|\alpha\rangle = |g_1, e_2\rangle$, $|\beta\rangle = |e_1, g_2\rangle$, then according to (15.33), the concurrence is given by

$$\begin{aligned} C &= \max(0, C_2), \\ C_2 &= 2(|\rho_{\alpha\beta}| - \sqrt{\rho_{gg}\rho_{ee}}). \end{aligned} \tag{16.9}$$

We need $C_2 > 0$, if the initial state is entangled. For the dephasing environment ρ_{ee} and ρ_{gg} do not evolve and $\rho_{\alpha\beta}$ evolves according to (16.7) and hence

$$C_2 = 2(|\rho_{\alpha\beta}| \exp\{-2t/T_2\} - \sqrt{\rho_{gg}\rho_{ee}}). \tag{16.10}$$

This gives us explicitly the amount of entanglement as a function of time t. If $\rho_{ee} = 0$, then we have monotonic decrease of entanglement. However, for nonzero ρ_{gg}, ρ_{ee} one has an interesting situation – the entanglement becomes zero at finite t determined by

$$t_D = \frac{T_2}{2} \log\left(\frac{|\rho_{\alpha\beta}|}{\sqrt{\rho_{gg}\rho_{ee}}}\right). \tag{16.11}$$

The term in the bracket is greater than unity since the state is entangled at $t = 0$. The vanishing of the entanglement for finite time has been called the sudden death of entanglement [14]. It may be added that this is the property of the initial entangled state rather than the dynamics. Needless to say, one can obtain similar results for other models of interaction with the environment. The concurrence can be obtained from experimental data by first doing quantum state tomography [15]. Almeida *et al.* [16] devised an alternate method by using the Bosonic representation of two-level systems (Eqs. (6.12), (6.13)). They thus used the polarization of photons to study the sudden death of entanglement in two-level systems.

16.3 Decoherence-free subspace

Sometimes it can happen that the interaction of, say, a system of spins with the environment is such that the spins interact with the environment via the collective operators $J_\pm = \sum_{j=1}^{N} S_\pm^{(j)}$, $J_z = \sum_{j=1}^{N} S_z^{(j)}$. In this case certain types of entangled states do not decay or do not undergo any decoherence. For example, consider the singlet state of two spins

$$|\psi_0\rangle = \frac{1}{\sqrt{2}}(|\uparrow\downarrow\rangle - |\downarrow\uparrow\rangle). \tag{16.12}$$

This state is such that

$$J_\mp|\psi_0\rangle = (S_\mp^{(1)} + S_\mp^{(2)})|\psi_0\rangle = 0, \qquad S_z|\psi_0\rangle = 0. \tag{16.13}$$

Consider for simplicity the master equation (9.56) for dephasing alone

$$\frac{\partial \rho}{\partial t} = -\mathrm{i}\omega[S_z, \rho] - \Gamma(S_z S_z \rho - 2S_z \rho S_z + \rho S_z S_z). \tag{16.14}$$

This equation was for a single spin or a two-level system. However, let us now assume that the two-atom system interacts with the dephasing environment via collective spins. Then Eq. (16.14) remains valid with S_z now representing the collective spin J_z. From the property (16.13), it is clear that

$$\frac{\mathrm{d}}{\mathrm{d}t}(|\psi_0\rangle\langle\psi_0|) = 0. \tag{16.15}$$

Thus for the dephasing model with collective interactions, the state $|\psi_0\rangle$ does not disentangle [17, 18], and one says that $|\psi_0\rangle$ lies in a decoherence-free subspace. Kwiat *et al.* [18] gave a very simple realization of the decoherence-free space by using photonic realization of spins. The decoherence is introduced via three quartz plates in each of the two paths for photons. Furthermore, the decoherence was made collective by choosing the relative orientation of the quartz plates.

Another very important case of environmental interaction is the one where the environment causes transitions between the states of the two-level system. Spontaneous emission is an example of this. On replacing the single two-level system operators with the collective operators and by setting $n_0 = 0$, $\Gamma = 0$, we obtain from (9.56),

$$\frac{\partial \rho}{\partial t} = -\mathrm{i}\omega[J_z, \rho] - \gamma(J_+ J_- \rho - 2J_- \rho J_+ + \rho J_+ J_-). \tag{16.16}$$

Using Eq. (16.13), we obtain from (16.16)

$$\frac{\partial}{\partial t}(|\psi_0\rangle\langle\psi_0|) = 0, \tag{16.17}$$

i.e. no disentanglement takes place. Note that, for two spins, the collective operators have triplet ($\frac{1}{\sqrt{2}}(|\uparrow, \downarrow\rangle + |\downarrow, \uparrow\rangle)$ and singlet (Eq. (16.12)) states. The singlet state is decoherence free against both dephasing (16.14) and radiative decay (16.16) models. This singlet state would then also be decoherence free under the combined influence of dephasing and state changing environmental interactions. Equation (16.16) allows dynamical evolution

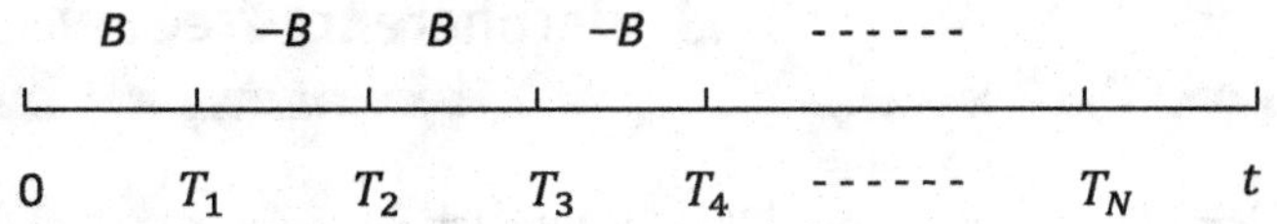

Fig. 16.1 A sequence of π-pulses is applied at times T_j.

in the triplet space and therefore the triplet state is not decoherence free under environmental perturbations given by (16.16).

The model (16.16) is the well-known description of the dynamics of Dicke superradiance from a small sample [19–21]. Its time-dependent solutions have been extensively studied for different number of atoms. We can realize a model like (16.16) in a cavity QED set up. We consider the case of a bad cavity and the atoms located at the antinodes of the cavity. Then the single-atom master equation becomes (16.16) if we replace the single-atom operators by the collective operators $J_{\pm}$.

16.4 Protection of decoherence due to dephasing via dynamical decoupling

In this section we show how a sequence of short, essentially instantaneous, π-pulses can protect against decoherence due to dephasing. In Section 9.8, we discussed a simple non-Markovian model of dephasing. The interaction of a two-level system (spin) with the bath is given by (9.107)

$$H = \hbar S_z \sum_i g_i (a_i e^{-i\omega_i t} + a_i^\dagger e^{i\omega_i t}) = \hbar S_z B(t), \tag{16.18}$$

where we work in the interaction picture. Consider the action of a π-pulse along the x axis. The operator S_z is then transformed to

$$e^{i\pi S_x} S_z e^{-i\pi S_x} = 2iS_x S_z(-2iS_x) = -4S_x S_x S_z = -S_z. \tag{16.19}$$

Thus the action of the π-pulse transforms H into $-H$, i.e. it reverses the direction of evolution and this is the reason why the action of pulse is called dynamic decoupling. Consider then the evolution over a time interval t and apply π-pulses at $T_1, T_2, \ldots, T_N$ as shown in Figure 16.1. Thus the evolution under the bath from $t = 0$ to $t = T_1$ is with (16.18), from $t = T_1$ to $t = T_2$ is with (16.18) with $B \to -B$. We showed in Section 9.8 that the off-diagonal element of ρ evolves as (Eqs. (9.111) and (9.112))

$$\begin{aligned} \rho_{eg}(t) &= \rho_{eg}(0)\zeta(t), \\ \zeta(t) &= \left\langle T \exp\left\{-i\int_0^t B(\tau)d\tau\right\}\right\rangle, \end{aligned} \tag{16.20}$$

where T stands for the time-ordering operator and the averaging is done over the state of the bath. We now need to modify (16.20) according to the scheme shown in Figure 16.1.

We need to replace $B(\tau)$ by $B(\tau)f(\tau)$ with

$$f(\tau) = \sum_{j=0}^{N-1} (-1)^j \theta(\tau - T_j)\theta(T_{j+1} - \tau), \qquad T_0 = 0. \tag{16.21}$$

Thus $f(\tau) = 1$ if $0 < \tau < T_1$, $f(\tau) = -1$ if $T_1 < \tau < T_2$, etc. Hence, under the influence of the π-pulses $\zeta(t)$ would be

$$\begin{aligned} \zeta(t) &= \langle w(t) \rangle, \\ w(t) &= T \exp\left\{ -\mathrm{i} \int_0^t B(\tau) f(\tau) \mathrm{d}\tau \right\}. \end{aligned} \tag{16.22}$$

The rest of the steps are similar to the ones in Section 9.8. By analogy to (9.123), we now obtain

$$\begin{aligned} \zeta(t) &= \exp\{\mathrm{i}\Phi(t)\} \exp\left\{ -\int \mathrm{d}\omega J(\omega) \cdot \left(n(\omega) + \frac{1}{2} \right) |f(\omega, t)|^2 \right\}, \\ f(\omega, t) &= -\int_0^t \mathrm{e}^{-\mathrm{i}\omega\tau} \mathrm{d}\tau f(\tau), \quad n(\omega) = \frac{1}{\mathrm{e}^{\hbar\omega/K_B T} - 1}, \\ \Phi(t) &= \int_0^t \mathrm{d}\tau_1 \int_0^{\tau_1} \mathrm{d}\tau_2 J(\omega) \sin[\omega(\tau_1 - \tau_2)] f(\tau_1) f(\tau_2), \end{aligned} \tag{16.23}$$

where $J(\omega)$ is the spectral function of the bath. For an ohmic bath it is given by (9.106)

$$J(\omega) = 2\alpha\omega\Theta(\omega_D - \omega) \text{ or } 2\alpha\omega \mathrm{e}^{-\omega/\omega_D}. \tag{16.24}$$

For a Lorentzian bath it is

$$J(\omega) = \frac{\Gamma/\pi}{\Gamma^2 + \omega^2} J_0, \tag{16.25}$$

where J_0 has the dimensions of frequency squared. The decay of the coherence $\rho_{eg}(t)$ is given by $\exp\{-\beta(t)\}$, where $\beta(t)$ is obtained from (16.20) and (16.23)

$$\beta(t) = \int \mathrm{d}\omega J(\omega) \left(n(\omega) + \frac{1}{2} \right) |f(\omega, t)|^2. \tag{16.26}$$

The function $f(\omega, t)$ can be evaluated to be

$$f(\omega, t) = \frac{\mathrm{i}}{\omega} \left[1 + (-1)^{N+1} \mathrm{e}^{-\mathrm{i}\omega t} + 2 \sum_{j=1}^{N} (-1)^j \mathrm{e}^{-\mathrm{i}\omega T_j} \right]. \tag{16.27}$$

For zero temperature $n(\omega) = 0$, and if no pulses are applied, then

$$\beta(t) \to 2 \int \mathrm{d}\omega J(\omega) \frac{\sin^2(\omega t/2)}{\omega^2}. \tag{16.28}$$

For large t, the function $\sin^2(\omega t/2)/\omega^2$ is sharply peaked and hence for a Lorentzian bath

$$\beta(t) \approx \pi J(0) t = \frac{t}{T_2}, \qquad T_2 = \frac{1}{\pi J(0)} = \frac{\Gamma}{J_0}. \tag{16.29}$$

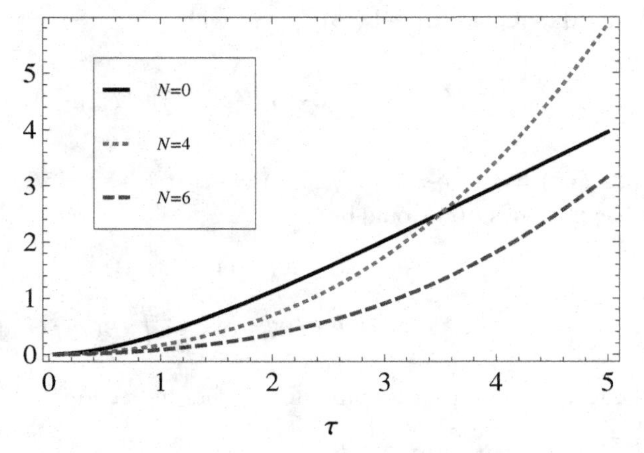

Fig. 16.2 The behavior of the parameter $\beta(t)$ (Eq. (16.26) with $n(\omega) = 0$) as a function of $\tau = \Gamma t$, when the CPMG sequence is applied. The values for $N = 0$ are one tenth of the actual values.

For a Lorentzian bath $\beta(t)$ (Eq. (16.28)) can also be evaluated in closed form for all t

$$\beta(t) = \frac{J_0}{\Gamma^2}[\Gamma t - 1 + \mathrm{e}^{-\Gamma t}]. \tag{16.30}$$

We next want to show how the application of two different types of pulse sequences makes the decay of coherence much slower.

16.4.1 Equidistant pulse sequence

One very well-known pulse sequence, first developed in the context of nuclear magnetic resonance, is the Carr–Purcell–Meiboom–Gill (CPMG) sequence. This consists of a set of N π-pulses uniformly distributed in the interval 0 to T, so that

$$T_j = \frac{jT}{N+1}, \qquad T_{N+1} = T. \tag{16.31}$$

The pulses being applied at T_1, T_2, $\ldots T_N$. The observation is done at time T. The series (16.27) can be evaluated using (16.31) with the result

$$\begin{aligned} |f(\omega, t)|^2 &= \frac{4}{\omega^2}\tan^2\left(\frac{\omega t}{2N+2}\right)\cos^2\left(\frac{\omega t}{2}\right), \quad N \text{ even} \\ &= \frac{4}{\omega^2}\tan^2\left(\frac{\omega t}{2N+2}\right)\sin^2\left(\frac{\omega t}{2}\right), \quad N \text{ odd.} \end{aligned} \tag{16.32}$$

A number of experiments [22–24], have established the great advantage of the CPMG sequence in slowing down decoherence particularly if the bath is described with Lorentzian spectral function (16.25). Using (16.26) and (16.32) we can evaluate $\beta(t)$. It can be shown that the integrals are well defined (see Exercise 16.5). We display $\Gamma T_2 \beta(t) = (\Gamma^2/J_0)\beta(t)$ as a function of $\tau = \Gamma t$ in Figure 16.2 for zero temperature. We show the behavior for two different numbers of pulses. Figure 16.2 shows how the CPMG helps in slowing down decoherence. Even for a small number of pulses, such as four or six, the scale for the

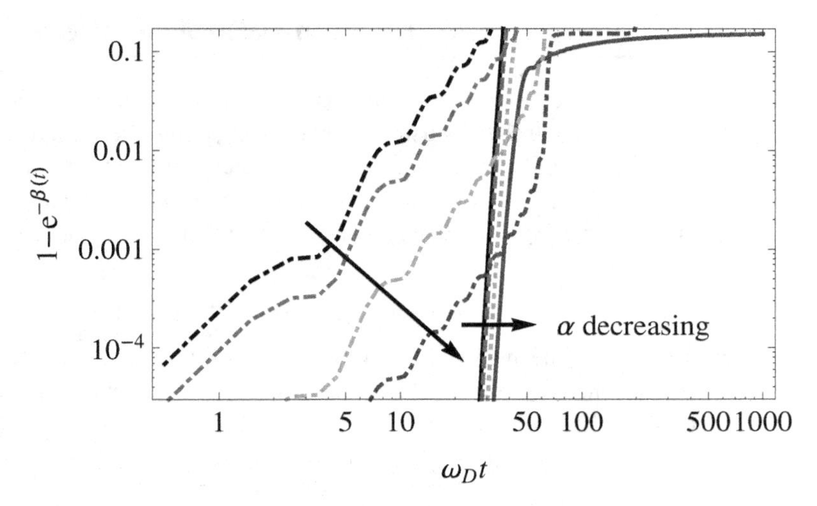

Fig. 16.3 Dotted: uniform CPMG sequence; solid: nonuniform Uhrig sequence for $\alpha = 0.25,\ 0.1,\ 0.01,\ 0.001$.

survival of coherence has increased by a factor of about ten. Sagi *et al.* [24] demonstrated the application of CMPG sequence in increasing the coherence time by a factor of 20 in a dense cold atomic ensemble of ^{87}Rb atoms. Other experiments have used solid-state systems [22, 23]. For example, Lange *et al.* [22] demonstrated enhancement in coherence time by a factor of 25 of a single nitrogen vacancy defect center in diamond coupled to a spin bath.

16.4.2 Uhrig pulse sequence (UDD)

Uhrig [25] proposed a pulse sequences in which pulses are applied at nonuniform intervals given by

$$T_j = T \sin^2\left(\frac{\pi j}{2(N+1)}\right). \tag{16.33}$$

It turns out that such a sequence is especially useful if the spectral density of the bath has a sharp cut off as in the ohmic case (16.24). In this case the function $f(\omega, t)$ can not be evaluated in analytic form. Uhrig [25] has given an asymptotic form valid for $\omega t/2(N+1) < 1$:

$$|f(\omega, t)|^2 \cong \frac{16}{\omega^2}(N+1)^2 J^2_{N+1}\left(\frac{\omega t}{2}\right). \tag{16.34}$$

The function $\beta(t)$ can be evaluated numerically. We show in Figure 16.3 the behavior of $1 - \exp\{-\beta(t)\}$ as a function of $\omega_D t$ for 20 pulses for both UDD and CPMG sequences. For ohmic baths, the advantage of UDD sequence is apparent. The UDD works even for strong coupling with the bath. Many experiments [26, 27] report the relative merits of the UDD and CPMG sequences. In experiments on an array of ions in a Penning trap, Blercuk *et al.* [27] could even manipulate the spectrum of noise and for different noise sources reported the performance of the UDD and CPMG sequences.

16.4.3 Protection against disentanglement

So far we have considered how the dephasing induced decoherence in a single two-level system can be controlled by using dynamic decoupling. We can extend the idea of dynamic decoupling to protect entanglement against deteriorating effects of environment. For EPR states like (16.6), the key quantity is the off-diagonal element $\rho_{eg,ge}$ (Eq. (16.7)). If the two spins are interacting with identical environments, then the decay of this quantity would be governed by

$$\rho_{eg,ge}(t) \sim \rho_{eg,ge}(0) \exp\{-2\beta(t)\}, \tag{16.35}$$

where $\beta(t)$ has been discussed in Sections 16.4.1 and 16.4.2. Clearly methods used to protect decoherence are equally valid to protect entanglement [28] as also confirmed by experiments [29, 30]. Even the p-particle GHZ state

$$|\psi\rangle = \frac{1}{\sqrt{2}}(|eee\ldots e\rangle - |ggg\ldots g\rangle) \tag{16.36}$$

can be protected. The disentanglement in presence of the pulses would be given by $\exp\{-p\beta(t)\}$. Incidentally this also shows the dependence of the disentanglement on the size (p) of the entangled state.

16.5 Control of the spectral density of environment for protection against decoherence

In Chapter 9 we showed that the relaxation times T_1, T_2 are determined by the structure of the correlation functions (Eq. (9.29)) of the bath. More specifically in the context of the Markovian theories, the decay rate is determined by the spectral density evaluated at system's frequency (Eq. (9.38)). Clearly if we can externally manipulate the spectral density such that its value at the system's frequency becomes small, then the environmental effects on the systems coherence and entanglement would be minimized. We consider this possibility by appropriately modulating the frequency of the system which effectively changes the characteristics of the relaxation induced by the bath [5–7].

Let us consider a modulation of the frequency of the two-level system

$$\omega_0 \to \omega_0 - m\nu \cos \nu t. \tag{16.37}$$

Let us consider the interaction of the two-level system with a bath of oscillators so that we write the Hamiltonian as

$$H = \hbar \sum_j \omega_j a_j^\dagger a_j + \hbar \sum_j (g_j a_j S_+ + H.c.) + \hbar(\omega_0 - m\nu \cos \nu t) S_z. \tag{16.38}$$

In the interaction picture (16.38) leads to

$$H_I(t) = \hbar \sum_j (g_j a_j \mathrm{e}^{-\mathrm{i}(\omega_j - \omega_0)t} \mathrm{e}^{-\mathrm{i}m \sin \nu t} S_+ + H.c.). \tag{16.39}$$

We would keep the analysis as simple as possible. We would take the bath of oscillators to be at zero temperature. Let the atom be in the state $|e\rangle$ at $t = 0$. We work in the Schrödinger picture. The wave function at any time can be written as

$$|\psi(t)\rangle = c_e(t)|e, \{0\}\rangle + \sum_j c_j(t)|g, \{1_j\}\rangle, \quad c_e(0) = 1, \quad c_j(0) = 0. \tag{16.40}$$

We also use the well-known expansion

$$\mathrm{e}^{-\mathrm{i}m\sin\nu t} = \sum_{q=-\infty}^{+\infty} J_q(m)\mathrm{e}^{-\mathrm{i}q\nu t}. \tag{16.41}$$

Using (16.38) and (16.41) and the Schrödinger equation, we obtain equations for the amplitudes

$$\dot{c}_e = -\mathrm{i}\sum_{j,q} g_j \mathrm{e}^{-\mathrm{i}(\omega_j-\omega_0)t-\mathrm{i}q\nu t} J_q(m) c_j(t), \quad \dot{c}_j = -\mathrm{i}\sum_q g_j^* \mathrm{e}^{\mathrm{i}(\omega_j-\omega_0)t} J_q(m)\mathrm{e}^{\mathrm{i}q\nu t} c_e(t). \tag{16.42}$$

We can now formally integrate (16.42) for $c_j(t)$ and substitute in the equation for c_e to obtain

$$\dot{c}_e = -\sum_{j,q,q'} |g_j|^2 \mathrm{e}^{-\mathrm{i}(\omega_j-\omega_0)t-\mathrm{i}q\nu t} J_q(m) J_{q'}(m) \int_0^t \mathrm{e}^{\mathrm{i}(\omega_j-\omega_0)(t-\tau)+\mathrm{i}q'\nu(t-\tau)} c_e(\tau)\mathrm{d}\tau. \tag{16.43}$$

We assume ν is large compared to the width of the bath so that we can do a time averaging over the scale of the order of ν^{-1}, then (16.43) reduces to

$$\dot{c}_e = -\sum_{j,q} |g_j|^2 J_q^2(m) \int_0^t \mathrm{e}^{-\mathrm{i}(\omega_j-\omega_0)\tau-\mathrm{i}q\nu\tau} c_e(\tau)\mathrm{d}\tau. \tag{16.44}$$

On taking the in Laplace transform, we obtain from (16.44)

$$p\hat{c}_e(p) - 1 + \sum_{jq} |g_j|^2 J_q^2(m)[p + \mathrm{i}(\omega_j - \omega_0) + \mathrm{i}q\nu]^{-1}\hat{c}_e(p) = 0. \tag{16.45}$$

On defining

$$\Sigma(p) = \sum_{jq} |g_j|^2 J_q^2(m)[p + \mathrm{i}(\omega_j - \omega_0) + \mathrm{i}q\nu]^{-1}, \tag{16.46}$$

the amplitude for finding the system in the state $|e\rangle$ is then

$$\hat{c}_e(p) = [p + \Sigma(p)]^{-1}. \tag{16.47}$$

The form of $\hat{c}_e(p)$ is similar to what we discussed in Chapter 7 (see, for example, Eq. (7.83)). The case discussed in Section 7.7.1 corresponds to $\nu = 0$. However, we can now use the modulation to fix the contribution coming from the bath. For example, we can choose modulation so that $J_0(m) = 0$. If we continue to assume that the coupling to bath is weak, then we can use the approximate expression

$$\hat{c}_e(p) = [p + \Sigma(0^+)]^{-1}, \tag{16.48}$$

$$\Sigma(0^+) = \lim_{p\to 0^+} \sum_{j,q\neq 0} |g_j|^2 J_q^2(m)[p + \mathrm{i}(\omega_j - \omega_0 + q\nu)]^{-1}. \tag{16.49}$$

Clearly the relaxation is now determined by the spectral density of the bath at the frequency $(\omega_0 - q\nu)$. For structured baths (nonwhite noise spectrum) with a width Γ (Eq. (9.100)), we can evaluate $\Sigma(p)$

$$\Sigma(p) = \sum_q \int J_q^2(m) \cdot \frac{\kappa\Gamma^2/\pi}{\Gamma^2 + (\omega - \bar{\omega})^2} \cdot \frac{\mathrm{d}\omega}{p + \mathrm{i}(\omega - \omega_0) + \mathrm{i}q\nu}, \tag{16.50}$$

which for $\bar{\omega} = \omega_0$ reduces to

$$\Sigma(p) = \sum_q J_q^2(m) \cdot \frac{\kappa\Gamma}{\Gamma + (p + \mathrm{i}q\nu)}. \tag{16.51}$$

Let us now compare the real part of $\Sigma(0^+)$ in the absence and presence of modulation

$$\begin{aligned} \mathrm{Re}\,\Sigma(0^+) &\to \kappa && \text{if } m = 0, \\ &\approx \kappa J_1^2(m) \cdot \frac{\Gamma^2}{\Gamma^2 + \nu^2} && \text{if } J_0(m) = 0, \\ &\ll \kappa && \text{if } \nu \gg \Gamma. \end{aligned} \tag{16.52}$$

The result (16.52) shows clear advantage of using modulation of the system's frequency in effectively reducing the decoherence effect of the bath [5]. Needless to say, given the form (16.51) of $\Sigma(p)$, we can evaluate the time dependence of c_e and we need not rely on the weak coupling approximation. The weak coupling is used to illustrate how modulation can substantially enhance relaxation times.

In the above we considered the survival of the population in the state $|e\rangle$ over longer periods. As discussed in Sections 16.1 and 16.2, the amplitude in the state $|e\rangle$ also determines the time over which coherence and entanglement would survive. Thus by using fast modulation both coherence and entanglement survive over much longer periods.

16.6 Modulation produced protection against disentanglement in cavity QED

Finally, we discuss the question of survival of entanglement in cavity QED. Consider two two-level systems in a bad cavity on resonance with the frequency of the two-level system. The interaction Hamiltonian in the interaction picture is

$$H = \hbar g[a(S_+^{(1)} + S_-^{(2)}) + H.c.]. \tag{16.53}$$

The effect of frequency modulation is to replace (16.53) by

$$H = \hbar g[a(S_+^{(1)} + S_-^{(2)})e^{-im\sin\nu t} + H.c.]. \tag{16.54}$$

The field in the cavity decays at the rate 2κ. Let us assume for simplicity that the initial state is a single-photon state

$$|\psi\rangle = \frac{1}{\sqrt{2}}(|e,g\rangle + |g,e\rangle)|0\rangle. \tag{16.55}$$

Then the state at time t would have the form

$$|\psi(t)\rangle = (\alpha_1(t)|e,g\rangle + \alpha_2(t)|g,e\rangle)|0\rangle + \alpha_d(t)|g,g\rangle|1\rangle + \alpha_f(t)|g,g\rangle|0\rangle. \tag{16.56}$$

The last part (α_f) represents the eventual decay of the excitation out of the cavity. The entanglement in the initial state and the state (16.56) can be calculated. The concurrence is

$$C(t) = 2\max\{0, |\alpha_1(t)||\alpha_2(t)|\}, \qquad C(0) = 1. \tag{16.57}$$

The equations for α_1, α_2 and α_d are

$$\begin{aligned} &\dot{\alpha}_1 = -\mathrm{i}g(t)\alpha_d, \quad \dot{\alpha}_2 = -\mathrm{i}g(t)\alpha_d, \quad \dot{\alpha}_d = -\mathrm{i}g^*(t)(\alpha_1 + \alpha_2) - \kappa\alpha_d, \\ &g(t) = g\mathrm{e}^{-im\sin\nu t}, \quad \alpha_1(0) = \alpha_2(0) = \frac{1}{\sqrt{2}}, \quad \alpha_d(0) = \alpha_f(0) = 0. \end{aligned} \tag{16.58}$$

These equations can be integrated numerically or by using the method of Section 16.5. As in Section 16.5, we use a value of m such that $J_0(m) = 0$. Furthermore, for large ν we expect retention of coherence and entanglement over much longer periods. For the case of a bad cavity $g \ll \kappa$ and for $m = 0$, we obtain approximate equations

$$\begin{aligned} &\alpha_d \approx -\mathrm{i}g^*(\alpha_1 + \alpha_2)/\kappa, \\ &\dot{\alpha}_1 = -g^2(\alpha_1 + \alpha_2)/\kappa, \qquad \dot{\alpha}_2 = -g^2(\alpha_1 + \alpha_2)/\kappa, \end{aligned} \tag{16.59}$$

and hence

$$\alpha_1(t) = \alpha_2(t) = \frac{1}{\sqrt{2}}\mathrm{e}^{-\frac{2g^2}{\kappa}t} \tag{16.60}$$

leading to time-dependent concurrence

$$C(t) = \exp\left\{-\frac{4g^2}{\kappa}t\right\}. \tag{16.61}$$

The loss of entanglement is given by twice the decay rate of the atom in the cavity. In presence of modulation we can integrate the set of Eqs. (16.58) by using the Runge–Kutta procedure. We show in Figure 16.4 the result of using fast modulation. Clearly the fast modulation considerably slows down the decay (16.61) of concurrence and thus entanglement can be made to survive over much longer times. This figure also shows that if m is not chosen to be a zero of $J_0(m)$ (case $m = 0.5$), then the modulation is not of much help. The bottom two curves in Figure 16.4 are hardly distinguishable from the decay given by (16.61).

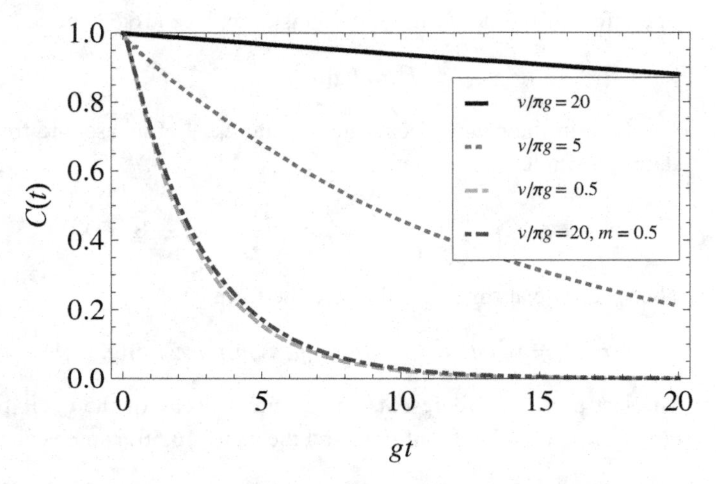

Fig. 16.4 The time evolution of the concurrence of the EPR state (16.55) for two atoms in a bad cavity. The figure shows how fast modulation can protect entanglement over much larger periods.

Exercises

16.1 Consider a two-atom system initially prepared in a Werner state defined by the mixed density matrix

$$\rho = F|\psi\rangle\langle\psi| + \frac{1-F}{4}I,$$

where $|\psi\rangle$ is the singlet state (16.6) and I is the unit operator in four-dimensional space. Note that $\text{Tr}\rho = 1$. The state is known to be entangled if $F > 1/3$ [31]. Assume that each two-level system decays independently according to (16.16) ($S_\pm$ spin-1/2 operators). Then find the density matrix at time t. Use the time dependence to find concurrence as a function of F. Can the concurrence vanish at finite t for all values of $F > 1/3$ [1].

16.2 Estimate T_D given by Eq. (16.11) for the Werner state of Exercise 16.1.

16.3 Using (16.4) find $1 - \text{Tr}\{\rho^2(t)\}$, which is a measure of how far the state deviates from a pure state.

16.4 Using contour integration prove the result (16.30).

16.5 Consider the integrand in (16.26) for Lorentzian bath and for CPMG sequence, Eq. (16.32). Show that there are no singularities on the real ω axis. It would be useful to express as $\cos px$, $\sin px$ for integer values of p in terms of powers of $\cos x$ and $\sin x$.

16.6 Consider the decay of the field in a cavity as described by the master equation (9.37) with $n_0 = 0$

$$\frac{\partial\rho}{\partial t} = -\kappa(a^\dagger a\rho - 2a\rho a^\dagger + \rho a^\dagger a).$$

One can use a feedback technique by measuring the field by a photo detector and by using the photocurrent to modulate the phase of the cavity field. In this case the master equation is modified to [10]

$$\frac{\partial \rho}{\partial t} = -\kappa(a^\dagger a\rho + \rho a^\dagger a - 2e^{i\pi a^\dagger a} a\rho a^\dagger e^{-i\pi a^\dagger a}).$$

Show that under this feedback the cat state $|\alpha\rangle + i|-\alpha\rangle$ will remain a cat state with $\alpha \to \alpha e^{-\kappa t}$. The general theory of feedback is developed in [32, 33].

References

[1] T. Yu and J. H. Eberly, *Phys. Rev. Lett.* **93**, 140404 (2004).
[2] L. Viola and S. Lloyd, *Phys. Rev. A* **58**, 2733 (1998).
[3] L. Viola, E. Knill, and S. Lloyd, *Phys. Rev. Lett.* **82**, 2417 (1999).
[4] P. Facchi, S. Tasaki, S. Pascazio, H. Nakazato, A. Tokuse, and D. A. Lidar, *Phys. Rev. A* **71**, 022302 (2005); and D. Vitali, and P. Tombesi, *Phys. Rev. A* **59**, 4178 (1999).
[5] G. S. Agarwal, *Phys. Rev. A* **61**, 013809 (1999).
[6] A. G. Kofman and G. Kurizki, *Phys. Rev. Lett.* **87**, 270405 (2001).
[7] A. G. Kofman and G. Kurizki, *Phys. Rev. Lett.* **93**, 130406 (2004).
[8] C. Search and P. R. Berman, *Phys. Rev. Lett.* **85**, 2272 (2000).
[9] G. S. Agarwal, M. O. Scully, and H. Walther, *Phys. Rev. Lett.* **86**, 4271 (2001).
[10] D. B. Horoshko and S. Ya. Kilin, *Phys. Rev. Lett.* **78**, 840 (1997).
[11] W. H. Zurek, *Rev. Mod. Phys.* **75**, 715 (2003).
[12] B. Bellomo, R. Lo Franco, and G. Compagno, *Phys. Rev. Lett.* **99**, 160502 (2007).
[13] S. Maniscalco, F. Francica, R. L. Zaffino, N. L. Gullo, and F. Plastina, *Phys. Rev. Lett.* **100**, 090503 (2008).
[14] T. Yu and J. H. Eberly, *Phys. Rev. Lett.* **97**, 140403 (2006).
[15] J. Laurat, K. S. Choi, H. Deng, C. W. Chou, and H. J. Kimble, *Phys. Rev. Lett.* **99**, 180504 (2007).
[16] M. P. Almeida, F. de Melo, M. Hor-Meyll *et al., Science*, **316**, 579 (2007).
[17] D. A. Lidar, I. L. Chuang, and K. B. Whaley, *Phys. Rev. Lett.* **81**, 2594 (1998).
[18] P. G. Kwiat, A. J. Berglund, J. B. Altepeter, and A. G. White, *Science* **290**, 498 (2000).
[19] G. S. Agarwal, *Phys. Rev. A* **2**, 2038 (1970).
[20] R. Bonifacio, P. Schwendimann, and F. Haake, *Phys. Rev. A* **4**, 302 (1971).
[21] R. Bonifacio, P. Schwendimann, and F. Haake, *Phys. Rev. A* **4**, 854 (1971).
[22] G. de Lange, Z. H. Wang, D. Riste, V. V. Dobrovitski, and R. Hanson, *Science* **330**, 60 (2010).
[23] C. A. Ryan, J. S. Hodges, and D. G. Cory, *Phys. Rev. Lett.* **105**, 200402 (2010).
[24] Y. Sagi, I. Almog, and N. Davidson, *Phys. Rev. Lett.* **105**, 053201 (2010).
[25] G. S. Uhrig, *Phys. Rev. Lett.* **98**, 100504 (2007).
[26] J. Du, X. Rong, N. Zhao, Y. Wang, J. Yang and R. B. Liu, *Nature (London)* **461**, 1265 (2009).

[27] M. J. Biercuk, H. Uys, A. P. VanDevender, N. Shiga, W. M. Itano and J. J. Bollinger, *Nature (London)* **458**, 996 (2009).
[28] G. S. Agarwal, *Phys. Scr*. **82**, 038103 (2010).
[29] Y. Wang, X. Rong, P. Feng *et al., Phys. Rev. Lett*. **106**, 040501 (2011).
[30] S. S. Roy, T. S. Mahesh, and G. S. Agarwal, *Phys. Rev. A* **83**, 062326 (2011).
[31] R. F. Werner, *Phys. Rev. A* **40**, 4277 (1989).
[32] H. M. Wiseman, *Phys. Rev. A* **49**, 2133 (1994).
[33] D. B. Horoshko and S. Ya. Kilin, *JETP* **79**, 691 (1994).

17 Coherent control of the optical properties

In Chapter 13 we saw how the optical properties of a two-level system can be modified by the application of an additional strong coherent field. For example, the absorption of light by a two-level system depends on the strength and frequency of the driving field. Figure 13.5 showed that in certain frequency regions we can amplify a probe beam. We assumed in Chapter 13 that the coherent light beam was acting on the same optical transition as the weak probe beam. However, the atomic/molecular systems have many energy levels and we can take advantage of this to produce a variety of ways of controlling the optical properties. This would offer much more flexibility as different optical transitions would have different frequencies and hence one could use a variety of sources. In this chapter we present results for the optical properties of a multilevel system. We show that coherent control can make an opaque medium transparent. We also show that the dispersive properties, which are important for the linear and nonlinear propagation of light, can be manipulated by light fields [1–3].

17.1 A simple model for coherent control

Let us consider the situation shown in Figure 17.1 where we have a plane wave travelling through a medium of length l. The medium is characterized by a complex optical susceptibility $\chi(\omega)$, which depends on the frequency of the electromagnetic wave. The output field is

$$E_{\text{out}}(\omega) = \exp\left[\mathrm{i}\frac{\omega}{c}n(\omega)l\right]E_{\text{in}}(\omega), \tag{17.1}$$

where $n(\omega)$ is the complex refractive index of the medium given by

$$n^2(\omega) = 1 + 4\pi\chi(\omega), \tag{17.2}$$

which for atomic vapors ($|\chi(\omega)| \ll 1$) reduces to

$$n(\omega) = 1 + 2\pi\chi(\omega). \tag{17.3}$$

The intensity I_{out} of the output is

$$I_{\text{out}}(\omega) = \exp[-\alpha(\omega)l]I_{\text{in}}(\omega), \tag{17.4}$$

where $\alpha(\omega)$ is the absorption coefficient of the medium

$$\alpha(\omega) = \frac{4\pi\omega}{c}\text{Im}(\chi(\omega)). \tag{17.5}$$

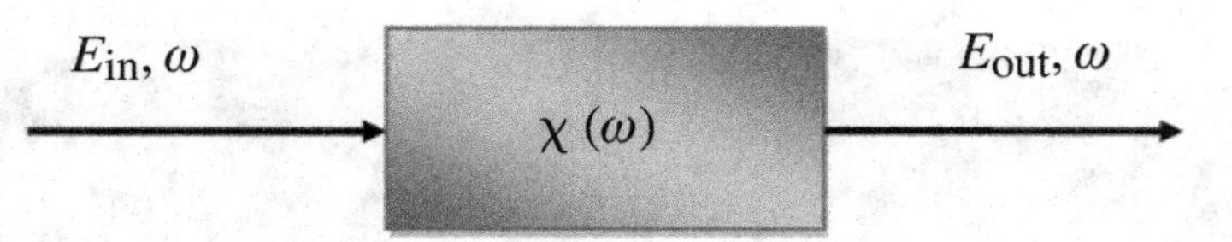

Fig. 17.1 A plane wave propagates through a medium.

The real part of $\chi(\omega)$ corresponds to the phase shift of the beam. The real and imaginary parts of $\chi(\omega)$ satisfy the Kramers–Kronig dispersion relations. The susceptibility $\chi(\omega)$ typically depends on the intrinsic properties of the medium, i.e. on the energy levels, life times, and various other relaxation processes (cf. result (13.14) for two-level systems). For two-level systems $\alpha(\omega)$ is maximum at resonance $\omega = \omega_0$

$$\alpha(\omega_0) = 4\pi \frac{T_2 |p|^2 n\omega_0}{\hbar c}, \tag{17.6}$$

where n is the density of atoms. This absorption coefficient is about 10 cm^{-1} for a density of 10^{10} of Na atoms and for the transition 3s $\rightarrow$ 3p (see the estimates after Eq. (13.16)). Thus the input beam would be attenuated over a distance of the order of 0.1 cm. In order to control the optical properties, the idea is to add another laser beam, which we call the control field and which would be acting on a different optical transition from the one on which the probe beam is acting. The application of the control field modifies the susceptibility $\chi(\omega)$ to $\chi(\omega, E_l)$, which can, in principle, depend on all powers of the control field E_l. We might be able to choose the parameters of the control field in such a way that $\alpha(\omega_0)$ becomes nearly zero and thus making the opaque medium transparent. Harris *et al.* [1] termed this the electromagnetically induced transparency. We now present explicit results of optical properties for a model system [1, 2].

Consider the three-level system shown in Figure 17.2. The optical transition of interest is $|a\rangle \rightarrow |b\rangle$. Many alkalis, such as sodium and rubidium, have this type of level structure. For example, for ^{87}Rb vapor, we can choose $|b\rangle = 5^2S_{1/2}$ $(F = 1)$, $|c\rangle = 5^2S_{1/2}$ $(F = 2)$, and $|a\rangle = 5^2P_{1/2}$ $(F = 2)$, where the hyperfine components $|b\rangle$ and $|c\rangle$ have an energy separation of 6.8 GHz. The probe and control lasers would be diode lasers. We consider the copropagating pump and probe fields

$$\vec{E} = \vec{E}_l e^{i\vec{k}_l \cdot \vec{r} - i\omega_l t} + \vec{E}_p e^{i\vec{k}_p \cdot \vec{r} - i\omega_p t} + c.c.. \tag{17.7}$$

The Hamiltonian of the three-level system, located at $\vec{r}$, in the rotating wave approximation is

$$\begin{aligned} H &= \hbar(\omega_a |a\rangle\langle a| + \omega_b |b\rangle\langle b| + \omega_c |c\rangle\langle c|) \\ &\quad - \hbar(g_p |a\rangle\langle b| e^{-i\omega_p t} + g_l |a\rangle\langle c| e^{-i\omega_l t} + H.c.), \\ g_p &= \frac{\vec{p}_{ab} \cdot \vec{E}_p}{\hbar} e^{i\vec{k}_p \cdot \vec{r}}, \qquad g_l = \frac{\vec{p}_{ac} \cdot \vec{E}_l}{\hbar} e^{i\vec{k}_l \cdot \vec{r}}. \end{aligned} \tag{17.8}$$

We remove all the optical frequencies by using the transformation

$$\begin{aligned} \tilde{\rho}_{ab} &= \rho_{ab} e^{i\omega_p t}, & \tilde{\rho}_{ac} &= \rho_{ac} e^{i\omega_l t}, \\ \tilde{\rho}_{bc} &= \rho_{bc} e^{-i\omega_p t + i\omega_l t}, & \tilde{\rho}_{aa} &= \rho_{aa}. \end{aligned} \tag{17.9}$$

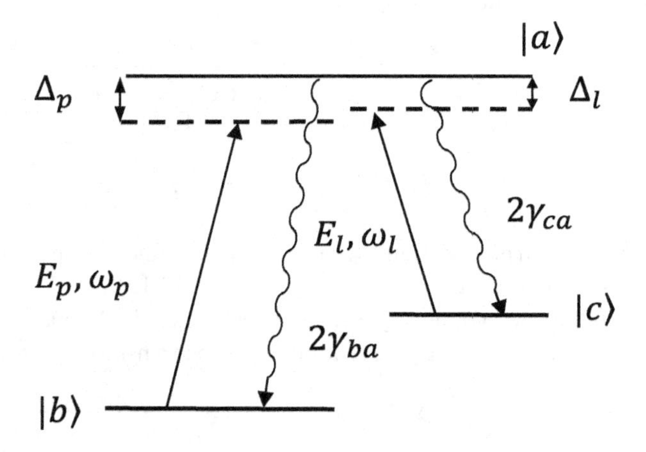

Fig. 17.2 A Λ-type three-level atomic system interacting with probe and control fields.

We include the radiative decay of the upper state $|a\rangle$ at the rate $2\gamma_{ba}$ $(2\gamma_{ca})$ to the state $|b\rangle$ $(|c\rangle)$. We also include the possibility of phase changing collisions. Thus each off-diagonal element $\rho_{\alpha\beta}$ of the density matrix decays at the rate $\Gamma_{\alpha\beta}$, which now includes contributions from radiative decay and phase changing collisions. We would thus use Eq. (9.67) to include all the decay processes. The working equations for the three-level model are obtained by using (9.67) and (17.8). These are given by

$$\begin{aligned}
\dot{\rho}_{ab} &= -\mathrm{i}\Delta_p\rho_{ab} - \Gamma_{ab}\rho_{ab} + \mathrm{i}g_p(\rho_{bb} - \rho_{aa}) + \mathrm{i}g_l\rho_{cb}, \\
\dot{\rho}_{ac} &= -\mathrm{i}\Delta_l\rho_{ac} - \Gamma_{ac}\rho_{ac} + \mathrm{i}g_p\rho_{bc} + \mathrm{i}g_l(\rho_{cc} - \rho_{aa}), \\
\dot{\rho}_{cb} &= \mathrm{i}(\Delta_l - \Delta_p)\rho_{cb} - \Gamma_{bc}\rho_{cb} - \mathrm{i}g_p\rho_{ca} + \mathrm{i}g_l^*\rho_{ab}, \\
\dot{\rho}_{aa} &= -2(\gamma_{ba} + \gamma_{ca})\rho_{aa} + \mathrm{i}g_p\rho_{ba} + \mathrm{i}g_l\rho_{ca} - \mathrm{i}g_p^*\rho_{ab} - \mathrm{i}g_l^*\rho_{ac}, \\
\dot{\rho}_{bb} &= 2\gamma_{ba}\rho_{aa} - \mathrm{i}g_p\rho_{ba} + \mathrm{i}g_p^*\rho_{ab}, \\
\dot{\rho}_{cc} &= 2\gamma_{ca}\rho_{aa} - \mathrm{i}g_l\rho_{ca} + \mathrm{i}g_l^*\rho_{ac},
\end{aligned} \tag{17.10}$$

where for brevity we have dropped the tilde from the density matrix elements and where

$$\Delta_l = \omega_a - \omega_c - \omega_l, \qquad \Delta_p = \omega_a - \omega_b - \omega_p. \tag{17.11}$$

17.1.1 The electromagnetically induced transparency

We assume that initially the atomic population is pumped into the state $|b\rangle$. The equations (17.10) can be solved to all orders in the fields g_l and g_p. However, in the spirit of the probe field, which, by definition, is a weak field, we calculate the coherence ρ_{ab} to lowest order in the weak field, i.e. to first order in g_p and to all orders in g_l. We first note that if $g_p = 0$, then the solution of (17.10) is $\rho_{bb} = 1$ and all the other elements of ρ are zero. To obtain ρ_{ab} we need to solve two coupled equations for ρ_{ab} and ρ_{cb} in the steady state with the results

$$\rho_{ab} = \frac{\mathrm{i}g_p}{(\Gamma_{ab} + \mathrm{i}\Delta_p) + \dfrac{|g_l|^2}{\Gamma_{bc} + \mathrm{i}(\Delta_p - \Delta_l)}}, \tag{17.12}$$

$$\rho_{cb} = -\frac{g_l^* g_p}{|g_l|^2 + (\Gamma_{ab} + \mathrm{i}\Delta_p)[\Gamma_{bc} + \mathrm{i}(\Delta_p - \Delta_l)]}. \tag{17.13}$$

The induced polarization ρ_{ab} can be used to define susceptibility $\chi^{(ab)}(\omega_p)$ for the transition $|a\rangle \leftrightarrow |b\rangle$ as we did in Section 13.1.1 (Eq. (13.12)).

Putting everything together and replacing dipole matrix elements by their rotational averages, the modified susceptibility for the transition $|a\rangle \leftrightarrow |b\rangle$ is

$$\chi^{(ab)}(\omega_p) = \frac{\chi_0 n\mathrm{i}}{(\Gamma_{ab} + \mathrm{i}\Delta_p) + \dfrac{|g_l|^2}{\Gamma_{bc} + \mathrm{i}(\Delta_p - \Delta_l)}}, \tag{17.14}$$

where n is the density of atoms. At two-photon resonance $\Delta_p = \Delta_l$, i.e., $\omega_p - \omega_l = \omega_{cb}$, the susceptibility reduces to

$$\chi^{(ab)}(\omega_p) = \frac{\chi_0 n\mathrm{i}\Gamma_{bc}}{\Gamma_{bc}(\Gamma_{ab} + \mathrm{i}\Delta_p) + |g_l|^2} \to 0 \quad \text{if } \Gamma_{bc} \to 0, \quad |g_l| \neq 0. \tag{17.15}$$

Thus, at two-photon resonance, both the real and imaginary parts of $\chi^{(ab)}$ become zero if the levels $|b\rangle$ and $|c\rangle$ are infinitely long lived. Furthermore, even if $\Gamma_{bc} \neq 0$, $\mathrm{Im}(\chi^{(ab)})$ can be made much smaller by increasing the power of the control laser. The vanishing of the imaginary part of $\chi^{(ab)}$ was called the electromagnetically induced transparency (EIT) by Harris and collaborators [1]. We show the behavior of the real and imaginary parts of the susceptibility (17.14) in Figures 17.3 and 17.4 for the case when the control field is at resonance, $\Delta_l = 0$ and $\Gamma_{bc} = 0$. We normalize all frequencies in units of Γ_{ab} and write $\chi' + \mathrm{i}\chi'' = \frac{\chi^{(ab)}\Gamma_{ab}}{n\chi_0}$, $G = |g_l|/\Gamma_{ab}$. The transparency window can be clearly seen. We now present an estimate of the width of the EIT window when the control field is at resonance, i.e. $\Delta_l = 0$ and when $\Gamma_{bc} = 0$. The susceptibility $\chi^{(ab)}(\omega_p)$ can be written in the neighborhood of $\Delta_p = 0$ as

$$\mathrm{Im}\big(\chi^{(ab)}(\omega_p)\big) \approx \frac{\chi_0 n}{\Gamma_{ab}} \cdot \frac{\Delta_p^2}{2\sigma^2}, \qquad \sigma = \frac{|g_l|^2}{\sqrt{2}\Gamma_{ab}}. \tag{17.16}$$

Note that $\chi_0 n/\Gamma_{ab}$ is the imaginary part of the susceptibility if there is no control field. The transmission (17.4) in the neighborhood of $\Delta_p \sim 0$ can now be expressed as

$$I_{\text{out}} \approx I_{\text{in}} \exp\left(-\frac{\alpha_0 l \Delta_p^2}{2\sigma^2}\right), \qquad \alpha_0 l = \frac{4\pi\omega_{ab}\chi_0 n}{c\Gamma_{ab}}. \tag{17.17}$$

The width of the window is governed by the power of the control laser and it increases linearly with the laser power, since σ is proportional to the laser power.

Another important consequence of the control field is the creation of atomic coherence $\rho_{bc} \neq 0$ between the two long-lived levels $|b\rangle$ and $|c\rangle$. This coherence oscillates at the

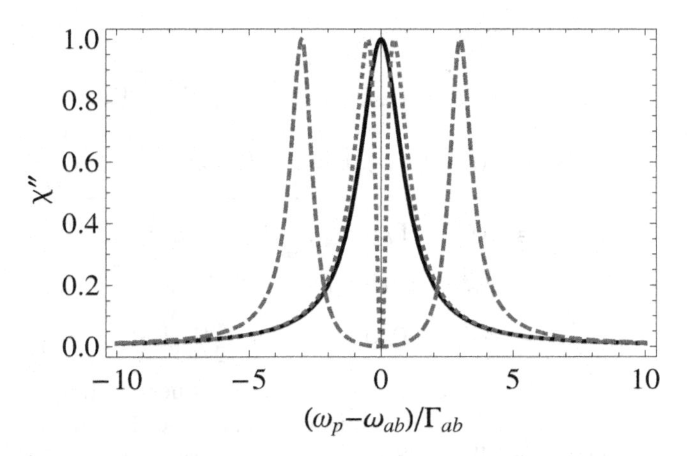

Fig. 17.3 The behavior of the absorption in the medium for $G = 0$ (solid), 0.5 (dotted), and 3 (dashed).

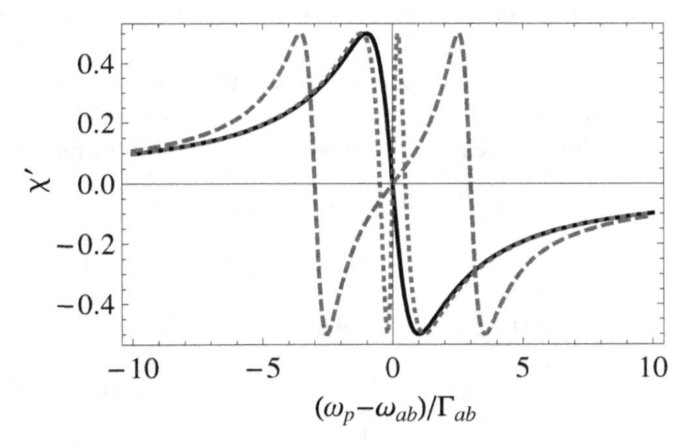

Fig. 17.4 The behavior of the dispersion in the medium for $G = 0$ (solid), 0.5 (dotted), and 3 (dashed).

frequency $(\omega_p - \omega_l)$. This coherence has many applications in storage, retrieval of pulses, and in creating memory devices for quantum information (Chapter 18).

Figure 17.4 for the behavior of the dispersion in the medium shows that in the region of line center the anomalous dispersion changes to normal dispersion. Furthermore, one can have steep dispersion depending on the power of the control field. The steep dispersion can be used for making the propagation of light ultraslow, as discussed in detail in Chapter 18. We also note that the nonlinear generation of radiation [3] can be enhanced by using both EIT and control of dispersion.

17.1.2 Nature of the quantum interference

We now discuss what leads to EIT. For this purpose, we examine the susceptibility (17.14). The behavior of the susceptibility depends on the roots of the denominator in (17.14), which for $\Delta_l = 0$ occur at

$$\Delta_p = \frac{i}{2}(\Gamma_{ab} + \Gamma_{bc}) \pm \sqrt{|g_l|^2 - \frac{1}{4}(\Gamma_{ab} - \Gamma_{bc})^2}, \tag{17.18}$$

$$\approx \frac{i}{2}(\Gamma_{ab} + \Gamma_{bc}) \pm |g_l|, \quad |g_l| \gg \frac{1}{2}|\Gamma_{ab} - \Gamma_{bc}|, \tag{17.19}$$

$$= \frac{i}{2}(\Gamma_{ab} + \Gamma_{bc}) \pm i\sqrt{\frac{1}{4}(\Gamma_{ab} - \Gamma_{bc})^2 - |g_l|^2} \quad \text{if } |g_l| < \frac{1}{2}|\Gamma_{ab} - \Gamma_{bc}|,$$

$$\approx i\Gamma_{ab},\ i\Gamma_{bc} + i|g_l|^2/(\Gamma_{ab} - \Gamma_{bc}),\ \Gamma_{ab} > \Gamma_{bc}. \tag{17.20}$$

$$= \frac{i}{2}(\Gamma_{ab} + \Gamma_{bc}), \quad \text{if } |g_l| = \frac{1}{2}|\Gamma_{ab} - \Gamma_{bc}|. \tag{17.21}$$

The case (17.19) is referred to as Autler–Townes splitting [4]. The two complex roots have distinct frequencies with identical width. In case (17.20) the two roots are purely imaginary and since generally $\Gamma_{ab} \gg \Gamma_{bc}$, the one root at $\Delta_p = 0$ has a very small line width. In case (17.21) we would have a double pole in the susceptibility. Note that the case $G = 0.5$ in Figures 17.3 and 17.4 corresponds to (17.21), where the dip is much narrower due to the double pole.

These roots enable us to understand the nature of the quantum interference that leads to the vanishing of absorption. The imaginary part of the susceptibility (17.14) for control field on resonance $\Delta_l = 0$ can be written by using the roots (17.19) in the form

$$\mathrm{Im}\big(\chi^{(ab)}(\omega_p)\big) = \frac{\pi\chi_0 n}{2}\left[L_w(\Delta_p - |g_l|) + L_w(\Delta_p + |g_l|)\right]$$
$$+ \frac{\beta}{|g_l|}\frac{\pi\chi_0 n}{2}\left[D_w(\Delta_p - |g_l|) - D_w(\Delta_p + |g_l|)\right], \tag{17.22}$$

where L and D are the Lorentzian and dispersive profiles defined by

$$L_w(x) = \frac{\omega/\pi}{x^2 + \omega^2}, \qquad D_w(x) = \frac{x/\pi}{x^2 + \omega^2}, \tag{17.23}$$

$$w = \frac{1}{2}(\Gamma_{ab} + \Gamma_{bc}), \qquad \beta = \frac{1}{2}(\Gamma_{ab} - \Gamma_{bc}). \tag{17.24}$$

At the line center $\Delta_p = 0$, which also corresponds to two-photon resonance since Δ_l is assumed to be zero, one has

$$\mathrm{Im}\big(\chi^{(ab)}(\Delta_p = 0)\big) = \pi\chi_0 n\left[\frac{w}{\pi(w^2 + |g_l|^2)} - \frac{\beta}{\pi(w^2 + |g_l|^2)}\right],$$
$$= \frac{\chi_0 n}{w^2 + |g_l|^2}(w - \beta) = \frac{\Gamma_{bc}\chi_0 n}{w^2 + |g_l|^2}. \tag{17.25}$$

On the other hand, if we had ignored the dispersive contributions (the D_w terms), then

$$\mathrm{Im}\big(\chi^{(ab)}(\Delta_p = 0)\big) = \frac{\chi_0 n}{\omega^2 + |g_l|^2}\frac{1}{2}(\Gamma_{ab} + \Gamma_{bc}), \tag{17.26}$$

which would never be zero. The absorption for $\Gamma_{bc} = 0$ goes to zero only due to the combined presence of both Lorentzian and dispersive contributions in (17.22) [5, 6]. We thus have the picture shown in Figure 17.5 in terms of the dressed states obtained by diagonalizing the part of the Hamiltonian (17.8) containing the states $|a\rangle$ and $|c\rangle$ (see Section 11.1.1).

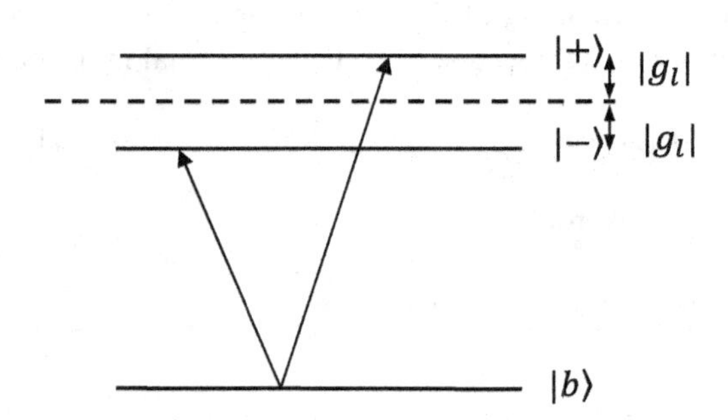

Fig. 17.5 Two channels of absorption starting from the state $|b\rangle$ to the dressed states $|\pm\rangle$.

A single channel of absorption becomes two-channel absorption. Furthermore, the absorption, which typically is Lorentzian, is now a combination of both Lorentzian and dispersive contributions. This is because the weight factor of each line is complex.

The first observation of the EIT was made by Boller *et al.* [7] using an autoionizing transition in neutral strontium. The probe transition was $5s5p^1P_1 \rightarrow 4d5d^1D_2$. The autoionizing level decays rather quickly and is coupled by a control laser to the level $4d5p^1D_2$. They observed the changes in transmission from e^{-20} to e^{-1}. These observations were extended to a medium with very strong collisional broadening by Field *et al.* [8]. Remarkably enough, EIT experiments at single-atom levels have also been reported by several groups using both high finesse cavities and ion traps [9–11]. The EIT has been experimentally studied when the probe is replaced by the squeezed vacuum [12]. Since at two-photon resonance the absorption is zero in the absence of collisions, one would expect that the degree of squeezing would not be affected by the medium. However, in practice there is degradation in squeezing as the squeezed light would have finite bandwidth, and different frequency components, other than at two-photon resonance, would suffer some absorption. In another experiment, Eisaman *et al.* [13] used the propagation of single-photon pulses in an EIT medium and demonstrated that the quantum nature of the single-photon pulses is nearly preserved.

In the discussion so far we have considered the Λ-scheme of energy levels which has the great advantage that the levels $|b\rangle$ and $|c\rangle$ which are not directly connected by optical fields are long lived. Many other schemes have been considered in literature. The ladder system has been discussed extensively [3, 14] and transparency in such systems also demonstrated experimentally [8,15,16]. Over the years a whole host of multilevel systems, including solid-state systems, even those with more than three levels, have been studied for uncovering newer and newer applications of coherent control. The double-lambda systems have turned out to be specially useful [17].

17.1.3 Subnatural and sub-Doppler line widths in absorption under conditions of EIT

The susceptibility (17.14) as a function of probe detuning (Δ_p) has a doublet structure. The line width of each doublet depends not only on the relaxation parameters Γ'^s, $|g_l|$ but also on the detuning of the control laser. The line width can be subnatural under certain conditions as we now discuss. The polynomial in the denominator in (17.14) is

$$P(\Delta_p) = \Delta_p^2 - i\Delta_p(\Gamma_{ab} + \Gamma_{bc} - i\Delta_l) - \Gamma_{ab}\Gamma_{bc} - |g_l|^2 + \mathrm{i}\Delta_l\Gamma_{ab}, \tag{17.27}$$

which has two roots

$$\Delta_p = \frac{\mathrm{i}(\Gamma_{ab} + \Gamma_{bc} - \mathrm{i}\Delta_l)}{2} \pm \sqrt{\Gamma_{ab}\Gamma_{bc} + |g_l|^2 - \left(\frac{\Gamma_{ab} + \Gamma_{bc} - \mathrm{i}\Delta_l}{2}\right)^2 - \mathrm{i}\Delta_l\Gamma_{ab}}. \tag{17.28}$$

Let $\Omega = \sqrt{4|g_l|^2 + \Delta_l^2}$, then in the limit of large Ω, the roots are approximately

$$\Delta_p = \frac{\Delta_l}{2} \pm \frac{\Omega}{2} + \frac{\mathrm{i}}{2}(\Gamma_{ab} + \Gamma_{bc}) \pm \frac{\mathrm{i}\Delta_l}{2\Omega}(\Gamma_{bc} - \Gamma_{ab}). \tag{17.29}$$

Thus the root $\Delta_p = \frac{\Delta_l}{2} + \frac{\Omega}{2}$ will have a full width

$$(\Gamma_{ab} + \Gamma_{bc}) + \frac{\Delta_l}{\Omega}(\Gamma_{bc} - \Gamma_{ab}) = \Gamma_{ab}\left(1 - \frac{\Delta_l}{\Omega}\right) + \Gamma_{bc}\left(1 + \frac{\Delta_l}{\Omega}\right), \tag{17.30}$$

which for large positive detuning $\Delta_l \approx \Omega$ would be subnatural since $\Gamma_{bc} \ll \Gamma_{ab}$. Rapol *et al.* [18] presented the demonstration of subnatural line width in room temperature Rb vapor. A key element in their experiment was the elimination of the first-order Doppler effect on the probe transition by using a counter-propagating saturating beam. Under these conditions the formula (17.30) applies.

The coherent control field can also produce sub-Doppler line widths as predicted in [19] and experimentally demonstrated in [20]. For copropagating fields, a moving atom with velocity v will see a frequency $\omega - kv$. Ignoring the difference between $k_l v$ and $k_p v$, this amounts to replacing detuning by $\Delta_p \to \Delta_p + x$, $\Delta_l \to \Delta_l + x$, $x = kv$. Let us set $\Gamma_{bc} = 0$ for simplicity. Then the root with upper sign in (17.28) in the limit of large Ω will be

$$\Delta_p = \frac{\Delta_l}{2} + \frac{\Omega}{2} + \frac{1}{2}\mathrm{i}\Gamma_{ab}\left(1 - \frac{\Delta_l}{\Omega}\right) - \frac{x}{2}\left(1 - \frac{\Delta_l}{\Omega}\right). \tag{17.31}$$

The parameter x is velocity dependent. It is fluctuating with zero mean value and dispersion equal to $\langle x^2 \rangle = \langle k^2 v^2 \rangle = D^2$. Thus the width of the line located at $\Delta_l/2 + \Omega/2$ would become

$$(\Gamma_{ab} + D)\left(1 - \frac{\Delta_l}{\sqrt{\Delta_l^2 + 4|g_l|^2}}\right), \tag{17.32}$$

showing the possibility of sub-Doppler line widths as the detuning Δ_l increases. The numerical simulations using (17.14) and Doppler averaging yield full line shapes [19]. The numerical solutions also confirm the approximate analytical result (17.32). Zhu and Wasserlauf [20] observed the behavior of the linewidth as predicted by (17.32) by performing EIT on the D_2 transition of Doppler broadened ^{87}Rb vapor.

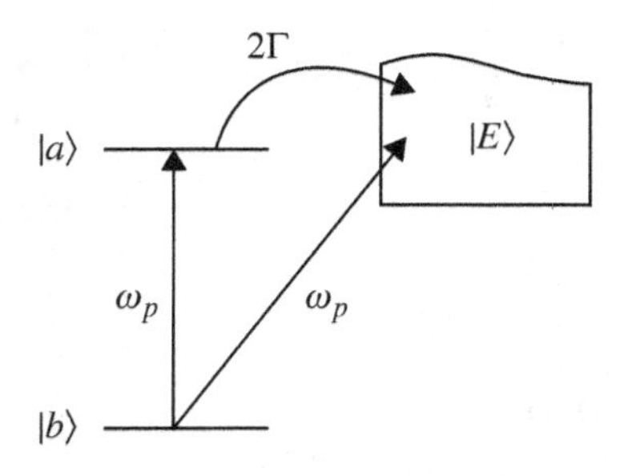

Fig. 17.6 Fano interference between the transitions $|b\rangle \to |E\rangle$ and $|b\rangle \to |a\rangle \to |E\rangle$.

17.1.4 Fano line shapes in EIT

In a classic paper Fano [21] discovered that the quantum interference of different transition amplitudes can produce absorption profiles which exhibited a zero, or more generally, a minimum. Let us consider the simplest possible situation as shown in Figure 17.6. Here the level $|b\rangle$ can be excited to the level $|a\rangle$ by a laser field. The level $|a\rangle$ is connected to a broad continuum of states $|E\rangle$. Furthermore, the population from the level $|b\rangle$ can go to the broad continuum via interaction with the laser field. Let 2Γ be the rate of decay of the state $|a\rangle$ to the continuum. The two transition amplitudes $|b\rangle \to |E\rangle$ and $|b\rangle \to |a\rangle \to |E\rangle$ interfere and lead to a non-Lorentzian line shape

$$S_F(\omega_l) = \frac{(\Delta_p/\Gamma + q)^2}{1 + (\Delta_p/\Gamma)^2}. \tag{17.33}$$

Here q is called the Fano asymmetry parameter, which is a measure of the strength of the interference between different channels of absorption. It is given by $q = V_{ba}/(\pi V_{bE} V_{Ea})$, where V is the interaction potential. The line shape has a zero at $\Delta_p/\Gamma = -q$. The interference effects are especially pronounced for small values of q, i.e. when the maximum and minimum are close to each other. Since the EIT results from the interference of different contributions, one would expect Fano-like line shapes under certain conditions. We examine (17.14) when the coupling laser is detuned from the transition $|c\rangle \leftrightarrow |a\rangle$ (Figure 17.2). We show a typical behavior of absorption in Figure 17.7 for $\Delta_l \neq 0$ and for different strengths of the coupling field. Note the appearance of Fano-like line shapes in the region $\Delta_p \approx \Delta_l$. This behavior can be understood if we examine (17.14) in the vicinity of $\Delta_p \sim \Delta_l$ and for $\Gamma_{bc} = 0$

$$\chi^{(ab)}(\omega_p) \approx \frac{-\chi_0 n(\Delta_p - \Delta_l)}{|g_l|^2 + (\Gamma_{ab} + \mathrm{i}\Delta_l)\mathrm{i}(\Delta_p - \Delta_l)}, \qquad \Delta_l > 0, \tag{17.34}$$

and hence the denominator has a pole at

$$\Delta_p - \Delta_l \cong \frac{|g_l|^2 \Delta_l}{\Delta_l^2 + \Gamma_{ab}^2} + \mathrm{i}\frac{|g_l|^2 \Gamma_{ab}}{\Delta_l^2 + \Gamma_{ab}^2}. \tag{17.35}$$

Thus the minimum and the maximum of the line shape are close to each other if $|g_l|^2\Delta_l/(\Delta_l^2 + \Gamma_{ab}^2) \lesssim \Gamma_{ab}$. This would then lead to Fano-like shapes in the region $\Delta_p \approx \Delta_l$.

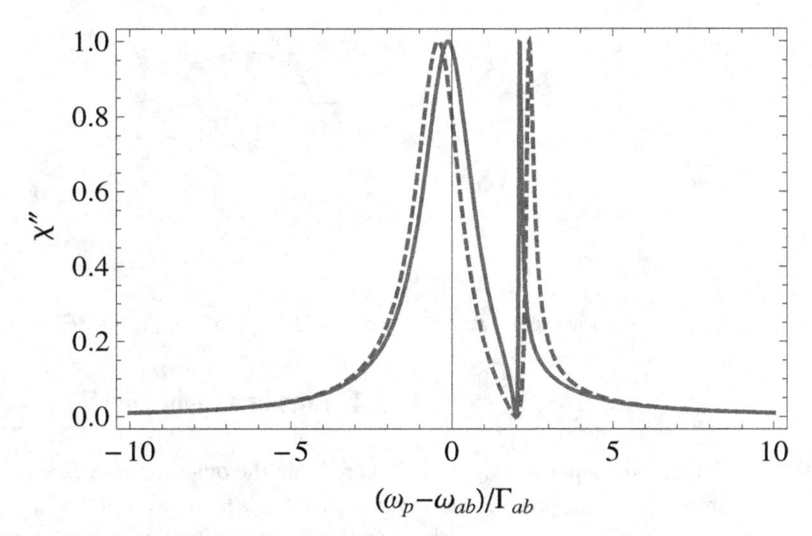

Fig. 17.7 Fano like line shapes in EIT, when the coupling laser is detuned $\Delta_l/\Gamma_{ab} = 2$ and for $G = 0.5$ (solid) and 1.0 (dashed).

17.2 Dark states and coherent population trapping

The Hamiltonian (17.8) for the Λ-system has some unusual features. After removing the optical frequencies, we have an effective Hamiltonian

$$H = \hbar(\Delta_p|a\rangle\langle a| + (\Delta_p - \Delta_l)|c\rangle\langle c|) - \hbar(g_p|a\rangle\langle b| + g_l|a\rangle\langle c| + H.c.), \quad (17.36)$$

which at two-photon resonance $\Delta_p = \Delta_l$ reduces to

$$H = \hbar\Delta_p|a\rangle\langle a| - \hbar(g_p|a\rangle\langle b| + g_l|a\rangle\langle c| + H.c.). \quad (17.37)$$

From (17.37) we see that the state $|\Psi_0\rangle$ defined by

$$|\psi_0\rangle = \frac{g_p|c\rangle - g_l|b\rangle}{\sqrt{|g_p|^2 + |g_l|^2}}, \quad (17.38)$$

is an eigenstate of H with zero eigenvalue $H|\psi_0\rangle = 0$. The state $|\psi_0\rangle$ is called a dark state as a system initially prepared in the state $|\psi_0\rangle$ would not evolve under the influence of (17.37). The state $|\psi_0\rangle$ can, of course, decay due to collisions. Such a dark state was first discovered experimentally [22, 23], and has found a very large number of applications in optical sciences and in laser cooling of atoms [24, 25]. It is remarkable to note that the dark state $|\psi_0\rangle$ is a coherent superposition of the states $|b\rangle$ and $|c\rangle$. This is known as coherent population trapping (CPT) since $|\psi_0\rangle$ does not evolve. The dark states have given rise to the technique of stimulated Raman adiabatic passage (STIRAP) [26] where one can transfer population adiabatically from one level to the other. The state (17.38) shows that if $g_l \to 0$, then up to a phase $|\psi_0\rangle \to |c\rangle$, and if $g_p \to 0$, then $|\psi_0\rangle \to |b\rangle$. If we use time-dependent

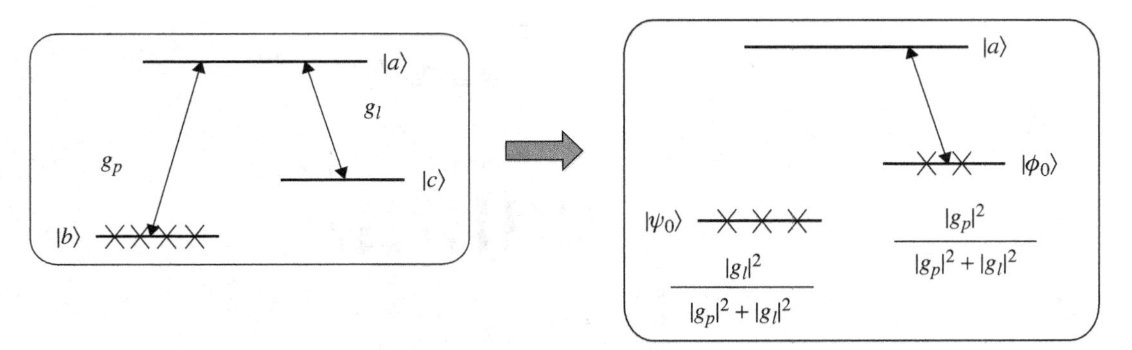

Fig. 17.8 Coupled and uncoupled states with population distribution as shown.

field envelopes $g_l(t)$ and $g_p(t)$ such that

$$\begin{aligned} g_p(t) &\to 0 \text{ as } t \to -\infty, \qquad & g_l(t) &\to 0 \text{ as } t \to +\infty, \\ g_l(t) &\to g_l \text{ as } t \to -\infty, \qquad & g_p(t) &\to g_p \text{ as } t \to +\infty, \end{aligned} \tag{17.39}$$

then the state $|b\rangle$ at $t = -\infty$ would have adiabatically evolved to the state $|c\rangle$ at $t = +\infty$. The pulse sequence (17.39) is called a counterintuitive sequence. The population transfer can be detected by using an auxiliary laser which would couple only $|c\rangle$ to a different excited state and by monitoring fluorescence from such a state. The technique of adiabatic passage has tremendous advantage in that the level $|a\rangle$ is never populated and thus the efficiency of adiabatic passage is not deteriorated by the effects of spontaneous emission. Needless to say, there are two other states $|a\rangle$ and $|\phi_0\rangle = \frac{g_l^*|c\rangle + g_p^*|b\rangle}{\sqrt{|g_l|^2+|g_p|^2}}$ which evolve in time. These are not the eigenstates of H. The state $|\phi_0\rangle$ is orthogonal to $|\psi_0\rangle$. The states $|\psi_0\rangle$ and $|\phi_0\rangle$ formed out of the two lower states are known as the uncoupled and coupled states respectively as shown in Figure 17.8.

17.2.1 Atom localization using CPT

We will now discuss a very interesting application of the coherent population trapping to atomic localization at the nanometric resolution level [27–31]. This method beats the well-known diffraction limit. Needless to say, there are many other methods to achieve the same goal [32, 33]. We consider the control beam to be a standing wave and consider the localization of the atom at the nodes of the control field as shown in Figure 17.9. The atomic system could be ^{87}Rb atoms in a MOT. The probe and coupling lasers could be resonant to the transitions $|F = 1\rangle \to |F' = 2\rangle$, $|F = 2\rangle \to |F' = 2\rangle$ of D_2 line of ^{87}Rb [30]. The field g_l of the standing wave would now be $g_l \sin kx$. The laser fields are in the xy plane so that the atom interacts with both at the same time. The dark state is then

$$|\psi_0\rangle = \frac{g_p|c\rangle - g_l \sin kx|b\rangle}{\sqrt{|g_p|^2 + |g_l|^2 \sin^2 kx}}. \tag{17.40}$$

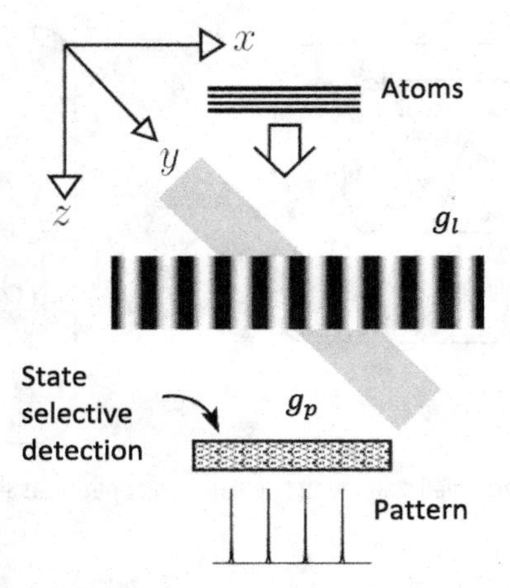

Fig. 17.9 Application of a strong standing-wave field, on the $|a\rangle \leftrightarrow |c\rangle$ transition, and a weak probe field, on the $|a\rangle \leftrightarrow |b\rangle$ transition, prepares the atom in a particular position-dependent superposition of the states $|b\rangle$ and $|c\rangle$ in steady state [28].

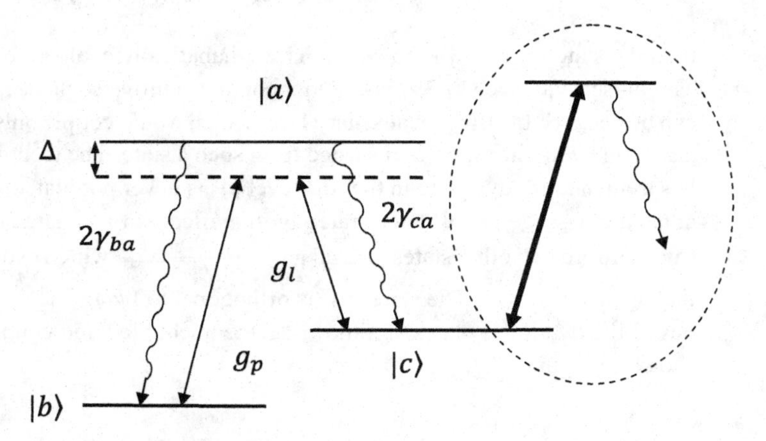

Fig. 17.10 Technique to monitor the atom localization. The CPT state of the atom could be probed by monitoring the population of the state $|c\rangle$, i.e. $p_c(x)$. Efficient measurement of $p_c(x)$ can be accomplished via the fluorescence shelving technique as depicted in the figure. Applying a strong shelving field and monitoring the resulting fluorescence leads to localization of the atom at the nodes of the standing-wave field. It is important that the atoms are detected in a state selective manner after their interaction with laser fields [28].

The population p_c of the state $|c\rangle$, as monitored by using an auxiliary laser, as shown in Figure 17.10, is

$$p_c(x) = \frac{1}{1 + R\sin^2 kx}, \qquad R = \frac{|g_l|^2}{|g_p|^2}. \tag{17.41}$$

It is interesting to note that $p_c(x)$ is similar to the transmission profile of a Fabry–Perot interferometer. The width of $p_c(x)$ narrows as the parameter R increases. The full width at

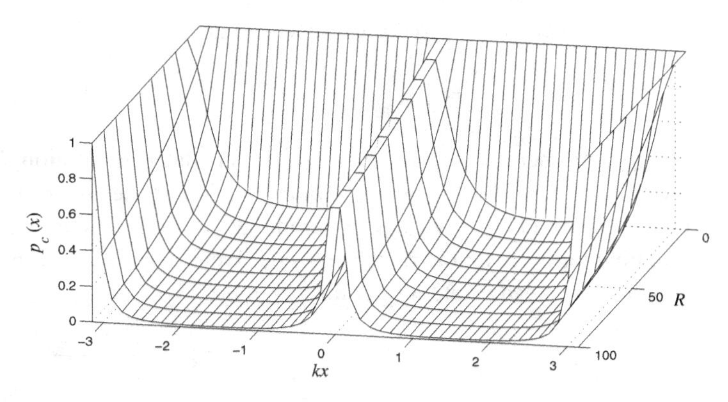

Fig. 17.11 Emergence of subwavelength localization as the ratio R is increased. Plot of $p_c(x)$ versus kx for various values of R. The range of kx, $\{-\pi, \pi\}$, covers one wavelength of the standing-wave field [28].

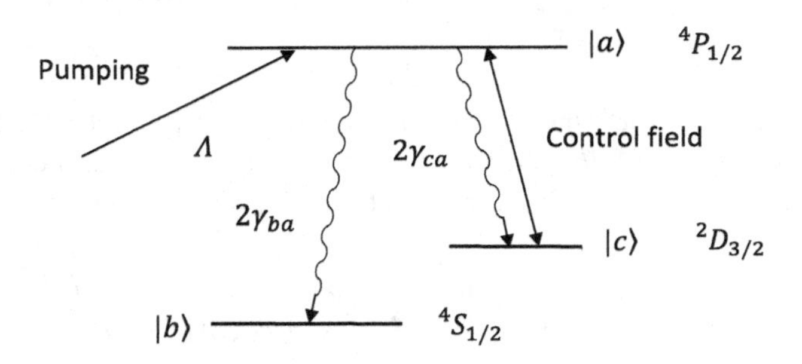

Fig. 17.12 A trapped ion, say $^{40}Ca^+$, which is pumped to the state $|a\rangle$ and is under the influence of the control field.

half maximum is $2/\sqrt{R}$. The resulting localization [28] of atoms at the nodes is clearly seen in Figure 17.11 as R increases. Proite *et al.* [30] present a beautiful observation of atomic localization along the above lines using about a billion Rb atoms in a magneto-optical trap at a temperature of 150 μK.

17.3 EIT in single-atom fluorescence

Just like the significant modifications of the absorption by a strong coherent field, we expect significant modifications of fluorescence. The latter is also a measure of the quantum fluctuations as we showed in Chapter 13. Thus one should expect an analog of EIT in fluorescence, which would be a method to detect EIT at the single-atom level [34]. Consider, for example, a single trapped ion such as $^{40}Ca^+$ with the energy levels as shown in Figure 17.12: the $|a\rangle \leftrightarrow |c\rangle$ transition wavelength is 866.21 nm, on which we apply the control field, the fluorescence is monitored on the transition $|a\rangle \rightarrow |b\rangle$, which would be around 396.85 nm. The ion is pumped to the excited state at the rate Λ. In the absence of the control field, the fluorescence spectrum is

$$S(\omega) = \left(\frac{\Lambda}{\Gamma_{ab}}\right) \frac{\gamma_{ba}/\pi}{(\omega - \omega_{ab})^2 + \Gamma_{ab}^2}, \qquad \Gamma_{ab} = \gamma_{ba} + \gamma_{ca}. \tag{17.42}$$

If $\gamma_{ca} = 0$, then (17.42) reduces to that calculated in Section 7.6. We next consider the modification of (17.42) by the control field. For the model of Figure 17.12, the basic equations are (17.10) with $g_p = 0$, and $\Delta_p = 0$. The equation for $\dot{\rho}_{aa}$ has an additional pumping term 2Λ on the right-hand side. We also set $\Delta_l = 0$ for simplicity. Solving these equations in steady state we get for $|g_l| \neq 0$

$$\rho_{aa} = \frac{\Lambda}{\gamma_{ba}}, \qquad \rho_{ac} = \frac{i g_l \gamma_{ca} \rho_{aa}}{|g_l|^2}. \tag{17.43}$$

By an argument similar to that leading to (13.51), we define the spectrum of emission as

$$S^{(ab)}(\omega) = \frac{1}{\pi}\left(\frac{\gamma_{ba}}{\Gamma_{ba}}\right) \cdot \mathrm{Re}\left[\int_0^{\infty} \mathrm{d}\tau\, \mathrm{e}^{-p\tau} \langle (|a\rangle\langle b|)_t (|b\rangle\langle a|)_{t+\tau}\rangle\right]_{p=-i\delta},$$
$$\delta = \omega - \omega_a + \omega_b. \tag{17.44}$$

The prefactor, which is the branching ratio γ_{ba}/Γ_{ba}, accounts for the fact that the state $|a\rangle$ decays to both $|b\rangle$ and $|c\rangle$. There is no collisional dephasing for the ion in the trap so that $\Gamma_{ba} = \gamma_{ba} + \gamma_{ca}$. Using the quantum regression theorem and the equations for ρ_{ab} and ρ_{cb} we find

$$\left(\frac{\mathrm{d}}{\mathrm{d}\tau} + \begin{pmatrix} \Gamma_{ab} & -ig_l \\ -ig_l^* & \Gamma_{bc} \end{pmatrix}\right) \begin{pmatrix} \langle (|a\rangle\langle b|)_t (|b\rangle\langle a|)_{t+\tau}\rangle \\ \langle (|a\rangle\langle b|)_t (|b\rangle\langle c|)_{t+\tau}\rangle \end{pmatrix} = 0, \tag{17.45}$$

which are to be solved subject to the initial conditions

$$\langle (|a\rangle\langle b|)_t (|b\rangle\langle a|)_t\rangle = \langle |a\rangle\langle a|\rangle = \rho_{aa},$$
$$\langle (|a\rangle\langle b|)_t (|b\rangle\langle c|)_t\rangle = \langle |a\rangle\langle c|\rangle = \rho_{ca}. \tag{17.46}$$

The required density matrix elements are given by (17.43). On using (17.44)–(17.46), we obtain the spectrum as [34]

$$S^{(ab)}(\omega) = \frac{1}{\pi}\left(\frac{\gamma_{ba}}{\Gamma_{ba}}\right) \rho_{aa} S(\omega),$$
$$S(\omega) = \mathrm{Re}\left[\frac{-i\delta + \gamma_{ca}}{(-i\delta + \Gamma_{ab})(-i\delta) + |g_l|^2}\right]. \tag{17.47}$$

We have set the collision term Γ_{bc} to zero as it is irrelevant. The structure of (17.47) is similar to the structure of (17.14) except for an addition term γ_{ca} in the numerator which makes $S^{(ab)}(\omega)$ nonzero at $\delta = 0$

$$S^{(ab)}(\delta = 0) = \frac{1}{\pi}\left(\frac{\gamma_{ba}}{\Gamma_{ba}}\right) \cdot \frac{\gamma_{ca}}{|g_l|^2} \rho_{aa} \to 0, \qquad \text{if } \gamma_{ca} \to 0. \tag{17.48}$$

We show the behavior of $S(\omega)$ for $\gamma_{ca}/\gamma_{ba} = 1/5$ for different values of $G = |g_l|/\Gamma_{ab}$ as a function of δ/Γ_{ab} in Figure 17.13. The quantum interference in the fluorescence spectrum is evident from this figure. It can be shown that the dip for $G = 0.5$ goes to zero in the limit $\gamma_{ca} \to 0$. Slodička *et al.* [11] detected EIT at a single-atom level by using an optically cooled

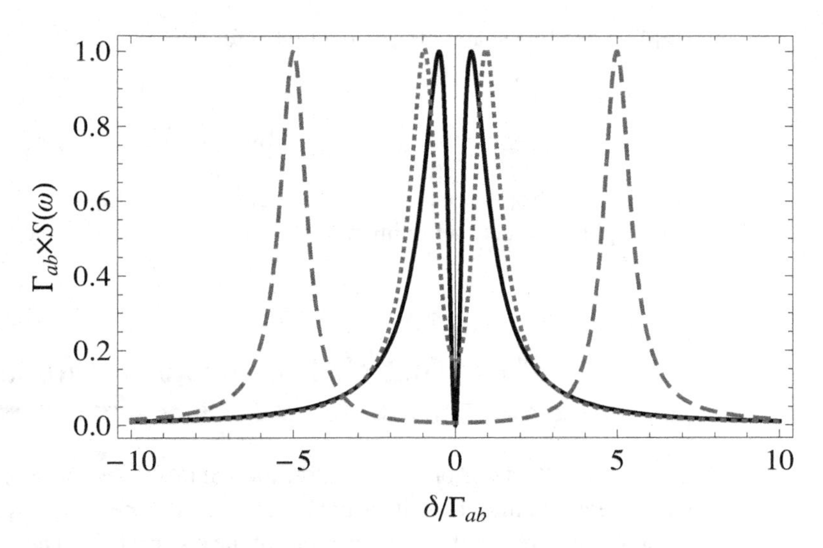

Fig. 17.13 Behavior of the fluorescence spectrum $S(\omega)$ as a function of δ/Γ_{ab} for different values of $G = |g_l|/\Gamma_{ab} = 1$ (dotted), 5 (dashed). The solid curve is for $G = 0.5$ and $\gamma_{ca} \to 0$.

$^{138}Ba^+$ ion in a trap. These authors also report interference to be minimum in fluorescence. In order to illustrate the idea of interference in fluorescence we have presented a very simple model [34]. The fluorescence spectrum for a Λ-system driven by two-strong fields under different relaxation mechanisms has been investigated extensively [35–37].

17.3.1 Initiation of higher-order poles of the S-matrix by the control field

In a classic paper Goldberger and Watson [38] considered the possibility that the decay law for an unstable particle can be more complex than a simple exponential. They showed the possibility of the existence of the poles of the S-matrix which were not necessarily simple poles. Since then higher-order poles have been extensively studied and several field theoretic models such as the Friedrich–Lee model, the cascade model, and their extensions have been examined [39–42]. One question is how the Breit–Wigner line shape formula is modified if the S-matrix posseses higher-order poles. Clearly, the decay law would not be simple exponential.

The experimental observation of the double poles does not seem to exist. There have been suggestions [43, 44] that such double poles can be tailor-made in the laboratory using laser fields. We now demonstrate how the model of Figure 17.12, i.e. how the control field can produce double poles and thus fluorescence from a single ion in a trap, would be a way to observe such double poles. From (17.47) we see that a double pole would occur if $|g_l| = \Gamma_{ab}/2$. The fluorescence spectrum is shown in Figure 17.13 (the curve marked $G = 0.5$). In the special case $\gamma_{ca} \to 0$ (which is equivalent to the extended Friedrich–Lee model [41]), the line shape $S(\omega)$ (Eq. (17.47) for $G = 0.5$) simplifies to

$$S(\omega) = \frac{\Gamma_{ab}\delta^2}{\left(\delta^2 + \frac{\Gamma_{ab}^2}{4}\right)^2}, \tag{17.49}$$

which is different from the usual Breit–Wigner line shape (Eq. (7.81)). Interestingly enough, the double pole produces a minimum at $\delta = 0$ rather than the usual maximum, which now occurs at $\delta = \pm\gamma_{ab}/2$.

17.4 Control of two-photon absorption

So far we have considered only the modification of the linear optical properties by a control field. In general one finds that all the nonlinear optical properties [1,3] undergo considerable modifications as a result of the applications of the control field. The explicit results to some extent, depend on the atomic level schemes. In this section we consider two examples.

17.4.1 Two-photon-induced transparency

We have already discussed in Chapter 8 that in two-photon absorption there is a possibility of interference arising from many intermediate states. The existence of the intermediate state depends on the specific atomic system. However, by applying coherent control fields, one can create new intermediate states. Such laser-created intermediate states can lead to two-photon transparency, i.e. can suppress two-photon absorption as predicted in [45] and demonstrated experimentally in [46, 47]. Let us consider the level scheme as shown in Figure 17.14 for two-photon absorption from two different laser fields of frequencies ω_1 and ω_2. For illustration we also give the relevant energy levels in ^{85}Rb. The efficiency of the two-photon absorption is determined by the population in the state $|a\rangle$. To bring out the idea of two-photon transparency, we apply a control laser on the transition $|b\rangle \leftrightarrow |d\rangle$. We can now examine the population of the level $|a\rangle$ for different strengths of the control laser and its detuning. After removing optical frequencies by using transformations similar to (17.9), we find the effective Hamiltonian (similar to (17.36)) for the two-photon absorption

$$\begin{aligned} H &= \hbar\Delta_b|b\rangle\langle b| + \hbar(\Delta_a + \Delta_b)|a\rangle\langle a| + \hbar(\Delta + \Delta_b)|d\rangle\langle d| \\ &\quad - \hbar(g_l|d\rangle\langle b| + g_a|a\rangle\langle b| + g_b|b\rangle\langle c| + H.c.), \end{aligned} \tag{17.50}$$

where detunings are defined as in Figure 17.14, $\Delta_b = \omega_b - \omega_c - \omega_2$, $\Delta_a = \omega_a - \omega_b - \omega_1$, $\Delta = \omega_d - \omega_b - \omega_l$, and the g's are defined as in (17.8). In addition, we have to account for the spontaneous emission from the levels $|a\rangle$, $|b\rangle$, and $|d\rangle$. The corresponding density matrix satisfies

$$\begin{aligned} \dot{\rho} &= -\frac{\mathrm{i}}{\hbar}[H, \rho] - \gamma_{ba}\{|a\rangle\langle a|, \rho\} - \gamma_{cb}\{|b\rangle\langle b|, \rho\} - \gamma_{bd}\{|d\rangle\langle d|, \rho\} \\ &\quad + 2\gamma_{ba}\rho_{aa}|b\rangle\langle b| + 2\gamma_{cb}\rho_{bb}|c\rangle\langle c| + 2\gamma_{bd}\rho_{dd}|b\rangle\langle b|, \end{aligned} \tag{17.51}$$

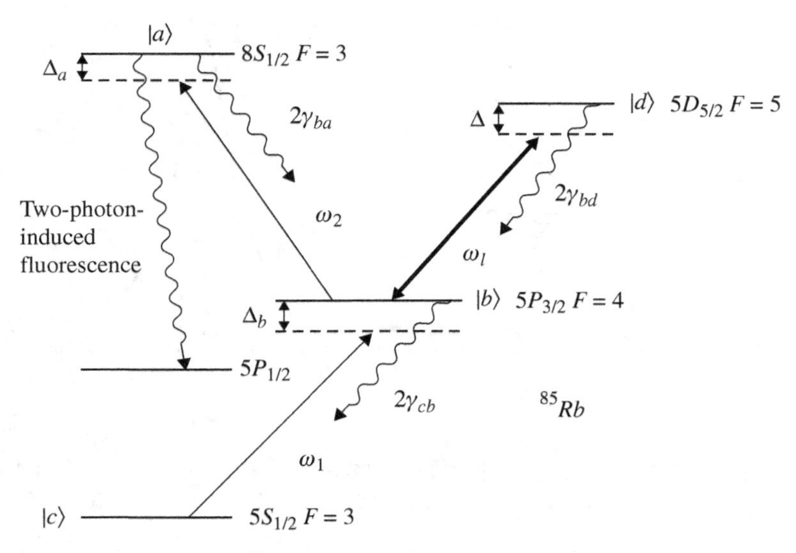

Fig. 17.14 Scheme of the relevant energy levels of ^{85}Rb. The γs and Δs are the corresponding spontaneous decay rates and detunings, respectively, after [46].

where H is given by (17.50). The resulting equations can be solved in the steady state. In order to illustrate the idea of two-photon transparency we find ρ_{aa} to the lowest order in g_a and g_b, i.e. to order $|g_a|^2|g_b|^2$. The two-photon excitation usually consists of two terms – one containing the two-photon resonant denominator $(\gamma_{ba} + \mathrm{i}\Delta_a + \mathrm{i}\Delta_b)$ and the other the two-step excitation. For large detuning from the intermediate state, the two-step excitation contribution is unimportant. A perturbative calculation then leads to

$$\rho_{aa} = \frac{|g_a|^2|g_b|^2}{\gamma_{ba}} \mathrm{Im} \frac{\mathrm{i}(\gamma_{bd} + \mathrm{i}\Delta + \mathrm{i}\Delta_b)}{(\gamma_{ba} + \mathrm{i}\Delta_a + \mathrm{i}\Delta_b)[(\gamma_{cb} + \mathrm{i}\Delta_b)(\gamma_{bd} + \mathrm{i}\Delta + \mathrm{i}\Delta_b) + |g_l|^2]}$$
$$\times \frac{\gamma_{ba} + \gamma_{bd} + \mathrm{i}\Delta_a - \mathrm{i}\Delta}{(\gamma_{ba} + \gamma_{cb} + \mathrm{i}\Delta_a)(\gamma_{ba} + \gamma_{bd} + \mathrm{i}\Delta_a - \mathrm{i}\Delta) + |g_l|^2}. \quad (17.52)$$

This expression shows the possibility of interference minimum when

$$\Delta = -\Delta_b, \qquad \Delta = \Delta_a, \qquad \Delta_a + \Delta_b = 0. \quad (17.53)$$

The existence of two-photon transparency can be explained in the same way as in Figure 17.5 as the control laser between $|b\rangle \leftrightarrow |d\rangle$ would produce the dressed state $|\psi_\pm\rangle$ and we would have two channels for two-photon transition $|c\rangle \rightarrow |\psi_+\rangle \rightarrow |a\rangle$, $|c\rangle \rightarrow |\psi_-\rangle \rightarrow |a\rangle$. The behavior of two-photon excitation at $\Delta_a + \Delta_b = 0$ is shown in Figure 17.15, which clearly shows the region of two-photon transparency [45]. In order to obtain realistic results one has to account for the residual Doppler width, details of which can be found in [45]. An unambiguous observation of two-photon transparency using the ladder scheme was reported by Wang *et al.* [46].

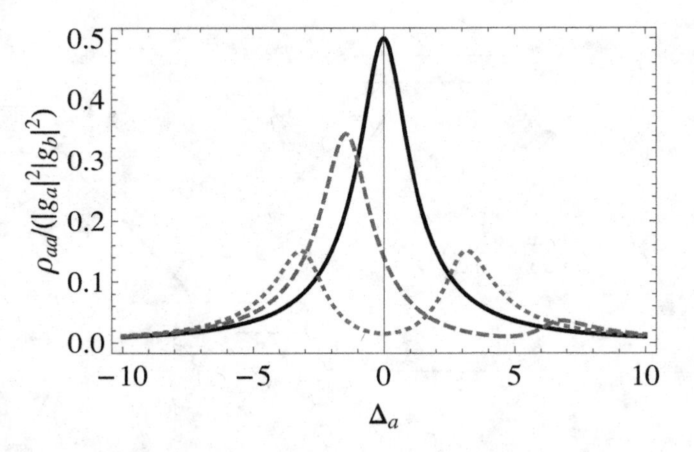

Fig. 17.15 The behavior of $\rho_{aa}/(|g_a|^2|g_b|^2)$ as a function of Δ_a at $\Delta_a + \Delta_b = 0$. All values of γ have been set to unity, and $g_l = 0$ (solid), $g_l = 3,\ \Delta = 0$ (dotted), $g_l = 3,\ \Delta = 5$ (dashed).

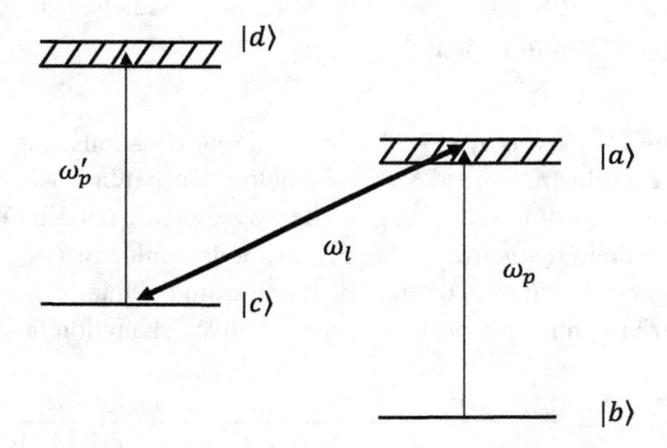

Fig. 17.16 Schematic representation of a four-state system. The line widths of states $|a\rangle$ and $|d\rangle$ are greater than all other characteristic frequencies.

17.4.2 Two-photon absorption without one-photon absorption

Harris and Yamamoto [48] proposed a four-level scheme which absorbs two photons but not one photon. The scheme shown in Figure 17.16, in principle, can operate as a single-photon switch. In the absence of ω_p', the photon of frequency ω_p is transmitted due to EIT, whereas in the presence of ω_p', the photon ω_p is not transmitted. The Hamiltonian for the system of Figure 17.16 is now equal to

$$H = \Delta_p|a\rangle\langle a| + (\Delta_p - \Delta_l)|c\rangle\langle c| + (\Delta_p - \Delta_l + \Delta_d)|d\rangle\langle d| \\ - \hbar(g_p|a\rangle\langle b| + g_l|a\rangle\langle c| + g_p'|d\rangle\langle c| + H.c.), \qquad \Delta_d = \omega_d - \omega_c - \omega_p'. \tag{17.54}$$

This is because $\tilde{\rho}_{db} = \rho_{db} \exp[\mathrm{i}(\omega_p' + \omega_p - \omega_l)t]$. All decays are to be included in the equations for the density matrix using the prescription of Section 9.4. The susceptibility for the probe transition $\chi^{(ab)}(\omega_p)$ can be calculated by evaluating the element ρ_{ab} in steady states and using the assumption of weak g_p. We obtain in place of (17.14)

$$\chi^{(ab)}(\omega_p) = \frac{\chi_0 n \mathrm{i}}{(\Gamma_{ab} + \mathrm{i}\Delta_p) + |g_l|^2/\mathcal{D}},$$
$$\mathcal{D} = [\Gamma_{bc} + \mathrm{i}(\Delta_p - \Delta_l)] + \frac{|g_p'|^2}{\Gamma_{da} + \mathrm{i}(\Delta_p - \Delta_l + \Delta_d)}. \tag{17.55}$$

Note that $\Delta_p - \Delta_l + \Delta_d$ is the three-photon detuning. It should be borne in mind that a simple structure emerges since the field g_p is weak and all the population is initially in the level $|b\rangle$. At two-photon resonance $\Delta_p = \Delta_l$, and at $\Gamma_{bc} = 0$

$$\chi^{(ab)}(\omega_p) = \frac{\chi_0 n \mathrm{i}}{(\Gamma_{ab} + \mathrm{i}\Delta_p) + \frac{|g_l|^2}{|g_p'|^2}(\Gamma_{da} + \mathrm{i}\Delta_d)}, \tag{17.56}$$

and hence on defining $\eta = |g_l|^2/|g_p'|^2$, the imaginary part of the susceptibility is

$$\mathrm{Im}\left(\chi^{(ab)}(\omega_p)\right) = \frac{\chi_0 n(\Gamma_{ab} + \eta\Gamma_{da})}{(\Gamma_{ab} + \eta\Gamma_{da})^2 + (\Delta_p + \eta\Delta_d)^2}. \tag{17.57}$$

Harris and Yamamoto [48] consider the case when the Γs are bigger than the detunings, in which case

$$\mathrm{Im}\left(\chi^{(ab)}(\omega_p)\right) = \frac{\chi_0 n}{\Gamma_{ab}\left(1 + \eta\frac{\Gamma_{da}}{\Gamma_{ab}}\right)}. \tag{17.58}$$

Note that the prefactor $\chi_0 n/\Gamma_{ab}$ is the factor when no control field is applied. Thus the ratio of the absorption in the presence and in the absence of the control field is

$$\frac{\mathrm{Im}\left(\chi^{(ab)}_{g_l \neq 0}(\omega_p)\right)}{\mathrm{Im}\left(\chi^{(ab)}_{g_l = 0}(\omega_p)\right)} = \frac{1}{1 + \eta\frac{\Gamma_{da}}{\Gamma_{ab}}}. \tag{17.59}$$

As long as the ratio $\eta\frac{\Gamma_{da}}{\Gamma_{ab}} \ll 1$, the absorption in the presence of ω_p and ω_p' is comparable to the absorption in the absence of the control field. Thus one can conclude that the probe photon is transmitted if $g_l \neq 0$, $\Gamma_{bc} = 0$ and $g_p' = 0$, whereas it is absorbed if $g_l \neq 0$, $\Gamma_{bc} = 0$ and $g_p' \neq 0$, this leads to the single-photon switch based on EIT. Höckel and Benson [49] demonstrated EIT in CS vapor using the D_1 line with probe pulses at the single-photon level.

The susceptibility (17.55) has another remarkable characteristic. It can lead to giant Kerr nonlinearities as discussed by Schmidt and Imamoğlu [50]. We calculate $\chi^{(ab)}(\omega_p)$ to order $|g_p'|^2$ and for large detunings and $\Delta_p = \Delta_l$, $\Gamma_{bc} = 0$,

$$\chi^{(ab)}(\omega_p) \approx \frac{\frac{\chi_0 n|g_p'|^2}{\Delta_d}}{\frac{\Delta_p|g_p'|^2}{\Delta_d} + |g_l|^2} \approx \frac{\frac{\chi_0 n|g_p'|^2}{\Delta_d}}{|g_l|^2} = \frac{\chi_0 n|g_p'|^2}{\Delta_d|g_l|^2}. \tag{17.60}$$

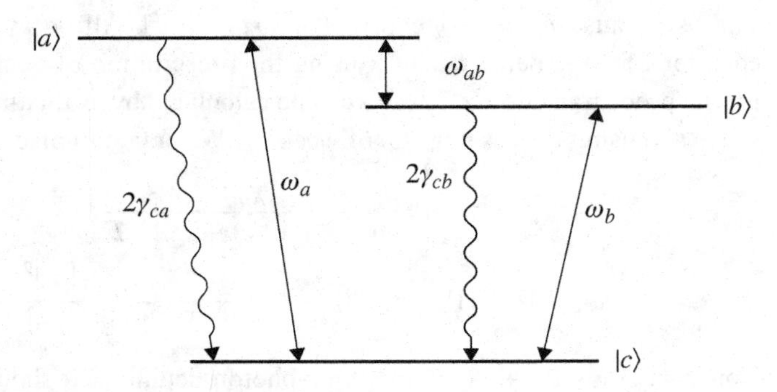

Fig. 17.17 The V system with two nearly degenerate levels.

Note that the nonlinearity for a two-level medium is $-4\chi_0 n|g|^2 T_1 T_2/(\Delta T_2)^3$ (Eq. (13.17)). On comparing this with (17.60), we see that the Kerr nonlinearity is enhanced by a factor of order $(\Delta/|g_l|)^2$, which could be of several orders.

17.5 Vacuum-induced coherence and interference

In the previous sections we studied in detail the interferences and coherences produced by using coherent control fields. However, much earlier it was found that the vacuum of the electromagnetic field can produce coherences in the process of spontaneous emission [51]. This is so even if no coherent fields are applied. Consider, for example, the V system shown in Figure 17.17 in which the excited states $|a\rangle$ and $|b\rangle$ decay to the lower state $|c\rangle$. According to the rate equations, we have

$$\dot{\rho}_{aa} = -2\gamma_{ca}\rho_{aa}, \qquad \dot{\rho}_{bb} = -2\gamma_{cb}\rho_{bb}. \tag{17.61}$$

The rate equations are derived by assuming that the spontaneous emission events $|a\rangle \to |c\rangle$, $|b\rangle \to |c\rangle$ are in no way related. However, the question is to what extent the two channels of emission $|a\rangle \to |c\rangle$, $|b\rangle \to |c\rangle$ are independent. This can be answered by a first principle derivation of the master equation. The basic master equation is given by (9.61) which was simplified to the standard form (9.67) by using the assumption (9.66). However, now (9.66) is to be relaxed. For the V system of Figure 17.17, one finds that the density matrix elements instead satisfy

$$\begin{aligned}
\dot{\rho}_{aa} &= -2\gamma_{ca}\rho_{aa} - \sqrt{\gamma_{ca}\gamma_{cb}}\cos\theta\left(\rho_{ab}e^{i\omega_{ab}t} + \rho_{ba}e^{-i\omega_{ab}t}\right),\\
\dot{\rho}_{bb} &= -2\gamma_{cb}\rho_{bb} - \sqrt{\gamma_{ca}\gamma_{cb}}\cos\theta\left(\rho_{ab}e^{i\omega_{ab}t} + \rho_{ba}e^{-i\omega_{ab}t}\right),\\
\dot{\rho}_{ab} &= -(\gamma_{ca} + \gamma_{cb})\rho_{ab} - \sqrt{\gamma_{ca}\gamma_{cb}}\cos\theta\,(\rho_{aa} + \rho_{bb})e^{-i\omega_{ab}t},
\end{aligned} \tag{17.62}$$

where $\cos\theta = (\vec{d}^{*}_{ac} \cdot \vec{d}_{bc})/(|\vec{d}_{ac}||\vec{d}_{bc}|)$. We now have a new feature where populations like ρ_{aa} are coupled to coherences like ρ_{ab} and coherences to populations. These couplings arise from the fact that the two channels of emission $|a\rangle \to |c\rangle$, $|b\rangle \to |c\rangle$ are not independent.

Due to the presence of the additional terms in (17.62), as compared to (17.61), the coherence ρ_{ab}, in steady state, is nonzero. We refer to this as the vacuum-induced coherence (VIC). Note that if ω_{ab} is large compared to other frequencies, then $e^{\pm i\omega_{ab}t}$ averages to zero, no interferences occur, and Eqs. (17.62) lead to (17.61). Thus interferences are most significant if $\omega_{ab} \to 0$ and $\cos\theta \to 1$, i.e. for degenerate levels $|a\rangle$ and $|b\rangle$ with two transition dipoles being parallel. In this case, the steady-state solution depends on the initial conditions. Thus, when interferences occur, the system possesses a memory of the initial state. From Eq. (17.62), it is clear that $\rho_{aa} + \rho_{bb} - \rho_{ab} - \rho_{ba} =$ constant, if $\gamma_{ca} = \gamma_{cb}$, $\theta = 0$, $\omega_{ab} = 0$. And in particular for $\rho_{aa}(0) = 1$, we find that in the steady state

$$\rho_{aa} = \rho_{bb} = 1/4, \qquad \rho_{ab} = -1/4. \tag{17.63}$$

Thus the population is trapped coherently in the excited state as $\rho_{ab} \neq 0$. The existence of the population trapping in an incoherent process follows from the nature of the interaction with the vacuum field $-\vec{p}\cdot\vec{E}$ with $\vec{p}$ expressed as

$$\begin{aligned}\vec{p} &= \vec{p}_{ac}(|a\rangle\langle c| + |b\rangle\langle c| + H.c.)\\ &= \sqrt{2}\vec{p}_{ac}\left[\left(\frac{|a\rangle + |b\rangle}{\sqrt{2}}\right)\langle c| + H.c.\right],\end{aligned} \tag{17.64}$$

where we assumed $\vec{p}_{ac} = \vec{p}_{bc}$. Thus the coupling to the vacuum field is via the state $|\Psi_c\rangle = (|a\rangle + |b\rangle)/\sqrt{2}$. The orthogonal state $|\Psi_{uc}\rangle = (|a\rangle - |b\rangle)/\sqrt{2}$ does not couple. Thus any population initially in the state $(|a\rangle - |b\rangle)/\sqrt{2}$ would not decay, whereas the population in $|\psi_c\rangle$ would decay to zero. This explains the result of Eq. (17.63). Clearly the result (17.63) can be thought of as quenching of spontaneous emission and this can have many applications in situations where one wants to reduce spontaneous emission noise, for example in the context of the preservation of quantum entanglement [52]. The nonorthogonality condition on dipole matrix elements $\vec{p}_{ac}$ and $\vec{p}_{bc}$ is not a serious one and is a consequence of the isotropy of the vacuum field. However, it was shown [53] that if the vacuum field is anisotropic, then one can even work with transitions with orthogonal dipole matrix elements. There are well-known cases when the vacuum is anisotropic, for example, when material bodies are present in the neighborhood of the radiating atom. Thus the vacuum-induced coherence effects can be enhanced in presence of metamaterials or nanostructures [54, 55] and photonic crystals [56]. Scully *et al.* [57] show how vacuum-induced coherence can lead to increase in the power of the quantum heat engines. For many varied applications of vacuum-induced coherence, we refer to the review articles [58, 59]. The vacuum-induced coherence effects can be monitored via both absorption and emission spectrum [60, 61].

17.5.1 VIC in the Λ system and the probe of VIC

In the above, we have discussed the vacuum-induced coherence in the context of the V scheme of transitions. Since Λ systems are extensively used in many applications and even some of the experimental confirmation is coming via Λ systems, we discuss the Λ scheme in detail [62–64]. We also discuss how the VIC can be probed in such systems.

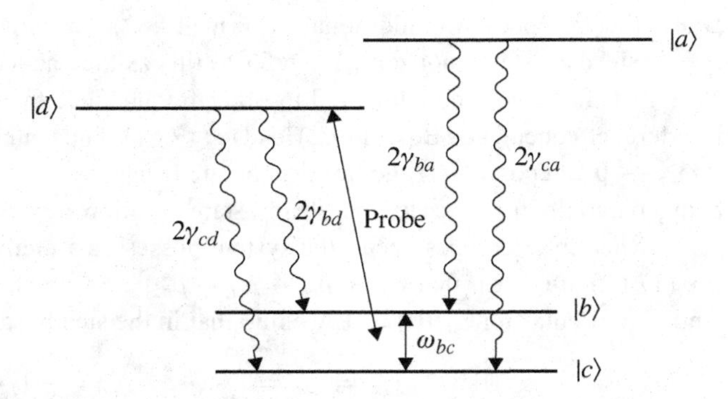

Fig. 17.18 Schematic diagram of a four-level model proposed for monitoring vacuum-induced coherence. The coherence created after spontaneous emission from $|a\rangle$ can be observed in the probe absorption [64].

In a spirit similar to our discussion of the probe of the CPT state, we will also discuss a probe of the VIC. We show the system in Figure 17.18. The idea is that the VIC would create a coherent superposition of populations p_b and p_a in the states $|b\rangle$ and $|c\rangle$. This coherence would be probed by a weak field which would excite the coherently trapped population to level $|d\rangle$. The population in the state $|d\rangle$ would show modulations at the frequency ω_{bc} due to vacuum-induced coherence among the levels $|b\rangle$ and $|c\rangle$. Thus the modulation of the population would be a direct measure of the coherence. The density matrix equations for this system, in the absence of the probe field, are

$$\begin{aligned}
\dot{\rho}_{aa} &= -2(\gamma_{ba}+\gamma_{ca})\rho_{aa}, & \dot{\rho}_{ab} &= -(\gamma_{ba}+\gamma_{ca}+\mathrm{i}\omega_{ab})\rho_{ab},\\
\dot{\rho}_{bb} &= 2\gamma_{ba}\rho_{aa}, & \dot{\rho}_{ac} &= -(\gamma_{ba}+\gamma_{ca}+\mathrm{i}\omega_{ac})\rho_{ac},\\
\dot{\rho}_{cc} &= 2\gamma_{ca}\rho_{aa}, & \dot{\rho}_{bc} &= -\mathrm{i}\omega_{bc}\rho_{bc}+2\sqrt{\gamma_{ba}\gamma_{ca}}\cos\theta\,\rho_{aa}.
\end{aligned} \tag{17.65}$$

In the long time limit a coherence between $|b\rangle$ and $|c\rangle$ is created

$$\rho_{bc} \to \frac{2\sqrt{\gamma_{ba}\gamma_{ca}}\cos\theta\,\mathrm{e}^{-\mathrm{i}\omega_{bc}t}}{2(\gamma_{ca}+\gamma_{ba})-\mathrm{i}\omega_{bc}}, \qquad t\to\infty, \tag{17.66}$$

which is especially significant if $\omega_{bc} < (\gamma_{ca}+\gamma_{ba})$ and if θ is close to zero. This coherence can again be understood in terms of the uncoupled and coupled states as in the case of the V system. Having created the coherence we now show how it can be monitored [64] by using a probe field coupling to the level $|d\rangle$ that is well separated from the level $|a\rangle$.

The Hamiltonian in the dipole approximation, with A's defined by (9.58), will be

$$H = \hbar\omega_{bc}A_{bb} + \hbar\omega_{ac}A_{aa} + \hbar\omega_{dc}A_{dd} - [(\vec{p}_{dc}A_{dc} + \vec{p}_{db}A_{db})\cdot\vec{E}_2\mathrm{e}^{-\mathrm{i}\omega_2 t} + H.c.], \tag{17.67}$$

where the counter-rotating terms in the probe field $\vec{E}_2$ have been dropped. The probe field is treated classically and has a frequency ω_2 and a complex amplitude $\vec{E}_2$. We use the master equation to derive equations for the reduced density matrix of the atomic system. We give the result of such a calculation

$$
\begin{aligned}
\dot{\rho}_{aa} &= -2(\gamma_{ba}+\gamma_{ca})\rho_{aa},\\
\dot{\rho}_{dd} &= -2(\gamma_{cd}+\gamma_{bd})\rho_{dd} + \mathrm{i}(g\rho_{bd}+f\rho_{cd})\mathrm{e}^{-\mathrm{i}\omega_2 t} - \mathrm{i}(g^*\rho_{db}+f^*\rho_{dc})\mathrm{e}^{\mathrm{i}\omega_2 t},\\
\dot{\rho}_{bb} &= 2\gamma_{ba}\rho_{aa} + 2\gamma_{bd}\rho_{dd} - \mathrm{i}g\mathrm{e}^{-\mathrm{i}\omega_2 t}\rho_{bd} + \mathrm{i}g^*\mathrm{e}^{\mathrm{i}\omega_2 t}\rho_{db},\\
\dot{\rho}_{bd} &= -(\gamma_{cd}+\gamma_{bd}-\mathrm{i}\omega_{db})\rho_{bd} - \mathrm{i}f^*\mathrm{e}^{\mathrm{i}\omega_2 t}\rho_{bc} + \mathrm{i}g^*\mathrm{e}^{\mathrm{i}\omega_2 t}(\rho_{dd}-\rho_{bb}),\\
\dot{\rho}_{cd} &= -(\gamma_{cd}+\gamma_{bd}-\mathrm{i}\omega_{dc})\rho_{cd} - \mathrm{i}g^*\mathrm{e}^{\mathrm{i}\omega_2 t}\rho_{cb} + \mathrm{i}f^*\mathrm{e}^{\mathrm{i}\omega_2 t}(2\rho_{dd}+\rho_{aa}+\rho_{bb}-1),\\
\dot{\rho}_{ba} &= -(\gamma_{ba}+\gamma_{ca}-\mathrm{i}\omega_{ab})\rho_{ba} + \mathrm{i}g^*\mathrm{e}^{\mathrm{i}\omega_2 t}\rho_{da},\\
\dot{\rho}_{ca} &= -(\gamma_{ba}+\gamma_{ca}-\mathrm{i}\omega_{ac})\rho_{ca} + \mathrm{i}f^*\mathrm{e}^{\mathrm{i}\omega_2 t}\rho_{da},\\
\dot{\rho}_{da} &= -[\gamma_{ba}+\gamma_{ca}+\gamma_{cd}+\gamma_{bd}-\mathrm{i}(\omega_{ac}-\omega_{dc})]\rho_{da} + \mathrm{i}(g\rho_{ba}+f\rho_{ca})\mathrm{e}^{-\mathrm{i}\omega_2 t},\\
\dot{\rho}_{bc} &= \eta_a\rho_{aa} + \eta_d\rho_{dd} - \mathrm{i}\omega_{bc}\rho_{bc} - \mathrm{i}f\mathrm{e}^{-\mathrm{i}\omega_2 t}\rho_{bd} + \mathrm{i}g^*\mathrm{e}^{\mathrm{i}\omega_2 t}\rho_{dc},
\end{aligned}
\tag{17.68}
$$

where we have used the trace condition $\sum_i \rho_{ii} = 1$ and where

$$
2\gamma_{bd} = \frac{4\omega_{db}^2|\vec{\wp}_{db}|^2}{3\hbar c^3}, \qquad 2\gamma_{cd} = \frac{4\omega_{dc}^2|\vec{\wp}_{dc}|^2}{3\hbar c^3}
\tag{17.69}
$$

define the spontaneous emission rates from the state $|d\rangle$ to states $|b\rangle$ and $|c\rangle$, respectively. The Rabi frequencies

$$
2g = 2\vec{E}_2\cdot\vec{\wp}_{db}/\hbar, \qquad 2f = 2\vec{E}_2\cdot\vec{\wp}_{dc}/\hbar
\tag{17.70}
$$

are for the probe field acting on transition $|d\rangle \leftrightarrow |b\rangle$ and $|d\rangle \leftrightarrow |c\rangle$, respectively. Furthermore, we can write $g = |g|\mathrm{e}^{-\mathrm{i}\phi_a}$ and $f = |f|\mathrm{e}^{-\mathrm{i}\phi_d}$, where the phase $\phi = \phi_a - \phi_d$ gives the relative phase between the complex dipole matrix elements $\vec{\wp}_{db}$ and $\vec{\wp}_{dc}$. The VIC parameters are

$$
\eta_a = 2\sqrt{\gamma_{ba}\gamma_{ca}}\cos\theta_a, \qquad \eta_d = 2\sqrt{\gamma_{bd}\gamma_{cd}}\cos\theta_d.
\tag{17.71}
$$

We thus include vacuum-induced coherence on all possible transitions.

In order to study probe absorption we solve Eqs. (17.68) perturbatively. We need to know $\rho_{dd}(t)$ to second order in the probe field, assuming that the atom was prepared in the state $|a\rangle$ at $t=0$. The calculations are straightforward. The result for the population of the level $|d\rangle$ in steady state is

$$
\begin{aligned}
&\rho_{dd}^{(2)}(t \gg (\gamma_{ba}+\gamma_{ca})^{-1}, (\gamma_{cd}+\gamma_{bd})^{-1})\\
&= \frac{\eta_a|f||g|\mathrm{e}^{-\mathrm{i}(\omega_{bc}t+\phi)}}{[2(\gamma_{cd}+\gamma_{bd})-\mathrm{i}\omega_{bc}][2(\gamma_{ba}+\gamma_{ca})-\mathrm{i}\omega_{bc}][(\gamma_{cd}+\gamma_{bd})-\mathrm{i}(\Delta_2+\omega_{bc}/2)]}\\
&\quad + \frac{\eta_a|f||g|\mathrm{e}^{\mathrm{i}(\omega_{bc}t+\phi)}}{[2(\gamma_{cd}+\gamma_{bd})+\mathrm{i}\omega_{bc}][2(\gamma_{ba}+\gamma_{ca})+\mathrm{i}\omega_{bc}][(\gamma_{cd}+\gamma_{bd})-\mathrm{i}(\Delta_2-\omega_{bc}/2)]}\\
&\quad + \frac{|g|^2}{2(\gamma_{cd}+\gamma_{bd})[\gamma_{cd}+\gamma_{bd}-\mathrm{i}(\Delta_2-\omega_{bc}/2)]}p_b\\
&\quad + \frac{|f|^2}{2(\gamma_{cd}+\gamma_{bd})[\gamma_{cd}+\gamma_{bd}-\mathrm{i}(\Delta_2+\omega_{bc}/2)]}p_c + c.c..
\end{aligned}
\tag{17.72}
$$

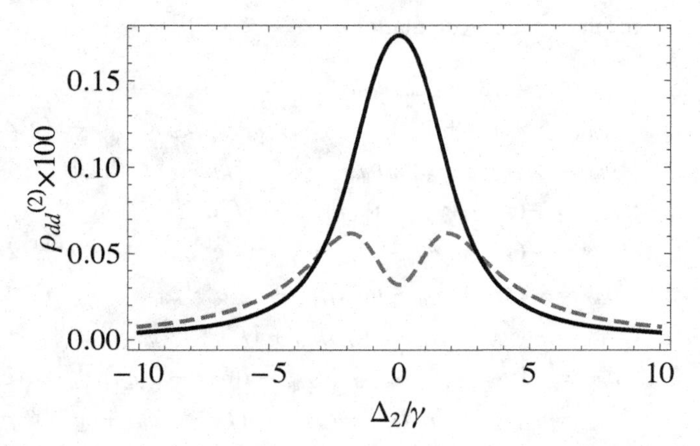

Fig. 17.19 The cosine (dashed) and sine (solid) components of the exited state population $\rho_{dd}^{(2)} \times 10^2$ as a function of Δ_2/γ. The parameters are $\gamma_{ba} = \gamma_{ca} = \gamma_{bd} = \gamma_{cd} = \gamma, \omega_{bc} = 2\gamma, |f| = |g| = 0.1\gamma, \theta_a = 0$, and $\phi = 0$.

In the absence of the vacuum-induced coherence $\eta_a \to 0$, the result (17.72) reduces to

$$\begin{aligned}
&\rho_{dd}^{(2)}(t \gg (\gamma_{ba} + \gamma_{ca})^{-1}, (\gamma_{cd} + \gamma_{bd})^{-1}) \\
&= \frac{|g|^2}{2(\gamma_{cd} + \gamma_{bd})[\gamma_{cd} + \gamma_{bd} - \mathrm{i}(\Delta_2 - \omega_{bc}/2)]} p_b \\
&+ \frac{|f|^2}{2(\gamma_{cd} + \gamma_{bd})[\gamma_{cd} + \gamma_{bd} - \mathrm{i}(\Delta_2 + \omega_{bc}/2)]} p_c + c.c.. \qquad (17.73)
\end{aligned}$$

Equation (17.73) is the expected result, which is the sum of the individual absorptions corresponding to the transitions $|b\rangle \to |d\rangle$, $|c\rangle \to |d\rangle$. The parameter $\frac{\omega_{db}+\omega_{dc}}{2} - \omega_2$ is the probe detuning defined with respect to the center of levels $|b\rangle$ and $|c\rangle$. The modulated term in probe absorption (17.72) is the result of VIC. This modulation is the signature of the VIC produced by the two paths of spontaneous emission $|a\rangle \to |b\rangle$, $|a\rangle \to |c\rangle$. Note the interesting phase dependence that arises in the probe absorption due to nonzero η_a. This phase dependence is another outcome of the presence of VIC in a system. Since the probe is treated to second order in its amplitude, the result is independent of the coherence parameter η_d for the transition $|d\rangle \to |b\rangle$, $|d\rangle \to |c\rangle$.

We next discuss the changes in absorption spectrum that can arise due to VIC. The modulated component of the population (17.72) can be written as

$$\begin{aligned}
\rho_{dd}^{(2)} &= \frac{2\eta_a|f||g|}{D}\{[2(\gamma_{ba} + \gamma_{ca})(\gamma_{cd} + \gamma_{bd})^2 + 2(\gamma_{ba} + \gamma_{ca})(\Delta_2^2 - \omega_{bc}^2/4) \\
&- (\gamma_{cd} + \gamma_{bd})\omega_{bc}^2] \cos(\omega_{bc}t + \phi) + \omega_{bc}[2(\gamma_{ba} + \gamma_{ca})(\gamma_{cd} + \gamma_{bd}) \\
&+ (\gamma_{cd} + \gamma_{bd})^2 + \Delta_2^2 - \omega_{bc}^2/4] \sin(\omega_{bc}t + \phi)\}, \qquad (17.74)
\end{aligned}$$

where

$$D = [4(\gamma_{ba}+\gamma_{ca})^2+\omega_{bc}^2][(\gamma_{cd}+\gamma_{bd})^2+(\Delta_2 + \omega_{bc}/2)^2][(\gamma_{cd} + \gamma_{bd})^2 + (\Delta_2 - \omega_{bc}/2)^2]. \qquad (17.75)$$

Since it is possible to separate the sine and cosine terms by the phase-sensitive detection we plot these in Figure 17.19 as a function of the probe detuning. These two components are the unambiguous signatures of the VIC.

The coherent control of the optical properties has many applications in both physics and chemistry. We have presented some applications in this chapter and some others will be presented in Chapter 18. Some of the other notable application areas of coherent control are bistability and more generally multistability [65, 66], tunneling, chemical reactions, photo dissociation [67]; chirality of the medium [68]; magneto-optical rotations [69–71]; slowing of decoherence produced by interactions with the environment [72, 73]; and suppression of the excited state absorption to produce ultraviolet tunable solid-state lasers [74].

Exercises

17.1 Consider a double Λ system as shown here. The wavy arrows give spontaneous emission transitions. The solid arrows are the interaction with the fields ω_p and ω_l. Write the Hamiltonian (generalization of (17.8)) for this system, make transformations like (17.9) to obtain an effective time independent Hamiltonian. Find the dark state for this system, if any.

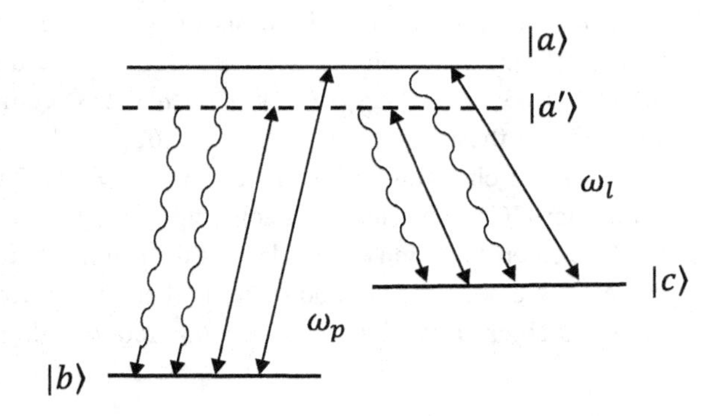

17.2 Show that (17.10) for $\Delta_p = \Delta_l = 0$, $\Gamma_{bc} = 0$ has an exact steady-state solution, $\rho_{aa} = \rho_{ab} = \rho_{ac} = 0$. In addition, show that ρ_{bb}, ρ_{cc}, ρ_{bc} values are consistent with the dark state (17.38).

17.3 Starting from Eqs. (17.68), prove the result (17.72).

17.4 Write all the density matrix equations explicitly using (17.51) and derive (17.52) to lowest order in $|g_a|^2|g_b|^2$.

17.5 Consider the multilevel scheme shown in the figure. Assuming that all the fields are resonant with their respective transitions. Then show that the dark state apart from a normalization constant is

$$|\Psi_0\rangle \approx G_1G_2|b_1\rangle - g_1G_2|b_2\rangle + g_1g_2|g_3\rangle.$$

For applications of this scheme to atomic localization see [31].

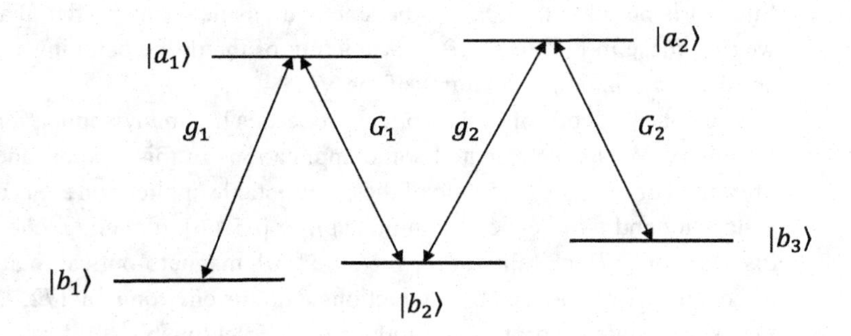

17.6 Starting with the Hamiltonian (17.37) and using the actual time-dependent profiles for $g_p(t)$ and $g_l(t)$ given by

$$g_l(t) = g_l \exp(-t^2/\tau^2),$$
$$g_p(t) = g_p \exp[-(t-T)^2/\tau^2], \qquad T = 1.5\tau,$$

and the Schrödinger equation, show that the state $|b\rangle$ at $t = -\infty$ would evolve adiabatically to the state $|c\rangle$ at $t = +\infty$. You can use the values $g_l\tau = g_p\tau = 12$. The Runge–Kutta method of integrating time dependent equations would be useful. Show further that the population in the upper state $|a\rangle$ remains negligibly small during the entire evolution.

17.7 Transform Eqs. (17.10) in the basis of the dressed states $|\Psi_+\rangle$, $|\Psi_-\rangle$ and $|b\rangle$ (cf. Figure 17.5). For simplicity, set $\Delta_p = \Delta_l = 0$ and $g_p = 0$. The states $|\Psi_\pm\rangle$ are the eigenstates of $-\hbar(g_l|a\rangle\langle c| + g_l^*|c\rangle\langle a|)$. Do retain the γs and Γs. Discuss the similarity of these equations to Eqs. (17.62) for the case of VIC, especially if there are no phase-changing collisions and if $\gamma_{ca} = 0$. This exercise shows how effects similar to VIC are produced by coherent driving fields.

17.8 Find the eigenstates and eigenvalues of the Hamiltonian (17.37) for arbitrary complex values of the couplings g_l and g_p. Find then the approximate form of the eigenfunctions and eigenvalues for Δ_p large compared to both g_l and g_p.

References

[1] S. E. Harris, J. E. Field, and A. Imamoğlu, *Phys. Rev. Lett.* **64**, 1107 (1990).
[2] O. A. Kocharovskaya and Ya. I. Khanin, *Zh. Eksp. Teor. Fiz.* **90**, 1610 (1986) [*Sov. Phys. JETP* **63**, 945 (1986)].
[3] S. P. Tewari and G. S. Agarwal, *Phys. Rev. Lett.* **56**, 1811 (1986).
[4] S. H. Autler and C. H. Townes, *Phys. Rev.* **100**, 703 (1955).
[5] G. S. Agarwal, *Phys. Rev. A* **55**, 2467 (1997).
[6] P. Anisimov and O. Kocharovskaya, *J. Mod. Opt.* **55**, 3159 (2008).
[7] K. Boller, A. Imamolu, and S. E. Harris, *Phys. Rev. Lett.* **66**, 2593 (1991).
[8] J. E. Field, K. H. Hahn, and S. E. Harris, *Phys. Rev. Lett.* **67**, 3062 (1991).

[9] M. Mücke, E. Figueroa, J. Bochmann *et al., Nature (London)* **465**, 755 (2010).
[10] T. Kampschulte, W. Alt, S. Brakhane *et al., Phys. Rev. Lett.* **105**, 153603 (2010).
[11] L. Slodička, G. Hétet, S. Gerber, M. Hennrich, and R. Blatt, *Phys. Rev. Lett.* **105**, 153604 (2010).
[12] D. Akamatsu, K. Akiba, and M. Kozuma, *Phys. Rev. Lett.* **92**, 203602 (2004).
[13] M. D. Eisaman, A. André, F. Massou, M. Fleischhauer, A. S. Zibrov, and M. D. Lukin, *Nature (London)* **438**, 837 (2005).
[14] R. M. Whitley and C. R. Stroud, *Jr. Phys. Rev. A* **14**, 1498 (1976).
[15] M. Xiao, Y.-Q. Li, S.-Z. Jin, and J. Gea-Banacloche, *Phys. Rev. Lett.* **74**, 666 (1995).
[16] F. S. Cataliotti, C. Fort, T. W. Hänsch, M. Inguscio, and M. Prevedelli, *Phys. Rev. A* **56**, 2221 (1997).
[17] O. Kocharovskaya and P. Mandel, *Phys. Rev. A* **42**, 523 (1990).
[18] U. D. Rapol, A. Wasan, and V. Natarajan, *Phys. Rev. A* **67**, 053802 (2003).
[19] G. Vemuri, G. S. Agarwal, and B. D. Nageswara Rao, *Phys. Rev. A* **53**, 2842 (1996).
[20] Y. Zhu and T. N. Wasserlauf, *Phys. Rev. A* **54**, 3653 (1996).
[21] U. Fano, *Phys. Rev.* **124**, 1866 (1961).
[22] E. Arimondo and G. Orriols, *Nuovo Cimento Lett.* **17**, 333 (1976).
[23] H. R. Gray, R. M. Whitley, and C. R. Stroud, *Jr., Opt. Lett.* **3**, 218 (1978).
[24] E. Arimondo, in *Progress in Optics*, edited by E. Wolf (Amsterdam: Elsevier, 1996), vol. **35**, p. 259.
[25] A. Aspect, E. Arimondo, R. Kaiser, N. Vansteenkiste, and C. Cohen-Tannoudji, *J. Opt. Soc. Am. B* **6**, 2112 (1989).
[26] K. Bergmann, H. Theuer, and B. W. Shore, *Rev. Mod. Phys.* **70**, 1003 (1998).
[27] E. Paspalakis, A. F. Terzis, and P. L. Knight, *J. Mod. Opt.* **52**, 1685 (2005).
[28] G. S. Agarwal and K. T. Kapale, *J. Phys. B* **39**, 3437 (2006).
[29] A. V. Gorshkov, L. Jiang, M. Greiner, P. Zoller, and M. D. Lukin, *Phys. Rev. Lett.* **100**, 093005 (2008).
[30] N. A. Proite, Z. J. Simmons, and D. D. Yavuz, *Phys. Rev. A* **83**, 041803(R) (2011).
[31] M. Kiffner, J. Evers, and M. S. Zubairy, *Phys. Rev. Lett.* **100**, 073602 (2008).
[32] S. W. Hell, *Science* **316**, 1153 (2007).
[33] Z. Liao, M. Al-Amri, and M. S. Zubairy, *Phys. Rev. Lett.* **105**, 183601 (2010).
[34] G. S. Agarwal, *Phys. Rev. A* **54**, R3734 (1996).
[35] C. Cohen-Tannoudji and S. Reynaud, *J. Phys. B* **10**, 2311 (1977).
[36] C. Cohen-Tannoudji and S. Reynaud, *J. Phys. B* **10**, 365 (1977).
[37] G. S. Agarwal and S. S. Jha, *J. Phys. B* **12**, 2655 (1979).
[38] M. L. Goldberger and K. M. Watson. *Phys. Rev. B* **136**, 1472 (1964).
[39] J. S. Bell and C. J. Goebel, *Phys. Rev. B* **138**, 1198 (1965).
[40] E. C. G. Sudarshan, *Phys. Rev. A* **50**, 2006 (1994).
[41] C. B. Chiu, E. C. G. Sudarshan, and G. Bhamathi, *Phys. Rev. D* **46**, 3508 (1992).
[42] A. Böhm, S. Maxson, M. Loewe, and M. Gadella, *Physica A* **236**, 485 (1997).
[43] N. J. Kylstra and C. J. Joachain, *Europhys. Lett.* **36**, 657 (1996).
[44] G. S. Agarwal, in *Frontiers of Quantum Optics and Laser Physics*, edited by S.-Y. Zhu, M. S. Zubairy, and M. O. Scully (Singapore: Springer, 1997), p. 155.
[45] G. S. Agarwal and W. Harshawardhan, *Phys. Rev. Lett.* **77**, 1039 (1996).

[46] D. Wang, J. Y. Gao, J. H Xu, G. C. La Rocca, and F. Bassani, *Europhys. Lett*. **54**, 456 (2001).
[47] M. Yan, E. G. Rickey, and Y. Zhu, *Phys. Rev. A* **64**, 043807 (2001).
[48] S. E. Harris and Y. Yamamoto, *Phys. Rev. Lett*. **81**, 3611 (1998).
[49] D. Höckel and O. Benson, *Phys. Rev. Lett*. **105**, 153605 (2010).
[50] H. Schmidt and A. Imamoğlu, *Opt. Lett*. **21**, 1936 (1996).
[51] G. S. Agarwal, *Quantum Optics* (Berlin: Springer-Verlag, 1974), p. 94.
[52] S. Das and G. S. Agarwal, *Phys. Rev. A* **81**, 052341 (2010).
[53] G. S. Agarwal, *Phys. Rev. Lett*. **84**, 5500 (2000).
[54] Y. Yang, J. Xu, H. Chen, and S. Zhu, *Phys. Rev. Lett*. **100**, 043601 (2008).
[55] V. Yannopapas, E. Paspalakis, and N. V. Vitanov, *Phys. Rev. Lett*. **103**, 063602 (2009).
[56] Q. Wang, S. Stobbe, H. Thyrrestrup *et al., Opt. Lett*. **35**, 2768 (2010).
[57] M. O. Scully, K. R. Chapin, K. E. Dorfman, M. B. Kim, and A. A. Svidzinsky, *PNAS* **108**, 15097 (2011).
[58] M. Kiffner, M. A. Macovei, J. Evers, and C. H. Keitel, in *Progress in Optics*, edited by E. Wolf (Amsterdom: Elsevier, 2010), vol. **55**, p. 85.
[59] Z. Ficek and S. Swain, *Quantum Interference and Coherence* (New York: Springer, 2005).
[60] E. Paspalakis and P. L. Knight, *Phys. Rev. Lett*. **81**, 293 (1998).
[61] P. Zhou and S. Swain, *Phys. Rev. Lett*. **77**, 3995 (1996).
[62] M. V. Gurudev Dutt, J. Cheng *et al., Phys. Rev. Lett*. **94**, 227403 (2005).
[63] J. Javanainen, *Europhys. Lett*. **17**, 407 (1992).
[64] S. Menon and G. S. Agarwal, *Laser Physics* **9**, 813 (1999), arXiv: quant-ph/9902021v1.
[65] W. Harshawardhan and G. S. Agarwal, *Phys. Rev. A* **53**, 1812 (1996).
[66] A. Joshi and M. Xiao, in *Progress in Optics*, edited by E. Wolf (Amsterdam: Elsevier, 2006), vol. **49**, p. 97.
[67] M. Shapiro and P. Brumer, *Principles of the Quantum Control of Molecular Processes* (Hoboken, NY: Wiley-Interscience, 2003).
[68] V. A. Sautenkov, Y. V. Rostovtsev, H. Chen, P. Hsu, G. S. Agarwal, and M. O. Scully, *Phys. Rev. Lett*. **94**, 233601 (2005).
[69] A. K. Patnaik and G. S. Agarwal, *Opt. Commun*. **179**, 97 (2000).
[70] V. A. Sautenkov, M. D. Lukin, C. J. Bednar *et al., Phys. Rev. A* **62**, 023810 (2000).
[71] F. S. Pavone, G. Bianchini, F. S. Cataliotti, T. W. Hänsch, and M. Inguscio, *Opt. Lett*. **22**, 736 (1997).
[72] G. S. Agarwal, *Phys. Rev. A* **61**, 013809 (1999).
[73] L. Viola and S. Lloyd, *Phys. Rev. A* **58**, 2733 (1998).
[74] E. Kuznetsova, R. Kolesov, and O. Kocharovskaya, *Phys. Rev. A* **70**, 043801 (2004).

18 Dispersion management and ultraslow light

In this chapter we discuss a variety of physical effects which primarily depend on the dispersive properties of the medium, i.e. how the real part of the refractive index depends on the frequency of light. For example, it is well known that the efficiency of nonlinear optical processes such as harmonic generation depends on the phase matching, which in turn depends on the refractive index at the fundamental and harmonic frequencies [1]. Thus a control of dispersion will enable us to obtain more efficient harmonic generation [2–5]. This in fact was the starting point of the work on control of dispersion [2]. Another subject where the dispersion is very important is in the propagation of the pulses which generally are distorted [6] by the dispersion of the medium and hence one needs to tailor the dispersion to obtain nearly distortionless propagation [7]. In Section 17.1, we have already shown how an appropriately chosen control field leads to a significant modification of the dispersion (Figure 17.4). We will now discuss some applications of this. We will also discuss how hole burning physics (Section 13.2) can be used to obtain very significant control of the dispersion.

18.1 Group velocity and propagation in a dispersive medium

Let us consider the one-dimensional propagation of an electromagnetic pulse in a dispersive medium characterized by susceptibility $\chi(\omega)$ and refractive index $n(\omega)$. For simplicity of the argument, we will assume real χ and n. Let us write a traveling wave pulse at the input face ($z = 0$) of the medium as

$$\vec{E}(z,t) = \hat{e}\int_{-\infty}^{+\infty} \mathrm{d}\omega \mathrm{e}^{-\mathrm{i}\omega(t-\frac{z}{c})} E(\omega) + c.c.. \tag{18.1}$$

The spectrum $E(\omega)$ has a narrow width and peaks around the frequency ω_p. The polarization of the input field is $\hat{e}$. The field at the output face $z = L$ is then

$$\begin{aligned} \vec{E}(L,t) &= \hat{e}T(t) + c.c., \\ T(t) &= \int_{-\infty}^{+\infty} \mathrm{d}\omega \mathrm{e}^{-\mathrm{i}\omega t + \mathrm{i}\frac{\omega n(\omega)L}{c}} E(\omega). \end{aligned} \tag{18.2}$$

In view of the assumed properties of $E(\omega)$, and if we further assume that the refractive index varies slowly around $\omega = \omega_p$, then

$$\omega n(\omega) = \omega_p n(\omega_p) + (\omega - \omega_p)\frac{\partial(\omega n)}{\partial \omega_p} + \frac{(\omega-\omega_p)^2}{2}\frac{\partial^2(\omega n)}{\partial \omega_p^2} + \cdots . \tag{18.3}$$

On substituting (18.3) in (18.2) and on retaining only the first two terms in (18.3), we get

$$T(t) = \mathrm{e}^{-\mathrm{i}\omega_p t + \frac{\mathrm{i}\omega_p n(\omega_p) L}{c}} \int_{-\infty}^{+\infty} \mathrm{d}\omega \mathrm{e}^{-\mathrm{i}\omega t + \frac{\mathrm{i}\omega L}{v_g}} E(\omega + \omega_p), \tag{18.4}$$

$$v_g = c \Bigg/ \frac{\partial(\omega_p n(\omega_p))}{\partial \omega_p}. \tag{18.5}$$

Note that $\int_{-\infty}^{+\infty} \mathrm{d}\omega \mathrm{e}^{-\mathrm{i}\omega t} E(\omega + \omega_p)$ defines the envelope $\mathcal{E}(t)$ of a slowly varying pulse $\vec{E}(t) = \hat{e}\, \mathcal{E}(t) \mathrm{e}^{-\mathrm{i}\omega_p t} + c.c.$ Thus (18.4) becomes

$$T(t) = \mathrm{e}^{-\mathrm{i}\omega_p t + \mathrm{i}\frac{\omega_p n(\omega_p) L}{c}} \mathcal{E}\left(t - \frac{L}{v_g}\right). \tag{18.6}$$

The quantity v_g is called the group velocity of the pulse and for positive v_g gives the group delay τ_g of the pulse, which is defined by

$$\tau_g = \frac{L}{v_g} - \frac{L}{c}. \tag{18.7}$$

The phase velocity is defined by

$$v_p = \frac{c}{n(\omega_p)}. \tag{18.8}$$

The group velocity clearly depends on the dispersion in the medium, because

$$v_g = \frac{c}{1 + \omega_p \frac{\partial}{\partial \omega_p} n(\omega_p)}, \tag{18.9}$$

which for a dilute gaseous medium reduces to

$$v_g \approx \frac{c}{1 + 2\pi \omega_p \frac{\partial \chi(\omega_p)}{\partial \omega_p}} < c, \tag{18.10}$$

since $n(\omega) \approx 1 + 2\pi \chi(\omega)$. The inequality in (18.10) is for the case when one is in the region of normal dispersion of the medium, i.e. when $\partial \chi / \partial \omega > 0$.

We will now show how coherent control can be used to slow down light. We specifically consider the model system of Figure 17.2 for which the susceptibility is given by Eq. (17.14). Figure 17.4 shows that the control field changes the anomalous dispersion region near the line center (solid curve) to normal dispersion (dashed curves) and when the dispersion becomes sharp, the group velocity could become much smaller than c [7]. Using (18.10), we define the group index n_g by

$$v_g = \frac{c}{n_g}, \qquad n_g \approx 2\pi \omega_p \frac{\partial \chi(\omega_p)}{\partial \omega_p}. \tag{18.11}$$

We use (17.14) to calculate $\partial \chi / \partial \omega_p = -\partial \chi / \partial \Delta_p$ at $\Delta_p = \Delta_l$ and at $\Gamma_{bc} = 0$. We find a simple result for the group index

$$n_g \approx 2\pi \omega_p \frac{\chi_0 n}{|g_l|^2}. \tag{18.12}$$

We also calculate $n_g^{(0)}$ in the absence of the control field $|g_l| = 0$ and when Δ_p is large compared to the homogeneous time width $2\Gamma_{ab}$ so that we are in the normal dispersion region of the two-level system. It is seen from (17.14) that

$$n_g^{(0)} \approx 2\pi\omega_p \frac{\chi_0 n}{\Delta_p^2}, \tag{18.13}$$

and thus the enhancement of n_g over $n_g^{(0)}$ is given by

$$\frac{n_g}{n_g^{(0)}} = \frac{\Delta_p^2}{|g_l|^2}. \tag{18.14}$$

We note that Δ_p could be about several GHz and $|g_l|$ in MHz range. The enhancement could be about 10^6–10^8 and then the group velocity could be only a few m/s. We can also express n_g in terms of the absorption parameter α_0 on resonance

$$n_g = \frac{(\alpha_0 L)(\frac{c}{L})}{2\Gamma_{ab}}\left(\frac{\Gamma_{ab}^2}{|g_l|^2}\right), \qquad \alpha_0 = \frac{4\pi\omega_p\chi_0 n}{c\Gamma_{ab}}, \tag{18.15}$$

$$= \frac{(\alpha_0 L)(\frac{c}{L})}{A}\left(\frac{A^2}{\Omega_l^2}\right), \tag{18.16}$$

where A is the Einstein $A[= 2\Gamma_{ab}]$ coefficient and $\Omega_l = 2|g_l|$ is the Rabi frequency of the control laser. The first experiment on ultraslow light, which made history, was performed by Hau *et al.* [8]. They considered the propagation of weak probe pulses in a Bose–Einstein condensate of sodium atoms in the presence of a coherent control field on the D_2 transition. They reported a group velocity of 17 m/s in the condensate at a temperature of about 50 nK. Even at a temperature of around the transition temperature of 435 nK, they reported v_g about 32.5 m/s, which is consistent with (18.16) for their experimental parameters $\alpha_0 L \sim 63$; $L = 229$ μm, $\Omega_l/A = 0.56$. The first observation of ultraslow light in a hot rubidium vapor cell at a temperature of 360 K was reported by Kash *et al.* [9]. Kash *et al.* observed group velocities of the order of 90 m/s. The group velocity is to be calculated by averaging the susceptibility (17.14) over the Doppler distribution. This is straightforward numerically although no simple analytical result is possible as the Doppler distribution is Gaussian. Interestingly, Kash *et al.* found that (18.12) is valid even for a Doppler broadened medium if the parameter $|g_l|^2$ is much bigger than the product of $\Gamma_{bc}D$, where D is the Doppler group width, and Γ_{bc}^{-1} is the life time of the atomic coherence between the levels $|b\rangle$ and $|c\rangle$. A later experiment with a much improved Rb cell reported an even lower velocity, about 8 m/s [10]. Slow-light experiments in solid-state media have also been repeated. Here Γ_{bc}^{-1} is rather short. However, a group velocity of about 45 m/s in Y_2SiO_5 doped with Pr at 5 K has been reported [11].

Next we examine the effect of the second derivative in (18.3). In order to understand its effect we examine the case of Gaussian pulses. Let us write the pulse shape as

$$\vec{E}(t) = \hat{e}\varepsilon \exp\left\{-\frac{t^2\Gamma^2}{4} - i\omega_p t\right\} + c.c., \tag{18.17}$$

which in the Fourier domain can be expressed as

$$\vec{E}(t) = \frac{2\varepsilon}{\Gamma}\sqrt{\pi}\hat{e}\int_{-\infty}^{+\infty} d\omega\, e^{-\frac{\omega^2}{\Gamma^2}-i\omega t-i\omega_p t} + c.c.$$

$$= \frac{2\varepsilon}{\Gamma}\sqrt{\pi}\hat{e}\int_{-\infty}^{+\infty} d\omega\, e^{-\frac{(\omega-\omega_p)^2}{\Gamma^2}-i\omega t} + c.c., \tag{18.18}$$

showing that for a Gaussian pulse $E(\omega)$ of (18.1) is $\frac{2\varepsilon}{\Gamma}\sqrt{\pi}\exp\left\{-\frac{(\omega-\omega_p)^2}{\Gamma^2}\right\}$. Using now (18.2) and (18.3) and the Gaussian shape of $E(\omega)$, the integral in (18.2) can be evaluated in closed form. The result is

$$T(t) = e^{-i\omega_p t+\frac{i\omega_p}{c}n(\omega_p)L}\frac{\varepsilon}{\sqrt{1-i\kappa L}}\exp\left\{-\frac{[\Gamma(t-\frac{L}{v_g})]^2}{4(1-i\kappa L)}\right\}, \tag{18.19}$$

where

$$\kappa = \left[\frac{\Gamma^2}{2c}\frac{\partial^2}{\partial\omega^2}(\omega n(\omega))\right]_{\omega=\omega_p}. \tag{18.20}$$

The result (18.19) shows that the pulse is broadened in time by a factor $(1+\kappa^2L^2)^{1/2}$ since

$$|T(t)|^2 = \frac{|\varepsilon|^2}{\sqrt{1+\kappa^2L^2}}\exp\left\{-\frac{[\Gamma(t-\frac{L}{v_g})]^2}{2(1+\kappa^2L^2)}\right\}. \tag{18.21}$$

Thus the second derivatives in (18.3) contribute to the broadening of the pulse. This is the reason that, for pulse propagation in optical fibers over long distances, one has to design a refractive index so that its second derivative remains small [6].

The result (18.21) has been derived using the reality of $n(\omega)$. For pulse propagation under EIT conditions the situation is different (see Exercise 18.1) where the refractive index is complex. In fact, for $\Delta_p = \Delta_l = 0$, $\Gamma_{bc} = 0$, $n \cong 1$; $\partial\chi/\partial\omega$ is real and $\partial^2\chi/\partial\omega^2$ is purely imaginary. The formula (18.19) still holds. We estimate the value of κ for $\Gamma_{bc} \to 0$,

$$\kappa L \approx \frac{2\pi\Gamma^2}{2c}\omega_p\frac{\partial^2\chi}{\partial\omega_p^2}L = i\frac{4\pi\chi_0 n\omega_p L}{c\Gamma_{ab}}\frac{\Gamma_{ab}^2}{|g_l|^4}\Gamma^2$$

$$= i(\alpha_0 L)\frac{\Gamma^2\Gamma_{ab}^2}{|g_l|^4} = i\alpha_0 L\left(\frac{\Gamma^2}{4\sigma^2}\right), \qquad \sigma^2 = \frac{|g_l|^4}{4\Gamma_{ab}^2}, \tag{18.22}$$

where σ defines the width of the transparency window. Note that κ is purely imaginary and positive, and the factor $(1-i\kappa L)$ becomes $(1+|\kappa|L)$. Thus in contrast to (18.21), the broadening is governed by $(1+|\kappa|L)^{1/2}$. For almost distortionless propagation of the pulse, we clearly require $|\kappa|L \ll 1$. This puts a condition $\Gamma \ll 2\sigma$ on the spectral width of the pulse in relation to the width of the transparency window.

Ultraslow light has many applications in interferometry, delay lines, and microwave photonics [12]. Leonhardt and Piwnicki [13] proposed an improvement in the sensitivity of gyroscopes by a factor n_g. Similar enhancements occur in the sensitivity of Mach–Zehnder interferometers [12]. Delay lines are needed in telecommunication applications as the problem of simultaneous arrival of data packets at an optical switch can be avoided. The slow light in an anisotropic medium [14] can be used to separate temporally the two components of a linearly polarized pulse by applying a coherent control field such that one component of the pulse becomes slow while the other component of the linearly polarized pulse undergoes only a small change in its group velocity. Finally, we mention that the dispersion management can also be used for controlling superluminal propagation [12, 15, 16].

18.2 Electromagnetically induced waveguides

As another application of coherent control, we consider the guiding or steering of a beam by another control beam [17, 18]. Let us consider the laser fields as beams with spatial structure for the envelopes. Assume the probe beam to be Gaussian and let the control field be a Laguerre–Gaussian beam (donut beam), so that the Rabi frequencies g_p and g_l become space dependent

$$g_p = \frac{g_{0p}}{w_p(z)}\sqrt{\frac{2}{\pi}}\exp\left(-\frac{\mathrm{i}k_p\rho^2}{2q_p}\right)\mathrm{e}^{\mathrm{i}k_pz-\mathrm{i}\omega_pt}, \tag{18.23}$$

$$g_l = \frac{g_{0l}}{w_l(z)}\frac{1}{\sqrt{3\pi}}\left(\frac{\sqrt{2}\rho}{w_2(z)}\right)^3\exp\left(-\frac{\mathrm{i}k_l\rho^2}{2q_l}-3\mathrm{i}\theta\right)\mathrm{e}^{\mathrm{i}k_lz-\mathrm{i}\omega_lt}, \tag{18.24}$$

$$\begin{aligned} &\rho^2 = x^2+y^2, \qquad \theta = \tan^{-1}\left(\frac{y}{x}\right), \qquad w(z) = w_0\sqrt{1+(\frac{z-z_0}{z_R})^2}, \\ &q = \mathrm{i}z_\mathrm{R} - z + z_0, \qquad z_\mathrm{R} = \frac{\pi w_0^2}{\lambda}. \end{aligned} \tag{18.25}$$

Here z_R is the Rayleigh range of the beam, w_0 is the waist radius, and z_0 is the location of the beam waist. The susceptibility (17.14) depends now on (x, y, z) coordinates as

$$|g_l|^2 = \frac{|g_{0l}|^2}{3\pi w_l^2(z)}\left(\frac{2\rho^2}{w_2^2(z)}\right)^3\exp\left\{-\frac{\mathrm{i}k_l\rho^2}{2}\left(\frac{1}{q_l}+\frac{1}{q_l^*}\right)\right\}. \tag{18.26}$$

From (17.14), the real part of $\chi^{(ab)}(\omega_p)$ for $\Delta_p = \Delta_l$ is

$$\mathrm{Re}[\chi^{(ab)}(\omega_p)] = \frac{\chi_0 n\Delta_p}{\Delta_p^2+\tilde{\Gamma}_{ab}^2}, \qquad \tilde{\Gamma}_{ab} = \Gamma_{ab} + \frac{|g_l|^2}{\Gamma_{bc}}. \tag{18.27}$$

For the donut beam with $|g_l|^2$ given by (18.26), $|g_l|^2 \to 0$ in the central region of the beam, whereas in the outer region $|g_l|^2$ is large. Furthermore, the smallness of the collisional parameter gives us

$$\tilde{\Gamma}_{ab}(\rho \sim \text{central}) \ll \tilde{\Gamma}_{ab}(\rho \sim \text{outer}).$$

Hence the refractive index $\mathrm{Re}[n(\omega)] \approx 1 + 2\pi\chi(\omega)$ has the property

$$\mathrm{Re}[n(\omega, \rho \sim \text{central})] > \mathrm{Re}[n(\omega, \rho \sim \text{outer})]. \tag{18.28}$$

The inequality in (18.28) holds if the probe beam is red detuned, i.e. if $\Delta_p > 0$. The above inequality reverses if the probe is blue detuned. The donut control beam has given a spatial structure to the refractive index of the probe beam. The property (18.28) is similar to what is used in waveguides where the core has a higher refractive index. Due to the property (18.28), a red detuned Gaussian probe beam would be guided into the center of the donut beam whereas a blue detuned probe beam would be guided into the ring region of the donut beam. If Γ_{bc} is nearly zero, then one has to keep nonzero two-photon detuning $\Delta_p - \Delta_l \neq 0$. In place of (18.27), we now have

$$\mathrm{Re}\,[\chi^{(ab)}(\omega_p)] = \frac{\chi_0 n \tilde{\Delta}_p}{\tilde{\Delta}_p^2 + \Gamma_{ab}^2}, \qquad \tilde{\Delta}_p = \Delta_p - \frac{|g_l|^2}{\Delta_p - \Delta_l}. \tag{18.29}$$

For the no control field case, one would generally take Δ_p to be large compared to Γ_{ab} and positive for red detuning. This is to avoid the effects of absorption. For waveguiding behavior $|g_l| \neq 0$ (outer region of the donut beam) we thus require that

$$\mathrm{Re}\,[\chi^{(ab)}(\omega_p, |g_l| \neq 0)] < \mathrm{Re}\,[\chi^{(ab)}(\omega_p, |g_l| = 0)].$$

Let us assume that we are working with Rabi frequencies and two-photon detunings such that $\tilde{\Delta}_p \gg \Gamma_{ab}$, then this condition reduces to $\Delta_p < \tilde{\Delta}_p$, $\Delta_p > 0$. This implies that we must choose the two-photon detuning $\Delta_p - \Delta_l$ to be negative. To summarize, the waveguiding would result if (a) probe is red detuned with $\tilde{\Delta}_p \gg \Gamma_{ab}$, (b) the two-photon detuning $\Delta_p - \Delta_l < 0$, and (c) $|g_l|$ is such that $\Delta_p - \frac{|g_l|^2}{\Delta_p - \Delta_l} \gg \Gamma_{ab}$.

In order to obtain the waveguiding behavior of the probe beam, we first have to average $\mathrm{Re}\,[\chi^{(ab)}(\omega_p)]$ over the Doppler distribution and then integrate the Maxwell equation for the slowly varying envelope g_p

$$\frac{\partial g_p}{\partial z} = \frac{\mathrm{i}c}{2\omega_p}\left(\frac{\partial^2}{\partial x^2} + \frac{\partial^2}{\partial y^2}\right) g_p + \frac{2\pi \mathrm{i}\omega_p}{c}\langle \mathrm{Re}\,[\chi^{(ab)}(\omega_p)]\rangle\, g_p, \tag{18.30}$$

where $\langle \cdots \rangle$ represents the Doppler average. Equation (18.30) can be integrated using standard numerical methods. We do not give the results of computations, which can be found in [18]. These computations show explicitly how the probe is guided in the central region of the donut beam for red detuning of the probe and is guided in the ring region for blue detuning of the probe. Truscott *et al.* [17] made the first observation of the waveguiding behavior using an Rb vapor cell where the probe was tuned near the D_2 line and the pump was tuned near the D_1 line. Their experiment corresponds to the V-scheme and the numerical calculations in [18] are for this scheme. The waveguiding using the Λ-scheme has been reported in [19], in which $\Gamma_{bc} \sim 0$, but the two-photon detuning was kept nonzero ($\Delta_p - \Delta_l < 0$) in conformity with our discussion after (18.29).

18.3 Storage and retrieval of optical pulses

In Chapter 17, specifically Eq. (17.13), we saw that the simultaneous application of the probe and control fields produces atomic coherence between the two long-lived levels $|b\rangle$ and $|c\rangle$. The life time of this coherence is determined by the collisional parameter Γ_{bc}. Thus when the control field is switched off after the creation of coherence, it will survive for a time $\sim \Gamma_{bc}^{-1}$. Fleischhauer and Lukin [20] extended the ideas of atomic coherence to the case when both control and probe fields are pulses. Assuming (i) adiabatic pulses and (ii) that the control field was a specified time-dependent field whose dynamics can be ignored on propagation, Fleischhauer and Lukin introduced the idea of a dark state polariton, which propagates without change in shape. Earlier Grobe *et al.* [21] presented most general solutions for both control and probe fields. Their solutions, called adiabatons, can be used to describe elegantly the storage and retrieval of pulses.

In order to demonstrate the storage and retrieval of optical pulses, we have to solve the density matrix equations (17.10) for the time-dependent field envelopes, i.e. when the fields, instead of (17.7), are of the form

$$\vec{E} = \vec{E}_l(z,t)\mathrm{e}^{-\mathrm{i}\omega_l t+\mathrm{i}k_l z} + \vec{E}_p(z,t)\mathrm{e}^{-\mathrm{i}\omega_p t+\mathrm{i}k_p z} + c.c.. \tag{18.31}$$

We will assume that the envelopes are slowly varying functions of space and time. We write the induced polarization in the form

$$\vec{P} = \vec{P}_l(z,t)\mathrm{e}^{-\mathrm{i}\omega_l t+\mathrm{i}k_l z} + \vec{P}_p(z,t)\mathrm{e}^{-\mathrm{i}\omega_p t+\mathrm{i}k_p z} + c.c., \tag{18.32}$$

where $\vec{P}_l(z,t) = n\vec{p}_{ac}^{\,*}\rho_{ac}\mathrm{e}^{-\mathrm{i}k_l z}$, $\vec{P}_p(z,t) = n\vec{p}_{ab}^{\,*}\rho_{ab}\mathrm{e}^{-\mathrm{i}k_p z}$. We now define g_p and g_l as

$$g_p = \frac{\vec{p}_{ab}\cdot\vec{E}_p}{\hbar}, \qquad g_l = \frac{\vec{p}_{ac}\cdot\vec{E}_l}{\hbar}, \tag{18.33}$$

which are dependent on space and time. The wave equation for the field is

$$\left(\frac{\partial^2}{\partial z^2} - \frac{1}{c^2}\frac{\partial^2}{\partial t^2}\right)\vec{E} = \frac{4\pi}{c^2}\frac{\partial^2\vec{P}}{\partial t^2}. \tag{18.34}$$

We use the forms (18.31) and (18.32) in (18.34), use slowly varying approximations to drop the second derivatives of $\vec{E}_l$, $\vec{E}_p$, $\vec{P}_l$, and $\vec{P}_p$, and the first derivatives of $\vec{P}_l$, $\vec{P}_p$. We can then convert the equations for $\vec{E}_l$ and $\vec{E}_p$ into equations for g_p and g_l using the definitions (18.33) with the results

$$\begin{aligned} \frac{\partial g_p}{\partial z} + \frac{\partial g_p}{\partial(ct)} &= \mathrm{i}\eta\rho_{ab}, \\ \frac{\partial g_l}{\partial z} + \frac{\partial g_l}{\partial(ct)} &= \mathrm{i}\eta\rho_{ac}, \end{aligned} \tag{18.35}$$

where we have set $\frac{2\pi\omega_p n}{c\hbar}|p_{ab}|^2 \approx \frac{2\pi\omega_l n}{c\hbar}|p_{ac}|^2 = \eta$. Note that η is related to the absorption coefficient $\alpha = 2\eta/\Gamma_{ab}$. To see this we write (18.35) for a linear medium. Using (17.12) for $g_l = 0$ and $\Delta_p = 0$, we get

$$\frac{\partial g_p}{\partial z} + \frac{\partial g_p}{\partial(ct)} = -\frac{\eta}{\Gamma_{ab}}g_p. \tag{18.36}$$

Note that since g_p and g_l are now space-time dependent, the density matrix elements $\rho_{\alpha\beta}$ also become space-time dependent. We now need to solve coupled set of equations (18.35) and (17.10) in the space-time domain. These set of equations are to be solved numerically. We will present results for $\Delta_l = \Delta_p = 0$, $\Gamma_{bc} = 0$. We also use the traveling coordinates

$$\tau = t - z/c, \qquad \zeta = z, \tag{18.37}$$

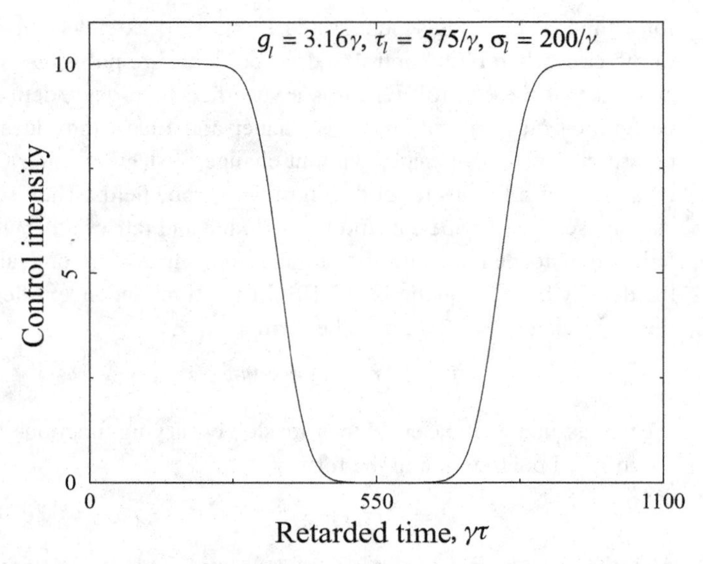

Fig. 18.1 The intensity $(g_l/\gamma)^2$ of the control field as a function of retarded time at the entry face of the medium at $\zeta = 0$; the super-Gaussian control field is switched off and on adiabatically.

so that

$$\frac{\partial}{\partial z} + \frac{\partial}{\partial (ct)} \to \frac{\partial}{\partial \zeta}, \qquad \frac{\partial}{\partial t} \to \frac{\partial}{\partial \tau}. \tag{18.38}$$

For the control field, we use a super-Gaussian profile (Figure 18.1) at the input face ($z = \zeta = 0$) of the medium

$$g_l(0, \tau) = g_l \left\{ 1 - \exp\left[- \left(\frac{\tau - \tau_l}{\sigma_l} \right)^{\beta} \right] \right\}. \tag{18.39}$$

For $\beta = 4$, we have adiabatic switching on and off of the control field. For higher values of β, we have nonadiabatic switching of the control field. For the probe field, we can take a Gaussian of the form

$$g_p(0, \tau) = g_p \exp\left(- (\tau - \tau_p)^2/\sigma_p^2 \right). \tag{18.40}$$

The set of equations (18.35) and (17.10) are numerically integrated using the profiles (18.39) and (18.40) and a set of numerical results, following [22], are shown in Figures 18.2 and 18.3. In Figure 18.1, we show the super-Gaussian (18.39) and how it gets switched off and then switched on. The scales are defined in terms of the parameter $\gamma = \gamma_{ba} = \gamma_{ca}$ (Figure 17.2). In Figure 18.2, we show how a weak probe pulse is stored in the medium when the control field is switched off and how it is retrieved when the control field is switched on. We also show in Figure 18.2(b) what happens to the storage and retrieval of the probe when the probe pulse is intense. We notice from Figure 18.2 that when the probe is weak, there is a complete retrieval of the original pulse, which is in accordance with the idea of a dark state polariton [20]. However, when the probe is strong, we have absorption and distortion by the medium. For the parameters used in Figure 18.2(a), the probe is within the

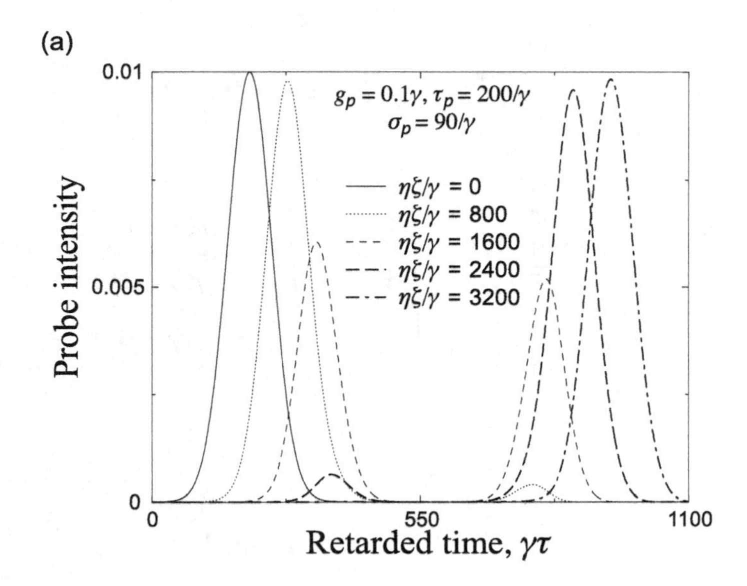

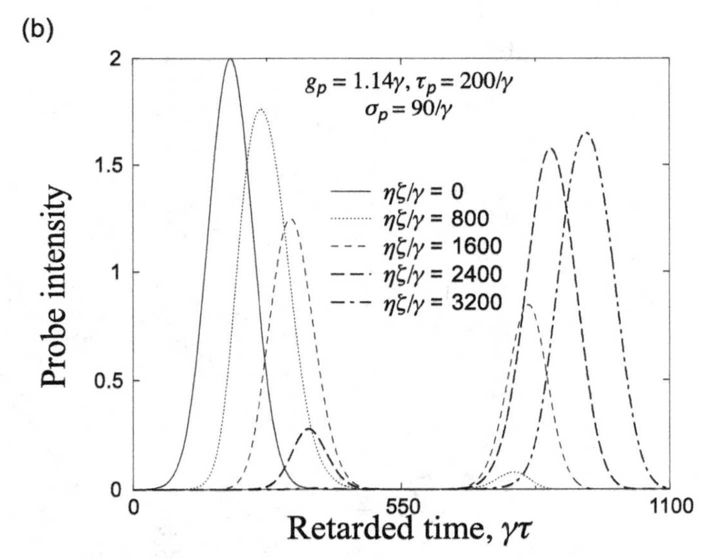

Fig. 18.2 Parts (a) and (b) show the time evolution of the weak and intense Gaussian probe pulses at different propagation distances $\eta\zeta/\gamma$, respectively. The peak position, width, and strength of the pulse are indicated in each panel and the y axes show $(g_p/\gamma)^2$. (Courtesy T. N. Dey.)

transparency window. For intense probes, however, the width of the transparency window goes down and the intense probe field suffers absorption and distortion. In order to study the generic nature of the storage and retrieval of probe pulses, one can examine other pulse shapes. In Figure 18.3, we show the results for sech pulses of the form

$$g_p(0, \tau) = g_p \left[\text{sech}\left(\frac{\tau - \tau_p}{\sigma_p}\right) + f\,\text{sech}\left(\frac{\tau - \tau_i}{\sigma_p}\right) \right]. \tag{18.41}$$

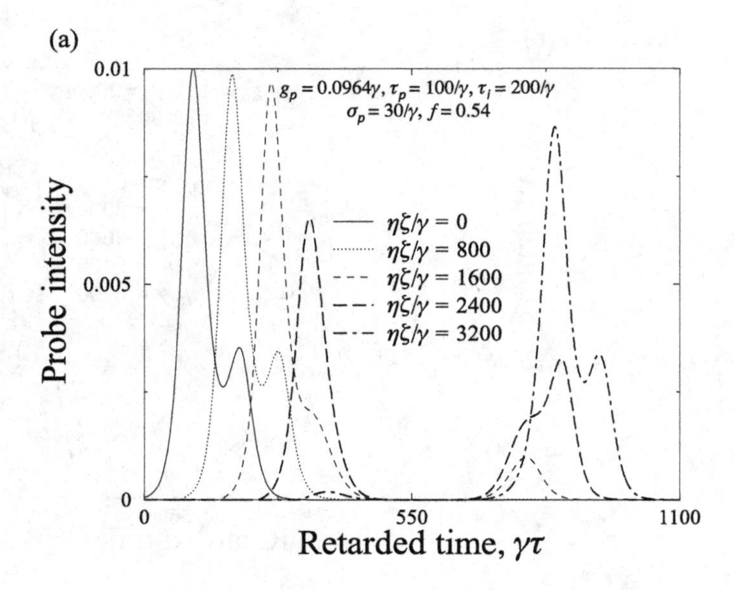

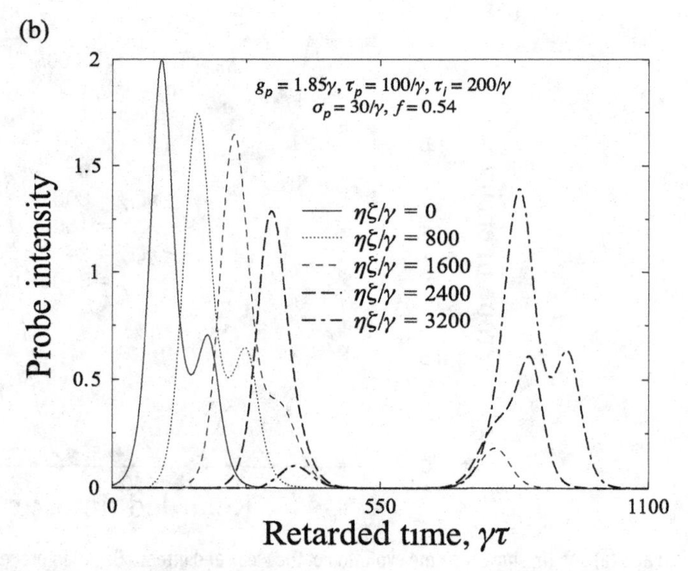

Fig. 18.3 Time evolution of the composite secant hyperbolic probe pulses at different propagation distances in (a) the weak and (b) the strong intensity limit, respectively. (Courtesy T. N. Dey.)

For more complex pulse shapes such as (18.41), we recover the pulse shapes if the pulse is weak. Even for intense probe pulses, however, we retrieve the information that the input pulse is a double humped pulse (Figure 18.3(b)). In the next section, we will use the theory of Grobe *et al.* [21] to understand these numerical results. Storage and retrieval of light pulses was observed in atomic vapors by Phillips *et al.* [23] and by Liu *et al.* [24]. Phillips *et al.* used atomic coherence stored in Zeeman levels to trap and retrieve light pulses in

Rb vapor. Storage times as long as 200 μs were reported by them. Liu *et al.* used a cold cloud of sodium atoms and storage times up to 1 ms were observed. Storage and retrieval experiments have been successfully repeated with the classical probe field replaced by the squeezed vacuum [25,26] and with single-photon pulses [27]. The experiments using input squeezed pulses could retrieve the squeezed vacuum after storing it for a time of about 1–3 μs, although the squeezing in the retrieved pulse was considerably degraded. Thus EIT and dark states could be very successfully utilized for applications in quantum memory elements.

18.4 Adiabatons and storage and retrieval of pulses

Grobe *et al.* [21] studied in general the possibility of shape-preserving solutions in a medium made of three-level atoms with two optically allowed transitions. Under the condition of adiabaticity, they found pulse pair solutions which propagated without a change of shape. In this section, we discuss the deep connection between the adiabatons and the problem of storage and retrieval of pulses. We will assume that the control field is switched on before the probe field, i.e. the pulses are applied in a counterintuitive sequence. This is essential to keep the system in the dark state (cf. Eq. (17.38)), which is required for the formation of the adiabatic pulse pair. Grobe *et al.* [21] introduced the two-pulse adiabaticity condition

$$g_l \frac{\partial g_p}{\partial \tau} - g_p \frac{\partial g_l}{\partial \tau} \ll G^3, \qquad G^2 = g_l^2 + g_p^2, \tag{18.42}$$

assuming the reality of g_l and g_p. Now in (17.10), we drop all the Γ's and set the Δ's to zero, then the density matrix elements can be written as

$$\rho_{\alpha\beta}(\tau) = \Psi_\alpha(\tau)\Psi_\beta^*(\tau),$$

$$\dot{\Psi}_a = \mathrm{i}g_l\Psi_c + \mathrm{i}g_p\Psi_b, \tag{18.43}$$

$$\dot{\Psi}_c = \mathrm{i}g_l\Psi_a, \qquad \dot{\Psi}_b = \mathrm{i}g_p\Psi_a. \tag{18.44}$$

Under the conditions of the counterintuitive sequence of g_p and g_l and (18.42), the solution of (18.43) is approximated by

$$\Psi_b \approx \frac{g_l}{G}, \qquad \Psi_c \approx -\frac{g_p}{G}, \tag{18.45}$$

which is like the dark state (17.38). Furthermore, the population in the level $|a\rangle$ is given by (18.44), i.e.

$$\Psi_a \cong \frac{\mathrm{i}}{g_l}\frac{\mathrm{d}}{\mathrm{d}\tau}\left(\frac{g_p}{G}\right) = -\frac{\mathrm{i}}{g_p}\frac{\mathrm{d}}{\mathrm{d}\tau}\left(\frac{g_l}{G}\right). \tag{18.46}$$

We can now use (18.45) and (18.46) in the Maxwell equations (18.35) for the field envelopes to obtain

$$\begin{aligned}\frac{\partial g_p}{\partial \zeta} &= -\frac{\eta}{G}\frac{\partial}{\partial \tau}\left(\frac{g_p}{G}\right),\\ \frac{\partial g_l}{\partial \zeta} &= -\frac{\eta}{G}\frac{\partial}{\partial \tau}\left(\frac{g_l}{G}\right).\end{aligned} \tag{18.47}$$

These two are coupled via the variable G. It can be seen that G does not depend on the space coordinate ζ

$$G(\zeta, \tau) = G(0, \tau), \tag{18.48}$$

which meas that any change in the probe pulse is compensated by the corresponding change in the control pulse. Thus G in (18.46) is a time-dependent function that is solely determined by the input fields. Analytical solutions can be obtained by changing the variable τ to

$$\xi = \frac{1}{\gamma^2}\int_{-\infty}^{\gamma\tau} G^2(0, \gamma\tau)\mathrm{d}(\gamma\tau) \tag{18.49}$$

with the results

$$\begin{aligned} g_p\left(\frac{\eta\zeta}{\gamma}, \gamma\tau\right) &= G(0, \gamma\tau)F_p\left[\xi - \frac{\eta\zeta}{\gamma}\right], \\ g_l\left(\frac{\eta\zeta}{\gamma}, \gamma\tau\right) &= G(0, \gamma\tau)F_l\left[\xi - \frac{\eta\zeta}{\gamma}\right], \end{aligned} \tag{18.50}$$

where $F_p[x] = g_p(0, \xi^{-1}(x))/G(0, \xi^{-1}(x))$ and likewise for $F_l[x]$. Here ξ^{-1} denotes the inverse function of ξ in the sense $\xi^{-1}[\xi(x)] = x$, $\xi^{-1}(x) \neq [\xi(x)]^{-1}$. We have chosen the initial fields strong enough to ensure the formation of an adiabatic pulse pair. The input fields g_l and g_p are chosen such that G is constant after a certain time T. Therefore, for $\tau \geq T$, the integral ξ can be analytically performed. For a continuous wave control field and a Gaussian probe pulse, we find the explicit results for the probe and control fields

$$\begin{aligned} g_p\left(\frac{\eta\zeta}{\gamma}, \gamma\tau\right) &= \frac{\sqrt{[g_p^2 \mathrm{e}^{-2(\gamma\tau-\gamma\tau_0)^2/(\gamma\sigma)^2} + g_l^2]}}{\sqrt{[g_p^2 \mathrm{e}^{-2(\gamma\tau-\gamma\tau_0-\gamma\eta\zeta/g_l^2)^2/(\gamma\sigma)^2} + g_l^2]}} g_p \mathrm{e}^{-(\gamma\tau-\gamma\tau_0-\gamma\eta\zeta/g_l^2)^2/(\gamma\sigma)^2}, \\ g_l\left(\frac{\eta\zeta}{\gamma}, \gamma\tau\right) &= \frac{\sqrt{[g_p^2 \mathrm{e}^{-2(\gamma\tau-\gamma\tau_0)^2/(\gamma\sigma)^2} + g_l^2]}}{\sqrt{[g_p^2 \mathrm{e}^{-2(\gamma\tau-\gamma\tau_0-\gamma\eta\zeta/g_l^2)^2/(\gamma\sigma)^2} + g_l^2]}} g_l, \\ & t \geq T. \end{aligned} \tag{18.51}$$

For the case when the control field is taken as a super-Gaussian pulse and the probe field is taken as a Gaussian pulse, it is not possible to evaluate the function ξ analytically. In Figure 18.4, the solution of Eq. (18.47) for both these cases is superimposed on the numerical results obtained from the density-matrix equations (17.10) and the Maxwell equations (18.35). It is remarkable that the solution of Eq. (18.47) obtained under the adiabatic approximation matches extremely well with the numerical solution of the complete set of coupled density-matrix Maxwell equations. It is evident from the temporal profiles of the control and probe fields at different propagation distances that a dip and a bump develop in the control field's intensity as it propagates through the medium. Figure 18.4 confirms unambiguously that the adiabatic-pulse pair (consisting of the dip in the pump and the broadened probe) travels loss-free over distances that exceed the weak probe absorption length (typical value of $\eta\zeta/\gamma = 2400$) by several orders of magnitude with an unaltered shape. In principle, G could be both space and time dependent. Within the adiabatic approximation, G does not depend on the space coordinate as shown in the inset of Figure 18.4. To keep G constant in the space domain, any change in the temporal shape of the control

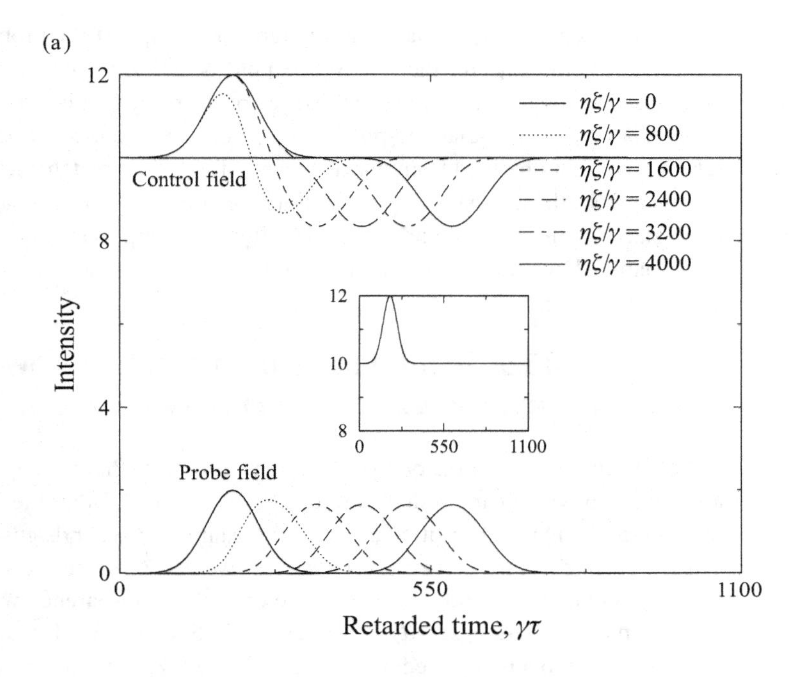

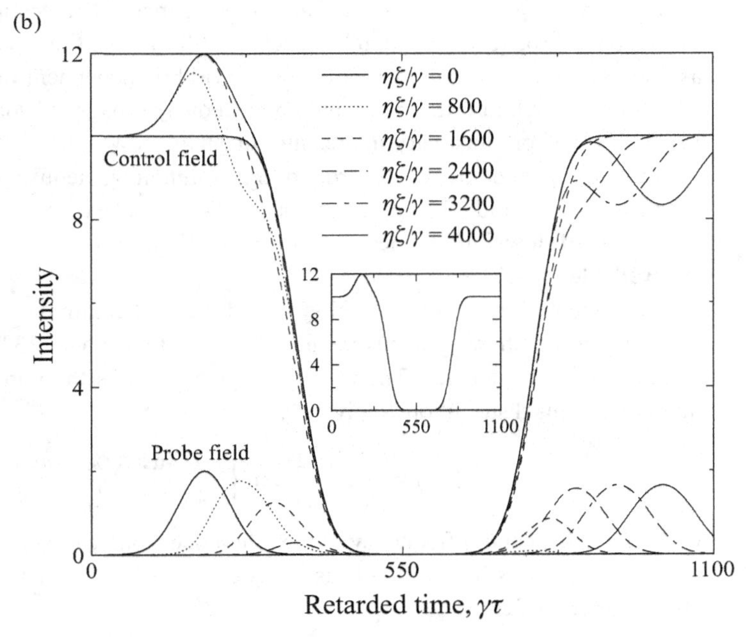

Fig. 18.4 Parts (a) and (b) show temporal profiles of the control $(g_l/\gamma)^2$ and probe field intensities $(g_p/\gamma)^2$ at different propagation distances within the medium. The temporal shape of $(G/\gamma)^2$ as shown in the inset does not depend on ζ. In (a) the input control field is a continuous wave field. In (b) the input control field is super-Gaussian with parameters $\tau_l = 575/\gamma, \sigma_l = 200/\gamma$. The common parameters of the above two graphs are chosen as: $g_l = 3.16\gamma, g_p = 1.414\gamma, \tau_p = 200/\gamma, \sigma_p = 90/\gamma$. The results of simulations using Maxwell–Bloch equations are indistinguishable from the results based on adiabaton theory. (Courtesy T. N. Dey.)

field is compensated by a change in the temporal shape of the probe field. When G^2 and the control field are zero, then the probe field is also zero, implying that the probe field gets stored inside the medium. The retrieved probe pulse that is a replica of the input probe pulse is a part of the adiabatic-pulse pair. Since the numerical results on the storage and retrieval of light obtained from density-matrix formalism match extremely well with those obtained from the adiabatic approximation, we conclude that adiabatons are important for understanding the storage and retrieval of light. An important aspect of the adiabatons is the dynamical evolution of the control field.

18.5 Non-EIT mechanisms for ultraslow light

From our discussions in the previous sections, it is clear that the key to slow light is to find a mechanism which can produce a transparency window. Other mechanisms are known to produce a hole in the absorption profile – for example, see our discussion in Section 13.2.1. Thus even in the case of two-level systems, we can produce a transparency window by using coherent pump fields. The existence of such a transparency window depends on the relaxation parameters T_1 and T_2. As discussed in Section 13.2.1, we need a medium where $T_2 \ll T_1$. Inhomogeneous media are especially useful as the inhomogeneous broadening is very large and typically results in a transparency window. For example, saturated absorption in a Doppler broadened medium has been used for a long time in a variety of applications as one now has a transparency hole whose width is dependent on T_2^{-1} and the power of the saturating field. Furthermore, inhomogeneously broadened solid state systems such as $Er^{3+} : Y_2SiO_5$ produce very interesting transparency windows. In fact, Baldit *et al.* [28] reported group velocities of the order of 3 m/s. Inhomogeneously broadened systems have been extensively studied both theoretically [29–32] and experimentally [28, 33–35] for the production of slow light. Reviews of such non-EIT methods for the production of slow light are available [36, 37].

The susceptibility of a two-level system driven coherently by a strong field is given by (13.29). For an inhomogeneous medium we need to average (13.29) over the distribution of the atomic frequencies. The atomic frequency enters through $\Delta = \omega_0 - \omega_l$. Let the inhomogeneous distribution be given by a Gaussian

$$p(\omega_0) = \frac{2\sqrt{\ln 2}}{\sqrt{\pi}\Gamma_{\text{inh}}} \exp\left\{-\frac{(4\ln 2)(\omega_0 - \bar{\omega}_0)^2}{\Gamma_{\text{inh}}^2}\right\}, \tag{18.52}$$

where $\bar{\omega}_0$ is the central frequency of the inhomogeneously broadened transition and Γ_{inh} is the inhomogeneous line width. Thus Δ dependence of (13.29) needs to be averaged using (18.52) or equivalently

$$p(\Delta) = \frac{2\sqrt{\ln 2}}{\sqrt{\pi}\Gamma_{\text{inh}}} \exp\left\{-\frac{4\ln 2}{\Gamma_{\text{inh}}^2}(\Delta - \bar{\Delta})^2\right\}, \qquad \bar{\Delta} = \bar{\omega}_0 - \omega_l. \tag{18.53}$$

In the following discussion, for simplicity we set $\bar{\Delta} = 0$.

We present the behavior of the real and imaginary parts of the susceptibility as a function of the detuning of the probe field in Figure 18.5. The real part of the susceptibility shows

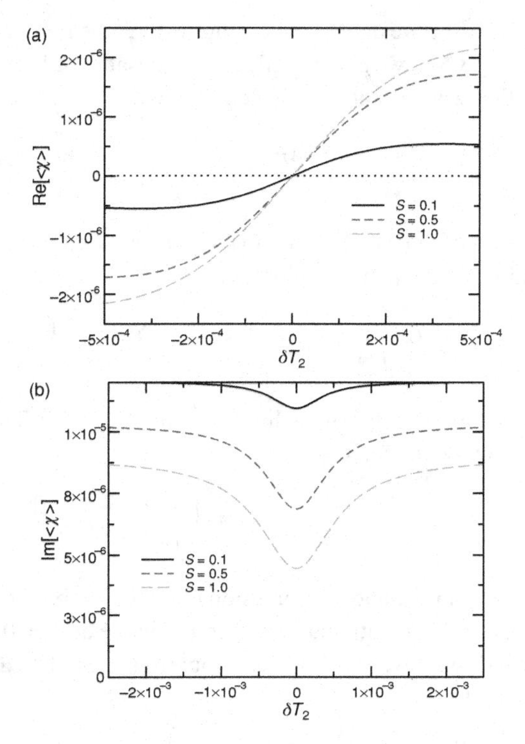

Fig. 18.5 Parts (a) and (b) show, respectively, the real and imaginary parts of the susceptibility $\langle \chi \rangle$ as a function of the probe frequency ω_p ($\delta = \omega_p - \omega_l$) in the presence of a co-propagating pump field g_l. The above plots are for an erbium-doped crystal, with the parameters: inhomogeneous absorption coefficient $\alpha_{\text{inh}} = 6.5$ cm^{-1}, inhomogeneous line width $\Gamma_{\text{inh}} = 1.3$ GHz, longitudinal relaxation time $T_1 = 8$ ms, and transverse relaxation time $T_2 = 3$ μs. (Courtesy T. N. Dey.)

normal dispersion. It is clear from Figure 18.5(a) that the slope of the normal dispersion attains its maximum when the saturation parameter $S = 4|g_l|^2 T_1 T_2 \sim 1$, which leads to ultraslow light. The imaginary part of $\langle \chi \rangle$ exhibits the absorption dip, which becomes deeper with the increase in the intensity of the pump field as shown in Figure 18.5(b). The spectral width of the absorption dip depends on the intensity of the pump field. This dip is associated with coherent population oscillations [38, 39].

Using the behavior of the averaged susceptibility, we can now discuss the possibility of ultraslow light in such a medium. We compute the group index n_g defined by

$$n_g = 1 + 2\pi\omega_p \frac{\partial}{\partial \omega_p} \text{Re}\langle \chi \rangle = 1 + 2\pi\omega_p \frac{\partial}{\partial \delta} \text{Re}\langle \chi \rangle = 1 - \frac{\alpha_{\text{inh}} c T_2}{2\pi} \left\langle \frac{F}{G} \right\rangle, \tag{18.54}$$

$$\begin{aligned} F &= \mathrm{i}(-\Delta + \mathrm{i})\big[S^2 + 2(\delta f + \mathrm{i})^2(\mathrm{i} + \delta + \Delta)^2(1 - \mathrm{i}\Delta) \\ &\quad + S(\mathrm{i} + \delta + \Delta)(-\mathrm{i} + \delta + 2f(\mathrm{i} + \delta - \mathrm{i}\delta^2 + \Delta) - \Delta)\big], \\ G &= 2(1 + S + \Delta^2)[S(\delta + \mathrm{i}) - (\mathrm{i} + \delta f)((\mathrm{i} + \delta)^2 - \Delta^2)]^2, \\ f &= \frac{T_1}{T_2}, \end{aligned} \tag{18.55}$$

where as before we denote the averaging over the inhomogeneous profile for the detuning Δ by $\langle\,\rangle$. In (18.54), we also introduced the unsaturated inhomogeneous absorption coefficient α_{inh} of the two-level atomic system defined by

$$\alpha_{\text{inh}} = \frac{4\pi\omega_p}{c}\langle \text{Im}[\chi]_{g_l=0}\rangle = \frac{8\pi^{3/2}\omega_p n|p_{eg}|^2\sqrt{\ln 2}}{c\hbar\Gamma_{\text{inh}}}. \tag{18.56}$$

In the limit of very large inhomogeneous broadening, we can replace the exponential term in (18.53) by unity. We then obtain

$$n_g = 1 - \frac{\alpha_{\text{inh}}cT_2}{2\pi}\int_{-\infty}^{+\infty}\frac{\text{i}(-\Delta+\text{i})[S^2+S(2f-1)(\Delta+\text{i})^2+2\text{i}(\Delta+\text{i})^3]}{(1+S+\Delta^2)^3}\text{d}\Delta, \tag{18.57}$$

where we have also taken the limit $\delta = \omega_p - \omega_l \to 0$. This integral is done by using contour integration, which leads to

$$n_g \cong c\alpha_{\text{inh}}T_1\left[\frac{S(4+S)}{16(1+S)^{5/2}}\right], \qquad \delta \to 0. \tag{18.58}$$

It is clear from the above expression that the group index varies as $S^{-1/2}$ for large values of S. The group index attains its maximum value at $S = 0.9$. In the case of a homogeneously broadened two-level system, the group index can be calculated using (18.54) and (18.55) with $\Delta = 0$

$$n_g \cong c\alpha_h T_1\left[\frac{S}{2(1+S)^3}\right], \qquad \delta \to 0, \tag{18.59}$$

where $\alpha_h = 4\pi\omega_p n|p_{eg}|^2T_2/c\hbar$ is the homogeneous absorption coefficient. Note that the ratio between the inhomogeneous and homogeneous unsaturated absortion coefficients is $\alpha_{in}/\alpha_h \approx \Gamma_{in}T_2$. For a homogeneously broadened two-level system the group index varies as S^{-2} at large S and peaks at $S = 0.5$. At large S, the group index for a two-level system falls much more slowly for an inhomogeneous medium as compared to the homogeneous case as shown in Figure 18.6. We thus find an important difference between inhomogeneously and homogeneouly broadened two-level systems. For nonzero values of δ, one has to evaluate (18.54) numerically and the results for the group index as a function of the pump intensity are shown in Figure 18.7. The results in Figures 18.6 and 18.7 are in agreement with the observations of Baldit *et al.* [28], who reported group velocities of the order of 3 m/s.

We note in passing that for a homogeneously broadened medium $n_g \approx cT_1\alpha_h/20$ if we take the square bracket in (18.59) as 0.05. Thus for a system like Ruby [40], $n_g \approx 10^7$, $v_g \sim 40$ m/s, for $\alpha_h \sim 1.17\ \text{cm}^{-1}$, $T_1 = 4.45$ ms. This is the value at $\delta = 0$. For nonzero δ, the group index is smaller or the v_g is larger. In fact, in the experiments on Ruby [40], the reported group velocity at $\delta \neq 0$ was 57 m/s. Finally, a good figure of merit for slow light experiments is the product of the group delay and bandwidth [41]. The bandwidth is determined by the transparency window. For an homogeneously broadened medium, we can give estimates as the half width of the transparency window is given by (13.40), i.e. $(1+S)/T_1$ and the delay bandwidth product becomes $\approx \frac{2(1+S)}{T_1}\cdot\frac{L}{v_g} = \frac{2(1+S)L}{T_1c}n_g = \frac{\alpha_h LS}{(1+S)^2}$, which is roughly close to $\alpha_h L/4$ for a saturation parameter equal to unity.

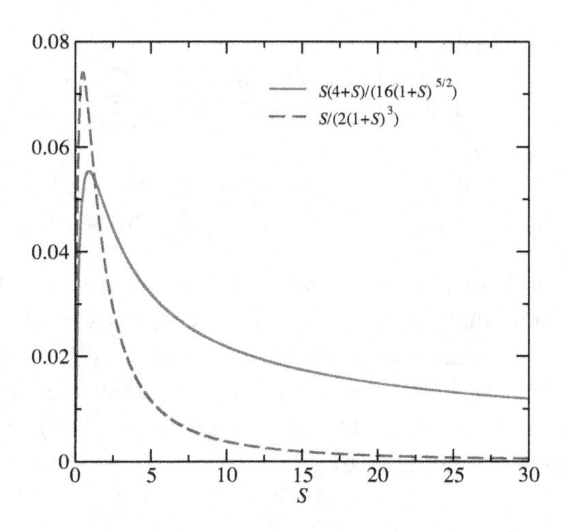

Fig. 18.6 Variation of the term in square brackets of Eqs. (18.58) and (18.59) as a function of the intensity of the pump field for inhomogeneous (solid) and homogeneous (dashed) cases of broadening. (Courtesy T. N. Dey.)

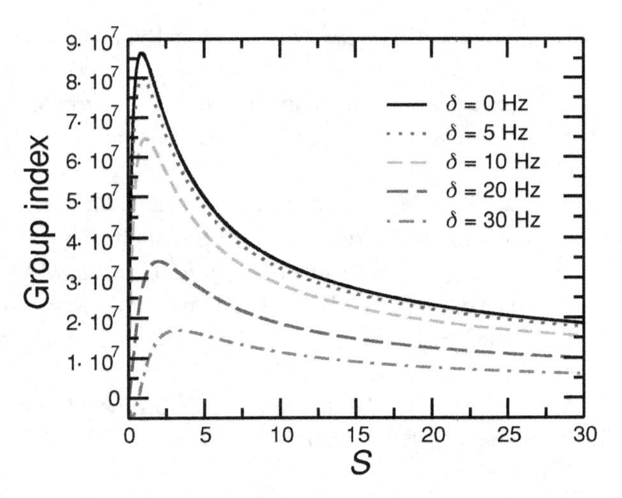

Fig. 18.7 The variation of the group index with the saturation parameter S. The parameters are chosen as $\alpha_{\text{inh}} = 6.5\ \text{cm}^{-1}$, $T_1 = 8$ ms, and $T_2 = 3\ \mu$s. (Courtesy T. N. Dey.)

Exercises

18.1 Use (17.14) to find the derivatives of $\chi^{(ab)}(\omega_p)$ at $\Delta_p = 0$, under the condition $\Delta_l = 0$. Show that $\left.\frac{\partial^2\chi^{(ab)}}{\partial\omega_p^2}\right|_{\Delta_p=0}$ is purely imaginary and $\frac{\partial\chi^{(ab)}}{\partial\omega_p}$ is purely real with values

$$\left.\frac{\partial \chi^{(ab)}}{\partial \omega_p}\right|_{\Delta_p=0} = \frac{\chi_0 n(|g_l|^2 - \Gamma_{bc}^2)}{(\Gamma_{ab}\Gamma_{bc} + |g_l|^2)^2} \xrightarrow{\Gamma_{bc}\to 0} \frac{\chi_0 n}{|g_l|^2},$$

$$\left.\frac{\partial^2 \chi^{(ab)}}{\partial \omega_p^2}\right|_{\Delta_p=0} = \frac{2\chi_0 n i[|g_l|^2(\Gamma_{ab} + 2\Gamma_{bc}) - \Gamma_{bc}^3]}{(\Gamma_{ab}\Gamma_{bc} + |g_l|^2)^3} \xrightarrow{\Gamma_{bc}\to 0} \frac{2\chi_0 n i\Gamma_{(ab)}}{|g_l|^4}.$$

For further discussions of the dispersive properties of EIT see [7].

18.2 Show that the formula (18.20) remains valid for the propagation of the Gaussian pulses under the conditions of Exercise 18.1.

18.3 By choosing an appropriate contour, show that (18.57) reduces to (18.58).

18.4 Derive the value of the α_{inh} as given by (18.56). Use (18.52) and the value of Im χ as given by Eq. (13.16).

18.5 Verify the solutions (18.50) and (18.51).

References

[1] R. W. Boyd, *Nonlinear Optics*, 3rd edition (Orlando, FL: Academic Press, 2008), p. 69.

[2] S. P. Tewari and G. S. Agarwal, *Phys. Rev. Lett.* **56**, 1811 (1986).

[3] M. Jain, H. Xia, G. Y. Yin, A. J. Merriam, and S. E. Harris, *Phys. Rev. Lett.* **77**, 4326 (1996).

[4] K. Hakuta, L. Marmet, and B. P. Stoicheff, *Phys. Rev. Lett.* **66**, 596 (1991).

[5] G. Z. Zhang, K. Hakuta, and B. P. Stoicheff, *Phys. Rev. Lett.* **71**, 3099 (1993).

[6] G. P. Agrawal, *Nonlinear Fiber Optics* (Boston, MA: Academic Press, 1989).

[7] S. E. Harris, J. E. Field, and A. Kasapi, *Phys. Rev. A* **46**, R29 (1992).

[8] L. V. Hau, S. E. Harris, Z. Dutton, and C. H. Behroozi, *Nature (London)* **397**, 594 (1999).

[9] M. M. Kash, V. A. Sautenkov, A. S. Zibrov *et al., Phys. Rev. Lett.* **82**, 5229 (1999).

[10] D. Budker, D. F. Kimball, S. M. Rochester, and V. V. Yashchuk, *Phys. Rev. Lett.* **83**, 1767 (1999).

[11] A. V. Turukhin, V. S. Sudarshanam, M. S. Shahriar, J. A. Musser, B. S. Ham, and P. R. Hemmer, *Phys. Rev. Lett.* **88**, 023602 (2001).

[12] R. W. Boyd and D. J. Gauthier, *Science* **20**, 1074 (2009).

[13] U. Leonhardt and P. Piwnicki, *Phys. Rev. A* **62**, 055801 (2000).

[14] G. S. Agarwal and S. Dasgupta, *Phys. Rev. A* **65**, 053811 (2002).

[15] L. J. Wang, A. Kuzmich, and A. Dogariu, *Nature (London)* **406**, 277 (2000).

[16] G. S. Agarwal and T. N. Dey, *Phys. Rev. Lett.* **92**, 203901 (2004).

[17] A. G. Truscott, M. E. J. Friese, N. R. Heckenberg, and H. Rubinsztein-Dunlop, *Phys. Rev. Lett.* **82**, 1438 (1999).

[18] R. Kapoor and G. S. Agarwal, *Phys. Rev. A* **61**, 053818 (2000).

[19] P. K. Vudyasetu, D. J. Starling, and J. C. Howell, *Phys. Rev. Lett.* **102**, 123602 (2009).

[20] M. Fleischhauer and M. D. Lukin, *Phys. Rev. Lett.* **84**, 5094 (2000).
[21] R. Grobe, F. T. Hioe, and J. H. Eberly, *Phys. Rev. Lett.* **73**, 3183 (1994).
[22] T. N. Dey and G. S. Agarwal, *Phys. Rev. A* **67**, 033813 (2003).
[23] D. F. Phillips, A. Fleischhauer, A. Mair, R. L. Walsworth, and M. D. Lukin, *Phys. Rev. Lett.* **86**, 783 (2001).
[24] C. Liu, Z. Dutton, C. H. Behroozi, and L. V. Hau, *Nature (London)* **409**, 490 (2007).
[25] K. Honda, D. Akamatsu, M. Arikawa *et al., Phys. Rev. Lett.* **100**, 093601 (2008).
[26] J. Appel, E. Figueroa, D. Korystov, M. Lobino, and A. I. Lvovsky, *Phys. Rev. Lett.* **100**, 093602 (2008).
[27] M. D. Eisaman, A. André, F. Massou, M. Fleischhauer, A. S. Zibrov, and M. D. Lukin, *Nature (London)* **438**, 837 (2005).
[28] E. Baldit, K. Bencheikh, P. Monnier, J. A. Levenson, and V. Rouget, *Phys. Rev. Lett.* **95**, 143601 (2005).
[29] G. S. Agarwal and T. N. Dey, *Phys. Rev. A* **68**, 063816 (2003).
[30] G. S. Agarwal and T. N. Dey, *Phys. Rev. A* **73**, 043809 (2006).
[31] R. N. Shakhmuratov, A. Rebane, P. Megret, and J. Odeurs, *Phys. Rev. A* **71**, 053811 (2005).
[32] A. Rebanea, R.N. Shakhmuratovb, P. Mégret, and J. Odeurs, *J. Luminescence* **127**, 22 (2007).
[33] R. M. Camacho, M. V. Pack, and J. C. Howell, *Phys. Rev. A* **74**, 033801 (2006).
[34] M. S. Bigelow, N. N. Lepeshkin, and R. W. Boyd, *Phys. Rev. Lett.* **90**, 113903 (2003).
[35] P. Wu and D. V. G. L. N. Rao, *Phys. Rev. Lett.* **95**, 253601 (2005).
[36] G. S. Agarwal and T. N. Dey, *Laser Photonics Rev*. **3**, 287 (2009).
[37] E. Baldit, S. Briaudeau, P. Monnier, K. Bencheikh, and A. Levenson, *C. R. Phys.* **10**, 927 (2009).
[38] R. W. Boyd, M. G. Raymer, P. Narum, and D. J. Harter, *Phys. Rev. A* **24**, 411 (1981).
[39] A. D. Wilson-Gordon, *Phys. Rev. A* **48**, 4639 (1993).
[40] M. S. Bigelow, N. N. Lepeshkin, and R. W. Boyd, *Phys. Rev. Lett.* **90**, 113903 (2003).
[41] J. Tidström, P. Jänes, and L. M. Andersson, *Phys. Rev. A* **75**, 053803 (2007).

19 Single photons and nonclassical light in integrated structures

In this chapter, we will study the behavior of single photons and entangled light in integrated structures – these could be coupled resonators or waveguide structures. These structures, which could be on a chip, enable one to study the transport of nonclassical light and quantum interferences in such transport. Typically such structures can be taken as coupled by the nearest neighbor Hamiltonian, which is similar to the tight binding Hamiltonian, well known in condensed matter physics. Thus integrated structures enable one to realize optically a number of condensed matter effects such as Anderson localization, Mott transition, etc. In addition, we can study a number of newer possibilities by considering the entangled photons in such structures. The integrated structures also allow the possibility of carrying out quantum logic operations [1].

19.1 Quantum optics in a coupled array of waveguides

We first investigate the transport of nonclassical light across an array of waveguides as shown in Figure 19.1. For simplicity, we will concentrate on the case of single-mode waveguides where the nearest neighbor waveguides are coupled by evanescent fields. Each waveguide could have its own refractive index n_j. Let us label the waveguide by the index j. We quantize the field in each waveguide. Let a_j and $a_j^\dagger$ be the annihilation and creation operators for the jth waveguide, these obey the commutation relations

$$[a_j, a_l^\dagger] = \delta_{jl}, \quad [a_j, a_l] = 0. \tag{19.1}$$

The Hamiltonian for the whole system can be written as

$$H = \hbar \sum_{j=1}^{N} \delta_j a_j^\dagger a_j + \hbar J \sum_{j=1}^{N-1} (a_j^\dagger a_{j+1} + a_j a_{j+1}^\dagger). \tag{19.2}$$

The first term describes the free propagation ($\delta_j \propto n_j$). We have assumed that the coupling constant J is independent of the index j, though one can fabricate devices such that J would be dependent on j. The numerical value of J is determined by the relative distance between the waveguides and J falls as the distance increases. Such an integrated structure has the flexibility that both δ_j and J are controllable. The Heisenberg equations of motion are

$$\dot{a}_j = -\mathrm{i}\delta_j a_j - \mathrm{i}J(a_{j+1} + a_{j-1}), \tag{19.3}$$

where t is related to the propagation distance along the direction of the waveguide. The Hamiltonian (19.2) is analogous to the tight binding Hamiltonian of condensed matter

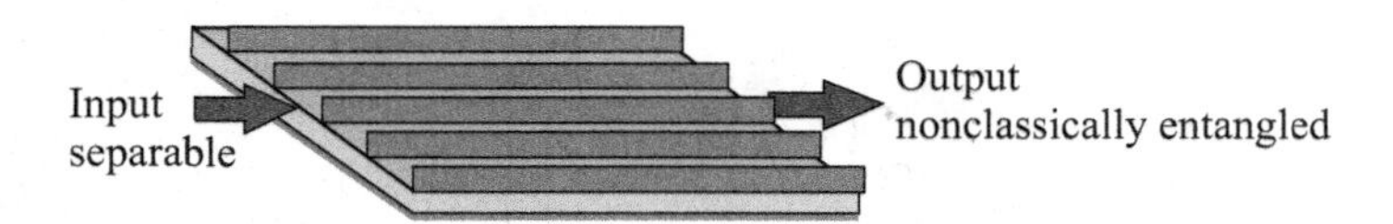

Fig. 19.1 Coupled waveguide array.

physics. The set of equations (19.3) is linear and hence one can write the formal solution in terms of the Green's function defined by

$$\dot{G}_{jl} = -\mathrm{i}\delta_j G_{jl} - \mathrm{i}J(G_{j+1,l} + G_{j-1,l}),$$

$$G_{jl}(t=0) = \delta_{jl}, \tag{19.4}$$

$$a_j(t) = \sum_l G_{jl}(t) a_l(0). \tag{19.5}$$

The δ_{jl} is the Kronecker delta. Thus the output from the waveguide can be calculated depending on the input to the waveguides. The input could be coherent, nonclassical, or entangled light. The average number of photons at the output of each waveguide is given by

$$N_j = \langle a_j^\dagger(t) a_j(t) \rangle = \sum_{ll'} G_{jl}^*(t) G_{jl'}(t) \langle a_l^\dagger a_{l'} \rangle$$

$$= \sum_l |G_{jl}(t)|^2 \langle a_l^\dagger a_l \rangle, \tag{19.6}$$

where the last line is for the uncorrelated input fields, i.e. $\langle a_l^\dagger a_{l'} \rangle \propto \delta_{ll'}$. Let p_{jl} be the joint probability of detecting a photon at the waveguide j and another photon at the waveguide l. This joint probability is proportional to

$$p_{jl} \propto \langle a_j^\dagger(t) a_j(t) a_l^\dagger(t) a_l(t) \rangle, \quad j \neq l. \tag{19.7}$$

For coherent inputs, all the outputs are coherent. This follows from the linearity of (19.5).

In case when all the waveguides have identical refractive index $\delta_j = \delta$, then (19.2) can be diagonalized [2] using

$$a_j(t) = \sum_{p=1}^{N} b_p S(j,p), \tag{19.8}$$

$$b_p(t) = \sum_{j=1}^{N} a_j S(j,p), \tag{19.9}$$

where the function $S(j,p)$ is defined as

$$S(j,p) = \sqrt{\frac{2}{N+1}} \sin\left(\frac{jp\pi}{N+1}\right). \tag{19.10}$$

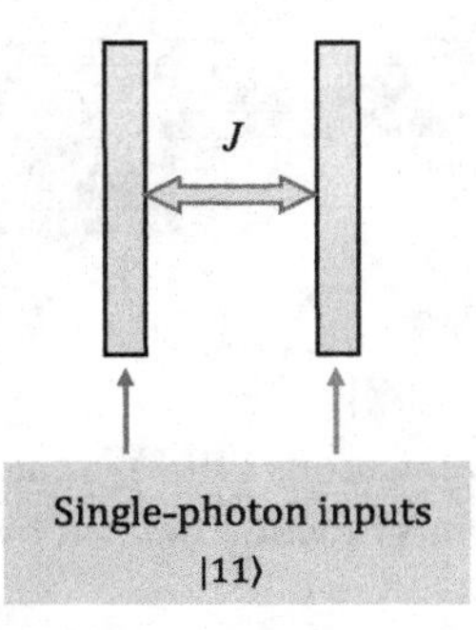

Fig. 19.2 Two evanescently coupled waveguides with single-photon inputs.

This function satisfies the orthonormality relations

$$\sum_{p=1}^{N} S(n,p)S(m,p) = \delta_{nm},$$
$$\sum_{p=1}^{N-1} [S(n,p)S(m,p+1) + S(n,p+1)S(m,p)] = 2\delta_{nm} \cos\left(\frac{n\pi}{N+1}\right). \tag{19.11}$$

Using (19.8) and (19.11), we obtain the diagonalized Hamiltonian

$$H = \hbar \sum_{p=1}^{N} (\delta + \beta_p) b_p^\dagger b_p, \qquad \beta_p \equiv 2J \cos\left(\frac{p\pi}{N+1}\right). \tag{19.12}$$

Note that for $N \to \infty$, β_p becomes the dispersion relation for an electron in periodic lattice. Using (19.12), the Green's function has the explicit form

$$G_{jl}(t) \equiv \sum_{p=1}^{N} \exp[-i(\delta + \beta_p)t] S(l,p) S(j,p). \tag{19.13}$$

The Green's function and (19.5) determine all the quantum properties of light in the coupled waveguide system.

19.2 The Hong–Ou–Mandel interference in a system of two coupled waveguides

We first consider the possibility of the Hong–Ou–Mandel interference [3] in integrated structures [1,2] (Figure 19.2). Let us consider a system of two identical waveguides coupled via evanescent fields. Let us further assume that each waveguide is fed with a single photon. However, the two photons are sent with a relative delay T. The effective Hamiltonian and the evolution operators are given by

$$H = \hbar J(a^\dagger b + a b^\dagger), \qquad U(t) = \exp\left[-iJt(a^\dagger b + a b^\dagger)\right], \tag{19.14}$$

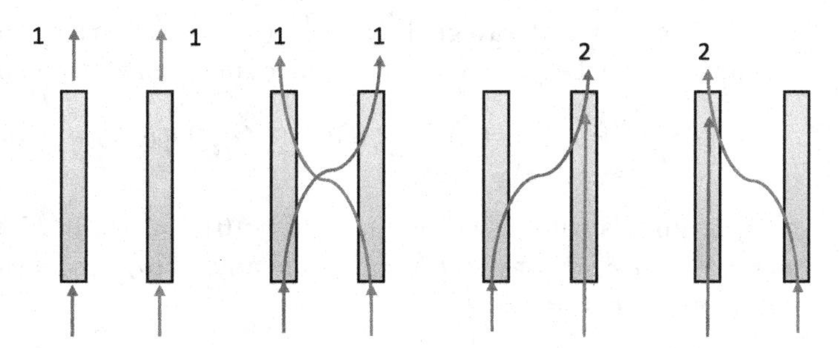

Fig. 19.3 Different paths for the two photons in two coupled waveguides leading to the Hong–Ou–Mandel interference.

where we have now set $a = a_1$, $b = a_2$. We assume that a photon is launched in waveguide 1 at $t = 0$. The waveguide fields evolve over an interval T given by (19.14), when the second photon is launched in waveguide 2. The output state $t > T$ is

$$|\psi(t)\rangle = \frac{U(t-T)b^\dagger U(T)a^\dagger|0,0\rangle}{\sqrt{\langle 1,0|U^\dagger(T)bb^\dagger U(T)|1,0\rangle}}. \tag{19.15}$$

In Eq. (19.15), $b^\dagger$ corresponds to the addition of a photon in the second waveguide at time T. Furthermore, the denominator in Eq. (19.15) arises as we have to ensure the normalization of the wave function $b^\dagger U(T)a^\dagger|0,0\rangle$ at time T. Using $U(t-T) = U(t)U^\dagger(T)$ and the definition of the Heisenberg operators $a(t) = U^\dagger(t)aU(t)$, the numerator in Eq. (19.15) simplifies to

$$\begin{aligned} U(t)b^\dagger(T)a^\dagger|0,0\rangle &= b^\dagger(T-t)a^\dagger(-t)U(t)|0,0\rangle \\ &= b^\dagger(T-t)a^\dagger(-t)|0,0\rangle. \end{aligned}$$

Using the solution of Heisenberg equations in this numerator and using Eq. (19.15) we find that the probability of finding simultaneously one photon at each output at time t is $(\theta = Jt, \theta_0 = JT)$

$$\begin{aligned} p(t,T) &= |\langle 1,1|\psi\rangle|^2 \\ &= \cos^2(2\theta - \theta_0)/(1+\sin^2\theta_0) \\ &= 0, \qquad \text{if } 2\theta - \theta_0 = \pi/2. \end{aligned} \tag{19.16}$$

This shows the two-photon interference dip at $2t - T = \pi/2J$ depends on the length (proportional to t) of the waveguides and the delay time. For a given structure such a dip can be scanned by varying the delay time T. Such an interference dip can be understood as an interference of different pathways that the two photons can take. This is displayed in Figure 19.3. The first two paths lead to destructive interference. In fact for $T = 0$, Eq. (19.15) leads to the generation of an entangled state

$$|\psi(t)\rangle = \mathrm{e}^{-\mathrm{i}\pi/2}(|2,0\rangle + |0,2\rangle)/\sqrt{2}, \qquad \text{if } Jt = \pi/4. \tag{19.17}$$

Politi *et al.* [1] observed the Hong–Ou–Mandel interference in coupled waveguides.

For a larger array of waveguides p_{jl} (Eq. (19.7)) can be calculated in terms of the Green's function (19.13). Let us label the waveguides in which each photon is sent as m and n, then

$$p_{jl} = \sum_{\alpha\beta\gamma\delta} G^*_{j\alpha}(t)G^*_{l\beta}(t)G_{j\gamma}(t)G_{l\delta}(t)\langle a^\dagger_\alpha a^\dagger_\beta a_\gamma a_\delta\rangle, \tag{19.18}$$

the averaging is with respect to the state $|\psi(0)\rangle = |1_m, 1_n\rangle$. Note that $a_\gamma a_\delta|\psi(0)\rangle = 0$ unless $\gamma = m, \delta = n$ or $\gamma = n, \delta = m$. Similarly, $\langle\psi(0)|a^\dagger_\alpha a^\dagger_\beta$ would be nonzero for $\alpha = m$, $\beta = n$ or $\alpha = n$, $\beta = m$, i.e.

$$\langle a^\dagger_\alpha a^\dagger_\beta a_\gamma a_\delta\rangle = (\delta_{\alpha m}\delta_{\beta n} + \delta_{\alpha n}\delta_{\beta m})(\delta_{\gamma m}\delta_{\delta n} + \delta_{\gamma n}\delta_{\delta m}),$$

and hence (19.18) reduces to

$$p_{jl} = |G_{jm}G_{ln} + G_{jn}G_{lm}|^2. \tag{19.19}$$

Note that (19.19) has four distinct contributions arising from the interference of two photon amplitudes corresponding to the photon from the mth guide (nth guide) reaching the jth guide (lth guide) and the process with m and n interchanged, i.e. $|1_m, 1_n\rangle \to |1_j, 1_l\rangle$, $|1_m, 1_n\rangle \to |1_l, 1_j\rangle$. A detailed study of p_{jl} is given by Bromberg *et al.* [4]. The behavior is shown in Figure 19.4 for two cases $m = 0, n = 1$; $m = -1, n = 1$ for an array consisting of 21 waveguides. In the first case photons exhibit bunching whereas in the second case there is no definite pattern.

19.2.1 Waveguide couplers – beam splitters on the chip

A very interesting consequence of (19.14) is the beam splitter-like action of two coupled waveguides. Thus coupled waveguides on a chip are important elements of optical circuits. From (19.14) we obtain solutions to the Heisenberg equations as

$$\begin{aligned} a(t) &= a(0)\cos Jt - ib(0)\sin Jt, \\ b(t) &= b(0)\cos Jt - ia(0)\sin Jt, \end{aligned} \tag{19.20}$$

which are the same transformation equations that we had for beam splitters (Eq (5.9)). The transmission of the beam splitter would be $\cos^2(Jt)$. Thus by changing the coupling constant or the length, we can produce beam splitters with varying transmission coefficients. Note that we need devices with varying transmission in any quantum architecture (cf. Exercises (5.5), (5.9), (5.10)). A typical waveguide coupler is shown in Figure 19.5.

19.3 Single-photon transport and coherent Bloch oscillations in a coupled array

Let us consider sending a single photon through the array. We can send a single photon through one of the waveguides or we can make a wave packet so that the single photon

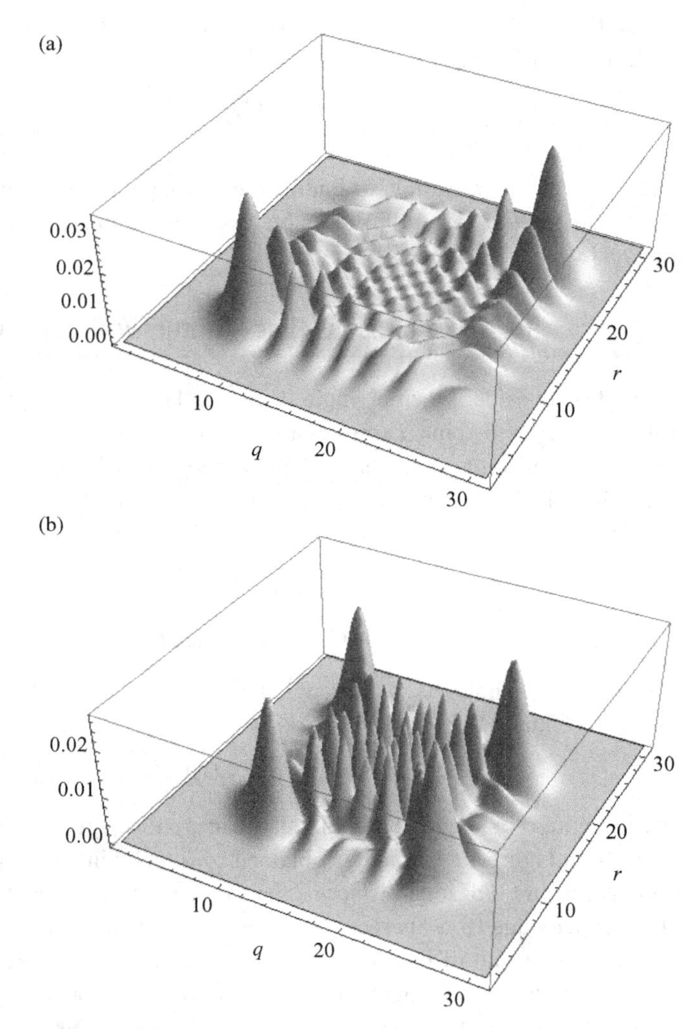

Fig. 19.4 The probability p_{qr} of (Eq. (19.19)) of finding one photon each at the sites q and r. (a) The photons are coupled to two adjacent waveguides, i.e. $|\psi(0)\rangle = a_0^\dagger a_1^\dagger|0\rangle$. The two photons exhibit bunching, and will emerge from the same side of the lattice. (b) The two photons are coupled to two waveguides separated by one waveguide, $|\psi(0)\rangle = a_{-1}^\dagger a_1^\dagger|0\rangle$. Here the two photons will both emerge either from the lobes or from the center. (Redrawn following [4].)

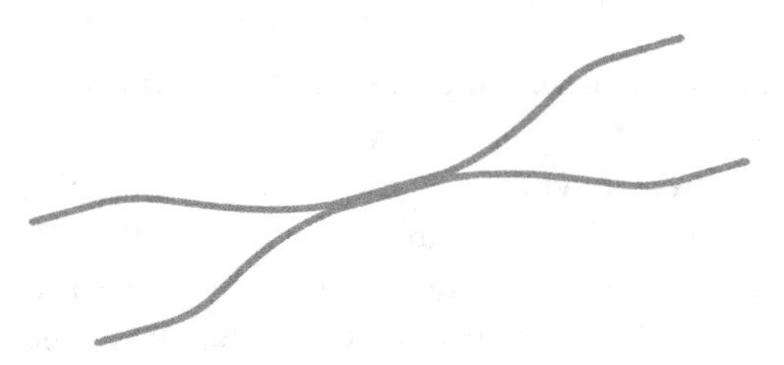

Fig. 19.5 A typical waveguide coupler on a substrate.

can be launched at each waveguide with some amplitude. The coherent wave packet has the form of a W state

$$|\psi\rangle = \sum_j c_j |1_j, \{0\}\rangle, \tag{19.21}$$

where $|c_j|^2$ is the probability of finding the single photon at the input of the jth waveguide. The W state of the form (19.21) can be produced by an arrangement of beam splitters and mirrors [5].

19.3.1 Continuous-time quantum walk of a single photon

Let us first consider when a single photon is launched at a given waveguide say zeroth waveguide $[c_j = \delta_{j0}]$ in an array running from $-N$ to $+N$. We will assume a large enough array so that boundary effects are unimportant. For identical waveguides δ is redundant. The solution to (19.4) can be written down by noting the recursion relations for Bessel functions $J_\nu(z)$

$$\begin{aligned} J_{\nu-1}(z) - J_{\nu+1}(z) = 2\frac{\mathrm{d}J_\nu(z)}{\mathrm{d}z}, \qquad J_\nu(0) = \delta_{\nu,0}, \\ G_{jl}(t) = J_{j-l}(2Jt)(-\mathrm{i})^{j-l}. \end{aligned} \tag{19.22}$$

Thus if a single photon is sent through the zeroth waveguide, then the intensity at output of the jth waveguide is

$$I_j(t) = |J_j(2Jt)|^2. \tag{19.23}$$

The behavior of a single photon in a large array for large and short propagation distances is shown in Figure 19.6, which is obtained by directly integrating (19.4) using the Runge–Kutta procedure. For long distances, the single photon tunnels through many waveguides. This happens due to evanescent coupling between the waveguides. The Bessel function (19.23) is the result of interference between all the quantum paths that the single photon can take. Perets *et al.* [6] comment that this kind of behavior is a hallmark of continuous-time quantum walks and thus photons in waveguides can be used to observe continuous-time quantum walk behavior. We have considered the propagation of a single photon. We note that the same result (19.23) holds for coherent fields except for a scaling factor.

19.3.2 Bloch oscillations with single photons

As already emphasized, the waveguide structures can be fabricated so as to model a very wide class of Hamiltonians. In particular, the well-known phenomenon of Bloch oscillations for an electron moving in a crystal lattice in the presence of electric field was studied experimentally using waveguide arrays [7, 8]. The effect of an electric field can be modeled by considering that the refractive index δ_j of the jth waveguide is proportional to j. Thus the basic equation for the Green's function (19.4) becomes

$$\dot{G}_{jl} = -\mathrm{i}\delta j G_{jl} - \mathrm{i}J(G_{j+1,l} + G_{j-1,l}), \qquad \delta_j = j\delta. \tag{19.24}$$

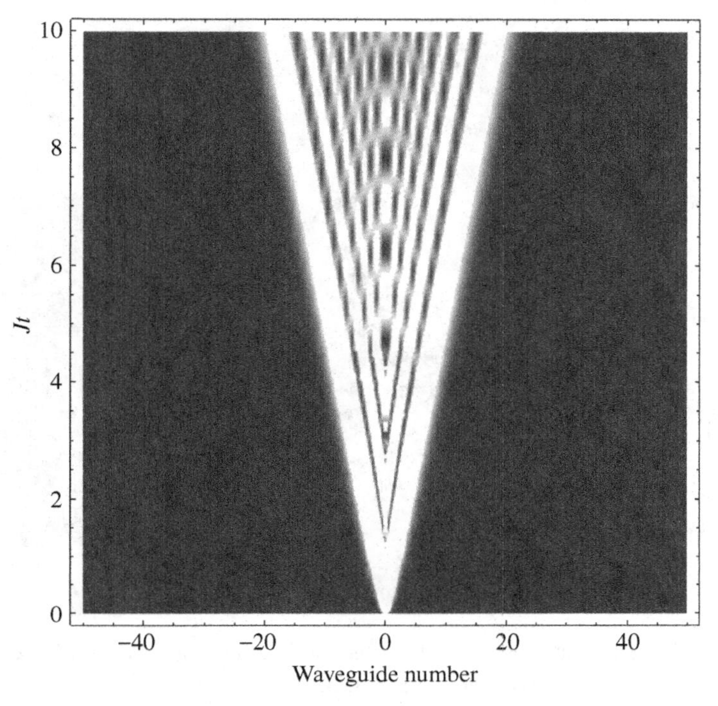

Fig. 19.6 Quantum walk of a single photon in a coupled waveguide array.

This equation is identical to the motion of an electron in a periodic lattice in the presence of an electric field.

We first study the transport of a single photon launched say at the middle (Nth) waveguide in an array consisting of $2N$ waveguides. The solution of (19.24) can be derived by using generating function or by Fourier techniques and is found to be [9]

$$G_{jl}(t) = \exp\left[-ilq\tau - \frac{\mathrm{i}(j-l)(\alpha\tau - \pi)}{2}\right] J_{l-j}\left[\frac{4}{\alpha}\sin\left(\frac{\alpha\tau}{2}\right)\right], \qquad \tau \equiv Jt, \qquad \alpha = \delta/J. \tag{19.25}$$

The output intensity distribution is then

$$I_j = \langle a_j^\dagger(t) a_j(t)\rangle = |G_{j,N}(t)|^2 = \left|J_{j-N}\left[\frac{4}{\alpha}\sin\left(\frac{\alpha\tau}{2}\right)\right]\right|^2. \tag{19.26}$$

This is shown in Figure 19.7. The behavior is determined by the zeroes of the Bessel function. The Bessel function in (19.26) has an argument with periodicity given by $\frac{\alpha\tau}{2} \to \frac{\alpha\tau}{2} + n\pi$. Thus the pattern repeats when $\frac{\alpha\tau}{2} \to \frac{\alpha\tau}{2} + n\pi$.

Next we consider the well-known coherent Bloch oscillation when the input to each waveguide is in a coherent state with amplitude α_j. In order to exhibit Bloch oscillations one needs a fairly wide distribution of fields at different inputs. As in the work of Peschel *et al.* [9] we assume a Gaussian distribution of α_j, i.e. we assume $\alpha_j \sim \exp[-(j-\bar{j})^2/2\sigma^2]$ up to a constant. The resulting Bloch oscillation is shown in Figure 19.8.

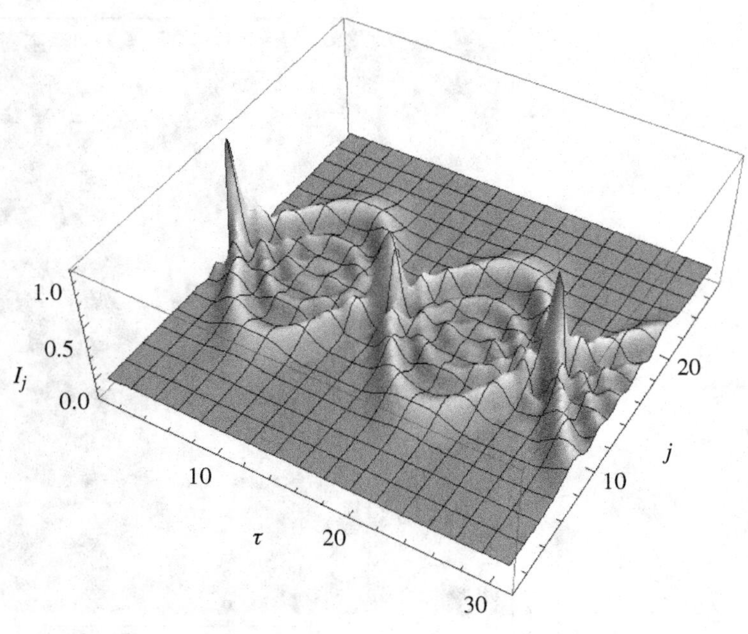

Fig. 19.7 The variation in output intensity distribution I_j for a single waveguide excitation as a function of τ for $j = 1, \ldots, 26$. The parameter α is $\alpha = 0.5$.

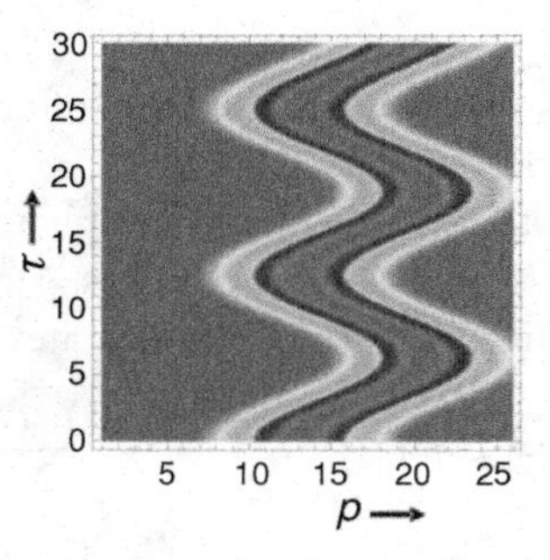

Fig. 19.8 Bloch oscillation for a Gaussian excitation. There are 26 waveguides in the system and the parameters for the Gaussian beam $\alpha_p \sim \exp\{-(p - \bar{p})^2/(2\sigma^2)\}$ are chosen as $\sigma = 3.6$ and $\bar{p} = 13$. The parameter $\alpha = \delta/J$ is 0.5.

We next discuss how it is possible to observe Bloch oscillations with single photons. We would consider a source of heralded single photons [10–12]. In order to do this we should be able to launch a single photon in any waveguide with a finite probability. Besides we need to launch it in a coherent manner. This is possible [5] if we prepare a single photon in the W state (Eq. (19.21)). Then the quantum correlations in the W state enable us to obtain coherent Bloch oscillations with single photons. For this purpose, we assume that the input to the waveguides is from the multiport device [13] of Figure 19.9. The state of the field at the input would be given by Eq. (19.21) with a distribution of c_js as shown in Figure 19.10.

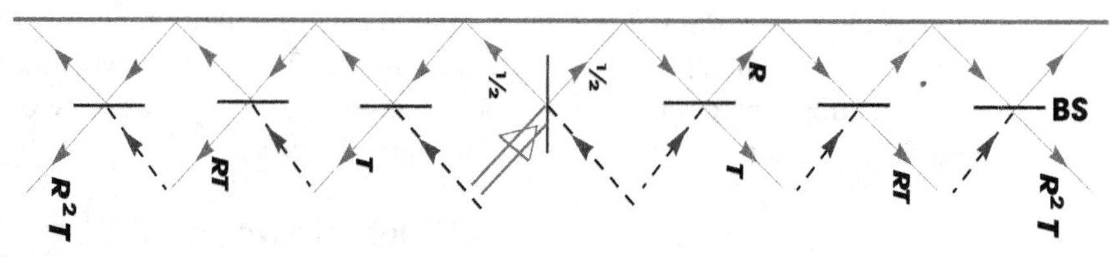

Fig. 19.9 The scheme for generating the required input W state. The long horizontal line is the mirror with 100% reflectivity. The double arrow indicates a heralded single photon from a source such as a parametric downconverter. The black lines show the vacuum fields at open ports. The transmissivity of the beam splitter is T. The output intensities at different ports are given by $TR^2/2$, $TR/2$, $T/2$, $T/2$, $TR/2$, $TR^2/2$.

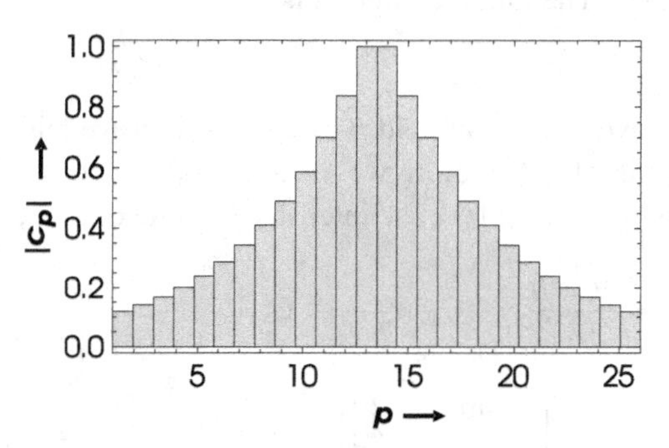

Fig. 19.10 Distribution of $|c_p|$ normalized to its maximum value as a function of p for $R = 1/\sqrt{2}$.

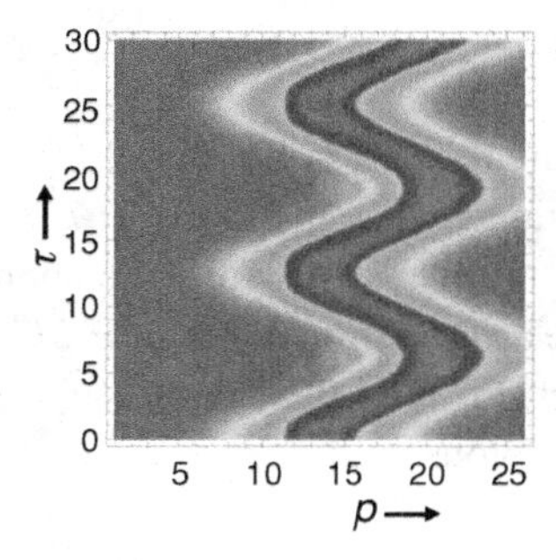

Fig. 19.11 The Bloch oscillation for a W state. The parameter α is chosen as $\alpha = 0.5$ and there are 26 waveguides in the system.

For a single photon in a W state we get

$$I_j = \left| \sum_p G_{j,p}(t) c_p \right|^2 , \tag{19.27}$$

where we use the c_ps from Figure 19.10. This distribution of the intensity is shown in Figure 19.11. In this case we recover the coherent Bloch oscillations even though we use incoherent single photons. This is possible due to the quantum correlations implicit in the W state of single photons.

Thus the observation of the single-photon Bloch oscillation would require (a) a heralded source of single photons of the type used in [1, 10–12]; (b) waveguide structures like, for example, the ones employed in [1, 4, 6]; and (c) a mirror assembly of the type shown in Figure 19.9. All these resources are currently available.

19.3.3 Optical ratchets

Finally, we discuss the possibility of phase-controlled photonic transport, i.e. the possibility of an optical ratchet [14]. Consider the feeding of two neighboring waveguides with coherent light but with different phases. The phase difference can be used for controlling the transport of photons. The initial condition is

$$|\psi\rangle = |\beta_0, \beta_1 \mathrm{e}^{\mathrm{i}\varphi}\rangle, \tag{19.28}$$

i.e. we have a coherent field β in the central waveguide "0" and a coherent field $\beta \mathrm{e}^{\mathrm{i}\varphi}$ in the waveguide "1". As before we assume a large enough waveguide array. Using Eq. (19.25) and the initial state $|\psi\rangle$, the intensity I_j at the output of the jth waveguide is then

$$\begin{aligned}\frac{I_j}{|\beta|^2} &= \frac{\langle a_j^\dagger a_j\rangle}{|\beta|^2} = |G_{j0}(t) + \mathrm{e}^{\mathrm{i}\varphi} G_{j1}|^2 \\ &= \left|J_{-j}\left[\frac{4}{\alpha}\sin\left(\frac{\alpha\tau}{2}\right)\right]\right|^2 + \left|J_{1-j}\left[\frac{4}{\alpha}\sin\left(\frac{\alpha\tau}{2}\right)\right]\right|^2 \\ &\quad - 2J_{-j}\left[\frac{4}{\alpha}\sin\left(\frac{\alpha\tau}{2}\right)\right] J_{1-j}\left[\frac{4}{\alpha}\sin\left(\frac{\alpha\tau}{2}\right)\right]\sin\left(\frac{\alpha\tau}{2} - \varphi\right), \quad -N \le j \le +N,\end{aligned} \tag{19.29}$$

where τ and α are defined in Eq. (19.25). We display the intensity patten in Figure 19.12 for the same parameters as in Figure 19.7. Figure 19.12(a) shows the intensity profile for $\varphi = 37°$. These profiles are asymmetric about the j axis indicating transport to the lower index side of the array. However, for $\varphi = 217°$, Figure 19.12(b) shows that the direction of transport is reversed. Thus the phase-displaced inputs can be used to control the direction of transport of photons.

19.4 The Anderson localization of quantum fields in coupled waveguide arrays

The Anderson localization has been extensively studied since its discovery more than 50 years ago. Extensive reviews and books on the subject are available [15, 16]. The newer studies now concentrate on systems where the parameters of the systems can be well controlled and such is the case with Bose condensates [16] and arrays of waveguides [17, 18]. We have seen beautiful demonstrations of the Anderson localization in such systems. In addition, waveguide structures offer the possibility of using strictly quantized radiation fields and thus newer aspects of the tight binding Hamiltonians can be uncovered

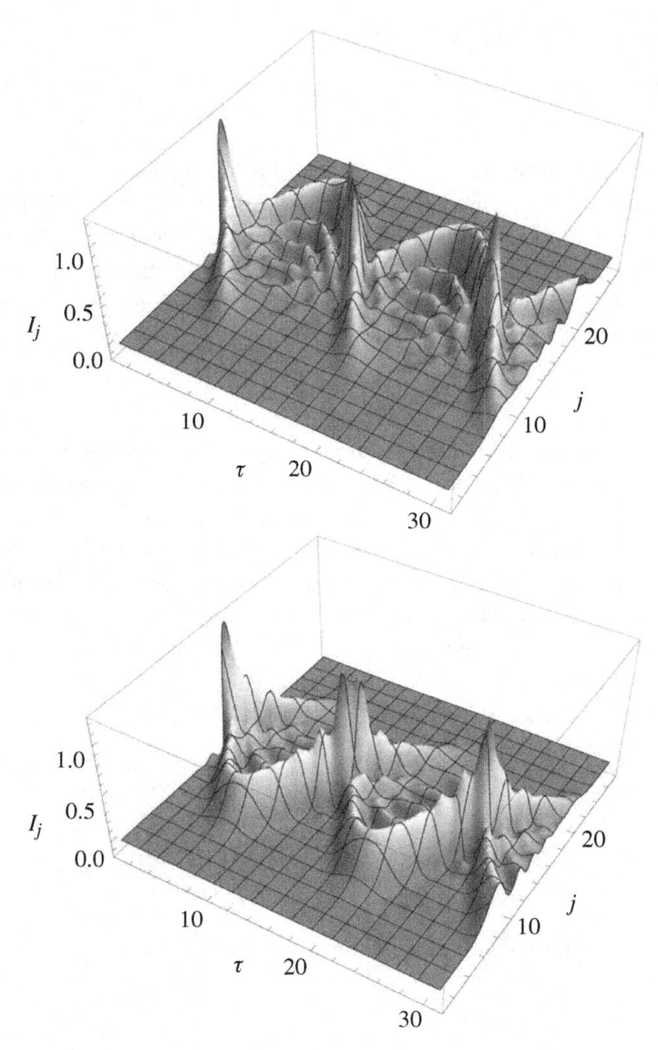

Fig. 19.12 Demonstration of the optical ratchet effect in the output intensity for different values of the relative phase φ in the input state (19.28) for $\alpha = 0.5$, after [14]. The waveguides are numbered as in Figure 19.7.

using nonclassical light. These, for instance, include issues such as quantum entanglement and photon–photon correlations in the fields at the output end of the waveguides. Thus from the perspective of quantum optics it is interesting to study the photon–photon correlations [17] in the Anderson localization of single photons as well as the Anderson localization with nonclassical light such as squeezed light or light with sub-Poissonian statistics [19]. The effects of the quantum statistics of the input light are seen in the intensity fluctuations and the intensity distribution of the output light.

We model the disorder in waveguides via the disorder in the refractive index of each waveguide. Thus the parameters δ_j in Eq. (19.2) for different j's are independent Gaussian variables with a distribution

$$P(\delta) = \frac{1}{\sqrt{2\pi\Delta^2}} \exp\left(-\frac{\delta^2}{2\Delta^2}\right), \tag{19.30}$$

where Δ^2 is the variance of the distribution and is a measure of the disorder in the medium. The final results are not very sensitive to the exact form of the distribution for the disorder. As a result of disorder, the Green's function becomes random and the physical results are obtained by the integration of the Green's function and its products over the distribution $P(\delta)$. Assume the input at the waveguide labeled as zero, Eq. (19.6) yields for the mean intensity at the jth waveguide

$$I_j = \langle |G_{j0}(t)|^2 \rangle \langle a_0^\dagger a_0 \rangle. \tag{19.31}$$

In the calculation of the mean intensity I_j, the quantum statistics of the field does not enter. We show in Figure 19.13 the results for I_j for increasing values of the disorder. As the disorder increases we get Anderson localization of the intensity in the waveguide in which the field was put in. Note that, after the localization has taken place, more than half of the input photons are found at the output of the waveguide through which the input field was sent. This characteristic property is essentially the reason for the enhanced radiation–matter interaction using localized modes (see, e.g., [20]).

The effect of the input photon statistics would appear at the level of fluctuations in the intensity, i.e. in the quantity

$$\tilde{g}^{(2)} = g^{(2)} - 1 = \frac{\langle a^{\dagger 2} a^2 \rangle}{\langle a^\dagger a \rangle^2} - 1, \tag{19.32}$$

$$= \frac{\langle |G_{j0}|^4 \rangle \langle a_0^{\dagger 2} a_0^2 \rangle}{\langle |G_{j0}|^2 \rangle^2 \langle a_0^\dagger a_0 \rangle^2} - 1, \tag{19.33}$$

and in the site-to-site correlations defined by

$$\langle I_l I_p \rangle = \langle |G_{l0}|^2 |G_{p0}|^2 \rangle \langle a_0^{\dagger 2} a_0^2 \rangle. \tag{19.34}$$

We can now consider different fields to illustrate the effect of quantum statistics. For a field in a coherent state $|\alpha\rangle$, we have $\langle a_0^\dagger a_0 \rangle = |\alpha|^2$, $\langle a_0^{\dagger 2} a_0^2 \rangle = |\alpha|^4$. For a field in a thermal state $\langle a_0^\dagger a_0 \rangle = \bar{n}$, $\langle a_0^{\dagger 2} a_0^2 \rangle = 2\bar{n}^2$. For a single-mode squeezed field, Eq. (2.41), $\langle a_0^\dagger a_0 \rangle = \sinh^2 r$, $\langle a_0^{\dagger 2} a_0^2 \rangle = \sinh^2 r(1+3\sinh^2 r)$, where r is the squeezing parameter. We exhibit the behavior of the intensity fluctuations in Figures 19.14 and 19.15. We first note that the fluctuations at the output are enhanced over the input via the disorder in waveguides. Figure 19.15 also shows how the fluctuations increase with an increase in the squeezing parameter. This happens as the input field has increasing super-Poissonian fluctuations. The site-to-sites correlations after localization has occurred are shown in Figure 19.16. The site-to-sites correlations can exhibit the spatial correlations within the localization peak if the two photons are launched in nonadjacent waveguides [18].

Let us now consider the case when light with sub-Poissonian statistics is launched into the disordered medium. Specifically, we take a two-photon Fock state $|2\rangle$ for which $\langle a_0^{\dagger 2} a_0^2 \rangle = 2$

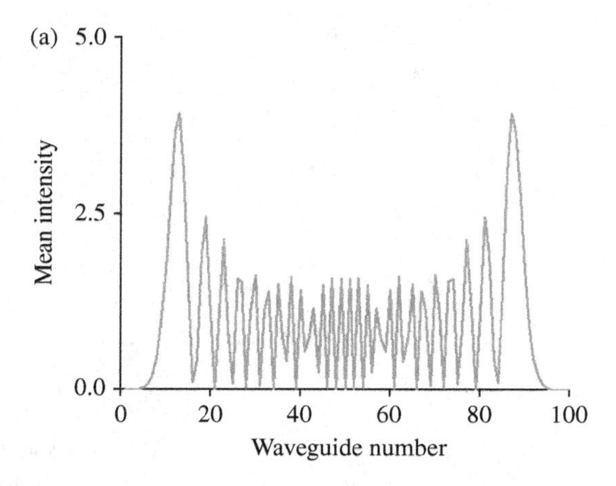

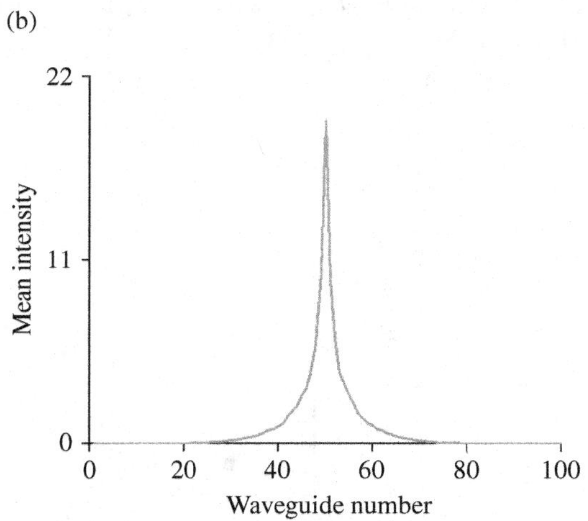

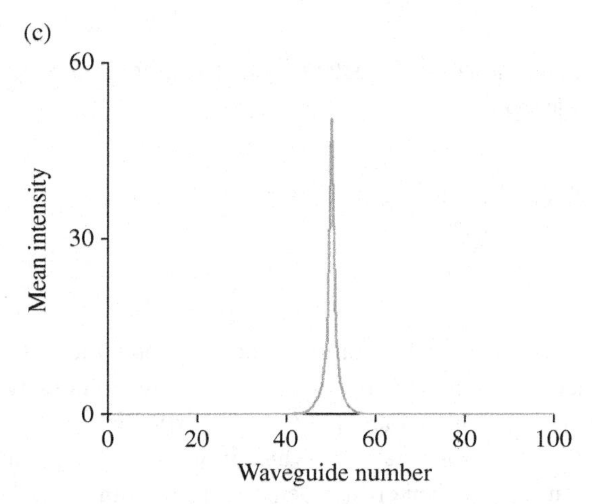

Fig. 19.13 Appearance of the Anderson localization in mean intensity as the disorder increases: (a) $\Delta/J = 0$, (b) $\Delta/J = 1$, and (c) $\Delta/J = 3$. Mean photon number for the input field in the waveguide numbered 50 is 100 [19].

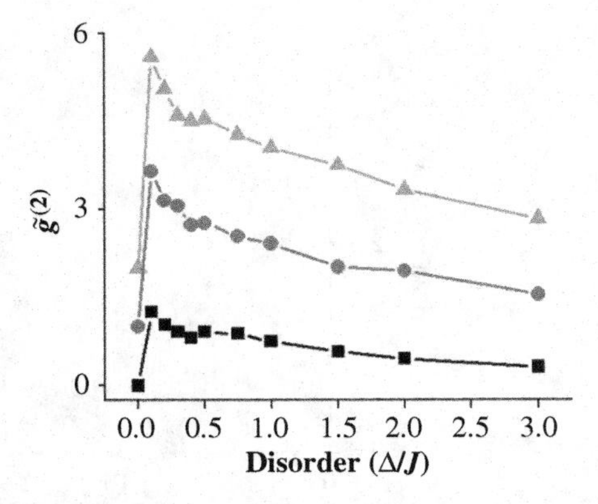

Fig. 19.14 Normalized variance versus disorder at the 50th waveguide for a Gaussian disorder. Mean photon number for all three input fields is 100. Curves shown are for coherent fields (squares), thermal fields (circles), and squeezed fields (triangles) [19].

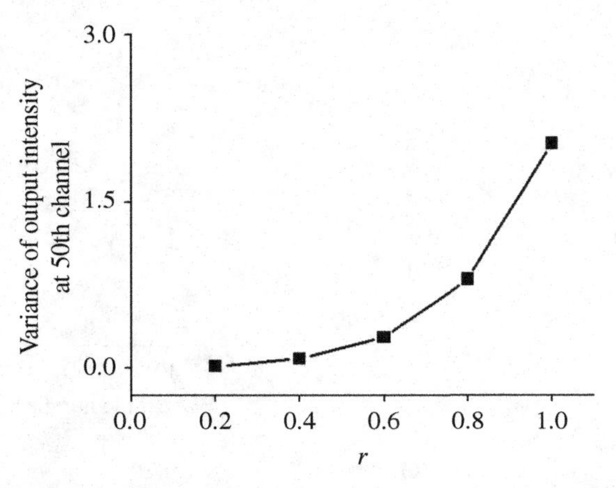

Fig. 19.15 Variance in output intensity at the 50th waveguide versus the squeezing parameter for a Gaussian disorder in the refractive index [19].

and $\langle a_0^\dagger a_0 \rangle = 2$. Since $g^{(2)}$ is given by

$$g^{(2)} = \frac{\langle |G_{j0}|^4 \rangle \langle a_0^{\dagger 2} a_0^2 \rangle}{\langle |G_{j0}|^2 \rangle^2 \langle a_0^\dagger a_0 \rangle^2}, \tag{19.35}$$

one can see that the first ratio (containing the Green's functions) provides information on the fluctuations induced by the disorder in the medium, whereas the second ratio is specific to the statistics of the input photons. In particular, this ratio is equal to 1 for a coherent state and $1/2$ for the Fock state $|2\rangle$, which immediately suggests that the output field is less noisy in the latter case. This result demonstrates an instance in which there is a suppression of the fluctuations (relative to the case where the input field has Poissonian statistics) due to the disorder of the medium by the nonclassical sub-Poissonian statistics of the input field.

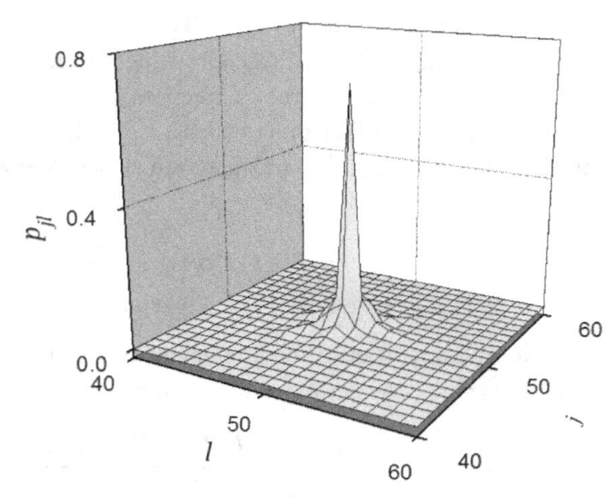

Fig. 19.16 Localization peak in site-to-site photon correlations for $r = 1$ and $\Delta/J = 3$ [19].

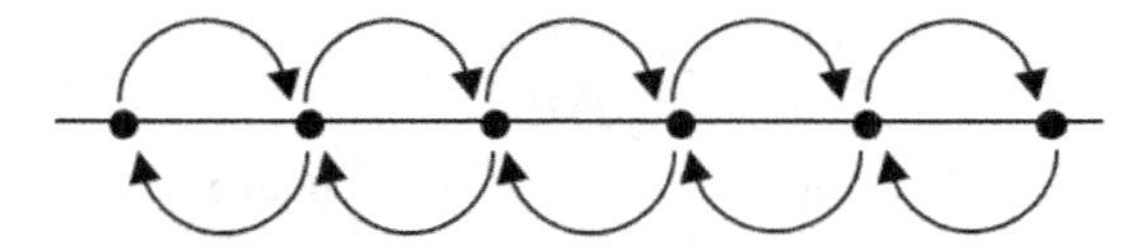

Fig. 19.17 Classical walker on a line.

A suppression of fluctuations has been reported in the multiple scattering of nonclassical light [21].

19.5 Discrete quantum walks via waveguide couplers on a chip

As another novel application of waveguide couplers on a chip, we consider quantum optical realization of discrete quantum walks. This is a new paradigm in quantum optics. We first present a discussion of such quantum walks and contrast these from the well-known classical walks.

19.5.1 The quantum walk on a line

Consider a classical walker moving on a lattice of points on a line. The walker can move with equal probability to the left or right as shown in Figure 19.17. It is well-known [22] that the probability of finding the walker at the site m in N steps is

$$P(N, m) = \frac{N!}{\left(\frac{N+m}{2}\right)!\left(\frac{N-m}{2}\right)!}\left(\frac{1}{2}\right)^N$$
$$\rightarrow \sqrt{\frac{2}{\pi N}}\exp\left(-\frac{m^2}{2N}\right), \quad \text{for large } N. \tag{19.36}$$

The probability is peaked around $m = 0$. Consider now a quantum walker such that it has additional quantum degree of freedom which decides whether the walker moves to the right or left. The additional degree of freedom could be spin. We can say that if the spin is up (down), then the walker moves to the right (left), i.e. we now represent by unitary transformation U the changes in the quantum state of the walker

$$\begin{aligned} U|\uparrow, l\rangle &\to |\uparrow, l+1\rangle, \\ U|\downarrow, l\rangle &\to |\downarrow, l-1\rangle, \\ U(\alpha|\uparrow\rangle + \beta|\downarrow\rangle)|l\rangle &\to \alpha|\uparrow, l+1\rangle + \beta|\downarrow, l-1\rangle. \end{aligned} \tag{19.37}$$

Such a transformation produces entanglement between the position and the spin degrees of freedom. Let us now consider a quantum walk represented by the following operations

$$|\psi_f\rangle = \underbrace{(UR)(UR)\cdots(UR)}_{N \text{ times}}|\psi_i\rangle, \tag{19.38}$$

where R is a rotation in the spin space

$$\begin{pmatrix}\alpha\\ \beta\end{pmatrix} \overset{\text{R}}{\to} \frac{1}{\sqrt{2}}\begin{pmatrix}1 & 1\\ -1 & 1\end{pmatrix}\begin{pmatrix}\alpha\\ \beta\end{pmatrix}. \tag{19.39}$$

The product UR can be written as a 2×2 matrix

$$UR = \frac{1}{\sqrt{2}}\begin{pmatrix}\mathrm{e}^{\mathrm{i}p} & \mathrm{e}^{\mathrm{i}p}\\ -\mathrm{e}^{-\mathrm{i}p} & \mathrm{e}^{-\mathrm{i}p}\end{pmatrix} \tag{19.40}$$

where $\mathrm{e}^{\mathrm{i}p}$ is defined by

$$\mathrm{e}^{\mathrm{i}p}|l\rangle \to |l \pm 1\rangle. \tag{19.41}$$

Thus we can write the Nth power of the matrix UR as

$$(UR)^N = \begin{pmatrix}A & B\\ C & D\end{pmatrix}, \quad A|0\rangle = \sum A_l|l\rangle, \text{ etc.}, \tag{19.42}$$

where A, B, C, and D can be obtained using (19.40). For example, the initial state $|\psi_i\rangle$ could be $|\downarrow\rangle|0\rangle$, i.e. in terms of a column matrix $|\psi_i\rangle = \binom{0}{1}|0\rangle$. In quantum computation, an operation represented by $\frac{1}{\sqrt{2}}\begin{pmatrix}1 & 1\\ 1 & -1\end{pmatrix}$ very similar to R is called the Hadamard transformation. In what follows, we use R as R is easily realizable in optics (see also Exercise 19.5). The final position of the quantum walker is obtained from

$$\rho_{\text{position}} = \text{tr}_{\text{spin}}(|\psi_f\rangle\langle\psi_f|). \tag{19.43}$$

In Table 19.1, we present the probability $P(N, m) = \langle m|\rho_{\text{position}}|m\rangle$ for the quantum walker and a comparison with the results for the classical walker [23]. The numbers for the classical walker, when these differ, are shown in the parenthesis. The results for the quantum walk are asymmetric and the probability distribution for $N = 50$ steps is shown in Figure 19.18. The quantum walk, unlike the classical walk, results from the interference of many pathways as implied by the series of unitary transformations leading from the initial state $|\psi_i\rangle$ to the final state $|\psi_f\rangle$. A proposal to realize quantum walks using a combination of beam splitters and phase shifters was made by Jeong *et al.* [24]. Do *et al.* [25] implemented

Table 19.1 Quantum walk with initial state $|\downarrow, 0\rangle$.

$N\backslash m$	-5	-4	-3	-2	-1	0	1	2	3	4	5
0						1					
1					$\frac{1}{2}$		$\frac{1}{2}$				
2				$\frac{1}{4}$		$\frac{1}{2}$		$\frac{1}{4}$			
3			$\frac{1}{8}$		$\frac{5}{8}\left(\frac{3}{8}\right)$		$\frac{1}{8}\left(\frac{3}{8}\right)$		$\frac{1}{8}$		
4		$\frac{1}{16}$		$\frac{5}{8}\left(\frac{1}{4}\right)$		$\frac{1}{8}\left(\frac{3}{8}\right)$		$\frac{1}{8}\left(\frac{1}{4}\right)$		$\frac{1}{16}$	
5	$\frac{1}{32}$		$\frac{17}{32}\left(\frac{5}{32}\right)$		$\frac{1}{8}\left(\frac{5}{16}\right)$		$\frac{1}{8}\left(\frac{5}{16}\right)$		$\frac{5}{32}$		$\frac{1}{32}$

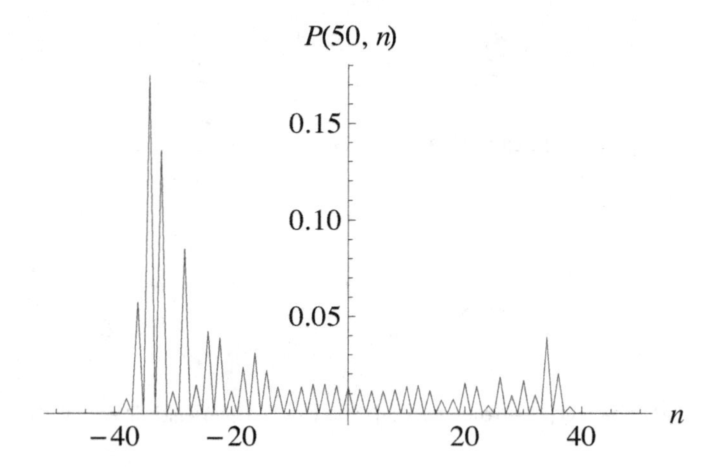

Fig. 19.18 The probability of finding the quantum walker at site n in 50 steps, starting from the initial state $|\downarrow, 0\rangle$.

quantum walks by using linear optical elements such as polarization beam splitters. Pathak and Agarwal [26] and Omar *et al.* [27] considered a two-dimensional walker so that the effects of quantum entanglement on the final states of the walker could be highlighted. Since it is possible to put many waveguide couplers on a chip, it is possible to realize quantum walks using waveguide couplers.

19.5.2 Realization of discrete quantum walks by waveguide couplers

We discussed in Chapter 5 the behavior of a single photon at a beam splitter (Eq. (5.2) and Figure 5.3). The transformation R is then realized by a single beam splitter. We can also use a waveguide coupler to realize R. We next have to introduce an analog of the different lattice sites. We can arrange a series of beam splitters or waveguide couplers as shown in Figure 19.19 for a five-step quantum walk. According to Table 19.1, the five-step quantum walker can be found only at the positions $-5, -3, -1, 1, 3, 5$. Note that the arrival of photon

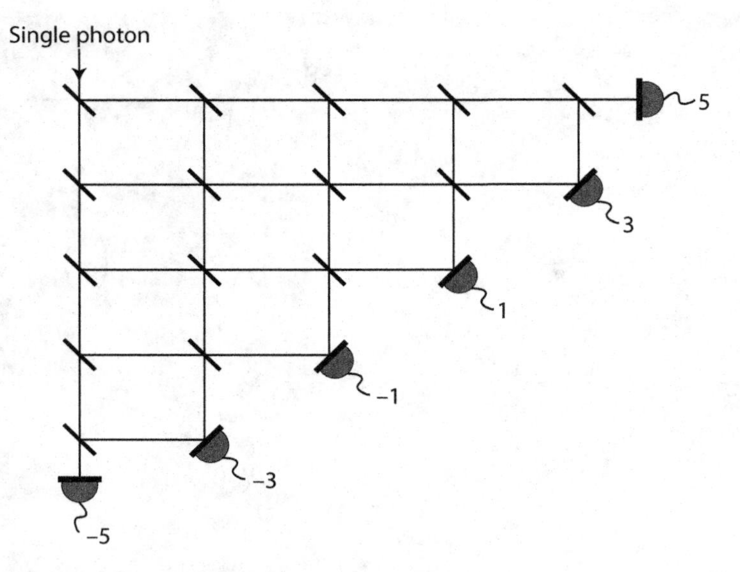

Fig. 19.19 Single photon five-step quantum walk realized by waveguide couplers.

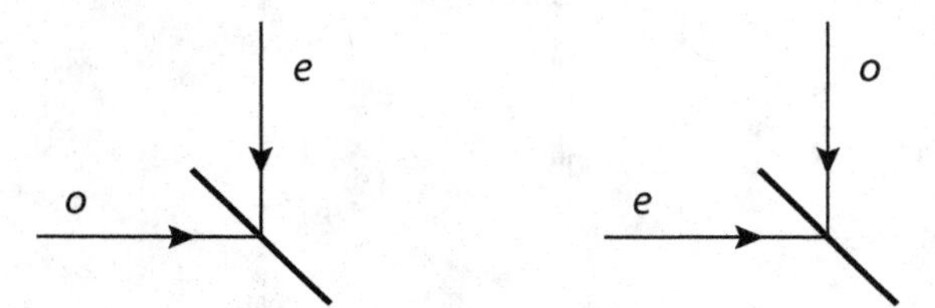

Fig. 19.20 The two parts of (19.44) corresponding to the direction from which different polarizations are launched in.

at either position ± 5 is via only one path. This explains that the probability of finding the quantum walker at the position ± 5 is the same as for a classical walker.

The quantum walk of two entangled photons [26, 27] can be considered by introducing polarization as the additional degree of freedom. For example, in the arrangement of Figure 19.19, we need to bring one photon from each side of the beam splitter. Each photon could have any of the two states e and o of orthogonal polarizations. Thus we can consider an entangled input state with arbitrary relative phase θ,

$$|\psi\rangle = \frac{1}{\sqrt{2}}(|e_v, o_h\rangle + \mathrm{e}^{\mathrm{i}\theta}|o_v, e_h\rangle), \tag{19.44}$$

where h and v represent the horizontal and vertical directions, respectively. Such a state is produced by a downconverter of type II [28]. This corresponds to the two configurations shown in Figure 19.20. Furthermore, we need beam splitters with reflection and transmissions independent of the polarization of light. Such beam splitters and waveguide couplers can be fabricated [29]. We now have optical elements that are independent of the polarization degrees of freedom. Thus given the input state (19.44) we can calculate the output from considerations of the passage of a single photon in the set-up of Figure 19.19.

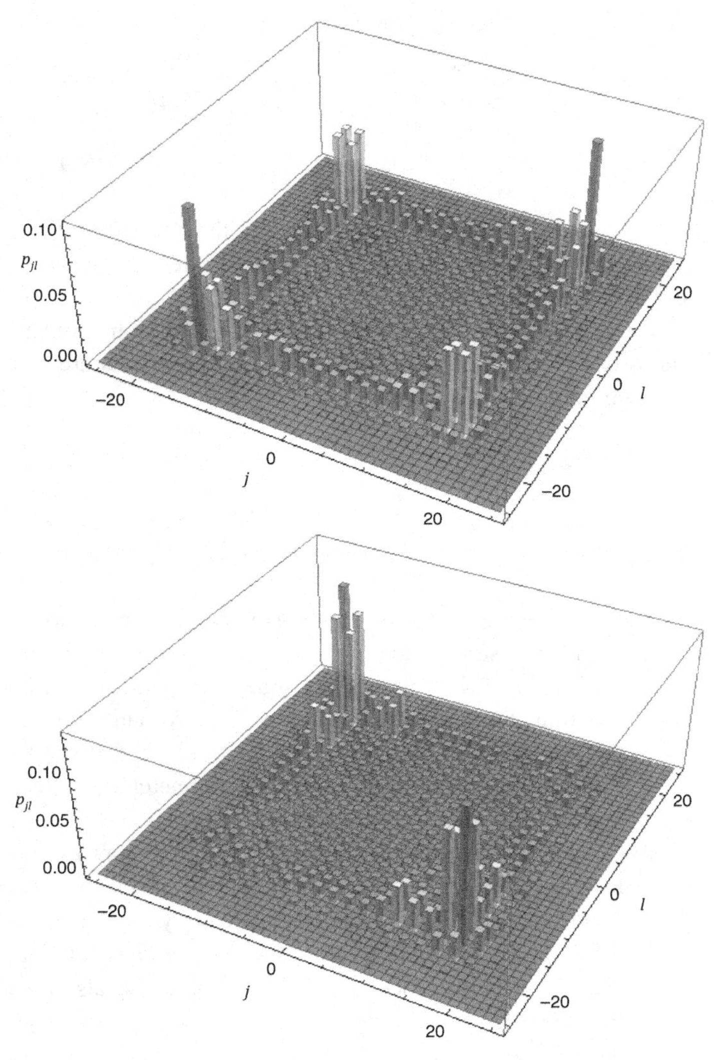

Fig. 19.21 Realization of quantum walk by two entangled walkers by waveguide couplers for the symmetric case $\theta = 0$ (top) and the antisymmetric case $\theta = \pi$ (bottom) in (19.44).

Thus a photon incident from either the horizontal or vertical direction from the input beam splitter would lead to

$$\begin{aligned} |x_h\rangle &\to \sum_l (B_l|l, x_v\rangle + D_l|l, x_h\rangle), \\ |x_v\rangle &\to \sum_l (A_l|l, x_v\rangle + C_l|l, x_h\rangle), \end{aligned} \tag{19.45}$$

where x stands for e or h. Using (19.44) and (19.45), the output state is calculated. The probability p_{jl} of finding a photon at the site j and one at the site l, irrespective of their polarizations, can be shown to be

$$p_{jl} = \frac{1}{2}|D_jC_l + D_lC_j e^{i\theta}|^2 + \frac{1}{2}|D_jA_l + B_lC_j e^{i\theta}|^2 \\ + \frac{1}{2}|B_jC_l + A_lD_j e^{i\theta}|^2 + \frac{1}{2}|B_jA_l + B_lA_j e^{i\theta}|^2 + (j \rightleftharpoons l). \tag{19.46}$$

The four terms in (19.46) correspond to the detection of combination of polarizations (e_h, o_h), (e_h, o_v), (e_v, o_h), (e_v, o_v). The interference terms arise from the entangled nature of the input state (19.44).

We show in Figure 19.21 the behavior of p_{jl} for symmetric and antisymmetric entangled states (19.44). The bunching and antibunching effects are evident in the symmetric and antisymmetric cases, respectively.

Exercises

19.1 For the model Hamiltonian (19.14) obtain solutions to the Heisenberg equations for a and b, assuming that the input state is a product of two single-mode squeezed states (Eq. (2.17)) with same squeezing parameter r. Then show that the Wigner function at time t is Gaussian. Find the covariance matrix defined by (3.95) and find the condition on the parameters Jt and r so that the state at time t is entangled. Calculate the logarithmic negativity parameter E_N defined by Eq. (3.100) and plot it as a function of Jt (for details see [30]).

19.2 Consider a waveguide array whose solution is approximately given by (19.23). Use it to study the discrete Talbot effect [31]. More specifically consider a binary coherent input $\{1, 0, 1, 0, \ldots\}$ to the waveguide array. Calculate the intensity pattern either by using the Bessel function solution or by directly integrating equations (19.4) using the Runge–Kutta procedure. Show periodic revivals of the intensity pattern.

19.3 Prove the solution of (19.25) for a large array satisfying Eq. (19.24). To prove this, write

$$G_{jl}(t) = \frac{1}{\sqrt{2\pi}} \int_{-\pi}^{+\pi} \tilde{G}_l(k,t) e^{ikj} dk,$$

$$\tilde{G}_l(k,t) = \frac{1}{\sqrt{2\pi}} \sum_j G_{jl}(t) e^{-ikj},$$

$$\tilde{G}_l(k,0) = \frac{1}{\sqrt{2\pi}} e^{-ikl},$$

then show that Eq. (19.24) has the solution

$$\tilde{G}_l(k,t) = \frac{1}{\sqrt{2\pi}} e^{-il(k+\alpha t)} \exp\left\{-\frac{2i}{\alpha}[\sin(k+\alpha t) - \sin k]\right\}.$$

The final result (19.25) is obtained by using the generating function for Bessel functions

$$e^{iz\sin\theta} = \sum e^{in\theta} J_n(z).$$

19.4 Obtain the normal modes of light propagation in an array of coupled nonlinear waveguides

$$H = \sum_{j=1}^{N} \hbar g(a_j^{\dagger 2} + a_j^2) + \hbar J \sum_{j=1}^{N-1} (a_j^\dagger a_{j+1} + a_{j+1}^\dagger a_j).$$

Show that the transformation, with S defined by

$$a_j = \sum_{p=1}^{N} b_p S(j, p), \qquad b_p = \sum_{j=1}^{N} a_j S(j, p),$$

yields diagonalization of the above H to

$$H = \sum_{p=1}^{N} \left[g b_p^{\dagger 2} + g b_p^2 + 2J \cos\left(\frac{p\pi}{N+1}\right) b_p^\dagger b_p \right].$$

(Source: Dr. J. Perk, private communication.)

19.5 Show that the results given in Table 19.1 and Figure 19.18 are also obtained if R in Eq. (19.38) is replace by the Hadamard transformation, i.e. the matrix (19.40) is replaced by $\frac{1}{\sqrt{2}}\begin{pmatrix} e^{ip} & e^{ip} \\ e^{-ip} & -e^{-ip} \end{pmatrix}$. Would the off-diagonal elements $\langle m|\rho_{\text{position}}|n\rangle$ $(m \neq n)$ also be identical?

19.6 Using (19.44) and (19.45) verify (19.46).

References

[1] A. Politi, M. J. Cryan, J. G. Rarity, S. Yu, and J. L. O'Brien, *Science* **320**, 646 (2008).

[2] A. Rai, G. S. Agarwal, and J. H. H. Perk, *Phys. Rev. A* **78**, 042304 (2008).

[3] C. K. Hong, Z. Y. Ou, and L. Mandel, *Phys. Rev. Lett.* **59**, 2044 (1987).

[4] Y. Bromberg, Y. Lahini, R. Morandotti, and Y. Silberberg, *Phys. Rev. Lett.* **102**, 253904 (2009).

[5] A. Rai and G. S. Agarwal, *Phys. Rev. A* **79**, 053849 (2009).

[6] H. B. Perets, Y. Lahini, F. Pozzi, M. Sorel, R. Morandotti, and Y. Silberberg, *Phys. Rev. Lett.* **100**, 170506 (2008).

[7] T. Pertsch, P. Dannberg, W. Elflein, A. Bräuer, and F. Lederer, *Phys. Rev. Lett.* **83**, 4752 (1999).

[8] R. Morandotti, U. Peschel, J. S. Aitchison, H. S. Eisenberg, and Y. Silberberg, *Phys. Rev. Lett.* **83**, 4756 (1999).

[9] U. Peschel, T. Pertsch, and F. Lederer, *Opt. Lett.* **23**, 1701 (1998).

[10] G. Bertocchi, O. Alibart, D. B. Ostrowsky, S. Tanzilli, and P. Baldi, *J. Phys. B* **39**, 1011 (2006).

[11] A. Zavatta, S. Viciani, and M. Bellini, *Science* **306**, 660 (2004).
[12] X. Shi, A. Valencia, M. Hendrych, and J. P. Torres, *Opt. Lett.* **33**, 875 (2008).
[13] M. Zukowski, A. Zeilinger, and M. A. Horne, *Phys. Rev. A* **55**, 2564 (1997).
[14] C. Thompson, G. Vemuri, and G. S. Agarwal, *Phys. Rev. B*, **84**, 214302 (2011).
[15] A. Lagendijk, B. van Tigglen, and D. S. Wiersma, *Phys. Today* **62**, 24 (2009).
[16] A. Aspect and M. Inguscio, *Phys. Today* **62**, 30 (2009).
[17] Y. Lahini, A. Avidan, F. Pozzi *et al., Phys. Rev. Lett.* **100**, 013906 (2008).
[18] Y. Lahini, Y. Bromberg, D. N. Christodoulides, and Y. Silberberg, *Phys. Rev. Lett.* **105**, 163905 (2010).
[19] C. Thompson, G. Vemuri, and G. S. Agarwal, *Phys. Rev. A* **82**, 053805 (2010).
[20] L. Sapienza, H. Thyrrestrup, S. Stobbe, P. D. Garacia, S. Smolka, and P. Lodahl, *Science* **327**, 1352 (2010).
[21] S. Smolka, A. Huck, U. L. Andersen, A. Lagendijk, and P. Lodahl, *Phys. Rev. Lett.* **102**, 193901 (2009); P. Lodahl, A. P. Mosk, and A. Lagendijk, *Phys. Rev. Lett.* **95**, 173901 (2005);
[22] S. Chandrasekhar, *Rev. Mod. Phys.* **15**, 1 (1943).
[23] J. Kempe, *Contemp. Phys.* **44**, 307 (2003).
[24] H. Jeong, M. Paternostro, and M. S. Kim, *Phys. Rev. A* **69**, 012310 (2004).
[25] B. Do, M. L. Stohler, S. Balasubramanian *et al., J. Opt. Soc. Am. B* **22**, 499 (2005).
[26] P. K. Pathak and G. S. Agarwal, *Phys. Rev. A* **75**, 032351 (2007).
[27] Y. Omar, N. Paunkovic, L. Sheridan, and S. Bose, *Phys. Rev. A* **74**, 042304 (2006).
[28] P. G. Kwiat, K. Mattle, H. Weinfurter, A. Zeilinger, A. V. Sergienko and Y. Shih, *Phys. Rev. Lett.* **75**, 4337 (1995).
[29] L. Sansoni, F. Sciarrino, G. Vallone *et al.*, *Phys. Rev. Lett.* **108**, 010502 (2012).
[30] A. Rai, S. Das, and G. S. Agarwal, *Opt. Express* **18**, 6241 (2010).
[31] R. Iwanow, D. A. May-Arrioja, D. N. Christodoulides, G. I. Stegeman, Y. Min, and W. Sohler, *Phys. Rev. Lett.* **95**, 053902 (2005).

20 Quantum optical effects in nano-mechanical systems

In this chapter we will show how many concepts from quantum optics, such as squeezing, nonclassicality, and quantum entanglement, can be applied to nano-mechanical systems leading to the possibility of realizing the quantized behavior of macroscopic systems [1]. Furthermore, nano-mechanical systems can exhibit a variety of rich nonlinear phenomena as the basic interaction between the nano-mechanical system and the radiation fields is via radiation pressure [2]. This interaction is nonlinear. Thus many nonlinear processes such as electromagnetically induced transparency, optical bistability, and up-conversion of radiation are expected to occur for nano-mechanical systems. Similarly, cavity QED effects such as vacuum Rabi splittings are also expected to occur provided one can design systems such that the interaction of a single photon with the nano-mechanical mirror is large. We note that the work on nano-mechanical systems originated with the discussion of Braginsky and collaborators [3] on how to measure small forces accurately. In this chapter, we will discuss only the fundamental quantum and nonlinear optical effects in nano-mechanical systems interacting with quantized and semiclassical fields.

20.1 The radiation pressure on the nano-mechanical mirror

We derive the radiation pressure on the nano-mechanical mirror via the Maxwell stress tensor. We consider a Fabry–Perot cavity with length L formed by a fixed partially transmitting mirror and a movable totally reflecting mirror, as shown in Figure 20.1. A plane electromagnetic wave with wave vector $\vec{k}$ ($k = |\vec{k}|$) at frequency ω_c in the cavity is propagating along the $\hat{x}$ direction. The electric field in the cavity is

$$\vec{E} = \sqrt{\frac{4\pi\hbar\omega_c}{V}}(\vec{\epsilon}a\mathrm{e}^{-\mathrm{i}\omega_c t} + \vec{\epsilon}^* a^\dagger \mathrm{e}^{\mathrm{i}\omega_c t})\sin(kx), \tag{20.1}$$

where V is the volume of the cavity and $V = AL$ (A is the surface area of the movable mirror), $\vec{\epsilon}$ is the polarization vector of the electric field: $\vec{\epsilon} = \epsilon_y\hat{y} + \epsilon_z\hat{z}$, $\vec{\epsilon}\cdot\vec{\epsilon}^* = 1$. By using the Maxwell equation $\vec{\nabla}\times\vec{E} = -\frac{1}{c}\frac{\partial\vec{B}}{\partial t}$, we can obtain the magnetic field in the cavity

$$\vec{B} = -\sqrt{\frac{4\pi\hbar\omega_c}{V}}\left(\mathrm{i}(\hat{x}\times\vec{\epsilon})a\mathrm{e}^{-\mathrm{i}\omega_c t} - \mathrm{i}(\hat{x}\times\vec{\epsilon}^*)a^\dagger\mathrm{e}^{\mathrm{i}\omega_c t}\right)\cos(kx). \tag{20.2}$$

Let the mirror in the volume $\tilde{V}$ be surrounded by a closed surface S, then the rate of the

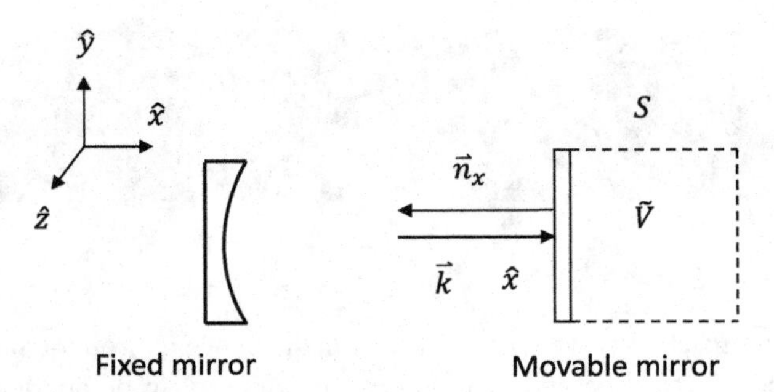

Fig. 20.1 A Fabry–Perot cavity with a moving mirror.

change of momentum would be

$$\frac{\mathrm{d}}{\mathrm{d}t}(\vec{p}_{\text{mech}} + \vec{p}_{\text{field}})_\alpha = \oint_S \sum_\beta T_{\alpha\beta} n_\beta \mathrm{d}S, \tag{20.3}$$

where $\vec{p}_{\text{mech}}$ and $\vec{p}_{\text{field}}$ are the mechanical and the electromagnetic momentum, respectively, of the mirror bounded by the closed surface S, $\vec{n}$ is the outward normal to the closed surface S, and $T_{\alpha\beta}$ is the Maxwell stress tensor [2]

$$T_{\alpha\beta} =: \frac{1}{4\pi}\left[E_\alpha E_\beta + B_\alpha B_\beta - \frac{1}{2}(\vec{E}\cdot\vec{E} + \vec{B}\cdot\vec{B})\delta_{\alpha\beta}\right]:, \tag{20.4}$$

where :: denotes normal ordering which is required when dealing with quantum fields. Note that inside the volume $\tilde{V}$ the electromagnetic field vanishes, hence $\vec{p}_{\text{field}} = 0$, and the force acting on the surface of the movable mirror is the only force acting on the system inside the volume $\tilde{V}$. Therefore the force acting on the mirror is given by

$$F_x = \frac{\mathrm{d}}{\mathrm{d}t}\vec{p}_{\text{mech}} = \int_A T_{xx} n_x \mathrm{d}S. \tag{20.5}$$

Since the outward normal to the surface of the movable mirror is in the $-\hat{x}$ direction, thus $n_x = -1$. Moreover, the Maxwell stress tensor T_{xx} can be obtained from Eq. (20.4) by noting that $E_x = B_x = 0$; $\vec{E}\cdot\vec{E} = 0$ at the mirror position which is a node of the field ($\sin(kL) = 0$). Furthermore, we drop terms a^2 and $a^{\dagger 2}$ which oscillate at optical frequencies to obtain

$$T_{xx} = -\frac{\hbar\omega_c}{V}a^\dagger a, \tag{20.6}$$

Thus the force on the movable mirror is

$$F_x = \frac{\hbar\omega_c}{L}a^\dagger a. \tag{20.7}$$

The contribution of the radiation pressure term to the Hamiltonian will be

$$H = -\frac{\hbar\omega_c}{L}a^\dagger a x, \tag{20.8}$$

where x is the displacement of the mirror.

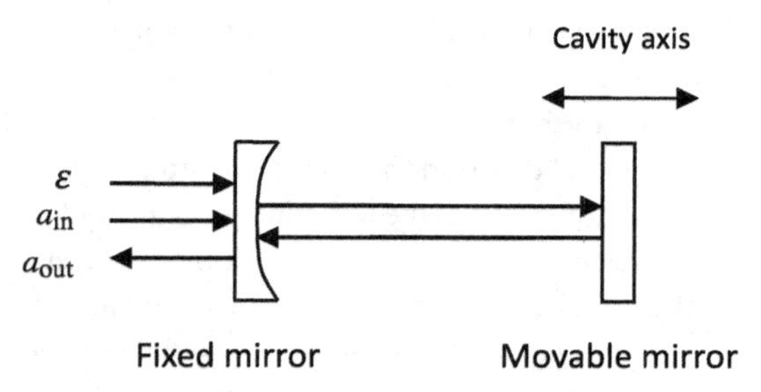

Fig. 20.2 Basic cavity optomechanical system. Here ε is a coherent field, and a_{in} is the quantum vacuum field.

There are other methods to derive the radiation pressure on the nano-mechanical mirror. For example, it also can be derived by the shift of the cavity resonance frequency. The photons in the cavity exert a radiation pressure force on the movable mirror so that the movable mirror makes small oscillations. In turn, the displacement of the movable mirror modifies the cavity resonance frequency. Thus the cavity resonance frequency depends on the displacement of the movable mirror, represented by $\omega_c(x) = \frac{n\pi c}{L+x}$, where n is the mode number in the cavity and c is the speed of light in a vacuum. If the the displacement of the movable mirror is very small, then $\omega_c(x) \approx \frac{n\pi c}{L}(1 - \frac{x}{L})$. The Hamiltonian of the cavity field becomes

$$\begin{aligned} H_c = \hbar\omega_c(x)a^\dagger a &= \hbar\frac{n\pi c}{L}\left(1 - \frac{x}{L}\right)a^\dagger a \\ &= \hbar\omega_c a^\dagger a - \frac{\hbar\omega_c}{L}a^\dagger a x, \end{aligned} \tag{20.9}$$

in which $\omega_c = \frac{n\pi c}{L}$ is the cavity resonance frequency in the absence of the optomechanical coupling. The first term in Eq. (20.9) is the bare energy of the cavity field, the second term is the interaction Hamiltonian H_{int} of the cavity field and the movable mirror. Thus the radiation pressure force on the movable mirror would be

$$\begin{aligned} F &= -\frac{\mathrm{d}}{\mathrm{d}x}H_{\text{int}} \\ &= \frac{\hbar\omega_c}{L}a^\dagger a, \end{aligned} \tag{20.10}$$

which is consistent with the result derived by the Maxwell stress tensor.

20.2 Basic quantum Langevin equations for the coupled system of cavity and NMO

We now develop the dynamical equations for the optomechanical system. The system is a Fabry–Perot cavity with one heavy, fixed partially transmitting mirror and one light, movable totally reflecting mirror of effective mass m (typically in the micro or nanogram range), as shown in Figure 20.2. The system is driven by an external laser at frequency ω_l, then the circulating photons in the cavity will exert a radiation pressure force on the movable mirror due to momentum transfer from the intracavity photons to the movable mirror. The force is proportional to the instantaneous photon number in the cavity. Moreover, the movable mirror is in thermal equilibrium with its environment at temperature T. Thus the mirror can move under the influence of the radiation pressure and at the same time undergoes Brownian motion as a result of its interaction with the environment. In turn, the movable mirror's small oscillation changes the length of the cavity and shifts the cavity resonance frequency so that the phase and amplitude of the cavity field are changed. Here, the movable mirror is modeled as a single-mode quantum harmonic oscillator with effective mass m, frequency ω_m and momentum decay rate γ_m. In the adiabatic limit, the mechanical frequency ω_m is much smaller than the cavity free spectral range $(c/2L)$, where L is the initial cavity length. The input laser drives only one cavity mode ω_c and scattering of photons from the driven mode into other cavity modes is negligible.

The Hamiltonian of the system is given by

$$H = \hbar\omega_c a^\dagger a - \hbar g a^\dagger a x + \frac{1}{2} m\omega_m^2 x^2 + \frac{p^2}{2m} + i\hbar\varepsilon(a^\dagger e^{-i\omega_l t} - a e^{i\omega_l t}). \tag{20.11}$$

In Eq. (20.11), the first term is the energy of the cavity field, a and $a^\dagger$ are the annihilation and creation operators for the cavity field satisfying the commutation relation $[a, a^\dagger] = 1$. The second term describes the interaction of the movable mirror with the cavity field, the parameter $g = \omega_c/L$ is the optomechanical coupling constant between the cavity and the movable mirror. The third and fourth terms give the energy of the movable mirror. The fifth term describes the cavity driven by a laser with power $\wp$ and $\varepsilon = \sqrt{\frac{2\kappa\wp}{\hbar\omega_l}}$, where κ is the photon decay rate due to the partial transmission of the fixed mirror. If $\mathcal{T}$ is the transmissivity of the mirror and $\mathcal{F}$ is the finesse [4], then

$$\kappa = \frac{\pi c}{2\mathcal{F}L}, \qquad \mathcal{F} = \frac{\pi(1-\mathcal{T})^{1/4}}{1-(1-\mathcal{T})^{1/2}}. \tag{20.12}$$

The Hamiltonian of the system in a rotating frame with respect to the laser frequency ω_l is defined by $H_{\text{rot}} = RHR^\dagger + i\hbar\frac{\partial R}{\partial t}R^\dagger$, where $R = e^{i\omega_l a^\dagger a t}$. By using $e^{\alpha A}Be^{-\alpha A} = B + \alpha[A, B] + \frac{\alpha^2}{2!}[A, [A, B]] + \cdots$, we can obtain the Hamiltonian of the system in a frame rotating at the laser frequency ω_l. For simplicity, we drop the subscript of H_{rot}, then we have

$$H = \hbar(\omega_c - \omega_l)a^\dagger a - \hbar g a^\dagger a x + \frac{1}{2} m\omega_m^2 x^2 + \frac{p^2}{2m} + i\hbar\varepsilon(a^\dagger - a). \tag{20.13}$$

For convenience, we represent the position and momentum operators x and p of the movable mirror in terms of the dimensionless position and momentum operators Q and P, which are defined in terms of the zero-point motion of the ground state. The zero-point motion is given by

$$\langle x^2 \rangle = \frac{\hbar}{2m\omega_m}, \qquad \langle p^2 \rangle = \frac{m\hbar\omega_m}{2}, \tag{20.14}$$

so that $Q = \sqrt{(m\omega_m)/\hbar}x$ and $P = \sqrt{1/(m\hbar\omega_m)}p$. Note that the scaled variables Q and P satisfy commutation relation $[Q, P] = i$ and for the ground state of the mirror

$\langle Q^2 \rangle = \langle P^2 \rangle = 1/2$. In terms of Q and P, the Hamiltonian of the system becomes

$$H = \hbar(\omega_c - \omega_l)a^\dagger a - \hbar\omega_m \chi a^\dagger a Q + \frac{\hbar\omega_m}{2}(Q^2 + P^2) + \mathrm{i}\hbar\varepsilon(a^\dagger - a), \tag{20.15}$$

where the dimensionless parameter $\chi = \frac{1}{\omega_m}\frac{\omega_c}{L}\sqrt{\frac{\hbar}{m\omega_m}}$ is the coupling constant.

The time evolution of the system operators can be derived by using the Heisenberg equations of motion and adding the corresponding damping and noise terms. We find a set of nonlinear quantum Langevin equations given by

$$\begin{aligned}
\dot{Q} &= \omega_m P, \\
\dot{P} &= \omega_m \chi a^\dagger a - \omega_m Q - \gamma_m P + \xi, \\
\dot{a} &= -\mathrm{i}(\omega_c - \omega_l - \omega_m \chi Q)a + \varepsilon - \kappa a + \sqrt{2\kappa}a_{\text{in}}, \\
\dot{a}^\dagger &= \mathrm{i}(\omega_c - \omega_l - \omega_m \chi Q)a^\dagger + \varepsilon - \kappa a^\dagger + \sqrt{2\kappa}a_{\text{in}}^\dagger.
\end{aligned} \tag{20.16}$$

Here a_{in} is the input vacuum noise operator with zero mean value and nonzero correlation function given by

$$\langle \delta a_{\text{in}}(t)\delta a_{\text{in}}^\dagger(t')\rangle = \delta(t - t'). \tag{20.17}$$

The force ξ is the Brownian noise operator associated with the mechanical damping, whose mean value is zero, and its correlation function reads

$$\langle \xi(t)\xi(t')\rangle = \frac{1}{2\pi}\frac{\gamma_m}{\omega_m}\int \omega e^{-\mathrm{i}\omega(t-t')}\left[1 + \coth\left(\frac{\hbar\omega}{2k_{\text{B}}T}\right)\right]\mathrm{d}\omega, \tag{20.18}$$

where k_{B} is the Boltzmann constant and T is the temperature of the environment of the mirror.

The output field a_{out} can be found by using the input–output relation [5],

$$a_{\text{out}} + a_{\text{in}} + \frac{\varepsilon}{\sqrt{2\kappa}} = \sqrt{2\kappa}a, \tag{20.19}$$

then the mean value of the output field would be

$$\langle a_{\text{out}}\rangle + \frac{\varepsilon}{\sqrt{2\kappa}} = \sqrt{2\kappa}\langle a\rangle, \tag{20.20}$$

since $\langle a_{\text{in}}\rangle = 0$. In the absence of the movable mirror $\chi = 0$, $\langle a\rangle = \varepsilon/[\kappa + \mathrm{i}(\omega_c - \omega_l)]$.

20.3 Steady-state solution of quantum Langevin equations in the mean field limit and bistability

From Eq. (20.16) we obtain equations for the mean values of the cavity field and the mirror's momentum and position. This set of equations is closed under the mean field approximation

$$\langle Qa\rangle \approx \langle Q\rangle\langle a\rangle, \qquad \langle Qa^\dagger\rangle \approx \langle Q\rangle\langle a^\dagger\rangle, \qquad \langle a^\dagger a\rangle \approx \langle a^\dagger\rangle\langle a\rangle. \tag{20.21}$$

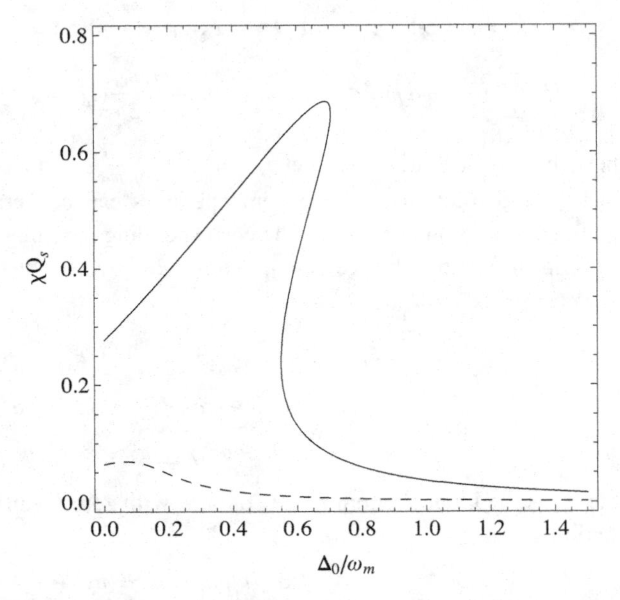

Fig. 20.3 The behavior of the mirror's displacement χQ_s in the steady state as a function of the normalized detuning Δ_0/ω_m for two different values of the input laser power: $\wp = 0.5$ mW (dashed), 5 mW (solid). Parameters used are: the wavelength of the laser $\lambda = 2\pi c/\omega_l = 1064$ nm, $L = 25$ mm, $m = 145$ ng, $\omega_m = 2\pi \times 947 \times 10^3$ Hz, $\kappa = 2\pi \times 215 \times 10^3$ Hz, the mechanical quality factor $Q' = \omega_m/\gamma_m = 6700$, and $\gamma_m = 2\pi \times 141$ Hz.

The equations for the mean values in the steady state then lead to

$$\langle P\rangle = P_s = 0, \qquad \langle Q\rangle = Q_s = \chi |a_s|^2, \qquad \langle a\rangle = a_s = \frac{\varepsilon}{\kappa + \mathrm{i}\Delta}, \tag{20.22}$$

where

$$\Delta = \omega_c - \omega_l - \omega_m \chi Q_s \tag{20.23}$$

is the effective cavity detuning, in which the term $-\omega_m \chi Q_s$ is the cavity resonance frequency shift due to radiation pressure. The variable Q_s denotes the new equilibrium position of the movable mirror with respect to that without the driving field. The variable a_s represents the steady-state amplitude of the cavity field. These equations for a_s and Q_s are coupled to each other and in principle can have multiple solutions, as shown in Figures 20.3 and 20.4. For the purpose of illustration we use the parameters from the experimental paper [6]. We note that there is nothing very special about these parameters. The results that follow are quite generic. From Figure 20.3, we can see that for $\wp = 5$ mW, the movable mirror can have three different equilibrium positions for a fixed detuning $\Delta_0 = \omega_c - \omega_l$. Two of these solutions would be stable, which indicates bistable behavior, as observed [7]. The stability of the solutions can be studied by linearization around the steady state (see discussion Section 20.4). Using (20.20) the mean value of the output field is

$$\langle a_{\text{out}}\rangle = \sqrt{2\kappa}\,\langle a\rangle - \frac{\varepsilon}{\sqrt{2\kappa}} = \frac{\kappa - \mathrm{i}\Delta}{\kappa + \mathrm{i}\Delta}\frac{\varepsilon}{\sqrt{2\kappa}}, \tag{20.24}$$

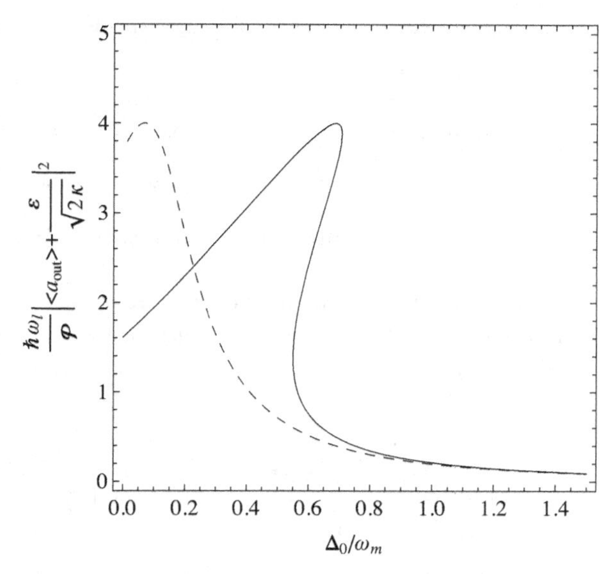

Fig. 20.4 The output photon number $\frac{\hbar\omega_l}{\wp}|\langle a_{\text{out}}\rangle + \frac{\varepsilon}{\sqrt{2\kappa}}|^2 = \frac{\hbar\omega_l}{\wp}2\kappa|\langle a\rangle|^2$ as a function of the normalized detuning $\Delta_0/\omega_m.\wp = 0.5$ mW (dashed), 5 mW (solid).

and hence $|\langle a_{\text{out}}\rangle|^2 = |\varepsilon|^2/(2\kappa)$. It should be noted that Δ depends on the cavity field. The photon number in the output field is constant for different detuning. But if we define the output field as the sum of the output field and the input coherent field, then these terms can interfere, which results in bistable behavior as shown in Figure 20.4. We can see that for $\wp = 5$ mW, the output photon number might have three different values, with two of them being stable. For the parameters of Figures 20.3 and 20.4, we find that the system exhibits bistable behavior when the pump power is greater than 2.9 mW.

20.4 Quantum fluctuations in optomechanical systems

The study of quantum fluctuations is complicated because of the nonlinear nature of the quantum Langevin equations (20.16). A standard quantum optical procedure is to assume that the quantum fluctuations around the steady-state mean values are small. We can then linearize the nonlinear equation (20.16) by writing each operator as the sum of its steady-state mean value and a small fluctuation with zero mean value,

$$Q = Q_s + \delta Q, \qquad P = P_s + \delta P, \qquad a = a_s + \delta a. \tag{20.25}$$

Inserting Eq. (20.25) into Eq. (20.16), assuming $|a_s| \gg 1$, and writing the product of two operators as $AB = A_sB_s + A_s\delta B + B_s\delta A + \delta A\delta B \approx A_sB_s + A_s\delta B + B_s\delta A$, the linearized quantum Langevin equations for the fluctuation operators take the form

$$\begin{aligned}
\delta\dot{Q} &= \omega_m \delta P, \\
\delta\dot{P} &= \omega_m \chi (a_s^* \delta a + a_s \delta a^\dagger) - \omega_m \delta Q - \gamma_m \delta P + \xi, \\
\delta\dot{a} &= -(\kappa + \mathrm{i}\Delta)\delta a + \mathrm{i}\omega_m \chi a_s \delta Q + \sqrt{2\kappa}\delta a_{\mathrm{in}}, \\
\delta\dot{a}^\dagger &= -(\kappa - \mathrm{i}\Delta)\delta a^\dagger - \mathrm{i}\omega_m \chi a_s^* \delta Q + \sqrt{2\kappa}\delta a_{\mathrm{in}}^\dagger.
\end{aligned} \tag{20.26}$$

Introducing the amplitude and phase quadratures of the cavity field $\delta x = (\delta a + \delta a^\dagger)/\sqrt{2}$, $\delta y = \mathrm{i}(\delta a^\dagger - \delta a)/\sqrt{2}$, and the input noise quadratures $\delta x_{\mathrm{in}} = (\delta a_{\mathrm{in}} + \delta a_{\mathrm{in}}^\dagger)/\sqrt{2}$ and $\delta y_{\mathrm{in}} = \mathrm{i}(\delta a_{\mathrm{in}}^\dagger - \delta a_{\mathrm{in}})/\sqrt{2}$, Eq. (20.26) can then be rewritten in matrix form as

$$\dot{f}(t) = Af(t) + \eta(t), \tag{20.27}$$

where $f(t)$ is the column vector of the fluctuations and $\eta(t)$ is the column vector of the noise sources. Their transposes are

$$\begin{aligned}
f(t)^{\mathrm{T}} &= (\delta Q, \delta P, \delta x, \delta y), \\
\eta(t)^{\mathrm{T}} &= (0, \xi, \sqrt{2\kappa}\delta x_{in}, \sqrt{2\kappa}\delta y_{in});
\end{aligned} \tag{20.28}$$

and the matrix A is given by

$$A = \begin{pmatrix}
0 & \omega_m & 0 & 0 \\
-\omega_m & -\gamma_m & \omega_m \chi (a_s + a_s^*)/\sqrt{2} & -\mathrm{i}\omega_m \chi (a_s - a_s^*)/\sqrt{2} \\
\mathrm{i}\omega_m \chi (a_s - a_s^*)/\sqrt{2} & 0 & -\kappa & \Delta \\
\omega_m \chi (a_s + a_s^*)/\sqrt{2} & 0 & -\Delta & -\kappa
\end{pmatrix}. \tag{20.29}$$

The linearized fluctuations can then be obtained by solving (20.27) using matrix methods.

While using linearized equations one has to make sure that the dynamical system is stable. The system is stable only if all the eigenvalues of the matrix A have negative real parts. The stability conditions for the system can be derived by applying the Routh–Hurwitz criterion [8]. This gives the stability conditions

$$\begin{aligned}
\omega_m^3 \chi^2 (2\kappa + \gamma_m)^2 |a_s|^2 \Delta + \kappa\gamma_m \{(\kappa^2 + \Delta^2)^2 + (2\kappa\gamma_m + \gamma_m^2)(\kappa^2 + \Delta^2) \\
+ \omega_m^2 [2(\kappa^2 - \Delta^2) + \omega_m^2 + 2\kappa\gamma_m]\} > 0, \\
\kappa^2 + \Delta^2 - 2\omega_m \chi^2 |a_s|^2 \Delta > 0.
\end{aligned} \tag{20.30}$$

All external parameters must be chosen so as to satisfy the stability conditions Eq. (20.30). These conditions also determine the regions of stability in Figures 20.3 and 20.4.

In subsequent sections we derive experimentally realized physical effects in opto-mechanical systems by studying the spectrum of fluctuations associated with the motion of the mirror and the field outside the cavity.

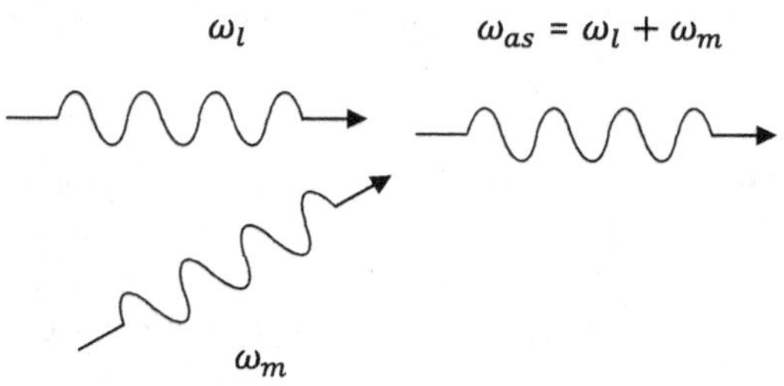

ω_l $\omega_{as} = \omega_l + \omega_m$

ω_m

Fig. 20.5 Sketch of the one-phonon process leading to the up-conversion of radiation.

20.5 Sideband cooling of the nano-mechanical mirror

The first series of experiments have demonstrated that the mechanical mirror can be cooled by the dynamical back-action of radiation pressure [9–11] (see Exercise 20.3). It was later shown theoretically [12, 13] that it is possible to cool the mechanical mirror to the quantum ground state by resolved sideband cooling. Sideband cooling was demonstrated experimentally by Schliesser *et al.* [14] and Park *et al.* [15]. For a mechanical mirror at frequency $\omega_m = 2\pi \times 100\,\text{MHz}$, the phonon number $\left(n = \left[\exp\left(\frac{\hbar\omega_m}{K_B T}\right) - 1\right]^{-1}\right)$ is around 312 at $T = 1.5\,\text{K}$, and about 1.6 at $T = 10\,\text{mK}$. If the phonon number in the mechanical oscillator is reduced to unity, the mirror should be cooled to 6.92 mK. Both these experiments started the system at about 1.5 K and showed cooling down to about 200 mK. The amount of cooling depends on the system parameters and the laser power. Interestingly enough, using the mechanism of references [9–11], Thompson *et al.* [16] showed that the lowest temperature achieved is 6.82 mK in an optical cavity with a vibrating membrane with a very low dissipation of about 1 Hz.

Before we give details of the theoretical discussion of sideband cooling, we discuss the physics which shows why sideband cooling results in cooling. When the pump field with frequency ω_l interacts with the mechanical mirror with frequency ω_m, absorption and emission of phonons creates the anti-Stokes field $(\omega_l + \omega_m)$ and the Stokes field $(\omega_l - \omega_m)$. During the anti-Stokes process, as shown in Figure 20.5, the pump field extracts a quantum of energy $\hbar\omega_m$ from the movable mirror, leading to the cooling of the movable mirror. While during the Stokes process, the pump field emits a quantum of energy $\hbar\omega_m$ to the movable mirror, leading to the heating of the movable mirror. If the pump frequency is tuned below the cavity resonance frequency by an amount ω_m, the amplitude of the anti-Stokes field is resonantly enhanced, since the frequency of the anti-Stokes field is close to the cavity resonance frequency ω_c; however, the Stokes field is suppressed since its frequency is far away from the cavity resonance frequency, thus the optomechanical coupling causes the cooling of the mirror. Furthermore, in the resolved sideband limit, the cavity amplitude decay rate κ is much less than the mechanical oscillation frequency ω_m. In this case, the linewidth κ of the cavity field is much smaller than the frequency spacing $2\omega_m$ between the anti-Stokes field and the Stokes field. Thus the amplitude of the Stokes field is close to zero, ground state cooling becomes possible.

In order to make estimates of the cooling of the mechanical oscillator, we need to calculate mean energy of the mechanical oscillator. This requires the fluctuations in Q and P to be calculated. Taking the Fourier transform of Eq. (20.26) to the frequency domain by using $f(t) = \frac{1}{2\pi}\int_{-\infty}^{+\infty} f(\omega)\mathrm{e}^{-\mathrm{i}\omega t}\mathrm{d}\omega$ and $f^{\dagger}(t) = \frac{1}{2\pi}\int_{-\infty}^{+\infty} f^{\dagger}(-\omega)\mathrm{e}^{-\mathrm{i}\omega t}\mathrm{d}\omega$, where $f^{\dagger}(-\omega) = [f(-\omega)]^{\dagger}$, and solving it, we obtain the position fluctuations of the movable mirror

$$\delta Q(\omega) = -\frac{\omega_m}{d(\omega)}\Big[\sqrt{2\kappa}\,\omega_m\chi\left\{[\kappa - \mathrm{i}(\Delta+\omega)]a_s^*\delta a_{in}(\omega) + [\kappa + \mathrm{i}(\Delta-\omega)]a_s\delta a_{in}^{\dagger}(-\omega)\right\}$$
$$+ [(\kappa - \mathrm{i}\omega)^2 + \Delta^2]\xi(\omega)\Big], \tag{20.31}$$

where

$$d(\omega) = 2\omega_m^3\chi^2\Delta|a_s|^2 + (\omega^2 - \omega_m^2 + \mathrm{i}\gamma_m\omega)[(\kappa - \mathrm{i}\omega)^2 + \Delta^2]. \tag{20.32}$$

In Eq. (20.31), the first term proportional to χ is the contribution of radiation pressure, while the second term involving $\xi(\omega)$ is the contribution of the thermal noise. In the absence of the cavity field, the movable mirror will make Brownian motion, $\delta Q(\omega) = \omega_m\xi(\omega)/(\omega_m^2 - \omega^2 - \mathrm{i}\gamma_m\omega)$, whose susceptibility has a Lorentzian shape centered at frequency ω_m with full width at half maximum γ_m.

The two-time correlation function of the fluctuations in position of the movable mirror is given by

$$\frac{1}{2}(\langle\delta Q(t)\delta Q(t+\tau)\rangle + \langle\delta Q(t+\tau)\delta Q(t)\rangle) = \frac{1}{2\pi}\int_{-\infty}^{+\infty}\mathrm{d}\omega S_Q(\omega)\mathrm{e}^{\mathrm{i}\omega\tau}, \tag{20.33}$$

in which $S_Q(\omega)$ is the spectrum of fluctuations in position of the movable mirror, defined by

$$\frac{1}{2}(\langle\delta Q(\omega)\delta Q(\Omega)\rangle + \langle\delta Q(\Omega)\delta Q(\omega)\rangle) = 2\pi S_Q(\omega)\delta(\omega+\Omega). \tag{20.34}$$

By using the correlation functions (20.17), (20.18) of the noise sources in the frequency domain,

$$\langle\delta a_{\mathrm{in}}(\omega)\delta a_{\mathrm{in}}^{\dagger}(-\Omega)\rangle = 2\pi\delta(\omega+\Omega),$$
$$\langle\xi(\omega)\xi(\Omega)\rangle = 2\pi\frac{\gamma_m}{\omega_m}\omega\left[1 + \coth\left(\frac{\hbar\omega}{2K_{\mathrm{B}}T}\right)\right]\delta(\omega+\Omega), \tag{20.35}$$

we obtain the spectrum of fluctuations in position of the movable mirror

$$S_Q(\omega) = \frac{\omega_m^2}{|d(\omega)|^2}\Big\{2\omega_m^2\chi^2\kappa(\kappa^2+\omega^2+\Delta^2)|a_s|^2$$
$$+\frac{\gamma_m}{\omega_m}\omega[(\Delta^2+\kappa^2-\omega^2)^2 + 4\kappa^2\omega^2]\coth\left(\frac{\hbar\omega}{2K_{\mathrm{B}}T}\right)\Big\}. \tag{20.36}$$

In Eq. (20.36), the first term involving χ arises from radiation pressure, while the second term originates from the thermal noise. So the spectrum $S_Q(\omega)$ of the movable mirror depends on radiation pressure and the thermal noise. In the absence of the optomechanical coupling ($\chi \to 0$), the spectrum $S_Q(\omega)$ of the movable mirror becomes

$$S_Q(\omega) = \frac{\gamma_m \omega_m \omega}{(\omega^2 - \omega_m^2)^2 + \gamma_m^2 \omega^2} \coth\left(\frac{\hbar\omega}{2K_B T}\right). \tag{20.37}$$

As $\omega \sim \omega_m$, Eq. (20.37) can be simplified to

$$S_Q(\omega) = \frac{\frac{\gamma_m}{4}}{(\omega - \omega_m)^2 + (\frac{\gamma_m}{2})^2} \coth\left(\frac{\hbar\omega}{2K_B T}\right). \tag{20.38}$$

We now find $S_P(\omega)$ by taking the Fourier transform of $\delta\dot{Q} = \omega_m \delta P$. This gives $\delta P(\omega) = -\frac{i\omega}{\omega_m}\delta Q(\omega)$, which leads to the spectrum of fluctuations in momentum of the movable mirror

$$S_P(\omega) = \frac{\omega^2}{\omega_m^2} S_Q(\omega). \tag{20.39}$$

The variances of position and momentum are then given by

$$\langle \delta Q(t)^2 \rangle = \frac{1}{2\pi} \int_{-\infty}^{+\infty} S_Q(\omega) \mathrm{d}\omega \quad \text{and} \quad \langle \delta P(t)^2 \rangle = \frac{1}{2\pi} \int_{-\infty}^{+\infty} S_P(\omega) \mathrm{d}\omega.$$

The phonon number n of the movable mirror can be calculated from the total energy of the movable mirror by using

$$\frac{\hbar\omega_m}{2}(\langle \delta Q(t)^2 \rangle + \langle \delta P(t)^2 \rangle) = \hbar\omega_m\left(n + \frac{1}{2}\right). \tag{20.40}$$

Then the effective temperature T_{eff} of the movable mirror can be determined from the phonon number n in the movable mirror from

$$T_{\text{eff}} = \frac{\hbar\omega_m}{K_B \ln(1 + \frac{1}{n})}, \quad n = \frac{1}{e^{\hbar\omega_m/(k_B T_{\text{eff}})} - 1}. \tag{20.41}$$

The parameters used are the same as those for Figure 20.3 except the effective detuning Δ and the pump power. Note that $\kappa/\omega_m \approx 0.23$, thus the system is operating in the resolved sideband regime. In the high-temperature limit $K_B T \gg \hbar\omega_m$, the approximation $\coth(\hbar\omega/2K_B T) \approx 2K_B T/\hbar\omega$ can be made. The laser is detuned below the cavity resonance frequency by an amount $\Delta = \omega_m$. All parameters are chosen to satisfy the stability condition (20.30). Figure 20.6 shows the variation of the effective temperature T_{eff} of the movable mirror with the laser power $\wp$. It is clear to see that the effective temperature T_{eff} of the movable mirror decreases with increases the laser power $\wp$. When $\wp = 100$ μW, the movable mirror can be cooled to about 50 mK, a factor of 20 below the starting temperature of 1 K [14, 15]. If the laser power is further increased to 1 mW, the movable mirror can be cooled to about 6 mK. Therefore the movable mirror can be effectively cooled in the resolved sideband limit.

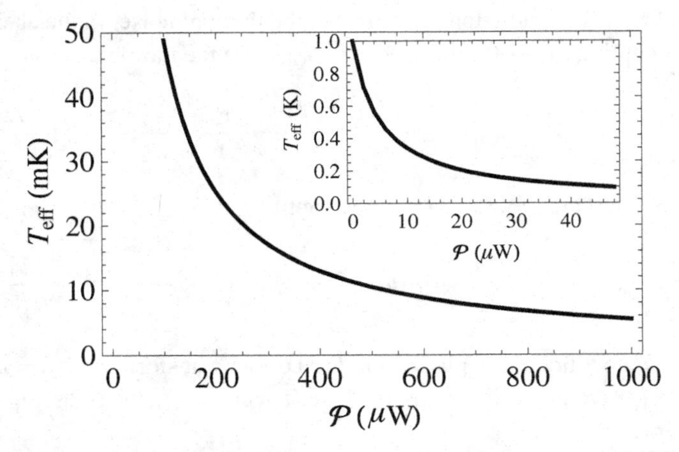

Fig. 20.6 The effective temperature T_{eff} (mK) of the movable mirror as a function of the laser power $\wp$ (μW). The initial temperature is taken to be 1 K.

20.6 Normal-mode splitting

20.6.1 Normal-mode splitting in mirror spectra

A phenomenon which is well known in the context of two coupled oscillators is the normal-mode splitting, which is a definite signature of the coupling between the oscillators. Consider for simplicity two oscillators a and b with degenerate frequencies such that their Hamiltonian is

$$H = \hbar\omega_a(a^\dagger a + b^\dagger b) + \hbar g(a^\dagger b + ab^\dagger). \tag{20.42}$$

This Hamiltonian can be diagonalized in straightforward manner with the result

$$\begin{gathered} H = \hbar(\omega_a + g)A^\dagger A + \hbar(\omega_a - g)B^\dagger B, \\ A = \frac{a+b}{\sqrt{2}}, \qquad B = \frac{a-b}{\sqrt{2}}, \\ [A, A^\dagger] = [B, B^\dagger] = 1, \qquad [A, B^\dagger] = 0. \end{gathered} \tag{20.43}$$

The normal-mode frequencies are $\omega_a \pm g$. The frequency splitting would be observable if the splitting g is bigger than the typical damping of the mode. For the optomechanical system, the two modes in question are (i) cavity mode with effective frequency Δ and (ii) the mirror with frequency ω_m. Clearly if the external field is detuned such that $\Delta \approx \omega_m$, then the normal-mode splitting [17] would occur if g exceeds the damping parameter, which itself would be determined by κ and ω_m. In addition, the strength of the coupling is determined by the driving field ε. In the previous section we have already calculated the spectrum of mirror's motion. We investigate $S_Q(\omega)$ given by (20.36) for different strengths of the driving field. We demonstrate the occurrence of the normal-mode splitting as the parameter ε increases. We choose the effective detuning $\Delta = \omega_m$, the pump power $\wp = 0.6$, 6.9, and

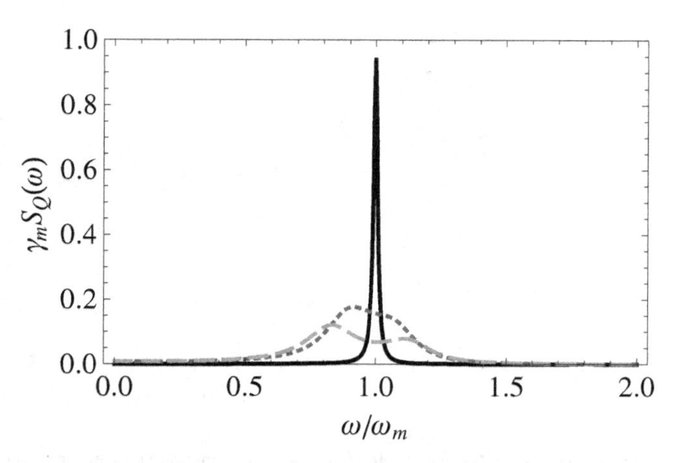

Fig. 20.7 The scaled spectrum $\gamma_m S_Q(\omega)$ as a function of the normalized frequency ω/ω_m for different pump power: $\wp = 0.6$ mW (solid curve), 6.9 mW (dotted curve), 10.7 mW (dashed curve). The curves for 6.9 mW and 10.7 mW are multiplied by a factor of 20.

10.7 mW, and the temperature of the environment $T = 300$ mK. The other parameters are the same as those mentioned for Figure 20.3.

The scaled spectrum $\gamma_m S_Q(\omega)$ as a function of the normalized frequency ω/ω_m for increasing the input laser power is shown in Figure 20.7. As we increase the laser power from 0.6 mW to 10.7 mW, the spectrum exhibits a doublet and the peak separation is proportional to the laser power. The doublet arises as the effective coupling between the movable mirror and the cavity field increases with increasing input laser power. Note that this coupling is determined by the steady-state photon number $|a_s|^2$ (Eq. (20.22)).

The structure of all the spectra is determined by the eigenvalues of $\mathrm{i}A$ (Eq. (20.29)) or the complex zeroes of the function $d(\omega)$ defined by Eq. (20.32). Clearly we need the eigenvalues of $\mathrm{i}A$ as the solution of Eq. (20.27) in the Fourier domain is $f(\omega) = \mathrm{i}(\omega - \mathrm{i}A)^{-1}\eta(\omega)$. Let us analyze the eigenvalues of Eq. (20.29). Note that in the absence of the coupling $\chi = 0$, the eigenvalues of $\mathrm{i}A$ are

$$\pm\sqrt{\omega_m^2 - \frac{\gamma_m^2}{4}} - \frac{\mathrm{i}\gamma_m}{2}; \qquad \pm\Delta - \mathrm{i}\kappa. \tag{20.44}$$

Thus the positive frequencies of the normal modes are given by Δ, $\sqrt{\omega_m^2 - \gamma_m^2/4}$ $(\omega_m > \gamma_m/2)$. The case that we consider in this section corresponds to

$$\omega_m \gg \frac{\gamma_m}{2}; \qquad \kappa \gg \gamma_m; \qquad \omega_m > \kappa. \tag{20.45}$$

The coupling between the normal modes would be most efficient in the degenerate case, i.e. when $\omega_m = \Delta$. It is known from cavity QED that the normal-mode splitting leads to symmetric (asymmetric) spectra in the degenerate (nondegenerate) case, provided that the dampings of the individual modes are much smaller than the coupling constant. Thus the mechanical oscillator is like the atomic oscillator, and the cavity mode in the rotating frame acquires the effective frequency Δ. All this applies provided that damping terms

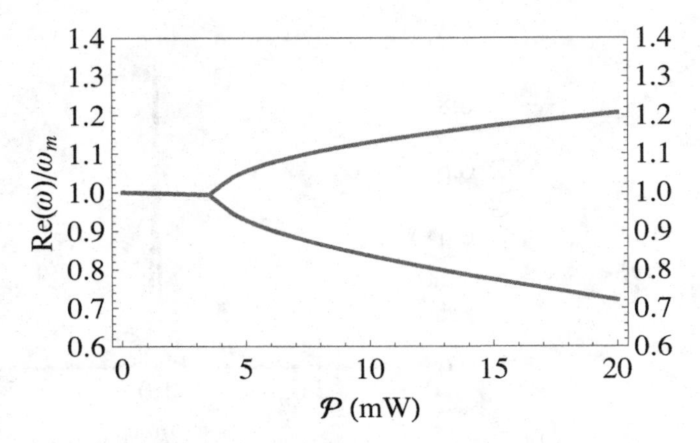

Fig. 20.8 The roots of $d(\omega)$ in the domain $\mathrm{Re}(\omega) > 0$ as a function of the pump power.

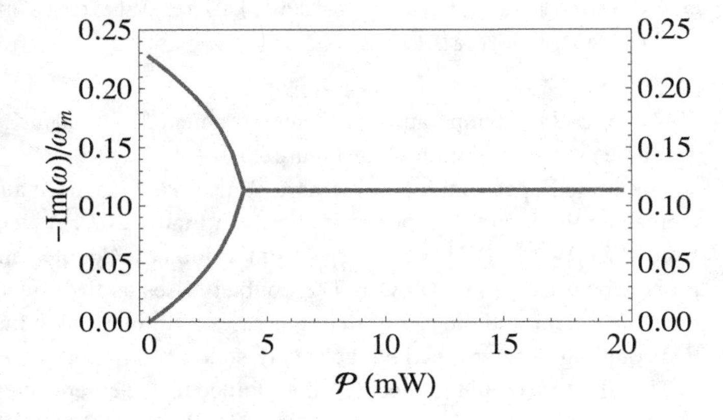

Fig. 20.9 The imaginary parts of the roots of $d(\omega)$ as a function of the pump power.

do not mix the modes significantly. An estimate of the splitting can be made by using the approximations given by Eq. (20.45) and the zeroes of $d(\omega)$. We find that the frequency splitting is given by

$$\omega_{\pm}^2 \cong \frac{\omega_m^2 + \Delta^2}{2} \pm \sqrt{\left(\frac{\omega_m^2 - \Delta^2}{2}\right)^2 + 2\omega_m^3 \chi^2 |a_s|^2 \Delta}. \tag{20.46}$$

It should be borne in mind that a_s is dependent on the pump power $\wp$. The splitting is determined by the pump power $\wp$.

Figure 20.8 shows the dependence of the real parts of the roots of $d(\omega)$ in the domain $\mathrm{Re}(\omega) > 0$ on the pump power. Figure 20.9 shows the dependence of the imaginary parts of the roots of $d(\omega)$ on the pump power. For small values of the pump power, the real parts of the roots of $d(\omega)$ have two equal values, so there is no splitting. However, there is lifetime splitting [18] as seen in Figure 20.9. If we increase the pump power beyond a threshold value, then the real parts of $d(\omega)$ in the domain begin to have two different values, and the difference between two real parts of the roots of $d(\omega)$ increases with increasing pump power.

20.6.2 Normal-mode splitting in the spectra of the output field

Next we examine the spectra of the output field. The output field carries the signature of the quantized motion of the mirror. The fluctuations $\delta a(\omega)$ of the cavity field can be obtained from Eq. (20.26). Furthermore, using the input–output relation (20.19), the fluctuations of the output field are given by

$$\delta a_{\text{out}}(\omega) = V(\omega)\xi(\omega) + E(\omega)\delta a_{\text{in}}(\omega) + F(\omega)\delta a_{\text{in}}^{\dagger}(-\omega), \tag{20.47}$$

where

$$\begin{aligned} V(\omega) &= -\frac{\sqrt{2\kappa}\omega_m^2\chi}{d(\omega)}\mathrm{i}[\kappa - \mathrm{i}(\Delta+\omega)]a_s, \\ E(\omega) &= \frac{2\kappa}{d(\omega)}\{-\omega_m^3\chi^2|a_s|^2\mathrm{i} + (\omega^2 - \omega_m^2 + \gamma_m\omega\mathrm{i})[\kappa - \mathrm{i}(\Delta+\omega)]\} - 1, \\ F(\omega) &= -\frac{2\kappa}{d(\omega)}\omega_m^3\chi^2 a_s^2\mathrm{i}. \end{aligned} \tag{20.48}$$

In Eq. (20.47), the first term associated with $\xi(\omega)$ stems from the thermal noise of the mechanical oscillator, while the other two terms are from the input vacuum noise. So the fluctuations of the output field are influenced by the thermal noise and the input vacuum noise.

The spectra of the output field are defined as

$$\begin{aligned} \langle \delta a_{\text{out}}^{\dagger}(-\Omega)\delta a_{\text{out}}(\omega)\rangle &= 2\pi S_{a\text{out}}(\omega)\delta(\omega+\Omega), \\ \langle \delta y_{\text{out}}(\Omega)\delta y_{\text{out}}(\omega)\rangle &= 2\pi S_{y\text{out}}(\omega)\delta(\omega+\Omega). \end{aligned} \tag{20.49}$$

where $\delta y_{\text{out}}(\omega)$ is the Fourier transform of the fluctuations $\delta y_{\text{out}}(t)$ of the output field, which is defined by $\delta y_{\text{out}}(t) = \mathrm{i}[\delta a_{\text{out}}^{\dagger}(t) - \delta a_{\text{out}}(t)]/\sqrt{2}$ [5]. Here $S_{a\text{out}}(\omega)$ denotes the spectral density of the output field, and $S_{y\text{out}}(\omega)$ is the spectrum of fluctuations in the y quadrature of the output field.

Combining Eq. (20.35), Eq. (20.47), and Eq. (20.49), we obtain the spectra of the output field

$$\begin{aligned} S_{a\text{out}}(\omega) &= V^*(\omega)V(\omega) \times \frac{\gamma_m}{\omega_m}\omega\left[-1 + \coth\left(\frac{\hbar\omega}{2k_{\text{B}}T}\right)\right] + F^*(\omega)F(\omega), \\ S_{y\text{out}}(\omega) &= -[V^*(\omega) - V(-\omega)][V^*(-\omega) - V(\omega)] \times \frac{\gamma_m}{2\omega_m}\omega\left[-1 + \coth\left(\frac{\hbar\omega}{2k_{\text{B}}T}\right)\right] \\ &\quad - [F^*(\omega) - E(-\omega)][E^*(-\omega) - F(\omega)]/2. \end{aligned} \tag{20.50}$$

In the following we will concentrate on discussing the dependence of the spectra on input laser power. The chosen parameters are the same as those for Figure 20.3, i.e. of the experiment by Gröblacher *et al.* [6]. Figures 20.10 and 20.11 show the spectra $S_{a\text{out}}(\omega)$ and $S_{y\text{out}}(\omega)$ as a function of the normalized frequency ω/ω_m for various values of the pump power. When $\wp = 0.6$ mW, the spectra do not exhibit normal-mode splitting. As the

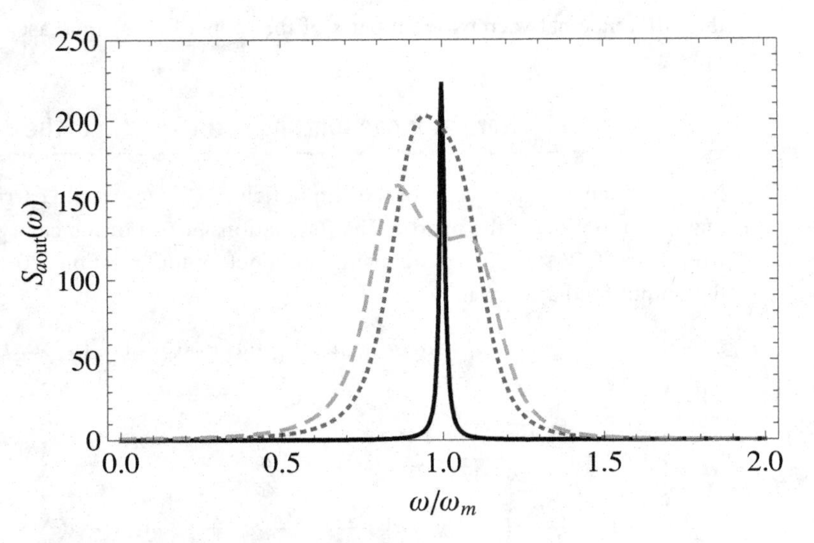

Fig. 20.10 The spectrum $S_{aout}(\omega)$ as a function of the normalized frequency ω/ω_m for different pump power: $\wp = 0.6$ mW (solid curve), 6.9 mW (dotted curve), 10.7 mW (dashed curve). The values in the last two cases are ten times the actual values.

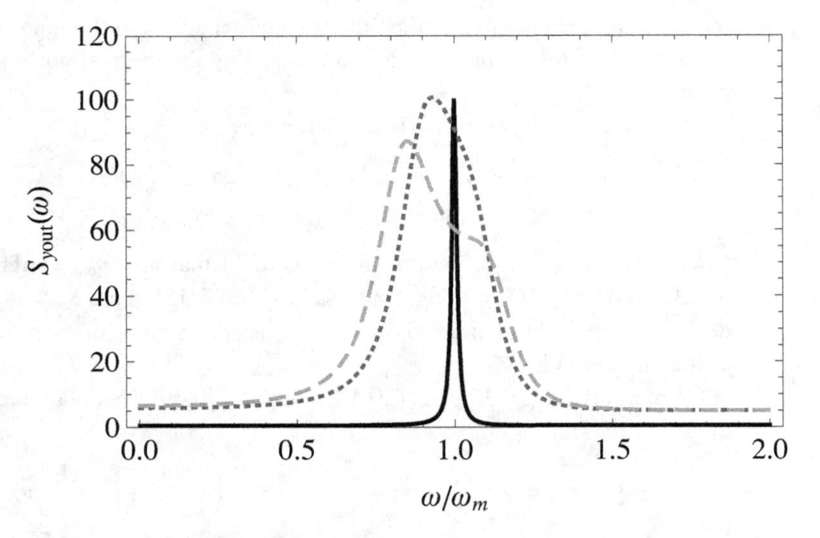

Fig. 20.11 The spectrum $S_{yout}(\omega)$ as a function of the normalized frequency ω/ω_m for different pump power: $\wp = 0.6$ mW (solid curve), 6.9 mW (dotted curve), 10.7 mW (dashed curve). The values in the last two cases are ten times the actual values.

pump power is increased, the normal-mode splitting becomes observable. This is due to significant changes in the line widths and their positions. Note that the separation between two peaks becomes larger as pump power increases. The reason is that increasing the pump power causes a stronger coupling between the movable mirror and the cavity field due to an

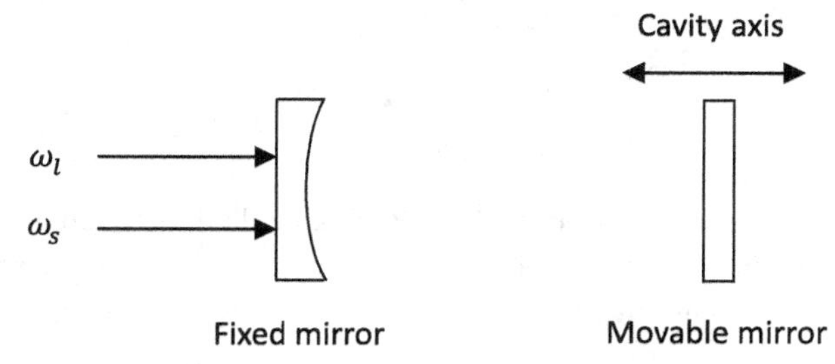

Fig. 20.12 Schematic sketch of the system for squeezing the mirror motion. A laser with frequency ω_l and squeezed vacuum light with frequency ω_s enter the cavity through the partially transmitting mirror, $\omega_s \approx \omega_c, \omega_l = \omega_c - \omega_m$.

increase in the photon number in the cavity. We have examined the contributions of various terms in Eq. (20.50) to the output spectrum. The dominant contribution comes from the mechanical oscillator. Also note the similarity [6] of the spectrum of the output quadrature y (Figure 20.11) to the spectrum of the mechanical oscillator (Figure 20.7).

Normal-mode splitting in the optomechanical system was demonstrated by Gröblacher *et al.* [6]. They showed that the splitting in the emission spectra of the driven cavity becomes clearly visible by increasing the optical pump power. The presence of a strong pump field leads to enhancement in the optomechanical coupling strength so that the optomechanical coupling strength exceeds both the photon decay rate κ and the mechanical damping rate γ_m. The normal-mode splitting effects [19] are quite generic, although for Figures 20.7–20.11 we use parameters from the experiment [6]. In this section, we studied the normal-mode splitting in the fluctuation spectra. An alternate method to study normal-mode splitting is via pump-probe spectroscopy of optomechanical systems as discussed in Section 20.8.

20.7 Squeezing of a nano-mechanical oscillator

In this section, we propose a scheme that is capable of generating squeezing of the movable mirror by feeding broadband squeezed vacuum light along with the laser light [20, 21]. The system shown in Figure 20.12 is similar to Figure 20.2, the only difference being that the cavity is also fed with squeezed light at frequency ω_s. Thus the input vacuum noise a_{in} in Eq. (20.16) is replaced by the input squeezed vacuum noise with frequency $\omega_s = \omega_l + \omega_m$. It has zero mean value, and nonzero time-domain correlation functions [22]

$$\begin{aligned}
\langle \delta a_{\text{in}}^\dagger(t)\delta a_{\text{in}}(t')\rangle &= N\delta(t-t'),\\
\langle \delta a_{\text{in}}(t)\delta a_{\text{in}}^\dagger(t')\rangle &= (N+1)\delta(t-t'),\\
\langle \delta a_{\text{in}}(t)\delta a_{\text{in}}(t')\rangle &= M\mathrm{e}^{-\mathrm{i}\omega_m(t+t')}\delta(t-t'),
\end{aligned} \tag{20.51}$$

where $N = \sinh^2(r)$, $M = \sinh(r)\cosh(r)\mathrm{e}^{\mathrm{i}\varphi}$, r is the squeezing parameter of the squeezed vacuum, and φ is the phase of the squeezed vacuum. It should be born in mind that a_{in} is in a frame rotating with the frequency ω_l. The fluctuations in position and momentum of the movable mirror are determined by

$$\langle \delta X(t)^2 \rangle = \frac{1}{4\pi^2} \iint_{-\infty}^{+\infty} \mathrm{d}\omega \mathrm{d}\Omega \mathrm{e}^{-\mathrm{i}(\omega+\Omega)t} \langle \delta X(\omega) \delta X(\Omega) \rangle, \tag{20.52}$$

where X stands for either Q or P. To calculate the fluctuations, we require the correlation functions of the noise sources in the frequency domain. These can be obtained using (20.51) and the definition $f(t) = \frac{1}{2\pi} \int f(\omega) \mathrm{e}^{-\mathrm{i}\omega t} \mathrm{d}\omega$ of the Fourier transforms

$$\begin{aligned} \langle \delta a_{\text{in}}^\dagger(-\omega) \delta a_{\text{in}}(\Omega) \rangle &= 2\pi N \delta(\omega + \Omega), \\ \langle \delta a_{\text{in}}(\omega) \delta a_{\text{in}}^\dagger(-\Omega) \rangle &= 2\pi (N+1) \delta(\omega + \Omega), \\ \langle \delta a_{\text{in}}(\omega) \delta a_{\text{in}}(\Omega) \rangle &= 2\pi M \delta(\omega + \Omega - 2\omega_m), \\ \langle \delta a_{\text{in}}^\dagger(-\omega) \delta a_{\text{in}}^\dagger(-\Omega) \rangle &= 2\pi M^* \delta(\omega + \Omega + 2\omega_m), \\ \langle \xi(\omega) \xi(\Omega) \rangle &= 2\pi \frac{\gamma_m}{\omega_m} \omega \left[1 + \coth\left(\frac{\hbar\omega}{2K_{\text{B}}T} \right) \right] \delta(\omega + \Omega). \end{aligned} \tag{20.53}$$

Combining Eqs. (20.31), (20.52), and (20.53), after some calculations, the fluctuations of Eq. (20.52) are written as

$$\begin{aligned} \langle \delta Q(t)^2 \rangle &= \frac{1}{2\pi} \int_{-\infty}^{+\infty} \omega_m^2 (A + B\mathrm{e}^{-2\mathrm{i}\omega_m t} + C\mathrm{e}^{2\mathrm{i}\omega_m t}) \mathrm{d}\omega, \\ \langle \delta P(t)^2 \rangle &= \frac{1}{2\pi} \int_{-\infty}^{+\infty} [\omega^2 A + \omega(\omega - 2\omega_m) B\mathrm{e}^{-2\mathrm{i}\omega_m t} + \omega(\omega + 2\omega_m) C\mathrm{e}^{2\mathrm{i}\omega_m t}] \mathrm{d}\omega, \end{aligned} \tag{20.54}$$

where

$$\begin{aligned} A &= \frac{1}{d(\omega)d(-\omega)} \Big(2\kappa\omega_m^2 \chi^2 |a_s|^2 \{ (N+1)[\kappa^2 + (\Delta+\omega)^2] + N[\kappa^2 + (\Delta-\omega)^2] \} \\ &\quad + \gamma_m \frac{\omega}{\omega_m} [(\Delta^2 + \kappa^2 - \omega^2)^2 + 4\kappa^2\omega^2] \Big[1 + \coth\left(\frac{\hbar\omega}{2K_{\text{B}}T} \right) \Big] \Big), \\ B &= \frac{2\kappa\omega_m^2 \chi^2 a_s^{*2} M}{d(\omega)d(2\omega_m - \omega)} [\kappa - \mathrm{i}(\Delta + \omega)][\kappa - \mathrm{i}(\Delta + 2\omega_m - \omega)], \\ C &= \frac{2\kappa\omega_m^2 \chi^2 a_s^2 M^*}{d(\omega)d(-2\omega_m - \omega)} [\kappa + \mathrm{i}(\Delta - \omega)][\kappa + \mathrm{i}(\Delta + 2\omega_m + \omega)], \end{aligned} \tag{20.55}$$

and $d(\omega)$ is defined by (20.32). In Eqs. (20.54) and (20.55), the term independent of χ is from the thermal noise; while those terms involving χ arise from the radiation pressure contribution, including the influence of the squeezed vacuum light. Moreover, either $\langle \delta Q(t)^2 \rangle$ or $\langle \delta P(t)^2 \rangle$ contains three terms, the first term being independent of time, but the second and third terms are time-dependent, which causes $\langle \delta Q(t)^2 \rangle$ and $\langle \delta P(t)^2 \rangle$ to vary with time. The complex exponential in Eq. (20.54) can be removed by working in the interaction picture. Let's define b ($b^\dagger$) and $\tilde{b}$ ($\tilde{b}^\dagger$) as the annihilation (creation) operators for the oscillator in the Schrödinger and interaction picture with $[b, b^\dagger] = 1$ and $[\tilde{b}, \tilde{b}^\dagger] = 1$. The relations between them are $b = \tilde{b}\mathrm{e}^{-\mathrm{i}\omega_m t}$ and $b^\dagger = \tilde{b}^\dagger \mathrm{e}^{\mathrm{i}\omega_m t}$. Then using $Q = (b + b^\dagger)/\sqrt{2}$, $P = \mathrm{i}(b^\dagger - b)/\sqrt{2}$,

$\tilde{Q} = (\tilde{b} + \tilde{b}^\dagger)/\sqrt{2}$, and $\tilde{P} = \mathrm{i}(\tilde{b}^\dagger - \tilde{b})/\sqrt{2}$, we get

$$\begin{aligned}\langle \delta \tilde{Q}^2 \rangle &= \frac{1}{2\pi}\int_{-\infty}^{+\infty} \omega_m^2 (A + B + C)\mathrm{d}\omega,\\ \langle \delta \tilde{P}^2 \rangle &= \frac{1}{2\pi}\int_{-\infty}^{+\infty} [\omega^2 A + \omega(\omega - 2\omega_m)B + \omega(\omega + 2\omega_m)C]\mathrm{d}\omega.\end{aligned} \tag{20.56}$$

Now according to the Heisenberg uncertainty principle,

$$\langle \delta \tilde{Q}^2 \rangle \langle \delta \tilde{P}^2 \rangle \geq \left| \frac{1}{2}[\tilde{Q}, \tilde{P}] \right|^2. \tag{20.57}$$

If either $\langle \delta \tilde{Q}^2 \rangle < 0.5$ or $\langle \delta \tilde{P}^2 \rangle < 0.5$, the movable mirror is said to be squeezed. From Eqs. (20.55) and (20.56), we find $\langle \delta \tilde{Q}^2 \rangle$ or $\langle \delta \tilde{P}^2 \rangle$ is determined by the effective detuning Δ, the squeezing parameter r, the laser power $\wp$, the cavity length L, and the temperature of the environment T. Here we focus on the dependence of $\langle \delta \tilde{Q}^2 \rangle$ and $\langle \delta \tilde{P}^2 \rangle$ on the squeezing parameter, the temperature of the environment, and the laser power. We numerically evaluate the mean square fluctuations in position and momentum of the movable mirror given by Eq. (20.56) to show squeezing of the movable mirror produced by feeding the squeezed vacuum light at the input mirror. For $T = 0$ and $\omega < 0\,(> 0)$, $\coth(\hbar\omega/(2k_{\mathrm{B}}T)) \simeq -1$ $(\simeq +1)$. Through numerical calculations, it is found that $\langle \delta \tilde{Q}^2 \rangle$ is not squeezed but $\langle \delta \tilde{P}^2 \rangle$ is squeezed. In the following we therefore concentrate on $\langle \delta \tilde{P}^2 \rangle$. Note that in the absence of the coupling to the cavity field, the movable mirror is in free space, and is coupled to the environment. Then the fluctuations are given by $\langle \delta \tilde{Q}^2 \rangle = \langle \delta \tilde{P}^2 \rangle = 0.5 + 1/(\mathrm{e}^{\hbar\omega_m/(k_{\mathrm{B}}T)} - 1)$ ≈ 440, for $T = 20\,\mathrm{mK}$, $\omega_m = 2\pi \times 947\,\mathrm{kHz}$. As is well known no squeezing of the movable mirror occurs.

Now we consider fluctuations in the presence of the coupling to the cavity field. If we choose $T = 1\,\mathrm{mK}$, and $\wp = 6.9\,\mathrm{mW}$, the fluctuation $\langle \delta \tilde{P}^2 \rangle$ is plotted as a function of the detuning Δ in Figure 20.13. Different graphs correspond to different values of the squeezing of the input light. It is seen that the fluctuation $\langle \delta \tilde{P}^2 \rangle$ falls below the standard quantum noise limit, so the momentum squeezing of the movable mirror occurs, and the maximum squeezing happens at about $r = 1$, the corresponding minimum value of $\langle \delta \tilde{P}^2 \rangle$ is 0.160, thus the maximum amount of squeezing is about 68%. So the injection of the squeezed vacuum light reduces, to a large extent, the fluctuation in the momentum of the mirror. Note that Eq. (20.26) shows how the fluctuations in momentum relate to the quadratures of the field in the cavity which in turn are determined by the input squeezed light. Clearly the squeezing of the mirror is quite sensitive to the temperature of the environment. Next we discuss how the temperature deteriorates the amount of achievable squeezing. We show in Figure 20.14 the effect of increasing temperature for a fixed laser power and for a fixed squeezing of the input. At $T = 5\,\mathrm{mK}$, the minimum value of $\langle \delta \tilde{P}^2 \rangle$ is 0.251, the corresponding amount of squeezing is about 50%. At $T = 20\,\mathrm{mK}$, the momentum squeezing of the movable mirror vanishes since the minimum value of $\langle \delta \tilde{P}^2 \rangle$ is 0.593, which is larger than 0.5. Note that the minimum value of $\langle \delta \tilde{P}^2 \rangle$ in the presence of the coupling to the cavity field is much less than that ($\langle \delta \tilde{P}^2 \rangle = 440$) in the absence of the coupling to the cavity field for $T = 20\,\mathrm{mK}$. So there is very large squeezing of the thermal fluctuations. The momentum fluctuations can be reduced by a factor more than seven hundred for $T = 20\,\mathrm{mK}$.

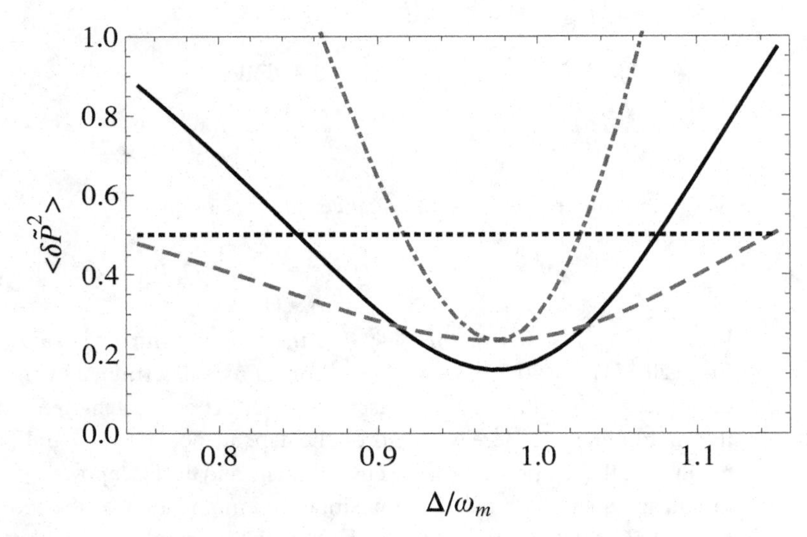

Fig. 20.13 The fluctuation $\langle \delta\tilde{P}^2 \rangle$ versus the detuning Δ/ω_m for different values of the squeezing of the input field: $r = 0.5$ (dashed line), $r = 1$ (solid curve), $r = 1.5$ (dot-dashed curve) for the temperature of the environment $T = 1$ mK and the laser power $\wp = 6.9$ mW. The minimum values, comparable to the ones at $T = 0$ mK, of $\langle \delta\tilde{P}^2 \rangle$ are 0.211 ($r = 0.5$), 0.138 ($r = 1$), 0.213 ($r = 1.5$). The flat dotted line represents the standard quantum noise limit ($\langle \delta\tilde{P}^2 \rangle = 0.5$), other parameters are the same as in Figure 20.3.

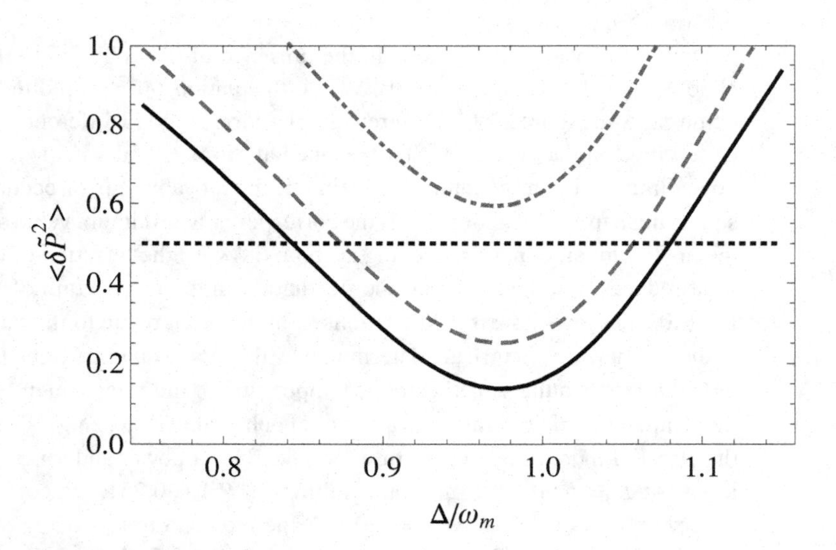

Fig. 20.14 The fluctuation $\langle \delta\tilde{P}^2 \rangle$ versus the detuning Δ/ω_m for different temperature of the environment: $T = 0$ K (solid curve), 5 mK (dashed curve), 20 mK (dotdashed curve) for the squeezing parameter $r = 1$ and the laser power $\wp = 6.9$ mW. The minimum values of $\langle \delta\tilde{P}^2 \rangle$ are 0.138 ($T = 0$ K), 0.251 ($T = 5$ mK), 0.593 ($T = 20$ mK). The flat dotted line represents the standard quantum noise limit ($\langle \delta\tilde{P}^2 \rangle = 0.5$).

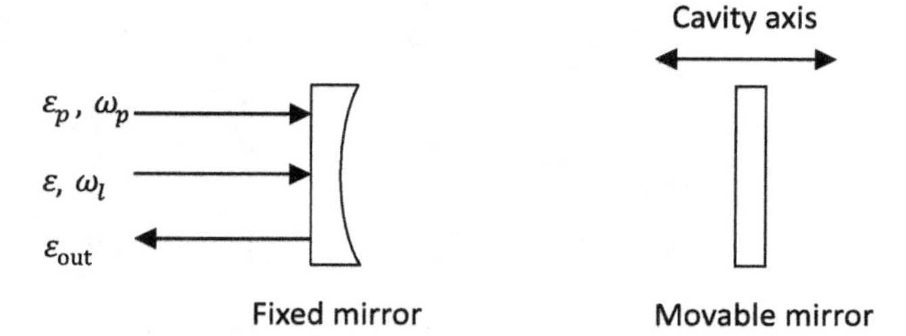

Fig. 20.15 Pump-probe response of the optomechanical system coupled to a high-quality cavity via radiation pressure effects.

The squeezing of the movable mirror can be considered as transfer of squeezing from one system to the other. The transfer process is quite efficient as Figure 20.13 shows. Furthermore, the feeding of squeezed light can be used to squeeze collective degrees of freedom for several mirrors inside the cavity. One can also use these techniques to entangle two mirrors [23].

20.8 Electromagnetically induced transparency (EIT) in the mechanical effects of light

In this section, we discuss the possibility of electromagnetically induced transparency (EIT) in the context of cavity optomechanics [19,24–26]. In Chapter 13 we studied how the pump probe spectroscopy of two level systems is an important tool to study the properties of two-level systems dressed by intense coherent fields. In Chapter 17 we studied the pump probe response of three-level systems and we showed how this response can lead to the phenomenon of EIT under certain conditions. Motivated by the discussions in Chapters 13 and 17, we consider the pump probe spectroscopy of optomechanical systems. In the context of EIT, the pump field is usually called the control or coupling field.

The radiation pressure interaction $\chi Q a^\dagger a$ is nonlinear and this nonlinearity was the reason for existence of bistability in optomechanical systems. If the average field $\langle a \rangle$ in the cavity had a modulation say at frequency δ, i.e. if $\langle a \rangle = \langle a \rangle_0 + \mathrm{e}^{-\mathrm{i}\delta t}\langle a \rangle_+ + \mathrm{e}^{\mathrm{i}\delta t}\langle a \rangle_-$, then the radiation matter interaction would have oscillating terms at $\pm\delta$, $\pm 2\delta$. In this case the mean field at the output would have such modulations. Clearly if the cavity (Figure 20.15) is driven by a second field ε_p at a frequency ω_p, then the cavity field would be modulated at $\delta = \omega_p - \omega_l$. Thus we expect that in this case the output fields would have frequency components ω_l, ω_p, and $2\omega_l - \omega_p$ in addition to a host of other components arising from the nonlinear mixing of the waves at frequencies ω_l and ω_p. The nonlinearity is due to the radiation pressure interaction. Furthermore, if we choose $\omega_p - \omega_l \cong \omega_m$, $\omega_c - \omega_l \cong \omega_m$, then $\omega_p \cong \omega_c$, and therefore the output field at frequency ω_p would be resonantly enhanced. The production of different frequency components can also be seen as Raman processes of different order as shown in Figure 20.16.

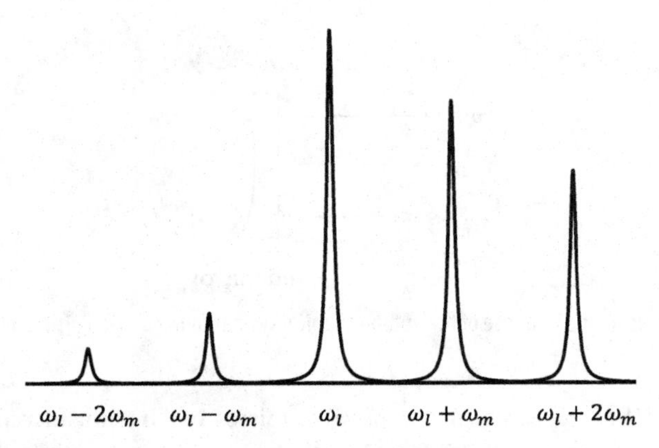

Fig. 20.16 Raman processes of different order arising from the radiation pressure interaction.

In this section we work with the mean field response which is obtained from (20.16) by taking expectation values and using the factorization $\langle AB\rangle \approx \langle A\rangle\langle B\rangle$, where A and B are any two operators, then (20.16) becomes

$$\begin{aligned}
\langle \dot{Q}\rangle &= \omega_m \langle P\rangle, \\
\langle \dot{P}\rangle &= \omega_m \chi \langle a^\dagger\rangle\langle a\rangle - \omega_m\langle Q\rangle - \gamma_m\langle P\rangle, \\
\langle \dot{a}\rangle &= -\mathrm{i}(\omega_c - \omega_l - \omega_m\chi\langle Q\rangle)\langle a\rangle - \kappa\langle a\rangle + \varepsilon + \varepsilon_p \mathrm{e}^{-\mathrm{i}\delta t}, \\
\langle \dot{a}^\dagger\rangle &= \mathrm{i}(\omega_c - \omega_l - \omega_m\chi\langle Q\rangle)\langle a^\dagger\rangle - \kappa\langle a^\dagger\rangle + \varepsilon + \varepsilon_p \mathrm{e}^{\mathrm{i}\delta t},
\end{aligned} \tag{20.58}$$

where $\delta = \omega_p - \omega_l$. In deriving (20.58), we have added the contribution from the probe field ε_p. The probe field ε_p is equal to $\sqrt{2\kappa\wp_p/(\hbar\omega_p)}$, where $\wp_p$ is the power of the probe field. Both ε and ε_p have dimensions of frequency. In the long time limit, all mean values would have the form

$$\langle A\rangle = \sum_{n=-\infty}^{+\infty} \mathrm{e}^{-\mathrm{i}n\delta t}\langle A\rangle_n. \tag{20.59}$$

The substitution of (20.59) in (20.58) leads to an hierarchy of coupled equations. We assume, as is standard in pump probe spectroscopy, that the probe is much weaker than pump $|\varepsilon_p| \ll |\varepsilon|$. In this case, we can terminate the series in (20.59) at $n = 1$. In what follows, we study quadratures of the cavity field at the frequency ω_p and at $2\omega_l - \omega_p$. We define dimensionless fields as

$$\varepsilon_{as} = 2\kappa\langle a\rangle_{n=1}/\varepsilon_p, \qquad \varepsilon_s = 2\kappa\langle a\rangle_{n=-1}/\varepsilon_p, \tag{20.60}$$

which directly give the output fields as well. The ε_{as} gives the output field at the frequency of the probe field and ε_s gives the field generated via the nonlinear mixing process. The field ε_s is at the frequency $2\omega_l - \omega_p$. On substituting (20.59) (with only $n = 0, \pm 1$ terms) in (20.58) and equating the coefficients of different Fourier components, we get algebraic equations which can be solved in a straightforward manner. We cite the results

$$\begin{aligned}\varepsilon_{as} &= \frac{2\kappa}{d(\delta)}\{(\delta^2-\omega_m^2+\mathrm{i}\gamma_m\delta)[\kappa-\mathrm{i}(\Delta+\delta)]-\mathrm{i}\omega_m\beta\},\\ \varepsilon_s &= -\frac{2\kappa}{d^*(\delta)}\mathrm{i}\omega_m^3\chi^2 a_s^2,\end{aligned} \tag{20.61}$$

where

$$\begin{aligned}d(\delta) &= 2\omega_m\Delta\beta+(\delta^2-\omega_m^2+\mathrm{i}\gamma_m\delta)[(\kappa-\mathrm{i}\delta)^2+\Delta^2],\\ \Delta &= \omega_c-\omega_l-\frac{\beta}{\omega_m}, \qquad \beta=\omega_m^2\chi^2|a_s|^2, \qquad a_s=\frac{\varepsilon}{\kappa+\mathrm{i}\Delta}.\end{aligned} \tag{20.62}$$

Note that if there were no control laser $\varepsilon=0$, then $a_s\to 0$, $\beta\to 0$, and

$$\varepsilon_{as}=\frac{2\kappa}{\kappa-\mathrm{i}(\omega_p-\omega_c)}, \qquad \varepsilon_s=0. \tag{20.63}$$

We define the two quadratures of the field as

$$\varepsilon_{as}=\upsilon_p-\mathrm{i}\tilde{\upsilon}_p. \tag{20.64}$$

When $\varepsilon=0$, $\upsilon_p=\frac{2\kappa^2}{\kappa^2+(\omega_p-\omega_c)^2}$, and $\tilde{\upsilon}_p=-\frac{2\kappa(\omega_p-\omega_c)}{\kappa^2+(\omega_p-\omega_c)^2}$. Clearly the two quadratures are like the imaginary and real parts of the susceptibility of a two-level atom in a weak field (Eq. (13.29)). Let us now examine the modifications in these quadratures as the control field is increased. It is also important to observe that the polynomial $d(\delta)$ appearing in the denominator in (20.61) is the same polynomial that appeared in the earlier section and that which determined the normal-mode splitting.

In order to understand the coupling field (ε) induced modification of the probe response ε_{as}, we make reasonable approximations. We work in the sideband resolved limit $\omega_m\gg\kappa$. This is the limit in which normal-mode splitting [6, 13, 17] has been discovered. Because it is known that the coupling between the nano-oscillator and the cavity is strongest whenever $\delta=\pm\omega_m$ or $\delta=\pm\Delta$, the case $\Delta\sim\omega_m$ is considered here. After some simplifications, we can write the field ε_{as} in an instructive form,

$$\varepsilon_{as}=\upsilon_p-\mathrm{i}\tilde{\upsilon}_p=\frac{2\kappa}{\kappa-\mathrm{i}x+\dfrac{\beta/2}{\frac{\gamma_m}{2}-\mathrm{i}x}}=\frac{A_+}{x-x_+}+\frac{A_-}{x-x_-}, \tag{20.65}$$

where $x=\delta-\omega_m$, which is the detuning from the line center of the mirror's frequency. Furthermore, it is seen that the denominator has two roots, which are

$$x_\pm=\frac{-\mathrm{i}(\kappa+\frac{\gamma_m}{2})\pm\sqrt{-(\kappa-\frac{\gamma_m}{2})^2+2\beta}}{2}, \tag{20.66}$$

whose nature depends on the power of the coupling laser. For coupling powers less than the critical power

$$\tilde{\wp}_c=\frac{\hbar\omega_l|a_s|^2(\kappa^2+\omega_m^2)(\kappa-\frac{\gamma_m}{2})^2}{4\kappa\beta}, \tag{20.67}$$

the two roots are purely imaginary. For $\wp>\tilde{\wp}_c$, the roots are complex conjugates of each other. The region $\wp>\tilde{\wp}_c$ corresponds to the region where normal-mode splitting [6, 13, 17]

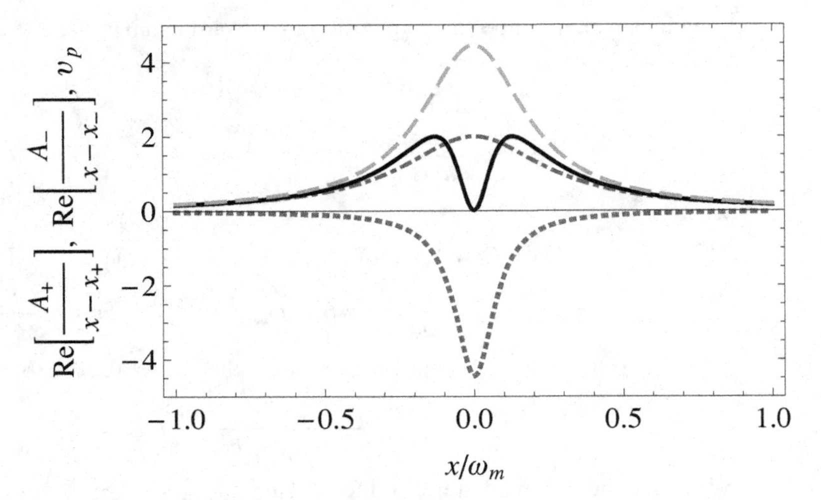

Fig. 20.17 Quadrature v_p of the field (solid curve) and the different contributions to it: the real parts of $A_+/(x-x_+)$ (dotted curve) and $A_-/(x-x_-)$ (dashed curve) as a function of the normalized frequency x/ω_m for input coupling laser power $\wp = 6$ mW. The dot-dashed curve is v_p in the absence of the coupling laser. Parameters used are the wavelength of the laser $\lambda = 2\pi c/\omega_l = 775$ nm, $\chi = 2\pi \times 12$ GHz/nm $\times \frac{1}{\omega_m}\sqrt{\frac{\hbar}{m\omega_m}}$, $m = 20$ ng, $\omega_m = 2\pi \times 51.8$ MHz, $\kappa = 2\pi \times 15$ MHz, $\gamma_m = 2\pi \times 41$ kHz, $\kappa/\omega_m = 0.289$, the mechanical quality factor $Q = \omega_m/\gamma_m = 1263$.

occurs and has been studied recently using a very different technique. In the context of optical physics, this is the region where Autler–Townes splitting [27] occurs. Furthermore, for EIT, it is important to have $\gamma_m \ll \kappa$.

In order to bring out prominently features like EIT [28–30], we specifically examine the case when the coupling power is less than the critical power. Here the relevant roots of $d(\delta)$ correspond to the parts of Figures 20.8 and 20.9 left to the bifurcation point. Note that $x_+ \to -\mathrm{i}\frac{\gamma_m}{2}$, $x_- \to -\mathrm{i}\kappa$ as $\beta \to 0$. Thus, the quadratures of the field have two distinct contributions in the limit of low values of the coupling laser strength. One contribution is extremely narrow as $\gamma_m \ll \kappa$. This characteristic property leads to the EIT dip. For numerical work, we use parameters from a recent experiment on the observation of optomechanically induced transparency [25]. We calculate the critical power $\tilde{\wp}_c$ to be 7.5 mW. In Figures 20.17 and 20.18, we show each contribution in Eq. (20.65) separately and also the total contribution. We observe that the narrow contribution is inverted relative to the broad contribution, and this leads to the typical EIT-like line shape for the quadrature v_p of the cavity field. The value at the dip is not exactly zero as $\gamma_m \neq 0$, although the value is very small as $\gamma_m \ll \kappa$. This is similar to what one has in the context of EIT in atomic systems where a strict zero is obtained if the ground-state atomic coherence has an infinite lifetime. In the absence of the coupling field, the narrow feature disappears (dot-dashed curve in Figure 20.17). The narrow feature's width has a contribution which depends on the coupling laser power. In leading order, the width is $\frac{\gamma_m}{2} + \frac{\beta}{\kappa}$. For the plot of Figure 20.17, the power-dependent contribution to the width in dimensionless units is $\beta/\kappa^2 \sim 0.398$. Thus

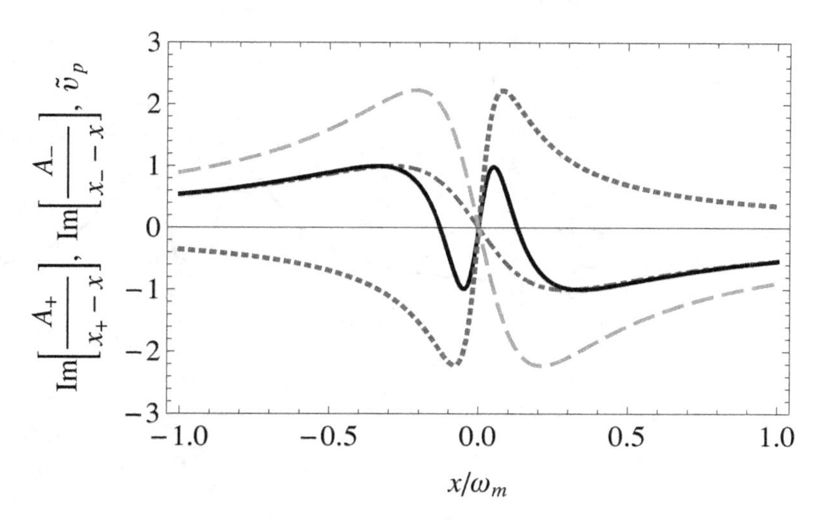

Fig. 20.18 Quadrature $\tilde{\upsilon}_p$ of the field (solid curve) and the different contributions to it: the imaginary parts of $A_+/(x_+ - x)$ (dotted curve) and $A_-/(x_- - x)$ (dashed curve) as a function of the normalized frequency x/ω_m for input coupling laser power $\wp = 6$ mW. The dot-dashed curve is $\tilde{\upsilon}_p$ in the absence of the coupling laser.

the width of the hole depends linearly on the power of the pump. This feature has been observed in experiments [19,25,26].

The quadrature $\tilde{\upsilon}_p$ exhibits dispersive behavior, and the coupling field changes the nature of dispersion from anomalous to normal in the region where quantum interferences are prominent. This behavior of dispersion is similar to the one found [31–33] in predictions of slow light in atomic systems.

We next present the nature of interferences in the region when $\wp > \tilde{\wp}_c$ in Figures 20.19 and 20.20. A typical behavior is shown in Figure 20.19 which clearly shows how the interference of the two contributions in Eq. (20.65) leads to the formation of the dip. The two contributions in Eq. (20.65) lead to asymmetric profiles. In the region of EIT, the tails from these contributions interfere. Unlike the case given by Figure 20.17, the two contributions have identical line widths. The roots correspond to the region to the right of the bifurcation points in Figures 20.8 and 20.9. From Figure 20.20, we also see how the dispersive behavior is changed by the coupling field from anomalous to normal in the region where quantum interferences are dominated. The inverted nature of the contribution A_+ should be noted, and it is this which changes the nature of dispersion.

In traditional EIT, say, in atomic vapors described by Λ systems (Chapter 17) the atomic coherence plays a very importance role. For the optomechanical system the quantity analogous to atomic coherence is the motion of the mirror which would have the structure (20.59). We show the component of Q oscillating as $e^{-i\delta t}$ in Figure 20.21. We use the normalization factor $(\varepsilon_p/\kappa)^2$ which is just the mean number of photons at the frequency $\omega_p = \omega_c$ if there is no pump.

We now explain the origin of the structure (20.65) for the probe response. Let us re-examine the Hamiltonian (20.15). Note that we drive the cavity with arbitrary pump

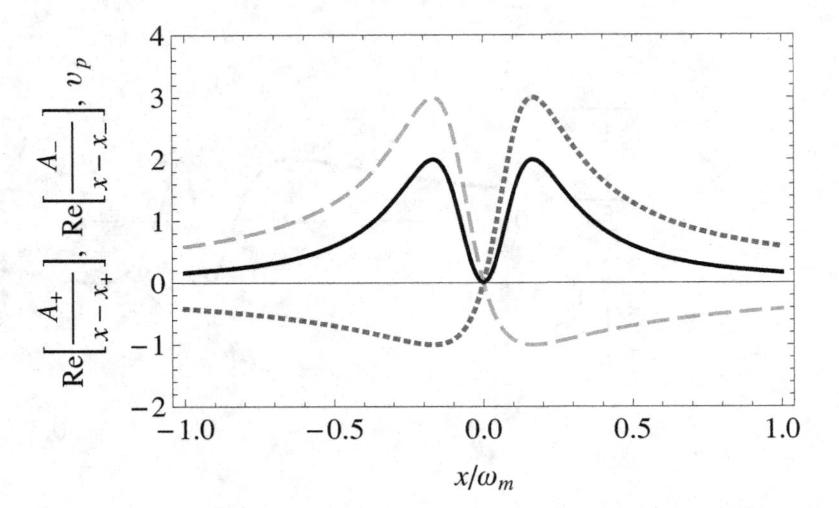

Fig. 20.19 Same as in Figure 20.17 except the input coupling laser power $\wp = 10$ mW and $\wp = 0$ case is not shown.

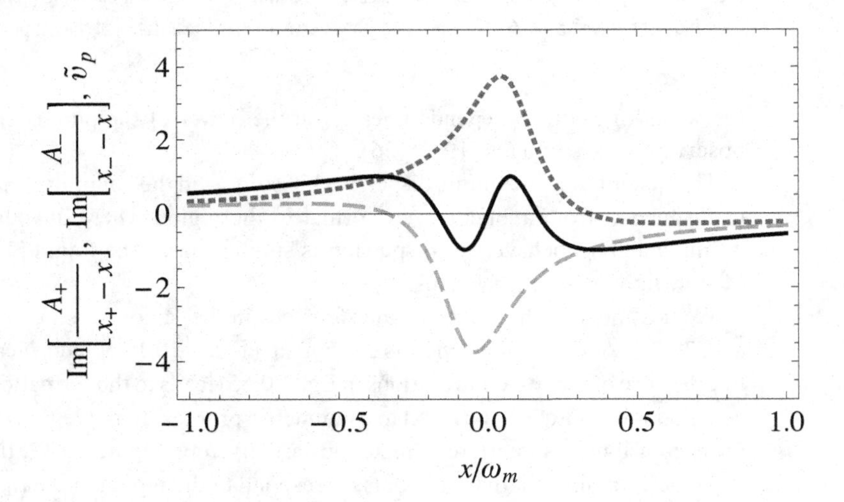

Fig. 20.20 Same as in Figure 20.18 except the input coupling laser power $\wp = 10$ mW and $\wp = 0$ case is not shown.

field ε_p. This effectively prepares the cavity in a coherent state with a value a_s if all the other interactions were zero. On writing the cavity operator a as $a_s + \delta a$, the trilinear interaction due to radiation pressure $\chi a^\dagger a Q$ can now be written as $\chi Q|a_s|^2 + \chi Q(a_s^* \delta a + a_s \delta a^\dagger)+$ higher-order terms. The pump thus has resulted in a bilinear interaction between the cavity oscillator and the mirror oscillator. The cavity oscillator is driven by the probe field, whereas the matter oscillator has no external drive. The cavity oscillator is damped at the rate κ, whereas the mirror is damped at the rate $\gamma_m \ll \kappa$. This situation typically results in line shapes such as (20.65). Thus for optomechanical systems, we have an exact analog of EIT in Λ systems (Chapter 17) provided the damping of the nanomechanical mirror is much smaller than the dissipation in the cavity. The EIT in optomechanical systems can

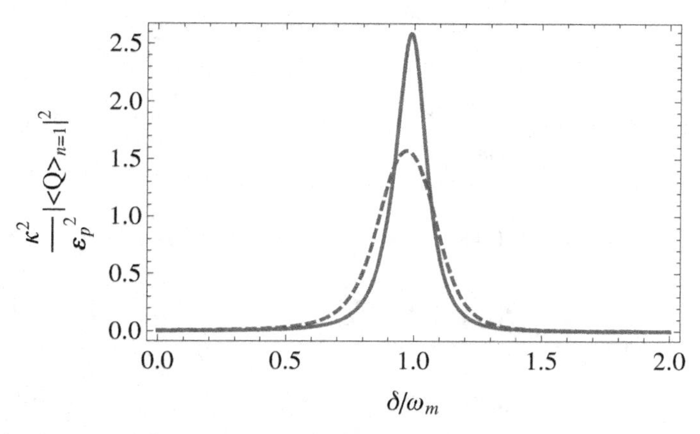

Fig. 20.21 The quantity $(\kappa^2/\varepsilon_p^2)|\langle Q\rangle_{n=1}|^2$ as a function of the normalized frequency δ/ω_m for two different values of the laser power: $\wp = 6$ mW (solid), 10 mW (dashed).

be understood as due to different interfering contributions as displayed, for example, in Figure 20.17.

Clearly, many of the applications of EIT in atomic vapors can be taken over in the context of optomechanical systems. For example light can be stored in the coherent motion of the mechanical oscillator [34]. The EIT in optomechanical systems with quantized fields is discussed in [35], and such systems can be used as quantum memory elements and single-photon routers [36].

20.8.1 Parametric generation of the Stokes field ε_s in optomechanics

We now give a short discussion of the field ε_s (Eq. (20.61)) generated at the frequency $2\omega_l - \omega_p$. Clearly this is generated via the nonlinear mixing of fields at ω_l and ω_p. No field is applied at the frequency $2\omega_l - \omega_p$. Note that ε_s is at least of order ε^2 as $a_s \propto \varepsilon$. Clearly it is a four-wave mixing contribution or it is produced by a parametric process. The fields at ω_l and ω_p produce the mirror's displacement at the frequencies $\omega_l - \omega_p$ and at $\omega_p - \omega_l$. These components beat with the field at ω_l to produce components at $\omega_l - \omega_p + \omega_l$, $\omega_p - \omega_l + \omega_l$, i.e. at $2\omega_l - \omega_p$ and ω_p. In Figure 20.22, we show the behavior of the generated field ε_s due to four-wave mixing.

20.9 Quantized states of the nano-mechanical mirror coupled to the cavity

If the mechanical oscillator is quantized, and its annihilation and creation operators b and $b^\dagger$ are defined by the dimensionless position and momentum operators,

$$b = \frac{1}{\sqrt{2}}(Q + iP), \qquad b^\dagger = \frac{1}{\sqrt{2}}(Q - iP), \tag{20.68}$$

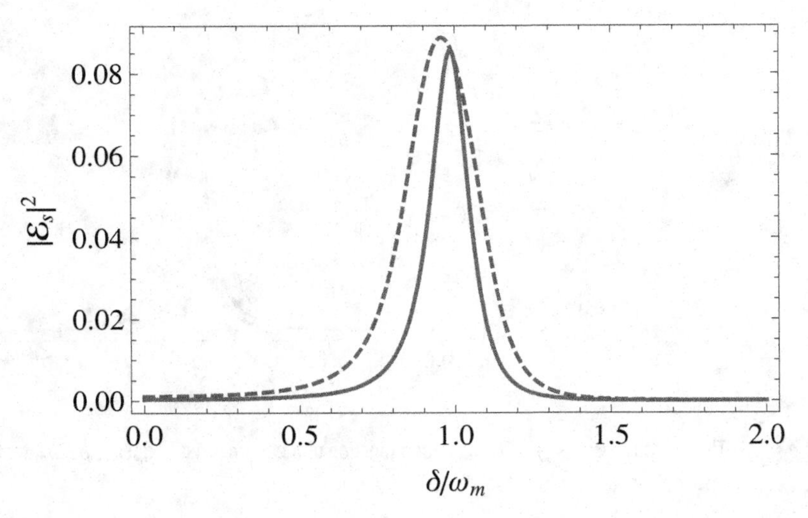

Fig. 20.22 The behavior of the generated field $|\varepsilon_s|^2$ as a function of the normalized frequency δ/ω_m for two different values of the laser power: $\wp = 6$ mW (solid), 10 mW (dashed).

then on dropping the zero point terms the first four terms in the Hamiltonian Eq. (20.11) of the system can be written as

$$H = \hbar\omega_c a^\dagger a + \hbar\omega_m b^\dagger b + \hbar G a^\dagger a(b + b^\dagger), \tag{20.69}$$

where $G = g\sqrt{\frac{\hbar}{2m\omega_m}}$. Assuming that the eigenfunctions and eigenvalues of H are $|\Psi\rangle$ and $\mathcal{E}$, i.e.

$$H|\Psi\rangle = \mathcal{E}|\Psi\rangle, \tag{20.70}$$

where

$$H = \hbar\omega_c a^\dagger a + \hbar\omega_m \left[b^\dagger b + \frac{G a^\dagger a}{\omega_m}(b + b^\dagger) \right]. \tag{20.71}$$

Let us work in the space with a fixed number of photons and thus we look for solution of the form

$$|\Psi\rangle = |n, \varphi\rangle, \qquad n = 0, 1, 2, \ldots, \tag{20.72}$$

in which n is the photon number of the cavity field. Hence

$$H|n, \varphi\rangle = \hbar \left\{ n\omega_c + \omega_m \left[b^\dagger b + \frac{Gn}{\omega_m}(b + b^\dagger) \right] \right\} |n, \varphi\rangle. \tag{20.73}$$

Let

$$h_n = \hbar\omega_m \left[b^\dagger b + \frac{Gn}{\omega_m}(b + b^\dagger) \right], \tag{20.74}$$

thus

$$h_n|n, \varphi\rangle = (\mathcal{E}_n - n\hbar\omega_c)|n, \varphi\rangle. \tag{20.75}$$

In addition, h_n can be written in a form

$$h_n = \hbar\omega_m B_n^\dagger B_n - \hbar\frac{G^2 n^2}{\omega_m}, \qquad B_n = b + \frac{Gn}{\omega_m}, \tag{20.76}$$

and B_n satisfies the commutation relation $[B_n, B_n^\dagger] = [b, b^\dagger] = 1$. Therefore B_n is like the annihilation operator for harmonic oscillator. Let us introduce the eigenstate of $B_n^\dagger B_n$ via

$$B_n^\dagger B_n |\Phi_\mu^{(n)}\rangle = \mu |\Phi_\mu^{(n)}\rangle, \tag{20.77}$$

and

$$|\Phi_\mu^{(n)}\rangle = \frac{(B_n^\dagger)^\mu}{\sqrt{\mu!}}|0\rangle, \tag{20.78}$$

where $|\Phi_\mu^{(n)}\rangle$ are the well-known harmonic oscillator states. Thus the eigenstate $|n, \varphi\rangle$ can be written as $|n, \Phi_\mu^{(n)}\rangle$, where $\mu = 0, 1, 2, \ldots, +\infty$, and combining Eqs. (20.75) and (20.76), we obtain

$$\mu\hbar\omega_m - \hbar\frac{G^2 n^2}{\omega_m} = \mathcal{E}_{n,\mu} - n\hbar\omega_c, \tag{20.79}$$

which leads to

$$\mathcal{E}_{n,\mu} = n\hbar\omega_c + \mu\hbar\omega_m - \hbar\frac{G^2 n^2}{\omega_m}. \tag{20.80}$$

Now we relate the eigenstates of $B_n^\dagger B_n$ to the eigenstates of $b^\dagger b$. This is done by using the properties of the displacement operator $\mathcal{D}(\beta) = \exp\left(\beta b^\dagger - \beta^* b\right)$ (Eq. (1.57)),

$$\mathcal{D}^\dagger(\beta) b \mathcal{D}(\beta) = b + \beta, \tag{20.81}$$

and hence

$$B_n = \mathcal{D}^\dagger\left(\frac{Gn}{\omega_m}\right) b \mathcal{D}\left(\frac{Gn}{\omega_m}\right). \tag{20.82}$$

Using (20.82) we can rewrite (20.77) as

$$\mathcal{D}^\dagger\left(\frac{Gn}{\omega_m}\right) b^\dagger b \mathcal{D}\left(\frac{Gn}{\omega_m}\right) |\Phi_\mu^{(n)}\rangle = \mu |\Phi_\mu^{(n)}\rangle, \tag{20.83}$$

or

$$b^\dagger b\left[\mathcal{D}\left(\frac{Gn}{\omega_m}\right)|\Phi_\mu^{(n)}\rangle\right] = \mu\left[\mathcal{D}\left(\frac{Gn}{\omega_m}\right)|\Phi_\mu^{(n)}\rangle\right]. \tag{20.84}$$

There

$$|\Phi_\mu^{(n)}\rangle = \mathcal{D}^\dagger\left(\frac{Gn}{\omega_m}\right)|\mu\rangle, \tag{20.85}$$

where

$$b^\dagger b|\mu\rangle = \mu|\mu\rangle, \qquad \mu = 0, 1, 2, \ldots. \tag{20.86}$$

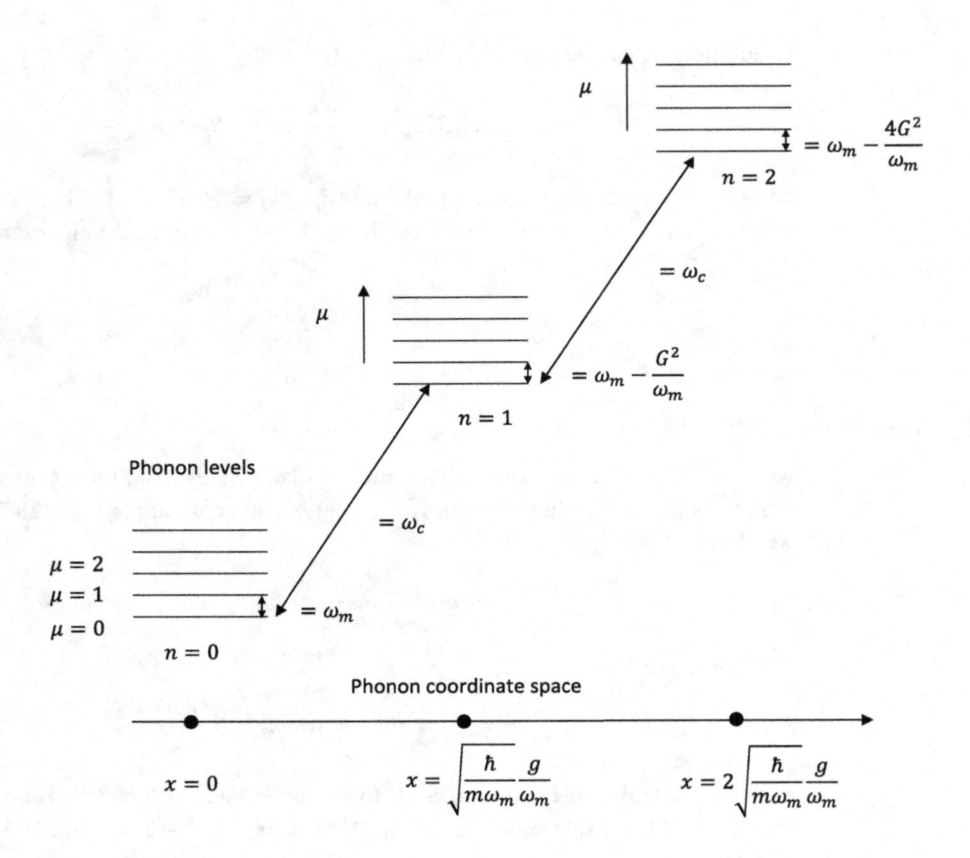

Fig. 20.23 Quantized states of the coupled cavity mirror system.

Thus to summarize the solution to the eigenvalue problem (20.70) is given by

$$|\Psi_\mu^{(n)}\rangle = \mathcal{D}^\dagger\left(\frac{Gn}{\omega_m}\right)|n, \mu\rangle,$$
$$\mathcal{E}_\mu^{(n)} = n\hbar\omega_c + \mu\hbar\omega_m - \hbar\frac{G^2n^2}{\omega_m}. \tag{20.87}$$

The eigenstates in the phonon space are obtained by displacing number states $|\mu\rangle$ for phonons. The displacement itself depends on the number of photons. These states are displayed in Figure 20.23. These states can be used in studying the time development of an initially prepared system.

We can now examine the possibility of transitions using the dressed states. Let us assume that the cavity is very weakly pumped. Thus effectively we can restrict to the one-photon space. Furthermore, the mirror can be at finite temperature and thus the occupation number of phonons is determined by the thermal distribution. The dissipation in the cavity leads the photon to leak out and thus one would have a transition to the zero-photon space. Even at zero temperature and negligible damping of the mirror, the initial state is a superposition of dressed states in one photon space, $|1,0\rangle \equiv \sum_\mu C_\mu |\Psi_\mu^{(1)}\rangle$, the coefficients C_μ are determined by using (20.87). We can ask the probability of making a transition from $|\Psi_\mu^{(1)}\rangle$ to $|\Psi_{\mu'}^{(0)}\rangle$ due to coupling of the cavity field to the environment outside. We can represent this coupling as $H_1 = aR^+ + a^\dagger R^-$, where $R^\pm$ are the environment operators (Chapter 9). This coupling leads to κ terms in the Langevin equations (20.16). We can now calculate using the perturbation theory the probability of making the transition from the state $|\Psi_\mu^{(1)}\rangle$ to the state $|\Psi_{\mu'}^{(0)}\rangle$. The relevant phonon dependent matrix element for this transition will be

$$|\langle\Psi_\mu^{(0)}|a|\Psi_{\mu'}^{(1)}\rangle|^2 = |\langle\mu|D(-G/\omega_m)|\mu'\rangle|^2, \tag{20.88}$$

where Eq. (20.87) has been used. The matrix element is given by the matrix element of the displacement operator in the phonon Fock state. This can be obtained from Eq. (8) of Table 1.1. The change in energy is $\mathcal{E}_{\mu'}^{(1)} - \mathcal{E}_\mu^{(0)} = \hbar\omega_c + (\mu' - \mu)\hbar\omega_m - \hbar G^2/\omega_m$. Thus the outgoing photon would have many sidebands determined by the multiples of phonon frequencies. The strength would be determined by the matrix element (20.88), which depends on the radiation pressure coupling coefficient G. A detailed discussion of the spectrum of the output photons is given in [37, 38].

Exercises

20.1 Using the exact eigenfunctions of H of Section 20.9, calculate the probability that the system would make a transition from $|1,0\rangle$ to $|1,\mu\rangle$ state, i.e. calculate

$$|\langle 1,\mu|e^{-iHt/\hbar}|1,0\rangle|^2.$$

Plot this probability as a function of $\omega_m t$ for different values of (G/ω_m) starting from the weak coupling case $(G/\omega_m < 1)$ to the strong coupling case $(G/\omega_m > 1)$.

20.2 A quadratically coupled membrane to the optical cavity is described by the Hamiltonian

$$H = \hbar\omega_c a^\dagger a + \hbar\omega_m b^\dagger b + gx^2 a^\dagger a.$$

Find the exact eigenstates of H in terms of the eigenstates $|n,\mu\rangle$ of $a^\dagger a$ and $b^\dagger b$. For details see [39].

20.3 The cooling of the mirror for the case when $\omega_m \ll \kappa$ can be understood in terms of the effective increase in damping of the mirror by radiation pressure. To see this treat Q, P, a, and $a^\dagger$ in (20.16) as numbers, and drop the ξ, a_{in} terms. Write the solution for a as

$$a \approx \left(\frac{\varepsilon}{\kappa + i\Delta_0}\right)\cdot\left[1 + i\omega_m\chi\int e^{-(\kappa+i\Delta_0)\tau}Q(t-\tau)d\tau\right],$$
$$\Delta_0 = \omega_c - \omega_l,$$

and use

$$Q(t-\tau) \approx Q(t) - \tau \frac{\partial Q}{\partial t},$$

to find an equation for P in which the effective frequency and damping of the mirror are modified. Find these modifications. In particular, show that

$$\gamma_m \to \gamma_m \left[1 + \frac{4\kappa \Delta_0 |\varepsilon|^2 \omega_m^3 \chi^2}{\gamma_m (\kappa^2 + \Delta_0^2)^3} \right].$$

The radiation pressure leads to increase in the damping of the mechanical oscillator if $\Delta_0 = \omega_c - \omega_l > 0$, i.e. for red detuning of the laser field. An increase in damping will lead to cooling of the mirror. Finally, justify the steps that led to the result for the effective damping. Note that the first successful experiments [9–11] on the cooling of the mechanical mirror are based on the idea of increasing damping.

20.4 Find the modes of a cavity with a dielectric bump described by a polarization term $4\pi\chi_0\delta(x - x_0)$, where x_0 is the location of the bump, as shown in the figure below. Use the solution of

$$\frac{d^2\varepsilon}{dx^2} + k^2[1 + 4\pi\chi_0\delta(x - x_0)]\varepsilon = 0, \qquad k = \omega/c,$$

and the boundary condition $\varepsilon = 0$ at $x = 0$ and $x = L$. Use the continuity of ε across

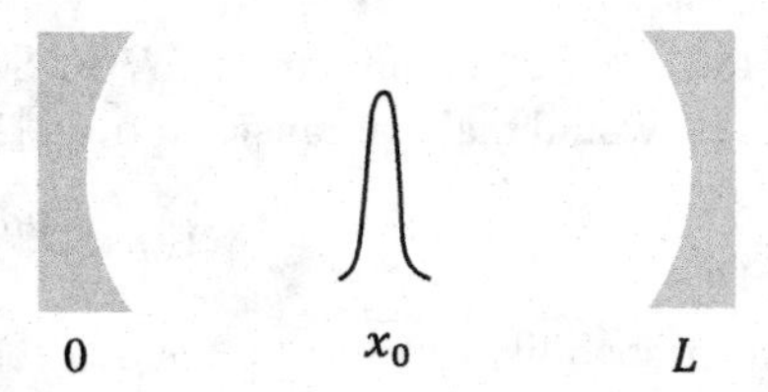

the bump. The discontinuity in the derivative of ε is determined by the differential equation for ε. Show that the eigenvalue equation is

$$4\pi\chi_0 k = \cot k\left(\frac{L}{2} + x\right) + \cot k\left(\frac{L}{2} - x\right),$$

$$x = x_0 - \frac{L}{2}.$$

Show further that, in the absence of the cavity, the transmission coefficient of the dielectric bump is $t = 1/(1 - 2\pi ik\chi_0)$ and hence $2\pi\chi_0 k = \sqrt{1-T}/\sqrt{T}$, $T = |t|^2$. Thus the above eigenvalue equation becomes [40]

$$\cot k\left(\frac{L}{2} + x\right) + \cot k\left(\frac{L}{2} - x\right) = 2\sqrt{\frac{1-T}{T}}.$$

A detailed numerical analysis of this eigenvalue equation can be found in [41].

References

[1] W. Marshall, C. Simon, R. Penrose, and D. Bouwmeester, *Phys. Rev. Lett.* **91**, 130401 (2003).

[2] J. D. Jackson, *Classical Electrodynamics*, 3rd edn. (New York: Wiley, 1999), p. 261.

[3] V. B. Braginsky and A. B. Manukin, *Measurement of Weak Forces in Physics Experiments* (Chicago, IL: University of Chicago Press, 1977).

[4] O. Svelto, *Principles of Lasers*, 4th edn. (Berlin: Springer, 1998).

[5] D. F. Walls and G. J. Milburn, *Quantum Optics* (Berlin: Springer, 1994), Chapter 7.

[6] S. Gröblacher, K. Hammerer, M. Vanner, and M. Aspelmeyer, *Nature (London)* **460**, 724 (2009).

[7] A. Dorsel, J. D. McCullen, P. Meystre, E. Vignes, and H. Walther, *Phys. Rev. Lett.* **51**, 1550 (1983).

[8] A. Hurwitz, *Selected Papers on Mathematical Trends in Control Theory*, edited by R. Bellman and R. Kalaba (NewYork: Dover, 1964).

[9] S. Gigan, H. R. Böhm, M. Paternostro *et al., Nature (London)* **444**, 67 (2006).

[10] O. Arcizet, P.-F. Cohadon, T. Briant, M. Pinard, and A. Heidmann, *Nature (London)* **444**, 71 (2006).

[11] D. Kleckner and D. Bouwmeester, *Nature (London)* **444**, 75 (2006).

[12] I. Wilson-Rae, N. Nooshi, W. Zwerger, and T. J. Kippenberg, *Phys. Rev. Lett.* **99**, 093901 (2007).

[13] F. Marquardt, J. P. Chen, A. A. Clerk, and S. M. Girvin, *Phys. Rev. Lett.* **99**, 093902 (2007).

[14] A. Schliesser, O. Arcizet, R. Rivière, G. Anetsberger, and T. J. Kippenberg, *Nature Physics* **5**, 509 (2009).

[15] Y. S Park and H. L. Wang, *Nature Physics* **5**, 489 (2009).

[16] J. D. Thompson, B. M. Zwickl, A. M. Jayich, F. Marquardt, S. M. Girvin, and J. G. E. Harris, *Nature (London)* **452**, 72 (2008).

[17] J. M. Dobrindt, I. Wilson-Rae, and T. J. Kippenberg, *Phys. Rev. Lett.* **101**, 263602 (2008).

[18] S. D. Gupta and G. S. Agarwal, *Opt. Commun.* **115**, 597 (1995).

[19] J. D. Teufel, D. Li, M. S. Allman *et al., Nature (London)* **471**, 204 (2011).

[20] S. Huang and G. S. Agarwal, arXiv:0905.4234.

[21] K. Jähne, C. Genes, K. Hammerer, M. Wallquist, E. S. Polzik, and P. Zoller, *Phys. Rev. A* **79**, 063819 (2009).

[22] C. W. Gardiner, *Phys. Rev. Lett.* **56**, 1917 (1986).

[23] M. J. Hartmann and M. B. Plenio, *Phys. Rev. Lett.* **101**, 200503 (2008).

[24] G. S. Agarwal and S. Huang, *Phys. Rev. A* **81**, 041803(R) (2010).

[25] S. Weis, R. Rivière. S. Deléglise *et al., Science* **330**, 1520 (2010).

[26] A. H. Safavi-Naeini, T. P. Mayer Alegre, J. Chan *et al., Nature (London)* **472**, 69 (2011).

[27] S. H. Autler and C. H. Townes, *Phys. Rev.* **100**, 703 (1955).

[28] S. E. Harris, *Phys. Today* **50**, 36 (1997).
[29] S. E. Harris, J. E. Field, and A. Imamoglu, *Phys. Rev. Lett*. **64**, 1107 (1990).
[30] K.-J. Boller, A. Imamoglu, and S. E. Harris, *Phys. Rev. Lett*. **66**, 2593 (1991).
[31] L. V. Hau, S. E. Harris, Z. Dutton, and C. H. Behroozi, *Nature (London)* **397**, 594 (1999).
[32] S. E. Harris, J. E. Field, and A. Kasapi, *Phys. Rev. A* **46**, R29 (1992).
[33] M. M. Kash, V. A. Sautenkov, A. S. Zibrov *et al., Phys. Rev. Lett*. **82**, 5229 (1999).
[34] V. Fiore, Y. Yang, M. C. Kuzyk, R. Barbour, L. Tian, and H. Wang, *Phys. Rev. Lett*. **107**, 133601 (2011).
[35] S. Huang and G. S. Agarwal, *Phys. Rev. A* **83**, 043826 (2011).
[36] G. S. Agarwal and S. Huang, *Phys. Rev. A* **85**, 021801 (R) (2012).
[37] A. Nunnenkamp, K. Børkje, and S. M. Girvin, *Phys. Rev. Lett*. **107**, 063602 (2011).
[38] P. Rabl, Phys. *Rev. Lett*. **107**, 063601 (2011).
[39] A. Rai and G. S. Agarwal, *Phys. Rev. A* **78**, 013831 (2008).
[40] W. J. Fader, *IEEE J. Quantum Electron*. **21**, 1838 (1985).
[41] M. Bhattacharya, H. Uys, and P. Meystre, *Phys. Rev. A* **77**, 033819 (2008).

Index

Printed in the United States
By Bookmasters